Nonlinear Evolution of Spatio-Temporal Structures in Dissipative Continuous Systems

NATO ASI Series

Advanced Science Institutes Series

A series presenting the results of activities sponsored by the NATO Science Committee, which aims at the dissemination of advanced scientific and technological knowledge, with a view to strengthening links between scientific communities.

The series is published by an international board of publishers in conjunction with the NATO Scientific Affairs Division

A	**Life Sciences**	Plenum Publishing Corporation
B	**Physics**	New York and London
C	**Mathematical and Physical Sciences**	Kluwer Academic Publishers Dordrecht, Boston, and London
D	**Behavioral and Social Sciences**	
E	**Applied Sciences**	
F	**Computer and Systems Sciences**	Springer-Verlag
G	**Ecological Sciences**	Berlin, Heidelberg, New York, London,
H	**Cell Biology**	Paris, and Tokyo

Recent Volumes in this Series

Volume 221—Guidelines for Mastering the Properties of Molecular Sieves: Relationship between the Physicochemical Properties of Zeolitic Systems and Their Low Dimensionality
edited by Denise Barthomeuf, Eric G. Derouane, and Wolfgang Hölderich

Volume 222—Relaxation in Complex Systems and Related Topics
edited by Ian A. Campbell and Carlo Giovannella

Volume 223—Particle Physics: *Cargèse 1989*
edited by Maurice Lévy, Jean-Louis Basdevant, Maurice Jacob, David Speiser, Jacques Weyers, and Raymond Gastmans

Volume 224—Probabilistic Methods in Quantum Field Theory and Quantum Gravity
edited by P. H. Damgaard, H. Hüffel, and A. Rosenblum

Volume 225—Nonlinear Evolution of Spatio-Temporal Structures in Dissipative Continuous Systems
edited by F. H. Busse and L. Kramer

Volume 226—Sixty-Two Years of Uncertainty: Historical, Philosophical, and Physical Inquiries into the Foundations of Quantum Mechanics
edited by Arthur I. Miller

Volume 227—Dynamics of Polyatomic Van der Waals Complexes
edited by Nadine Halberstadt and Kenneth C. Janda

Volume 228—Hadrons and Hadronic Matter
edited by Dominique Vautherin, F. Lenz, and J. W. Negele

Series B: Physics

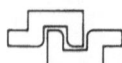

Nonlinear Evolution of Spatio-Temporal Structures in Dissipative Continuous Systems

Edited by

F. H. Busse and L. Kramer

University of Bayreuth
Bayreuth, Federal Republic of Germany

Plenum Press
New York and London
Published in cooperation with NATO Scientific Affairs Division

Proceedings of a NATO Advanced Research Workshop on
Nonlinear Evolution of Spatio-Temporal Structures
in Dissipative Continuous Systems,
held September 24-29, 1989,
in Streitberg (near Bayreuth), Federal Republic of Germany

Library of Congress Cataloging-in-Publication Data

NATO Advanced Research Workshop on Nonlinear Evolution of Spatio-Temporal Structures in Dissipative Continuous Systems (1989 : Streitberg, Wiesenttal, Germany)
Nonlinear evolution of spatio-temporal structures in dissipative continuous systems / edited by F.H. Busse and L. Kramer.
p. cm. -- (NATO ASI series. Series B, Physics ; v. 225)
"Proceedings of a NATO Advanced Research Workshop on Nonlinear Evolution of Spatio-Temporal Structures in Dissipative Continuous Systems, held September 24-29, 1989, in Streitberg (near Bayreuth), Federal Republic of Germany"--T.p. verso.
"Published in cooperation with NATO Scientific Affairs Division."

ISBN-13: 978-1-4684-5795-7 e-ISBN-13: 978-1-4684-5793-3

DOI: 10.1007/978-1-4684-5793-3

1. Fluid dynamics--Congresses. 2. Nonlinear theories--Congresses. 3. Mathematical physics--Congresses. I. Busse, F. H. II. Kramer, L. III. North Atlantic Treaty Organization. Scientific Affairs Division. IV. Title. V. Series.
QC151.N37 1989
532'.05--dc20 90-7591
CIP

Softcover reprint of the hardcover 1st edition 1990

A Division of Plenum Publishing Corporation
233 Spring Street, New York, N.Y. 10013

SPECIAL PROGRAM ON CHAOS, ORDER, AND PATTERNS

This book contains the proceedings of a NATO Advanced Research Workshop held within the program of activities of the NATO Special Program on Chaos, Order, and Patterns.

Volume 208—MEASURES OF COMPLEXITY AND CHAOS
edited by Neal B. Abraham, Alfonso M. Albano, Anthony Passamante, and Paul E. Rapp

Volume 225—NONLINEAR EVOLUTION OF SPATIO-TEMPORAL STRUCTURES IN DISSIPATIVE CONTINUOUS SYSTEMS
edited by F. H. Busse and L. Kramer

PREFACE

This volume contains papers contributed to the NATO Advanced Research Workshop "Nonlinear Evolution of Spatio-Temporal Structures in Dissipative Continuous Systems" held in Streitberg, Fed. Rep. Germany, Sept. 24 through 30, 1989. The purpose of the rather long title has been to focus attention on a particularly fruitful direction of research within the broad field covered by terms like Nonlinear Dynamics or Non-Equilibrium Systems.

After physicists have been occupied for several decades mainly with the microscopic structure of matter, recent years have witnessed a resurgence of interest in macroscopic patterns and dynamics. Research on these latter phenomena has not been dormant, of course, since fluid dynamicists interested in the origin of turbulence, meteorologists studying weather patterns and numerous other scientists have continued to advance the understanding of the structures relevant to their disciplines. The recent progress in the dynamics of nonlinear systems with few degrees of freedom and the discovery of universal laws such as the Feigenbaum scaling of period-doubling cascades has given rise to new hopes for the understanding of common principles underlying the spontaneous formation of structures in extended continuous systems.

The goal of the Workshop has been to bring together scientists from diverse fields ranging from chemical engineering to astrophysics to discuss the mechanisms of structure formation and the mathematical tools for their description in various continuous media. Fluids and liquid crystals offer the largest variety of spatio-temporal patterns, but chemical oscillations, solidification processes and defect ordering in solids also provide important examples for structure formation.

In arranging the contributions we have started with classical simple fluid systems exhibiting a transition to turbulence through a sequence of bifurcations. They continue to offer new and sometimes surprising dynamic phenomena, and the possibilities for novel solutions of the Navier-Stokes equations are far from being exhausted. Thermal convection in binary fluids has received much attention in recent years because it offers an experimentally easy, accessible Hopf-bifurcation and the puzzles of confined, blinking and other states have been a major topic of discussion at the Workshop. In spite of recent progress the nonlinear theory is still far from a satisfactory state. Systems with two separated phases offer even more degrees of freedom and it is not surprising that the experimentally realised phenomena far exceed the theoretical models for their interpretation.

Papers on the theoretical description of localized structures and the dynamics of defects of patterns are collected in the fourth part of the proceedings. Stimulated by the observations of confined states in binary fluids and by the importance of defects for the induction of chaotic states this area is an especially active field of research. Chemically reacting fluids and liquid crystals exhibit a number of fascinating patterns, a selection of which is presented in this volume. Again, the experimental results dominate this area, but the theoretical understanding is catching up rapidly. The following section emphasises the role of symmetries and resonances in the mathematical description of nonlinear phenomena. Most of the papers have applications to several different physical systems. The crystallisation of a solid from its melt may be considered as a problem of front propagation. But this topic encompasses a much wider variety of phenomena as becomes evident from the papers of the seventh section. Finally a number of papers related to the subject of fully developed turbulence have been collected which demonstrate the utility of new approaches to problems of high Reynolds number flows.

The coherence in the motivations, goals and methods of research and the collaborative spirit which, in spite of some controversial points, were evident in the discussions at the Workshop may not have been captured as well as we would have liked in the diverse contributions of this volume. But we believe that the book offers a timely state-of-the-art account of a most fascinating field of research and we trust that it will be useful to all students and scientists interested in problems of structure formation.

We are grateful to our colleagues on the Organising Committee, Profs. H. Haken, A. C. Newell and Y. Pomeau for their advice and encouragement and to the NATO Scientific Affairs Division for its generous support of the Workshop. Additional financial support has been received from the University of Bayreuth, from the Emil-Warburg-Foundation, Bayreuth, and from the Deutsche Forschungsgemeinschaft through Sonderforschungsbereich 213. Last, but not least, we are indebted to our collaborators at the chairs Theoretical Physics II and IV, University of Bayreuth, without whose dedicated help the organisation of the Workshop would not have been possible.

Bayreuth, December 1989

F.H. Busse
L. Kramer

CONTENTS

NONLINEAR DYNAMICS IN SIMPLE FLUIDS

BINARY FLUID CONVECTION

TWO-PHASE FLOWS

LOCALIZED STRUCTURES AND DEFECTS

NONLINEAR DYNAMICS IN REACTING FLUIDS, NEMATICS, AND SOLIDS

SYMMETRY, RESONANCE, AND CHAOS

NONLINEAR PROBLEMS OF CRYSTAL GROWTH AND FRONT PROPAGATION

STRUCTURES IN TURBULENT FLOWS

SPATIAL AND DYNAMICAL PROPERTIES OF 1 – D RAYLEIGH-BENARD CONVECTION

M. Dubois, F. Daviaud, M. Bonetti

Service de Physique du Solide et de Résonance Magnétique
CEN-Saclay
91191 Gif-sur-Yvette Cedex, France

The studies of one-dimensional systems have gained a lot of interest in the last few years, as these particular geometries can provide intermediate situations between pure dynamical behaviours and very complex spatio-temporal behaviours. Rayleigh-Bénard convection is used to study the specific behaviours in such $1 - D$ geometries and we report here the main observed features.

Rayleigh-Bénard convection[1] is the instability undergone by an horizontal fluid layer submitted to a distabilizing vertical temperature gradient $\Delta T/d$, with ΔT the temperature difference across the layer and d the depth. The evolution of the convective state is controlled by the Rayleigh number Ra, an adimensionless measure of the temperature difference. The dynamics itself is related to the Prandtl number Pr, but the aspect ratios, defined as the ratio of the horizontal extensions of the layer to its depth play also a fundamental role. Observations of patterns instaured in different geometries with different aspect ratios have shown that a good approximation of a $1 - D$ system[2] is obtained when one of the aspect ratio is lower than 0.6. This means, that under this condition, the convective behaviour in these narrow channels is only dependent on one space variable, namely X for example, along the greatest extension of the layer. In the following, the results concern rectangular and annular cells with such aspect ratios ($\Gamma_y <$ 0.6). A general description of the experimental set-up has been given elsewhere[3]. The measurements are performed through shadowgraphic images and the light intensity is recorded by photodiodes or by image processing.

When the Rayleigh number is varied, three domains can be evidenced. A first Ra domain corresponds to stationnary patterns and wave-number selection. The second domain is related to the first time dependences, initiated by the appearance of oscillators in the boundary layers. Then, spatial defects appear in the roll pattern. These defects can propagate and lead to spatio-temporal intermittencies.

As it was already pointed out, the first striking feature observed in narrow channels is the presence of very small wavelengths[2]. Their selection mechanism is not still entirely understood. A more systematic study in a rectangular channel ($\Gamma_x = L_x/d \simeq 26$, $\Gamma_y = L_y/d = 0.29$, $Pr \simeq 7.5$) has shown that by increasing or decreasing very slowly the temperature difference, the wavenumber selection takes place in a very efficient manner when $r = Ra/Rac < 20$. The observed stability domain is shown in Fig.1. It is situated between the "Busse balloon", calculated in the case of infinite geometry[4], and the stability domain calculated by Kvernvold for Hele-Shaw cells[5]. Wavelengths greater than $0.7\lambda_c$ ($\lambda_c \simeq 2d$) have never been observed and the minimum observable wavelength λ_m is such as $\lambda_m/2 = L_y$ (smallest extension of the layer). This wavelength is always observed

Nonlinear Evolution of Spatio-Temporal Structures in Dissipative Continuous Systems
Edited by F.H. Busse and L. Kramer
Plenum Press, New York, 1990

near $r = 10$, if this r value is reached by very slow Ra increase or decrease. The stability domain then seems very narrow around $r = 10$. For higher Ra values ($r > 20$), a relatively broad domain of stability is recovered. Up to day, no theoretical explanation has been proposed for this particular property of the patterns in narrow channels.

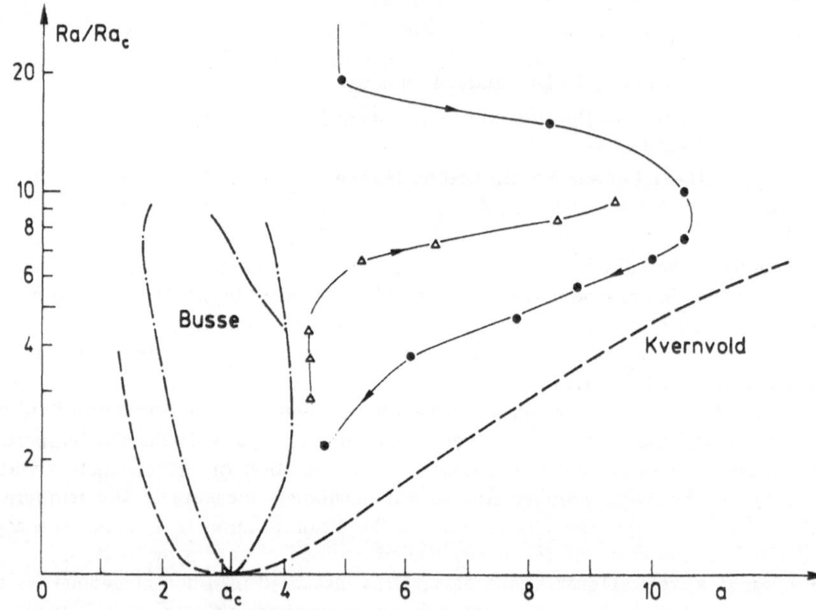

Figure 1

Evolution of the wave number $a = 2\pi d/\lambda$ versus Ra/Rac, observed in a narrow rectangular channel (the lateral boundaries are made of plexiglass). $\Gamma_x = 26$, $\Gamma_y = 0.29$ with $d = 7mm$. The fluid under study is silicon oil with $Pr \simeq 7.5$.

Δ – *Ra increasing* • – *Ra decreasing*

–. – .– *Stability domain as calculated by Busse for infinite geometries*

– – – *Stability domain, calculated by K Kvernvold for Hele-Shaw cells.*

Rac is the critical convective onset for the studied geometry.

When the Rayleigh number is increased up to a value r_0, which depends on the transverse aspect ratio and on the actual wavelength, oscillators appear in the rolls, as hot plumes for example. The convective layer then acts as a complex dynamical system, i.e. a chain of coupled non linear oscillators. When $\lambda = \lambda_m$ a specific oscillatory mechanism takes place: the rolls oscillate periodically around their mean position with phase opposition between the hot and cold main streams. When the pattern is quasi-perfect (same wavelength everywhere), the amplitude of the motion is spatially modulated. This behaviour can be described by the presence of a standing wave[6], with wavelength around 5 to 6 λ_0 (λ_0 is the wavelength of the pattern). When *Ra* is increased, the behaviour which may become chaotic, remains unchanged both in time and in space at fixed *Ra*, until the appearance of new events, leading to a spatial symmetry breaking.

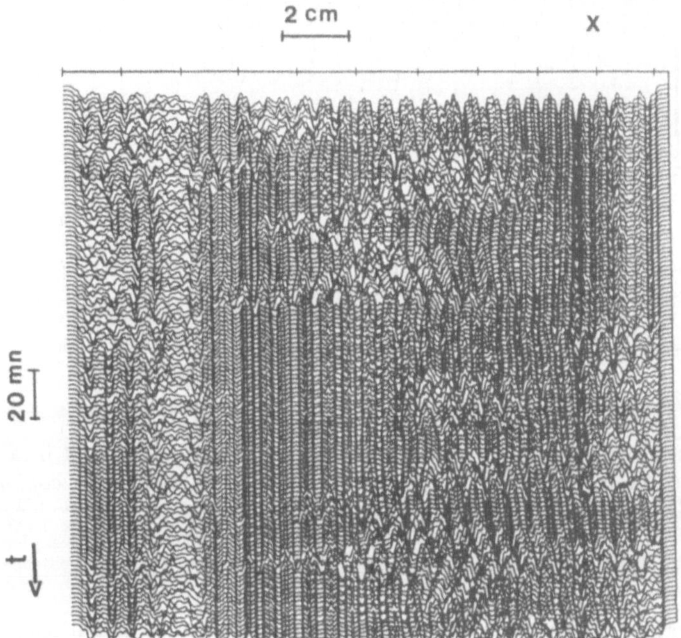

Figure 2

Spatio-temporal evolution of the pattern in a rectangular channel from shadowgraphic image processing (120s between each horizontal "line"). $\Gamma_x \simeq 26$, $\Gamma_y \simeq 0.43$, $Pr \simeq 7.5$, $r = 363$).

Each horizontal curve gives a snapshot of the intensity versus X, *at a given height of the image, then at a given height in the fluid layer. The solitary waves are reflected by robust oscillating domains (oscillation period* $\simeq 1s$).

The first manifestation of this symmetry breaking is given by the presence of solitary waves. In the studied narrow channels, these waves correspond to the propagation of a local spatial defect, generally a wavelength greater than its neighbours. They are not easy seen, because their velocity is very low (from 10^{-2} to 5 $10^{-2}\lambda_0 s^{-1}$, see Figs.2 and 3). Nevertheless they have always been observed as preceding the appearance of local turbulent states and the development of spatio-temporal intermittencies. As a matter of fact, the interaction of these solitary waves with present oscillators, or the interaction of two of them with opposite direction can lead to the formation of local large cells as shown in Fig.3. These local abnormal wavelengths are germs of turbulent patches which may develop. The coexistence of these turbulent patches with laminar domains and their mutual exchange versus time correspond then to what has been defined as spatio-temporal intermittencies[7,8].

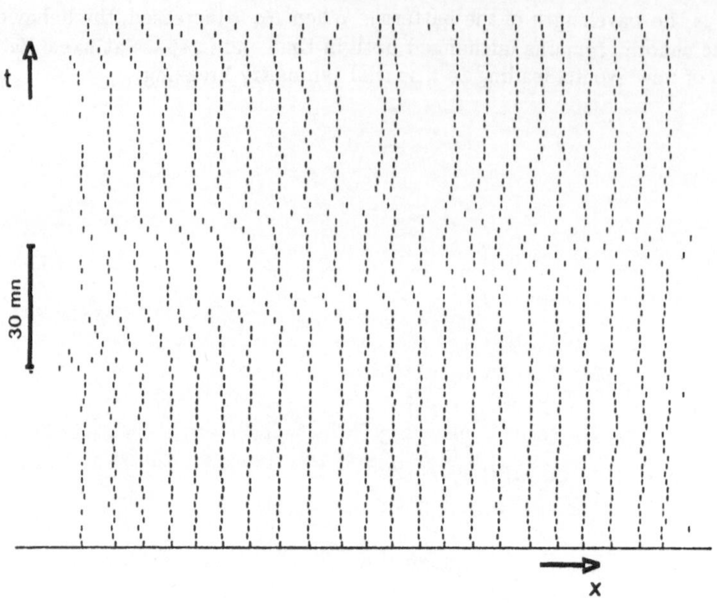

Figure 3
Time variation of the position of the intensity minima (hot streams) in a rectangular channel $\Gamma_x \simeq 21$, $\Gamma_y \simeq 0.35$, $Pr \simeq 7.5$, $r = 230$. (Here the fluid is confined between horizontal glass plates). Two solitary waves collapse and give rise to two local large cells.

To perform quantitative analysis of these spatio-temporal intermittencies (S.T.I.), it is fundamental to find a relevant criterion in the qualification of the local state as "laminar or turbulent". In narrow channels, the local spatial properties are very different in the two states. The spatial order with a defined wavelength is dominant in laminar domains meanwhile complete spatial disorganization prevails in turbulent regions. So a spatial criterion has been used to perform statistical analysis[9].

The first global behaviour we can look at is the evolution of the turbulent fraction Ft versus r, for $r > r_S$, r_S being the threshold for the appearance of the first turbulent events. The two curves shown in Fig.4 correspond to different geometries, one rectangular ($\Gamma_x \simeq 26$, $\Gamma_y \simeq 0.43$, $Pr = 7.5$), the other one annular ($\Gamma_p \simeq 35$ along the circumference, $\Gamma_y \simeq 0.29$, $Pr = 7.5$). In both geometries, we are in presence of a critical transition in analogy with directed percolation process[10,11]. This transition looks quasi-perfect in the rectangular geometry, in contrast to the results observed in the annular geometry. Nevertheless these behaviours have to be related to the results of other data analysis by studying, for example, the statistics of the lengths of the laminar domains.

We do not want to enter into the details which are reported elsewhere[9,12]. We want just to note that in the sustained S.T.I. regime, the number $N(L)$ of laminar domains of length L decreases with L according to an exponential decay. This introduces a characteristic length ρ which varies as

$$m = 1/\rho \div (r - r'_S)^{1/2} .$$

in the two kinds of geometry and which has been also observed in an other experimental situation[13]. Nevertheless, if $r'_S = r_S$, within the experimental error, in the case of the rectangular channel, r'_S differs from r_S in the annulus. More generally, all the statistical measurements confirm that, at least in the case of our experiments, the transition to

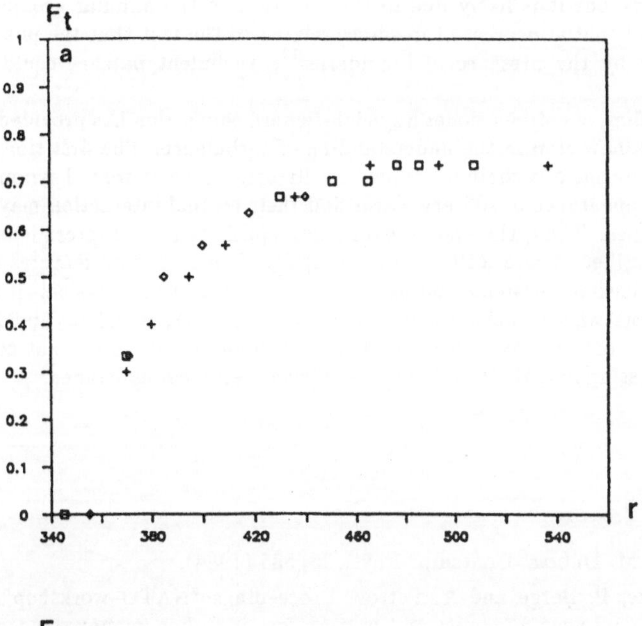

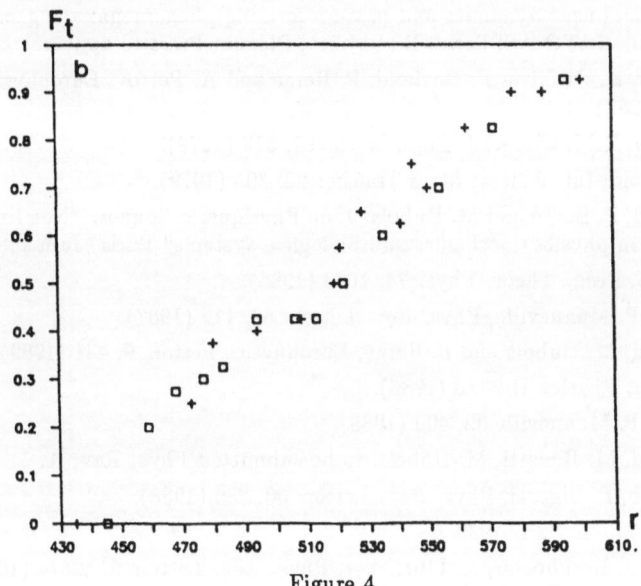

Figure 4

Evolution of the mean turbulent fraction F_t versus the reduced Rayleigh number $r = Ra/Rac$. F_t refers to the number of turbulent cells divided by the total number of cells in the container.

a) rectangular geometry $r_S \simeq r'_S \simeq 360$

b) annular geometry $r_S = 450$.

turbulence via S.T.I. is quasi-perfect in rectangular channels and is not in annulus though the critical character of the transition remains. The explanation of this difference is not clear to day but it is likely due to the fact that in the annular geometry, there are some additional spatial degrees of freedom, related to the fact that the phase of the rolls is not blocked by the presence of boundaries[12]. Turbulent patches could then appear spontaneously.

In conclusion, one-dimensional Rayleigh-Bénard convection has provided informations which can be a new step in the understanding of turbulence. The first time dependences are similar to the one of a chain of coupled oscillators. The first spatial symmetry breaking arises by the appearance of solitary waves and their mutual interaction may give birth to turbulent patches. Then, the system enters into spatio-temporal intermittencies, as they have been described theoretically and numerically. Some of these features seems generic of the one-dimensional systems and has been observed in other physical situations[15,16].

The authors wish to acknowledge P. Bergé, G. Balzer, H. Chaté and P. Manneville for constructive discussions. They thank also P. Hède for his important contribution in the data processing and B. Ozenda for his efficient technical assistance.

REFERENCES

1. P. Bergé, M. Dubois, Contemp. Phys. 25, 535 (1984).
2. M. Dubois, P. Bergé and A. Petrov, Proceedings "NATO workshop" (1988): new trends in nonlinear dynamics and pattern forming phenomena: the geometry of non equilibrium. NATO ASI Series B (Physics) Plenum Press to appear.
3. M. Dubois, R. da Silva, F. Daviaud, P. Bergé and A. Petrov, Europhysics Letters 8, 135-139 (1989).
4. F.H. Busse, R.M. Clever, J. Fluid Mech. 91, 319 (1979).
5. O. Kvernvold, Int. J. Heat Mass Transfer 22, 395 (1979).
6. F. Daviaud, P. Bergé and M. Dubois, J. de Physique; colloques. "Non linear coherent structures in physics, mechanics and biological systems" Paris (Juin 1988).
7. K. Kaneko, Prog. Theor. Phys. 74, 1033 (1985).
8. H. Chaté, P. Manneville, Phys. Rev. Letters 58, 112 (1987).
9. F. Daviaud, M. Dubois and P. Bergé, Europhysics Letters 9, 441 (1989).
10. Y. Pomeau, Physica D 23, 3 (1986).
11. H. Chaté, P. Manneville 32, 409 (1988).
12. F. Daviaud, M. Bonetti, M. Dubois, to be submitted Phys. Rev. A.
13. S. Ciliberto, P. Bigazzi, Phys. Rev. Letters 60, 286 (1988).
14. H. Chaté, Thèsis Paris 1989.
15. A. Simon, J. Bechhoefer, A. Libchaber, Phys. Rev. Letters 61, 2574 (1988).
16. M. Rabaud, S. Michalland, Y. Couder, to be published.

DIMENSIONS AND LYAPUNOV SPECTRA FROM MEASURED TIME SERIES OF TAYLOR-COUETTE FLOW

Thorsten Buzug, Torsten Reimers and Gerd Pfister

Institut für Angewandte Physik
Universität Kiel
West Germany

1. INTRODUCTION

The extraction of Lyapunov spectra and fractal dimensions of chaotic attractors from experimental time series has become an important instrument in analyzing the dynamic properties of non linear systems [1]. Though the principle ideas are clear the practical handling of the estimation of dynamical properties depends sensitively on the quality of the reconstructed phase space of the attractor. Additionally all experimental data are covered with noise so that the algorithms used should be insensitive to external noise.

The first step in analyzing the experimental data is a reconstruction of the phase space of the attractor. Using delay time coordinates with a proper time delay and a sufficient embedding dimension we can get a topologically similar map of the attractor [2]. The most significant parameter describing chaos is the Lyapunov spectrum of an attractor. It yields a measure of the dynamical behaviour of the coordinates in the phase space of adjacent points averaged over the attractor. For practical considerations we estimated the correlation dimension for homogenious attractors and the distribution of pointwise dimensions when the averaging makes no sense. We present here three kinds of scenarios of Taylor-Couette flow which is only a small fraction of possible ways to chaotic states in this system. Firstly we investigated quasiperiodic states which show stable 3-mode-states [3]. Secondly a route to chaotic motion is shown which passes a periodic 3-window [4]. Thirdly we want to analyze the onset of chaos nearby a homoclinic orbit [5].

2. THE TAYLOR APPARATUS AND MEASURING TECHNIQUES

The inner cylinder of our Taylor-Couette apparatus was made from stainless steel with a radius of r_1 = 12.5 mm, the outer cylinder was a Perspex cylinder with a radius r_2 = 25 mm giving a radius ratio of $\eta = r_1 / r_2 = 0.5$. The gap length could be adjusted from 0 to 600 mm continously. The fluid confined in the gap was silicon oil with various viscosities. The control

Nonlinear Evolution of Spatio-Temporal Structures in Dissipative Continuous Systems
Edited by F.H. Busse and L. Kramer
Plenum Press, New York, 1990

parameter is the Reynolds number defined as $Re = \Omega r_1 d / \nu$, where Ω is the angular rotation frequency, $d = r_2 - r_1$ the gap width and ν the kinematic viscosity. The entire cylinder was surrounded by a square box which contained silicon oil which acts as a thermostating fluid as well as an index matching for the Laser Doppler velocimetry (LDV). With this LDV the local velocities could be measured on-line using a phase-locked-loop tracker. A more detailed description of the measuring techniques can be found in [6,7].

3. METHODS

Until recently power spectra and correlation functions were used to characterize states of dynamical systems. But with these methods one can hardly distinguish between very complicated time series. Therefore new mathematical tools like estimating dimensions and Lyapunov exponents from reconstructed attractors are provided. We have tested these tools with the usual model attractors like Lorenz, Rössler and Hénon attractor. We used the x-coordinate of each system and normalized those data sets to [0,1023] for simulating a 10-bit A/D-converter.

3.1 RECONSTRUCTION OF PHASE SPACE

In order to estimate the Lyapunov spectra and the dimensions from the measured time series, one has to reconstruct the phase space using delay time coordinates from an observable ξ [2]:

$$x(t) = \left\{ \xi(t), \xi(t+\tau), \ldots, \xi(t+(dim_E-1)\tau) \right\} .$$

Though the delay time can in principle be chosen arbitrarily, in practice we are limited to a finite resolution of our experimental data and the finite number of data points. The choice of the correct delay time τ and the embedding dimension dim_E is discussed in detail elsewhere [8,9]. The influence of the filtering process of the experimental time series on the dynamics will be published [10]. For the states which we will present the cut-off frequency of the low pass is far away from frequencies which are relevant in the system.

3.2 THE LYAPUNOV EXPONENTS (LE)

The best tool to detect chaotic motion and to distinguish different chaotic states is calculating LE. These LE yield a dynamic characterization, e.g. the stability of solutions of the flow.

a) <u>The definition</u>

The local deformation of an infinitesimal unit n-sphere to a n-ellipsoid by the flow defines the LE:

$$\lambda_k = \lim_{t_{ev} \to \infty} \frac{1}{t_{ev}} \log \frac{p_k(t_{ev})}{p_k(0)} .$$

$p_k(0)$ is the radius of the n-sphere at the starting point and

$p_k(t_{ev})$ are the principal axes of the n-ellipsoid after some evolution time t_{ev}. We can calculate the λ_k using the linearized flow map $T_x^t \equiv D_x\varphi_t$ which is the Jacobi matrix of the flow:

$$\lambda_k = \lim_{m\to\infty} \frac{1}{m * T_a} \log \left\| T_x^{jt_{ev}} e_j^k \right\| .$$

Fig.3.1 shows how an orthonormal base is deformed by the flow.

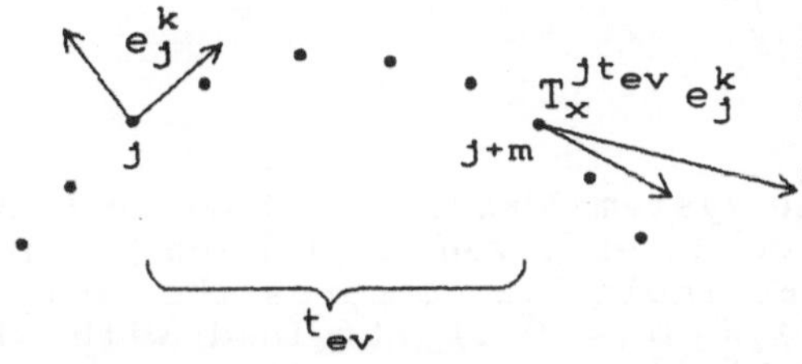

Fig. 3.1.
The base e_j^k $(k=1,\ldots,n)$ is mapped on $T_x^{jt_{ev}} e_j^k$ after the evolution time $t_{ev} = m T_a$ (n: dimension of phase space).

Due to the properties of nonlinear dynamical systems the base will collapse in the direction of strongest growth. To avoid numerical difficulties we normalize our base periodically after the evolution time t_{ev} (fig. 3.2).

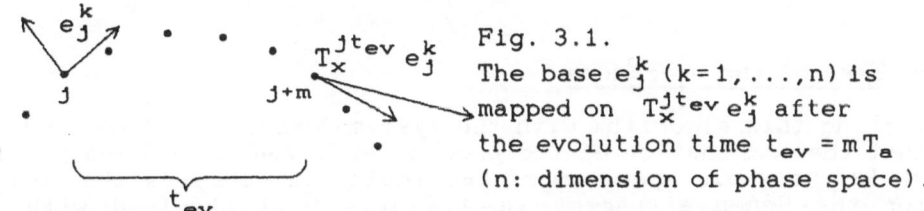

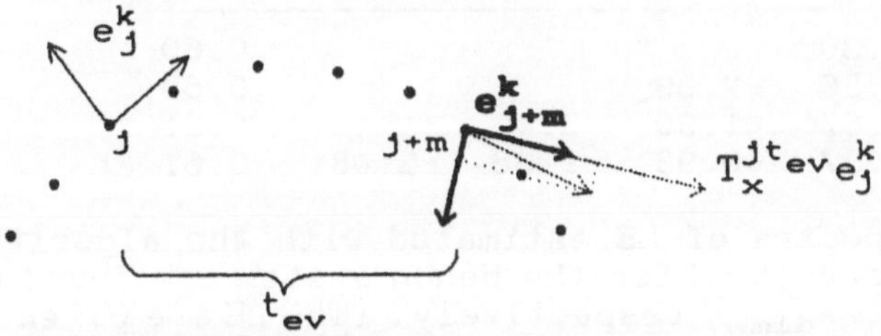

Fig. 3.2 . Normalization of our base.

Normalizing the base with the QR-decomposition of the linearized flow map one gets the LE from the following successive procedure [11]:

$$T_1 E = Q_1 R_1$$

$$T_2 T_1 E = T_2 Q_1 R_1 = Q_2 R_2 R_1$$

$$\vdots$$

$$\prod_{k=1}^{M} T_k E = Q_M \prod_{k=1}^{M} R_k$$

Averaging over the corresponding diagonal elements r_{jj} of R_k yields the Lyapunov exponents:

$$\lambda_j = \frac{1}{M * t_{ev}} \log \prod_{i=1}^{M} r_{jj}^i ,$$

$$\lambda_j = \frac{1}{M * t_{ev}} \sum_{i=1}^{M} \log r_{jj}^i$$

b) The experimental situation

We have to approximate the linearized flow map. The displacement vectors to the neighbouring points of the reference point span the tangential space [12] (fig. 3.3).

The evolution of the $\{y^i\}$ to the $\{z^i\}$ is represented by $T_j \equiv T_x^{jt_{ev}}$: $z^i \approx T_j y^i$. We get the T_j by a least squares fit:

$$\min S = \min \frac{1}{N_{nb}} \sum_{i=1}^{N_{nb}} \left| z^i - T_j y^i \right|^2 .$$

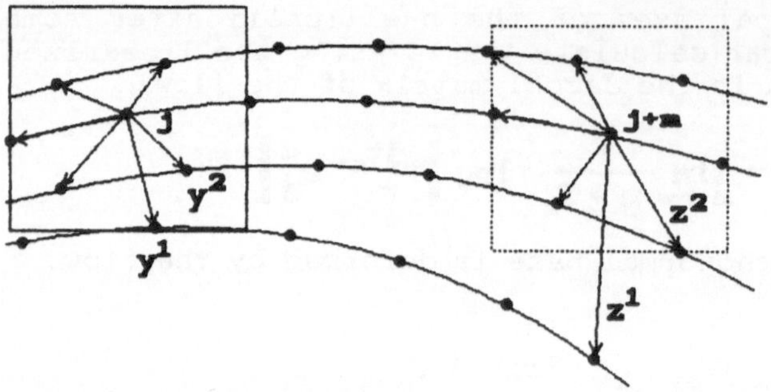

Fig. 3.3. Approximation of the tangential space by the sets of displacement vectors $\{y^i\}$ and $\{z^i\}$, respectively.

c) Test of the procedure

Testing this algorithm with the systems described above we found that the combination of the procedures given in [12] and [13] may yield the whole spectrum of LE. Table 3.1 compares the results for the Hénon attractor ($a = 1.4$, $b = 0.3$) obtained with the algorithms developed by [13] and [14].

Table 3.1. (published by [15])

dim_E	λ_1	λ_2	λ_3	λ_4	λ_5	λ_1^{Wolf}
2	0.61	-2.30				0.60
3	1.18	0.56	-2.39			0.60
4	1.24	0.57	-1.90	-2.46		0.61
5	1.24	0.57	-0.93	-1.66	-2.48	0.61

Table 3.2 shows the spectra of LE estimated with the algorithm explained above. The correct LE for the Hénon system are found by averaging the λ_1 and λ_{dim_E}, respectively. (The LE are given in units of bits/iteration.)

Table 3.2

dim_E	λ_1	λ_2	λ_3	λ_4	λ_5	λ_6	λ_7
2	0.62	-2.57					
3	0.61	-1.17	-2.34				
4	0.61	-0.64	-1.14	-2.20			
5	0.61	-0.24	-0.65	-1.08	-2.19		
6	0.61	0.07	-0.34	-0.68	-1.03	-2.14	
7	0.60	0.23	-0.19	-0.40	-0.62	-0.90	-2.20

The question of the physical relevance of the 'parasitical' LE which arise if we increase the embedding dimension is still unsolved. But one of the most interesting problems is how to extract the correct negative LE from the parasitical in the general case. For discrete dynamics the situation is as above. For continuous systems the relevant negative LE do not appear on the diagonal of a triangle shown in table 3.2, but as a column, which means fixed index, if the embedding dimension is increased. The discussion of this extraction is controverse [16,17]. Unfortunately one cannot get all exponents in every situation. If the absolute value of the true negative LE is much higher than the positive LE, we will get flat manifolds. The dimensionality of the manifold is nearly an integer and smaller than the embedding dimension. So, the displacement vectors of fig.3.3 do not span the tangential space. This is the case for the Lorenz and the Rössler attractor.

In our experimental situation we have to test the following condition:

$$\dim_E - 1 \ll D_2 \le \dim_E ,$$

where $\dim_E$ is the smallest correct embedding dimension.

If this condition is satisfied, like for the Hénon attractor or Arnold's Cat-Map, we get the whole spectrum of LE though the experimental measurements were done with a 10-bit resolution.

The results of this algorithm are strongly influenced by the choice of the embedding parameters, e.g. the sampling time T_a, the reconstruction delay τ and the embedding dimension $\dim_E$, like all other actual methods.
To extract useful LE one has to choose the following parameters very carefully: the radius of the clouds of neighbouring points, the number of neighbouring points N_{nb} , the number of data points N_{dat}, the level of noise, the number of averaging intervals M and the evolution time t_{ev}.

3.3 THE DIMENSION

In contrast to the LE the dimension yields a more static property of our attractor. Before we calculate the dimensions we define the generalized dimension D_q [1].

a) <u>The generalized dimension D_q</u>

To determine the generalized dimension we subdivide our $\dim_E$ -dimensional phase space into cells of size $r^{\dim_E}$. The cells have the numeration i=1 to i=M(r). The probability of finding a point of the attractor in one of the M(r) cells is p_i. So we can write the generalized dimension D_q as

$$D_q = \lim_{r \to 0} \frac{1}{q-1} \frac{\log\left(\sum_{i=1}^{M(r)} p_i^q\right)}{\log r}$$

We prefer the following dimensions, because they are easy to implement on a small computer.

b) <u>The correlation dimension [18]</u>

For q = 2 we get the correlation dimension

$$D_2 \approx \lim_{r \to 0} \frac{\log(C(r))}{\log(r)} ,$$

where C(r) is the correlation integral, defined as

$$C(r) \approx \frac{1}{N_{Ref}} \sum_{j=1}^{N_{Ref}} \frac{1}{N} \sum_{i=1}^{N} \sigma(r - |x_i - x_j|),$$

σ is the Heaviside function, $N = N_{dat} - (\tau / T_a)(\dim_E - 1)$: the number of points in the phase space and N_{ref}: the number of reference points. In our calculations we use N_{dat} = 32768 data-points and N_{ref} = 1000 reference points.

c) <u>The pointwise dimension</u>

$$D_{pw} = \lim_{r \to 0} \lim_{N \to \infty} \frac{\log(\ C^j(r)\ )}{\log(\ r\ )} .$$

$C^j(r)$ is the correlation integral for the jth reference point.

d) <u>The Kaplan-Yorke-Dimension [19]</u>

The n-dimensional Lyapunov spectra has to be sorted in the form

$$\lambda_1 \geq \lambda_2 \geq \ldots \geq \lambda_n .$$

Then the Kaplan-Yorke-Dimension is defined as:

$$D_{KY} := \begin{cases} j + \dfrac{\sum_{k=1}^{j} \lambda_k}{|\lambda_{j+1}|} , & \sum_{k=1}^{n} \lambda_k < 0 \text{ where} \\ & \sum_{k=1}^{j} \lambda_k \geq 0 \text{ and } \sum_{k=1}^{j+1} \lambda_k < 0 \end{cases} .$$

4. Experimental results

4.1 <u>Scenario of a three torus</u>

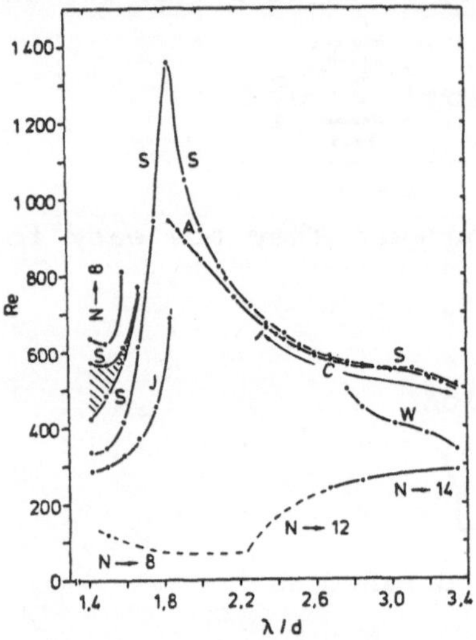

Fig. 4.1.
Stability diagram of the 10-vortex state (The modes: Antijet (A), Core (C), Jet (J), Wavy (W) and Snaily (S); N→..: change in number of vortices; λ/d: normalized vortex length) [3].

The type and the critical Reynolds numbers of oscillatory states in rotational Taylor-Couette flow depend sensitively on aspect ratio and vortex length, which is shown for a 10-vortex state in fig. 4.1. The scenario described below is observed at an aspect ratio of $\lambda/d = 3.2$, where the Wavy-, the Core- and the Snaily mode successively appear when the Reynolds number is increased. Fig. 4.2 shows the Poincaré sections. At Re =543 one sees the section of the Wavy mode. At Re = 553, 562, 578 the Core mode growth and a torus with a frequency ratio of $\omega_c/\omega_w \approx 7$ arise. The second row of sections shows tori with thick walls caused by a third mode, the Snaily mode ($\omega_s/\omega_w \approx 0.5$). At Re =583 the frequency ratio ω_s/ω_w is exactly 0.5. It is not a re-stabilization as mentioned in [20]. The spectra of LE for the last four states are (0,0,0,-).

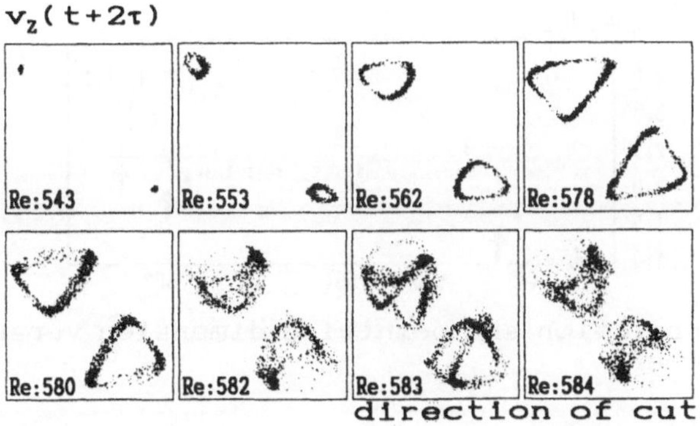

Fig. 4.2. Poincaré sections for successive increased Reynolds numbers

Careless embedding leads to wrong estimated LE as we will show in [9]. As an example for the estimation of the correlation dimension the local slopes from the double logarithmic plot of the correlation integral versus radius are shown for various embedding dimensions. For small values of R the influence of noise dominates and the phase space of any embedding dimension dim_E is entirely filled up giving for the slope the value of dim_E. For too large values of R the points of the attractor used in the algorithm come from the other wall or even from the other side of the torus. The right part of fig. 4.3 shows the proper choice of R, where for embedding dimensions larger than 4 the value of the slopes converges.

Fig. 4.4 displays the expected behaviour of the correlation and mean pointwise dimension. The arrows at Re ≈ 450 and Re ≈ 580 show the onset of the core mode and snaily mode, respectively, and the expected step in the value of dimensions. The third arrow marks a Reynolds number where the value of D_2 suggests a kind of restabilization, e.g. the disappearance of one oscillatory mode.

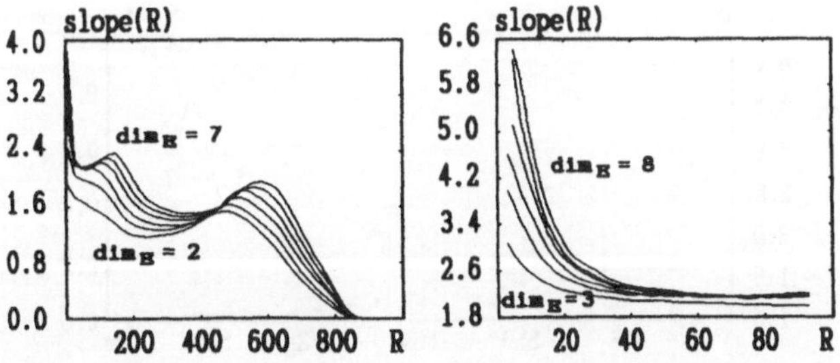

Fig. 4.3. Local slopes from the double logarithmic correlation integral (Re = 555). The left picture shows the scaling behaviour of the whole attractor and on the right picture the relevant region is blown up.

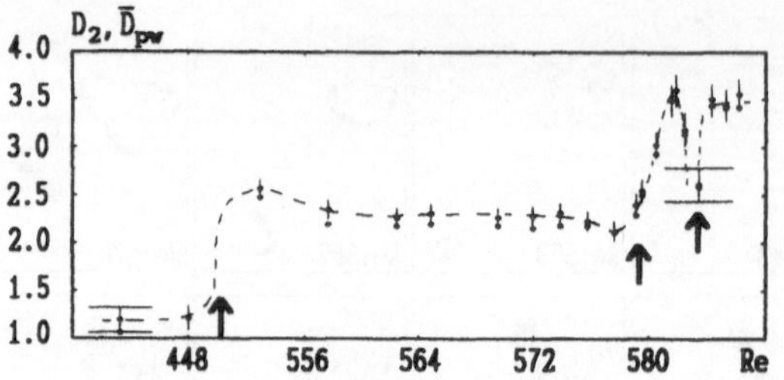

Fig. 4.4. Correlation and pointwise dimension versus Reynolds number.

Ploting the frequency ratios of the modes

$$v_1=(\omega_c/\omega_w) \bmod 1 \quad \text{and} \quad v_2=(\omega_w/\omega_s) \bmod 1$$

versus Reynolds numbers (fig. 4.5), it is obvious that the condition of incommensurability of the three frequencies is violated at this point. Strictly spoken, those commensurate frequencies appear infinitely often but only the smaller rational numbers can be resolved with our measurements. We conclude for this example, that the knowledge of dimensions alone leads to misinterpretations of the results.

4.2 Period-3-window

The period doubling cascade in Taylor-Couette flow has already been described in [4] and so we focus on the appearance of the period-3-window above the onset of chaos. Fig. 4.6 gives an idea of the attractor at three Reynolds numbers. For this example the various kinds of dimensions defined above and the LE display the expected dynamics very well, even at a negative LE (fig. 4.7 and 4.8). For completeness we want to stress the fact, that in the region of the period-3-window, contrary to what one would expect from the simple logistic equation, the period 6 does not appear. Instead we observe another period trebling, so we found at least one ninth of the fundamental frequency.

Fig. 4.9 shows the convergence of the estimated LE as a function of increasing embedding dimension for the period-3-window.

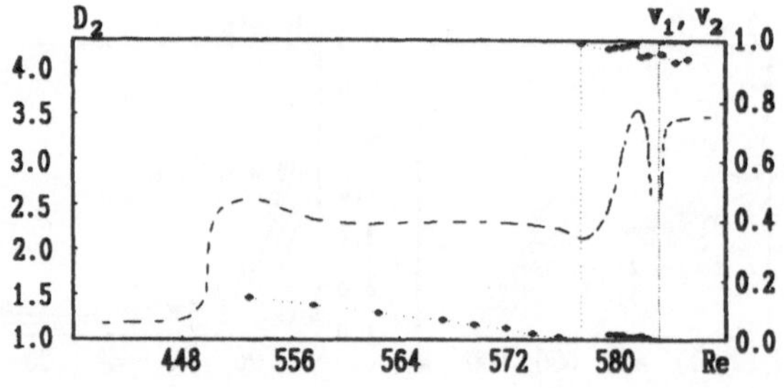

Fig. 4.5. Correlation dimension and frequency ratios versus Reynolds number.
• : v_1 ,
∘ : v_2 .

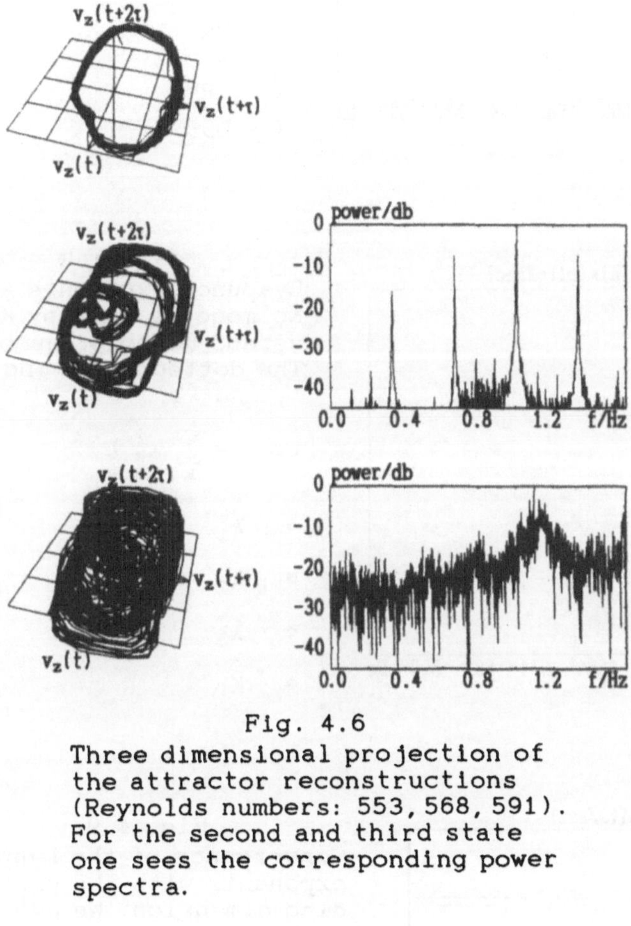

Fig. 4.6
Three dimensional projection of the attractor reconstructions (Reynolds numbers: 553, 568, 591). For the second and third state one sees the corresponding power spectra.

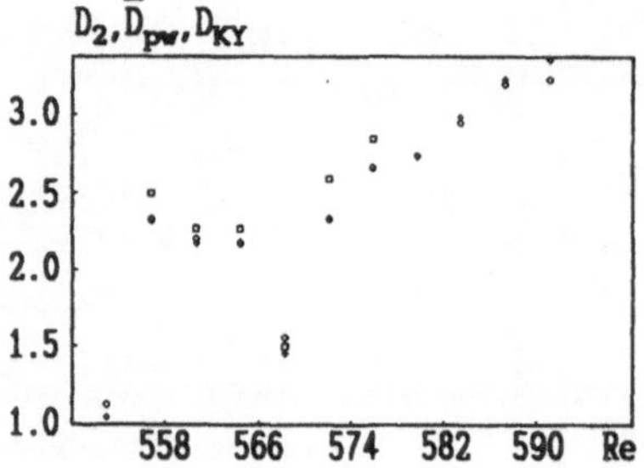

Fig. 4.7
Correlation dimension, averaged pointwise dimension and Kaplan-Yorke dimension versus Reynolds number.

○ : D_2

□ : $\bar{D}_{pw}$

◇ : D_{KY}

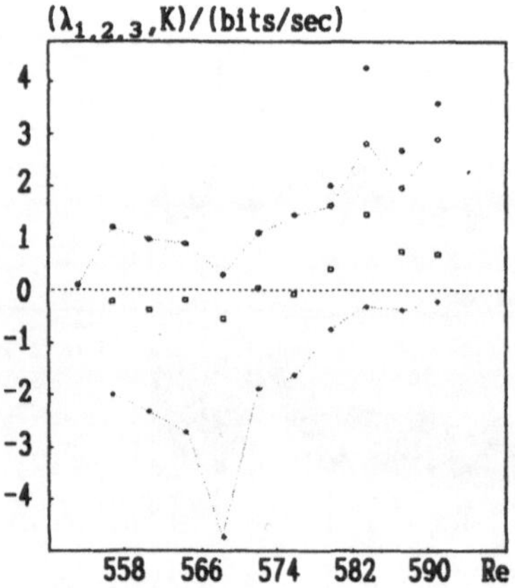

Fig. 4.8
Lyapunov exponents and Kolmogorov entropy K versus Reynolds number. The dotted lines indicate λ_1 and λ_3.

$(K = \sum_k \lambda_k^+)$

○ : λ_1

□ : λ_2

◇ : λ_3

• : K

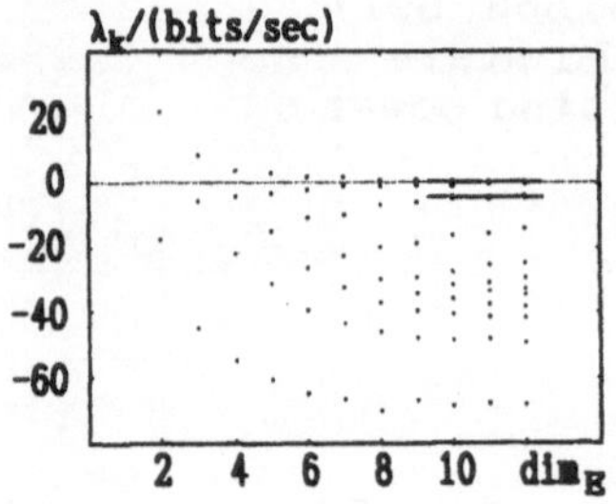

Fig. 4.9
Convergence of the Lyapunov exponents with the embedding dimension. We averaged the relevant exponents in the interval where the convergence is achieved (indicated by the solid lines), Re = 568.

4.3 Intermittent route nearby a homoclinic orbit

As a last example we applied the analysis on an intermittent route to chaos. Beside many other locations in the control parameter space this kind of scenario has been found for a one vortex state at aspect ratio $\Gamma = 1.2$ for Reynolds numbers of order 1600 (fig. 4.10). This region seems very interesting because we found two lines of homoclinic orbits, marked with HO1 and HO2, respectively, separated by a stable region. Details of this bifurcations will be published elsewhere [5].

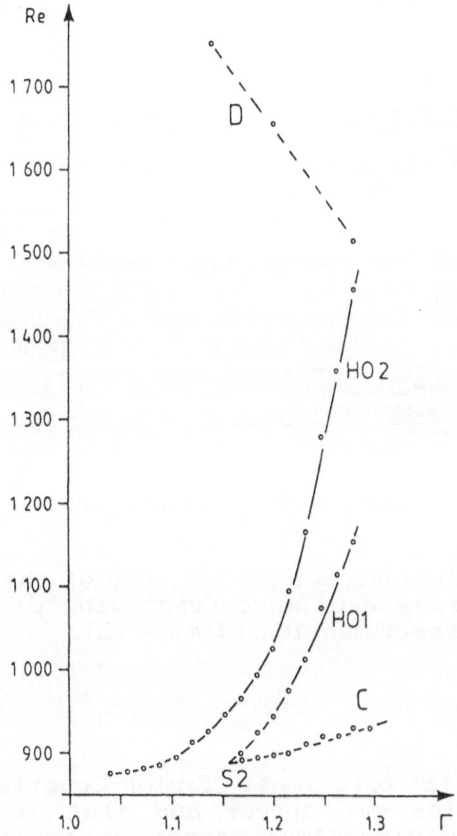

Fig. 4.10
Stability diagram of a m = 2 mode. HO1 and HO2 indicate homoclinic orbits. At line C one passes a Hopf bifurcation. Line D is the boundary to chaotic regime. (Γ: ratio of cylinder height and gap width) [5].

Fig. 4.11 shows the reconstructed phase spaces and the distributions of the pointwise dimensions calculated with $dim_E = 12$. The three figures show a noisy limit cycle, the intermittent region and the purely chaotic state. For a better understanding of the dynamics it makes no sense to plot a mean value of the pointwise dimensions or to give a value of the correlation dimension. For an inhomogenious attractor as in the intermittent region the distribution of the pointwise dimensions is the proper way to describe this data set.

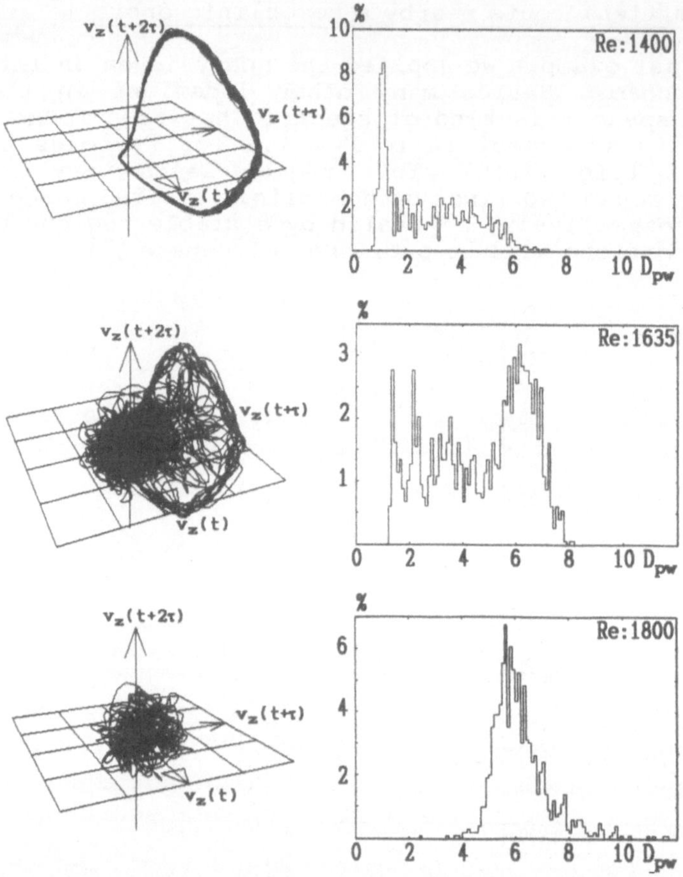

Fig. 4.11. Three dimensional projections of the attractor reconstructions and the corresponding distributions of the pointwise dimension ($dim_E = 12$).

5. CONCLUSION

We showed that in rotational Taylor-Couette flow extremely different scenarios can appear and that the algorithms for estimating LE and dimensions cannot be applied automatically without knowledge of the structure of the attractor. One must keep in mind that even the reconstruction is part of the algorithms and that a non proper treatment of the raw experimental data cannot be compensated later.

REFERENCES

[1] Schuster, H.G., Deterministic Chaos, An Introduction, VCH-Verlag, Weinheim (1988).

[2] Takens, F., in: Dynamical Systems and Turbulence, Lecture Notes in Math. **898**, D.A. Rand and L.-S. Young (eds.), Spinger Verlag, Warwick (1980).

[3] Gerdts, U., Dissertation, Universität Kiel (1985).

[4] Pfister, G., in: G.E.A. Meier and F. Obermeier (eds.), Flow of Real Fluids, Lecture Notes in Physics **235**, Springer Verlag (1985).
[5] Pfister, G. and Lensch, B., to be published.
[6] Pfister, G., Schmidt, H., Cliffe, K.A. and Mullin, T., J. Fluid Mech., **191**, 1 (1988).
[7] Pfister, G., Gerdts, U., Lorenzen, A. and Schätzel, K., in: Photon Correlation Techniques in Fluid Mech., Springer Verlag (1983).
[8] Liebert, W., Pawelzik, K. and Schuster, H.G., preprint, to be published in Phys. Rev. Lett. (1989).
[9] Buzug, Th., Reimers, T. and Pfister, G., in preparation.
[10] Chennaoui, A., Pawelzik, K., Liebert, W., Schuster, H.G. and Pfister, G., preprint (1989).
[11] Eckmann, J.P. and Ruelle, D., Rev. Mod. Phys., **57**, 617 (1985).
[12] Sano, M and Sawada, Y., Phys. Rev. Lett., **55**, 1082 (1985).
[13] Eckmann, J.P., Kamphorst, S.O., Ruelle, D. and Ciliberto, S., Phys. Rev., **34A**, 4971 (1986).
[14] Wolf, A., Swift, J.B., Swinney, H.L. and Vastano, J.A., Physica **16D**, 285 (1985).
[15] Vastano, J.A. and Kostelich, E.J., in: G. Mayer-Kress (ed.), Dimensions and Entropies in Chaotic Systems, Springer Verlag (1986).
[16] Stoop, R. and Meier, P.F., J. Opt. Soc. Am. **5B**, 1037 (1988).
[17] Holzfuss, J. and Lauterborn, W., Phys. Rev. **39A**, 2146 (1989).
[18] Grassberger, P. and Procaccia, I., Phys. Rev. Lett. **50**, 346 (1983).
[19] Frederickson, P., Kaplan, J.L., Yorke, E.D. and Yorke, J.A., J. Diff. Eqns. **49**, 185 (1983).
[20] Buzug, Th. , Reimers, T., Wilkening, V. and Pfister, G., Poster 41 at the Dynamics Days Düsseldorf (1989).

DYNAMICAL BEHAVIOUR OF TAYLOR VORTICES WITH SUPERIMPOSED AXIAL FLOW

Karl Bühler and Norbert Polifke

Institut für Strömungslehre und Strömungsmaschinen
Universität Karlsruhe
D-7500 Karlsruhe 1, Kaiserstr.12, West-Germany

We present theoretical and experimental results on the stability and time-behaviour of instabilities in circular Couette flow with superimposed axial flow. Linear stability theory is used within the small gap approximation to explain the stability and dynamics of the instabilities in form of ring and spiral vortices. Ring vortices can travel only in the direction of throughflow. In contrast, spiral vortices can be obtained either in a steady state or time-dependent travelling in the direction or in opposite direction of the throughflow. The travelling direction depends on the ratio of the Taylor number to the throughflow Reynolds number. With throughflow as an initial condition a new secondary instability is found at high Taylor numbers.

1. INTRODUCTION

The stability of the flow between two concentric cylinders has received considerable attention in nature and technology. This problem was first treated by Taylor [1] and extended by Goldstein [2] to the case with throughflow. Many different aspects of the spiral flow were reported in the papers [3,4,5]. Some references concerning the laminar-turbulent transition are given in a recent review [6]. Investigations of spherical gap flows with superimposed mass flux in meridional direction have shown, that spiral vortices exhibit a special time-behaviour [7]. It was therefore an open question, whether this behaviour is also true for spiral vortices in cylindrical gaps. In a recent investigation of the stability of spiral Poiseuille flow [5] such a behaviour was not observed. The present work is concerned with an analytical and experimental investigation of the stability and time-behaviour of spiral Couette flow.

2. LINEAR STABILITY THEORY

We developed the linear theory for the small gap limit to explain the stability and time-behaviour of the torodial and spiral Taylor vortices. The analysis is based on the idea of Oswatitsch[8] and further developments by Bühler[9,10]. The problem is described with local cartesian coordinates as shown in Fig.1 and Fig.2 with the centrifugal force incorporated. The Coriolis force can be neglected in the small gap limit. The nondimensional perturbation equations leads to the following parameters:

$$Re = \frac{R_1 \omega_1 s}{\nu} \quad , \quad Re_D = \frac{Ws}{\nu} \quad , \quad Ta = Re \cdot \delta^{\frac{1}{2}} \tag{2.1}$$

Nonlinear Evolution of Spatio-Temporal Structures in Dissipative Continuous Systems
Edited by F.H. Busse and L. Kramer
Plenum Press, New York, 1990

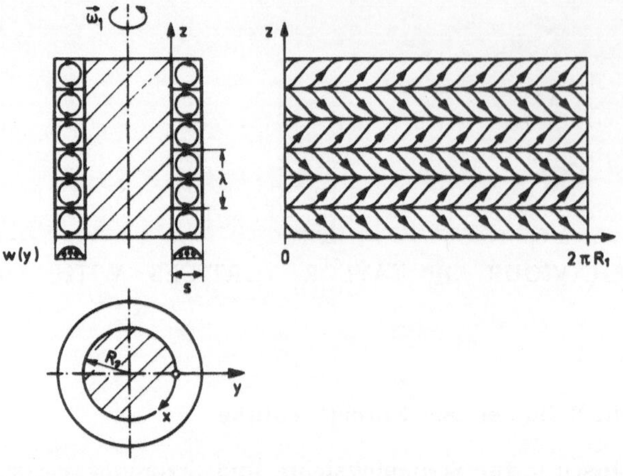

Fig.1. Toroidal structure of the Taylor vortices with and without axial flow.

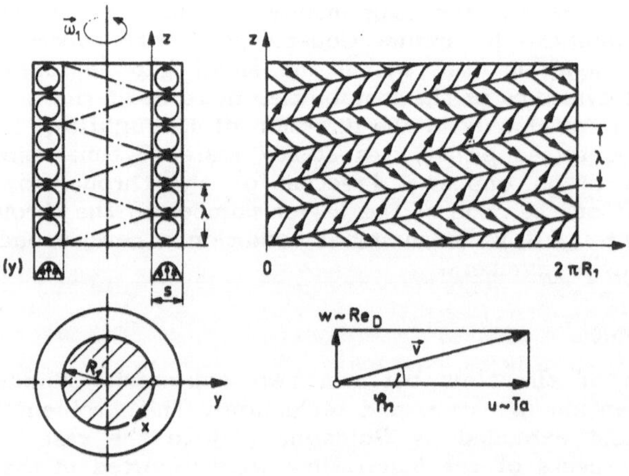

Fig.2. Spiral structure of the Taylor vortices with superimposed axial flow, small aspect ratio.

with $\delta = s/R_1$ as nondimensional gap width. W is the volumetric mean axial velocity. The perturbation equations allow a solution in normal mode form:

$$(u'', v'', w'', p'') = \left[\bar{u}(y'), \bar{v}(y'), \bar{w}(y'), \bar{p}(y')\right] \exp\left(i\alpha_1 x' + i\alpha_2 z' + \sigma t'\right) \qquad (2.2)$$

The wave numbers α_1 and α_2 are responsible for the periodicity in x' and y' direction. The wave numbers are given by $\alpha_1 = n \cdot s/R_1 = n \cdot \delta$ in the circumferential direction and $\alpha_2 = 2\pi \cdot s/\lambda$ in the axial direction. The inclination angle φ_n of the spiral vortices in Fig.2 follows as

$$\varphi_n \simeq \tan \varphi_n = n \frac{\lambda}{2\pi R_1} = \frac{\alpha_1}{\alpha_2} \qquad (2.3)$$

The existence condition for nontrivial solutions of the pertubation equations leads to the characteristic equation:

$$F(Ta, Re_D, \delta, \alpha_1, \alpha_2, \sigma) = 0 \qquad (2.4)$$

In the case of neutral - stability the complex eigenvalue $\sigma = \beta + i\gamma$ has a vanishing real part $\beta = 0$. From the condition of real solutions of the equation (2.4) we obtained the characteristic eigenvalue relations. An analytical solution of (2.4) is possible for "stress - free" boundary conditions:

$$Ta^2 = \frac{(\pi^2 + \alpha_2^2)^3}{\alpha_2^2} \quad , \; Ta_c = 25.642 \; , \; \alpha_{2c} = \frac{\pi}{\sqrt{2}} \tag{2.5}$$

$$-\gamma = \frac{1}{2}\alpha_1 \, Ta \, \delta^{-\frac{1}{2}} + \frac{3}{2}\alpha_2 \, Re_D \tag{2.6}$$

In the assumption of vanishing gap width the stability behaviour (2.5) is independent of the wave number α_1 and therefore independent of the inclination. This fact is confirmed by symmetry arguments, which show that the stability behaviour must be independent of the direction of the rotation and of the throughflow. In contrast to this, the equation (2.6) displays the fact, that steady solutions with $\gamma = 0$ are possible for some Re_D depending on the wave numbers α_1 for a finite gap width. For the case of "rigid" boundary conditions we obtain the critical eigenvalues for the stability behaviour:

$$Ta_c = 41.325 \; , \quad \alpha_{2c} = 3.117 \tag{2.7}$$

The eigenvalue spectrum $Ta^2 = Ta^2(\alpha_2)$ is given in Fig. 3. The result for the eigenvalue γ is identical with the expression (2.6). In Fig. 4 the eigenvalues $-\gamma$ are plotted as function of the throughflow Reynolds number Re_D for different inclinations, given by the integer n. Steady solutions with $\gamma = 0$ at the critical values (2.7) and the gap width $\delta = s/R_1 = 0.25$ used in the experiments are possible for the throughflow Reynolds numbers corresponding to the inclinations n:

n	0	1	2	3
Re_D	0	2.210	4.419	6.629

(2.8)

Besides these steady states, the solutions in the neighbourhood are time-dependent. The eigenvalue $-\gamma$ is proportional to the velocity w_c of the travelling vortices in z' - direction. It follows, that the ring vortices with n=0 have a positive axial velocity $w_c \sim -\gamma \sim Re_D$, which is proportional to the throughflow Reynolds

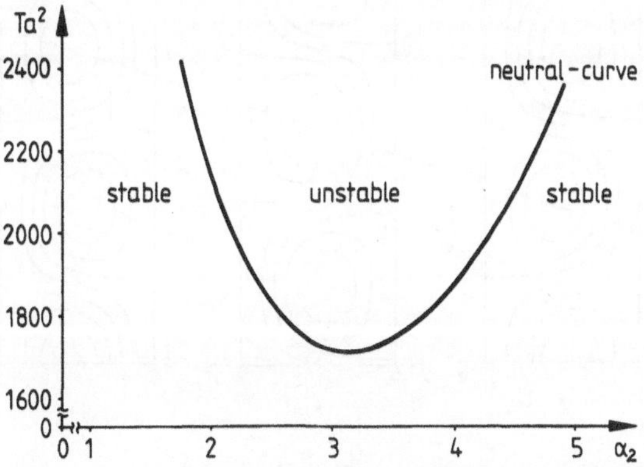

Fig. 3. Eigenvalue spectra for cylindrical gap flows. Critical Taylor number Ta^2 as function of the axial wave number α_2.

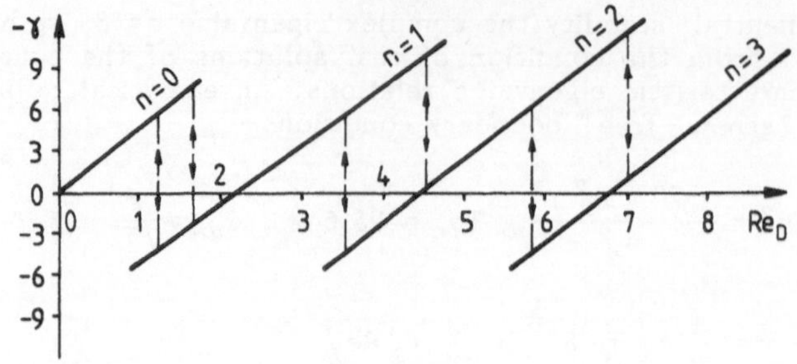

Fig.4. Eigenvalue $-\gamma$ as function of the throughflow Reynolds number Re_D for different inclination values n, the arrows mark possible transitions between the modes with different inclinations.

number. Starting with steady spiral vortices $n=1$ and the corresponding Re_{D1} we find with increasing $Re_D > Re_{D1}$ spiral vortices travelling in the direction of throughflow with $w_c > 0$. For decreasing Reynolds numbers $Re_D < Re_{D1}$ the spiral vortices travel in the opposite direction of the throughflow. The same behaviour is possible for all spiral vortices with different inclinations. Transitions between the different modes are possible and marked by arrows in Fig.4. The values of Re_D at which the transition occurs, cannot determined within the linear stability theory. The flow structure of the time-dependent ring vortices in the meridional plane is plotted in Fig.5 at different times of one period. The streamfunction is obtained from the eigenfunctions of the linear stability analysis.

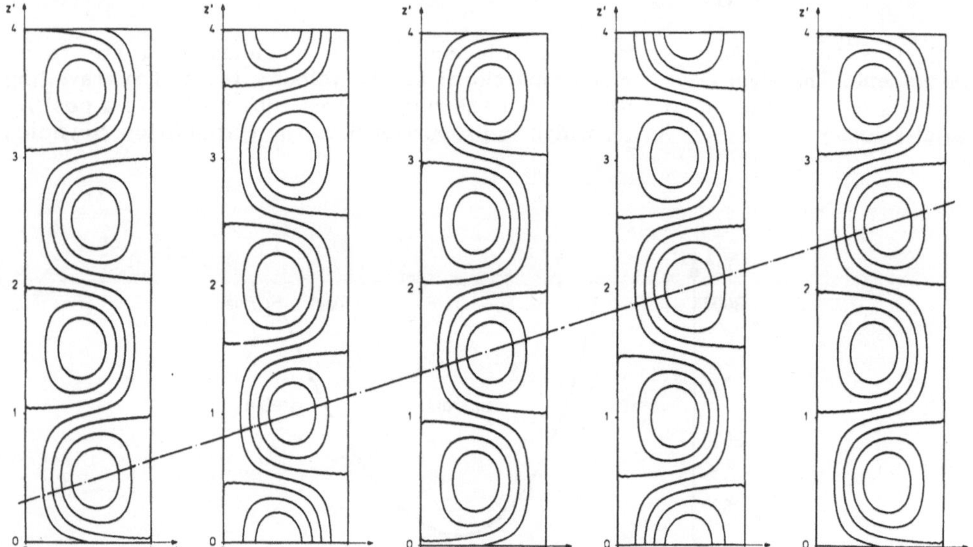

Fig.5. Meridional flow structure with lines of constant streamfunction in the z', y'- plane, travelling ring vortices at different times of one period, the axial wave length is $\alpha_2 = 3.117$, $\alpha_1 = 0$, $\delta = 0.1$.

3. EXPERIMENTS

The experimental apparatus is shown schematically in Fig.6. The inner cylinder rotates while the outer transparent one is at rest. The inlet is at the lower and the outlet at the upper part. The mass flux through the gap in axial direction is realized by an external circulation, driven by a pump. Silicon oil with a kinematic viscosity $\nu = 40 \cdot 10^{-6}\ m^2/s$ was used as test fluid. Small aluminium flakes were served into the fluid to visualize the flow structure.

The experiments are focused on the stability of the spiral basic flow, the structure of the instabilities and their dynamical behaviour. We obtain toroidal ring vortices $n=0$ with a structure as shown principally in Fig.1. In the presence of throughflow, the ring vortices travel in the direction of the axial velocity. At higher throughflow rates, spiral vortices with different inclinations proportional to the integer n are observed. These spiral vortices can either be steady or can move in the direction of throughflow ($\uparrow$) or in opposite direction of the throughflow ($\downarrow$). The existence regions of the vortex configurations are given in Fig.7. The arrows in Fig. 7 indicate the direction of the travelling vortices. The Taylor number Ta is increased with fixed throughflow Reynolds number Re_D. Steady ring vortices as shown in Fig. 8 occur above the critical Taylor number $Ta \geq Ta_c = 47.5$ without throughflow. Time-dependent ring vortices $n = 0$ are observed in the range $0 < Re_D < 2.7$. Fig. 9 shows, that in this case the size of the vortices is changed alternately in larger and smaller ones. At higher Taylor numbers the wavy mode in circumferential direction is superimposed on the travelling vortices as can be seen in Fig.10. In the range $2.7 < Re_D < 4.6$ we find steady spiral vortices $n=1$ as shown in Fig. 11 and 12. The time-dependence of this spiral mode at different Taylor numbers is caused by the variation of the axial wavelength λ. If we decrease the throughflow rate, the spiral vortices $n = 1$ are travelling in opposite direction as can be seen in Fig.13. Spiral vortices $n=2$ are observed in the range $6 < Re_D < 13$. A steady state is shown in Fig.14, while the spiral vortices in Fig.15 are time-dependent. Spiral vortices with higher inclination $n=3$ are observed for $Re_D > 13$. In this region the spiral basic flow becomes first unstable with travelling ring vortices, which change into the spiral mode at higher Taylor

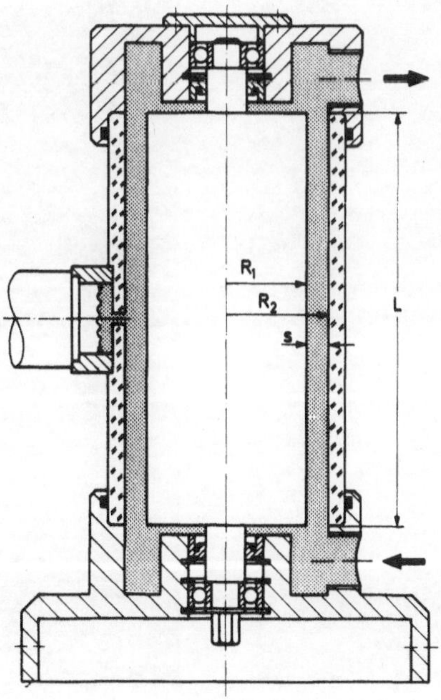

Fig.6. Principal sketch of the experimental apparatus, geometrical data $R_1 = 24$ mm, $s = R_2 - R_1 = 6$ mm, $L = 120$ mm.

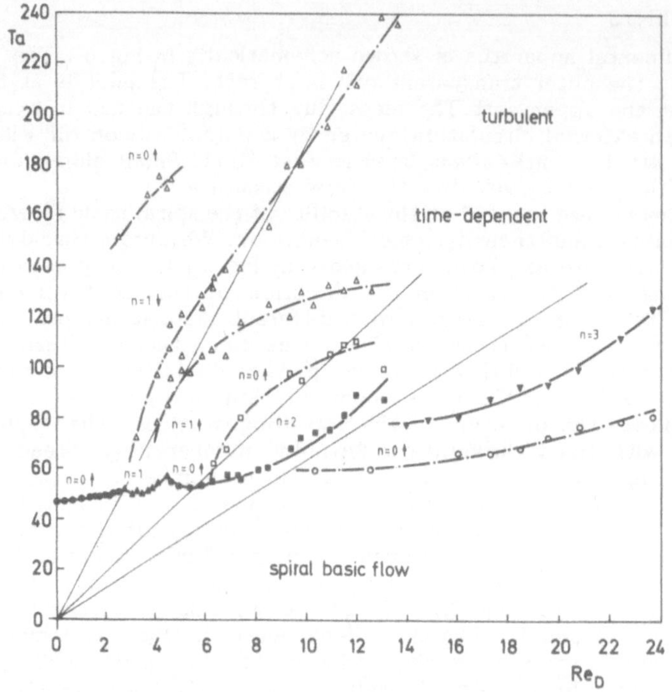

Fig.7. Stability diagram for the spiral flow in cylindrical gaps and existence regions for steady and time-dependent supercritical modes, $\delta = s / R_1 = 0.25$.

Fig.8. Steady toroidal ring vortices
$Ta = 51$, $Re_D = 0$

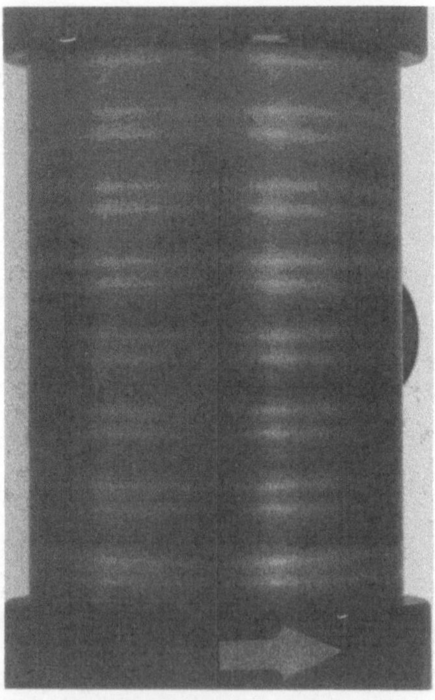

Fig.9. Time-dependent ring vortices
$n = 0$↑, $Ta = 67.6$, $Re_D = 1.3$

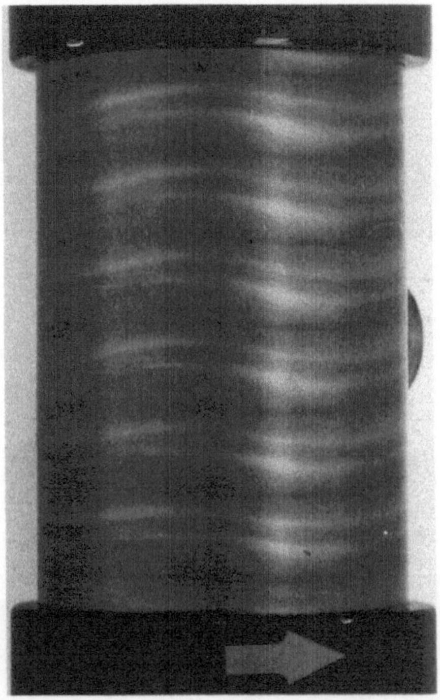

Fig.10. Travelling ring vortices with superimposed wavy mode $n=0\uparrow$, $Ta=98.2$, $Re_D=1.3$

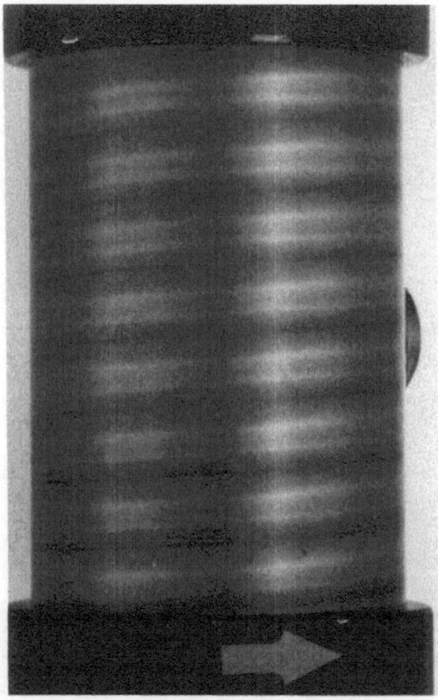

Fig.11. Steady spiral vortices $n=1$, $Ta=61.5$, $Re_D=3.6$

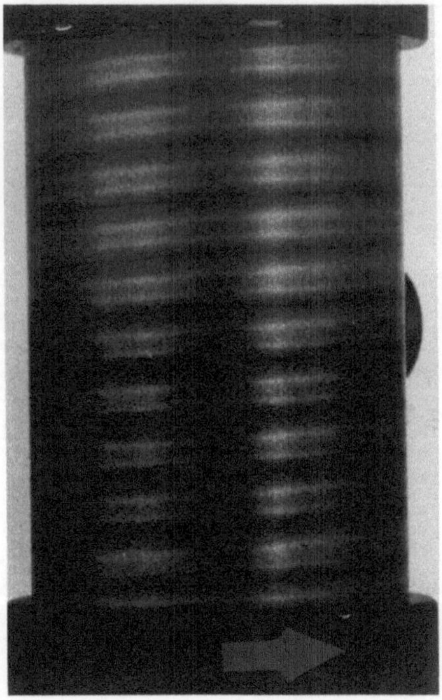

Fig.12. Steady spiral vortices $n=1$, $Ta=84.7$, $Re_D=3.8$

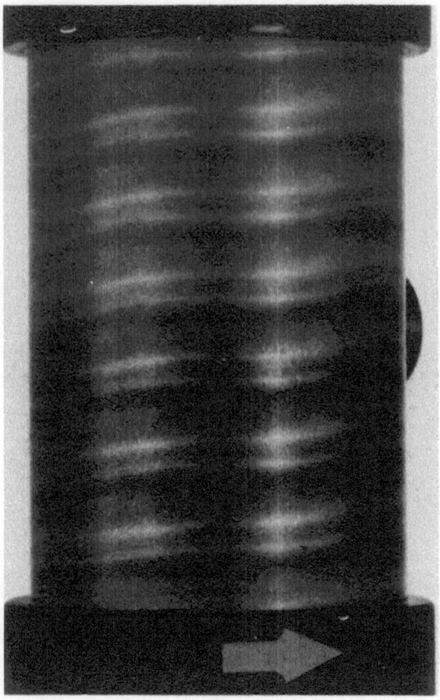

Fig.13. Travelling spiral vortices $n=1\downarrow$, $Ta=85$, $Re_D=2.2$

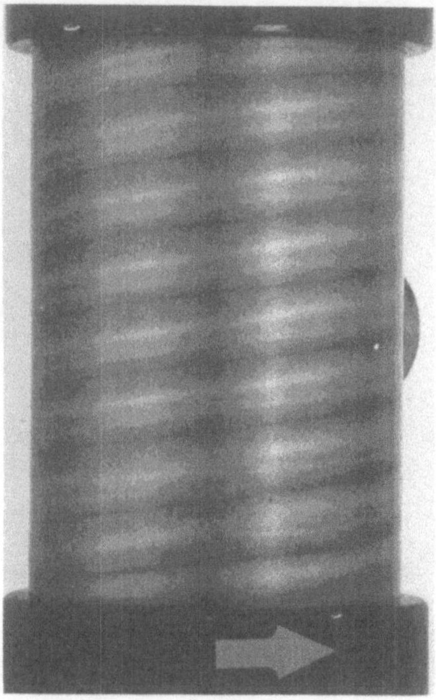

Fig.14. Steady spiral vortices n = 2 , Ta = 66 , Re_D = 8.0

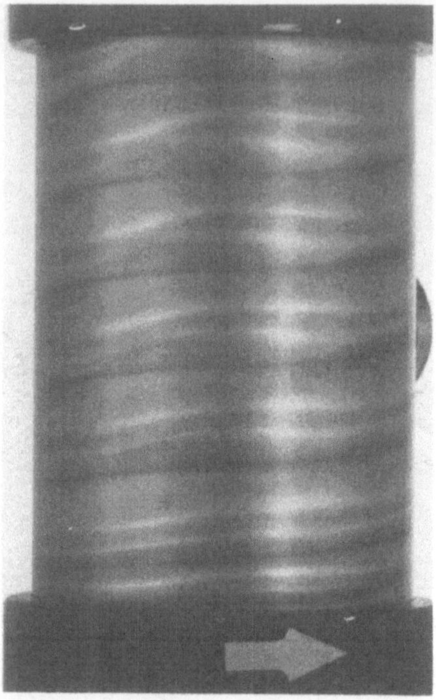

Fig.15. Travelling spiral vortices n = 2↓ , Ta = 102 , Re_D = 7.8

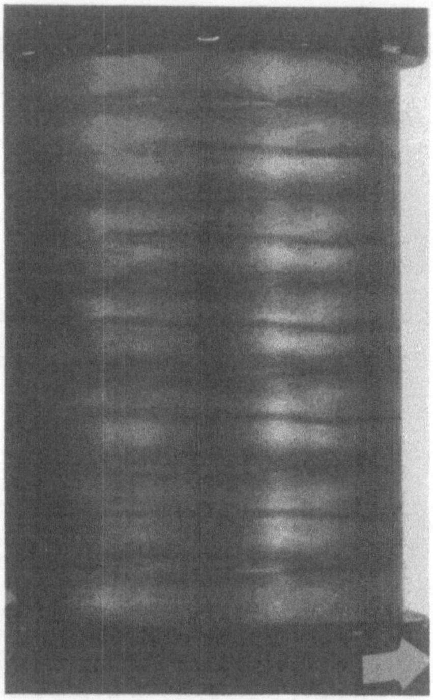

Fig.16. Ring vortices with secondary instabilities, small axial wavelength, Ta = 660, Re_D = 0

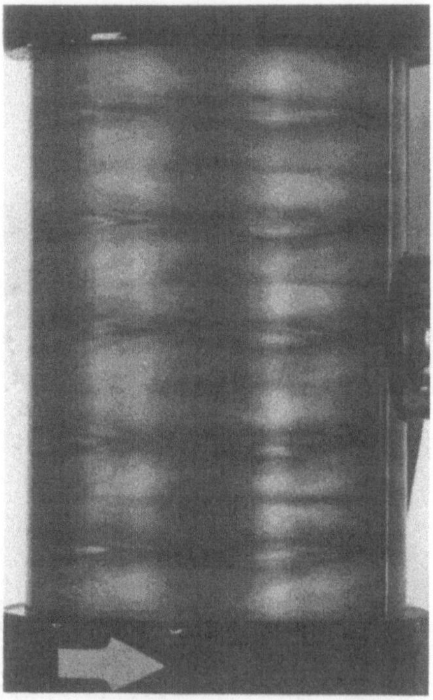

Fig.17. Ring vortices with secondary instabilities, large axial wavelength, Ta = 668, Re_D = 0

numbers. The Ta-Re_D relation for steady spiral vortices obtained from heuristic considerations agrees with the relation (2.6) for steady solutions with $\gamma=0$ and is plotted in the stability diagram of Fig.7 for three inclinations. The spiral vortices are steady, if their inclination is parallel to the velocity vector in the middle of the gap. The lines in Fig.7 coming from the origin cross the regions of the steady spiral modes. The variation of the axial wavelength is responsible for the fact that steady spiral modes occur in a region around these lines. An apparently new secondary instability is obtained for high Taylor numbers. The structure of these instability is shown in Fig.16 and 17 between the Taylor vortices at different axial wavelength. Their structure is clearly indicated by pressure measurements in [11].

4. CONCLUSIONS

The stability and time-behaviour of the flow in a cylindrical gap with superimposed axial mass flux is studied both theoretically and experimentally. Interesting experimental facts with respect to the stability and time-behaviour can be explained with the linear theory. These include the increase of the critical Taylor number with variing wavelength as well as the dynamical behaviour of the toroidal and spiral vortices. The spiral vortices can either be steady or can travel in direction of the throughflow or in opposite direction. The new instability at high Taylor numbers needs further investigations to explain their physical nature.

5. REFERENCES

[1] Taylor G.I., Stability of a viscous liquid contained between two rotating cylinders. Proc.Roy.Soc. (A) 223, 289 - 343 (1923)

[2] Goldstein S., The Stability of viscous fluid flow between rotating cylinders. Camb.Phil.Soc.33, 41-61 (1937)

[3] Snyder H.A., Experiments on the stability of spiral flow at low axial Reynolds numbers. Proc.Roy.Soc. (A) 265, 198-214 (1962)

[4] Schwarz K.W., Springett B.E., Donnelly R.J., Modes of instability in spiral flow between rotating cylinders. J.Fluid Mech. 20, 2, 281-289 (1964)

[5] Takeuchi D.I., Jankowski D.F., A numerical and experimental investigation of the stability of spiral Poiseuille flow. J.Fluid Mech. 102, 101-126 (1981)

[6] Swinney H.L., Gollub J.P., Hydrodynamic instabilities and the transition to turbulence. Topics in Appl.Physics Vol.45, Berlin, Springer, 1981

[7] Bühler K., Strömungsmechanische Instabilitäten zäher Medien im Kugelspalt. Fortschritt-Berichte VDI, Reihe 7, Nr.96, Düsseldorf 1985

[8] Oswatitsch K., Physikalische Grundlagen der Strömungslehre. Handbuch der Physik, Bd. VIII/1, Hrsg. S. Flügge, Berlin, Springer, 1959

[9] Bühler K., Ein Beitrag zum Stabilitätsverhalten der Zylinderspaltströmung mit Rotation und Durchfluß. Strömungsmechanik und Strömungsmaschinen 32, 35-44 (1982)

[10] Bühler K., Der Einfluß einer Grundströmung auf das Einsetzen thermischer Instabilitäten in horizontalen Fluidschichten und die Analogie zum Taylor-Problem. Strömungsmechanik und Strömungsmaschinen 34, 67 - 76 (1984)

[11] Bühler K., Instabilitäten spiralförmiger Strömungen im Zylinderspalt. ZAMM 64, T180-T184 (1984)

EXTERNALLY MODULATED HYDRODYNAMIC SYSTEMS

Russell J. Donnelly

Department of Physics
University of Oregon
Eugene, Oregon 97403

ABSTRACT

This report covers a series of experiments on hydrodynamic stability of systems subject to modulation of one parameter. Examples are Taylor-Couette flow with the speed of one cylinder periodically varied and Bénard convection with the temperature of one boundary periodically modulated. Correspondence with theory and outstanding difficulties are discussed.

I. INTRODUCTION

The stability of mechanical systems subject to parametric oscillations has a long history. Dynamic stabilization and destabilization can lead to dramatic modifications of behavior depending on conditions of tuning of the amplitude and frequency of modulation. The first examination of the stability of a hydrodynamic system subject to parametric modulation appears to have been an experiment by Donnelly, Reif and Suhl (1962), and Donnelly (1964), who examined the stability of Taylor-Couette flow with the inner cylinder modulated with some amplitude and frequency about a mean angular velocity $\bar{\Omega}$. The response of the system as determined from the radial component of disturbance u_r, appeared to be the production of a low amplitude secondary flow which occurred before the unmodulated onset of instability Ω_c, followed by a switch from a linear dependence of u_r on $\bar{\Omega}$ to a Landau-law dependence with u_r proportional to $(\bar{\Omega}^2 - \bar{\Omega}_c^2)^{1/2}$. Since we had just discovered the square root dependence in normal Taylor-Couette flow (Donnelly and Schwarz, 1963), I was drawn to the idea that the onset of the square-root law should signal the onset of instability. It appears in retrospect that a better choice would have been the location of the *first* departure from Couette flow; which suggest modulation *destabilizes* rather than *stabilizes* the flow, as was soon found by theorists. In reviewing the status of the subject Davis (1976) suggested that further experiments were needed, and our group has now performed a number of experiments in both Taylor-Couette and Rayleigh-Bénard flows. This conference seems an appropriate time to take an overview of the situation to see where we stand today. I shall confine my remarks mostly to our own experiments, recognizing that other groups have made important and significant contributions as well.

Nonlinear Evolution of Spatio-Temporal Structures in Dissipative Continuous Systems
Edited by F.H. Busse and L. Kramer
Plenum Press, New York, 1990

One common notion among all the experiments I shall discuss is the Stokes layer which characterizes the penetration depth of viscous waves near an oscillating plane, or thermal fluctuations near a periodically heated plane. The characteristic lengths involved are defined in Figure 1.

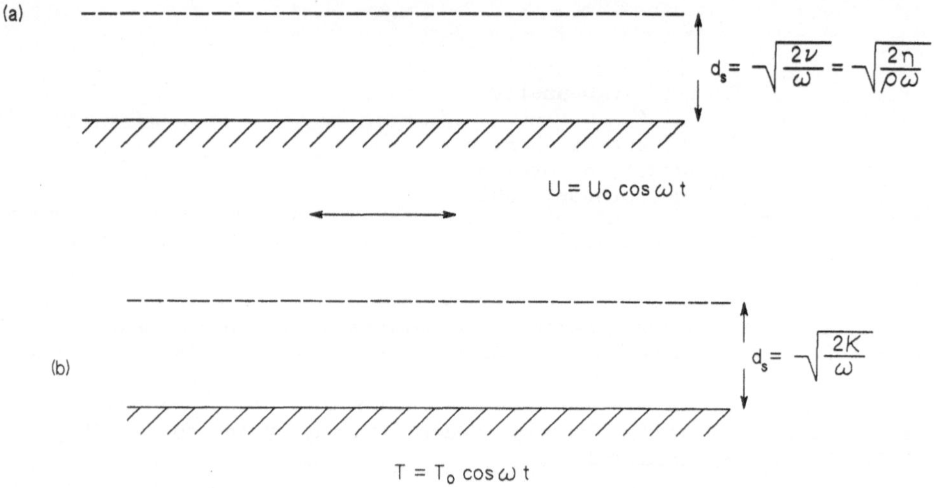

Figure 1. Stokes Layers. (a) Motion near an oscillating plane. (b) Temperature variations near a periodically heated surface.

II. MODULATED TAYLOR-COUETTE SYSTEMS

Figure 2 shows three different types of experiments which we have performed. Given the confusion in interpreting the earliest experiment, we decided to try the very simplest experiment we could think of--namely the Stokes flow near an oscillating cylinder in an unbounded fluid (Park, Barenghi and Donnelly, 1980), for which the stability theory had been done by Seminara and Hall (1976).

The Taylor number for this system is defined by Seminara and Hall (see Figure 2) as

$$T = \frac{2\Delta^2}{R} \sqrt{\frac{\omega}{\nu}}$$

providing the thickness of the Stokes layer is much less than R. Seminara and Hall obtained a critical value of $T_c = 232.5$ and a critical wavenumber $k_c = 0.86$, where k is defined in terms of the wavelength of the vortices λ by $k = 2\pi\delta/\lambda$.

Our apparatus consisted of cylinders 2.40, 2.22 and 1.00 cm in radius, all 27.5 cm long and placed vertically in a glass fish tank 40 cm x 20 cm x 22 cm in height. Motion was imparted by a computer-controlled stepping motor and the flow visualized by an indicator dye technique. Using different cylinders and frequencies, we obtained the data of Figure 3. At large R/δ, where the analysis is expected to be valid, we obtained $k_c = 0.88$ and $T_c = 246 \pm 5$ in reasonable agreement with Seminara and Hall. There was, however, a surprise.

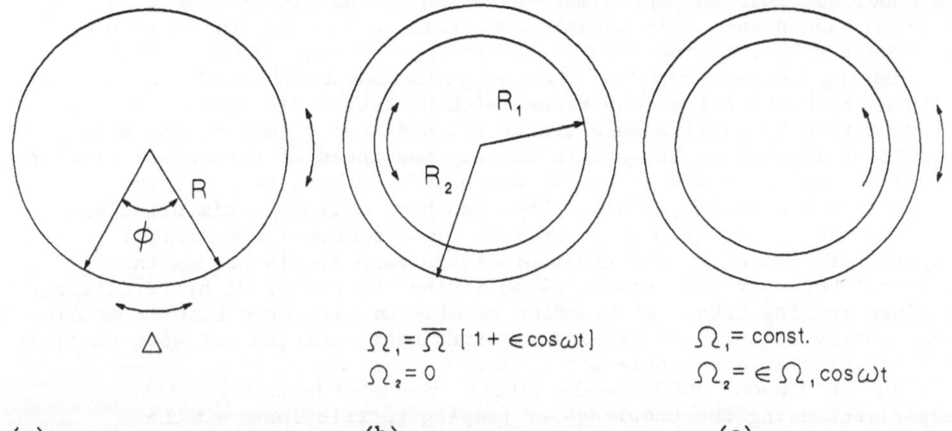

Figure 2. (a) Cylinder in infinite fluid.
(b) Modulated inner cylinder, oscillating about a mean speed $\bar{\Omega}$.
(c) Steady rotation of inner cylinder, outer cylinder rotating about zero mean.

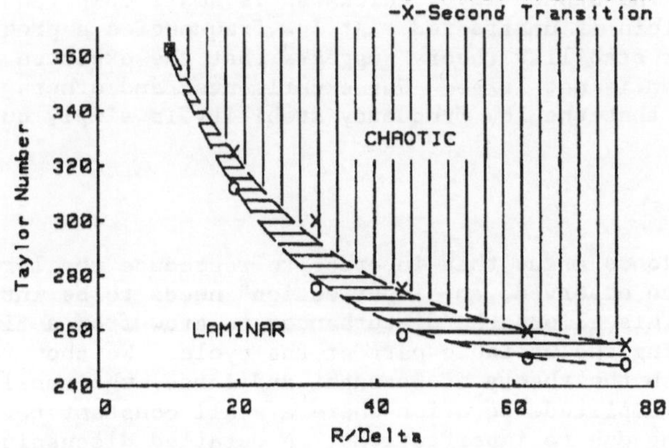

Figure 3. Two transitions in the flow near an oscillating cylinder. The first is from laminar flow to vortices in the boundary layer; the second is a snap-through instability with a subharmonic spatial bifurcation.

The flow on increasing Δ beyond its critical value is remarkable. Just beyond the first transition the vortices are regular and comparable in scale with the Stokes thickness. After a while, however, the vortex array is modulated by some long wavelength and eventually the vortices are destroyed in pairs--a spatial subharmonic destabilization. This stimulated Philip Hall to examine the stability of the vortex flow and he concluded that an approximate solution of the linear stability problem showed the cells becoming unstable at $T_c = 260$ in remarkable agreement with the experimental result $T_{c2} = 256 \pm 5$. (Hall, 1981)

During the next several years we conducted a series of experiments with K. Park and gained experience with photoelectric measurements of flows marked by Kalliroscope tracer (fish flakes). One of the most important lessons we learned is to ramp the speed of the inner cylinder slowly enough to avoid an acceleration-induced hysteresis (Park, Crawford and Donnelly, 1981). These authors defined a dimensionless ramping rate a^* which we have used to characterize our modulated experiments as well. The procedure is to ramp slowly across the critical boundary and return, establishing the degree of hysteresis for a given ramping rate. a^* is often so slow in real time that we do not necessarily believe the results of experiments carried out without first establishing an acceptable a^*.

Walsh, Wagner and Donnelly (1987), repeated Donnelly's 1964 experiment using the knowledge of ramping restrictions and the convenience of computer controlled stepping motors and data acquisition systems. The results were plots of critical Reynolds number R_c as a function of $\gamma = d/\delta = (d^2\omega/2\upsilon)^{1/2}$ the dimensionless modulation frequency where d is the width of the gap between the cylinders. Figure 4 shows the results for three radius ratios $\eta = R_1/R_2 = 0.95$, 0.88 and 0.719 plotted as reduced Reynolds number

$$\Delta = (R_c - R_{co})/R_{co}$$

vs dimensionless frequency γ. At high γ, modulation has little effect on the flow. The Stokes layer thickness is small compared to d and the bulk of the fluid is unaffected. At low frequencies a problem arises because linear stability theory suggests that the expected destabilization is not large. Our experiments, and others (Thompson, 1969) suggest that the low frequency stability is simply quasi-static, i.e.

$$R_c \to R_c(1+\epsilon)$$

Barenghi and Jones argue that in order to reproduce the large destabilization observed, an "imperfection" needs to be introduced into the theory. This allows the disturbances to grow from a finite amplitude during the unstable part of the cycle. We show in Figure 5 a comparison with the theory of Barenghi and Jones, who consider an inhomogeneous amplitude equation where a small constant term c accounts for the forcing due to imperfections. A detailed discussion of various theories and comparison with experiments is contained in Barenghi and Jones (1989) and Kuhlmann, Roth and Lücke (1989).

Walsh and Donnelly (1988) report an experiment where the inner cylinder rotation is steady, but the outer cylinder is oscillated about zero mean (see Figure 2(c)). This flow was examined theoretically by Carmi and Tustaniskyi (1981) who predicted that the oscillating outer cylinder should destabilize the flow. The very first qualitative experiments with this apparatus showed that instead, oscillation of the outer cylinder stabilizes the flow, especially at low frequencies. This is not surprising: as Taylor (1923) first showed, the flow is

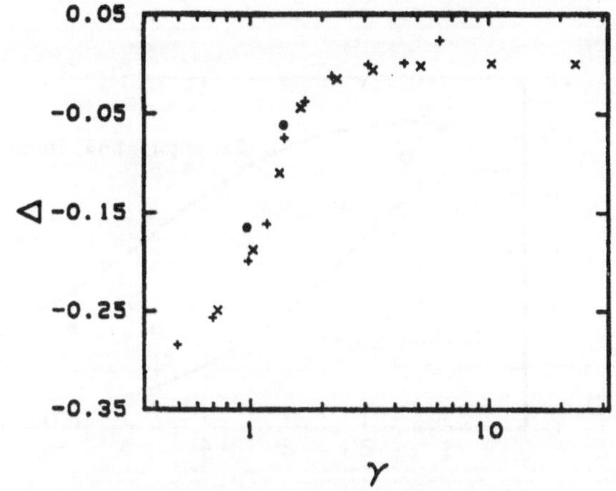

Figure 4. Reduced Reynolds numbers Δ as a function of dimensionless frequency of modulation for three radius ratios, $\eta = 0.95$ (circles), $\eta = 0.88$ (plusses) and $\eta = 0.719$ (crosses).

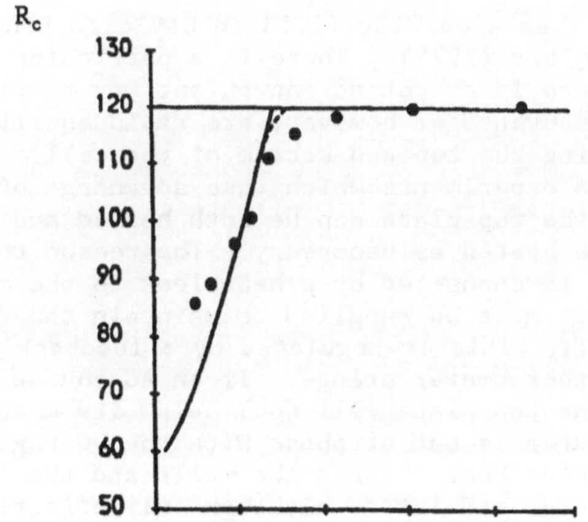

Figure 5. Comparison of destabilization at low frequencies with the theory of Barenghi and Jones with an imperfection term $c = 0.001$.

stabilized with respect to the inner cylinder rotating alone, if the outer cylinder is rotated in either direction. The theory has been studied by Barenghi and Jones (1989) and by Wu and Swift (1989). Both capture the essential trend and are shown in Figure 6.

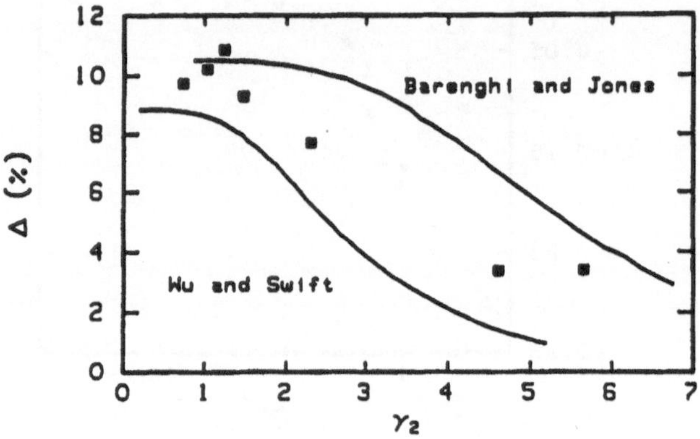

Figure 6. Comparison of the fractional stabilization Δ produced by modulation of the outer cylinder around zero mean, at dimensionless frequency γ_2 (solid squares), compared with recent calculations of Wu and Swift, and Barenghi and Jones.

III. MODULATED RAYLEIGH-BÉNARD SYSTEMS

Liquid helium I as a working fluid in Rayleigh-Bénard convection was pioneered by Ahlers (1975). There is a particular disadvantage to using helium I--there is as yet no convenient way to visualize the flow. Countering this disadvantage, however, are rapid equilibrium times for metal plates defining the top and bottom of the cell. We show in Figure 7 a class of Bénard experiments which take advantage of the cryogenics in a unique way: the top plate can be both heated and cooled, while the bottom plate can be heated as necessary. The reason this can be done is that the top plate is connected by a heat leak to the main bath which is below 2K. Thus heat must be supplied to maintain the top plate at some temperature above T_λ. This is regulated by a feedback circuit working from a resistance thermometer bridge. If an AC source is put in series with the bridge, the top plate will cool below its mean temperature whenever the AC source is out of phase with the DC regulating current. The bottom plate loses heat through the walls and the fluid to the top plate, and can also be modulated, although only effectively at low frequencies.

With the apparatus, Niemela and Donnelly were able to investigate the stability of a Stokes layer near the top plate at a high enough frequency that the characteristic thickness $\delta \ll d$, where d=0.41 cm is the plate spacing.

The convective onset is observed experimentally by plotting $\Delta T = T_L - T_u$, the measure of convective contributions to the heat flux, against the Rayleigh number, Ra, as the latter is increased in value. The Rayleigh number is here defined as

$$Ra = g\alpha T_o d_s^3/\kappa\nu$$

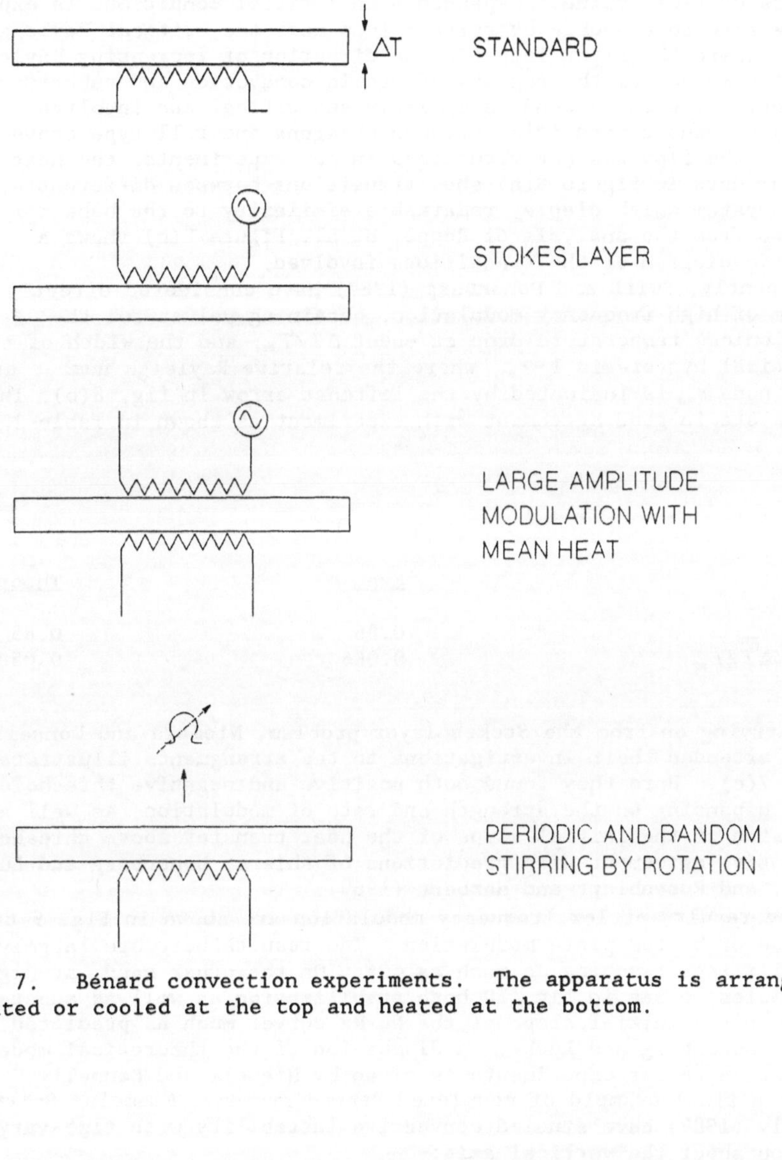

Figure 7. Bénard convection experiments. The apparatus is arranged to be heated or cooled at the top and heated at the bottom.

where g is the acceleration of gravity, α, κ, and ν are the thermal expansion coefficient, thermal diffusity, and kinematic viscosity of the fluid. Additionally, the Prandtl number is defined as $\sigma = \nu/\kappa$. It turns out that the heat transfer ΔT can be normalized to T_{oc} the amplitude of modulation at the onset of convection. Then experiments done at different frequencies then fall on one universal curve.

As we were reporting these results at a meeting, Steve Davis suggested that we look for hysteresis about the convective onset. Roppo, Davis and Rosenblat (1984) had predicted a finite-amplitude onset to a hexagonal convective planform as the Rayleigh number is increased past its critical value. Dependence on intitial conditions is expected to give rise to a double hysteresis loop near the critical Rayleigh number, where the first loop (in the direction of increasing Rayleigh number) transverses the regimes of stable conduction and subcritical hexagons, while the second is entirely subcritical and involves a history-dependent transition between hexagons and roll-type convection. Although the flow was not visualized in our experiments, the heat transfer data in figure 8(a) show transitions between different states of the system which display remarkable similarity to the behavior expected from the analysis of Roppo, et al. Figure 8(b) shows a schematic diagram of the transitions involved.

Recently, Swift and Hohenberg (1987) have considered directly the problem of high frequency modulation, obtaining values for the dimensionless temperature drop at onset $\Delta T/T_{oc}$, and the width of the subcritical hysteresis $1-r_{sn}$, where the relative Rayleigh number at the saddle node r_{sn} is indicated by the leftmost arrow in fig. 8(b). These results are in good agreement with experiment as shown in Table 1.

Table 1

	Exp.	Theory
r_{sn}	0.86	0.85
$\Delta T/T_{oc}$	0.066	0.058

Carrying on from the Stokes layer problem, Niemela and Donnelly (1987) extended their investigations to the arrangments illustrated in Figure 7(c). Here they found both positive and negative threshold shifts depending on the strength and rate of modulation, as well as a depression of the initial slope of the heat transfer above threshold, in qualitative agreement with predictions of Ahlers, Hohenberg and Lücke (1985), and Rosenblatt and Herbert (1969).

The results of low frequency modulation are shown in Fig. 9 taken with use of bottom-plate modulation. The results here are largely destabilizing, reaching as much as 40%. On the other hand, at higher frequencies we see in Fig. 10 both stabilization as well as substantial change in the initial slope of the Nu-Ra curve, much as predicted by Ahlers, Hohenberg and Lücke. A discussion of the theoretical models and application to our experiments is given by Niemela and Donnelly.

As a final example of modulated Bénard systems, Niemela, Smith and Donnelly (1989) have studied convective instability with time-varying rotation about the vertical axis:

$$\Omega(t) = \Omega_s + \Omega_o \cos(\tilde{\omega} t)$$

where $\tilde{\omega} = \omega\tau_v$, with ω the dimensioned modulation frequency and $\tau_v = d^2/\nu$ a

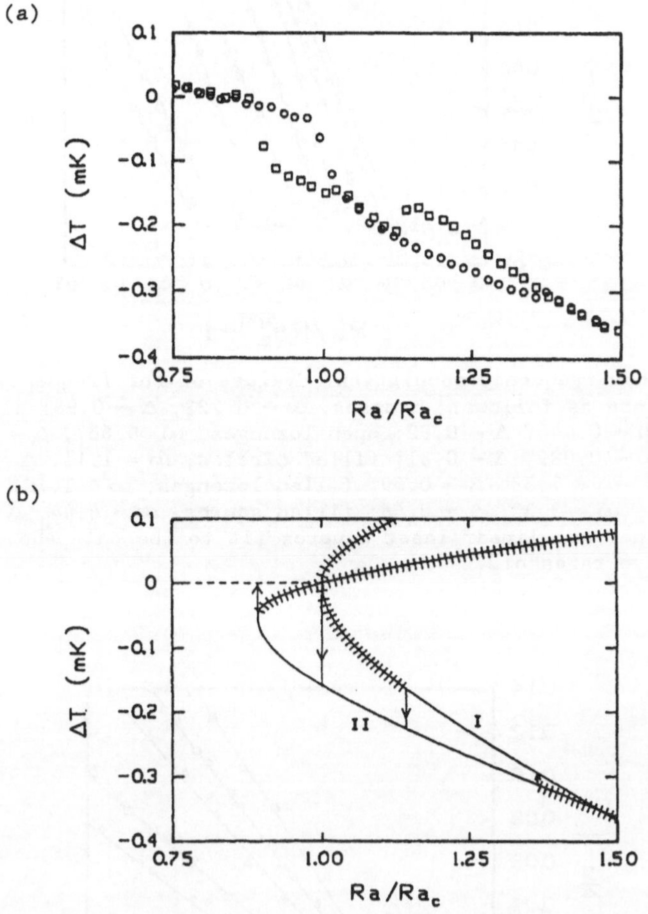

Figure 8. Results of a hysteresis run of the Stokes experiment of Figure 7(a) for Prandtl number $\sigma = 0.49$ and frequency $f = 0.048$ Hz.
(a) The data corresponding to increasing Rayleigh numbers are circles, and decreasing Rayleigh numbers are squares.
(b) A schematic picture of the results assuming transitions between two different planforms. The unhatched portions of the curves were plotted to overlay the data in (a), where the arrows indicate the location and direction of the transitions observed between planforms I and II.

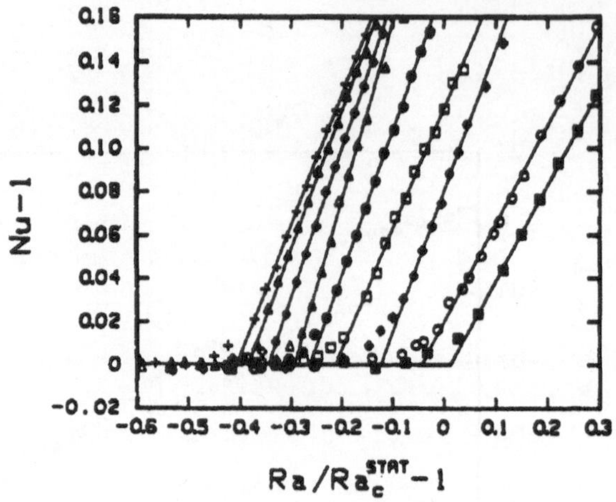

Figure 9. Low-frequency modulation. Values of the frequency ω and amplitude Δ are as follows: crosses, ω = 0.222, Δ = 0.98; filled triangles, ω = 0.444, Δ = 0.92; open lozenges, ω =0.667, Δ = 0.90; open triangles, ω = 0.889, Δ = 0.81; filled circles, ω = 1.11, Δ = 0.75; open squares, ω = 1.33, Δ = 0.69; filled lozenges, ω = 1.78, Δ = 0.58; open circles, ω = 3.33, Δ = 0.68; filled squres, ω = 4.44, Δ = 0.68. The solid lines are linear least squares fit to the data above and below the convective threshold.

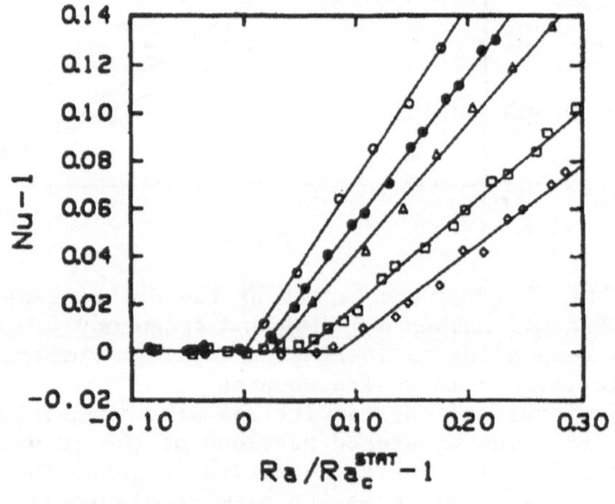

Figure 10. Moderately high-frequency modulation: ω = 10.7. Open circles, Δ = 0; filled circles, Δ = 0.30; triangles, Δ = 0.45; squares Δ = 0.68; lozenges, Δ = 0.76. The data were obtained for a top-plate modulation. The lines represent fits to the data as in Figure 9.

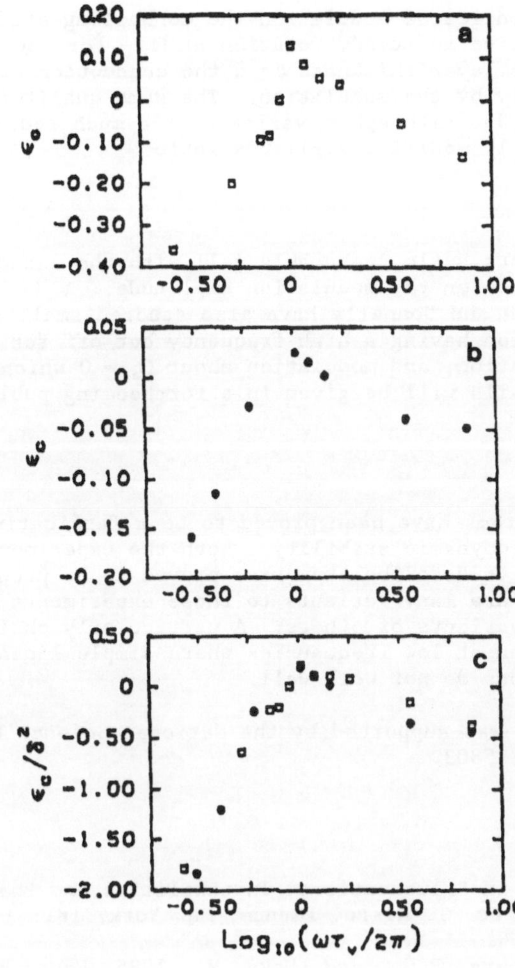

Figure 11. Convective thresholds for (a)Δ = .70 (open squares), and (b)Δ = .35 (filled circles). Dependence on the modulation amplitude of the results in (a) and (b) is suggested in (c), where $\delta = (1 + \epsilon_c)\Delta$.

characteristic vertical viscous diffusion time. In a first set of experiments, $\tilde{\omega}$ was varied over a wide range of values with a fixed mean rotation $\Omega_s = 37.6$ and two fixed values of the amplitude ratio $\Delta = \Omega_o/\Omega_s = 0.35$ and 0.70. The convective thresholds for $\Delta = 0.70$ are shown in Figure 11(a) where ϵ_c is a reduced critical Rayleigh number

$$\epsilon_c = \frac{Ra_c(\Omega)}{Ra_c(\Omega_s)} - 1$$

At very low frequencies of modulation the conducting state is severely destabilized relative to steady rotation at Ω_s. For modulation periods producing a Stokes layer thickness $\delta \approx d$ the conduction state is stabilized slightly by the modulation. The same qualitative behavior is evident for $\Delta = 0.35$, although ϵ_c varies over a much reduced range. If we define a slightly modified amplitude ratio

$$\delta = (1 + \epsilon_c)\Delta$$

we see that the data scale reasonably well with the assumption of a quadratic dependence on the modulation amplitude.

Niemela, Smith and Donnelly have also studied small amplitude broadband modulation having a high frequency cut off for which they observed stabilization, and modulation about $\Omega_s = 0$ which is also stabilizing. Details will be given in a forthcoming publication.

IV. CONCLUSIONS

Modulated systems have been proved to be a fascinating branch of the theory of hydrodynamic stability. Both the experiments and the theories invented to describe them have proved to be prone to unwelcome surprises. There are many variants to these experiments, but few can be taken as trivial variants of others. A particularly challenging area has been modulation at low frequencies where simple linear theories with perfect bifurcations do not work well.

This research was supported by the National Science Foundation under grant DMR 8815803.

REFERENCES

Ahlers, G., 1975, in "Fluctuations, Instabilities and Phase Transitions," ed. T. Riste, Plenum, New York, 181.

Ahlers, G., Hohenberg, P.C., and Lücke, M., 1985, Phys. Rev. Lett., 32A, 3493.

Barenghi, C.F., and Jones, C.A., 1989, J. Fluid Mech. (in press).

Carmi, S., and Tustaniwskyi, J.I.., 1981, J. Fluid Mech., **108**, 19.

Davis, S.H., 1976, Ann. Rev. Fluid Mech., **8**, 57.

Donnelly, R.J., 1964, Proc. Roy. Soc. A, **281**, 130.

Donnelly, R.J., Reif, F., and Suhl, H., 1962, Phys. Rev. Lett., **9**, 363.

Donnelly, R.J., and Schwarz, K.W., 1963, Phys. Lett., **5**, 322.

Hall, P., 1981, J. Fluid Mech, **105**, 523.

Kuhlmann, H., Roth, D., and Lücke, M., 1989, Phys. Rev. A, **39**, 745.

Niemela, J.J., and Donnelly, R.J., 1986, Phys. Rev. Lett., **57**, 583.

Niemela, J.J., and Donnelly, R.J., 1986, Phys. Rev. Lett., **57**, 2524.

Niemela, J.J., and Donnelly, R.J., 1987, Phys. Rev. Lett., **59**, 2431.

Niemela, J.J., Smith, M.R., and Donnelly, R.J., 'Convective Instability with Time-Varying Rotation', preprint.

Park, K., Barenghi, C., and Donnelly, R.J., 1980, Phys. Lett., **78A**, 152.

Park, K., Crawford, G.L., and Donnelly, R.J., 1981, Phys. Rev. Lett., **47**, 1448.

Roppo, M.N., Davis, S.H., and Rosenblat, S., 1984, Phys. Fluids, **27**, 796.

Rosenblat, S., and Herbert, D.M., 1969, J. Fluid Mech., **43**, 385.

Seminara, G., and Hall, P., 1976, Proc. Roy. Soc. A, **350**, 299.

Swift, J.B., Hohenberg, P.C., 1987, Phys. Rev. A, **36**, 4870.

Taylor, G.I., 1923, Phil. Trans. Roy. Soc. A, **223**, 289.

Thompson, R., 1968, "Instabilities of Some Time Dependent Flows," D. Phil. Thesis, MIT.

Walsh, T.J., Wagner, W.T., and Donnelly, R.J., 1987, Phys. Rev. Lett., **58**, 2543.

Walsh, T.J., and Donnelly, R.J., 1988, Phys. Rev. Lett., **60**, 700.

Wu, X., and Swift, J.B., 1989, Phys. Rev. A, **40**, 7197.

SIDEBAND INSTABILITY OF WAVES

WITH PERIODIC BOUNDARY CONDITIONS

B. Janiaud, E. Guyon, D. Bensimon and V. Croquette

LPS-ENS, 24 rue Lhomond, 75231 Paris cedex 05

ABSTRACT

The sideband instability for traveling waves is studied experimentally under periodic boundary conditions. Such an instability, known as the Eckhaus instability [1], is subcritical in the case of stationary patterns. In contrast, for traveling waves, we show that the instability occurs via a forward bifurcation. This enables the study of the development of the phase instability and the change in wavenumber. We analyze the phase equation describing this instability. We show that the bifurcation is indeed supercritical. It is characterized by soliton solutions at onset and a transition to phase turbulence away from onset.

INTRODUCTION

Wave patterns have received an extensive interest during these last years since their phase instabilities are predicted to lead to chaotic behavior. However, experimental studies of the mechanism of the phase instabilities are not common. We have investigated a wave pattern obtained in the oscillatory instability of Rayleigh-Bénard convective rolls. We have chosen to impose periodic boundary conditions to be able to study the pure mode evolution. We find that the longitudinal phase instability, called Eckhaus in the case of *stationary* pattern, is significantly different in the case of *wave*. In particular the spontaneous wavenumber modulation which appears at the instability threshold, saturates nonlinearly and exhibits a pulse-like shape.

EXPERIMENTAL APPARATUS

The experimental apparatus is very similar to the one described in [2]. The cell of thickness $d = 1mm$ is sandwiched between a sapphire and a copper mirror. The container is of annular geometry, a 4 mm in diameter plastic disc constitutes the inner boundary and a 11.5 mm plastic ring defines the outer boundary. In order to stabilize azimuthal rolls, we have placed a small heating wire along the sidewalls [3].

The experimental control parameter is the Rayleigh number $R = \Delta T \alpha g d^3/\nu\kappa$ where α, ν and κ are respectively the thermal expansion coefficient, the viscosity and the thermal diffusivity of the fluid, g is the gravitational field and ΔT the temperature difference applied to the fluid layer. The convective fluid is argon under 70 atm, with a Prandtl number of $Pr = \nu/\kappa = .7$. The convection threshold R_C is reached when $\Delta T = 2.5°K$. A second control parameter in this experiment is the current flowing in the sidewall wire which produces a lateral temperature gradient stabilizing the roll parallel to the sidewall. The width of the annulus is chosen such that the wavenumber of a two roll pattern is smaller than 2. Thus this pattern is stable versus the skewed-varicose instability [4] at high R. Our measuring technique consists of projecting a shadowgraphic image of the convective pattern on an annular row of 64 CCD photodiodes. The scanning electronics and the amplifier are connected to a PC which collects the data.

Nonlinear Evolution of Spatio-Temporal Structures in Dissipative Continuous Systems
Edited by F.H. Busse and L. Kramer
Plenum Press, New York, 1990

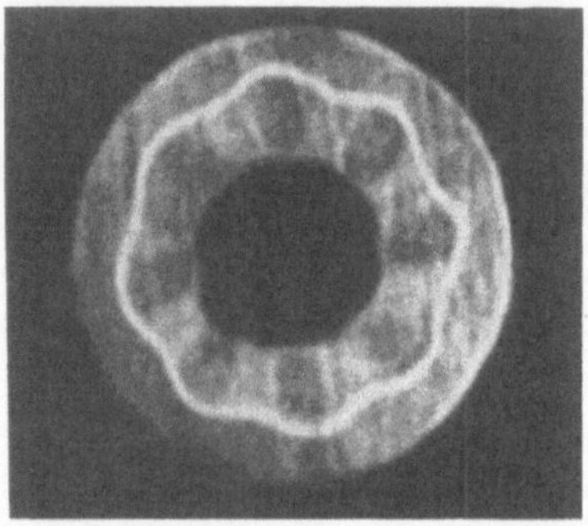

Figure 1.Shadowgraphic picture of the annular convective pattern, the wiggling white circle is the frontier between the two concentric rolls. The wave shape of this frontier corresponds to the 8 wavelength pattern observed at $R = 2R_{C0}$.

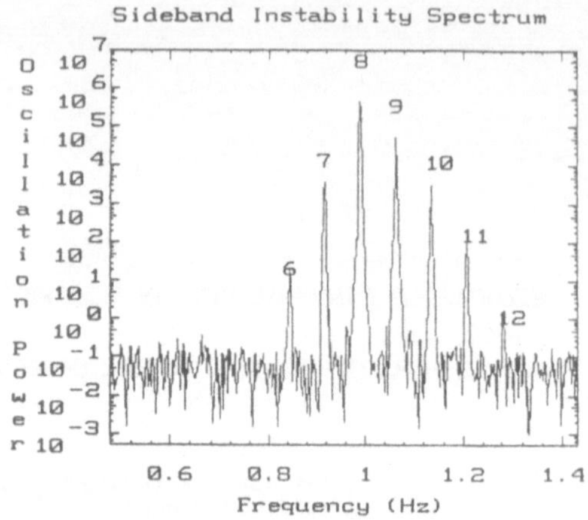

Figure 2. Time spectrum of the local position of the frontier between rolls. This spectrum presents a dominant peak ω_8 surrounded by a series of sidebands corresponding to the wavenumber modulation, $R = 1.04R_{C0}$.

RESULTS

We have studied the evolution of the oscillatory waves as a function of the Rayleigh number. The frontier between the two rolls where the oscillatory instability develops is a circle 24.3 mm in perimeter. Turning on this circle the waves experience periodic boundary conditions. We have imposed a constant ratio between the lateral temperature gradient and the Rayleigh number.

When slowly increasing the control parameters, the oscillatory instability appears via a forward bifurcation at $R_{C0} \approx 6 \times R_C$. After a transient, a wave turning either clockwise or counterclockwise settles along the circular boundary between the two rolls as may be seen in Fig. 1. The amplitude of this wave is homogeneous along this circle. The number of wavelengths at onset is an integer: 9 in this experiment. Assuming that the two rolls are equal in size, the corresponding wavenumber equals 2.34. This is very close to the critical wavenumber predicted theoreticaly [4] and to the one observed in a rectangular container [2]. A local measurement shows a sinusoidal signal with a frequency $\omega_9 \approx 1Hz$.

In a large domain of control parameters, this wave is stable and leads to a spatio-temporal Fourier transform with a single frequency and its harmonics at the spatial wavenumber 9. Until $R_{C1} = 1.5 \times R_{C0}$, the square of the peak amplitude grows linearly with R, as does the frequency ω_9. At $R > R_{C2} = 2 \times R_{C0}$ the pure 9 wavelengths mode exhibits an amplitude modulation which leads to a biperiodic spectrum, the new frequency being nearly $\omega_9/5$. By increasing further R, subharmonics appear in this biperiodic state, and a transition to chaos occurs at $R = 3R_{C0}$. In the chaotic regime, several spatial modes are present simultaneously with time dependent amplitudes. If we lower R from such an erratic state to a value below the threshold of biperiodism, we regain a regular wave patterns but with a wavenumber which may be different from the critical one. Following this procedure, we have often observed an 8 wavelength pattern, and have once obtained a 7 wavelength one.

After preparing such a pattern, we study its stability by decreasing R. For a large range of R these modes remain stable and display the same behavior as the 9 wavelengths pattern, except that they have smaller amplitude and lower frequency. However as the wave amplitude reaches low values, we observe an Eckhaus type of instability. The original instability predicted by Eckhaus [1], applies to *stationary* patterns and consists of a *subcritical* phase instability which ultimately produces a change of wavenumber. In contrast, we observe a *supercritical* phase instability which ultimately leads to a change of wavenumber. The Fourier spectrum performed at a single point shows that first a few sidebands appear on each side of the main peak (corresponding to the dominant mode). When we lower R the amplitude of these sidebands increases significantly as well as their numbers as may be seen on Fig. 2. When the sidebands are strong enough, subharmonic of the beating frequency appear (1/3 and 1/2). Finally lowering further R leads to a change of wavenumber, the spectrum displays again a single peak but at the frequency of what was before the

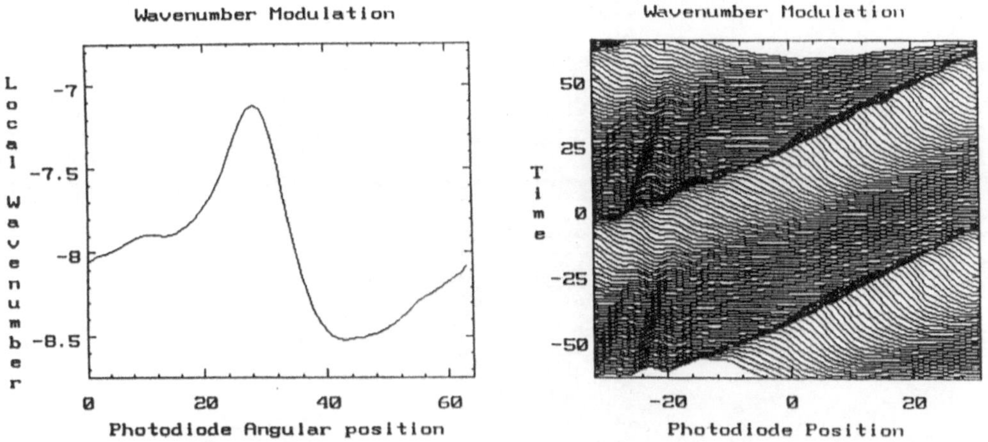

Figure 3. Shape of the wavenumber modulation along the cell, corresponding to the spectrum of Fig. 2. The first graph was obtained by measuring the local wavenumber along the cell at a precise time. This pulse propagate along the cell with a velocity close to the group velocity as may be seen in the second graph.

larger sideband. Using a 2D-FFT, it is easy to see that the sidebands correspond to modes having nearby but integer wavenumbers. Furthermore, by measuring the local wavenumber, we show that the instability encountered is a phase instability. The local wavenumber displays a modulation having one wavelength along the annulus (see Fig. 3) and propagating nearly at the group velocity.

The way the pattern changes modes is not easy to track. In the case of stationary patterns the wavenumber modulation drives the amplitude to a local zero, where a new wavelength may appear or disappear [5]. In the case of traveling wave we suggest that another scenario may take place: in a transient recording during the phase instability, we have observed that a standing wave appears over a few wavelengths as may be seen in Fig. 4. A possible explanation could be that the wavenumber modulation drives an amplitude modulation. When the amplitude minimum is close to zero, the wave propagating in the reverse direction is free to grow, leading to a small standing wave. Such a competition between the two waves could rapidly lead to a change of wavenumber of the dominant wave. As expected and as may be seen on Fig. 5, this instability occurs for the mode 7 at a higher R than it does with the mode 8 but leads in both cases to the same scenario.

DISCUSSION

In a previous experiment [2], we have shown that the amplitude A of the oscillatory instability waves $A = \exp -i(kx - \omega t)$ can be described by a complex Landau-Ginsburg amplitude equation:

$$\tau_0 \left(\frac{\partial A}{\partial t} + v \frac{\partial A}{\partial x} \right) = \epsilon(1 + ic_0)A + \xi_0^2(1 + ic_1)\frac{\partial^2 A}{\partial x^2} - \gamma(1 + ic_2)|A|^2 A. \qquad (1)$$

Here $\tau_0 = .163\ d^2/\kappa$, $\xi_0 = .52\ d$, $v = 5.02\ \kappa/d$ and $\gamma = 2.3$ are respectively the characteristic time, the correlation length, the group velocity and the nonlinear saturation parameter, $\epsilon = (R - R_{CO})/R_{CO}$. The imaginary parts $c_0 = 1.57$, $c_1 = -.826$, $c_2 = -1.15$ represent the frequency change as a function of respectively ϵ, the wavenumber shift Q and the amplitude A . Eq. (1) describes a wave traveling to the right, a similar equation having the opposite group velocity would be needed to treat the left wave.

The solutions of Eq. (1) have the form $A(x,t) = A_Q \exp -i(Qx - \Omega t)$. The stability analysis of these solutions to perturbations of the form $a(x,t) \exp i[\Omega t - Qx + \varphi(x,t)]$ leads to the following equation for the phase φ :

$$\frac{\partial \varphi}{\partial t} = a_1 \frac{\partial \varphi}{\partial x} + D_{\parallel} \frac{\partial^2 \varphi}{\partial x^2} + a_3 \frac{\partial^3 \varphi}{\partial x^3} + D_4 \frac{\partial^4 \varphi}{\partial x^4} + a_2 (\frac{\partial \varphi}{\partial x})^2 + d(\frac{\partial \varphi}{\partial x})(\frac{\partial^2 \varphi}{\partial x^2}) \qquad (2).$$

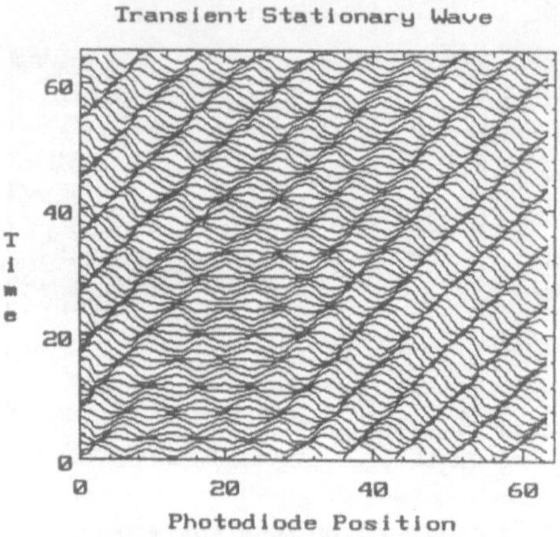

Figure 4. Transient space-time recording, obtained just after a change of wavenumber. Notice that the dominant right going wave presents a low amplitude where a standing wave has developed.

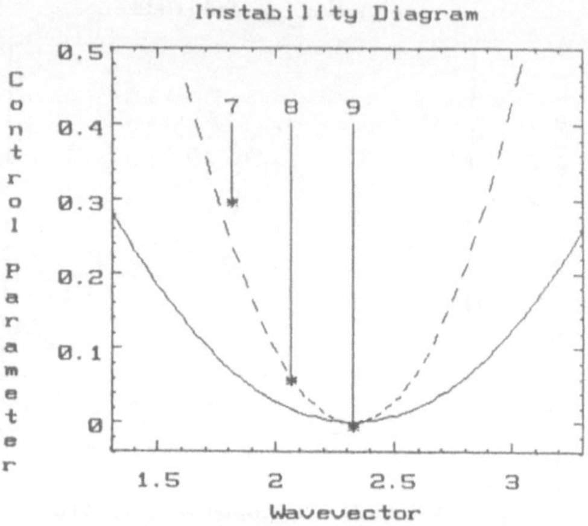

Figure 5. Stability domain of pure modes in the annular cell: the full line corresponds to the marginal stability of the oscillatory modes, the dashed curve represents the Eckhaus instability prediction based upon the value of $c_{1,2}$ obtained in [2]; the tars indicate the loci of the instability found in the experiment.

The coefficients of this equation may be found in [6]. For stationary patterns, the a_i are all null since $c_{1,2} = 0$ and the phase equation becomes a diffusion equation with a nonlinear term. As stressed by Pomeau and Manneville [7], the usual Eckhaus instability is observed when the diffusion coefficient $D_{\|} = (\xi_0^2/\tau_0)(\epsilon - 3q^2)/(\epsilon - q^2)$ becomes negative. As the representative point of the pattern (q, ϵ) is located below the parabola $\epsilon = 3q^2$, a phase modulation spontaneously appears. The dominant nonlinear term $(\partial\varphi/\partial x)(\partial^2\varphi/\partial x^2)$ does not saturate, and leads to a subcritical bifurcation which ultimately produces a change of the wavenumber Q $(q = Q\xi_0)$.

In our situation, the coefficients a_i are non zero and alter this picture. In the long wavelength limit the dominant non linear term is $(\partial\varphi/\partial x)^2$ which is saturating and leads to a forward bifurcation. Furthermore the diffusion coefficient $D_{\|}$ itself is modified and becomes :

$$D_{\|} = \frac{\xi_0^2}{\tau_0}\left[1 + c_1c_2 - 2q^2.\frac{1 + c_2^2}{\epsilon - q^2}\right]. \tag{3}$$

The curvature of the parabola in the $q - \epsilon$ plane, where $D_{\|} = 0$, now depends on c_1 and c_2 and is not 3 any more as for stationary patterns.

Using the values of c_i found in [2] enables us to make a semi-quantitative comparison. Notice that we observe a phase instability in a range of R below R_{C1}, where we believe the amplitude equation (1) to be valid. The first test concerns the stability domain of a pure mode, $D_{\|} > 0$. We have observed only two transitions from 7 to 8 wavelengths and 8 to 9, which leads to two points (three with the threshold) to determine a parabola. However, these points are in good agreement with the curve $D_{\|} = 0$ obtained with the values of c_1 and c_2 found in [2]. The slight upward shift of ϵ may be expected from finite size effects [5][8]. Clearly a stability domain exists for pure modes confirming that the Benjamin-Feir criterion [9] is not met, in other words $c_1c_2 > -1$. The spectra of Fig. 2 demonstrate the mechanisms involved in sideband instability : the sidebands have integer wavenumbers and their frequencies are given, in a first approximation, by the dispersion relation $\omega_i = \omega_9 + v(k_i - k_9)$. Since the instability is accompagnied by a frequency shift, the frequency of the sidebands are also shifted from the value given by the dispersion relation. However, this shift is small in the experiment and the $\delta\omega/\delta k$ has a value very similar to the group velocity determined in [2].

We have established the criticality of the bifurcation in two ways: first we obtain the instability by lowering R and we measure a regular increase of the magnitude of the sidebands. But the domain of existence for the instability is small and we have not yet got enough points to draw a convincing bifurcation diagram. However we may analyze directly the wavenumber modulation during the instability and extracts its second harmonic. It turns out that the major contribution to the second harmonic comes from the $(\partial\varphi/\partial x)^2$ term which constitutes an indirect proof that the bifurcation is forward. As a matter of fact, it is easy to verify that if the phase modulation fundamental is of the form of $\varphi_1(x) = b_1 \cos(2\pi x/L)$, the $(\partial\varphi/\partial x)^2$ term leads to a second harmonic $\varphi_2(x) = b_2 \cos(4\pi x/L)$ whereas $(\partial\varphi/\partial x)(\partial^2\varphi/\partial x^2)$ leads to a second harmonic which is out of phase: $\varphi_2(x) = b_3 \sin(4\pi x/L)$.

The forward bifurcation that we observe was predicted by Fauve in [10], where he studied the phase instability of Benjamin-Feir unstable waves at the critical wavenumber. In this situation he derived a phase equation where the $(\partial\varphi/\partial x)^2$ nonlinearity appeared and showed that a forward bifurcation was expected. It is easy to show that driving the system unstable through the Eckhaus instability at $k \neq k_c$ is similar to the Benjamin-Feir instability at $k = k_c$ treated by Fauve. However, since $c_{1,2} \neq 0$, Eq. (2) contains terms $\partial\varphi/\partial x$ and $\partial^3\varphi/\partial x^3$,and they create an interesting difference. While the first term can be absorbed by going to a moving frame, the second is dispersive and has a drastic importance. Considering Eq. (2) in the neighbourhood of the phase instability, the dissipative terms are very small in comparison with the dispersive term and may be neglected in a first approximation, the dominant nonlinear term is $(\partial\varphi/\partial x)^2$, thus we are left with the following equation:

$$\frac{\partial\varphi}{\partial t} = a_3\frac{\partial^3\varphi}{\partial x^3} + a_2\left(\frac{\partial\varphi}{\partial x}\right)^2 \tag{4}$$

which is the Korteweg de Vries equation well-known to have a family of solitons solutions. Adding the dissipative terms leads to a *selection* of one of these solutions [11], thus, close to the phase instability, the wavenumber modulation is expected to have a pulse-like shape as indeed suggested by the experimental results of Fig. 3.

Furthermore, notice that Eq. (2) with the dominant nonlinearity $(\partial\varphi/\partial x)^2$ and away from the onset of the phase instabilitie, when the dispersive terms can be neglected in respects with the dissipatives ones, is the Kuramoto-Shivasinsky equation, well-known to have chaotic solutions [12].Thus we infer that Eq. (2) should display an interesting transition from well ordered pulse-like solutions to a disordered and chaotic state, as the control parameter drives the phase instability more and more unstable.

CONCLUSION

We have studied the longitudinal phase instability mechanism experienced by the waves of the oscillatory instability. We have evidenced the sidebands characteristic of this instability and shown that the wavenumber modulation indicates a forward bifurcation. We also demonstrated that this wavenumber modulation has a pulse shape which may be explained considering the solitary solutions of the phase equation obtained for waves.

ACKNOWLEDGEMENTS

We wish to acknowledge stimulating discussions with A. Pumir, S. Fauve, P. Tabeling, O. Cardoso, M. Rabaud, Y. Couder, V. Hakim, S. Zaleski, L. Kramer, P. Manneville and H. Chaté. We thank DRET for its financial support through the contract 88/1343/DRET/DS/SR.

REFERENCES

[1] W. Eckhaus, "Studies in Nonlinear Stability Theory", Springer, New-York (1965).

[2] V. Croquette and H. Williams, *Nonlinear Competition Between Waves in Convective Rolls* Phys. Rev. **A 39** , 2765 (1989).

[3] V. Croquette, M. Mory and F. Schosseler, *Rayleigh-Bénard Convective Structures in a Cylindrical Container*, J. de Physique **44**, 293-301 (1983).

[4] R.M. Clever and F.H. Busse, *Transition to Time-Dependent Convection*, J. Fluid Mech. **65**, 625 (1974).

[5] L. Kramer and W. Zimmerman, *On the Eckhaus Instability for Spatially Periodic Patterns*, Physica **D 16** , 221 (1985).

[6] J. Lega, *Défauts Topologiques Associés à la Brisure de l'Invariance de Translation dans le Temps*, Université de Nice, *Thèse d' état* (1989).
[7] Y. Pomeau and P. Manneville, *Phase Diffusion in Rayleigh-Bénard Convection* J. Phys. Lett. **47**, 835 (1981).
[8] G. Ahlers, D.S. Cannel, M.A. Dominguez-Lerma, R. Heinrichs, *Wavenumber Selection and Eckhaus Instability in Couette Taylor Flow*, Physica **D 23** , 202 (1986); M. Lowe and J.P. Gollub, *Solitons and the Commensurate-Incommensurate Transition in a Convecting Nematic Fluid* Phys. Rev. Lett. **55** , 2575 (1985).
[9] T.B Benjamin and J.E. Feir,*The Disintegration of Wave Train on Deep Water* J. Fluid Mech. **27** , 417 (1966); J.T. Stuart and R.C. DiPrima,*The Eckhaus Instability and Benjamin-Feir Resonance Mechanisms* Proc. Roy. Soc. London, Ser. **A 362** , 27 (1978).
[10] S. Fauve *Large Scale Instabilities of Cellular Flows*, in: "Instabilities and Nonequilibrium Structures", , 63-88 Ed. Riedel Publishing Company (1987).
[11] T. Kawahara, Phys. Rev. Lett. **51** , 381 (1983).
[12] H. Chaté and P. Manneville, *Transition to Turbulence via Spatiotemporal Intermittency* Phys. Rev. Lett. **58** ,112 (1988).

TRANSITION TO CHAOS IN NON-PARALLEL TWO-DIMENSIONAL FLOW IN A CHANNEL

P.G. Drazin, W.H.H. Banks and M.B. Zaturska

School of Mathematics
University Walk
Bristol BS8 1TW, UK

INTRODUCTION

This paper is a review of some new work on an old problem. We shall first describe the origin of the problem, and then present a few snapshots of some of the solutions. Please forgive us for taking the snapshots from a parochial point of view, because this review is mostly of work done at Bristol.

The problem concerns two-dimensional flow of an incompressible fluid of kinematic viscosity ν. In his famous work on steady flow near a stagnation point of attachment, Hiemenz (1911) assumed that the stream function is of the similarity form

$$\psi(x, y) = xF(y), \tag{1}$$

where the stagnation point is at the origin O on a rigid wall $y = 0$ and the flow is in the half-plane $y > 0$. Some have attributed this similarity form to Newell (1906), a paper we have been unable to obtain: whilst a sober examination of the original papers may reveal Newell's prophetic vision, we are convinced that it would also confirm that Hiemenz was the first to give the form explicitly. Now, on substituting the stream function (1), the Navier-Stokes equations, in the form of the vorticity equation, reduce to the ordinary differential equation,

$$\nu F^{iv} + FF''' - F'F'' = 0. \tag{2}$$

In a fertile paper on two-dimensional *un*steady flow near a stagnation point of separation at O, Proudman and Johnson (1962) assumed the similarity form

$$\psi(x, y, t) = xf(y, t). \tag{3}$$

Then the x- and y-components of the velocity of the fluid are

$$u = xf_y, \quad v = -f,$$

respectively. The similarity form and the vorticity equation now lead to the partial differential equation

Nonlinear Evolution of Spatio-Temporal Structures in Dissipative Continuous Systems
Edited by F.H. Busse and L. Kramer
Plenum Press, New York, 1990

$$f_{yyt} = \nu f_{yyyy} + f f_{yyy} - f_y f_{yy}, \tag{4}$$

although Proudman and Johnson in fact stated only the integral of (4) with respect to y rather than (4) explicitly.

Hiemenz's similarity solution has been generalized also by Homann (1936) for axisymmetric steady flow, Howarth (1951) for three-dimensional steady flow and by Taylor *et al.* (1990) for three-dimensional unsteady flow. They all show how the work below may be applied to three-dimensional flows, albeit still to rather special exact solutions of the Navier-Stokes equations. However, we have not enough space here to describe these generalizations.

The nonlinear diffusion equation (4) is reminiscent of the canonical nonlinear evolution equations, e.g. the Kuramoto-Sivashinsky equation, so fashionable today; although its origin is rather different, not coming from wave modulation, it will be seen to have, like those canonical equations, a rich variety of solutions. It has been applied not only to flow near a stagnation point but also to various other flow problems, each with its own set of boundary conditions, as summarized below.

(i) Flow near a stagnation point O (Hiemenz 1911, Proudman and Johnson 1962, Childress *et al.* 1989, etc.):

$$f = f_y = 0 \quad \text{at } y = 0, \quad |f_y| < \infty \quad \text{as } y \to \infty. \tag{5}$$

(ii) Flow in a channel with equal and uniform steady suction on both walls (Berman 1953, etc.):

$$f = \mp V, \; f_y = 0 \quad \text{at } y = \pm h, \tag{6}$$

for some constants V, h. Berman first used the flow with these boundary conditions to model gaseous diffusion through the porous walls of a tube. The configuration of the channel for this flow is sketched in Fig. 1. The resulting boundary-value problem is symmetric under reflection in the centre line $y = 0$ of the channel, so it admits symmetric solutions with f as an odd function of y.

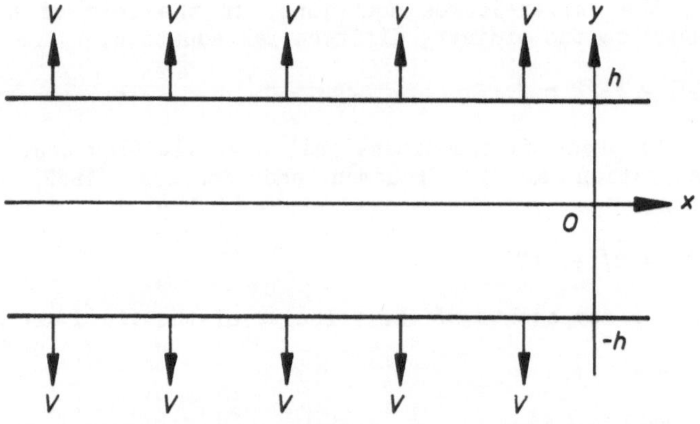

Fig. 1. A sketch of the configuration of the channel for the Berman problem.

(iii) More general flow in a channel with accelerated walls and suction (Brady and Acrivos 1981, E.B.B. Watson *et al.* 1990, Cox 1989, P. Watson 1989, etc.):

$$f = V_{\pm},\ f_y = E_{\pm} \quad \text{at } y = \pm h, \tag{7}$$

for some constants V_+, V_-, E_+, E_-.

(iv) Unbounded flow (unpublished work of the authors):

f has period $2h$ in y, $f - y$ has period $2h$ in y,

or $|f_y| < \infty$ as $y \longrightarrow \infty$.

(v) Unsteady suction and acceleration of channel walls (unpublished work of the authors):

$f = V_{\pm}(t)$, $f_y = E_{\pm}(t)$ at $y = \pm h$.

With each set of boundary conditions, a dimensionless form of the problem may be taken, and then the bifurcations of the solutions examined as the dimensionless parameters vary. For example, with boundary conditions (6), the Reynolds number $R = Vh/\nu$ is increased. For conditions (7) there are four independent dimensionless parameters to vary. In this way the transition to spatio-temporal chaos has been found (although it should be remembered that in a real flow other types of instability, perhaps three-dimensional, also may occur). Next we present a few snapshots of the results, assuming a dimensionless form of each problem.

THE BERMAN PROBLEM

Since the original paper of Berman (1953), there has been much work on equation (2) with the symmetric boundary conditions (6). For example, Terrill (1964) and others explored thoroughly the steady symmetric solutions as R varies. Over the years the custom grew of presenting the results for steady flows in a bifurcation diagram with R as the abscissa and $F''(1)$ as the ordinate, because the skin friction on the wall at $y = 1$ is proportional to $F''(1)$. Zaturska *et al.* (1988) initiated the exploration of the asymmetric and unsteady solutions of the problem of (4) subject to (6). They found many bifurcations en route to chaos. There is a unique steady flow, which is symmetric, for $-\infty < R \leq R_1$, ≈ 6; this solution, and its continuation for $R > R_1$, is said to be of type I. (Note that negative values of R correspond to blowing rather than suction through the walls.) Some of the bifurcations for $R \geq R_1$ are presented in the diagram of Fig. 2. Two asymmetric steady solutions, said to be of types I_1 and I_1', can be seen to arise at a supercritical pitchfork bifurcation at $R = R_1$. Also two stable asymmetric periodic solutions arise at supercritical Hopf bifurcations at $R = R_{11}$, ≈ 13. (There are also solutions of types II and III, which are steady, symmetric and unstable, and are not mentioned further here, because they do not appear to be important in the interpretation of the chaotic dynamics. Further unstable asymmetric steady solutions bifurcate from the solutions of type III.)

Zaturska *et al.* found that as R increases there are not only the pitchfork and Hopf bifurcations presented in Fig. 2 but also at least one period doubling, and a homoclinic explosion at $R = R_c$, ≈ 19.7. It appears, from numerical integrations of the nonlinear initial-value problem, that at $R = R_c$ there are two homoclinic connections of a fixed point in phase space; this fixed point is a saddle point which corresponds to the steady solution of type I. (There are two connections rather than one because of the symmetry of the mathematical problem corresponding to the symmetry of the flow about the centre line of the channel.) The fixed points corresponding to asymmetric flows of types I_1 and I_1' are saddle foci. They, and the periodic orbits they shed at the Hopf bifurcations, also seem important in the chaotic dynamics.

It appears plausible that the essential nonlinear dynamics of the chaos, which ensues as R increases above R_c, occurs in a three-dimensional submanifold of the infinite-dimensional phase space. An impression of this is given in Fig. 3 of a conjectured (X, Y, Z)-space. The evidence in support of Fig. 3 is made up of (i) analysis of the linear stability of the fixed points I, I_1 and I_1', (ii) the symmetry of the system about the line $Y = 0$ corresponding to the symmetry of the configuration of the flow about the centre line of the channel, (iii) many numerical integrations of

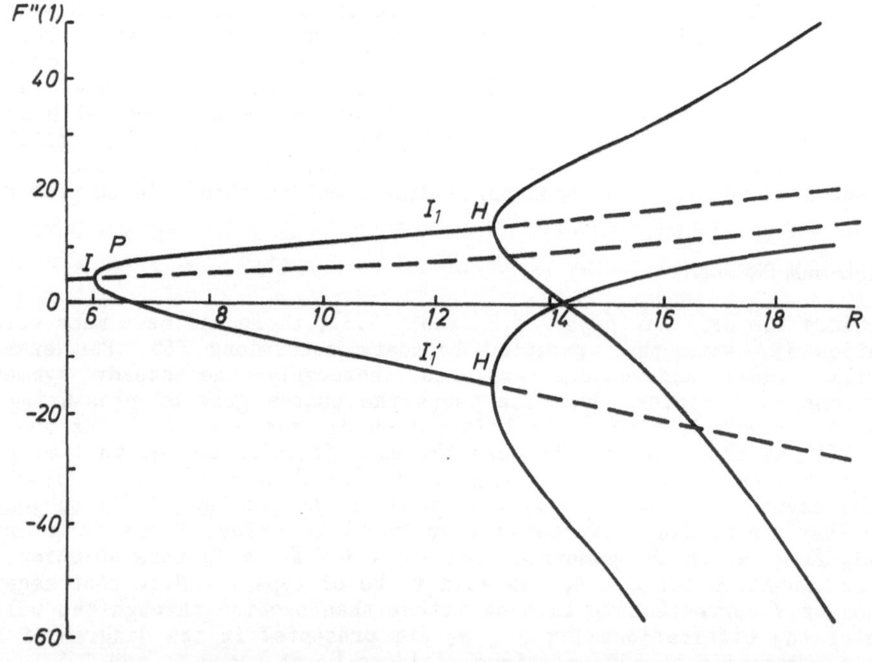

Fig. 2. The bifurcation diagram for the Berman problem. Values of $F''(1)$ for symmetric steady solutions of type I, asymmetric steady solutions of types I_1 and I_1', and periodic solutions are plotted as a function of R. The points marked P at $R = R_1$ and H at $R = R_{11}$ respectively represent pitchfork and Hopf bifurcations. The continuous curves represent stable steady, periodic, or quasi-periodic solutions, and the broken curves unstable solutions. The asymmetric solution I_1 is represented by *both* the values $\pm F''(\pm 1)$ and each one of the two periodic solutions by the *four* values $\max_t\{\pm f_{yy}(\pm 1, t)\}$ and $\min_t\{\pm f_{yy}(\pm 1, t)\}$. These quantities give the quantities for the other solution by reflection in the centre line $y = 0$ of the channel.

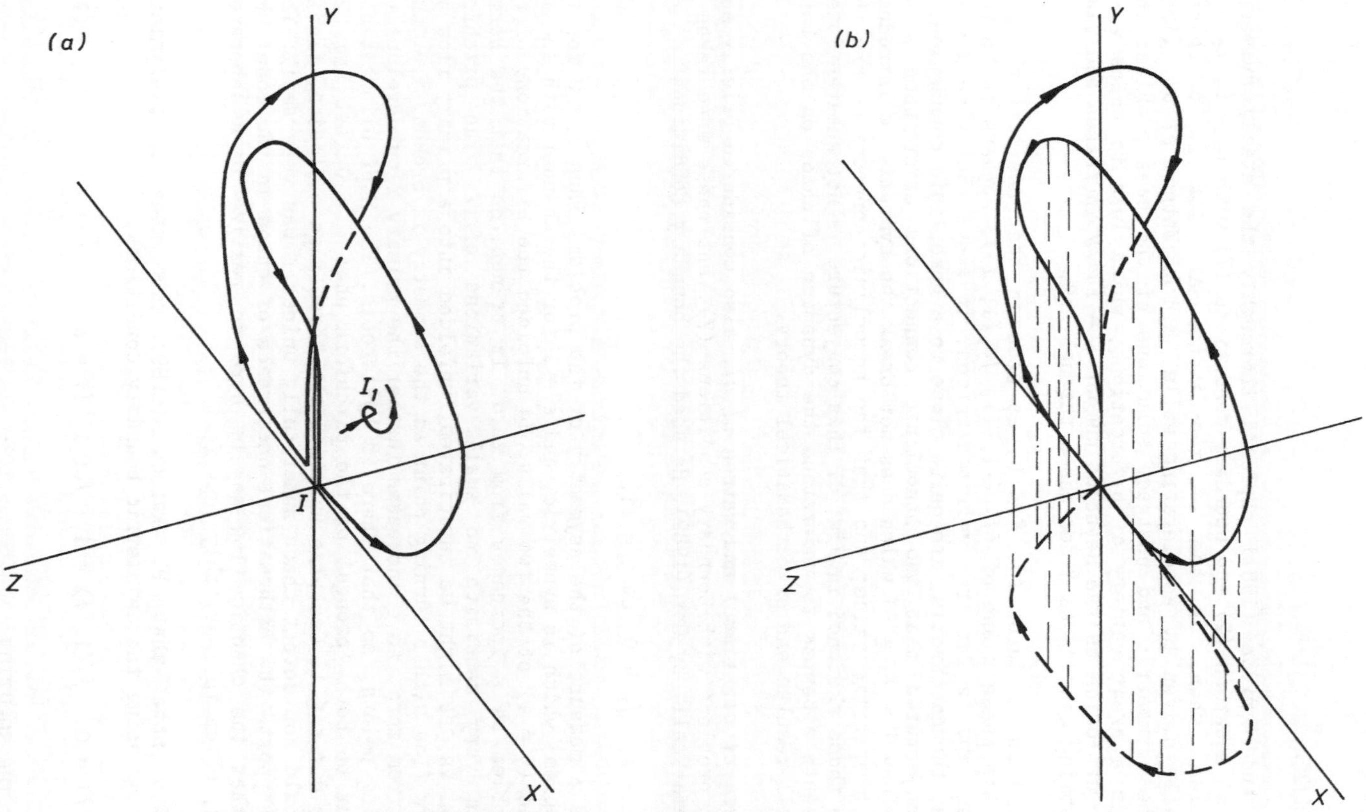

Fig. 3. Perspective sketches of some orbits of the Berman problem as if they were in a three-dimensional space of (X, Y, Z). (a) $0 < R_c - R \ll 1$. (b) Homoclinic connections at $R = R_c$.

phase plots, chiefly in the plane of $(f_{yy}(-1, t), f_{yy}(1, t))$, and (iv) examination of the theory of the explosion of homoclinic connections (cf. Glendinning 1988).

OTHER PROBLEMS

Brady and Acrivos (1981) explored thoroughly the steady symmetric solutions of equation (2) and boundary conditions (7) with $V_\pm = 0$, $E_\pm = E$. This problem, then, is one of two-dimensional flow in a channel symmetrically driven by accelerating walls. E.B.B. Watson *et al.* (1990) studied the asymmetric and unsteady solutions of the same problem. They found again a great variety of bifurcations, which lead to chaos via a homoclinic explosion as the Reynolds number $R = Eh^2/\nu$ increases and via a period doubling cascade as R decreases through -79.

The pattern of chaos for $R < 0$, as seen in the numerical plots of orbits in the phase plane of $(f_{yy}(-1, t), f_{yy}(1, t))$, appears to be very complicated, and is far from well understood. It seems that when $R \approx -85$ the orbits, though chaotic, are quite close to a homoclinic connection. It may be conjectured that two homoclinic connections exist when a new parameter (say $V = V_\pm \neq 0$) which does not break the symmetry is introduced and takes a specific value; and that the homoclinic connections are then similar to those sketched in Fig. 4. This conjecture is not substantiated, but represents attempts to understand the dynamics of chaos on the basis of numerical results and of mathematical theory.

Zaturska *et al.* (1988) and Watson *et al.* also considered briefly some asymmetric problems with boundary conditions (7), but these were taken up more systematically by Cox (1989). He used the boundary conditions

$$f = \mp 1 - \epsilon, \quad f_y = 0 \quad \text{at } y = \pm 1, \tag{8}$$

where ϵ is a measure of the asymmetry of the problem. Thus $\epsilon = 0$ for the Berman problem, which is symmetric, and $\epsilon = 1$ for the channel with suction on only one ($y = 1$) of the two walls. He unfolded the bifurcations of the Berman problem as ϵ increases from zero. It transpires that the Berman problem is very sensitive to small variations of ϵ. The pitchfork bifurcation is, as might be anticipated, unfolded into a primary flow and a secondary flow with a turning point as the symmetry is broken, i.e. as ϵ increases from zero. As ϵ increases further the primary flow develops two more turning points, so that there is hysteresis. The Hopf bifurcations, which occur on both 'prongs' of the pitchfork when $\epsilon = 0$ (see Fig. 2), remain for $0 < \epsilon \lesssim 1$. Cox also unfolded the homoclinic explosion, finding that he could not detect chaos numerically unless ϵ was very small. Thus his work supports the mathematical occurrence of chaos in the model, but suggests that the chaotic flow may be hard to realize in a laboratory experiment, which is inevitably imperfect.

We also note that P. Watson (1989) has made a preliminary investigation using the asymmetric boundary conditions,

$$f(\pm 1, t) = 0, \quad f_y(1, t) = 1, \quad f_y(-1, t) = \alpha. \tag{9}$$

BLOW-UP OF THE SOLUTION

By direct numerical integration of the initial value-problem for the partial differential equation (4) and boundary conditions (8) for $\epsilon = 1$, Cox (1989) examined the stable periodic solutions following the first Hopf bifurcation, and found that the amplitude of the limit cycle grows rapidly

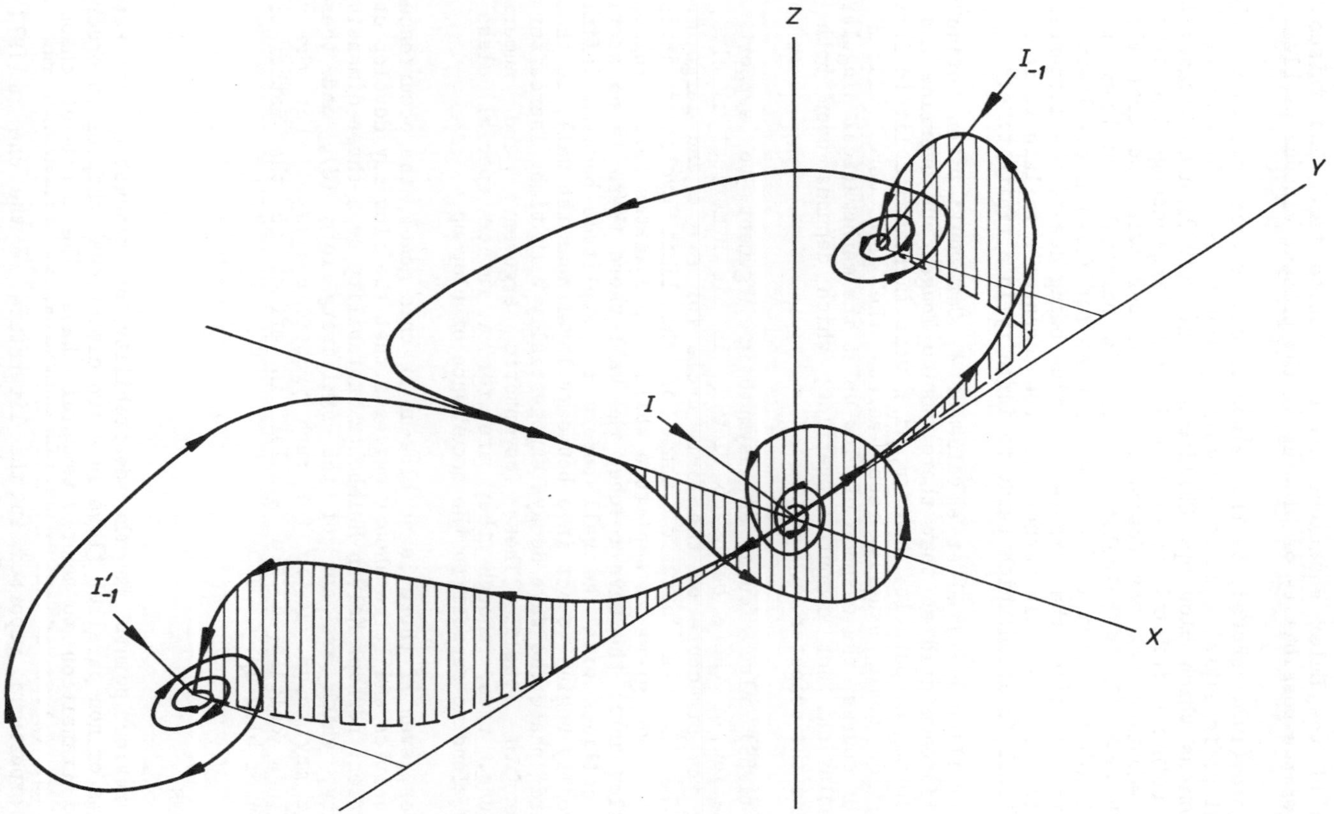

Fig. 4. A perspective sketch of two homoclinic connections in a three-dimensional phase space which seem *close* to orbits computed by E.B.B. Watson *et al.* (1990) for $R \approx -85$.

as R increases. For example, with R = 37.5 he found that $f_{yy}(-1, t) \approx$ -12,000 and $f_{yy}(1, t) \approx$ 8,000 at one stage of the cycle. As R increases further the amplitude of the limit cycle seems to grow like R^4, so that soon the solution grows so large that the computer stops because of overflow. This is reminiscent of the 'blow-up' of two-dimensional solutions of the Euler equations after a finite time, and raises the question of the possibility of blow-up in the present viscous problem.

Calogero (1984) showed, by the method of characteristics, that a class of partial differential equations including the *inviscid* form of (4) have some solutions which blow up. Childress *et al.* (1989) demonstrated that once the solution for unbounded flow becomes large enough the inviscid terms in equation (4) may dominate the viscous term, so that blow-up occurs. However, Cox's calculations for his particular problem suggest that, although the limit cycle grows with R, for fixed R the limit cycle persists for all time, the computer overflow being due to the largeness of the amplitude of the cycle rather than the computer's failure to represent a mathematical solution which tends to infinity in a finite time.

When overflow occurred at a value of R, Cox could often refine the finite-difference grid so that there was no longer any overflow and the limit cycle was computable. However, if R were increased a little further then overflow would usually recur unless the grid were refined yet further. Of course, the essence of blow-up is the mathematical singularity of the solution, not computer overflow, which depends upon both the program and the computer.

Cox (1989) also found an asymptotic argument to support his interpretation of there being no blow-up in his numerical results, by examining the structure of the limit cycle for ϵ = 1 and large R. It appears that in the 'outward' phase of the limit cycle an inviscid sinusoidal 'wave' grows in amplitude while it propagates into a region of uniform flow until the wave reaches the wall where there is no suction. The wave collides with the wall before its amplitude becomes infinite. Then viscosity begins to act in a boundary layer near the wall, so that in the 'inward' phase the wave decays exponentially with time. Thereafter the cycle repeats itself. These asymptotic arguments and numerical calculations, then, though they are for a rather special class of solutions, lend no support to the occurrence of blow-up.

However, even if there is no blow-up in this model, the occurrence of large velocities and large shears suggests that the flow may develop other instabilities, perhaps Kelvin-Helmholtz instability or a three-dimensional instability, which are not of the similarity form (3), and thereby turbulence may ensue. It is for such practical reasons that the very strong flows calculated are similar in effect to the mathematical phenomenon of blow-up.

CONCLUSIONS

This review points to the desirability of conducting laboratory experiments of non-parallel flows in a two-dimensional channel in order to understand transition to spatio-temporal chaos. The onset of chaos at quite modest values of the Reynolds number looks promising. The only relevant experiment reported in the literature is the one in 1974 by Raithby and Knudsen on the Berman flow with symmetric uniform suction at the two walls of the channel. However, such an experiment is not easy, and the theory current in 1974 seems not to have led the experimentalists to look in the right direction, so their results are not very conclusive. It should be remembered that the best of experiments have imperfections, such as irregularities in the suction on the walls, so that no experiment can be perfectly symmetrical. It is unfortunate, then, that the quantitative effects of small asymmetries are so large in some of the problems discussed above. Also in a laboratory experiment there is no restriction to ensure that only flows of the similarity form (3) occur. Tollmien-Schlichting waves may occur where $|x| \gg h$ because the local longitudinal velocity u, and hence the local Reynolds number, is large there.

The problem of Cox (1989) with $\epsilon = 1$ seems the easiest to realize in the laboratory - indeed, that was the original motivation for his calculations - but his not finding chaos makes the problem less interesting for an experimental study of transition.

We have mentioned the historical origin of the work in the theory of the problem of flow near a stagnation point, and perhaps the work reviewed above is most important for its relevance to that fundamental problem of fluid mechanics (see, for example, Stuart 1988).

For lack of space, we have in this article described more of the temporal properties than the spatial properties of the flows. The spatial properties need a lot of pictures to be described clearly. Some pictures can be found in the papers we have cited. However, it can be readily appreciated that the motion of a point in its phase space corresponds to the evolution of a solution of equation (4), and thence corresponds to spatial as well as temporal variations of the velocity components of the fluid, even though the assumption of the similarity solution (3) precludes a description of a real turbulent flow.

Lastly, let us commend the problems we have reviewed for their didactic value. They exhibit, in a relatively simple fashion, a wide range of the bifurcations found in the early stages of transition to turbulence.

ACKNOWLEDGEMENT

We thank S.M. Cox for the suggestion which was the genesis of Fig. 4.

REFERENCES

Berman, A.S., 1953, Laminar flow in channels with porous walls, *J. Appl. Phys.*, 24:1232.

Brady, J.F. and Acrivos, A., 1981, Steady flow in a channel or tube with an accelerating surface velocity. An exact solution to the Navier-Stokes equations with reverse flow, *J. Fluid Mech.*, 112: 127.

Calogero, F., 1984, A solvable nonlinear wave equation, *Studies Appl. Math.*, 70:189.

Childress, S., Ierley, G.R., Spiegel, E.A. and Young, W.R., 1989, Blow-up of unsteady two-dimensional Euler and Navier-Stokes solutions having stagnation-point form, *J. Fluid Mech.*, 203:1.

Cox, S.M., 1989, A similarity solution of the Navier-Stokes equations for two-dimensional flow in a porous-walled channel, Ph.D. dissertation, University of Bristol.

Glendinning, P., 1988, Global bifurcations in flows, *in* "New Directions in Dynamical Systems", T. Bedford and J. Swift ed., Cambridge University Press.

Hiemenz, K., 1911, Die Grenzschicht an einem in den gleichförmigen Flüssigkeitsstrom eingetauchten geraden Kreiszylinder, *Dinglers J.*, 326:321.

Homann, F., 1936, Der Einfluss grosser Zähigkeit bei der Strömung um den Zylinder und um die Kugel, *Z. angew. Math. Mech.*, 16:153.
Howarth, L., 1951, The boundary layer in three-dimensional flow. Part II: The flow near a stagnation point. *Phil. Mag.*, (7) 42:1433.
Newell, A.C., 1906, Side-roll instability – a new solution of the Guiness-Löwenbräu problem, *Proc. Irish Narrative Soc.*, 8:1.
Proudman, I. and Johnson, K., 1962, Boundary-layer growth near a stagnation point, *J. Fluid Mech.*, 12:161.
Raithby, G.D. and Knudsen, D.C., 1974, Hydrodynamic development in a duct with suction and blowing, *A.S.M.E. J. Appl. Mech.* 41:896.
Stuart, J.T., 1988, Nonlinear Euler partial differential equations: singularities in their solution, *in* "A Symposium to Honor C.C. Lin", D.J. Benney, F.H. Shu and C. Yuan ed., World Scientific Publishing, Singapore.
Taylor, C.L., Banks, W.H.H., Zaturska, M.B. and Drazin, P.G., 1990, Three-dimensional flow in a porous channel (to be published).
Terrill, R.M., 1964, Laminar flow in a uniformly porous channel, *Aero. Quart.*, 15:299.
Watson, E.B.B., Banks, W.H.H., Zaturska, M.B. and Drazin, P.G., 1990, On transition to chaos in a two-dimensional channel flow driven symmetrically by accelerating walls, *J. Fluid Mech.* 212: (in the press).
Watson, P., 1989, Symmetry breaking in a laminar channel flow driven by accelerating walls, M.Sc. dissertation, University of Bristol.
Zaturska, M.B., Drazin, P.G. and Banks, W.H.H., 1988, On the flow of a viscous fluid driven along a channel by suction at porous walls, *Fluid Dynamics Res.*, 4:151.

DRIFTING CONVECTION ROLLS INDUCED BY SPATIAL MODULATION

G. Hartung, F.H. Busse and I. Rehberg

Institute of Physics
University of Bayreuth
8580 Bayreuth, FRG

ABSTRACT

A sinusoidal variation of the height of a convection channel gives rise to the onset of time dependent convection. The drift rate of the convection rolls is proportional to the amplitude of the modulation, to the sine of the phase difference between the modulation at the upper and lower boundary and to the amount of asymmetric properties of the convecting fluid such as the curvature in the temperature dependence of the density. These results have been obtained experimentally through the measurement of convection in an annular channel heated from below.

INTRODUCTION

Rayleigh-Bénard convection is usually studied in the case of horizontally uniform external conditions. For many applications in geophysics or in engineering problems inhomogeneous conditions in the horizontal dimensions are encountered, however, and the role played by spatial modulations of external parameters becomes an important question. The problem of sinusoidal modulations of the height of the convection layer or of the temperatures at the boundaries was considered by Kelly and Pal (1978) and Pal and Kelly (1978). In this paper we report experimental observations of the onset of convection in a channel heated from below with a sinusoidally varying height.

Since the local Rayleigh number varies in a spatially modulated system, the onset of convection occurs inhomogeneously. Because of the spatial order imposed by the local variations of the Rayleigh number one might expect even less variations in time of the convection flow than in the case of a homogeneous layer. Indeed, the analysis of Kelly and Pal predicts a stationary onset of convection even in cases when the sinusoidal modulations at the lower and upper boundaries are out of phase. The experimental observations described in the following demonstrate the unexpected phenomenon of time dependent convection near threshold. A detailed analysis of the problem to be published elsewhere traces the origin of the discrepancy between existing theory and laboratory measurements to the neglection of non-Boussinesq effects by Kelly and Pal (1978).

Nonlinear Evolution of Spatio-Temporal Structures in Dissipative Continuous Systems
Edited by F.H. Busse and L. Kramer
Plenum Press, New York, 1990

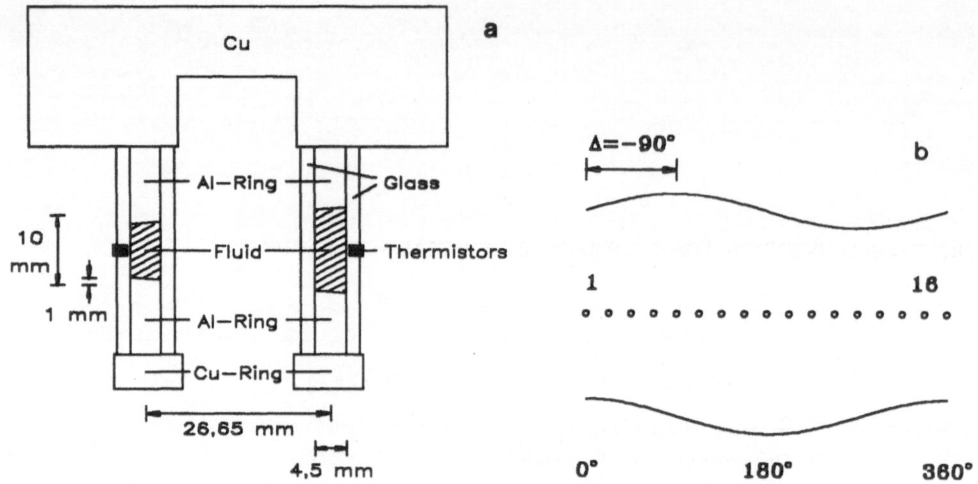

Fig. 1. a) Cross section of the convection apparatus. Domain of convecting fluid is hatched.

b) Postition of the thermistors and definition of the phase shift angle Δ.

THE ANNULAR CONVECTION CHANNEL

A cross-section of the cell used in the experiments is shown in figure 1a. In order to achieve periodic boundary conditions in one of the horizontal dimensions the cell was constructed as an annulus. Two anodized aluminium rings form the upper and lower boundaries. The inner and outer cylindrical boundaries are made out of plexiglas®. The upper copper block is kept at a constant temperature by a thermostatically controlled water bath, while the temperature difference is produced by a current flowing through a constantan wire which is winded around the lower copper ring. Copper constantan thermoelements are used to measure the temperature difference. The dimensions of the fluid container are also shown in figure 1a. The mean height, thickness and diameter of the annular fluid channel are 10, 4.5 and 25.65 mm, respectively. The variation of the upper and lower boundary with an amplitude of 1 mm was produced by oblique cuts of the aluminum rings which yield a sinusoidal variation of height along the circumference. Since the phase of modulation at the upper and lower boundaries is arbitrary, a general variation of height as shown in figure 1b is obtained. 16 thermistors are placed equidistantly in the outer plexiglas ring at the equatorial plane of the annular channel cell. Thermistor 1 is placed above the maximum height of the lower boundary and defines the origin of the coordinate around the circumference.

EXPERIMENTAL OBSERVATIONS

In order to realize a convecting fluid with asymmetric material properties, the annular channel was filled with water at a mean temperature near 4° C. The onset of convection is demonstrated in figure 2. To obtain optimal sensitivity for the onset of convection the thermistor signal at a temperature difference below the critical value has been subtracted from the measured signal. For better presentation of the data, the data for the interval $0° < \varphi < 180°$ are repeated for $360° < \varphi < 540°$.

The lines through the data are obtained by a Fourier series interpolation. In figures 3 and 4 only these interpolating functions are shown for simplicity . As must be expected the onset of convection occurs first at φ=135° where the maximum height of the channel is located. With increasing ΔT the entire channel is filled with convection rolls.

In order to visualize the time dependence of convection the thermistor signals must be recorded at different times for a fixed value of ΔT. Figure 3 shows typical results for opposite values of the phase shifts, Δ=-90° and Δ=+90°. At the onset of convection the rolls drift in the direction in which the channel is bent upwards (see figure 1b). Once the entire channel is occupied by convectiton rolls the drift is reversed. In both cases the average drift is roughly proportional to sinΔ. At Δ=180° the opposite drift is observed on the two sides of the maximal height. Occasionally a drift is observed even at Δ=0° due to imperfections of the channel. Because of the two different mechanisms for the drift apparent in the data it is not surprising that steady convection rolls are sometimes observed in parts of the channel as shown in figure 4. It has also been found that states with different numbers of rolls can be realized and that elimination of a roll pair can give rise to hysteresis phenomena when the Rayleigh number is increased first and decreased subsequently.

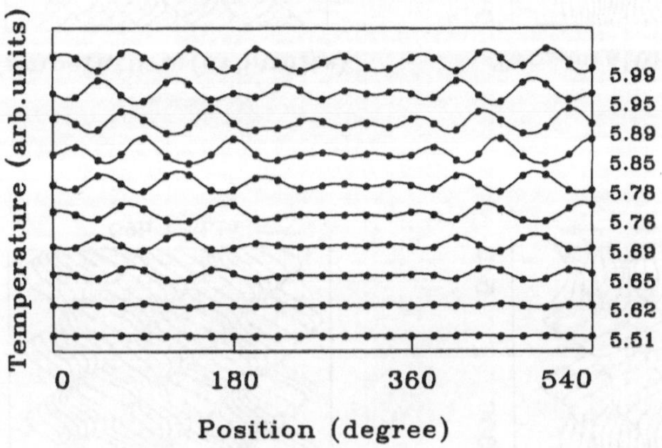

Fig. 2. Onset of convection for Δ=-90°. The numbers at the right side are the applied temperature differences. The mean temperature varies between 3.6° C and 3.8°C. The abscissa gives the angle φ measured in degrees.

DISCUSSION

In the presence of a modulation the basic state before onset of convection is characterized by a circulation with the spatial periodicity of the modulation. The advection by this circulation of the packets of the convection rolls appearing at the onset of convection seems to be responsible for the time dependence shown in figure 3a,c. As the convection becomes distributed more uniformly throughout the channel, the effects of the advection by the basic circulation cells tend to cancel in

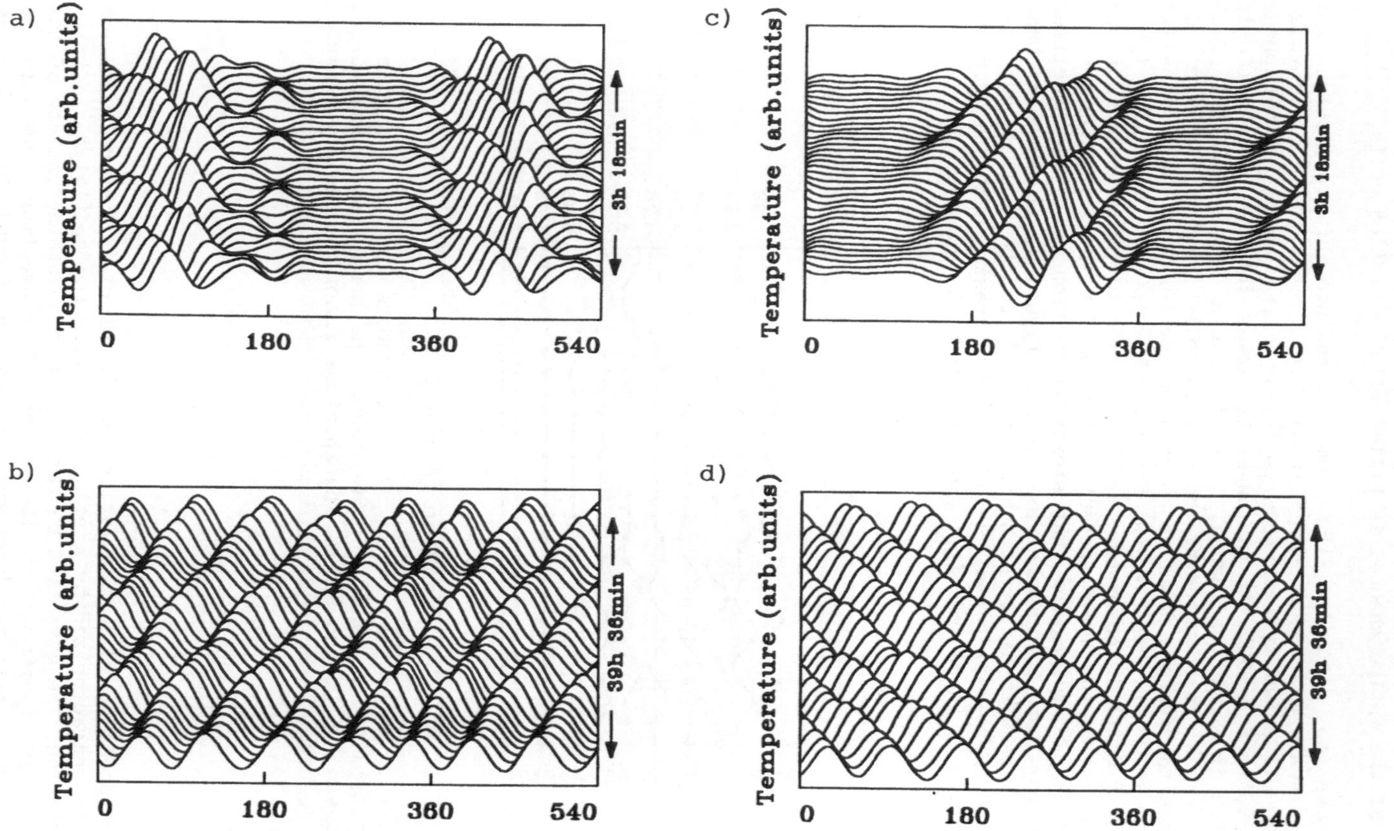

Fig. 3. Time dependence of convection. The lines represent measurements at equidistant time interval. The entire duration of the measurements is indicated on the right side of each graph.

a) $R = 1.02\ R_c$, $\Delta = -90°$, $T_{mean} = 3.7°$

b) $R = 1.21\ R_c$, $\Delta = -90°$, $T_{mean} = 4.3°$

c) $R = 1.01\ R_c$, $\Delta = +90°$, $T_{mean} = 4.3°$

d) $R = 1.26\ R_c$, $\Delta = +90°$, $T_{mean} = 4.9°$

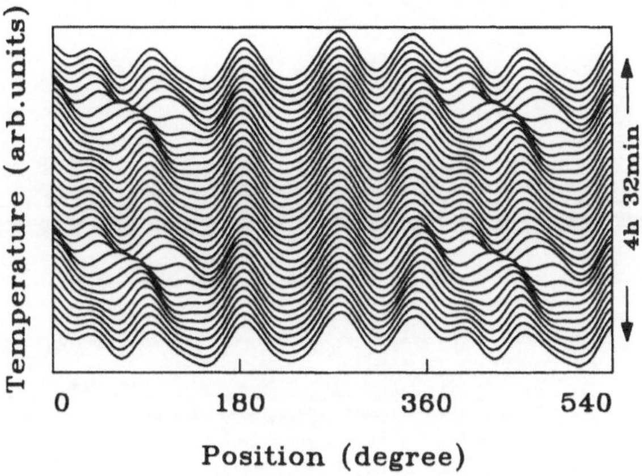

Fig. 4. Convection in the channel for $R = 1.09\ R_c$, $\Delta = +90°$, $T_{mean} = 4.5°$ at equidistant times. This figure indicates the coextistence of steady and drifting convection roles.

different regions of the channel. On the other hand a weaker nonlinear influence of the basic circulation cells becomes noticeable. Through Reynolds stresses a mean azimuthal flow is generated in the channel which causes the advection of the convection rolls in the opposite direction. Both types of advection depend on the asymmetry of the fluid channel in the vertical direction. Without this asymmetry the effects of advection vanish in agreement with the theoretical model of Pal and Kelly (1978).

It is not appropriate to outline the theory of the phenomena in more detail at this time, since not all numerical computations have yet been completed. Characterizing the amplitude of modulation by δ and the amount of asymmetry by γ we find that the drift near onset is of the order $\gamma\delta\sin\Delta$ while at higher Rayleigh number a drift of the order $\gamma\delta^2\sin\Delta$ is found. A detailed comparison between theoretical predictions and experimental measurements will be given in a future paper.

ACKNOWLEDGEMENT

The support by the Volkswagenstiftung for the research reported in this paper is gratefully acknowledged.

REFERENCES

Kelly, R.E., and Pal, D., 1978: Thermal Convection With Spatially Periodic Boundary Conditions: Resonant Wavelength Excitation, J. Fluid Mech. 86, 433-456

Pal, D., and Kelly, R.E., 1978: Thermal Convection With Spatially Periodic Nonuniform Heating: Nonresonant Wavelength Excitation, pp. 235-238 in "Heat Transfer 1978" (Proc. Int. Heat Transfer Conf.), Vol. 2.

ACKNOWLEDGMENT

REFERENCES

CONVECTION IN A ROTATING CYLINDRICAL ANNULUS WITH RIGID BOUNDARIES

M. Schnaubelt and F.H. Busse

Institute of Physics
University of Bayreuth
8580 Bayreuth, FRG

ABSTRACT

Convection driven by centrifugal buoyancy in a rotating cylindrical annulus exhibits dynamical behavior quite different from that of Rayleigh-Bénard convection in a layer heated from below. The present analysis extends the work of Or and Busse (1987) for stress-free cylindrical walls to the experimentally relevant case of no-slip boundaries. While the major bifurcations of convection flows are preserved, the details of the stability regions are changed and new instabilities are found.

1. INTRODUCTION

Convection driven by centrifugal buoyancy in a rotating cylindrical annulus cooled from within and heated from the outside represents the simplest realisation of a basic mechanism for heat transport that is responsible for motions in the atmospheres of the major planets, in the Earth's core and in rotating stars. The essential feature of a horizontal orientation of the axis of rotation is typical for the equatorial regions of the celestial bodies and can be modeled in the laboratory through use of the centrifugal force as effective gravity (Busse and Carrigan, 1976). Besides the planetary and stellar application the problem is of interest in own right, because the novel dynamics of convection in the presence of conical end surfaces of the annular region leads to motions that are quite different from those of ordinary Rayleigh-Bénard convection. Because the height of the annular region in the direction of the axis of rotation varies with distance from the axis, the onset of convection occurs in the form of traveling waves akin to the Rossby waves studied in the context of atmospheric dynamics. Inspite of the fact that the Proudman-Taylor condition remains approximately satisfied and the convection flow preserves its nearly two-dimensional character in the form of columns aligned parallel to the axis of rotation, there is a rich variety of instabilities and transitions that ultimately lead to the appearance of chaotic types of flow.

The possibility of the two-dimensional formulation of the problem leads to a considerable simplification of the numerical analysis of finite amplitude convection flows. Previous investigations (Busse and Or, 1985; Or and Busse, 1987) have taken advantage of this property and have

Nonlinear Evolution of Spatio-Temporal Structures in Dissipative Continuous Systems
Edited by F.H. Busse and L. Kramer
Plenum Press, New York, 1990

explored some representative parts of the parameter space. At the cylindrical walls stress-free boundary conditions have been assumed in that work which can be justified as applicable even in the case of rigid boundaries if the rotation rate is sufficiently high such that thin Stewartson layers with a thickness of the order $E^{1/3}$ are formed (Busse, 1970). For a comparison with experiments, however, the range of asymptotically small Ekman numbers E is not appropriate and the no-slip boundary condition at the cylindrical walls must be taken into account in a direct way. This problem is addressed by the present paper. Because the emphasis of the analysis is on qualitative and quantitative differences from the previous work the formulation of the problem and the numerical methods will be discussed only briefly in section 2. The principal results are described in section 3 and concluding remarks are added in section 4.

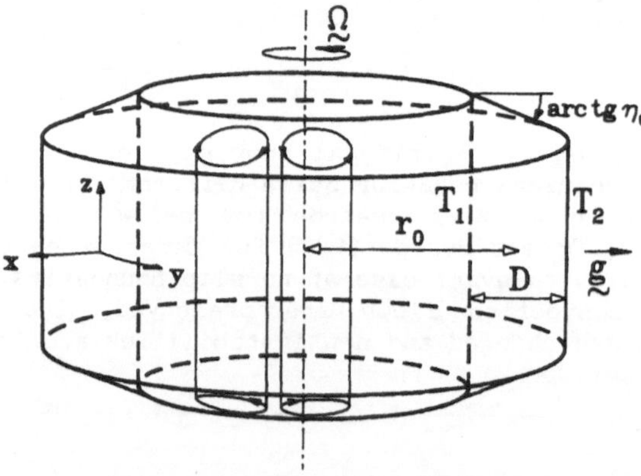

Fig. 1. Geometrical configuration of the Rotating Cylindrical Annulus

2. MATHEMATICAL FORMULATION OF THE PROBLEM

We consider the cylindrical annulus shown in figure 1 and introduce a dimensionless description of the problem by using the gap width D as length scale, D^2/ν as time scale where ν is the kinematic viscosity, and $(T_2-T_1)\cdot P$ as scale of the temperature. In the limit of high rotation rates the velocity field $\underset{\sim}{u}$ can be seen as a modification of a geostrophic velocity field

$$\underset{\sim}{u} = \nabla\times\underset{\sim}{k}\psi\ (x,y) + \underset{\sim}{u}' \tag{1}$$

where $\underset{\sim}{u}$ is of the order η_0 smaller than the geostrophic component on the right end side of (1). For the orientation of the Cartesian system of coordinates and unit vectors $\underset{\sim}{i}$, $\underset{\sim}{j}$, $\underset{\sim}{k}$ see figure 1. η_0 is the tangens of the small angle of the conical boundaries with the equatorial plane of the system. In order to determine the streamfunction ψ which is independent of z according to the geostrophic balance we consider the averaged z-component of the curl of the equation of motion

$$\left(\frac{\partial}{\partial t}+ \frac{\partial}{\partial y}\psi\frac{\partial}{\partial x} - \frac{\partial}{\partial x}\psi\frac{\partial}{\partial y}\right) \Delta_2\psi - \eta^* \frac{\partial}{\partial y}\psi = -R \frac{\partial}{\partial y}\Theta + \Delta_2{}^2\psi \tag{2a}$$

This equation together with the heat equation for the deviation Θ of the temperature from the state of pure conduction

$$P\left(\frac{\partial}{\partial t}+\frac{\partial}{\partial y}\psi\frac{\partial}{\partial x}-\frac{\partial}{\partial x}\psi\frac{\partial}{\partial y}\right)\Theta+\frac{\partial}{\partial y}\psi=\Delta_2\Theta \tag{2b}$$

provides the basis for the numerical analysis. The two-dimensional Laplacian Δ_2 is defined by $\Delta_2 \equiv \nabla^2-(\underset{\sim}{k}\cdot\nabla)^2$. The rotation parameter η^*, the Rayleigh number R, and the Prandtl number P are defined in the usual way,

$$\eta^*=\frac{4\eta\cdot\Omega D^3}{L\nu},\quad R=\frac{\gamma g(T_2-T_1)D^3}{\nu\kappa},\quad P=\frac{\nu}{\kappa} \tag{3}$$

For further details we refer to Busse (1970, 1986). Although η_0 is regarded as a small parameter of the problem, η^* is large compared to unity in typical applications. The influence of Ekman layers at the conical boundaries at $z=\pm L/2D$ will be neglected in the following, but the no-slip conditions at $x=\pm\frac{1}{2}$ will be taken into account,

$$\psi\pm a=\frac{\partial}{\partial x}\psi=\Theta=0 \qquad \text{at } x=\pm\tfrac{1}{2} \tag{4}$$

The constant a in these conditions refers to the possibility of a mean flow in the azimuthal direction with the integrated flux 2a. The solution of equations (2) together with the boundary conditions (4) is obtained in the form of a Galerkin expansion

$$\psi=\sum_{n,m}\left[\hat{a}_{nm}\sin m\alpha(y-ct)+a_{nm}\cos m\alpha(y-ct)\right]g_n(x)+a(4x^3-3x) \tag{5a}$$

$$\Theta=\sum_{n,m}\left[\hat{b}_{nm}\sin m\alpha(y-ct)+b_{nm}\cos m\alpha(y-ct)\right]\sin n\pi(x+\tfrac{1}{2}) \tag{5b}$$

where the functions $g_n(x)$ have been introduced by Chandrasekhar (1961, p.635). They vanish with their derivatives at $x=\pm\frac{1}{2}$. In order to obtain a finite system of algebraic equations for the coefficients we assume that the coefficients vanish for

$$n+m>N_T \tag{6}$$

where the truncation parameter can be adjusted to check the quality of the approximation.

Once a solution of the form (5) has been determined by a Newton-Raphson iteration method its stability can be investigated by the superposition of infinitesimal disturbances of the form

$$\tilde{\psi}=\sum_{n,m}\tilde{a}_{nm}\,g_n(x)\exp\{i(m\alpha+d)(y-ct)+\sigma t\} \tag{7a}$$

$$\tilde{\Theta}=\sum_{n,m}\tilde{b}_{nm}\sin n\pi(x+\tfrac{1}{2})\exp\{i(m\alpha+d)(y-ct)+\sigma t\} \tag{7b}$$

The linear homogeneous equations for the coefficients $\tilde{a}_{nm}$, $\tilde{b}_{nm}$ represent an eigenvalue problem for the growth rate σ. Whenever the maximum of the real part σ_r of σ becomes positive as a function of d the stationary solution (5) is unstable; otherwise it is regarded stable.

3. STABILITY REGIONS OF CONVECTION FLOWS

When figure 2a is compared with the corresponding figure 1 of Or and

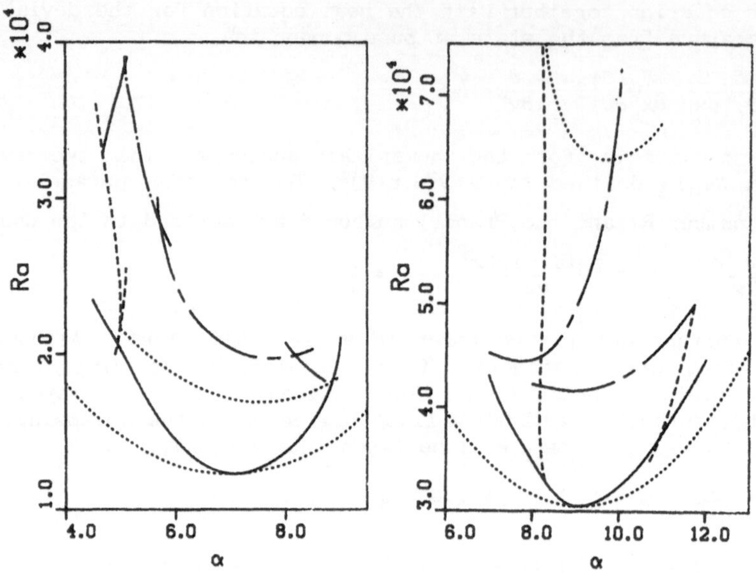

Fig. 2. Stability regions of convection flows for $\eta^* = 1600$ in the case $P = 0.7$ (left) and $P = 7$ (right). Convection in the form of symmetric thermal Rossby waves sets in at the lower dotted line. Transition to mean flow convection occurs at the upper dotted line. The solid lines denote the onset of side band instabilities. The mean flow solution becomes unstable to vacillations at the long dash-short dash boundary for $P = 0.7$. For $P = 7$ there is a range of intermediate instability bounded by the long-dash-short dash curves. The dash-dotted line (left) describes the onset of the instability IV referred to in the text.

Busse (1987) a general similarity can be noticed. But the details show significant differences. The solution setting in at the critical Rayleigh number R_c and describing symmetric rolls in the form of thermal Rossby waves remains stable until the mean flow instability occurs about 40 % above R_c. At still higher Rayleigh numbers the mean flow solution is replaced by the onset of vacillations as in the case of Or and Busse (1987). Towards small wavenumbers α the mean flow solution remains stable up to relatively high Rayleigh numbers and the stability boundary is no longer given by the vacillating instability but instead by various instabilities with finite values of the Floquet wavenumber d. While the imaginary part of the growth rate σ does not vanish in general for finite values of d it vanishes with d for instability IV as the latter parameter approaches zero in contrast to the vacillating instability. Because most of the new instabilities occur as instabilities of the mean flow solution, there are no symmetry properties left by which they can be distinguished. A further exploration of the manifold of convection solutions is planned by time integrations of the basic equations in those cases where the stability boundaries are characterized by small or vanishing values of d, such that finite growth rates appear for d=0 in the region beyond the stability boundary.

In the case of figure 2b the mean flow instability is shifted to high Rayleigh numbers and the stability region of the symmetric roll solution

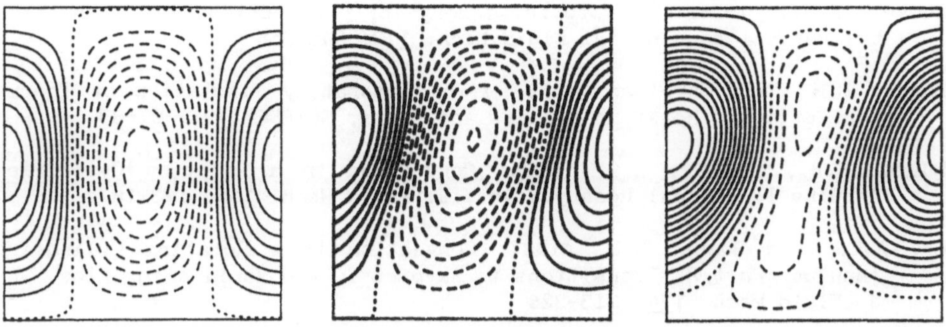

Fig. 3. Streamlines, ψ=const, of symmetric thermal Rossby waves (left, R=17000), of the mean flow solution without y-average of ψ (middle, R = 21000) and with y-average of ψ included (right, R = 21000). All graphs are for η^{*} = 1600, P = 0.7.

is intersected by the appearance and disappearance of an instability characterized by a vanishing imaginary part but a finite real part of σ at d=0. Only at considerably higher values of R does the mean flow instability set in. The unusual intermediate region of instability requires further study which be attempted in the special case of d=0.

The stream line pattern of the symmetric convection solution and of the mean flow solution are shown in figure 3. They resemble comparable graphs in the paper of Or and Busse and clearly demonstrate the significant change that occurs after the onset of the mean flow instability.

4. CONCLUSION

The problem of convection driven by centrifugal buoyancy in a rotating cylindrical annulus is the representative model for a typical dynamical mechanism operating the deep atmospheres of celestial bodies and in the Earth's core. In spite of the restriction to two spatial dimensions, the basic equations (2) exhibit at least two different types of solutions which are steady with respect to a drifting frame of reference and numerous other solutions with more complicated time dependence. The variety of instabilities in the presence of no-slip cylindrical walls appears to be even richer than in the case of stress-free boundaries. Since there are few examples of realistic fluid dynamical systems confined to two dimensions which exhibit such interesting features, further theoretical studies and eventual comparisons with experimental observations are warranted.

The support under grant Bu589/2 by the Deutsche Forschungsgemeinschaft for the research reported in this paper is gratefully acknowledged.

REFERENCES

Busse, F.H., 1970: Thermal Instabilities in Rapidly Rotating Systems, J. Fluid Mech. 44, 441-460

Busse, F.H., 1986: Asymptotic Theory of Convection in a Rotating Cylindrical Annulus, J. Fluid Mech. 173, 545-556

Busse, F.H., and Carrigan, C.R., 1976: Laboratory Simulation of Thermal Convection in Rotating Planets and Stars, SCIENCE 191, 81-83

Busse, F.H., and Or, A.C., 1986: Convection in a Rotating Cylindrical Annulus I: Thermal Rossby Waves, J. Fluid Mech. 166, 173-187

Or, A.C., and Busse, F.H., 1987: Convection in a Rotating Cylindrical Annulus. Part 2. Transitions to Asymmetric and Vacillating Flow, J. Fluid Mech. 174, 313-326

COMPLEX DEMODULATION TECHNIQUES FOR EXPERIMENTS ON TRAVELING-WAVE CONVECTION

Paul Kolodner and Hugh Williams

AT&T Bell Laboratories
Murray Hill, NJ 07974 USA

ABSTRACT

Near onset, convection in binary fluid mixtures in a one-dimensional geometry is dominated by the interaction between oppositely-propagating traveling waves. In this paper, we explore the use of shadowgraphic flow visualization and complex demodulation techniques to extract the spatiotemporal behavior of these waves. These techniques can allow a quantitative comparison with theories based on coupled complex Ginzburg-Landau equations.

INTRODUCTION

Experimental systems which allow the study of the dynamics of nonlinear traveling waves have lately been the subject of intense interest. Examples of such experiments include spiral waves in Taylor-Couette flow [1], electrohydrodynamic convection in nematics [2], transverse waves propagating along convection rolls in low-Prandtl-number fluids [3], and traveling-wave convection in binary fluid mixtures [4,5]. Each of these four systems exhibits a linear instability to a state of traveling waves, and each exhibits dynamical behavior due to the nonlinear interaction between different traveling-wave components. Experiments on convection in binary fluids suffer from the fact that time scales are determined by diffusion, making them quite time-consuming. However, this system also has some compelling advantages. Convection experiments can be performed with extreme precision and stability, and the relevant parameters of the fluid, the apparatus, and the equations governing the physics can be accurately measured [6,7] and calculated [8,9]. In a narrow rectangular [4,5,10] or annular [11] geometry, the waves are essentially one-dimensional. Furthermore, with appropriate choice of the separation ratio ψ, the experiment can be operated in the "weakly nonlinear" regime [4,5], by which is meant that the nonlinear traveling waves are truly a minor perturbation on the underlying quiescent state - the basic requirement for perturbation-type approaches to understanding such systems. Thus, our goal in pursuing these experiments is to provide a quantitative test of one-dimensional theories of weakly nonlinear traveling waves.

In constructing such a theory, the assumption is made that the flow fields can be written as a sum of left- and right-going waves whose complex amplitudes

Nonlinear Evolution of Spatio-Temporal Structures in Dissipative Continuous Systems
Edited by F.H. Busse and L. Kramer
Plenum Press, New York, 1990

$A_L(x,t)$ and $A_R(x,t)$ vary only slowly in time and space. One then writes a systematic expansion of the fluid equations in powers of the wave amplitudes and their derivatives. The lowest-order nonlinear truncation of these coupled complex Ginzburg-Landau equations has the following form [12]:

$$\tau_0\left(\frac{\partial A_R}{\partial t}+s\frac{\partial A_R}{\partial x}\right) = \epsilon(1+ic_0)A_R+\xi_0^2(1+ic_1)\frac{\partial^2 A_R}{\partial x^2}-g_1(1+ic_2)|A_R|^2A_R$$
$$-g_2(1+ic_3)|A_L|^2A_R; \tag{1a}$$

$$\tau_0\left(\frac{\partial A_L}{\partial t}-s\frac{\partial A_L}{\partial x}\right) = \epsilon(1+ic_0)A_L+\xi_0^2(1+ic_1)\frac{\partial^2 A_L}{\partial x^2}-g_1(1+ic_2)|A_L|^2A_L$$
$$-g_2(1+ic_3)|A_R|^2A_L. \tag{1b}$$

In a finite geometry, these equations must be supplemented by boundary conditions at the endwalls. The study of such systems of equations has produced a great deal of understanding of experiments on convection in binary fluids [8,9,12-14]. However, there remain several open questions which might be resolved by accurate experimental measurements of the complex wave amplitudes A_L and A_R :

1.) The transient linear waves seen at onset [6] can be quantitatively explained by setting the nonlinear coefficients $g_1 = g_2 = 0$ in Eqs. (1), and one aspect of this is that the measured [6] and calculated [8] values of the linear coefficients agree quite well. However, the details of the boundary conditions that the wave amplitudes A_L and A_R must satisfy are not well understood [12,15]. These boundary conditions determine with exquisite sensitivity the shape of the wave amplitudes near the endwalls [12] and the nature of the selected modes in the linear [6] as well as nonlinear [4] states. Thus, an important lack of understanding exists, even in the linear regime.

2.) While some understanding of the lowest-order nonlinearity in (1) exists [9], it is clear that higher-order nonlinear terms must be added to these equations in order to describe the experimentally-observed subcritical bifurcation to the first nonlinear state [4,5,16]. However, it is not obvious which terms are the important ones, and calculations of the relevant coefficients have not yet been performed. It is hoped that quantitative measurements of the wave amplitudes will illuminate this theoretical problem. In some sense, we hope to experimentally answer the question, what are the higher-order analogs of (1) that the experimental wave fields actually satisfy?

3.) A more qualitative question concerning the nonlinear parts of Eqs. (1) remains open. Cross has shown [12] that solutions of Eqs. (1) with all $c_i = 0$; $g_1 = g_2 > 0$ can exhibit time-dependent nonlinear behaviors which are remarkably similar to experimental observations [4,5]. However, it is known that, with the non-zero values of c_i exhibited by the fluids used in these experiments [7,9], these equations are subject to modulational instabilities [17]. These are known theoretically [18] and experimentally [19] to be the cause of strongly time-dependent dynamical behavior. Furthermore, neither of these two mechanisms has been explored for the experimentally relevant case of a subcritical bifurcation. So the relative importance of modulational instabilities and the mechanism in Ref. 12 remains to be determined. Clarifying the role of modulational instabilities in traveling-wave convection will require an experimental understanding of the full, complex wave amplitudes.

APPARATUS AND EXPERIMENTAL PROCEDURE

A diagram of the experimental apparatus is shown in Fig. 1. The convection cell consists of a mirror-polished, rhodium-plated copper bottom plate, a sapphire top plate, and sidewalls formed by a plastic frame of height d = 0.35 cm. The lateral dimensions of the cell, in units of d, are $3.0 \times \Gamma$, where the long aspect ratio Γ can be continuously over the range 12.2 to 20.9 by means of a screw driven by a stepper motor. The bottom plate is heated with Peltier cells, and temperature-regulated water flows over the top plate. The vertical temperature difference applied across the cell is $\Delta T_c = 2.94\,^\circ$ C at onset, and this is regulated with a stability of $\pm$ 0.1 mK. A surrounding radiation shield and vacuum box provide thermal isolation from the external environment. The fluid used in these experiments is a 0.30 % by weight solution of ethanol in water, with a mean temperature of 21.43 ° C, a separation ratio $\psi = -0.021 \pm 0.001$, Prandtl number $P = 6.97$, and Lewis number $L = 0.0079$ [7].

The shadowgraphic flow-visualization system sketched in Fig. 1 is modeled after the design of Croquette [20] and is essentially an inverted telescope. A point source of white light illuminates the cell through a collimating lens f_1, and, after reflection from the bottom plate of the cell and refraction by the lateral temperature gradients in the convecting fluid, is recollimated by a smaller lens f_2. This system forms an inverted, demagnified image one focal length f_2 behind the recollimating lens. It can easily be shown [20] that this image exhibits no contrast

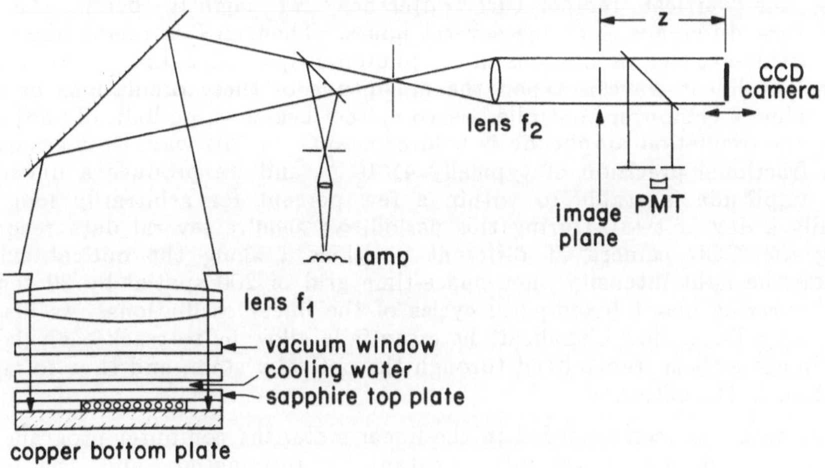

FIGURE 1. Sketch of experimental apparatus. The convection cell, which consists of a copper bottom plate and a sapphire top plate with cooling water flowing over it, sits inside a radiation shield and a vacuum box, neither of which are shown. The optical system consists of the inverting telescope formed by lenses f_1 and f_2 and is illuminated through a beamsplitter by the lamp. The CCD camera and photomultiplier tube PMT view the cell image which is formed a distance f_2 behind the small lens. The focalization distance z to the camera plane is varied between 30 mm and 60 mm in these experiments.

due to the convection. Rather, a ray of light which emerges from a given point in the cell at an angle θ' with respect to the optical axis will pass through the corresponding point in the image plane, but now at an angle $\theta = \theta' f_1/f_2$. Thus, a camera placed a distance z behind the image plane records a shadowgrahic visualization of the *image* which is equivalent to that which would be seen at a distance $z' = z(f_1/f_2)^2$ from the *cell* if we merely illuminated it with collimated light. Since $f_1/f_2 \approx 10$, the demagnification allows the use of an extremely compact optical system. We typically use a focalization distance of only $z = 30$ to 60 mm.

In the camera plane of the optical system is placed a Thompson-CSF CCD camera chip, which is an array of 384$\times$576 pixels. Data is acquired from a row of 200 patches of 4$\times$32 pixels. The long sides of the patches are parallel to the short side of the image of the cell and cover the central 80% of its width. Approximately 15 patches cover one wavelength of the convection pattern. Thus, we measure the light intensity averaged along the short direction of the cell - this is the y direction - as a function of position in the direction of propagation of the waves parallel to the long side of the cell - this is the x direction. With an integration time of 12.75 sec (oscillation periods are typically 200 sec), the noise level in the data is dominated by shot noise, which appears in the fourteenth bit of the 16-bit digitized signal. At an equivalent position viewed through a beamsplitter, a photomultiplier tube measures the intensity of the light transmitted by a narrow slit which is parallel to the convective rolls in the image. This signal is extremely useful for monitoring and controlling the convective state (see below).

The procedure employed in these experiments is quite straightforward. We begin by setting the long aspect ratio Γ of the cell at some desired value and holding the vertical temperature difference ΔT slightly below the onset temperature difference ΔT_c for several hours. Then, ΔT is raised above ΔT_c, and smoothly-growing oscillations in the photomultiplier signal reveal the presence of transient linear waves. When the magnitude of these oscillations exceeds a preset value, a computer-controlled servo system begins to periodically adjust ΔT so that the oscillation amplitude is held constant. In this way, we measure ΔT_c with a fractional precision of typically 4×10^{-5}, and we produce a linear state whose amplitude is stable to within a few percent for arbitrarily long times (typically a day or two). During this period, we acquire several data records by placing the CCD camera at different positions z along the optical axis and sampling the light intensity on a space-time grid of 200 spatial by 32 temporal points, covering about 5 temporal cycles of the linear oscillations. As discussed below, these linear flow visualizations essentially allow us to track each light ray in the image as it is transmitted through the optical system, and thus to measure distortions in the optics.

Once data has been recorded in the linear state, the computer program which holds the oscillation amplitude constant is terminated, and the applied temperature difference is increased above the measured onset temperature by a small amount. Setting $\epsilon \equiv (\Delta T - \Delta T_c)/\Delta T_c = 3\times10^{-4}$ allows a nonlinear state to grow up and stabilize over the course of a day or two. Then, another CCD-camera run is recorded. The nonlinear states in this system are characterized in the time domain by slow modulations, with a modulation period of typically 40 linear oscillation periods at onset [4,5]. Thus, to record two such cycles, we must acquire a substantial amount of data - sometimes as much as 200 spatial points by 800 temporal points. Using calibration information from the linear data taken previously, complex demodulation techniques are then applied to extract information about the left- and right-going wave amplitudes.

COMPLEX DEMODULATION OF SHADOWGRAPH DATA

The physics of shadowgraphs is well understood [20,21]. However, a quantitative measurement requires that we also understand distortions in the optical system. For this purpose, we make reference to Fig. 2, which shows a typical light ray traveling from the image plane to the camera. A ray emerging at a distance x from one end of the image of the cell is refracted by the convection through an angle $\theta(x,t)$ which is proportional to the horizontal gradient of the temperature field in the cell. The function $\theta(x,t)$ is the quantity we wish to recover from our data for the purpose of demodulation. Distortions in the optical system are represented in this sketch by the wavy surface behind the image, and they cause a further deflection through a time-independent vector angle $\vec{\alpha}(x,y)$; since the data are averaged along the direction y parallel to the short side of the cell, Fig. 2 shows the average projection $\alpha_x(x)$ in the x direction. At the camera plane a distance z behind the image plane, this ray is mapped to a time-averaged position

$$x'(x,z) = x + z\alpha_x(x). \tag{2}$$

Thus, the light intensity at x' is given by [21]

$$I(x',z,t) = I_0(x)[1+z(\frac{d}{dx}\theta(x,t)+\nabla\cdot\vec{\alpha}(x,y))]^{-1}. \tag{3}$$

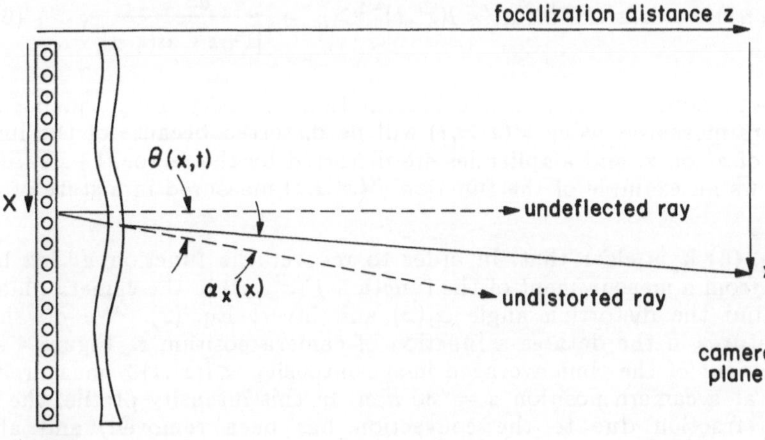

FIGURE 2. Definitions of the symbols used in the discussion of the back end of the optical system. An image of the cell is formed in the image plane, where counterpropagating waves travel in the x direction. A light ray which emerges from this image at a distance x from the end of the image is refracted by the convection pattern by an angle $\theta(x,t)$ and travels a distance z to the camera plane, where it strikes the CCD camera at a position $x'(x,t)$. In the camera, a row of patches is arrayed along x', and the light intensity is averaged along the transverse direction y. The effects of distortions in the optical system are represented by the wavy surface behind the image of the cell. They contribute an additional, time-independent deflection angle $\vec{\alpha}(x,y)$ whose projection $\alpha_x(x)$ is shown.

In this equation, $I_o(x)$ is the intensity of illumination provided by the lamp, and it is understood that the right side of the equation is averaged along y. Part of the problem in interpreting this equation is that the measurement made at the position x' in the camera plane must be mapped back the corresponding position x in the image plane - in other words, we must find a way to invert Eq. (2). More on this below.

Equation (3) contains both the signal from the time-dependent convection pattern, in a narrow band of temporal frequencies centered near the relatively high linear oscillation frequency ω_o, and drifts at very low frequencies, typically associated with changes in the laboratory temperature. These can be effectively separated by filtering in the time domain. Lopass filtering of the *inverse* of the light-intensity signal separates out slow drifts and steady distortions:

$$<I(x',z,t)^{-1}>_{lo} = I_o(x)^{-1}[1+z\nabla\cdot\vec{\alpha}], \tag{4}$$

while hipass filtering isolates the signal component due to convection:

$$<I(x',z,t)^{-1}>_{hi} = I_o(x)^{-1}z\frac{d\theta}{dx}(x,t). \tag{5}$$

Thus, in the ratio of these two components,

$$f'(x',z,t) \equiv <I(x',t)^{-1}>_{hi}/<I(x',t)^{-1}>_{lo} = \frac{z\frac{d\theta}{dx}(x,t)}{[1+z\nabla\cdot\vec{\alpha}(x,y)]}, \tag{6}$$

slow drifts and nonuniformity in the illumination intensity are normalized out. Wavenumbers measured using $f'(x',z,t)$ will be distorted because of the implicit dependence of x' on x, and amplitudes are distorted by the factor $[1+z\nabla\cdot\vec{\alpha}(x,y)]$. Figure 3 shows an example of the function $f'(x',z,t)$ measured in a state of linear waves.

From Eq. (6) it is clear that, in order to recover the function $d\theta/dx$ in the image plane from a measurement of the function $f'(x',z,t)$ in the camera plane, we must first find the distortion angle $\alpha_x(x)$ and invert Eq. (2). We do this by following features in the data as a function of camera position z. Figure 4 shows one measurement of the time-averaged image intensity $<I(x',t)>$ measured in a linear state, at a camera position $z =$ 30 mm. In this intensity profile, the time-dependent refraction due to the convection has been removed, and all the remaining structure is due to small scratches in the bottom plate of the convection cell. These scratches are too fine to have a noticeable effect on the convection, but they scatter light out of the optical system and thus show up with high contrast in the camera signal. To use these features as a length calibration scale, we compute time-averaged light intensity profiles for three different focalization distances z. For each profile, we map the abscissas in a small domain around the ith feature by applying the transformation

$$x'_i(z) = a_i + b_i z. \tag{7}$$

The three shifted profiles are plotted on the same graph, and the parameters a_i and b_i are varied until the mapped features best overlap. We can then identify b_i as the distortion angle α_x at the image-plane position $x_i = a_i$. Figure 5 shows a set of such measurements, along with a fit to a smooth function of x. With this function $\alpha_x(x)$, we can associate each point x' in the camera plane at z with a

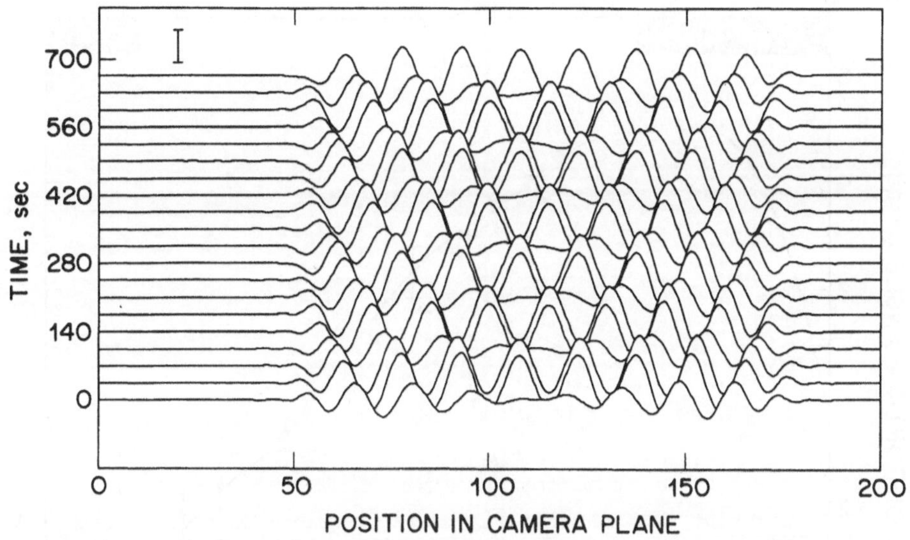

FIGURE 3. The filtered data function $f'(x',z,t)$ of Eq. (6) shown as a function of time and camera-plane coordinate x' for a state of linear counterpropagating waves in a cell of length $\Gamma = 17.243$, at a focalization distance $z = 30$ mm. The horizontal coordinate is the number of the patch of pixels on the CCD chip. The cell extends from patch 53 to patch 174, so that, for data recorded outside this interval, $f'(x',z,t) = 0$. The vertical bar in the upper left corner represents the magnitude of the signal caused by a 1% change in light intensity. Near the edges of the cell, outwardly-propagating waves are dominant. In the center, their interference produces a standing wave.

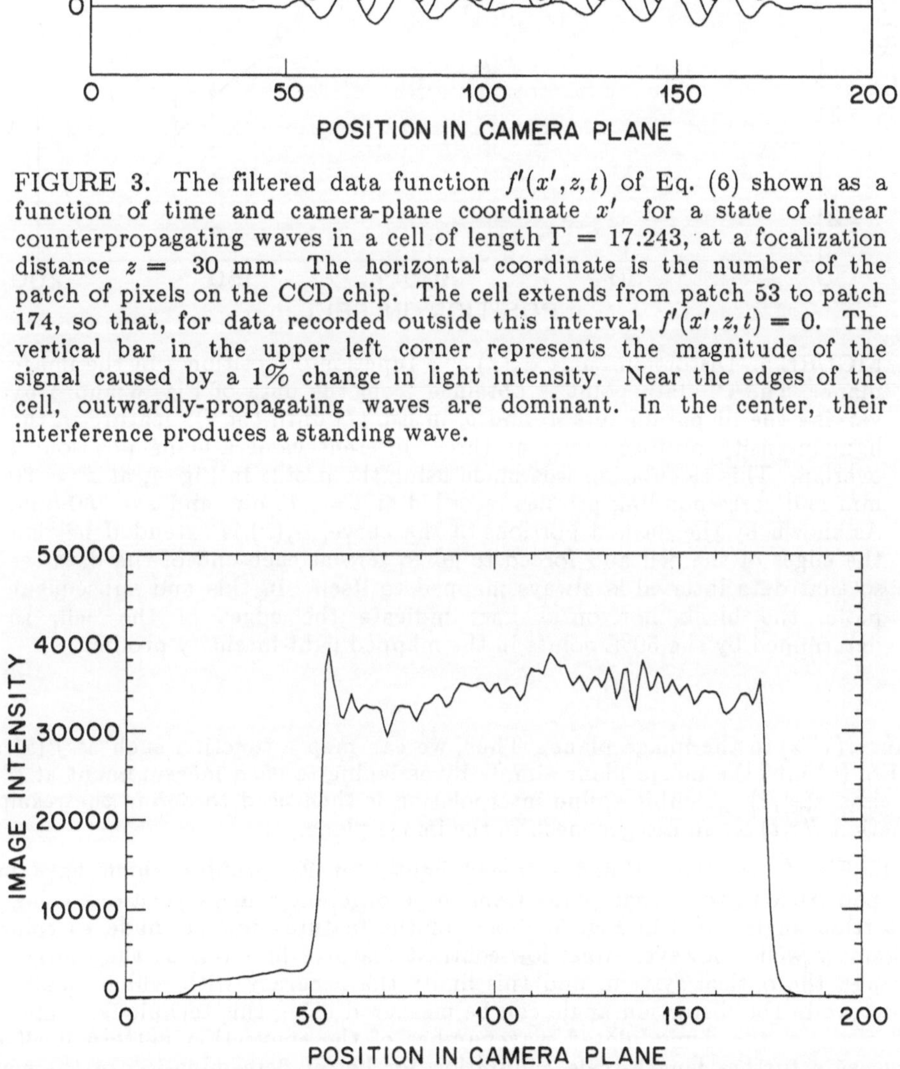

FIGURE 4. Time-averaged light intensity profile $<I(x',t)>$ calculated from the linear data used to make Fig. 3. The sharp features are caused by scattering of light out of the optical image by fine scratches on the bottom plate of the convection cell and conveniently serve as fiducial marks for length calibration of the optics.

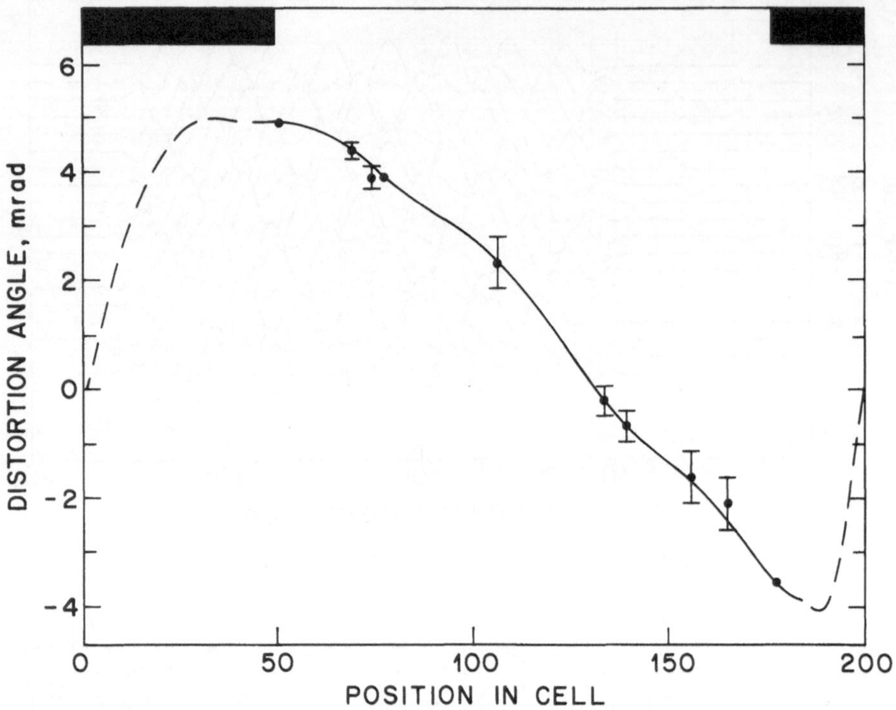

FIGURE 5. Distortion angle $\alpha_x(x)$ as a function of position x in the image plane. The ith data point is obtained from the data of Figs 3 and 4 by varying the fit parameters a_i and b_i in Eq. (7) until the ith feature in the light-intensity profiles made at three different camera-plane positions z overlap. This calculation was made using the profile in Fig. 4, at $z =$ 30 mm and corresponding profiles recorded at $z =$ 45 mm and $z =$ 60 mm. As shown by the dashed portions of the curve, $\alpha_x(x)$ is extended beyond the edges of the cell and forced to go to zero at each end of the data set so that data interval is always mapped to itself. In this and subsequent plots, the black horizontal bars indicate the edges of the cell, as determined by the 50% points in the mapped light-intensity profile.

point $x(x',z)$ in the image plane. Thus, we can map a function such as $f'(x',z,t)$ in Eq. (6) into the image plane simply by assigning to each measurement at x' an abscissa $x(x',z)$. A cubic spline interpolation is then used to define the resulting function $f(x,t)$ on an integer mesh in the image plane.

In Fig. 6, we show timed-averaged light-intensity profiles which have been mapped back to the image plane from three different camera positions using the distortion angle $\alpha_x(x)$ in Fig. 5. Some of the features can be made to coincide extremely well. However, fine, low-contrast features blur out as they propagate through the optical system, and this limits the accuracy with which small-scale structure in the distortion angle can be measured using this technique. Later, we will see that the demodulated wavenumber of the convection pattern itself can serve as a further length-scale calibration for better determination of the small-scale structure of the distortion angle. For the moment, let us notice that, since the ends of the profiles can be mapped together with great accuracy, and since the length of the actual cell is known quite well, the precision of the overall length calibration given by this technique is quite high - of the order of 1%.

Notice that, while the *positions* of small features in the intensity profiles in Fig. 6 can be made to coincide by mapping, the actual *intensities* of the three profiles do not coincide at every point. This is caused by the amplitude distortion factor $[1+z\nabla\cdot\vec{\alpha}(x,y)]$ in Eq. (6). As yet, we do not have any information on the dependence of the distortion angle $\vec{\alpha}$ on the transverse position y, and so we are not yet able to calculate this correction. Below, we will use the dependence of the demodulated wave amplitudes on the focalization distance z for this purpose.

With optical distortions removed from the data, we can now perform the actual complex demodulation. The assumption underlying this technique is that the mapped data $f(x,t)$ can be written in the form

$$f(x,t) = A_L(x,t)\cos(k_L x+\omega t)+A_R(x,t)\cos(k_R x-\omega t), \tag{8}$$

where the amplitudes $A_{L,R}$ and the wavenumbers $k_{L,R}$ vary slowly in time and space. By computing spatial Fourier spectra at several times, we indeed find that there is a mean wavenumber k_d and a bandwidth $\delta k << k_d$ such that essentially all the wave energy lies in the interval $[k_d-\delta k, k_d+\delta k]$. We then define

$$\Delta k_L(x,t) \equiv k_L(x,t)-k_d; \tag{9a}$$

$$\Delta k_R(x,t) \equiv k_R(x,t)-k_d. \tag{9b}$$

Similarly, we can define a mean temporal frequency ω_d and a bandwidth $\delta\omega << \omega_d$.

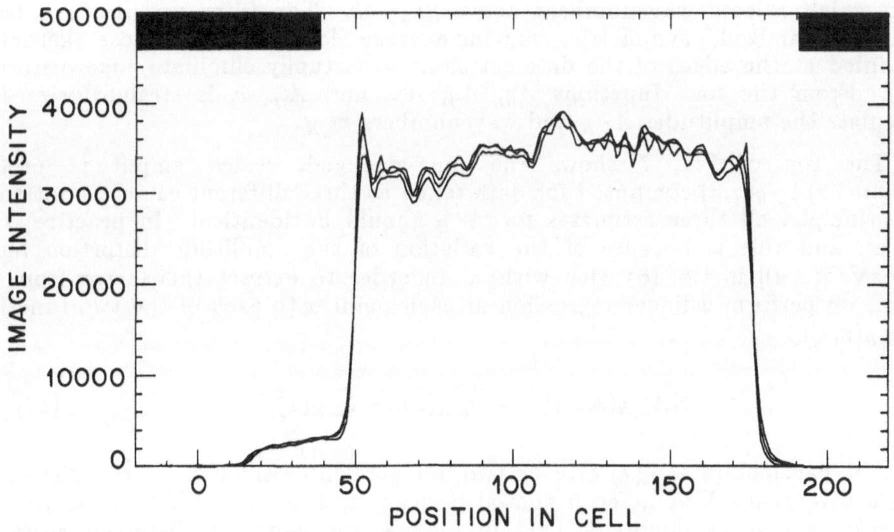

FIGURE 6. The three light-intensity profiles used to construct Fig. 5 are shown mapped into the image plane using the smooth function $\alpha_x(x)$. This mapping is successful at constraining some well-defined features to coincide, but fine features which are blurred out at large focalization distances z are harder to match. The 50% points, which are used to define the edges of the cell, can be very accurately located and are always forced to coincide.

In a state of one-dimensional traveling waves, each wave component corresponds to a different peak in a two-dimensional Fourier spectrum of the function $f(x,t)$. Thus, in principle, complex demodulation can be efficiently performed by computing a two-dimensional fast-Fourier-transform of $f(x,t)$, separately shifting each peak to the origin, and performing an inverse fft. However, because the data only comprise a small number of spatial wavelengths - typically 9 - this technique will perform poorly at the edges of the cell. Instead, we resort to direct demodulation by multiplying $f(x,t)$ by sines and cosines at k_d and ω_d and stripping off the resulting second harmonics by lopass filtering. We use a lopass filter which correctly handles the edges of the data set.

We first perform a temporal demodulation by computing the functions

$$C(x,t) = \langle f(x,t)\cos\omega_d t\rangle_t = A_L(x,t)\cos k_L x + A_R(x,t)\cos k_R x; \quad (10a)$$

$$S(x,t) = \langle f(x,t)\sin\omega_d t\rangle_t = A_L(x,t)\sin k_L x + A_R(x,t)\sin k_R x. \quad (10b)$$

Here, $\langle\ \rangle_t$ denotes a temporal lopass filter which passes frequencies below $\delta\omega$ but which rejects frequencies above $2\omega_d - \delta\omega$. We then perform a spatial demodulation by forming the four functions

$$A_{il} = \langle C(x,t)\sin k_d x + S(x,t)\cos k_d x\rangle_x = -2A_L\sin\Delta k_L x \quad (11a)$$

$$A_{rl} = \langle C(x,t)\cos k_d x - S(x,t)\sin k_d x\rangle_x = 2A_L\cos\Delta k_L x \quad (11b)$$

$$A_{ir} = \langle C(x,t)\sin k_d x - S(x,t)\cos k_d x\rangle_x = -2A_R\sin\Delta k_R x \quad (11c)$$

$$A_{rr} = \langle C(x,t)\cos k_d x + S(x,t)\sin k_d x\rangle_x = 2A_R\cos\Delta k_R x \quad (11d)$$

Again, $\langle\ \rangle_x$ denotes a spatial lopass filter which passes wavenumbers below δk but which rejects wavenumbers above $2k_d - \delta k$. The filter we use is a least-squares-optimized, symmetric, moving-average lopass filter whose kernel is modified at the edges of the data set so as to virtually eliminate edge distortion [22]. From the four functions A_{il}, A_{rl}, A_{ir}, and A_{rr}, it is straightforward to calculate the amplitudes $A_{L,R}$ and wavenumbers $k_{L,R}$.

The top of Fig. 7 shows the time-averaged, scaled amplitude profiles $(45mm/z)A_{L,R}(x,z)$ computed for data taken at three different camera positions z. In principle, all three estimates for $A_{L,R}$ should be identical. In practice, they differ, and this is because of the variation of the amplitude distortion factor $[1+z\nabla\cdot\vec{\alpha}(x,y)]$ in Eq. (6) with with z. In order to extract this factor from the data, we perform a linear regression at each point x to each of the two functions $zA_{L,R}(x,z)$:

$$z(A_{L,R}(x,z))^{-1} = a_{L,R}(x)[1+b_{L,R}(x)z]. \quad (12)$$

The fit parameters $b_{L,R}(x)$ give us two independent estimates for the distortion-angle divergence $\nabla\cdot\vec{\alpha}$ at each spatial position x, and we set $\nabla\cdot\vec{\alpha}$ equal to their average, which we denote $b_{av}(x)$. In Fig. 8, we show the distortion-corrected functions $(45mm/z)[1+b_{av}(x)z]A_{L,R}(x)$ computed from the data in the top frame of Fig. 7. These curves represent our three estimates for the true wave amplitudes in the linear state, and they are consistent and reproducible at the level of a few percent.

There are several notable features in the profiles in Fig. 8. Each of the two waves exhibits approximately exponential growth in space in the direction of its propagation. This form has been observed previously [4-6], and Cross [12] has

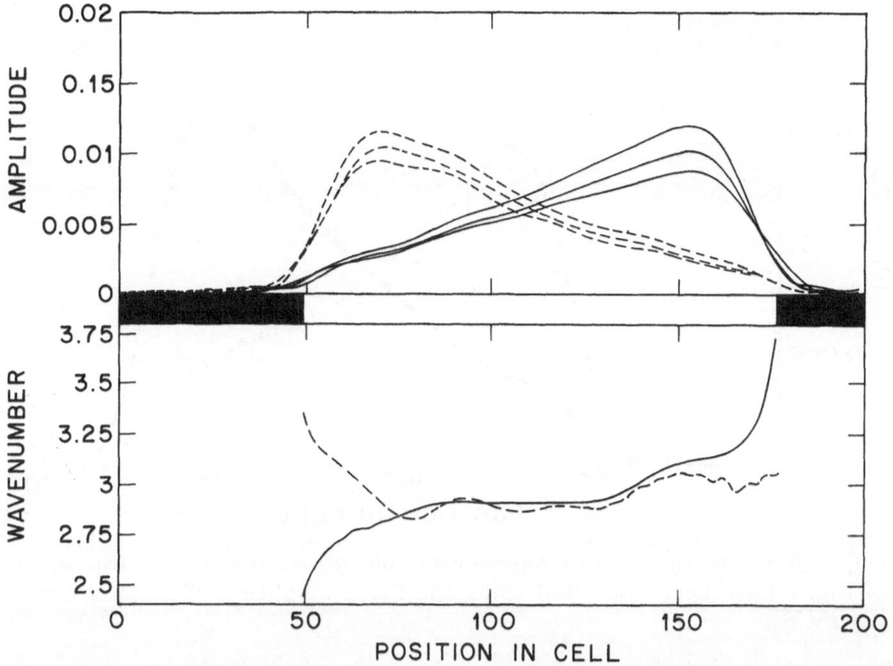

FIGURE 7. Top: time-averaged demodulated wave-amplitude profiles $(45mm/z)A_{L,R}(x)$ for the state of linear waves represented in Fig. 3. The focalization distances z corresponding to the top, middle, and bottom profiles are 30 mm, 45 mm, and 60 mm, respectively. In such plots, we always represent the left-going waves by dashed curves and the right-going waves by solid curves. Bottom: time-averaged, distortion-corrected wavenumber profiles $k_{L,R}(x)$ for this state, computed from data at focalization distance $z = 30$ mm.

pointed out that it is the result of linear growth in time coupled with drift in space. The waves are quenched within a distance of approximately one wavelength of the endwalls towards which they propagate. This is the result of the boundary conditions imposed on the solutions of Eq. (1) [12]; however, integrations of the linear parts of Eqs (1) exhibit a much shorter healing length [23] than is seen in Fig. 8. This discrepancy apparently results from the lack of detailed understanding of the boundary conditions at the endwalls, and we hope to use our data to resolve this issue. Finally, there is a marked asymmetry in the peak amplitudes of the left- and right-going waves. In measurements on five linear states in cells of lengths ranging from 17.243 to 17.625, the ratio of the peak right- to left-wave amplitude was found to be 1.16 ± 0.03. Such an asymmetry can be produced in numerical integration of linear parts of the Ginzburg-Landau equations (1) if we introduce a spatial gradient in the control parameter ϵ which is of a magnitude that is consistent with the levels of stray gradients in our experiment [23]. In fact, by inserting the demodulated wave-amplitude profiles into the Ginzburg-Landau equations, we can extract the spatial dependence of ϵ, for use as an input to integration of the full equations for the purpose of modeling our nonlinear data. It is worth pointing out that the profiles obtained in Fig. 8 - in particular, the shape of the healing regions at the ends - are insensitive to the details of our demodulation processing and are preserved if we form artificial data by multiplying the demodulated profiles by pure sine waves and demodulating again.

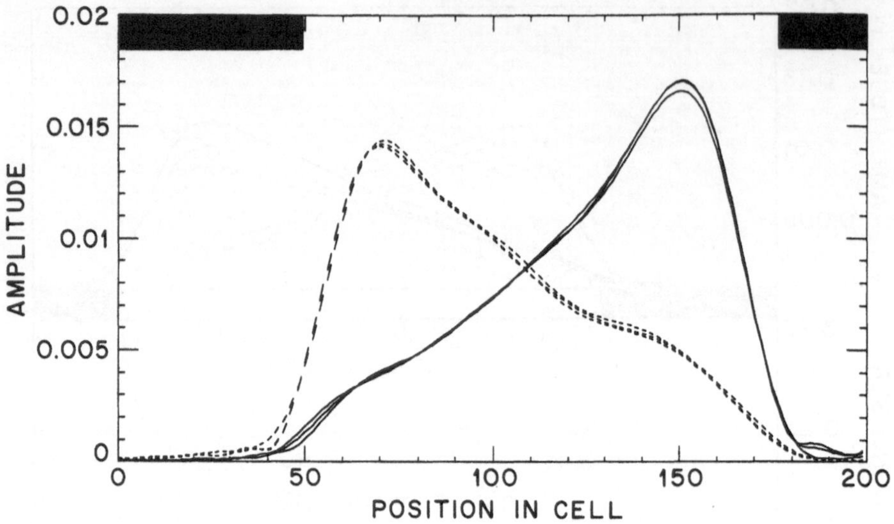

FIGURE 8. In this Figure, the wave-amplitude profiles in the top frame of Fig. 7 have been corrected using the fitted amplitude distortion factor $[1+z\nabla\cdot\vec{\alpha}]$.

There are several features worth noting in the linear wavenumber profiles plotted in the bottom frame of Fig. 7. First, the left and right wavenumbers $k_{L,R}$ are identical in the center of the cell. It would not be physically reasonable for the two wavenumbers to be different in a region where the gradients of the amplitudes $A_{L,R}$ are weak [23]. On the other hand, the wavenumbers develop strong gradients at the edges of the cell. This divergence is not seen when artificial data with flat wavenumber profiles is demodulated and is thus unrelated to the strong amplitude modulation and the ends of the cell. In numerical integrations of the linear parts of Eq. (1), a sharp divergence of the wavenumber is seen within a very short distance from the edges of the cell [23]. This difference may be a further manifestation of the uncertainty in the boundary conditions used. Third, there is a net gradient in the wavenumbers from one end of the cell to the other; we are not yet sure whether this is a physically reasonable consequence of the gradient in ϵ which results in the asymmetry in wave amplitudes. Finally, there are some weak wiggles in the wavenumber profiles which appear to be caused by residual optical distortions of short scale length which are within the error bars in Fig. 5.

If we postulate that the average wavenumber in the linear state is actually a constant k_a throughout the cell, then we can use the demodulated wavenumber profiles in Fig. 7 to calculate a further distortion-angle correction to apply to the data. For purposes of uniform notation, let us say that the profiles in Fig. 7 have been measured in a distorted plane whose coordinate is again x', and let us denote by $k_m(x')$ the average measured wavenumber $(k_L(x')+k_R(x'))/2$. Since the data is known in the primed plane, it is useful to rewrite Eq. (2) in terms of primed coordinates:

$$x(x') = x' - z\beta(x'). \tag{13}$$

If $z\beta(x')<<1$, then $\beta\approx\alpha$. The fundamental *Ansatz* here is that the measured phase whose derivative is $k_m(x')$ is equal to $k_a x(x')$. If we denote that phase by $\phi_m(x')$, then we can write

$$k_m(x') = \frac{d}{dx'}\phi_m(x') = \frac{d}{dx'}(k_a(x'-z\beta(x'))) = k_a(1-z\frac{d\beta}{dx'}). \tag{14}$$

So

$$z\beta(x') = \int_{x_c}^{x'} (1-\frac{k_m(x'')}{k_a})dx''. \tag{15}$$

Thus, the additional distortion correction necessary to impose a flat average wavenumber profile is determined by the demodulated data to within the two constants x_c and k_a, which are set by requiring that the ends of the cell map to themselves. Figure 9 shows the result of applying this calculation to a linear data set. This procedure has little effect on the wave amplitudes, but it renders the wavenumber profiles quite flat. It may well be completely unjustified to impose a flat wavenumber profile in this way. In particular, we have as yet little real knowledge of the details of the shape of the wavenumber profiles at the cell edges, and yet this is imposed by our *Ansatz*. However, the usefulness of this second correction is that, if we apply it to subsequent nonlinear data, then we are essentially subtracting out any structure in the wavenumber profiles inherent in the linear states, and details which are due to nonlinearities are easy to see. This is done below.

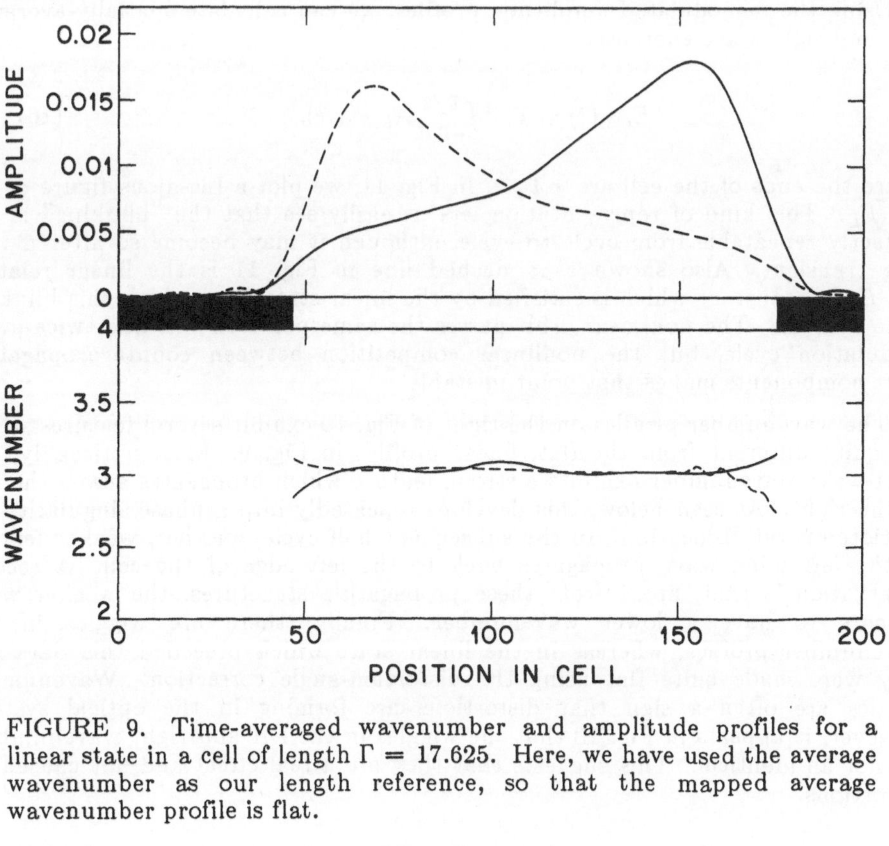

FIGURE 9. Time-averaged wavenumber and amplitude profiles for a linear state in a cell of length Γ = 17.625. Here, we have used the average wavenumber as our length reference, so that the mapped average wavenumber profile is flat.

DEMODULATION OF NONLINEAR WAVE STATES

Once we have acquired the three linear data sets necessary to calculate the distortion angle $\alpha_x(x)$ and the amplitude distortion factor $[1+z\nabla\cdot\vec{\alpha}(x,y)]$, we can turn up the temperature difference applied across the cell to produce a nonlinear convective state. Our plan is to use the linear corrections to remove distortions from new nonlinear data. In practice, during the day or two that is required for the new state to stabilize, the optical system continues to drift. Thus, mapping with the distortion angle $\alpha_x(x)$ results in a data set which is still slightly distorted. Now, however, the time-averaged light-intensity profile mapped from a particular camera plane z in the new state is very similar to the corresponding profile in Fig. 6, and that profile has already been very carefully distortion-corrected. Since these two profiles have nearly identical features, and since we generally use the shortest focalization distance z to obtain the sharpest features, we can construct a new distortion-angle correction which maps the new profile onto the old one quite accurately. This distortion-angle correction is then applied to the data and to the amplitude distortion factor $[1+z\nabla\cdot\vec{\alpha}(x,y)]$. The additional distortion is quite effectively removed in this way.

This procedure has been applied in Figure 10, which shows the amplitude and wavenumber profiles calculated for one half of a modulation cycle in a "blinking" state. This state evolved at $\epsilon = 3\times10^{-4}$ from the linear state in Fig. 9; those data provided the distortion correction. The amplitude profiles on the left of Fig. 10 clearly show the wave energy moving from right-going waves on the right of the cell to left-going waves on the left. In the next half-cycle, the waves will return to the right of the cell. Note that the modulation is asymmetric - the left-going waves are noticeably weaker at their peak than the right-going waves. This asymmetry may or may not be due to a gradient in the control parameter ϵ.

Using the demodulated amplitude profiles, we can calculate spatially-averaged left- and right-wave energies

$$E_{L,R}(t) = \Gamma^{-1}\int_{-\Gamma/2}^{\Gamma/2} A_{L,R}(x,t), \tag{16}$$

where the ends of the cell are $\pm\,\Gamma/2$. In Fig. 11, we plot a Lissajous figure of E_R *vs.* E_L. This kind of representation lets us easily see that the "blinking" is not perfectly repeatable from cycle to cycle, although it may become so after a very long transient. Also shown as a dashed line in Fig. 11 is the linear relation $E_R/E_L =$ constant which is satisfied by the linear state from which this blinking state evolved. The nonlinear orbit passes the same point on this line twice every modulation cycle, but the nonlinear competition between counterpropagating wave components makes that point unstable.

The wavenumber profiles on the right of Fig. 10 exhibit several features which are quite different from the flat, linear profiles in Fig. 9. Most noticeably, the right-wave wavenumber exhibits a strong feature which propagates across the cell to the right. As seen below, this develops repeatedly into a phase singularity - a spatiotemporal dislocation. In the subsequent half-cycle, another, weaker feature in the left-going wave propagates back to the left edge of the cell. A second observation is that, apart from these propagating structures, the weaker wave appears to have a lower wavenumber. Finally, there are wiggles in the wavenumber profiles, whereas in the linear state which preceded this data set, they were made quite flat using the distortion-angle correction. Wavenumber wiggles are often a sign that distortions are forming in the optical system. However, it appears in Fig. 10 that the wiggles in the left and right wavenumbers are not all identical. This suggests that they are a real effect and not caused by distortions.

In Fig. 12, we plot the spatially-averaged left and right wavenumbers as a function of time. The propagating features seen in the right-wavenumber profiles in Fig. 10 show up here as pulses which correspond to the gain and then loss of exactly one wavelength every modulation cycle. By contrast, the left wavenumber merely exhibits a weak periodic modulation.

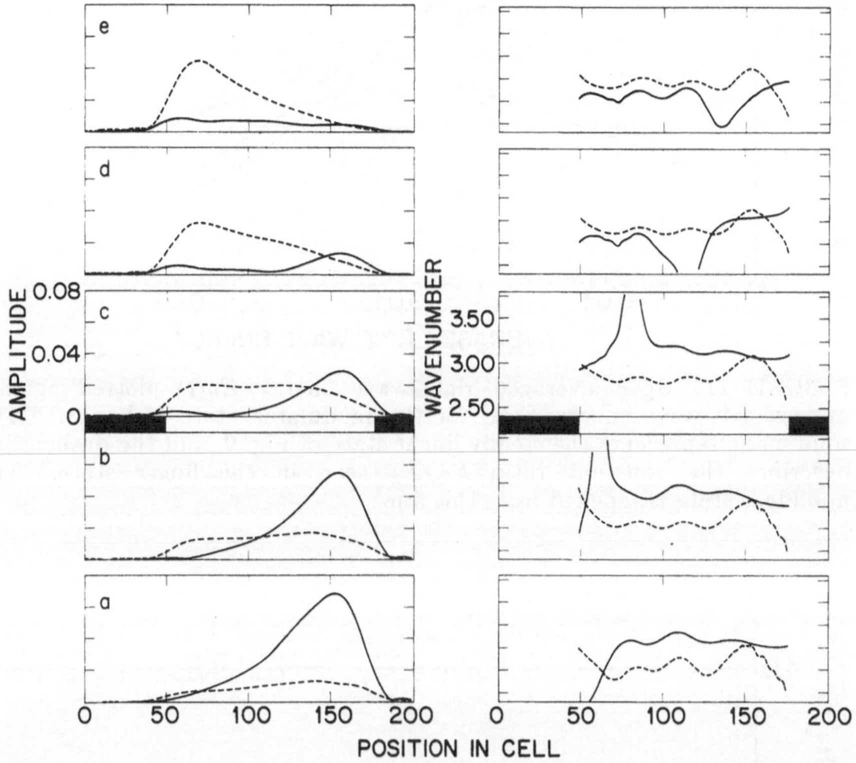

FIGURE 10. Wavenumber and amplitude profiles for a time-dependent "blinking" state produced at $\epsilon = 3\times10^{-4}$ in a cell of length $\Gamma = 17.625$. These data have been mapped using distortion corrections derived from the data of Fig. 9. Each horizontal pair of frames represents the convective state at a particular time in the modulation cycle. Time starts in the bottom frame, made at the peak of the right-wave amplitude, and progresses upwards. One half of a modulation cycle is shown. On the left, the wave energy is seen to "blink" from the right of the cell to the left. On the right, a structure in the right wavenumber is seen to propagate across the cell. These curves are samples from a larger data set which is used to construct all the subsequent figures.

The relationship between the propagating phase structures and the "blinking" wave-amplitude profiles shows up clearly if we make a space-time plot of both features on the same graph. In Fig. 13, which shows data from two modulation cycles, the closed solid curves represent equal-amplitude contours in spacetime, while the near-vertical lines of dots are equal-phase points. These are constructed by marking the horizontal line corresponding to each time step with a dot at

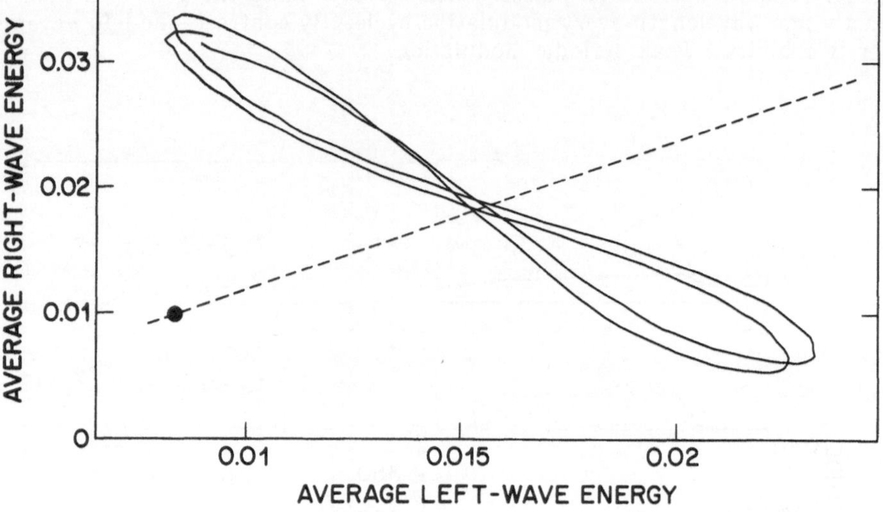

FIGURE 11. Space-averaged right-wave energy $E_R(t)$ plotted against average left-wave energy $E_L(t)$ for the modulated state of Fig. 10. The solid circle represents the steady linear state of Fig. 9, and the dashed line describes the constant ratio E_R/E_L seen in the linear state. The nonlinear state is repelled from this line.

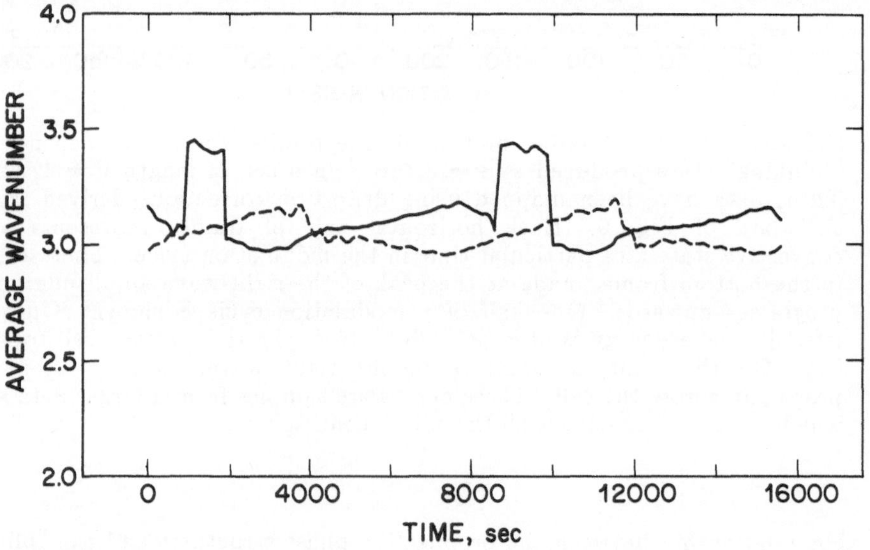

FIGURE 12. Space-averaged left- and right wavenumbers $k_{L,R}(t)$ *vs.* time in the modulated state of Figs. 10 and 11. The right wavenumber exhibits a pulse every modulation cycle, while the left wavenumber merely describes a smooth oscillation.

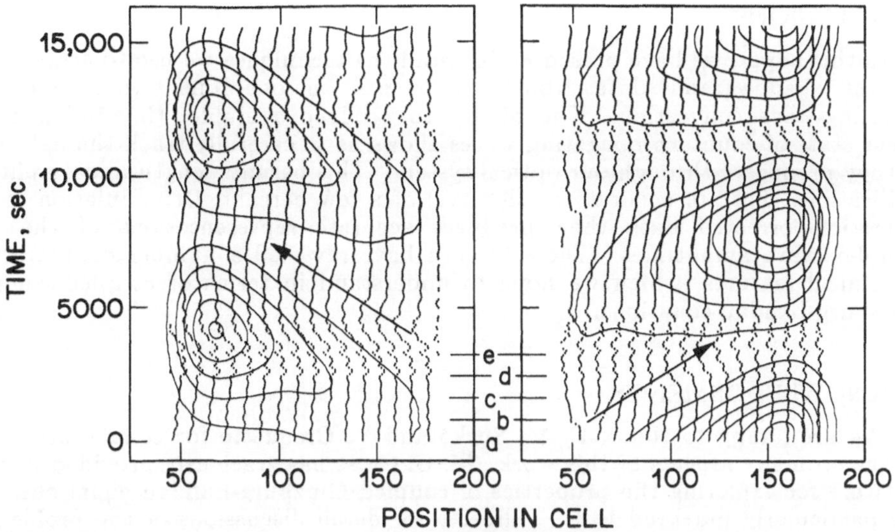

FIGURE 13. Space-time plots of the full, complex wave amplitudes for the left- and right-going wave components in the state of Figs. 10 - 12 are shown in the left and right frames, respectively. The letters a through e mark the times at which the profiles in Fig. 10 were computed. The solid, closed curves are constant-amplitude contours. The vertical curves are made up of points whose phase differs by multiples of π. Once per modulation cycle - at times 2500 - 4700 sec and again at times 10000 - 12000 sec - a "shimmer" or global, nonpropagating phase oscillation appears. Propagating phase distortions are also seen and are marked by arrows. In the right-wave phase, this feature develops into a pair of spatiotemporal dislocations.

every position that the phase of the wave equals a multiple of π. The actual curves of constant phase in these plots are nearly horizontal. However, since the temporal sampling is commensurate with the average oscillation frequency, points with phase differences of multiples of π line up approximately vertically. This allows us to clearly see the topological structure of the phase. Two repetitive phase disturbances are revealed in this representation. The first is a "shimmer": a nonpropagating, global oscillation of the equal-phase lines seen in both wave components once per cycle. This "shimmer" appears in several different dynamical states and may represent the early growth of a modulational instability. The second type of phase disturbance is the propagating defect seen earlier in Fig. 10. It shows up clearly as a pair of spatiotemporal dislocations which appear in the right wave once per modulation cycle. Their trajectory is shown in the bottom cycle of the right frame of Fig. 13 by an arrow which indicates their propagation velocity: two cell lengths per modulation period. A similar arrow in the left frame indicates the phase distortion in the left-wave phase which subsequently propagates back across the cell at the same velocity. The left-wave distortion appears as a bending of the equal-phase lines and does not develop into an actual defect. Observations of similar data in several different dynamical states reveal that the propagating phase distortions always appear just after the amplitude of the wave reaches its peak in the modulation cycle and propagate in spacetime parallel to the subsequent equal-amplitude contours. Whether these propagating structures develop into a full-fledged defect or not seems to depend only on the strength of the corresponding wave amplitude. We intend to investigate whether these "hidden spatiotemporal defects" are indications of the presence of modulational instabilities.

CONCLUSIONS

In this paper, we have presented in detail the techniques we use to acquire and process shadowgraph data from experiments on one-dimensional, nonlinear traveling-wave convection. The ability to produce a stable, time-independent linear state of counterpropagating waves allows us to carefully track the light rays as they propagate through our optical system. This permits length and amplitude calibrations that are accurate at the level of a few percent. Demodulation of the distortion-corrected data then produces accurate representations of the full, complex wave amplitudes. These in turn have revealed a number of interesting dynamical features, which we hope to understand in terms of coupled complex Ginzburg-Landau equations.

ACKNOWLEDGEMENTS

We are happy to thank C. M. Surko and V. Croquette for collaboration and advice on many aspects of this work. M. C. Cross has graciously provided us with software for exploring the properties of coupled Ginzburg-Landau equations. We are particularly indebted to S. Ciliberto for detail discussions of the problem of optical distortions.

REFERENCES

[1] R. Tagg, W. S. Edwards, H. L. Swinney, and P. S. Marcus Phys. Rev. **A 39,** 3734 (1989)

[2] I. Rehberg, B. L. Winkler, M. de la Torre Juarez, S. Rasenat, and W. Schoepf, Advances in Solid State Physics **29,** 35 (1989); A. Joets and R. Ribotta, to be published in J. de Physique **C3** (1989)

[3] V. Croquette and H. Williams, Phys. Rev. **A 39,** 2765 (1989); V. Croquette and H. Williams, Physica **D37,** 300 (1989); A. Chiffaudel, B. Perrin, and S. Fauve, Phys. Rev. **A 39,** 2761 (1989)

[4] P. Kolodner, C. M. Surko, and H. Williams, Physica **D37,** 319 (1989); V. Steinberg, J. Fineberg, E. Moses, and I. Rehberg, Physica **D37,** 359 (1989)

[5] P. Kolodner and C. M. Surko, Phys. Rev. Lett. **61,** 842 (1988); J. Fineberg, E. Moses, and V. Steinberg, Phys. Rev. Lett. **61,** 838 (1988)

[6] C. M. Surko and P. Kolodner, Phys. Rev. Lett. **58,** 2055 (1987) ; P. Kolodner, C. M. Surko, H. L. Williams, and A. Passner, in *Propagation in Systems Far from Equilibrium,* J. E. Wesfreid *et al,* eds., Springer, Berlin (1988), p. 282

[7] P. Kolodner, H. Williams, and C. Moe, J. Chem. Phys. **88,** 6512 (1988)

[8] M. C. Cross and K. Kim, Phys. Rev. **A 37,** 3909 (1988); S. J. Linz and M. Lucke, Phys. Rev. **A 35,** 3997 (1987); B. J. A. Zielinska and H. R. Brand, Phys. Rev. **A 35,** 4349 (1987); E. Knobloch and D. R. Moore, Phys. Rev. **A 37,** 860 (1988)

[9] W. Schoepf and W. Zimmerman, Europhys. Lett. **8,** 41 (1989)

[10] In Ref 4, we pointed out several examples of strongly two-dimensional flow in a rectangular cell whose width was 4.9 times its height. In the present work, at transverse aspect ratio 3.0, such effects are absent

[11] P. Kolodner, D. Bensimon, and C. M. Surko, Phys. Rev. Lett. **60,** 1723 (1988); D. Bensimon, P. Kolodner, C. M. Surko, H. Williams, and V. Croquette, submitted to J. Fluid Mech.

[12] M. C. Cross, Phys. Rev. **A 38,** 3593 (1988)

[13] M. Bestehorn, R. Friedrich, and H. Haken, Z. Phys. **B 75** 265 (1989)

[14] R. J. Deissler, J. Stat. Phys. **40,** 371 (1985); R. J. Deissler and H. R. Brand, Phys. Lett. **130A,** 293 (1988)

[15] M. C. Cross, private communication

[16] T. S. Sullivan and R. J. Deissler, submitted to Phys. Rev. **A** (1989)

[17] J. T. Stuart and R. C. DiPrima, Proc. Roy. Soc. (London) **A 362,** 27 (1978)
[18] C. S. Bretherton and E. A. Spiegel, Phys. Lett. **96A,** 152 (1983); H. R. Brand, P. S. Lomdahl, and A. C. Newell, Physica **D23,** 345 (1986)
[19] P. Kolodner, H. Williams, and J. A. Glazier, unpublished
[20] V. Croquette, thesis, Université de Paris, 1986 (unpublished)
[21] S. Rasenat, G. Hartung, B. L. Winkler, and I. Rehberg, Experiments in Fluids **7,** 412 (1989)
[22] P. Bloomfield, *Fourier Analysis of Time Series: an Introduction,* Wiley, New York (1976), chapter 6
[23] P. Kolodner and H. Williams, unpublished

CONVECTION IN BINARY LIQUIDS WITH SORET EFFECT: WHAT WE CAN LEARN FROM LASER DOPPLER VELOCIMETRY EXPERIMENTS

J.K. Platten and O. Lhost

State university of Mons
21 avenue Maistriau
7000 Mons – Belgium

1. INTRODUCTION

It has been recognized since a long time that Rayleigh–Bénard convection in binary mixtures possesses a rich variety of dynamical behaviors due to the coupling between the temperature and the concentration fields (since a more recent review does not exist, see e.g. Platten & Legros 1984). For example, the oscillatory onset of convection was not only theoretically predicted almost 20 years ago (Platten 1971 – Shteinberg 1971 – Hurle & Jakeman 1971 – Legros et al. 1972) but also experimentally observed by inserting temperature probes (thermocouple, thermistance, ...) inside the liquid layer (Platten & Chavepeyer 1973). In this last paper, far away from the critical point, very regular and sustained temperature oscillations were recorded in a circular container of large aspect ratio (diameter/height $\simeq$ 13 cm/ 0.32 cm) were a regular roll pattern is even not expected. Also, an hysteretic effect in Schmidt–Milverton plots was discovered (Platten & Chavepeyer 1973). However the exact nature of the "oscillator" was not resolved. More recently, a lot of dynamical behaviors has been experimentally described: steady overturning convection (SOC), traveling waves (TW) (Walden et al. 1985 – Surko et al. 1986 – Moses & Steinberg 1986a – Lhost & Platten 1988 – Kolodner et al. 1988 – Lhost & Platten 1989a), modulated TW (Heinrichs et al. 1987 – Lhost & Platten 1988), localized TW (Moses et al. 1987 – Kolodner & Surko 1988), counter–propagating TW (Kolodner et al. 1986), zipper (Walden et al. 1985), ... and theoretically analysed (Cross 1986a and 1986b – Ahlers & Lücke 1987 – Linz & Lücke 1987a and 1987b – Deane et al. 1988 – Cross 1988 – Linz et al. 1988 – Lücke 1988 – Linz & Lücke 1988 – Schöpf & Zimmermann 1989 – Barten et al. 1989 – Bensimon et al. 1989) together with a few predictions on the transitions between these convective states. Such a progress in the knowledge of the different states has been made possible essentially by the visualisation of the whole convective field using transparent plates (like sapphire plates) instead of the usual copper plates inadequate for optical access (Walden et al. 1985 – Moses & Steinberg 1986a – Kolodner et al. 1988 – Moses & Steinberg 1988). This decisive step being made, little attention has been given to a quantitative study of the velocity field itself in the spirit of the work done by Berge and coworkers (1978) in the more standard Rayleigh–Bénard convection in a pure fluid. Therefore, we believe that a quantitative determination of the velocity components would be a valuable contribution to this problem and laser Doppler velocimetry (hereafter called LDV) seems the appropriate tool since no material disturbing probe has to be inserted in the liquid. And sometimes, when the convective amplitude and the wave number are small, it is not sure that any method based on the variation of the index of refraction will be sufficient to reveal the convective structure. The aim of the present paper is to summarize the works performed by the present authors during the last three years in binary mixtures using LDV.

Nonlinear Evolution of Spatio-Temporal Structures in Dissipative Continuous Systems
Edited by F.H. Busse and L. Kramer
Plenum Press, New York, 1990

2. BINARY FLUIDS WITH NEGATIVE SEPARATION RATIO HEATED FROM BELOW

The separation ratio ψ is defined by $\psi = \frac{D_T}{D} N_1 N_2 \frac{\beta}{\alpha}$ where D_T is the thermal diffusion coefficient, D the isothermal diffusion coefficient, N_1 the mass fraction of the denser component, $N_2 = 1 - N_1$, β the mass expansion coefficient and α the thermal expansion coefficient. When ψ is negative (e.g. in a mixture 90% water – 10% alcohol), the denser component (water) migrates towards the hot lower plate. The resulting concentration gradient is stabilizing and convection is delayed. Generally, the new state is not a state of steady convection. In that case, TW are observed and by increasing the Rayleigh number, a new transition to SOC occurs. Galerkin type techniques have failed to give a correct picture of this transition, probably because of the difficulty to represent, by a few Fourier modes, the concentration field, profoundly modified by the velocity field (Lhost et al. 1989). Recently, two different theories were proposed for the transition from TW to SOC and presented at this conference. Bensimon et al. (1989) have developed a "boundary layer theory", performing an expansion around the pure fluid convecting state, and not, as in the Galerkin techniques, an expansion around the binary mixture conducting state. Thus, they consider the transition SOC → TW as an instability of the concentration boundary layers in the convective flow. The main result of these authors is to predict a critical Peclet number P for the transition:

$$P_{crit} = \frac{V_{crit}\, h}{D} = \left[\sqrt{16.98}\; |\psi|^{1/2} L^{-1}\right]^{8/7} \qquad (1)$$

where L is the Lewis number. Clearly, one could check not only the 4/7 exponent of ψ for this transition, but also the value of V_{crit} itself by LDV measurements.

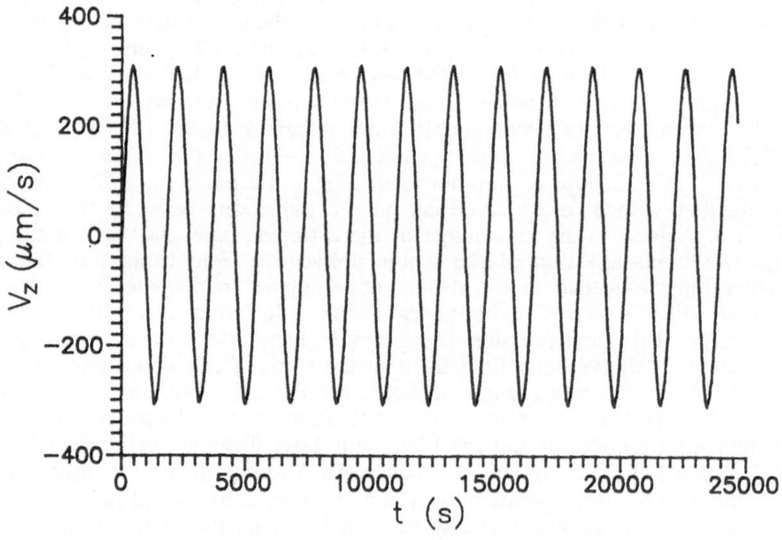

Figure 1. Time evolution of the vertical component of the velocity in a mixture water 90% – isopropanol 10% at $\Delta T = 3.10$ K (the onset of free convection occcurs at $\Delta T = 2.975$ K).

The strategy adopted by Barten et al. (1989) is completely different: they solved the full hydrodynamical equations by numerical techniques based on finite differences. They succeeded to produce TW and the transition to SOC for $\psi = -0.25$, $\sigma = 10$ (Prandtl number) and L = 0.01 (Lücke 1989). When $|\psi|$ increases, e.g. for $\psi = -0.6$, only TW are reported (Barten et al. 1989).

As a contribution to the study of Rayleigh–Bénard convection in water–alcohol solution, we have measured the velocity field in a rectangular container of aspect ratio 1 – 3.6 – 28 (all the dimensions are reduced by the height h = 4.15 mm). The container is filled with a mixture 90% water – 10% isopropanol. More details about the experimental protocole are given elsewhere (Platten et al. 1988). The Doppler bursts are Fourier analysed by a HP 3651A dynamical signal analyser, the detected frequency being proportional to the velocity at the measuring point. The signal analyser is interfaced with a computer and it is a standard job to record the velocity each 2 seconds during several days when necessary. A typical plot of the vertical component of the velocity V_z as a function of time is given in figure 1 for $\Delta T = 3.10$ K ($\Delta T^{hopf}_{crit} = 2.975$ K). The maximum value of V_z is 310.3 µm/s. The corresponding frequency is determined to be $5.37 * 10^{-4}$ Hz (or a period of 1862 s). Let us emphasize that for this mixture, $\Delta T = 3.10$ K is the largest temperature difference that TW can support (TW are observed in the range $2.975\text{ K} < \Delta T < 3.10\text{ K}$ but also below 2.975 K showing the hysteretic effect). An increase of ΔT from 3.10 K to 3.12 K produces a transition to SOC (figure 2) and the new stable state of SOC is typically reached after one day by oscillations of decreasing amplitude: the wave stops traveling (at t = 44120 s as indicated by the vertical dashed line on figure 2) and starts to oscillate around its new equilibrium position. In figure 2, since the new velocity amplitude is rather small, the measurement probe was nearly at the center of a fixed roll. And in order to record the maximum value of V_z in the SOC state, the measurement probe is moved along the horizontal

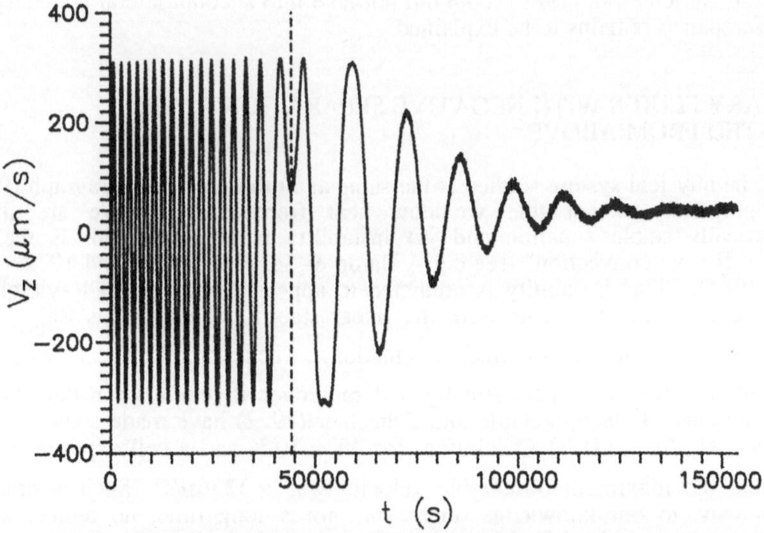

Figure 2. Time evolution of the vertical component of the velocity during the transition from TW to SOC in a mixture water 90% – isopropanol 10% when increasing the temperature difference from $\Delta T = 3.10$ K to 3.12 K.

direction. In some other experiment (Lhost & Platten 1988 figure 5), the new amplitude is already (by chance) maximum.

In any case, a precise determination of V_z is obtained just before and after the transition; the experimental values are listed below:

ΔT (K)	V_z (μm/s)	f (Hz)
3.10	310.3	$5.37 * 10^{-4}$
3.12	317.9	0

Thus, the mean value of 314 μm/s probably is predicted with less than 1% error (3 μm/s). We are thus able to estimate $P_{crit} \simeq 1230$. The calculated value $P_{crit} = 760$ shows that the order of magnitude is preserved. A possible explanation of the origin of this difference is that the theory is valid for small $|\psi|$ which is not met in our experiment at $\psi = -0.41$. However, the same experiment repeated with 80% water – 20% isopropanol ($\psi = -0.16$) reveals the same discrepancy.

The numerical results presented by Lücke (1989) are difficult to use since we only dispose of the transition TW → SOC at $\psi = -0.25$, $\sigma = 10$ and $L = 0.01$. Let us however point out that in terms of the reduced quantity $\varepsilon^{TW \to SOC} = \frac{Ra^{TW \to SOC}}{1708} - 1$, the transition occurs at $\varepsilon^{TW \to SOC} \simeq 0.65$ when $\psi = -0.25$. It is clear from Barten's results that $\varepsilon^{TW \to SOC} \to \infty$ at $\psi = -0.6$. Thus at $\psi = -0.4$, $\varepsilon^{TW \to SOC}$ could not be smaller than 0.65, a value to be compared with our experimental finding ($\varepsilon^{TW \to SOC} = 0.79$). Let us however emphasize that, in their numerical work, Barten et al. (1989) use periodic boundary conditions. Rigid lateral walls could stabilize SOC and explain our small ε value. However, comparing the theoretical results of Bensimon et al. (1989) and Barten et al. (1989) (both did not take into account lateral boundary effects) a huge discrepancy remains to be explained.

3. BINARY FLUIDS WITH NEGATIVE SEPARATION RATIO HEATED FROM ABOVE

The physical system studied is the same as in the previous paragraph (90% water - 10% isopropanol) except that we now heat from above. We are thus in an hydrostatically stable situation and yet instability is predicted: this is the so–called "double diffusive convection" (see e.g. Velarde & Schechter 1971 and 1972 – Platten & Legros 1984). This instability is predicted to appear at very small Rayleigh numbers. For the mixture that we have used, the linear stability theory gives $Ra_{crit} \simeq -13$ (i.e. $\Delta T_{crit} \simeq -0.013$ K in our experimental situation). Such a small critical Rayleigh number is difficult to measure experimentally and moreover it requires the detection of very small velocities. Indeed, Velarde and Schechter (1972) have made some estimations at $\psi < 0$ (namely for a LiI–H_2O solution) for $\Delta T = 10$ K and a cell depth of 1 cm. They found that the maximum observable velocity was $\simeq 32$ μm/s. Such a small velocity explains why, to our knowledge, there was, for a long time, no direct experimental evidence of this double diffusive convection. And since the wavenumber goes to zero (large scale convection with almost no vertical motion , except near the lateral boundaries) this is thus a typical situation where LDV is the appropriate tool, since shadowgraphy or other refractive index variation based techniques will probably not

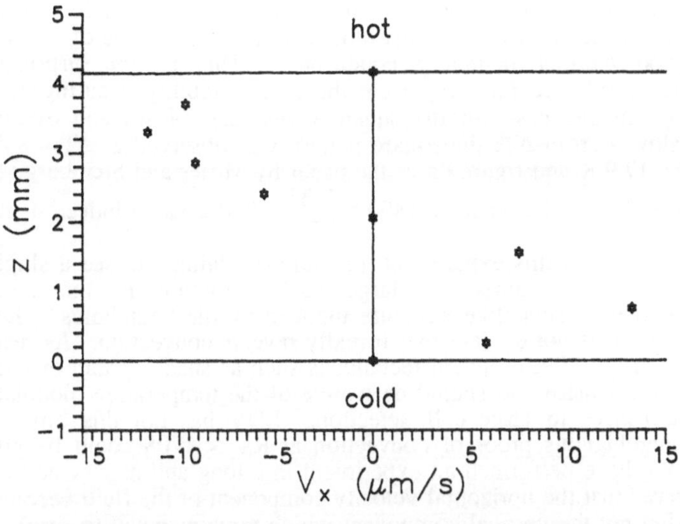

Figure 3. Measurement of the horizontal component of the velocity as a function of the elevation at $\Delta T = -5.64$ K in a mixture water 90% – isopropanol 10%.

suffice to reveal the flow structure. Figure 3 reproduces the measurement of the horizontal velocity component as a function of the elevation at $\Delta T = -5.64$ K at a given value of x. Similar plots may be obtained at different x values and the experimental results suggest the existence of only two large convective cells in the container (Lhost & Platten 1989b).

4. BINARY FLUIDS WITH POSITIVE SEPARATION RATIO HEATED FROM BELOW

In this situation, both gradients (of temperature and concentration) are destabilizing and the linear theory predicts a decrease of the critical Rayleigh number and of the wavenumber as well (Legros et al. 1972 – Linz & Lücke 1987a – Knobloch & Moore 1988 – Cross & Kim 1988a and 1988b). Some authors have pointed out the analogy with the case of a pure fluid with insulating boundaries, where $Ra_{crit} = 720$ and $k_{crit} = 0$ (Velarde & Schechter 1972 – Moses & Steinberg 1986b – Bigazzi et al. 1990). For the mixture that we have used in the experiment described below (58% water – 42% isopropanol), $\psi = 0.38$ and $Ra_{crit} \simeq 20$ together with $k_{crit} = 0$. It is clear that in such a case, convection is unable to transport heat and that a plot of the Nusselt number versus the Rayleigh number will not reveal an instability at small Rayleigh numbers (see e.g. the experiments performed by Legros et al. (1968a, 1968b and 1968c)). However, there is a second instability near $Ra \simeq 1708$. Nonlinear theory (Müller & Lücke 1988 – Knobloch 1989) and some experiments (Moses & Steinberg 1986b) revealed the existence of a square pattern below $Ra \simeq 1708$ and a corresponding wavenumber $k \simeq 3$, as well as a transition to the usual roll pattern when ΔT is increased. Moses and Steinberg (1986b) have used different containers, namely a cylindrical one with aspect ratio 20, square containers with aspect ratio 1:8.9:8.9 or 1:24:24 and a rectangular cell 1:4:12 of almost the same aspect ratio as ours (1:3.6:28). In their original paper, using the rectangular container, they observed a regular squarelike pattern at $\Delta T = 1.96$ K where convection is driven by the concentration gradient and also a regular roll pattern at $\Delta T = 2.20$ K corresponding to the usual Rayleigh–Bénard structure. The nature of the square pattern and its competition with the roll structure (leading to oscillations) is examined in cylindrical and square containers of large aspect ratio. In all cases, the wavevector was of the order of ≃ 3, very similar to the usual value of 3.117 observed for rolls with good thermal boundary conditions. This is not surprising since the experiments are conducted far away from the first instability occuring at Ra << 1708 (say Ra ≃ 20). In one case (in the square container), Moses and Steinberg (1986b) conducted a slow scan in ΔT: the square pattern was observed at $\Delta T = 8.47$ K, the roll pattern at $\Delta T \simeq 17.9$ K and figure 4a of the paper by Moses and Steinberg (1986b) shows a result at $\Delta T = 2.31$ K (i.e. at $Ra \simeq 1800 * \frac{2.31}{17.9} \simeq 250$ a value indeed a little bit above the first instability). In this experiment, the authors claimed to see a single convective cell in their large container, a "large scale structure in the interior of the cell ... modulated by a spokelike structure induced by the boundaries". Looking at the published picture, it is not evident that it really reveals convection. As recently pointed out by Bigazzi et al. (1990), optical techniques such as shadowgraph or index refraction measurements are sensitive to spatial derivative of the temperature modulation and thus are rather unsensitive to large cell detection. LDV has not this limitation. In our opinion, an unambiguous proof of convection at Ra << 1708 could be given by LDV experiment. We have performed an experiment in a long and narrow cell (1:3.6:28) and we have observed that the <u>horizontal</u> velocity component of the fluid becomes significant at small ΔT (but not the vertical component which remains equal to zero). By scanning the measuring point, we are able to show the existence of a large roll extending all along the cell. A typical horizontal component of the velocity as a function of the height is given in figure 4. The same profile can be recorded at every x position: there is thus no x periodicity. Also, we have verified that there is no y periodicity corresponding to square cells. This experiment is thus a direct proof of convection at small ΔT. Increasing ΔT, we have observed the usual transition near Ra ≃ 1708 showing indeed a x periodicity with k ≃ 3 and a vertical velocity component of the same order of magnitude as the horizontal one. However, we have not yet looked at the convective pattern a little bit below Ra ≃ 1708 and specially the square to roll transition. To be very clear, there is no contradiction with the findings of Moses and Steinberg (1986b). We have just demonstrated the existence of large scale convection with velocity amplitudes of ≃ 3 µm/s at small ΔT (e.g. $\Delta T = 0.5$ K) with no x or y periodicity, i.e. the non existence of square cells at $\Delta T = 0.5$ K. This does not mean that at $\Delta T = 1.16$ K, when $V_x \simeq 31$ µm/s (see table 2 of our original paper (1989b)), a square pattern is not the preferred mode of convection. We should look at this particular point in the future. At $\Delta T = 1.24$ K, when $V_z \simeq V_x \simeq 117$ µm/s, the usual roll pattern is observed.

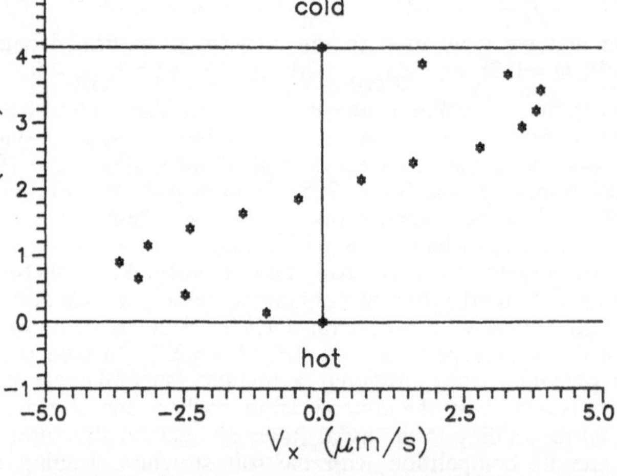

Figure 4. Measurement of the horizontal component of the velocity as a function of the height in the mixture water 58% – isopropanol 42% at $\Delta T = 0{,}88$ K.

5. CONCLUSION

LDV seems an interesting tool to study free convection problems. It is of course a local measurement technique but that can be successfully used in characterizing the convective pattern especially in the time independant regime. Of course, the price we have to pay is to scan in both horizontal directions, and this can take some time. This technique has been used in another context (Ouazzani et al. 1990) namely Rayleigh–Bénard convection with a superimposed Poiseuille flow and we were able to distinguish between transverse and longitudinal rolls. The main advantage of LDV over index gradient measurements is to provide a quantitative study of the velocity field and to make possible comparisons with theoretical predictions. Indeed, Galerkin type techniques or finite differences produce non–dimensional velocity components, and since the velocity scale is known, dimensional values are obtained. Such a comparison for a lot of data is in progress (Lhost 1990).

REFERENCES

Ahlers G. & Lücke M. 1987, *Phys. Rev. A*, **35**, 470.
Barten W., Lücke M., Hort W. & Kamps M. 1989, *Phys. Rev. Lett.*, **63**, 376.
Bensimon D., Pumir A. & Shraiman B.I. 1989, *J. de Phys.* (to appear).
Berge P. & Dubois M. 1978, in "*Lecture Notes in Physics*" (Springer–Verlag), 72, 133.
Bigazzi A., Ciliberto S. & Croquette V. 1990, prepint, (to appear).
Cross M.C. 1986a, *Phys. Lett. A*, **119**, 21.
Cross M.C. 1986b, *Phys. Rev. Lett.*, **57**, 2935.
Cross M.C. & Kim K. 1988a, *Phys. Rev. A*, **37**, 3909.
Cross M.C. & Kim K. 1988b, *Phys. Rev. A*, 38, 529.
Cross M.C. 1988, *Phys. Rev. A*, **38**, 3593.
Deane A.E., Knobloch E. & Toomre J. 1988, *Phys. Rev. A*, **37**, 1817.
Heinrichs R., Ahlers G. & Cannell D.S. 1987, *Phys. Rev. A*, **35**, 2761.
Hurle D.T.J. & Jakeman E. 1971, *J. Fluid Mech.*, **47**, 667.
Knobloch E. & Moore D.R. 1988, *Phys. Rev. A*, 37, 860.
Knobloch E. 1989, *Phys. Rev. A*, **40**, 1549.
Kolodner P., Bensimon D. & Surko C.M. 1988, *Phys. Rev. Lett.*, **60**, 1723.
Kolodner P., Passner A., Surko C.M. & Walden R. 1986, *Phys. Rev. Lett.*, **56**, 2621.
Kolodner P. & Surko C.M. 1988, *Phys. Rev. Lett.*, **61**, 842.
Legros J.C., Platten J.K. & Poty P.G. 1972, *Phys. Fluids*, 15, 1383.
Legros J.C., Van Hook W.A. & Thomaes G. 1968a, *Chem. Phys. Lett.*, 1, 696.
Legros J.C., Van Hook W.A. & Thomaes G. 1968b, *Chem. Phys. Lett.*, **2**, 249.
Legros J.C., Van Hook W.A. & Thomaes G. 1968c, *Chem. Phys. Lett.*, 2, 251.
Lhost O. & Platten J.K. 1988, *Phys. Rev. A*, 38, 3143.
Lhost O. & Platten J.K. 1989a, *Phys. Rev. A*, **40**, 4552.
Lhost O. & Platten J.K. 1989b, *Phys. Rev. A*, 40 (to appear).
Lhost O., Lücke M., Linz S.J., Müller H.W. & Niederländer J. 1989, (unpublished).
Lhost O. 1990, Ph.D. thesis, University of Mons, (to appear).
Linz S.J. & Lücke M. 1987a, *Phys. Rev. A*, **35**, 3997.
Linz S.J. & Lücke M. 1987b, *Phys. Rev. A*, **36**, 3505.
Linz S.J., Lücke M., Müller H.W. & Niederländer J. 1988, *Phys. Rev. A*, **38**, 5727.
Linz S.J. & Lücke M. 1988, in "*Propagation in Nonequilibrium Systems*" Springer Series in Synergetics (eds J.E. Wesfreid et al.), 41, 292.
Lücke M. 1988, in "*Nonequilibrium Phase Transitions*", Lecture Notes in Physics, (ed. Garrido), Springer 1988/1989).
Lücke M. 1989, private communication.
Moses E., Fineberg J. & Steinberg V. 1987, *Phys. Rev. A*, **35**, 2757
Moses E. & Steinberg V. 1986a, *Phys. Rev. A*, **34**, 693.
Moses E. & Steinberg V. 1986b, *Phys. Rev. Lett.*, **57**, 2018.
Moses E. & Steinberg V. 1988, *Phys. Rev. Lett.*, **60**, 2030.
Müller H.W. & Lücke M. 1988, *Phys. Rev. A*, 38, 2965.
Platten J.K. 1971, *Bull. Classe Sci. Acad. Roy. Belg.* **57**, 669.
Platten J.K. & Chavepeyer G. 1973, *J. Fluid Mech.* **60**, 305.

Platten J.K. & Legros J.C. 1984, Convection in liquids, chap. IX (Springer–Verlag – Berlin).
Platten J.K., Villers D. & Lhost O. 1988, in "*Laser Doppler Anemometry in fluid mechanics*" Vol III, 245 (eds: R.J. Adrian et al. – Ladoan – Instituto Superior Technico – Lisbon).
Ouazzani M.T., Platten J.K. & Mojtabi A. 1990, *Int. J. Heat Mass Transfer*, (to appear).
Schöpf W. & Zimmermann W. 1989, *Europhys. Lett.*, **8**, 41.
Shteinberg V.A. 1971, *PMM* 35, 335.
Surko C.M., Kolodner P., Passner A. & Walden R.W. 1986, *Physica D*, **23**, 220.
Velarde M.G. & Schechter R.S. 1971, *Chem. Phys. Lett.*, **12**, 312.
Velarde M.G. & Schechter R.S. 1972, *Phys. Fluids*, **15**, 1707.
Walden R.W., Kolodner P., Passner A. & Surko C.M. 1985, *Phys. Rev. Lett.* **55**, 496.
Wesfreid J., Pomeau Y., Dubois M., Normand C. & Berge P. 1978, *J. de Phys.* **39**, 725.

BOUNDARY LAYER ANALYSIS OF TRAVELING WAVES IN BINARY CONVECTION

David Bensimon[1], Alain Pumir[1] and Boris I. Shraiman[2]

[1] Laboratoire de Physique Statistique, Ecole Normale Supérieure
24, rue Lhomond, 75231 Paris Cedex, France

[2] AT&T Bell Laboratories, Murray Hill, NJ 07974, USA

Introduction

A systematic study of the phenomenon of convection in binary mixtures has revealed a rich variety of behaviors. Many interesting dynamical states have been found experimentally[1-6]. The situation is very different from what happens in pure fluid convection, due to the fact that the conducting state gets destabilized at a finite frequency through a Hopf bifurcation[7]. Propagative rolls are often observed, bifurcating subcritically when the control parameter is increased. States of confined traveling wave patterns have also been found experimentally, and their interactions induce rather complex and interesting phenomena which let one ti think that binary mixture convection might an appropriate system for testing theoretical ideas on spatio-temporal disorder and phase turbulence. Indeed, traveling waves seem to be a crucial ingredient for the understanding of many interesting and complex dynamical properties of quasi-1 dimensional systems. This paper is devoted to a thorough analysis of the emergence of nonlinear traveling waves in binary mixtures convection.

The linear stability of the conducting state is very well understood[7,8,9]. In this problem, the temperature stratification that leads to a destabilization of the conducting state of a pure fluid, when heated from below, competes with a concentration stratification, due to the Soret effect. The latter may either tend to bring light fluid in the hot region, thus making the conducting state even more unstable, and lowering the threshold of instability, or, on the contrary, tend to bring heavy fluid in the hot region, thus diminishing the mechanism of instability, and increasing the convection threshold. The separation ratio, $\psi = -(\alpha'/\alpha)S_T\, c_0(1-c_0)$, where α' is the solutal expansion, α the thermal expansion, S_T the Soret coefficient and c_0 the average concentration measures the importance of concentration stratification. When $\psi < 0$, the conducting state is destabilised through a Hopf bifurcation, corresponding to traveling waves. The subcriticality is due to the fact that at small Lewis numbers, the concentration gradients are expelled from the zones with closed streamlines (the Lewis number is the ratio D/κ, where D is the molecular diffusivity and κ is the thermal diffusivity). Therefore, once the convection regime is started, the concentration stratification is destroyed, except very near the walls, and one is left with an (almost) pure fluid, above threshold. Numerical solutions of the full Boussinesq equations confirm these expectations[10]. The formation of concentration boundary layers near the walls invalidates any weakly nonlinear analysis [11] and requires a different approach, which is the subject of this paper. Because of the expulsion of the concentration gradients from the rolls, a good starting point for doing perturbation theory is the convective state for a pure fluid. When $\psi = 0$, the equations for the velocity and temperature fields decouple from the equation for the concentration field. By treating ψ as a small parameter, together with the convection amplitude ε, one has to solve the equation for the concentration field, γ, and then to study the effect of γ on the velocity and temperature fields. From the solvability condition at first order in perturbation, one obtains the traveling wave velocity as a function of ε and ψ[12].

In this paper, we briefly recall the results of the perturbation analysis. We then sketch the boundary layer solution for the concentration field and present some numerical results. The full analysis can be found elsewhere[13].

Nonlinear Evolution of Spatio-Temporal Structures in Dissipative Continuous Systems
Edited by F.H. Busse and L. Kramer
Plenum Press, New York, 1990

Basic equations

A theoretically convenient limit for studying convection is the infinite Prandtl number limit: $Pr = \nu/\kappa \to \infty$ (where ν is the viscosity of the fluid and κ the heat diffusion coefficient). We denote the horizontal direction by x and the vertical direction by z. Let us introduce the streamfunction φ, defined by : $u = (-\partial_z\varphi, 0, \partial_x\varphi)$. The variables in the equations of motion are rescaled using the height of the cell, d, as the unit of length, the diffusion time $\tau_d = d^2/\kappa$ as the unit of time, δT as the unit of temperature and $\delta T\, S_T\, c_0(1-c_0)$ as the unit of concentration. The temperature fluctuation θ is also defined by $T = -z + \theta$. In the Boussinesq limit, the equations are :

$$\nabla^4 \varphi + R\, \partial_x \theta = R\, \psi\, \partial_x \gamma \qquad (1a)$$

$$\nabla^2 \theta + \partial_x \varphi = \partial_t \theta + u.\nabla \theta \qquad (1b)$$

$$L\nabla^2 \gamma = \partial_t \gamma + u.\nabla\gamma \qquad (1c)$$

R is the usual Rayleigh number:

$$R = \frac{g\alpha d^3 \delta T}{\nu\kappa}$$

Here, we have slightly changed the definition of γ in Eqs (1a,b,c). In the literature, one rather uses a variable of concentration Γ related to ours by $\Gamma = \gamma + LT/(1-L)$. As it has already been stated before, when $\psi = 0$, the equations for the velocity and temperature field completely decouple from the equation for the concentration field (1c). This is the starting point of our analysis. We will concentrate on the realistic case of rigid-rigid-impermeable boundary conditions on $z = 0,1$:

$$\varphi = \partial_z \varphi = 0 \qquad (1d)$$

$$\partial_z \gamma = 1 - \partial_z \theta$$

We are looking for traveling wave solutions :

$$\zeta(x,z,t) = \varepsilon\, \zeta_1(x-v_0t,z) + \varepsilon^2\, \gamma_2(x-v_0t,z) + \ldots$$

$$\gamma(x,z,t) = \gamma_0(x-v_0t,z) + \varepsilon\, \gamma_1(x-v_0t,z) + \ldots$$

where ζ stands for φ and θ. Notice that γ is not expanded around the perturbation state, *i.e.* we do *not* assume that γ_0 is linear in z. We are thus able to correctly describe the concentration boundary layers existing for convection at small Lewis numbers. We define the distance to the critical Rayleigh number R_c by $\varepsilon^2 = (R-R_c)/R_c$, where R_c is the critical Rayleigh number for a pure fluid with the same boundary conditions. It is suggested by experimental data as well as by the linear stability results that, in the domain of existence of traveling wave solutions, $\varepsilon \approx (-\psi)^{1/2}$. The velocity of the traveling wave, v_0 is also expected to be of the same order as ε. Therefore, one can define $\alpha = v_0/\varepsilon$ and $\beta = \psi/\varepsilon^2$. Expanding the fields φ, θ and γ in power series of ε, and substituting into equations (1a,b,c), one obtains, at $O(\varepsilon)$:

$$R_c^{-1} \nabla^4 \varphi_1 + \partial_x \theta_1 = 0 \qquad (2a)$$

$$\nabla^2 \theta_1 + \partial_x \varphi_1 = 0 \qquad (2b)$$

$$p_0^{-1} \nabla^2 \gamma_0 + \partial_x \gamma_0 - \alpha^{-1} u_1.\nabla\gamma_0 = 0 \qquad (2c)$$

with $p_0 = v_0/L$. Equations (2a,b) correspond to the Rayleigh Bénard problem for a pure fluid with infinite Prandtl number at the onset of convection, and can be easily solved:

$$\varphi_1(x,z) = A \sin(kx)\, w(z)$$

$$\theta_1(x,z) = A \cos(kx)\, \theta(z)$$

where w(z) and θ(z) are well known functions[14]. Equation (2c) is more difficult and in the limit that interests us here : $\varepsilon|v_1| \gg v_0$, necessitates the use of boundary layers techniques (see below). At order ε^2, one gets :

$$R_c^{-1} \nabla^4 \varphi_2 + \partial_x \theta_2 = \beta \partial_x \gamma_0 \qquad (3a)$$

$$\nabla^2 \theta_2 + \partial_x \varphi_2 = -\alpha\, \partial_x \theta_1 + u_1.\nabla\theta_1 \qquad (3b)$$

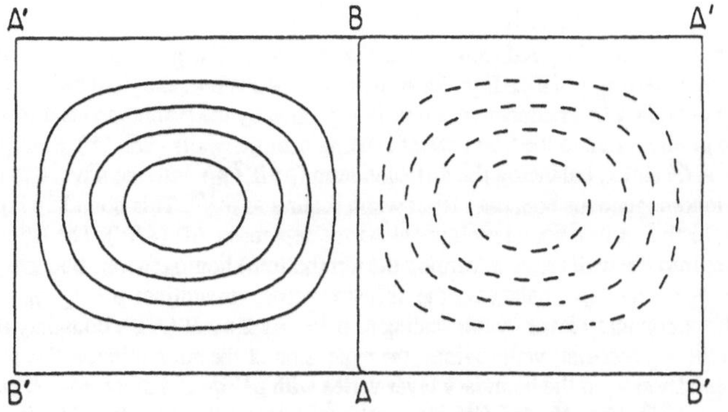

Fig. 1. Steady flow patterns($v_0 = 0$).

In order to solve Eqns (3a,b) for the velocity and for the temperature field, one has to solve a linear system, with a nontrivial kernel. One thus has to insist that the right hand side is orthogonal to the zero modes of the adjoint. These zero modes are $(\varphi_1,-\theta_1)$ and $(\partial_x\varphi_1,-\partial_x\theta_1)$. The first orthogonality condition is automatically satisfied, whereas the second gives rise to a non trivial condition :

$$v_0 = -\psi \frac{<\partial_x\gamma_0 \, , \, \partial_x\varphi_1>}{<(\partial_x\theta_1)^2>} \qquad (4)$$

This equation determines the velocity of the traveling wave, which is therefore related to the parity symmetry (x -> -x) of the concentration field. A zero velocity solution ($v_0 = 0$) always exists, since in this case, the parity of γ ($\gamma(x) = \gamma(-x)$) is opposite to the parity of φ_1 ($\varphi_1(-x) = -\varphi_1(x)$), and the term $<\partial_x\gamma_0,\partial_x\varphi_1>$ is thus zero by symmetry. The existence of a non zero velocity is related to a breaking of the parity symmetry : $\gamma(x) \neq \gamma(-x)$. The antisymmetric part of γ : $\gamma_A(x) = 1/2\,(\gamma(x) - \gamma(-x))$ can be expressed in all generality as a power series in v_0. It turns out that this expansion contains only odd terms in v_0:

$$\gamma_A = v_0\gamma_{A,1} + v_0^3\gamma_{A,3} + v_0^5\gamma_{A,5} + \ldots \qquad (5)$$

Inserting this expansion in Eqn 4, one obtains :

$$v_0 = (-\psi)v_0\,(a_1 + a_3\,v_0^2 + a_4 v_0^4 + \ldots) \qquad (6)$$

The transition from stationary convection to traveling waves happens for a value of ε where $a_1(-\psi) = 1$. Note in passing that if v_0 is a solution of Eqn. 6, $-v_0$ is also a solution, as it should. Numerically, we find the transition to be second order. Notice that the connection between the existence of traveling wave solutions and breaking of the parity symmetry is a well established feature, that has been noticed many times[15]. In the following we will see how to calculate the concentration field γ (and the coefficient a_1) using boundary layer techniques.

Boundary layer solution for the concentration field

In this paragraph we consider the case where the phase velocities v_0 of the waves is slow. Physically, the Péclet number p = |u|/L, based on the convection velocity is large and the mixing due to the flow has a much stronger effect on the concentration field than on the temperature. While the uniform vertical temperature gradient established in the quiescent conducting regime is slightly perturbed and modulated by the flow, the uniform vertical concentration gradient is completely destroyed in the limit L-> 0. Instead, the concentration gradients, imposed by the coupling to the temperature field are confined to the boundary layersnear the walls (where the velocity vanishes), and to the free boundary layers along the vertical separatrices of the flow (see Fig. 1).

In order to explicitly solve the problem, we first consider the case of steady state convection ($v_0 = 0$) in the large Péclet number limit, that is Eqn. (2c) rewritten as :

$$p^{-1}\nabla^2\gamma_0 + g\partial_x\gamma_0 - \hat{u}_1.\nabla\gamma_0 = 0 \qquad (7a)$$

with $g \equiv v_0/\varepsilon A = 0$ and where $p = \varepsilon A/L$ is the Péclet number and the flow field ($u_1 \equiv A\hat{u}_1$) is defined by the streamfunction φ_1 ($\equiv A\Phi_1$) corresponding to the single mode rigid boundary conditions (Eqn 1d). The boundary conditions for the concentration field are:

$$\partial_z \gamma_0 |_{z=0,1} = 1 - \varepsilon\partial_z\theta_1 \approx 1 \qquad (7b)$$

The treatment of this equation is inspired from references 16 and 17. The general solution of Eqn. 7a can be written as $\gamma_0 = \gamma_h + \gamma_p$ where γ_h satisfies Eqn. 7a with homogeneous boundary conditions : $\partial_z\gamma_h |_{z=0,1} = 0$. The inhomogeneous boundary layer concentration γ_p is generated by the simultaneous diffusion and advection of the concentration gradient due to the Soret effect existing near the walls (Eqn 7b). Since at distances $\zeta << 1$ from the wall, $\varphi_1 \approx A\,\zeta^2 \sin kx$, balancing the diffusion term $(p^{-1}\partial_z^2\gamma_0)$ with the advection term $(u_1.\nabla\gamma_0)$ in Eqn. (2c) yields an inhomogeneous boundary layer width scaling as $p^{-1/3}$. This boundary layer detaches at points A (A') giving birth to a free boundary layer along the separatrix AB (A'B'). The latter in turn extends along the streamlines into the wall regions forming the top (bottom) homogeneous boundary layers. Since along the separatrix $\varphi_1 \approx kAw(z)x$, balancing the diffusion across streamlines $p^{-1}\,\partial_x^2\,\gamma_h$ with advection along the streamlines yields a boundary layer width scaling as $p^{-1/2}$. As the AB (A'B') boundary layer extends along the streamlines into the top (bottom) wall regions, the expansion of the streamlineswill change the width of the layer. The concentration γ_h in the boundary layer varies with $p^{1/2}\varphi$, and since $\varphi \approx z^2$, we infer that the concentration scales as $p^{1/4}$. The effect of diffusion $(p^{-1}\partial_z^2\gamma_h)$ across this broadened layer is of $O(p^{-1/2})$ and can thus be neglected in comparison with advection parallel to the wall $(\partial_z\varphi\partial_x\gamma_h \approx z\partial_x\gamma_h)$ which is of order $O(p^{-1/4})$. Thus the homogeneous boundary layers along the walls are dominated by advection and the variation of γ_h along streamlines in these regions can be neglected.

It is important to notice that the boundary layer (of width $p^{-1/4}$) is much thicker than the inhomogeneous boundary layer (of width $p^{-1/3}$). Thus, the inhomogeneous boundary layer can be accounted for by including a source of strength Γ $(-\Gamma)$ at the stagnation point A (A') in the equation for γ_h. By conservation of flux the strength of this source is $\Gamma = \pi k^{-1}p^{1/2}$ which is obtained by integrating $\partial_z\gamma_p|_{z=0,1}$ along AB' (A'B) and rescaling with $p^{1/2}$ as apropriate for the AB (A'B') boundary layer.

We now solve for the concentration field along the separatrices. Introducing streamlines coordinates :

$$\sigma = p^{1/2}\Phi_1 \qquad (8a)$$

$$\tau(z) = k\int_0^z w(z')\,dz' \qquad (8b)$$

with $\Phi_1 \approx kxw(z)$ one derives from Eqn. 7a the equation for the AB boundary layer :

$$\partial_\sigma^2\gamma_h = \partial_\tau\gamma_h \qquad (9)$$

whose solution is :

$$\gamma_h(\sigma,\tau) = \int\left[-\Gamma\delta(\sigma') + \gamma_h(\sigma',0)\right] G(\sigma-\sigma',\tau)\,d\sigma' \qquad (10)$$

$G(\sigma,\tau)$ is the diffusion propagator :

$$G(\sigma,\tau) = \frac{\exp(-\sigma^2/4\tau)}{2\sqrt{\pi\tau}} \qquad (11)$$

To obtain an equation for $\gamma_h(\sigma,0)$ we follow γ_h around the closed loop ABA'B' and remember that diffusion has no effect on γ_h in the wall region BA', B'A. We find :

$$\gamma_h(\sigma,0) = \Gamma\left[G(\sigma,\tau_0) - G(\sigma,2\tau_0)\right] + \int \gamma_h(\sigma',0)\,G(\sigma-\sigma',2\tau_0)\,d\sigma' \qquad (12)$$

where $\tau_0 = \tau(1)$. Introducing the Fourier transform

$$\tilde{\gamma}(k,\tau) = \int \gamma_h(\sigma,\tau)\,e^{ik\sigma}\,d\sigma \qquad (13)$$

and Fourier transforming Eqn. 12 yields :

$$\tilde{\gamma}(k,0) = \frac{\Gamma}{1+e^{k^2\tau_0}} \qquad (14)$$

Substituting into Eqn. 10 yields :

$$\tilde{\gamma}(k,\tau) = -\frac{\Gamma e^{-k^2\tau}}{1+e^{-k^2\tau_0}} = -\Gamma\sum_{n=0}^{\infty}(-1)^n e^{-k^2(\tau+n\tau_0)} \qquad (15)$$

which back in real space has the form :

$$\gamma_0(x,z) = -\Gamma\,p^{1/2}\sum_{n=0}^{\infty}(-1)^n\,\frac{\exp\left[-\Phi_1(x,z)/2\eta_n^2\right]}{\sqrt{2\pi}\eta_n} \qquad (16)$$

with : $\eta_n^2 = 2(\tau(z) + n\tau_0)/p$.

Having found the concentration profile for the "passive" impurity we can turn back the coupling to the flow, $|\psi| \neq 0$, and calculate the corrections to the flow perturbatively. As has been shown earlier, the perturbation theory will involve the solvability condition Eqn 5. Provided that $\gamma_0(x,y)$ is symmetric under x -> -x, the solvability condition is trivially satisfied. However, if the boundary layers were asymmetric, the solvability condition would require a non-vanishing translational velocity of the rolls -that the TW velocity $v_0 \neq 0$. We can look for such a non-trivial self consistent solution by simply replacing the streamfunction Φ_1 in Eqn. 16 by $\mathit{\Phi}_1 \equiv g z + \Phi_1$ which corresponds to the TW convective flow as seen in the comoving frame (with $g \equiv v_0/\varepsilon A$ being the rescaled TW velocity). As long as g is small, i.e. smaller than all the other parameters in the problem, our analysis for the stationary convection applies without change and the concentration field is given by Eqn. 16 with $\mathit{\Phi}_1$ replacing Φ_1 . The self consistent value of g is determined by the solvability condition of Eqn 5 :

$$g = \frac{\psi}{p^2 L^2} \frac{\langle \partial_x \gamma_0, \partial_x \Phi_1 \rangle}{\langle (\partial_x \vartheta)^2 \rangle} \tag{17}$$

with $\vartheta \equiv A^{-1}\,\theta_1$. The inner product : $\langle \partial_x\gamma_0, \partial_x\varphi_1 \rangle = 2\langle \partial_x\gamma_{AB}, \partial_x\varphi_1 \rangle$ (the factor 2 comes from the contribution of the A'B' separatrix), can be evaluated from Eqn. 16 (see ref. 13). The result is :

$$g = \frac{\psi}{p^2 L^2} \kappa_1 p^{1/4} g + O(g^3) \tag{18}$$

with $k_1 \approx 17$ is a positive constant. The nontrivial ($g \neq 0$) solution appears for $p \leq p_c$ with p_c (g -> 0 as p -> p_c from below) :

$$p_c = (\sqrt{\kappa_1}\, |\psi|^{1/2} / L)^{8/7} \tag{19}$$

The $p = p_c$ point corresponds to the transition from stationary to the TW convection which, in terms of the reduced Rayleigh number is therefore predicted at $\varepsilon_{TW} \approx p_c L \approx |\psi|^{4/7}$. Eqn. 19 is confirmed by a numerical solution of the equations deduced from the perturbative treatment of our problem (see fig. 2,3) which also shows the bifurcation from the steady to TW convection to be critical. Concentration boundary layers are also obvious from fig. 4 that shows the isoconcentration lines for the parameter values $L = 10^{-2}$, $\psi = -1/4$, $\varepsilon^2 = .175$. Notice that the convection amplitude is smooth through this transition. This is in partial agreement with the experimental results, where a continuous transition from TW to steady convection was observed on heating. On cooling though, the transition from the steady convective state to TW convection was hysteretic. However the experiments were done in a rectangular geometry where the influence of the lateral wall is known to be important. These walls may stabilize the steady convecting state state against a TW perturbation, thus leading to hysteretic behavior.

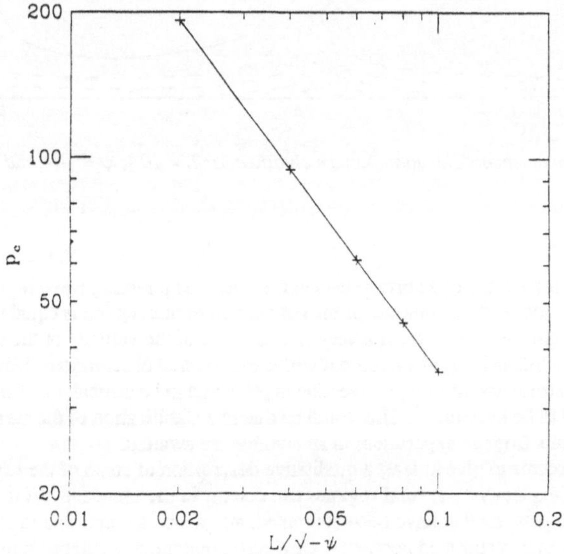

Fig. 2. The critical Péclet number p_c at the transition from TW to stationary convection as a function of $l \equiv L/\sqrt{-\psi}$.

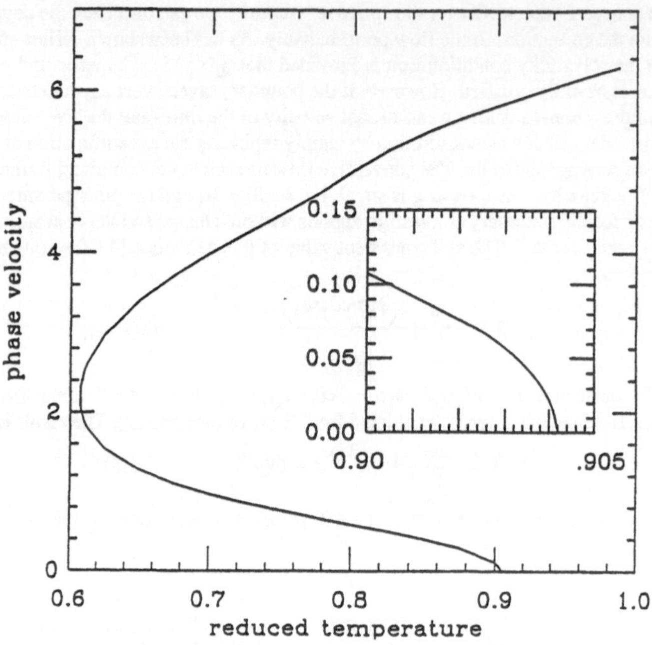

Fig. 3 . Velocity of the traveling wave v_0 as a function of ε^2/ψ for a value of the L = 10^{-2} and $\psi = 0.25$.

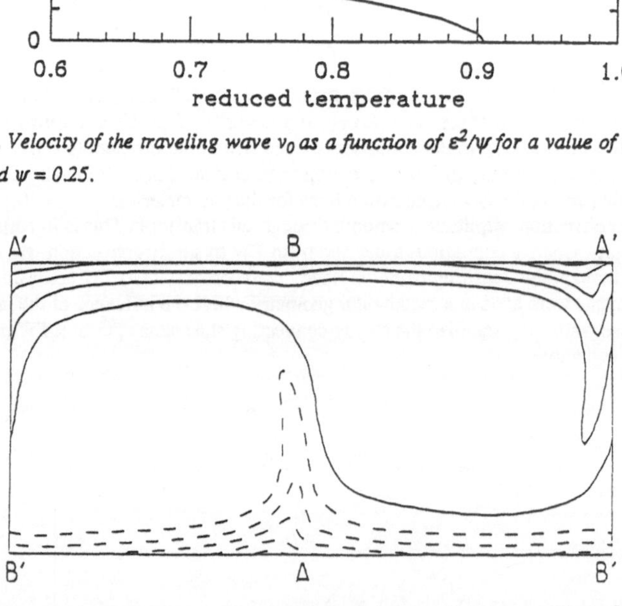

Fig. 4. A typical isoconcentration pattern obtained for $L = 10^{-2}$, $\psi = -1/4$ and $\varepsilon^2 = .175$

Discussion

Physically, the lower branch of fig.3 corresponds to the observed traveling wave regime[1-4]. In agreement with experimental observations, the amplitude of the convection in this regime is equal to the amplitude of the pure fluid convection. From fig. 3, we expect a very fast decrease of the velocity of the waves, as ε varies by no more than 10%. Again this is in good agreement with experimental observations. However, the transition from traveling waves to steady overturning convection is predicted to be critical, *i.e.*, forward (fig. 3). It is experimentally observed to be hysteretic[1]. This could be due to a stabilization of the steady overturning state by the lateral walls. Results from an experiment in an annulus are awaited.

Although our theory seems to give at least a qualitative description of some of the experimental observations, it does not explain why several regimes can coexist in the same cell[4]. It is conceivable that some of the fast traveling wave patterns that have been computed, and which are unstable in an infinite geometry, can be partially restabilized in a confined geometry. Obviously, quantitative comparisons between the present theory and experimental data are needed.

References

1) E. Moses and V. Steinberg, Phys. Rev. **A34**, 693 (1986)
2) E. Moses, J. Fineberg and V. Steinberg, Phys. Rev. **A35**, 2757 (1987)
3) R.W. Walden, P. Kolodner, A. Passner and C. Surko, Phys. Rev. Lett. **61**, 2030 (1985)
4) P. Kolodner, D. Bensimon and C. Surko, Phys. Rev. Lett. **60**, 1723 (1988)
5) T.S. Sullivan and G. Ahlers, Phys. Rev. Lett. **61**, 78 (1988)
6) O. Lhost and J. K. Platten, these Proceedings
7) D.T.J. Hurle and F. Jakeman, J. Fluid Mech. **47**, 667 (1971)
8) B.J.A. Zielinska and H.R. Brand, Phys. Rev. **A35**,4349 (1987)
9) M.C. Cross and K. Kim, Phys. Rev. **A37**, 3909 (1988)
10) W. Barten, M. Lucke, W. Hart, M. Kamps, Phys. Rev. Lett. **63**, 376 (1989)
11) E. Knobloch, Phys. Rev. **A34**, 1536 (1986)
12) Using similar approaches, M. Proctor (J. Fluid Mech. **105**, 507 (1981)) has considered the problem of steady thermohaline convection
13) D. Bensimon, A. Pumir and B. Shraiman, J. Physique **50**,3039 (1989)
14) S. Chandrasekhar, Hydrodynamic and Hydromagnetic stability (Dover, New-York, 1981)
15) B.A. Malomed and M.I. Tribelsky, Physica **14D**, 67 (1984), see also S. Douady, S. Fauve and O. Thual, Europhys. Lett. **10**, 309 (1989), P. Coullet et al. Phys. Rev. Lett. **63**,1954 (1989)
16) B. Shraiman, Phys. Rev. **A36**, 261 (1987)
17) M.N. Rosenbluth, H.L. Berk, I. Doxas and W. Horton, Phys. Fluids **30**, 2636 (1987)

NONLINEAR CONVECTION IN BINARY MIXTURES

E. Knobloch

Department of Physics
University of California
Berkeley, CA 94720, USA

D. R. Moore

Department of Mathematics
Imperial College
London SW7 2BZ, UK

I. Introduction

The experimental observation of travelling wave convection [1] has stimulated much of the recent work, both experimental and theoretical, on convection in binary fluids. Two systems have been extensively studied experimentally: ^{3}He–^{4}He mixtures above the λ-point [2], and water-ethanol mixtures [3,4]. Of these the former cannot be visualized, and the dynamics has to be inferred from point measurements. The experiments reveal the existence of both time-independent patterns [5], and a variety of time-dependent travelling waves [3,4]. The latter are small amplitude states that come into existence near the Hopf bifurcation from the pure conduction state. This bifurcation occurs in binary fluid mixtures characterized by a sufficiently negative separation ratio S. This ratio provides a measure of the stabilizing concentration gradient set up in response to a destabilizing temperature gradient by the (negative) Soret effect. With increasing Rayleigh number the conduction state loses stability to growing oscillations provided the restoring force due to the concentration gradient is sufficiently strong to overcome viscous dissipation. The instability occurs because heat diffuses faster than concentration, setting up a phase difference between the concentration and temperature fields, which persists into the nonlinear regime, regardless of whether the instability evolves into a travelling or standing pattern [6].

We focus in this article primarily on two-dimensional spatially periodic patterns. This idealization, commonly made in both theoretical and numerical treatments of large-aspect-ratio systems, is based on the assumption that in such systems the sidewalls manifest themselves as a small perturbation on the dynamics taking place in the interior. Indeed, depending on the parameters and the geometry of the system, approximately spatially periodic two-dimensional states are readily observed [7]. Recent experiments in small-gap annular regions are designed to approximate this situation even more closely [8]. The most important point about the assumption of periodic boundary conditions in a continuous reflection-symmetric system is that it introduces the symmetry $O(2)$, the group of rotations and reflections of a circle, into the problem. The presence of this symmetry changes the bifurcation behavior from that expected in the absence of the symmetry. We discuss below these changes by focusing on the neighborhood of particular bifurcation points, and then employ a model of binary fluid convection to study the extent to which the resulting behavior persists to parameter values away from their

Nonlinear Evolution of Spatio-Temporal Structures in Dissipative Continuous Systems
Edited by F.H. Busse and L. Kramer
Plenum Press, New York, 1990

bifurcation values. The model system is derived for idealized boundary conditions and we discuss how it compares with fully nonlinear solutions of the partial differential equations with realistic boundary conditions. Towards the end of the article we discuss how these ideas generalize to three-dimensional patterns.

II. Two-dimensional convection with periodic boundary conditions

The bifurcation to convection, be it steady or oscillatory, breaks the translation symmetry of the conduction state. Consequently the symmetry $O(2)$ acts non-trivially, and the number of critical eigenvalues is doubled. Below we summarize the essential results.

(a) Steady state bifurcation

If we write the vertical velocity $w(x,z,t)$ at the bifurcation point $R = R_{SS}$ in the form

$$w(x,z,t) \;=\; \Re e\{v(t)\,e^{ikx}\}\,f(z)\ , \tag{1}$$

where k is the wave number minimizing the Rayleigh number R_{SS} for the onset of the instability and $f(z)$ is the vertical eigenfunction, then the group $O(2)$ acts on v by

$$\text{translation:} \qquad x \;\rightarrow\; x+\ell\ , \qquad\qquad v \;\rightarrow\; e^{ik\ell}\,v \tag{2a}$$

$$\text{reflection:} \qquad x \;\rightarrow\; -x\ , \qquad\qquad v \;\rightarrow\; \bar{v}\ . \tag{2b}$$

Since the partial differential equations are equivariant with respect to translations and reflections it follows that the equation for v when $0 \leq |\lambda| \ll 1$ must be equivariant with respect to the action (2) of $O(2)$. Thus

$$\dot{v} \;=\; g(\lambda, |v|^2)\,v\ , \qquad\qquad g(0,0) \;=\; 0\ . \tag{3}$$

where $\lambda \propto R - R_{SS}$, and g is a real-valued function. In real variables, $v = r\exp(i\theta)$, (3) becomes

$$\dot{r} \;=\; (\lambda + ar^2)\,r \,+\, \mathcal{O}(r^5, \lambda r^3)\ , \tag{4a}$$

$$\dot{\theta} \;=\; 0\ . \tag{4b}$$

Observe that at bifurcation ($\lambda = 0$) there are in fact two zero eigenvalues. The second one (eq. (4b)) describes neutral stability of the roll pattern with respect to translation. Note that (4) imply that the bifurcation is a pitchfork of revolution.

(b) Hopf bifurcation

The general solution of the linear problem of $R = R_{\mathrm{Hopf}}$ is an arbitrary superposition of left- and right-travelling waves,

$$w(x,z,t) \;=\; \Re e\{v(t)\,e^{ikx} + w(t)\,e^{ikx}\}\,f(z)\ , \tag{5a}$$

where

$$\begin{pmatrix}\dot{v}\\ \dot{w}\end{pmatrix} \;=\; \begin{pmatrix} i\omega_0 & 0\\ 0 & -i\omega_0\end{pmatrix}\begin{pmatrix} v\\ w\end{pmatrix}\ . \tag{5b}$$

Since v, w are complex the center eigenspace is four-dimensional. To construct nonlinear equations for (v,w) note that $O(2)$ acts by

$$\text{translation:} \qquad x \;\rightarrow\; x+\ell\ , \qquad\qquad (v,w) \;\rightarrow\; e^{ik\ell}\,(v,w)\ , \tag{6a}$$

$$\text{reflection:} \qquad x \;\rightarrow\; -x\ , \qquad\qquad (v,w) \;\rightarrow\; (\overline{w}, \bar{v})\ . \tag{6b}$$

Consequently (v, w) satisfy

$$\begin{pmatrix} \dot{v} \\ \dot{w} \end{pmatrix} = \begin{pmatrix} g_1 & g_2 \\ \bar{g}_2 & \bar{g}_1 \end{pmatrix} \begin{pmatrix} v \\ w \end{pmatrix}, \tag{7}$$

where g_j, $j = 1, 2$, are complex-valued functions of $\lambda \propto R - R_{\text{Hopf}}$, $r^2 \equiv |v|^2 + |w|^2$, and of vw^* and v^*w. Additional near-identity nonlinear coordinate changes can be performed at $\lambda = 0$ to simplify the nonlinear terms. One finds that (7) are replaced by the normal form [9]

$$\begin{pmatrix} \dot{v} \\ \dot{w} \end{pmatrix} = \begin{pmatrix} h(\lambda, |w|^2, r^2) & 0 \\ 0 & \bar{h}(\lambda, |v|^2, r^2) \end{pmatrix} \begin{pmatrix} v \\ w \end{pmatrix}. \tag{8}$$

Note that the normal form has the additional symmetry,

$$\text{phase-shift:} \qquad t \to t + \phi/\omega_0, \qquad (v, w) \to (e^{i\phi}v, e^{-i\phi}w). \tag{6c}$$

To determine the small amplitude solutions of (8) we can expand h in powers of v, w. In terms of the variables $(v, w) = (x_1 e^{i\theta_1}, x_2 e^{i\theta_2})$ we obtain

$$\dot{x}_1 = [\lambda + ax_2^2 + br^2]\,x_1 + \mathcal{O}(5) + \lambda\mathcal{O}(3) \tag{9a}$$

$$\dot{x}_2 = [\lambda + ax_1^2 + br^2]\,x_2 + \mathcal{O}(5) + \lambda\mathcal{O}(3) \tag{9b}$$

with two decoupled equations for the phases θ_1, θ_2. Here $\omega - \omega_0 = \mathcal{O}(\lambda)$, a and b are real coefficients, and $\mathcal{O}(n)$ denotes the terms of order n in (x_1, x_2).

When the nondegeneracy conditions

$$a \neq 0, \qquad b \neq 0, \qquad a + 2b \neq 0 \tag{10}$$

hold the higher order terms can be omitted, and equations (9) admit the following three types of small amplitude solutions (x_1, x_2): (i) conduction $(0,0)$, (ii) left- and right-travelling wave $(x, 0)$, $(0, x)$, respectively, and (iii) standing waves (x, x). We summarize the properties of these solutions in fig. 1. Observe that the branches of travelling (TW) and standing (SW) waves bifurcate simultaneously, a consequence of the doubling of the eigenvalues. Stable solutions are found only if both branches bifurcate supercritically, and the stable branch has larger amplitude.

Of interest in the following is the degeneracy $b = 0$. If $|b| \ll 1$ then fifth order terms must be retained:

$$\dot{x}_1 = [\lambda + ax_2^2 + br^2 + cx_2^4 + dx_2^2r^2 + er^4]\,x_1 + \mathcal{O}(7) + (\lambda, b)\,\mathcal{O}(5) \tag{11a}$$

$$\dot{x}_2 = [\lambda + ax_1^2 + br^2 + cx_1^4 + dx_1^2r^2 + er^4]\,x_2 + \mathcal{O}(7) + (\lambda, b)\,\mathcal{O}(5) \tag{11b}$$

with (λ, b) as the two unfolding parameters. As in (9) the truncation of (11) can be rigorously justified [10] provided $a \neq 0$, $e \neq 0$. If $0 < b \ll 1$, $e < 0$, the TW branch is initially unstable (cf. fig. 1) but acquires stability with increasing amplitude at a secondary saddle-node bifurcation. The SW branch is unstable.

Similarly, one finds that the degeneracy $a + 2b = 0$ gives rise to a saddle-node bifurcation of the SW branch, while the degeneracy $a = 0$ leads to a secondary branch of modulated travelling waves. The stability properties of this branch depend on the coefficients of seventh order terms [11].

(c) Takens-Bogdanov bifurcation

The codimension-two bifurcation occurs when $R_{SS} = R_{\text{Hopf}} = R_{CT}$, say, a condition that requires that a second parameter S, say, takes a particular value S_{CT}. The study of this bifurcation has long

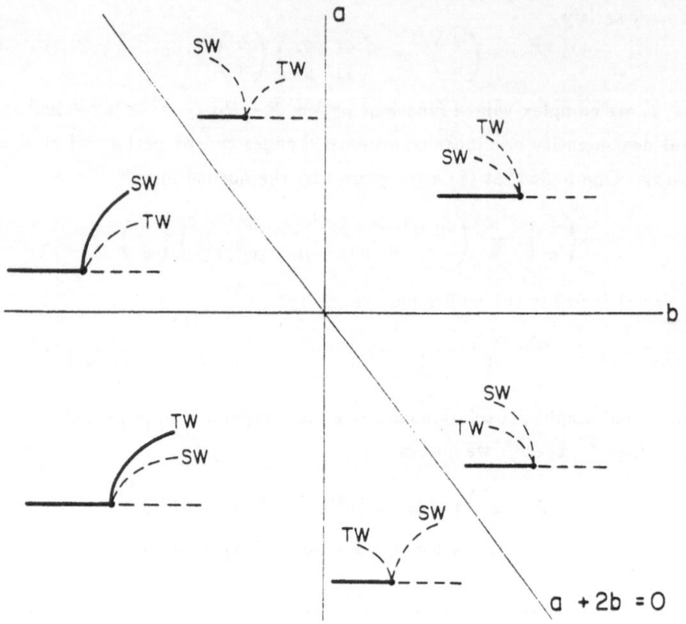

Fig. 1. Bifurcation diagrams $r(\lambda)$ in the (a,b) plane. Stable (unstable) branches are indicated by solid (broken) lines.

been recognized as providing the key to the interaction between steady and oscillatory convection [12]. At $R = R_{CT}$, $S = S_{CT}$ the Hopf frequency ω_0 vanishes, and owing to the $O(2)$ symmetry, there are four zero eigenvalues [13]. Consequently

$$w(x,z,t) = \Re e\{v(t)\, e^{ikx}\}\, f(z)\,, \tag{12a}$$

where

$$\begin{pmatrix} \dot{v} \\ \dot{w} \end{pmatrix} = \begin{pmatrix} 0 & 1 \\ 0 & 0 \end{pmatrix} \begin{pmatrix} v \\ w \end{pmatrix}. \tag{12b}$$

The group $O(2)$ acts on both v, w as in (2), and implies that the nonlinear equation for (v, w) at $R = R_C$, $S = S_C$ takes the form

$$\begin{pmatrix} \dot{v} \\ \dot{w} \end{pmatrix} = \begin{pmatrix} g_1 & g_2 \\ g_3 & g_4 \end{pmatrix} \begin{pmatrix} v \\ w \end{pmatrix}, \tag{13}$$

where g_j, $j = 1,2,3,4$, are real-valued functions of $|v|^2$, $|w|^2$ and $v\overline{w} + \overline{v}w$. Near-identity coordinate changes enable one to write (13) in the simpler form [13]

$$\begin{aligned} \dot{v} &= w \\ \dot{w} &= [A|v|^2 + B|w|^2 + C(v\overline{w} + \overline{v}w)]\, v + D|v|^2 w\,, \end{aligned} \tag{14a}$$

neglecting $\mathcal{O}(5)$ terms. Under appropiate nondegeneracy conditions on the coefficients A, D and $M \equiv 2C + D$ two unfolding parameters suffice to unfold this equation. If one introduces a small parameter ϵ ($0 < \epsilon \ll 1$) such that $R - R_{CT} = \mathcal{O}(\epsilon^2)$, $S - S_{CT} = \mathcal{O}(\epsilon^2)$, and a slow time $t' = \epsilon t$, then the scalled v ($v \to \epsilon v$) satisfies [13]

$$v'' = (\mu + A\,|v|^2)\,v + \epsilon\left[\nu\, v' + C\,(v\,\bar{v}' + \bar{v}\,v')\,v + D\,|v|^2\,v'\right] + \mathcal{O}(\epsilon^2)\,, \tag{15}$$

where the prime denotes differentiation with respect to t'.

Equation (15) has five solution types most easily described in terms of the variables (r, θ) with $v = r\exp(i\theta)$: (i) conduction ($r = 0$), (ii) steady convection (hereafter SS) given by ($r' \neq 0$, $r \neq 0$, $\theta' = 0$), (iii) SW ($r' \neq 0$, $\theta' = 0$), (iv) TW ($r' = 0$, $\theta' \neq 0$) and (v) modulated travelling waves (hereafter MW) given by ($r' = 0$, $\theta' \neq 0$). These solutions and their stability properties are discussed as a function of the unfolding parameters μ, ν in ref. 13. When $|D| \ll 1$, the degeneracy corresponding to that discussed in (11), a fifth order term $E|v|^4 w$ must be added to (15). Although the resulting equation has not been fully analyzed, it is clear that $\operatorname{sgn} D$ determines whether the TW branch bifurcates sub- or supercritically, while its larger amplitude behavior is determined by $\operatorname{sgn} E$, and resembles that in (15) with $D = \mathcal{O}(1)$, $\operatorname{sgn} D = \operatorname{sgn} E$. The predictions for $A > 0$, $C < 0$, $D = 0$, $E < 0$ and $E > 0$ are summarized in fig. 2.

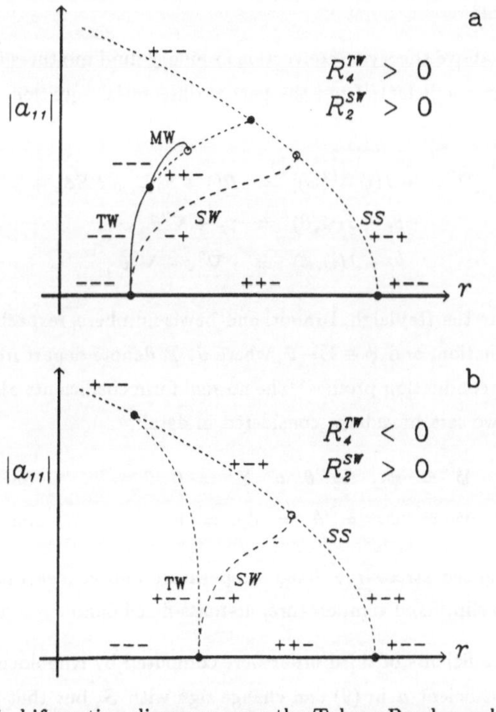

Fig. 2. Schematic bifurcation diagrams near the Takens-Bogdanov bifurcation point with $\mathcal{O}(2)$ symmetry for $A > 0$, $C < 0$, $D = 0$ and (a) $E < 0$, (b) $E > 0$. Stable (unstable) portions of the branches are indicated by solid (broken) curves. Local bifurcations are marked by filled circles, global ones by open circles.

There are three noteworthy features. First, near the steady state bifurcation ($\mu = 0$) and the Hopf bifurcation ($\nu = 0$, $\mu < 0$) equation (15) reduces to equations (3) and (9), respectively. For the latter the correspondence is: $\operatorname{sgn} M = \operatorname{sgn}(a + 2b)$, $\operatorname{sgn} D = \operatorname{sgn} b$. Second, in order for the TW branch to terminate on the (unstable) SS branch it must be unstable. To see this, note that the bifurcation to TW from SS is a steady state bifurcation from a group orbit of (nontrivial) equilibria. At this bifurcation a positive (unstable) eigenvalue remains unstable. In fig. 2a the bifurcation from SS to TW is backwards (i.e., in the direction of increased stability). By the well-known exchange of stability rules, this requires that the TW branch has two unstable eigenvalues, one of which reaches zero when the branch terminates. In fig. 2a we see that the TW branch accomplishes this by shedding an MW branch in a secondary Hopf bifurcation. The MW branch terminates on a global bifurcation that occurs when MW becomes homoclinic to TW. In contrast in fig. 2b the steady state bifurcation at the end of the TW branch is forward and so requires only one unstable eigenvalue on the TW branch. Since there already is one such eigenvalue because the TW branch is subcritical, no secondary bifurcations are required before the TW branch terminates. Since the bifurcation from SS to TW is a pitchfork (the waves can travel in either direction) the phase velocity of the TW varies as $(R_{TW} - R)^{1/2}$ near the end of the TW branch. Third, the SW branch also terminates on the SS branch but does so in a global bifurcation. This bifurcation does not therefore require a change in the stability of the SS branch provided it is a (generalized) saddle. Near the end of the SW branch the oscillation period varies as $-\ln(R_{SW} - R)$ [12].

(d) Coefficient calculations

In order to apply the above theory to convection in binary fluid mixtures it is necessary to calculate the essential normal form coefficients from the partial differential equations. Suitably nondimensionalized these are [9]

$$\frac{1}{\sigma}[\nabla^2\psi_t + J(\psi, \nabla^2\psi)] = R(1+S)\theta_x + RS\phi_x + \nabla^4\psi \tag{16a}$$

$$\theta_t + J(\psi, \theta) = \psi_x + \nabla^2\theta \tag{16b}$$

$$\phi_t + J(\psi, \phi) = \tau\nabla^2\phi - \nabla^2\theta\,, \tag{16c}$$

where R, σ and τ denote the Rayleigh, Prandtl and Lewis numbers, respectively. S is the separation ratio, ϕ is the stream function, and $\phi \equiv \Sigma - \theta$, where θ, Σ denote departures of the temperature and concentration from their conduction profiles. The normal form coefficients also depend on the adopted boundary conditions. Two sets have been considered in detail,

$$\text{(i)}\qquad \psi = \psi_{zz} = \theta = \Sigma = 0 \qquad \text{on} \quad z = 0, 1 \tag{17a}$$

$$\text{(ii)}\qquad \psi = \psi_z = \theta = \phi_z = 0 \qquad \text{on} \quad z = 0, 1\,, \tag{17b}$$

corresponding to (i) idealized stress-free, fixed temperature and concentration boundary conditions, and (ii) experimental no-slip, fixed temperature, no-mass-flux boundary conditions.

For case (i) all the coefficients of third order were computed by Knobloch [9]. The calculation shows in particular that the coefficient a in (9) can change sign with S, but that $b \equiv 0$ identically [14]. This unfortunate degeneracy necessitates the calculation of the fifth order coefficient e; this coefficient can also change sign with S [15]. We find that unless the Prandtl number is very small $e < 0$ and the TW branch bifurcates supercritically; for $-1 < S < S_{CT} < 0$ the SW branch also bifurcates supercritically but is unstable. The SS branch can also change direction of bifurcation but is typically subcritical. For case (ii) the coefficient b no longer vanishes identically. In fact the TW bifurcate supercritically for S very close to S_{CT}, and subcritically otherwise [16]. The fifth order terms necessary to describe the vicinity of the various degenerate bifurcations have not been calculated. The coefficients A, M and D in the Takens-Bogdanov normal form (15) follow on evaluating the coefficients a, b in (4) and (9) at $S = S_{CT}$. In this region the calculations predict a bifurcation diagram of the form shown in fig. 2a. In both cases the theory therefore predicts a secondary bifurcation to MW at small amplitudes for S close to S_{CT}, at least for parameters typical of the experiments ($\tau = 0.03$, $\sigma = 0.6$ for ^{3}He$-^4$He mixtures, and $\tau = 0.02$, $\sigma = 10.0$ for water-ethanol mixtures). It should be noted, however, that in case (ii) the codimension-two point is shielded, in the sense that steady state instabilities with a different wavenumber set in for R slightly less than R_{CT} [17]. Consequently, the relevance of these predictions to the experiments remains to be elucidated.

(e) Numerical simulations

Numerical simulations of the closely related doubly diffusive convection with boundary conditions (17a) show that convection takes the form of TW for most parameter values [14]. This is because in this system $b \equiv 0$ also [18]. Indeed stable SW are found only in the vicinity of the degeneracy $a = b = 0$ [6]. These studies show that the heat transported by a TW is independent of time, in contrast to SW, as predicted theoretically [13], and reveal the physical mechanism responsible for propagation [6]. Simulations of binary fluid convection with the boundary conditions (17b) have been made only recently [19]. For the parameter values typical of water/ethanol mixtures these reveal the presence of a thin boundary layer in the concentration that in a TW snakes between top and bottom of the layer. In contrast, in SW or SS convection the boundary layers remain attached to the top and bottom, and do not weave between the cells. Consequently to calculate the TW comparable numerical resolution is required in the horizontal as in the vertical. This is not so for SW or SS.

In all these calculations care must be taken to choose the spatial period $2\pi/k$ appropiately. It is well-known that an incorrect (i.e., artificial) choice of $2\pi/k$ typically causes spurious secondary bifurcations [20]. This fact raises the possibility that solutions which do not allow k to evolve with amplitude (or Rayleigh number) may also experience spurious bifurcations, or even spurious chaos.

(f) The minimal system

In this section we describe a model of binary fluid convection with the boundary conditions (17a) which allows us to follow relatively easily to larger amplitudes the various branches whose existence is revealed by the local bifurcation analyses described above [15]. This model is constructed to reproduce exactly the bifurcation results near the Hopf, steady state and Takens-Bogdanov bifurcations from the conduction solution, and in particular the coefficient e in (11). An examination of the perturbation

TABLE 1. Modes required by the bifurcation analysis of equations (16,17a).

Order	ψ	θ	Σ
ε	$e^{ikx}\sin\pi z$	$e^{ikx}\sin\pi z$	$e^{ikx}\sin\pi z$
ε^2	—	$\sin 2\pi z$	$\sin 2\pi z$
ε^3	$e^{ikx}\sin 3\pi z$	$e^{ikx}\sin 3\pi z$	$e^{ikx}\sin 3\pi z$
ε^4	$\sin 2\pi z$	—	—

calculations required to calculate these coefficients shows that they depend only on the spatial modes listed in table 1. If we write

$$\psi = \Re e\left\{\frac{2\sqrt{2p}}{ik} e^{ikx}\sin\pi z\, a_{11}(t') - \frac{p}{\pi ik}\sin 2\pi z\, a_{02}(t') + \frac{2\sqrt{2p}}{ik} e^{ikx}\sin 3\pi z\, a_{13}(t')\right\} \qquad (18a)$$

$$\theta = \Re e\left\{2\sqrt{\frac{2}{p}}\, e^{ikx} \sin \pi z\; b_{11}(t') - \frac{1}{\pi} \sin 2\pi z\; b_{02}(t') + 2\sqrt{\frac{2}{p}}\, e^{ikx} \sin 3\pi z\; b_{13}(t')\right\} \tag{18b}$$

$$\Sigma = \Re e\left\{2\sqrt{\frac{2}{p}}\, e^{ikx} \sin \pi z\; d_{11}(t') - \frac{1}{\pi} \sin 2\pi z\; d_{02}(t') + 2\sqrt{\frac{2}{p}}\, e^{ikx} \sin 3\pi z\; d_{13}(t')\right\}, \tag{18c}$$

where $t' = pt$, $p \equiv k^2 + \pi^2$, and project equations (16) onto these modes we obtain the following set of coupled ordinary differential equations for the mode amplitudes [15]

$$a'_{11} + (\varpi - 1)a_{11}a_{02} + (\varpi + 1)a_{13}a_{02} = \sigma(rb_{11} + rSd_{11} - a_{11}) \tag{19a}$$

$$a'_{02} - \varpi(a_{11}a^*_{13} - a^*_{11}a_{13}) = -\sigma\varpi a_{02} \tag{19b}$$

$$a'_{13} + \left(\frac{1-\varpi}{1+2\varpi}\right)a_{11}a_{02} = \frac{\sigma}{1+2\varpi}(rb_{13} + rSd_{13}) - \sigma(1+2\varpi)a_{13} \tag{19c}$$

$$b'_{11} + a_{11}b_{02} - b_{11}a_{02} - a_{13}b_{02} + b_{13}a_{02} = a_{11} - b_{11} \tag{19d}$$

$$b'_{02} - \frac{1}{2}\varpi\{a_{11}b^*_{11} + a^*_{11}b_{11} - b_{11}a^*_{13} - b^*_{11}a_{13} - a^*_{11}b_{13} - a_{11}b^*_{13}\} = -\varpi b_{02} \tag{19e}$$

$$b'_{13} - a_{11}b_{02} + b_{11}a_{02} = a_{13} - (1+2\varpi)b_{13} \tag{19f}$$

$$d'_{11} + a_{11}d_{02} - d_{11}a_{02} - a_{13}d_{02} + d_{13}a_{02} = a_{11} - \tau d_{11} + \tau b_{11} \tag{19g}$$

$$d'_{02} - \frac{1}{2}\varpi\{a_{11}d^*_{11} + a^*_{11}d_{11} - d_{11}a^*_{13} - d^*_{11}a_{13} - a^*_{11}d_{13} - a_{11}d^*_{13}\} = -\varpi\tau d_{02} + \varpi\tau b_{02} \tag{19h}$$

$$d'_{13} - a_{11}d_{02} + d_{11}a_{02} = a_{13} - (1+2\varpi)\tau d_{13} + (1+2\varpi)\tau b_{13}, \tag{19i}$$

Here $\varpi \equiv 4\pi^2/p$ $(0 < \varpi < 4)$, and $r \equiv R/R_0$, $R_0 \equiv p^3/k^2$.

Equations (19) can be readily solved for SS convection by setting $d/dt' = 0$, while TW take the form

$$(a_{11}, a_{13}, b_{11}, b_{13}, d_{11}, d_{13}) = e^{i\omega_1 t}(\hat{a}_{11}, \hat{a}_{13}, \hat{b}_{11}, \hat{b}_{13}, \hat{d}_{11}, \hat{d}_{13}), \tag{20a}$$

$$(a_{02}, b_{02}, d_{02}) = (\hat{a}_{02}, \hat{b}_{02}, \hat{d}_{02}), \tag{20b}$$

where the hatted quantities are time-independent. For given $|a_{11}|$ this yields a complex eigenvalue problem for $\omega = \omega_0 + \mathcal{O}(|a_{11}|^2)$. Finally, SW can be determined by restricting the mode amplitudes to be real. Note that this implies that $a_{02} = 0$. Stability of the solutions is also readily determined; all three solutions have a zero eigenvalue owing to translation equivariance. Figs. 3 summarize typical results. In Fig. 3a the supercritical TW branch terminates on the lower (unstable) part of the SS branch. For reasons already explained, the TW branch sheds an MW branch in a secondary Hopf bifurcation before doing so. This Hopf bifurcation is supercritical, and consequently the MW branch can be followed by integrating equations (19) in time. Fig. 3b shows a case in which the supercritical TW branch terminates on the upper SS branch. In this case the pitchfork bifurcation at the end of the TW branch is forward, and no change of stability along the TW branch is required; the branch remains stable throughout. Note that a second TW branch bifurcates supercritically from the SS branch at larger r. In both cases the SW branch is unstable.

Figs. 3 indicate that the qualitative behavior revealed by the analysis of the Takens-Bogdanov bifurcation persists for r and S substantially far from r_{CT}, S_{CT}. In addition, the figures indicate that the MW branch disappears as S decreases through S_{SN} and the end point of the TW branch moves past the saddle-node on the SS branch. In ref. 15 we have studied in detail the unfolding of the codimension-two bifurcation that occurs when the TW branch terminates at the saddle-node. At this point the SS branch has three zero eigenvalues, one from translation equivariance, the second because it is a saddle-node bifurcation and the third to allow the TW branch to terminate there. The motion of the dominant eigenvalues along the TW branch is indicated for $S \simeq S_{SN}$ in fig. 4. It shows that in addition to the zero eigenvalue from phase-shift equivariance, a pair of complex eigenvalues coalesces

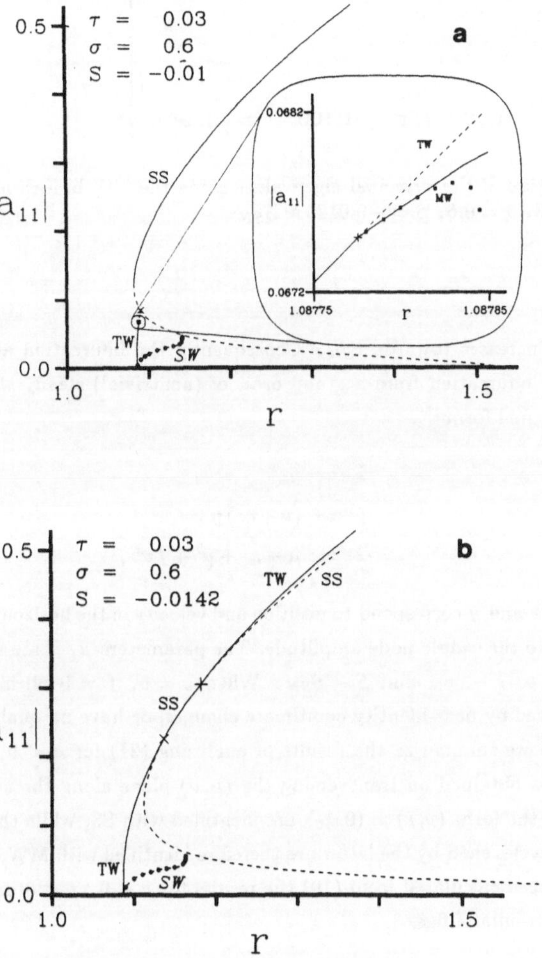

Fig. 3. Bifurcation diagrams for a $^3He-^4He$ mixture with boundary conditions (17a) and (a) $S = -0.01$, (b) $S = -0.0142$.

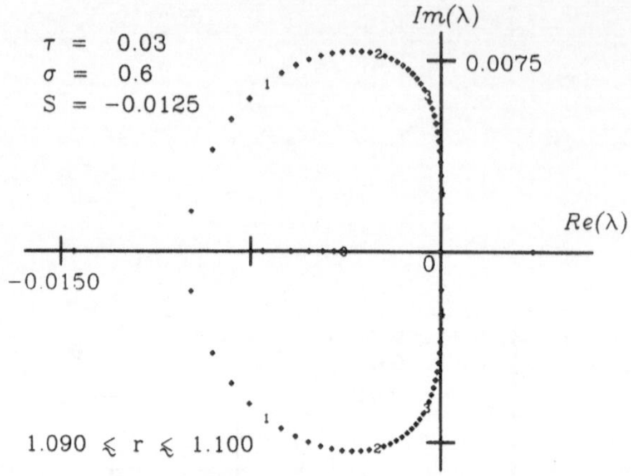

Fig. 4. Motion of the principal eigenvalues along the TW branch as a function of r for $\tau = 0.03$, $\sigma = 0.6$, $S = -0.0125 \approx S_{SN}$.

at the origin as r increases towards r_{SN}. Consequently the bifurcation at $r = r_{SN}$, $S = S_{SN}$ is a Takens-Bogdanov bifurcation from a group orbit of (nontrivial) steady states. This bifurcation is described by the normal form [15]

$$\dot{x} = y \qquad (21a)$$

$$\dot{y} = (\nu + cz)\, y \qquad (21b)$$

$$\dot{z} = \mu - z^2 + y^2 + fz^3 , \qquad (21c)$$

where the variables x and y correspond to position and velocity in the horizontal, and z is an amplitude coordinate relative to the saddle-node amplitude. The parameters μ, ν are unfolding parameters and are linearly related to $r - r_{SN}$ and $S - S_{SN}$. When $c \neq 0$, $f \neq 0$ all higher order terms in (21) can either be removed by near-identity coordinate changes, or have no qualitative effect on the local dynamics. In fig. 5 we summarize the results of analyzing (21) for $c < 0$, $f < 0$, and exhibit the bifurcation diagrams obtained on transversing the (μ, ν)-plane along the two lines indicated. From (21) fixed points of the form $(y, z) = (0, z_0)$ are identified with SS, while those with $(y, z) = (y_0, z_0)$ are TW. The limit cycles shed by the latter are therefore identified with MW. Although the coefficients c and f have not been calculated from (19) the results for $c < 0$, $f < 0$ are in complete agreement with the numerical results of figs. 3.

Fig. 5 shows that the MW branch must disappear in a global bifurcation along some line Γ, as indicated in the figure. This bifurcation is not, however, described by the normal form (21). This is because within (21) the limit cycles grow without bound as one traverses from region 5 into region 1, and the local analysis breaks down. What typically happens is that the limit cycle runs into fixed points that are not captured by the normal form (21), such as the $\mathcal{O}(1)$ fixed point $(y, z) \simeq (0, f^{-1})$ discarded in fig. 5. Consequently to study the nature of this global bifurcation we have integrated equations (19) along the MW branch, using an eigth order Runge-Kutta scheme with 128 bit precision. Figs. 6 show the MW very near its end point; observe that the wave is highly nonlinear, and the modulation period P (most easily seen in fig. 6b) is long. Fig. 7 shows P as a function of $-\ln(r_{MW} - r)$, showing behavior expected from an approach to a global bifurcation. In figs. 8 we show these solutions in a

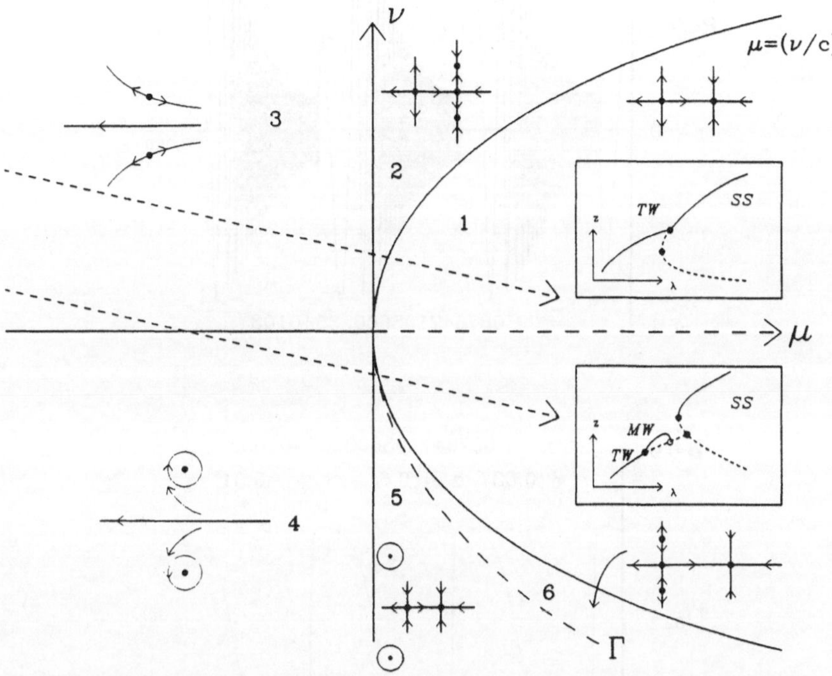

Fig. 5. The (μ, ν) plane for the normal form (21) with $c < 0$, $f < 0$ showing the different phase portraits $(z(t), y(t))$ characteristic of the various regions. The line Γ is a codimension-one surface of homoclinic bifurcations. Bifurcation diagrams $z(\lambda)$ obtained along typical sections through the (μ, ν) plane are shown as insets.

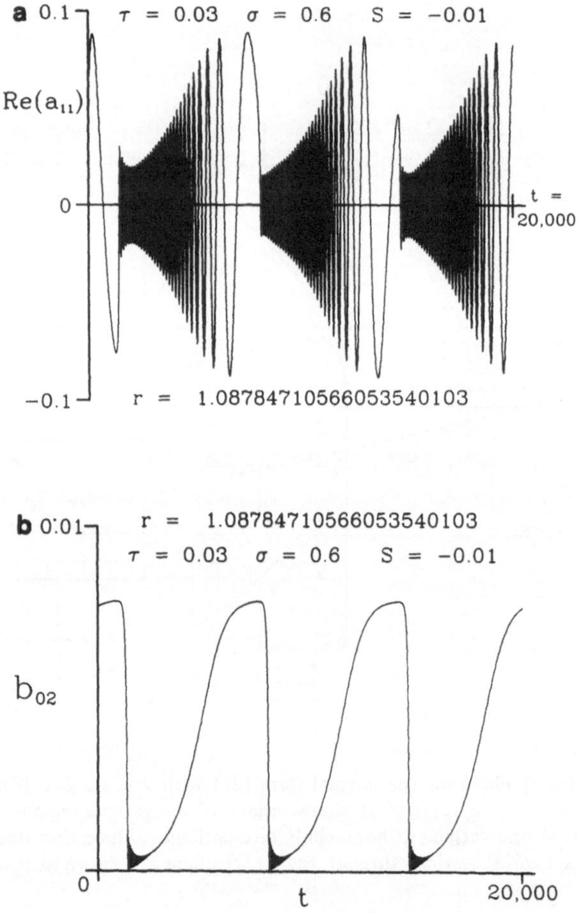

Fig. 6. Time series (a) $\Re e\, a_{11}(t)$ and (b) $b_{02}(t)$ for parameter values near the end of the MW branch.

different form. The rich structure in these figures arises from the phase lags between the different modes which are present in all travelling waves. As is evident from figs. 8 the global bifurcation differs from that predicted by the Takens-Bogdanov analysis. Instead of becoming homoclinic to the TW (indicated by + in figs. 8) the MW appears to become homoclinic to the origin. At these parameter values the origin has an 11-dimensional stable manifold and a 4-dimensional unstable manifold. The homoclinic two-torus is most easily described using fig. 8b. It has a slow phase, starting at the origin, looping around the TW fixed point and reaching $\Re e\, d_{11} \simeq 0$ at the point marked A. The slow phase is followed by a fast phase along $\Re e\, d_{11} \simeq 0$. The fast phase is present when the fast phase of the modulation cycle (fig. 6b) is faster than the TW period $2\pi/\omega$.

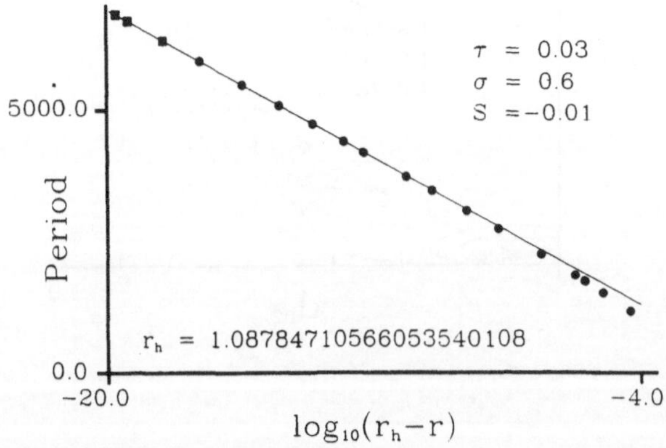

Fig. 7. The modulation period $P(r)$ near the end of the MW branch. The solid line is a fit of the theory to the three longest periods, denoted by squares, and determines r_h to 20 decimal places.

(g) Eulerian and Lagrangian mean flows

From table 1 (see also eq. (18)) we see that there is one mode in ψ that is independent of x, and that contributes to the direction of branching of the TW. Another x-independent $\mathcal{O}(\epsilon^4)$ term, proportional to $\sin 4\pi z$, does not and has been omitted from the table. Together these two terms describe the Eulerian mean flow given by $u = -\psi_z$. Fig. 9 shows the resulting mean flow profile $u(z)/|a_{11}|^4$. The mean flow is in the direction of propagation of the wave at the top and bottom, and in the opposite direction in the center. With experimental boundary conditions a mean Eulerian drift will appear already at $\mathcal{O}(\epsilon^2)$, and will vanish at the top and bottom [21,22]. At larger amplitudes such flows were computed by Barten et al. [19]

The Eulerian mean flow must be distinguished from the Lagrangian mean flow [21,22]. With idealized boundary conditions the Eulerian mean flow contributes an $\mathcal{O}(\epsilon^4)$ Lagrangian drift. This drift is swamped, however, by the $\mathcal{O}(\epsilon^2)$ Lagrangian drift arising simply from the usual Stokes' drift in waves of finite amplitude, even in the absence of Eulerian mean flow [23]. From (18) the Lagrangian mean flow is

$$\bar{u}_L \;=\; \frac{A^2}{2c}(1+\cos 2\pi z) + \mathcal{O}(A^4)\,, \tag{22}$$

where $A \equiv 2\sqrt{2p}\,(\pi/k)\,\epsilon\,|a_{11}|$ and c is the phase speed of the wave. Note that this flow vanishes in the midplane, and once again is largest at the walls and in the direction of c. The velocity (22) describes the mean drift speed of the particles in the fluid. Note that in the present case these drifts are usually larger than the Eulerian, i.e., laboratory frame, mean flow.

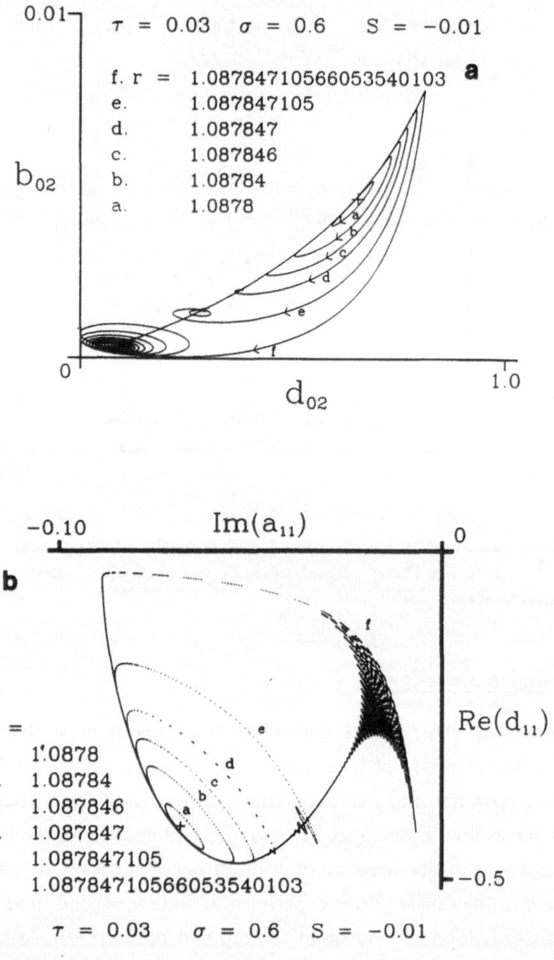

Fig. 8. (a) limit cycles in the (b_{02}, d_{02}) plane for several values of r and (b) Poincaré section through the MW two-torus corresponding to (a). $\Im m\, a_{11}$ is plotted against $\Re e\, d_{11}$ every time $\Re e\, a_{11} = 0$ with $\Re e\, \dot{a}_{11} > 0$.

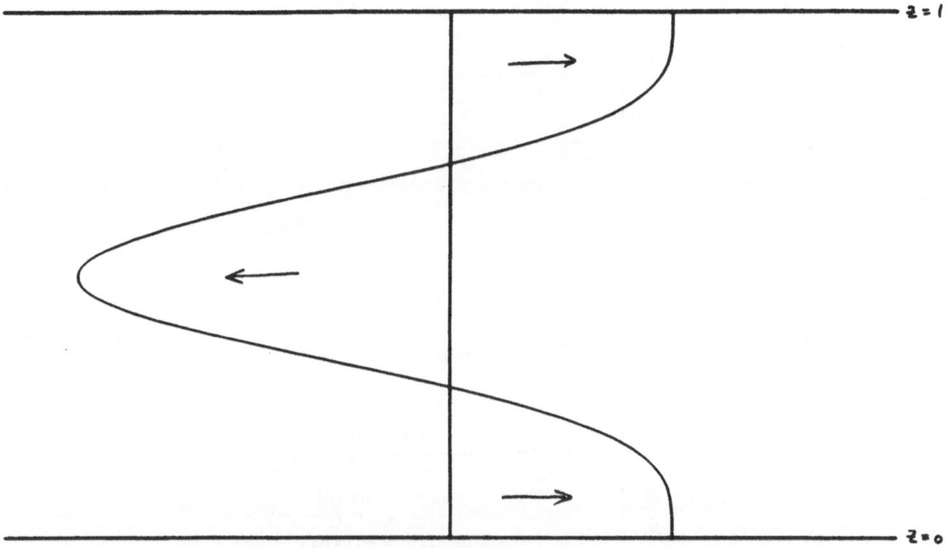

Fig. 9. Profile of mean Eulearian flow at $\mathcal{O}(\epsilon^4)$ for a right-travelling wave with idealized boundary conditions. Its mean vanishes.

(h) Experimental boundary conditions

We have solved equations (16) with the boundary conditions (17b) for parameter values typical of $^3He/^4He$ mixtures using a Fourier decomposition in the horizontal and in time [24]. This procedure yields a coupled set of ordinary differential equations for the vertical structure of the mode amplitudes. These are solved as a two-point boundary value problem using an eighth order deferred correction scheme. Ten equal intervals covered the half region $0 \leq z \leq 1/2$ with appropiate reflection used to complete the solution. Doubling the number of intervals in the vertical changes the result in the sixth decimal place only. Typically four horizontal modes were used, in addition to the $k = 0$ mode. This procedure restricts the results to small negative values of the separation ratio. For the SS and TW patterns this translates into 36 and 72 coupled first order equations, respectively. The accuracy of the resulting solutions was checked in most cases by including as many as eight horizontal modes. For the cases illustrated here the resulting bifurcation diagrams show no discernible differences. Additional checks against a finite difference code with the same boundary conditions and horizontal period were made, and the present method found to be much more efficient. All calculations were carried out with $2\pi/k = 2.0$, close to the value 2.0162 for a pure fluid. This procedure ignores a weak parameter dependence [17] of the linear theory k and, as already mentioned, a possible Rayleigh number dependence in the nonlinear regime.

In fig.11 we show bifurcation diagrams for two negative values of S: (a) $S = -0.001$ and (b) $S = -0.004$. The TW branch bifurcates subcritically for $S < S_{tc}^{TW} \approx -7.87 \times 10^{-4}$ and supercritically for $S_{tc}^{TW} < S < S_{CT} \simeq -0.00056$ (cf. the experimental value $S_{CT} \approx -0.0044$[25]). In case (a) S is just less than S_{tc}^{TW} and the branch exhibits a secondary saddle-node bifurcation. Since the calculations do not explicitly determine the stability of the solutions we use the theoretical results summarized in the preceding sections to conclude that at the saddle-node bifurcation the TW branch acquires stability

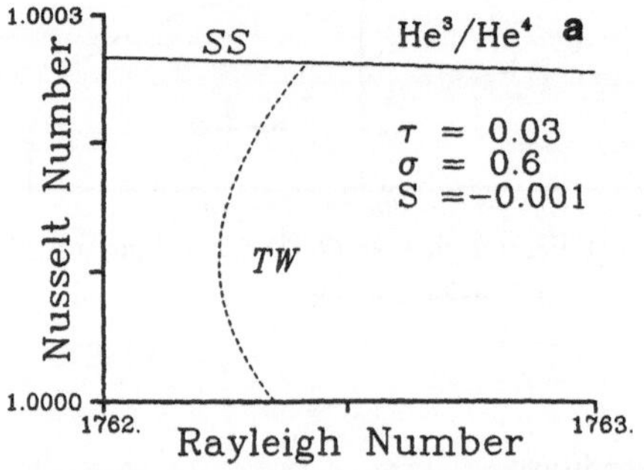

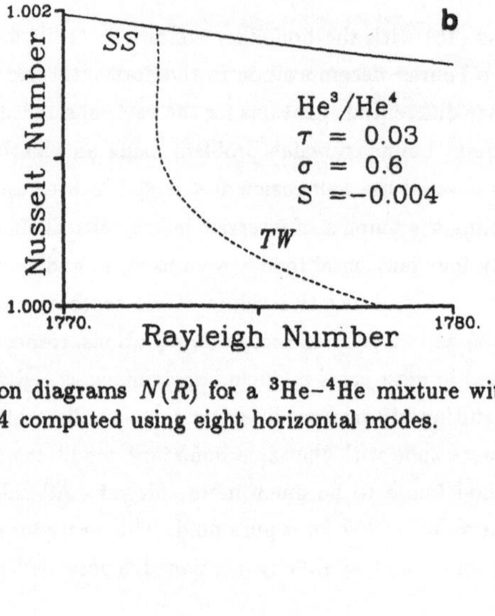

Fig. 10. Bifurcation diagrams $N(R)$ for a ^{3}He$-^4$He mixture with (a) $S = -0.001$ and (b) $S = -0.004$ computed using eight horizontal modes.

with respect to amplitude perturbations, but loses it again to an MW branch before terminating on the SS branch. The methods used do not allow us to locate this bifurcation or the MW branch it produces. In case (b) the TW branch exhibits two secondary saddle-node bifurcations in rapid succession, with the stability interval dramatically reduced. In this case the secondary bifurcation to MW is no longer necessary. An important feature of case (b) is that the stability interval occurs for Rayleigh numbers preceding the Hopf bifurcation from the conduction solution. In this case the initial Hopf bifurcation *cannot* develop into a periodic TW. In both cases the SW branch is supercritical but unstable to TW. The SS branch also exhibits a tricritical point, at $S \equiv S_{tc}^{SS} \approx -4.7 \times 10^{-6}$ in contrast to the experimental value $S_{tc}^{SS} \simeq 0.0035$[25]. This discrepancy may perhaps be attributed to finite size effects or to uncertainties in the determination of S. In fig.11 we show a typical bifurcation diagram for $S > 0$, showing a clear signature of the transition from the Soret régime, $R_{SS} < R \lesssim R_0$, to the Rayleigh-Bénard régime $R \gtrsim R_0{}^2$. Here $R_0 \simeq 1708$ the Rayleigh number for the onset of convection in a pure fluid. The results described above do not address the possibility of localized TW or Benjamin-Feir unstable patterns.

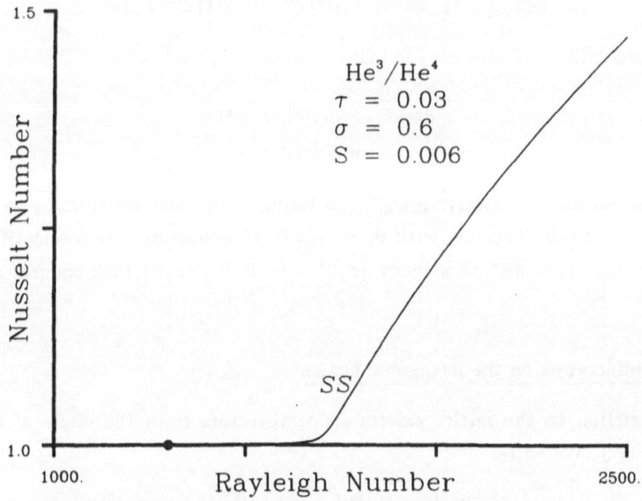

Fig. 11. Bifurcation diagram $N(R)$ for a ^{3}He–^{4}He mixture with $S = 0.006$.

III. Three-dimensional convection with periodic boundary conditions

In this section we summarize existing results on pattern selection among doubly periodic patterns in the plane. These results provide a natural extension of the results of the preceding section to three dimensions. There are three lattices that isotropically tile the plane: rhombic, square and hexagonal. When the partial differential equations are restricted to functions periodic on these lattices the symmetries of the unit cell together with translations in two horizontal directions reduce the group of symmetries from the Euclidean group to $D_2 \times T^2$, $D_4 \times T^2$ and $D_6 \times T^2$, respectively. When the boundary conditions at the top and bottom are identical and non-Boussinesq effects absent, an additional midplane reflection symmetry Z_2 is also present. These facts guarantee that the center eigenspace is finite-dimensional and that all other eigenvalues are bounded away from the imaginary axis. Consequently the partial differential equations can be reduced near bifurcation from a trivial solution to a finite-dimensional system of the form

$$\dot{z} = g(z, \lambda) , \tag{23a}$$

where z is the vector of the critical mode amplitudes, and λ the bifurcation parameter. The vector field g commutes with the symmetry group Γ:

$$\gamma g(z, \lambda) = g(\gamma z, \lambda) , \qquad \gamma \in \Gamma . \tag{23b}$$

This condition restricts the form of g, and assists in the determination of the stability properties of the various branches. We do not consider the rhombic lattice further.

(a) Steady state bifurcation on the square lattice

On the square lattice at $\lambda = 0$ we may write

$$w(x, y, z, t) = \Re e\{z_1(t) e^{ikx} + z_2(t) e^{iky}\} f(z) , \tag{24}$$

where z_1, z_2 satisfy [26]

$$\dot{z}_1 = g(\lambda, |z_2|^2, r^2) z_1 \tag{25a}$$
$$\dot{z}_2 = g(\lambda, |z_1|^2, r^2) z_2 , \tag{25b}$$

with g is real-valued, and $r^2 = |z_1|^2 + |z_2|^2$. In terms of the real variables $z_j = x_j e^{i\theta_j}$, $j = 1, 2$, equations (25) reduce to (9), together with $\dot{\theta}_1 = \dot{\theta}_2 = 0$. Consequently the results of section IIb apply to the present problem provided we identify $(x, 0)$ with Rolls, (x, x) with squares and (x_1, x_2) with cross-rolls.

(b) Steady state bifurcation on the hexagonal lattice

Here the restriction to the lattice selects six wavevectors from the circle of marginally stable wavevectors at $\lambda = 0$. We write

$$w(x, y, z, t) = \Re e\{z_1(t) e^{i\vec{k}_1 \cdot \vec{x}} + z_2(t) e^{i\vec{k}_2 \cdot \vec{x}} + z_3(t) e^{i\vec{k}_3 \cdot \vec{x}}\} f(z) , \tag{26}$$

where $\vec{k}_1 = (k, 0)$, $\vec{k}_{2,3} = k(-1, \pm\sqrt{3})/2$ and $\vec{x} = (x, y)$. When only the symmetry $D_6 \times T^2$ is present it is well-known that there are three primary branches, two of hexagons ($H^{\pm}$) and one of rolls. All are unstable [27–29]. With the extra midplane reflection symmetry there are generally four primary branches. Writing $z_j = x_j + iy_j$, $j = 1, 2, 3$, these are [29]

$$\text{Rolls (R):} \qquad (z_1, z_2, z_3) = x(1, 0, 0) \tag{27a}$$
$$\text{Hexagons (H}^{\pm}\text{):} \qquad (z_1, z_2, z_3) = x(1, 1, 1) \tag{27b}$$
$$\text{Regular Triangles (RT):} \qquad (z_1, z_2, z_3) = y(i, i, i) \tag{27c}$$
$$\text{Patchwork Quilt (PQ):} \qquad (z_1, z_2, z_3) = x(0, 1, 1) . \tag{27d}$$

Here $H^{\pm}$ are distinguished by sgn x but are taken into one another by the reflection in Z_2. To distinguish between H and RT a fifth order term must be computed. The fourth pattern (PQ) can never be stable.

(c) Hopf bifurcation on the square lattice

This problem, partially analyzed by Swift[30] and completed by Silber[31], gives rise to an eight-dimensional system of amplitude equations. In the generic case there are five primary solution branches, corresponding to travelling (TR) and standing (SR) rolls, travelling (TS) and standing (SS) squares, and alternating rolls (AR). In some cases a sixth solution, called standing cross-rolls (SCR), may also bifurcate simultaneously with the other branches. It is found that TR and TS cannot both be stable. In addition, while it is possible to have two coexisting stable branches when the first five branches all bifurcate supercritically, it is also possible to have all five bifurcate supercritically with none being stable. This observation has potentially interesting dynamical consequences.

(d) Hopf bifurcation on the hexagonal lattice

This case, also analyzed in ref. 30 and completed by Roberts et al.[32], leads to a twelve-dimensional system of equations. There are eleven primary branches whose stability properties are known. In this case the extra midplane symmetry yields no additional restriction on the equations and the theory applies equally to systems with identical or distinct boundary conditions at top and bottom.

(e) Applications

Explicit coefficient calculations for the idealized boundary conditions (17a) show that when steady rolls are supercritical they are also stable with respect to all perturbations lying on the square and hexagonal lattices[26], a result already known for doubly diffusive convection[30,33].

With the realistic boundary conditions (17b) only partial results are available[34]. These results are exact but are restricted to long wavelength patterns found when the conduction solution loses stability for $0 < S_\infty - S \ll S_\infty$, where S_∞ is the value of the separation ratio at which k first vanishes[17]. The theory shows that on the square lattice squares are stable, unless the asymmetry in the boundary conditions at top and bottom is too large when rolls become stable. By examining a particular degeneracy in (9) the existence of a stable branch of cross-rolls connecting rolls and squares was also established. As mentioned in section IIb such a calculation depends on coefficients of seventh order terms in $g(\lambda, z)$. The result that squares are typically stable in this system lends support to the suggestion[5] that with realistic boundary conditions squares should be stable. In the parameter regime the reflection-symmetric problem on the hexagonal lattice is degenerate; in this singular case hexagons are stable, but when the midplane reflection symmetry is broken rolls become stable. Finally, with non-Boussinesq terms included, squares bifurcate subcritically if these are large enough, and no stable solutions are found on the hexagonal lattice near onset[35].

IV. Conclusions

In this article we have summarized the results of analyzing a number of simple codimension-one and codimension-two bifurcations in the presence of symmetry, and showed how these can be used to make predictions about binary fluid convection. These techniques show clearly which aspects of the problem are due to the symmetries, and which explicit calculations need to be performed to apply the theory to specific experimental systems. These techniques are ideally suited to the study of the existence and stability properties of spatially periodic patterns. Although such patterns are by no means all of the states that are of interest, many experiments do reveal the existence of approximately spatially periodic states. Much work, particularly studies of modulational instabilities, remains to be done.

Finally, it must be emphasized that particularly for travelling patterns the effect of distant sidewalls, though small, is not unimportant. A discussion of this issue for two-dimensional convection can be found elsewhere in this volume[36].

Acknowledgment: one of us (EK) wishes to acknowledge support under NSF/DARPA grant DMS-8814702 during the preparation of this article.

References

1 R. W. WALDEN, P. KOLODNER, A. PASSNER AND C. M. SURKO: "Traveling waves and chaos in convection in binary fluid mixtures." Phys. Rev. Lett. **55**: 496 (1985)

2 G. AHLERS AND I. REHBERG: "Convection in a binary mixture heated from below." Phys. Rev. Lett. **56**: 1373 (1986)

3 V. STEINBERG, E. MOSES AND J. FINEBERG: "Spatio-temporal complexity at the onset of convection in a binary fluid." Nuclear Phys. B (Proc. Suppl.) **2**: 109 (1987)

4 P. KOLODNER, C. M. SURKO AND H. WILLIAMS: "Dynamics of traveling waves near the onset of convection in binary fluid mixtures." Physica D **37**: 319 (1989)

5 E. MOSES AND V. STEINBERG: "Competing patterns in a convective binary mixture." Phys. Rev. Lett. **57**: 2018 (1986)

6 A. E. DEANE, E. KNOBLOCH AND J. TOOMRE: "Traveling waves and chaos in thermosolutal convection." Phys. Rev. A **36**: 2862 (1987)

7 E. MOSES AND V. STEINBERG: "Flow patterns and nonlinear behavior of traveling waves in a convective binary fluid." Phys. Rev. A **34**: 693 (1986)

8 P. KOLODNER, D. BENSIMON AND C. M. SURKO: "Traveling wave convection in an annulus." Phys. Rev. Lett. **60**: 1723 (1988)

9 E. KNOBLOCH: "Oscillatory convection in binary mixtures." Phys. Rev. A **34**: 1538 (1986)

10 M. GOLUBITSKY AND M. ROBERTS: "A classification of degenerate Hopf bifurcations with $O(2)$ symmetry." J. Diff. Eq. **69**: 216 (1987)

11 E. KNOBLOCH: "On the degenerate Hopf bifurcation with $O(2)$ symmetry." Contemp. Math. **56**: 193 (1986)

12 E. KNOBLOCH AND M. R. E. PROCTOR: "Nonlinear periodic convection in double-diffusive systems." J. Fluid Mech. **108**: 291 (1981)

13 G. DANGELMAYR AND E. KNOBLOCH: "The Takens-Bogdanov bifurcation with $O(2)$ symmetry." Phil. Trans. Roy. Soc. London **322**: 243 (1987)

14 E. KNOBLOCH, A. E. DEANE, J. TOOMRE AND D. R. MOORE: "Doubly diffusive waves." Contemp. Math. **56**: 203 (1986)

15 E. KNOBLOCH AND D. R. MOORE: "A minimal model of binary fluid convection." Preprint (1989)

16 W. SCHÖPF AND W. ZIMMERMANN: "Multicritical behavior in binary fluid convection." Europhys. Lett. **8**: 41 (1989)

17 E. KNOBLOCH AND D. R. MOORE: "Linear stability of experimental Soret convection." Phys. Rev. A **37**: 860 (1988)

18 C. BRETHERTON AND E. A. SPIEGEL: "Intermittency through modulational instability." Phys. Lett. **96A**: 152 (1983)

19 W. BARTEN, M. LÜCKE, W. HORT AND M. KAMPS: "Fully developed traveling-wave convection in binary fluid mixtures." Phys. Rev. Lett. **63**: 376 (1989)

20 G. DANGELMAYR AND E. KNOBLOCH: "Interaction between standing and travelling waves and steady states in magnetoconvection." Phys. Lett. A **117**: 394 (1986)

[21] E. Knobloch and D. R. Moore: "Particle drifts associated with travelling wave convection in binary fluid mixtures." Bull. Amer. Phys. Soc. **33**: 370 (1988)

[22] S. J. Linz, M. Lücke, H. W. Müller and J. Niederländer: "Convection in binary fluid mixtures: traveling waves and lateral currents." Phys. Rev. A **38**: 5727 (1988)

[23] E. Knobloch and J. B. Weiss: Mass transport by wave motion, in "The Internal Solar Angular Velocity," B. R. Durney and S. Sofia, eds., Reidel, Dordrecht (1987)

[24] D. R. Moore and E. Knobloch: "Nonlinear convection in binary mixtures." Bull. Amer. Phys. Soc. **33**: 2285 (1988) and preprint (1989)

[25] T. S. Sullivan and G. Ahlers: "Hopf bifurcation to convection near the codimension-two point in a ^{3}He$-^4$He mixture." Phys. Rev. Lett. **61**: 78 (1988)

[26] M. Silber and E. Knobloch: "Pattern selection in steady binary fluid convection." Phys. Rev. A **38**: 1468 (1988)

[27] F. H. Busse: "Nonlinear properties of thermal convection." Rep. Prog. Phys. **41**: 1929 (1978)

[28] E. Ihrig and M. Golubitsky: "Pattern selection with $O(3)$ symmetry." Physica D **13**: 1 (1984)

[29] M. Golubitsky, J. W. Swift and E. Knobloch: "Symmetries and pattern selection in Rayleigh-Bénard convection." Physica D **10**: 249 (1984)

[30] J. W. Swift: "Bifurcation and symmetry in convection." Ph.D. Thesis, University of California, Berkeley (1984)

[31] M. Silber: "Bifurcations with D_4 symmetry and spatial pattern selection." Ph.D. Thesis, University of California, Berkeley (1989)

[32] M. Roberts, J. W. Swift and D. H. Wagner: "The Hopf bifurcation on a hexagonal lattice." Contemp. Math. **56**: 283 (1986)

[33] W. Nagata and J. W. Thomas: "Bifurcation in doubly diffusive systems I. Equilibrium solutions." SIAM. J. Math. Anal. **17**: 91 (1986)

[34] E. Knobloch: "Pattern selection in binary-fluid convection at positive separation ratios." Phys. Rev. A **40**: 1549 (1989)

[35] E. Knobloch: Nonlinear binary fluid convection at positive separation ratios, in "Cooperative Dynamics in Complex Systems," H. Takayama, ed., Springer-Verlag, Berlin (1989)

[36] G. Dangelmayr and E. Knobloch: On the Hopf bifurcation with broken $O(2)$ symmetry, in "The Physics of Structure Formation: Theory and Simulation," W. Güttinger and G. Dangelmayr, eds., Springer-Verlag, Berlin (1987); Dynamics of slowly varying wave trains in finite geometry, this volume

STRUCTURE AND DYNAMICS OF NONLINEAR CONVECTIVE STATES IN BINARY FLUID MIXTURES

W. Barten[1], M. Lücke[1], and M. Kamps[2]

[1]*Institut für Theoretische Physik, Universität des Saarlandes, D–6600 Saarbrücken*
[2]*Institut für Festkörperforschung, Kernforschungsanlage, D–5170 Jülich*

Various properties of traveling wave (TW) and stationary overturning convection (SOC) are determined for ethanol–water parameters by finite–differences numerical solutions of the basic hydrodynamic field equations subject to realistic horizontal boundary conditions. Bifurcation– and phase diagrams for TW and SOC solutions are presented. Unstable SOC patterns that decay into a stable TW or the conductive state can be stabilized by phase pinning lateral boundaries. The structural changes at the transition TW ⟷ SOC are shown. The mean flow, the lateral currents of heat and concentration, and the particle motion associated with a TW are elucidated.

I INTRODUCTION

Convection in horizontal layers of binary fluid mixtures heated from below has attracted in recent years a great deal of experimental and theoretical activities[1-20]. Of particular interest has been the primary pattern forming instability of the motionless conductive state and the structure, dynamics, and bifurcation behaviour of the convective states that bifurcate out of the basic conductive state. In a pure fluid the instability that occurs first on increasing the Rayleigh number

$$R = \frac{\alpha g d^3}{\kappa\nu} \Delta T \qquad (1.1)$$

beyond the threshold $R_c^0 = 1707.8$ is a stationary one. The conductive state looses stability in a forwards bifurcation to a stable pattern of stationary overturning convection (SOC) rolls. However, in a mixture the primary instability is either oscillatory or stationary depending on the separation ratio

$$\psi = -\frac{\beta}{\alpha T_0} k_T \, . \qquad (1.2)$$

Here d is the vertical thickness of the layer, κ the thermal diffusivity, ν the kinematic viscosity, T_0 the mean temperature, $\alpha = -(1/\rho)\partial\rho(T,p,C)/\partial T$ the thermal expansion coefficient, $\beta = -(1/\rho)\partial\rho(T,p,C)/\partial C$ the solutal expansion coefficent, and k_T the

Nonlinear Evolution of Spatio-Temporal Structures in Dissipative Continuous Systems
Edited by F.H. Busse and L. Kramer
Plenum Press, New York, 1990

thermodiffusion ratio of the fluid mixture. Furthermore, g is the gravitational constant and ΔT the vertical temperature difference across the fluid layer.

In Fig. 1 we show *schematically* the reduced Rayleigh numbers

$$r = R/R_c^0 \tag{1.3}$$

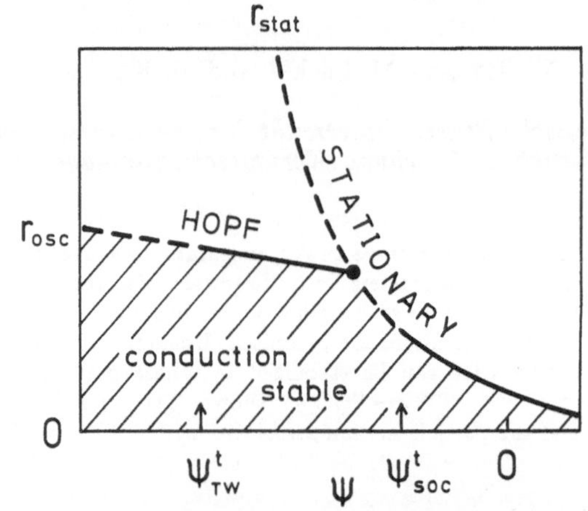

Fig. 1. Schematic bifurcation lines r_{osc} and r_{stat} of oscillatory and stationary convection out of the conductive state. Plotting the vertical convective heat current in an N vs r diagram TW (SOC) convection bifurcates forwards — $\partial N/\partial r > 0$ — out of the conductive state across the full part of the $r_{osc}(r_{stat})$ line and backwards — $\partial N/\partial r < 0$ — across the dashed part. At the tricritical values ψ^t the initial slope diverges: $N - 1 \propto (r - r^t)^{1/2}$. Bifurcating SW solutions are not considered here. For $\psi = 0$ one has the behaviour of a pure fluid. The diagram is not to scale: For room temperature ethanol–water mixtures $\psi^t_{SOC} = \mathcal{O}(-10^{-6})$, $\psi^t_{TW} = \mathcal{O}(-10^{-4})$[12].

where convective solutions branch off the conductive state. From experiments[1-5], linear stability analysis[6,7] of the conductive state, few–mode Galerkin approximations[8-11], and weakly nonlinear perturbation analysis[12] of the bifurcating convective states[13,14] has emerged the following picture: (1) In the r–ψ plane the stability range of the conductive state is bounded by an oscillatory bifurcation threshold, r_{osc}, and a stationary threshold, r_{stat}. (2) On the stationary branch, r_{stat}, of the stability boundary there is at ψ^t_{SOC} a tricritical point in the sense of Landau's meanfield classification of phase transitions. There the vertical convective heat current flowing through the layer grows with divergent initial slope: $N-1 \propto (r-r^t_{SOC})^{1/2}$ with N being the Nusselt number. Above (below) ψ^t_{SOC} there is a forwards nonhysteretic (backwards hysteretic) transition at r_{stat} and the convective heat current of the bifurcating SOC solution grows close to threshold with positive (negative) slope: $N-1 \propto |r-r_{stat}|$. (3) There is a Hopf bifurcation threshold, r_{osc}, at which convective roll pattern solutions in the form of traveling waves (TW) and standing waves (SW) bifurcate out of the conductive state. (4) Hopf frequency ω_H and critical wave numbers k^c_{osc} and k^c_{stat} of the bifurcating oscillatory and stationary patterns decrease with increasing ψ. The wave numbers differ also at the "codimension–two point" where r_{stat} and r_{osc} intersect. (5) For TW and SW solutions there exist tricritical points, ψ^t_{TW} and ψ^t_{SW}, respectively, on r_{osc} such that the slope $\partial N/\partial r$ of the convective heat current at the threshold r_{osc} diverges there. (6) For $\psi > \psi^t_{TW}$ ($\psi < \psi^t_{TW}$) a stable (unstable) TW branches off the conductive state in a forwards (backwards) bifurcation. Close to threshold the squared frequency of the TW decreases proportional to the distance from onset: $\omega^2_H - \omega^2 \propto |r-r_{osc}|$. (7) Weakly nonlinear SW solutions close to threshold are always unstable. Nonlinear SW's were neither found in experiments nor in numerical simulations.

We shall consider henceforth only SOC and TW convection and we restrict ourselves in this work to negative ψ values to the left of ψ^t_{TW} where the TW solution bifurcates subcritically.

II THE SYSTEM

Here we specify the system, boundary conditions, and the numerical procedure to solve the hydrodynamical field equations.

A The equations

Consider a binary fluid layer of vertical height d in the homogeneous gravitational field $-g\,\mathbf{e}_z$. Perfectly heat conducting horizontal plates impose a temperature difference ΔT between the warm bottom and the cool top plate. Convection is described by balance equations of mass, momentum, heat, and concentration which lead in Oberbeck–Boussinesq approximation to the equations

$$0 = \nabla \cdot \mathbf{u} \tag{2.1}$$

$$\partial_t \mathbf{u} = -\nabla \cdot (\mathbf{u} : \mathbf{u}) - \nabla p - \sigma \nabla \times (\nabla \times \mathbf{u}) + b\,\mathbf{e}_z \tag{2.2}$$

$$\partial_t T = -\nabla \cdot (\mathbf{u}\,T) + \nabla^2 T \tag{2.3}$$

$$\partial_t C = -\nabla \cdot (\mathbf{u}\,C) + L\nabla^2 (C - \psi T) \tag{2.4}$$

for the fields of velocity $\mathbf{u}$, pressure p, temperature T, and concentration C. For small deviations

$$\delta F = F - F_0 \qquad (2.5)$$

from the mean values $F_0 = T_0, p_0, C_0$ the vertical buoyancy force caused by temperature– and concentration–induced density variations is

$$b = \sigma R (\delta T + \delta C) . \qquad (2.6)$$

Making use of incompressibility (2.1) we have written the field equations into a form that in its discretized version is advantageous to integrate numerically.

We measure lengths in units of d, times by the vertical thermal diffusion time d^2/κ, temperature by ΔT, and concentration by $\Delta T\ \alpha/\beta$. In addition to the Rayleigh number R and the separation ratio ψ two material parameters enter into the reduced field equations: the Prandtl number $\sigma = \nu/\kappa$ and the Lewis number $L = D/\kappa$ where D is the concentration diffusion constant. Throughout this work we use $\sigma = 10$ and $L = 0.01$ which is appropriate[4, 5, 21] for ethanol in water around room temperature.

We impose no slip, impermeable boundary conditions on the fluid at the horizontal plates thus enforcing the vertical concentration current to vanish there. This causes via the Soret effect a coupling of concentration– and temperature gradient, $\partial_z C = \psi\, \partial_z T$, at $z = 0, 1$ the presence of which significantly influences[8] the convective behaviour of the mixture. The Soret coupling also generates in the conductive state

$$\mathbf{u}_{cond} = 0 \ ; \quad T_{cond}(z) = T_0 + \frac{1}{2} - z \qquad (2.7a)$$

a concentration gradient

$$C_{cond}(z) = C_0 + \psi \left(\frac{1}{2} - z\right) . \qquad (2.7b)$$

Consequently the buoyancy force

$$b_{cond}(z) = \sigma R \left(\frac{1}{2} - z\right)(1 + \psi) \qquad (2.8)$$

is for $\psi < 0$ reduced in comparison to the pure fluid, $\psi = 0$, thus stabilizing via the Soret effect the conductive state.

B Numerical simulation

We have numerically solved the governing equations (2.1–2.6) in a vertical crossection of the layer, i. e. in an x–z plane perpendicular to the axes of the convective rolls. We suppress any y–dependence of the fields along the roll axes and we assume the velocity field

$$\mathbf{u} = u\,(x,z;\,t)\,\mathbf{e}_x + w\,(x,z;\,t)\,\mathbf{e}_z \tag{2.9}$$

to have the lateral component u and the vertical component w. Furthermore, we enforce convective patterns of wave number $k = \pi$ with lateral periodicity $\lambda = 2$ by imposing periodic boundary conditions on all fields: $F\,(x,z;\,t) = F\,(x + \lambda,z;\,t)$. The boundary conditions at the bottom and top of the layer are the experimental ones described in Sec. II.A. In some runs we *in addition* enforced the lateral velocity field u to vanish at the lateral boundaries $x = 0,\lambda$. The associated phase–pinning prevents TW solutions in our short system and stabilizes certain SOC solutions that otherwise would be unstable towards a TW or the conductive state.

We used the MAC[22] method to discretize the fields on three staggered grids with uniform spacing $\Delta x = \Delta z = 0.05$. Spatial derivatives are replaced by central differences and time derivatives by forward differences. Pressure and velocity fields are iteratively adapted[23] to each other using (2.1) and (2.2). Typical relaxation times towards a TW (SOC) state were about 100 (10) vertical diffusion times. The critical Rayleigh number R_c^0 for onset of convection resulting from the discretized system for the pure fluid case, $\psi = 0$, is about 1 % below the threshold of the continuous system. We use henceforth R_c^0 of our discretized system to reduce Rayleigh numbers according to $r = R/R_c^0$.

III RESULTS

A Bifurcation properties

As a representative example for the bifurcation behaviour at not too small negative ψ we show in Fig. 2b the Nusselt number of nonlinear SOC and TW convective states vs r for $\psi = -0.25$ in comparison with the SOC Nusselt number for a pure fluid, $\psi = 0$. Our method — as experiments — is not suited to determine unstable states (open triangles in Fig. 2b on the upper SOC branch are states that have been stabilized by phase–pinning lateral boundary conditions). Thus the lower unstable TW and SOC solution branches are not shown. The former bifurcates at r_{osc} subcritically with finite negative slope out of the conductive state and connects to the shown upper branch (full dots in Fig. 2b) at the saddle r^s_{TW}. The lower SOC branch lies almost parallel to the r–axis — the stationary threshold r_{stat} has moved to ∞ already at much smaller negative ψ. This topology of the TW (SOC) solution branch is supported by calculations of Bensimon et al.[14] (the Galerkin model of Linz et al.[8, 9]).

The frequency of the TW monotonously decreases[14] from its Hopf value at r_{osc} as one moves along the lower unstable branch to the saddle, r^s_{TW}. Fig. 2a shows the ensuing monotonous drop in ω on the upper stable TW branch until the TW merges at r* nonhysteretically with vanishing frequency with the SOC branch. Close to r* we found a square–root variation[9, 24] $\omega \simeq 1.3\sqrt{r^*-r}$. The nonhysteretic character[9, 13] of

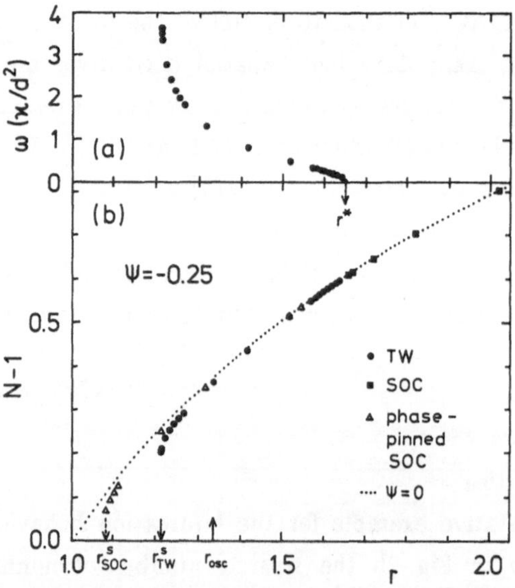

Fig. 2. Numerically obtained bifurcation properties of nonlinear convective states. a) Frequency of stable TW and b) Nusselt numbers vs r. Full symbols (thin dots) refer to states in the mixture (pure fluid) that are stable under lateral periodic boundary conditions. At r^* the TW merges nonhysteretically with zero frequency with the SOC branch. Open triangles denote unstable SOC states in the mixture that have been stabilized by phase pinning lateral boundaries. Within our numerical accuracy we identify the position of the lowest TW (SOC) state with the saddle r^s_{TW} (r^s_{SOC}) of the TW (SOC) solution branch. L=0.01, σ=10, ψ= −0.25, $r_{osc}(k=\pi)=1.33$[7], $\omega_H(k=\pi)=11.2$[7], $r^s_{TW} \simeq 1.21$, $r^s_{SOC} \simeq 1.08$, $r^* \simeq 1.65$.

the transition TW ⟷ SOC at r^* was also obtained by Bensimon et al.[14], however, their r^* is below r_{osc}. On the other hand, in experiments[3] hysteresis has been observed that might be due to phase pinning forces exerted e. g. by lateral sidewalls.

If we do not allow TW solutions in our short system by applying the phase pinning lateral boundary conditions described in Sec. II.B. then the stable SOC branch (squares in Fig. 2b) extends all the way down (triangles) to the saddle r^s_{SOC}. Upon lifting the stabilizing phase pinning these SOC states (triangles) decay into the TW state for $r^s_{TW} < r < r^*$ and into the conductive state below the TW saddle. In the latter case the decay proceeds via a transient TW in which the frequency grows while the amplitude decays to zero. Note finally how close the upper branches of the Nusselt number curves for TW and SOC states in the mixture are to the forwards bifurcating SOC solution in a pure fluid, $\psi = 0$.

B Phase diagram

In Fig. 3 we elucidate the existence range of stable conductive and nonlinear convective states in our system without lateral phase pinning in the ψ–r plane. The double logarithmic manner was chosen to resolve details at small ψ, r–1. The conductive state is stable below r_{osc}.

In the ψ range of Fig. 3 TW's bifurcate subcritically at r_{osc}. The TW's on the upper solution branch are stable in the dashed area between the TW saddle and r^* where the TW solution branch ends on the SOC branch. For large negative ψ the upper existence boundary r^* moves to larger and larger values. For less negative ψ–values the existence range of stable TW's shrinks to zero when the TW saddle r^s_{TW}

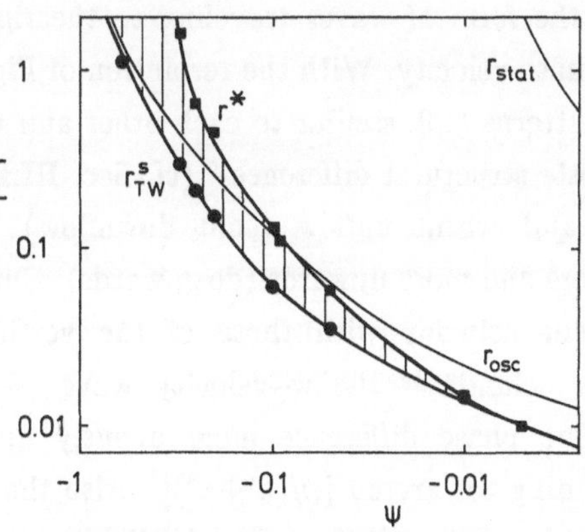

Fig. 3. Boundaries of stable states in a double–logarithmic r–1 vs ψ phase diagram (parameters as in Fig. 2). Symbols come from simulation and full lines are guides to the eye. Bifurcation thresholds r_{stat}, r_{osc} were obtained[7] with a shooting method. The conductive state is stable below r_{osc}. TW's are stable in the dashed region between the saddle, r^s_{TW}, and r^* where the TW solution branch ends on the SOC branch. SOC's are stable above r^* or the saddle, r^s_{SOC}. See text for details.

and r^* come together i. e. when the TW branch ends with divergent slope on the SOC branch (at $\psi \simeq -0.01$; for $\psi = -0.005$ we did not find stable TW's). Furthermore, we find within our numerical accuracy that this pinching of the stable TW's existence range happens close to the SOC saddle, r^s_{SOC}. Whenever r^* is on the upper SOC branch then SOC's are stable (full squares in Fig. 2b) above r^* and unstable (triangles in Fig. 2b) below[24]. In case r^* is on the lower SOC branch then SOC's are stable above the saddle r^s_{SOC}.

The phase diagram in Fig. 3 shows that there are three cases with different experimental consequences to be distinguished. Case (i): $r^* > r_{osc}$ is realized for large negative ψ (cf. Fig. 2b). Then an experiment shows upon heating a first order transition to the stable TW state on the upper solution branch when r increases beyond r_{osc}. Reducing now r there is a transition back to the conductive state from the TW saddle at r^s_{TW}. Case (ii): $r^s_{TW} < r^* < r_{osc}$ is realized for smaller negative ψ. Then there is on heating beyond r_{osc} a transition (with an oscillatory transient) towards a final stable SOC state on the upper SOC branch. Upon decreasing r there will first be the smooth transition at r^* to a stable TW and finally on decreasing r further below r^s_{TW} the transition back to the conductive state. And finally in case (iii) for $\psi \gtrsim -0.01$ we saw no more stable nonlinear TW state.

C Structure of TW and SOC states

Having discussed the bifurcation behaviour of the convective states we now elucidate structural properties.

1. Fields In Fig. 4 we show structural changes as one moves with increasing r along the upper TW solution branch of Fig. 2b from the saddle r^s_{TW} to the endpoint r^* and beyond onto the SOC branch. After transients have died out the final TW convective fields have the form of waves traveling to the right: $F = F(x - v_p t, z)$ where $v_p = \omega/k$ is the phase velocity. With the resolution of Fig. 4a the velocity fields of TW and SOC roll patterns look similar to each other and to that in a pure fluid. However there is a subtle structural difference[19] (cf. Sec. III.D.). With increasing r, convection intensifies and warm upflow (cold downflow) bends the conductive horizontal isotherms more and more upwards (downwards). But in a TW the extrema of the isotherms do not coincide with those of the vertical velocity field: the temperature wave lags behind[9, 19] the w–velocity wave — cf. the lateral wave profiles at $r = 1.22$. The phase difference monotonously decreases with the TW frequency roughly according to[9] $\arctan[\omega/(k^2+\pi^2)]$. Also the concentration wave is phase shifted relative to the w–wave[9, 19] (cf. Sec. III.D.).

Note that the concentration wave (squares in Fig. 4e) is highly anharmonic. It is the TW concentration field that shows the most pronounced structure. Furthermore its changes when approaching the SOC state at r^* are most conspicuous. To understand these properties one has to keep in mind that convection mixes the fluid

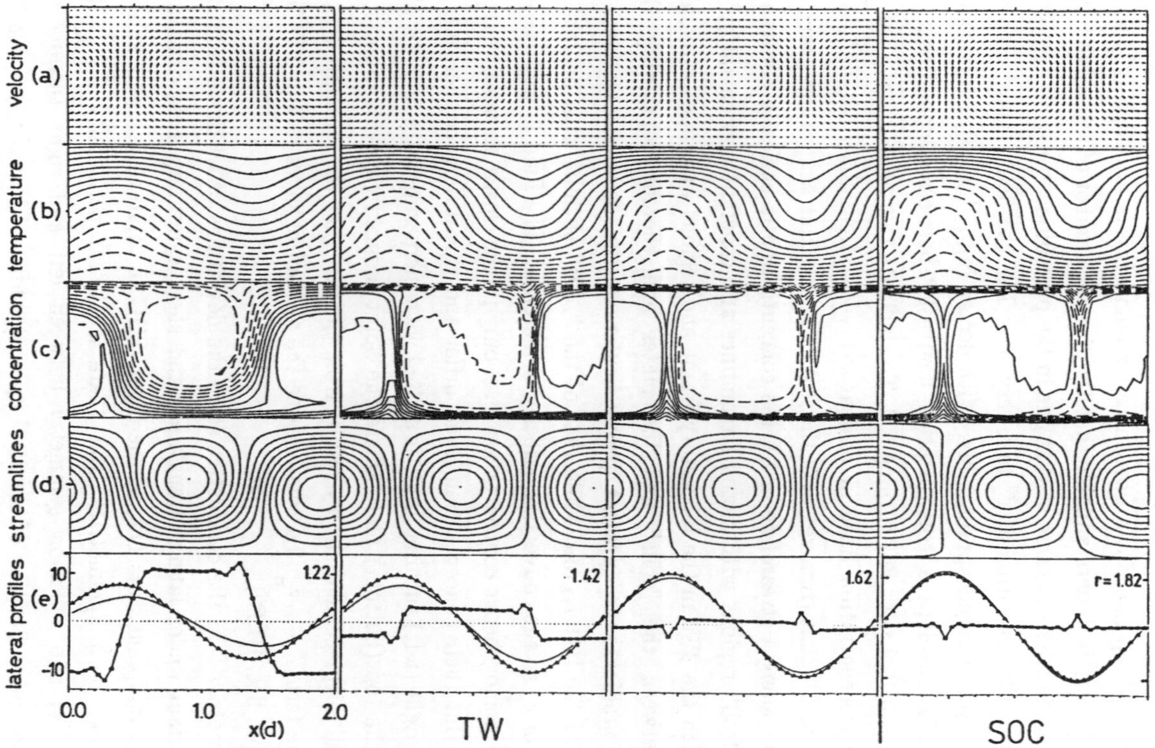

Fig. 4. Structural properties of TW and SOC convective states of Fig. 2b on the upper solution branch: velocity field (a), equidistant isolines of the temperature (b) and concentration (c) field, streamlines (d) (for TW's in the frame comoving with the phase velocity to the right), and lateral profiles (e) at midheight, $z = 1/2$, of w (thin line), 40 $(T-T_0)$ (triangles), and 400 $(C-C_0)$ (squares). The first three columns represent TW's traveling to the right. The first one at $r = 1.22$ with frequency $\omega \simeq 0.3\ \omega_H$ is close to the saddle, $r^S_{TW} \simeq 1.21$. For the second one $\omega \simeq 0.07\ \omega_H$. The last one at $r = 1.62$ is close to $r^* \simeq 1.65$ with $\omega \simeq 0.02\ \omega_H$. The SOC state at $r = 1.82$ is well beyond the transition point r^*.

and destroys[19] the conductive state's Soret induced concentration gradient between top and bottom of the layer. In fact in the SOC where the streamlines of the velocity field are closed the alcohol is practically homogeneously distributed at the mean concentration level, C_0, over the bulk of the fluid. Only in the two narrow top and bottom boundary layers the width of which decreases with increasing convection intensity there is an almost linear concentration variation with z. And at the positions of maximal up and down flow these boundary layers are slightly deformed into the bulk. Thus it is understandable that the Nusselt number of intensive SOC in the mixture is practically the same as that one in the pure fluid.

Just below the transition at r^* to a very slow TW the fields (third column in Fig. 4) have not yet changed appreaciably. With decreasing r, i. e. with increasing ω, however, the TW develops a concentration contrast[19] between alcohol rich right turning roll and alcohol poor left turning roll. To elucidate this phenomenon we show snapshots of isoconcentration lines in comparison with streamlines of the velocity field in the frame Σ' that comoves with the TW. The two line shapes are practically identical so that along a streamline in Σ' the concentration is basically constant. In Σ' there are "roll" regions with closed streamlines and constant concentration that travel along with the TW in the laboratory frame and regions with open streamlines meandering between the "rolls" and the plates and thereby separating them. Concentration gradients are present perpendicular to the open streamlines. The spatial extent of the latter grow while that of the "rolls" shrink[9] with decreasing r since the ratio of v_p to the convective velocity w increases. The fluid is well mixed in Σ' to a locally homogeneous concentration level only in the closed streamline regions of the "rolls". But their concentration levels differ and this difference increases as r decreases: The right (left) turning "roll" is shifted with increasing ratio of v_p/w closer and closer to the top (bottom) plate where the Soret induced concentration is high (low). The "roll" displacement simply results from the fact that the streamfunction in Σ' contains the term $-z\,v_p$. To some degree the alcohol behaves in the TW like a passive scalar (cf. Sec. III.E.).

2. Symmetries In the conductive state the deviations of all fields from their mean are invariant under lateral translation and show symmetry behaviour under reflexion at the midplane, $z = 1/2$, of the layer. SOC in mixtures as well as in the pure fluid breaks both symmetries. But there is lateral periodicity $\delta F_{SOC}(x,z) = \delta F_{SOC}(x+\lambda,z)$ and reflexion symmetry between left and right turning rolls: Taking $x = 0$ to be the position of maximal up or downflow $\delta F_{SOC}(x,z) = \pm\,\delta F_{SOC}(-x,z)$ with + for w, T, C, p and -1 for u. Furthermore the SOC state shows point symmetry, $\delta F_{SOC}(x+\frac{\lambda}{4}, z) = \pm\,\delta F_{SOC}(\frac{\lambda}{4}-x, 1-z)$, around the roll center $x = \lambda/4$, $z = 1/2$ with + for p and − for u,w,T,C. Combining these symmetries one obtains $\delta F_{SOC}(x,z) = \pm\,\delta F_{SOC}(x+\lambda/2, 1-z)$ with + for u,p and − for w,T,C.

While a TW has lateral periodicity the reflexion symmetries of SOC are broken. But within our numerical accuracy the TW fields still show the combination symmetry

$$\delta F_{TW}(x,z;t) = \pm\, \delta F_{TW}(x + \lambda/2, 1 - z; t) \tag{3.1}$$

that has first been observed in our numerical simulation[19]. As an aside we mention that our code was written such as to guarantee the absence of symmetry breaking roundoff errors[25]. Thus we can as an alternative to phase pinning generate an unstable SOC state also by starting from initial conditions with SOC reflexion symmetry.

D Lateral flow and lateral currents in a TW

1. Mean lateral flow The TW velocity field generates a mean Reynoldsstress, <wu>. The brackets imply a lateral average over one wavelength, which for a TW is equivalent to a time average over one oscillation period $\tau = 2\pi/\omega$

$$<\phi> = \frac{1}{\lambda}\int_0^{\lambda} dx\, \phi\,(x - \tfrac{\omega}{k}t, z) = \frac{1}{\tau}\int_0^{\tau} dt\, \phi\,(x - \tfrac{\omega}{k}t, z)\,. \tag{3.2}$$

This averaging procedure projects out the stationary, large scale component of any field ϕ. In particular it is the correct way to determine any global current or flow (cf. ref. 26 for a different opinion).

The mean Reynoldsstress arises[9] because in particular the first lateral Fourier mode of the vertical velocity field has a z–dependent phase $\varphi_w(z)$ as in Fig. 5a of ref. 19. Thus, in a laterally periodic system without adverse mean lateral pressure gradient the above stress drives[9] a mean flow

$$U(z) = < u\,(x,z,t) > \tag{3.3}$$

that is directed opposite to the TW with a symmetric vertical profile[19] that has a maximum at midheight, z = 1/2. We expect this mean flow to occur in annular experimental set–ups while the situation in rectangular geometry with lateral sidewalls is unclear. It should be present also in a weakly nonlinear TW close to threshold since the phase $\varphi_w(z)$ of the linear mode of w is z–dependent[6,7]. In Fig. 5c we show the variation of the maximum of U with r on the upper TW branch. The mean flow vanishes when the TW ends at r* with zero phase velocity on the SOC branch and it seems to change direction close to the TW saddle. Note that the mean flow is rather small ($U/v_p \simeq -0.006$ for r = 1.5, $\psi = -0.25$).

2. Mean lateral currents Let us consider the lateral current densities of heat and concentration

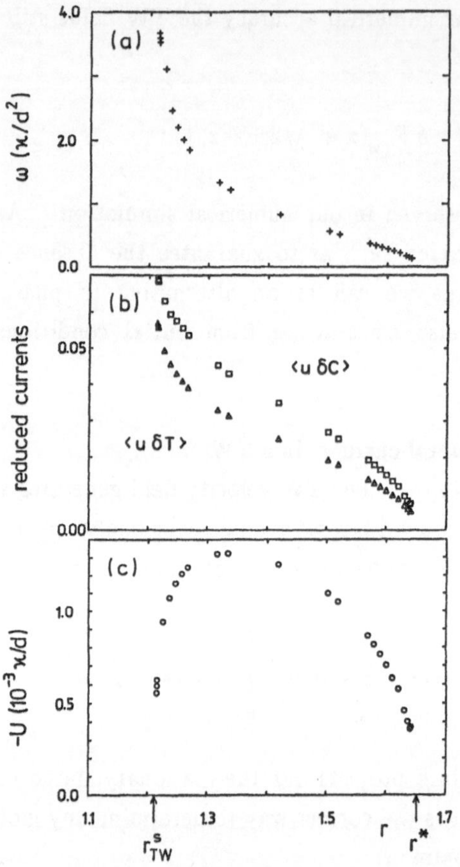

Fig. 5. (a) Frequency, maximal values of (b) mean lateral currents of heat <uδT> and concentration <uδC>, and of (c) mean lateral flow for the TW's of Fig. 2. For the shape of the vertical profiles of the currents and the flow see Fig. 5 of ref. 19.

$$Q_x = u\,T - \partial_x T \quad ; \quad J_x = u\,C - L\,\partial_x\,(C - \psi\,T)\,. \tag{3.4}$$

A TW generates stationary, global, z–dependent, lateral mean currents of heat and concentration

$$\langle Q_x \rangle = \langle u\,T \rangle = U\,T_0 + \langle u\,\delta T \rangle \tag{3.5a}$$

$$\langle J_x \rangle = \langle u\,C \rangle = U\,C_0 + \langle u\,\delta C \rangle\,. \tag{3.5b}$$

The mean lateral currents are convective since diffusive parts average out in an extended TW state.

The first contribution, U F_0, to (3.5) describing transport of the mean F_0 by the mean flow U depends on the magnitude of F_0 and on the existence of a mean flow. We estimate[27] $|U\,F_0/\langle u\,\delta F \rangle| \simeq 0.7$ for temperature and $\simeq 0.08$ for concen–

tration. The second contribution, $<u\,\delta F>$, arising also without mean flow[9] and in confined TW states[28] is more important. It is dominated by the convective part f of the fields

$$\delta F = F - F_0 = F_{cond} + f - F_0 \,. \tag{3.6}$$

In fact the vertical profiles of $<u\,\delta F>$ and of $<u\,f>$ are practically identical and have the form shown in refs. 9, 19: They are antisymmetric around the midplane so that the net current $\int_0^1 dz <u\,\delta F>$ vanishes. But in the upper half of the layer there is a heat current $<u\,\delta T>$ (concentration current $<u\,\delta C>$) flowing parallel to **k** (− **k**) and vice versa in the lower half. The current maxima are located at $z \simeq 1/4$ and $3/4$. Their magnitudes are plotted in Fig. 5b vs r. One sees roughly a proportionality $<u\,\delta C> \propto <u\,\delta T> \propto \omega$ which holds also for the weakly nonlinear TW of the Galerkin model[9]. The lateral heat current near the TW saddle is roughly one third of the convective heat current, N−1, flowing vertically through the layer.

It is instructive to express the current $<u\,f>$ in terms of lateral Fourier modes of f and the w–velocity field

$$<u\,f> = \frac{1}{2k}\, \mathrm{Im} \sum_{n=1} \frac{1}{n} \hat{f}_n\, \partial_z \hat{w}_n^* \,. \tag{3.7}$$

Taking only the first lateral modes and ignoring a small contribution $\sim \partial_z \varphi_w$ one finds

$$<u\,f> \simeq \frac{1}{2k}\, \hat{f}_1(z)\, \partial_z |\hat{w}_1(z)|\, \sin(\varphi_f - \varphi_w) \,. \tag{3.8}$$

The product $\hat{f}_1\, \partial_z |\hat{w}_1|$ explains[9, 19] the z–variation of the lateral current profiles and the sine shows that the current is driven by the phase difference between the f–wave and the w–wave. Since also the linear modes $\hat{f}_1(z)$ and $\hat{w}_1(z)$ at threshold are phase shifted[7, 12] the currents (3.7) are generated[9] also in weakly nonlinear TW's.

3. Mass currents Here we compare the lateral mass current

$$u\,\rho_{ethanol} = u\,\underline{C}\,\rho \tag{3.9}$$

of ethanol, $\rho_{ethanol} = \underline{C}\,\rho$, that is associated with the concentration current $u\underline{C}$ with the net current $u\rho$ of total mass, $\rho = \rho_{ethanol} + \rho_{water}$ Note that

$$\underline{C} = C\, \Delta T\, \frac{\alpha}{\beta} \tag{3.10}$$

is the unreduced concentration — $0<\underline{C}<1$ — while C is the reduced concentration used throughout this paper. We keep track of the contributions from the mean flow $U = \langle u \rangle$ and the fluctuating part $\tilde{u} = u - U$. With

$$\rho = 1 + \delta\rho = 1 - \alpha\,\Delta T\,(\delta T + \delta C) \tag{3.11}$$

one gets for the total mass current

$$\langle u\,\rho \rangle = U + \langle u\,\delta\rho \rangle \simeq U + \langle \tilde{u}\,\delta\rho \rangle\,. \tag{3.12}$$

We hasten to add that a current like $\langle u\,\delta\rho \rangle$ should be interpreted with caution (and even more so $\langle w\,\delta\rho \rangle$) when as in our case the velocity field is constraint by incompressibility, i. e., $\delta\rho = 0$. We obtain[27] $|\langle u\,\delta\rho \rangle| \simeq 2\cdot 10^{-5}$ which is small compared to the mean mass current $|U| \sim 10^{-3}$ generated by the mean flow when laterally periodic boundary conditions apply.

From (3.9) one finds that the mean lateral mass current of ethanol

$$\langle u\,\rho_{\text{ethanol}} \rangle = \langle u\,(\underline{C}_0 + \delta\,\underline{C})\,(1 + \delta\rho) \rangle \simeq U\,\underline{C}_0 + \langle u\,\delta\,\underline{C} \rangle \tag{3.13}$$

is dominated by two contributions since terms involving $\delta\rho \ll 1$ are negligible. Here[27] the ethanol mass current sustained by the mean flow, $|U|\underline{C}_0 \simeq 8\cdot 10^{-5}$, is significantly smaller than the current $|\langle u\,\delta\,\underline{C} \rangle| \simeq |\langle \tilde{u}\,\delta\,\underline{C} \rangle| \simeq 10^{-3}$ that is generated by the phase difference between C– and w–field in a TW of Fig. 5 close to the saddle.

Comparing (3.12) with (3.13) one finds: In the case of periodic lateral boundary conditions allowing a finite mean flow U the magnitude of the total mass current $\langle u\,\rho \rangle \simeq U$ is comparable with the ethanol mass current $\langle u\,\underline{C}\,\rho \rangle$ associated with the concentration current. However, in the case that the mean flow U is suppressed then $|\langle u\,\rho \rangle / \langle u\,\underline{C}\,\rho \rangle| \simeq 0.02$. In such a situation the mass currents of ethanol, $\rho_{\text{ethanol}} = \underline{C}\,\rho$, and of water, $\rho_{\text{water}} = (1 - \underline{C})\,\rho$ compensate each other almost perfectly — $\langle u\,\rho \rangle$ is almost zero — while the concentration current $\langle u\,\delta\,C \rangle$ is important. This result on the significance and magnitude of a mean mass transport in a TW with suppressed mean flow has previously been derived[9] within the Galerkin model for weakly nonlinear TW's. The result is at odds with the interpretation of the Lagrangian motion[9, 10, 29] of passive marked particles in a TW suggested in refs. 26, 29.

E Motion of passive particles in a TW

In this section we review some properties of the Lagrangian dynamics of particles that are passively advected in the time–dependent TW velocity field and the associated mixing behaviour[30] of the TW (see refs. 9,10 for details). Such an investigation is of interest in view of the fact (cf. Fig. 4) that the alcohol concentration field in the TW state shows some characteristics of a passive scalar. Furthermore, an experiment[26, 29] monitoring the motion of passive markers that were initially in a small spatial section of the fluid layer has recently been performed partly with the aim to elucidate large scale flow properties of a TW.

For the sake of simplicity we consider instead of our numerically obtained velocity field a model TW velocity field[9, 10, 29]

$$w = \cos\,[\pi(x - vt)]\ \mathscr{C}_1(z - \tfrac{1}{2})\,. \tag{3.14a}$$

$$u = -\frac{1}{\pi} \sin\,[\pi(x - vt)]\ \partial_z\ \mathscr{C}_1\,(z - \tfrac{1}{2})\,. \tag{3.14b}$$

It is purely harmonic with wave number $k = \pi$ and phase velocity v and has by construction no meanflow

$$U = \langle u \rangle = 0\,. \tag{3.15}$$

Here $\mathscr{C}_1$ is the first Chandrasekhar function which looks like $\sin^2(\pi z)$ with a maximum of about 1.5. The Lagrangian mixing dynamics is most easily understood in the frame Σ' comoving with the TW, $x' = x - vt$, where the velocity field $w' = -\partial\phi/\partial x'$, $u' = \partial\phi/\partial z$ is stationary. Here

$$\phi\,(x',z) = -\frac{1}{\pi} \sin\,(\pi x')\ \mathscr{C}_1(z - \tfrac{1}{2}) - vz \tag{3.16}$$

is the stream function in Σ' along the isolines of which passive particles move in Σ'. In Fig. 6 we show streamlines of ϕ for a relatively fast TW ($v = 0.4$) and the time evolution of passive non diffusing particles that were at $t = 0$ distributed evenly over a pair of rolls, $1 \lesssim x \lesssim 3$. Particles that are in the regions of closed streamlines circulate around the elliptical fixed points in Σ',i. e., they comove in the laboratory frame with the TW to the right. Those on open streamlines move in Σ' to the left and, depending on the magnitude of their mean velocity, they do so also in the laboratory frame. The region of closed streamlines grows (shrinks) with decreasing (increasing) ratio v/w_{max}. But for any $v>0$ there are open streamlines in Σ' and also backwards motion in the laboratory frame. For $v>(1/\pi)\max \partial_z \mathscr{C}_1 \simeq 1.6$ all streamlines are open.

Fig. 5b shows that the TW velocity field efficiently mixes the particles of the fluid — those which were initially in different parts of one roll end up in vastly different spatial domains in the laboratory frame. But one cannot infer[26, 29] from this large scale backwards and forwards Lagrangian motion of individual particles that the TW (3.14) sustains a mean stationary large scale mass transport. Here for passive

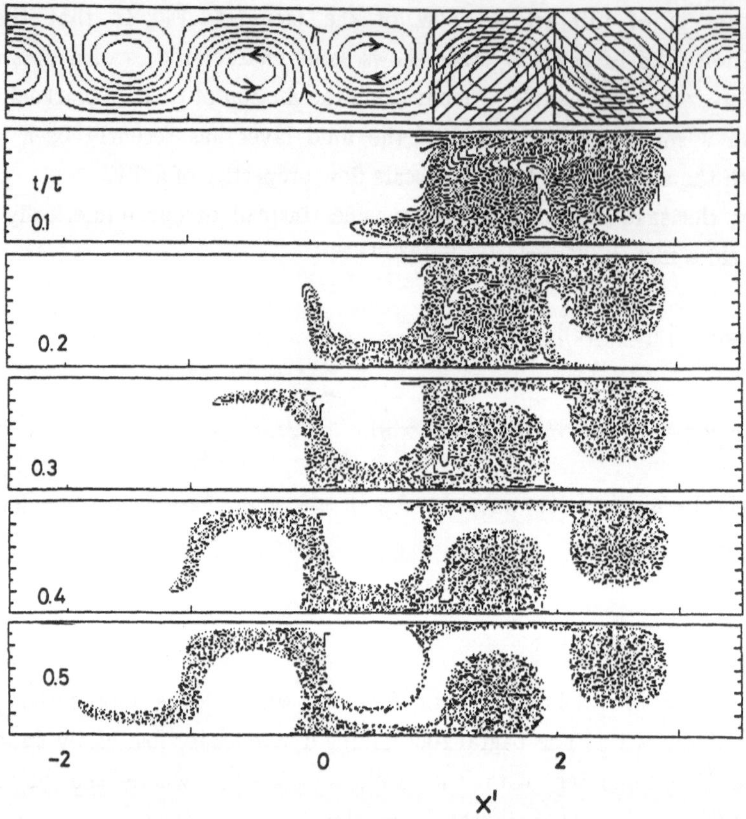

Fig. 6. Lagrangian particle motion in the model TW velocity field. Top row shows streamlines in the frame Σ' comoving with the phase velocity $v = 0.4$ of the TW to the right. Positions of 5000 particles are shown in Σ' that were initially distributed homogeneously over the shaded regions of a left ($1 \leq x' \leq 2$) and right ($2 \leq x' \leq 3$) turning roll.

particles that are for simplicity distributed at $t = 0$ with constant density ρ_0 all over the system the mass being transported by (3.14) through any area element $d\mathbf{s}$ in the laboratory frame during time τ is

$$\rho_0 \int_0^\tau dt\ \mathbf{u}(\mathbf{x} - \mathbf{v}t, z) \cdot d\mathbf{s} = 0. \tag{3.17}$$

In fact, during the oscillation period τ as many passive particles move through $d\mathbf{s}$ in positive as in negative direction. In the TW state of a real binary fluid where the total density ρ or that of one of the components, say $\rho_{ethanol}$, is neither constant nor a passive scalar one has to integrate $\rho\mathbf{u}$ or $\rho_{ethanol}\mathbf{u}$ over one period in order to determine the associated transport through a given area element whithin τ, i. e., one has to evaluate the mean currents of Sec. III.D.

This work was supported by Deutsche Forschungsgemeinschaft.

REFERENCES

1. For an early review see J. K. Platten and J. C. Legros, *Convection in Liquids* (Springer, Berlin, 1984). For later work we refer to the references in recent experimental (Refs. 2–5,26,29) and theoretical (Refs. 6–20) papers.
2. R. Heinrichs, G. Ahlers, and D. S. Cannell, Phys. Rev. A **35**, 2761 (1987); T. S. Sullivan and G. Ahlers, Phys. Rev. Lett. **61**, 78 (1988).
3. E. Moses and V. Steinberg, Phys. Rev. A **34**, 693 (1986); J. Fineberg, E. Moses, and V. Steinberg, Phys. Rev. A **38**, 4939 (1988).
4. P. Kolodner, D. Bensimon, and C. M. Surko, Phys. Rev. Lett. **60**, 1723 (1988); P. Kolodner and C. M. Surko, Phys. Rev. Lett. **61**, 842 (1988).
5. O. Lhost and J. K. Platten, Phys. Rev. A **38**, 3147 (1988); Phys. Rev. A **40**, 4552 (1989).
6. E. Knobloch and D. R. Moore, Phys. Rev. A **37**, 860 (1988); M. C. Cross and K. Kim, Phys. Rev. A **37**, 3909 (1988).
7. W. Hort, Diplomarbeit, Universität Saarbrücken, 1990 (unpublished).
8. S. J. Linz and M. Lücke, Phys. Rev A **35**, 3997 (1987); in *Propagation in Systems Far from Equilibrium*, edited by J. E. Wesfreid, H. R. Brand, P. Manneville, G. Albinet, and N. Boccara (Springer, Berlin, 1988), p. 292.
9. S. J. Linz, M. Lücke, H. W. Müller, and J. Niederländer, Phys. Rev. A **38**, 5727 (1988).
10. M. Lücke, in *Far from Equilibrium Phase Transitions*, in Lecture Notes in Physics, vol. 319, edited by L. Garrido (Springer, Berlin, 1988), p. 195.
11. S. J. Linz, Ph. D. thesis, Universität Saarbrücken, 1989 (unpublished).
12. W. Schöpf and W. Zimmermann, Europhys. Lett. **8**, 41 (1989); W. Schöpf, Diplomarbeit, Universität Bayreuth, 1988 (unpublished).
13. E. Knobloch, Phys. Rev. A **34**, 1538 (1986).
14. D. Bensimon, A. Pumir, and B. I. Shraiman, J. Phys. France **50**, 3089 (1989).
15. H. R. Brand, P. C. Hohenberg, and V. Steinberg, Phys. Rev. A **30**, 2548 (1984); H. R. Brand, P. S. Lomdahl, and A. C. Newell, Physica **23D**, 345 (1986).
16. M. C. Cross, Phys. Rev. Lett. **57**, 2935 (1986).
17. A. E. Deane, E. Knobloch, and J. Toomre, Phys. Rev. A **37**, 1817 (1988).
18. M. Bestehorn, R. Friedrichs, and H. Haken, Z. Phys. **B75**, 265 (1989).
19. W. Barten, M. Lücke, W. Hort, and M. Kamps, Phys. Rev. Lett. **63**, 376 (1989). In this paper the correct ordinate labels of Fig. 5a are $U(10^{-2}\,\kappa/d)$ and $\varphi_w(10^{-3}\,2\pi)$.
20. H. Yahata (unpublished).
21. P. Kolodner, H. Williams, and C. Moe, J. Chem. Phys. **88**, 6512 (1988).
22. J. E. Welch, F. H. Harlow, J. P. Shannon, and B. J. Daly, Los Alamos Scientific Laboratory Report No. LA–3425, 1966.

23. C. W. Hirt, B. D. Nichols, and N. C. Romero, Los Alamos Scientific Laboratory Report No. LA–5652, 1975.

24. G. Dangelmayr and E. Knobloch, Phil. Trans. R. Soc. Lond. **A322**, 243 (1987).

25. W. Barten, M. Lücke, and M. Kamps, J. Comp. Phys. (in press).

26. E. Moses and V. Steinberg, Physica D **37**, 341 (1989).

27. For our estimates we use $T_0 = 300$ K, $\Delta T = 7$ K, $\alpha = 3 \cdot 10^{-4}\,K^{-1}$, $\beta = 0.15$. Thus for $\underline{C}_0 = 8$ weight % ethanol in water $C_0 = 5.7$ in reduced units. Furthermore we take a TW at $\psi = -0.25$ near the saddle with extrema $\langle u\delta T\rangle \simeq 0.06$, $\langle u\delta C\rangle \simeq -0.07$ at $z = 1/4$ and $U \simeq -0.001$ at $z = 1/2$.

28. W. Barten, M. Lücke, and M. Kamps, unpublished.

29. E. Moses and V. Steinberg, Phys. Rev. Lett. **60**, 2030 (1988).

30. J. M. Ottino, *The kinematics of mixing: stretching, chaos, and transport* (Cambridge University Press, 1989).

ONSET OF SORET AND DUFOUR DRIVEN CONVECTION IN BINARY FLUID MIXTURES

W. Hort, S. J. Linz, and M. Lücke

Institut für Theoretische Physik, Universität des Saarlandes
D–6600 Saarbrücken, West–Germany

I INTRODUCTION

In the last few years a lot of progress has been made to understand convection in binary fluid layers in an external temperature gradient from the experimental[1] as well as from the theoretical[2-4] side. All this work[1-4] was concentrated on *liquid* mixtures. Recently one of us has shown[5] for a layer in an external concentration gradient that the onset of convection is significantly influenced by the different physical properties of liquid and gas mixtures:

(i) the Lewis number L, i. e. the ratio of mass diffusivity D and thermal diffusivity κ, in liquid mixtures being of the order 10^{-2}, is of order 1 in gaseous mixtures and

(ii) the Dufour coupling of the mass diffusion current into the temperature field equation, is in gases about 10^4 times larger than in liquid mixtures and can no longer be ignored.

Here we present results on the *onset of convection* in binary mixtures heated from below and in particular how the inclusion of the Dufour effect changes the stability properties of the conductive state. A more detailed account will be given elsewhere[6]. For the treatment of small Dufour coupling we refer to Ref. 7–9.

II BASIC EQUATIONS

Consider a layer of height d of a binary fluid mixture of mean concentration C_m between two parallel horizontal, perfectly heat conducting plates at z=0, and z=d subject to the vertical gravitational field $\mathbf{g}=-g\mathbf{e}_z$ and to a vertical temperature gradient ΔT between the plates (upper plate at temperature T_0, lower plate at $T_0+\Delta T$). In Oberbeck–Boussinesq approximation[7] the conductive profile for the temperature and concentration

$$T_{cond}(z)-T_0=R\,(1-z) \quad \text{and} \quad C_{cond}(z)-C_m=R\psi\,(\tfrac{1}{2}-z) \tag{1}$$

are unchanged by the Dufour effect. The linearized equations for the vertical velocity w and the deviations from the conductive profiles of temperature, $\theta=T-T_{cond}$, and concentration, $c=C-C_{cond}$, read[7]

$$(\partial_t-\sigma\nabla^2)\nabla^2 w=\sigma\partial_x^2(\theta+c) \tag{2a}$$

$$\partial_t\theta=Rw+(1+LQ\psi^2)\nabla^2\theta-LQ\psi\nabla^2 c \tag{2b}$$

$$\partial_t c=R\psi w+L\nabla^2(c-\psi\theta). \tag{2c}$$

We have nondimensionalized length by d, time by d^2/κ, temperature by $\kappa\nu/\alpha g d^3$, and

Nonlinear Evolution of Spatio-Temporal Structures in Dissipative Continuous Systems
Edited by F.H. Busse and L. Kramer
Plenum Press, New York, 1990

concentration by $\kappa\nu/\beta g d^3$. Here α is the thermal expansion coefficient, β the solutal expansion coefficient, ν the kinematic viscosity, $\sigma=\nu/\kappa$ the Prandtl number, and R the Rayleigh number i. e. the nondimensionalized temperature difference between the plates and ψ the separation ratio. We have introduced the Dufour number Q as in Ref. 5

$$Q=(T_0\alpha^2/c_p\beta^2)\ \partial\mu/\partial c\ , \tag{3}$$

where c_p is the specific heat at constant pressure and μ the chemical potential of the mixture. Since $\partial\mu/\partial c>0$ also Q is always positive. Setting Q=0 the Soret driven convection problem of binary fluids is recovered. Setting $\psi=0$ leads to the stability problem in a one–component fluid. The off diagonal Dufour coupling $\sim LQ\psi$ and the diagonal Dufour coupling $\sim LQ\psi^2$ are small in *liquid* mixtures where $Q \cong 0.2$, L=0.01, and $|\psi| \lesssim 0.5$. There the Dufour effect has no perceptible effect[8]. But in *gaseous* mixtures where L is of the order 1 and Q typically[6] about 10 or 20 the Dufour effect cannot be ignored. The main reason why Q is so large in gases seems to be that the thermal expansion coefficient α entering quadratically in Q is about 10 times larger than in liquids. The combination of the other constants entering in Q has nearly the same size in liquids as in gases. Our estimate of Q coincides roughly with calculations[10] for a gaseous mixture used in a recent experiment.

III ONSET OF CONVECTION FOR FREE SLIP, PERMEABLE BOUNDARIES

While the impermeability of the plates has significant influence[2-4] on the stability properties many *changes* when varying Q can be understood with idealized free slip, permeable boundaries, ($w=\partial^2_z w=\theta=c=0$ at z=0 and z=1) which allow an analytical treatment[7].

One finds for the critical Rayleigh number for onset of stationary convection reduced by the value in a pure fluid $R^0_c=R_c(\psi=0)=27\pi^4/4$

$$r^c_{stat}= [(1+\psi)(1+Q\psi^2)+\psi/L]^{-1}. \tag{4a}$$

The corresponding critical wave number, $\pi/\sqrt{2}$ is independent of Q. While r^c_{stat} (Q=0) decreases monotonously from $+\infty$ at $\psi=-L/(L+1)$ to zero at $\psi\to\infty$, this is no longer the case if $Q>3(1+1/L)$: Then r^c_{stat} develops a maximum and a minimum at $\psi_{1,2}=-(1/3)[1\pm\{1-3(1+L)/QL\}^{1/2}]$ as shown in Fig. 1. With growing Q the location of the maximum (minimum) approaches $\psi=0$ ($\psi=-2/3$) connected with a strong de–stabilization of the conductive state in that ψ–range. For large Q r^c_{stat} diverges at

$$\psi=-1+L^{-1}Q^{-1}-(L-1)L^{-2}Q^{-2}+(L^2-4L+2)L^{-3}Q^{-3}+0(Q^{-4}). \tag{4b}$$

Enhancing Q shifts the divergence of r^c_{stat} to much smaller ψ values.

For the reduced critical value of the onset of oscillatory convection we find

$$r^c_{osc}=\frac{1}{\sigma}\,\frac{L(1+Q\psi^2)[\sigma(\sigma+2)+L+L\sigma(1+Q\psi^2)]+\sigma^2+\sigma+L}{1+(1+\psi)\sigma}\ , \tag{4c}$$

whenever the corresponding square of the Hopf frequency ω^2 given by

$$\omega^2=-\frac{9}{4}\pi^4\,\frac{L[1+\sigma(1+Q\psi^2)][\psi+L(1+\psi)(1+Q\psi^2)]+\sigma\psi[1+LQ\psi(1+\psi)]}{1+(1+\psi)\sigma}, \tag{4d}$$

is non negative. Also here the critical wave number is $\pi/\sqrt{2}$. The ψ–value where ω^2 and r^c_{osc} diverges, $\psi=-(1+1/\sigma)$, being the lower bound of the existence range of an oscillatory instability is not changed by incorporating of the Dufour effect with this kind of boundary conditions. The codimension–two (CT) points are given by the solution of $\omega^2=0$. If Q=0, there is only one CT point. For Q>0 the CT points are determined by the real zeros of a fifth–order polynomial in ψ, which leads for appropriately chosen L,Q and σ to more than one CT point, e. g. three. Eq. (4c) shows that any nonzero Q shifts the oscillatory threshold upwards relative to the case Q=0 (cf.Fig.1). For large Q and not too small $|\psi|$ one finds $r^c_{osc}(1+\sigma+\sigma\psi)=L^2Q^2\psi^4+0(Q)$ being totally different in comparison to $r^c_{osc}(1+\sigma+\sigma\psi)=(1+L)(1+L+\sigma+L/\sigma)$

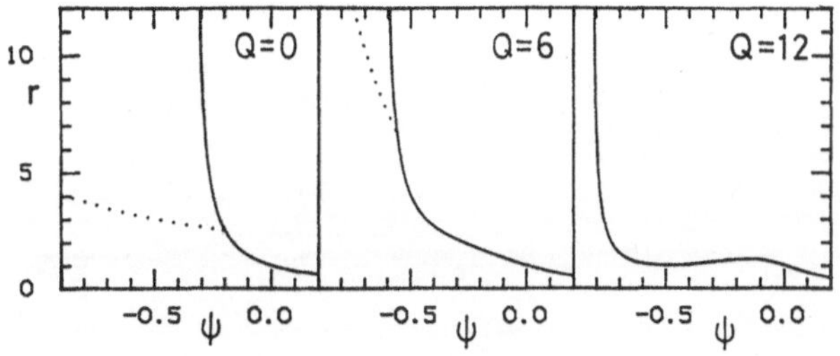

Fig. 1. **Free slip, permeable boundary conditions: Reduced stability thresholds r^c_{stat} (full lines) for stationary and r^c_{osc} (dots) for oscillatory onset of convection for different Dufour numbers Q. The parameters L=1/2, σ=1 are typical for gaseous mixtures.**

in the case Q=0. Furthermore the ψ–range where there is an oscillatory instability is shifted to the left in Fig. 1. For example, at large Q the CT point marking the right endpoint of r_{osc} moves to

$$\psi_{ct}= -1+L^{-1}Q^{-1} - (L-1)L^{-2}Q^{-2}+(L^2 - 3L+2)L^{-3}Q^{-3}+0(Q^{-4}) \quad (4e)$$

$$r_{ct}=L^2Q^2+0(Q), \quad (4f)$$

while without Dufour effect, Q=0, $\psi_{ct} = -0.2$ and $r_{ct} = 2.5$ for L=1/2, σ=1. A comparison of (4b) and (4e) shows that for large Q the CT point is always located at larger ψ values as the point where r^c_{stat} diverges. This implies the existence of a range of oscillatory instabilities for every finite Q.

Let us finally mention that the changes in r^c_{stat} and r^c_{osc} arise in both cases from the diagonal *and* the off diagonal Dufour couplings.

IV ONSET OF CONVECTION FOR NO SLIP, IMPERMEABLE BOUNDARIES

Let us now turn to realistic no slip, impermeable boundary conditions, i. e. $w=\partial_z w=\theta=\partial_z(c-\psi\theta)=0$ at the plates. By using a shooting method[11] we have determined numerically for a representative gaseous mixture with L=1/2 and σ=1 the reduced Rayleigh number of the stationary and oscillatory instability r^c_{stat} and r^c_{osc}, the reduced critical stationary and oscillatory wave numbers and the Hopf frequency ω taken at the critical oscillatory wave number. They are shown in Fig. 2 as function of ψ for five different Dufour numbers. With impermeable boundaries the critical wave numbers for the onset of stationary and oscillatory convection vary with ψ and this variation changes with increasing Dufour number.

For Q=0 the stability thresholds of the conductive state and the Hopf frequency look similar to those in Ref. 3,4. Here, however, because of the large Lewis number (L=1/2) the CT point where $r^c_{osc}=r^c_{stat}$ is shifted to a more negative ψ (ψ_{ct} =–0.065) and furthermore the difference between the stationary and oscillatory wave number at ψ_{ct} is much larger, about 0.2. Increasing Q to Q=8 shifts the divergence of r^c_{stat} in Fig. 2 to larger negative ψ. Simultaneously r^c_{osc} is bent strongly upwards leading to a stabilization of the conductive state below r^c_{osc}. Furthermore the critical oscillatory wave number k^c_{osc} does no longer increase monotonously, with decreasing ψ, but decreases first for not too negative ψ. Increasing the Dufour number further to Q=10 the oscillatory instability curve crosses three times the stationary one.

Due to the impermeable boundaries the Hopf frequency ω does not vanish at the CT points. For Q=16 there are two ranges of onset of oscillatory instabilities, one for $-0.33 \leq \psi \leq -0.08$ and the other for $\psi<-0.63$. At Q=20 the smaller interval of oscillatory instabilities has shrunk to zero. Simultaneously a pair of a minimum and a maximum on the stationary instability is created similar as in Sec. III which destabilizes the conductive state in the range $-0.7<\psi<0.2$. The remaining CT point is shifted to much more negative ψ, $\psi \simeq -0.83$ which also leads to a strong stabilization below the oscillatory branch. For Q=20 in the considered ψ range the oscillatory instability has almost entirely vanished (the CT point is located at ψ_{ct}=–0.87 and r_{ct}=13.3), the minimum at ψ=–0.56, r=0.624 and maximum at ψ=–

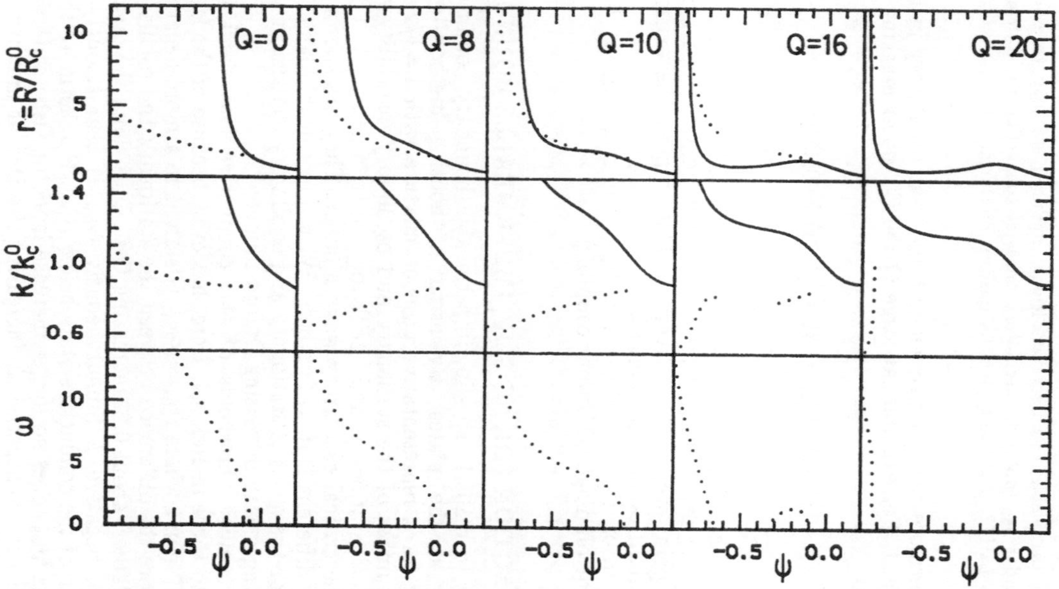

Fig. 2. Stability properties for no slip, impermeable boundary conditions vs ψ for different Dufour numbers Q and L=1/2, σ=1. Top row shows critical reduced Rayleigh numbers r^c_{stat} (full lines) and r^c_{osc} (dots). Second row contains the critical reduced wave numbers for the onset of stationary convection (full lines) and oscillatory convection (dots). R^0_c=1707.8 and k^0_c=3.116 are the reference values for a pure fluid. Bottom row shows the Hopf frequencies at the critical oscillatory wave number.

0.12, r=1.287 of r^c_{stat} are more pronounced as for Q=16. The stability diagram now resembles very closely the right most one in Fig. 1.

V CONCLUSIONS AND OUTLOOK

We have shown that for gaseous mixtures where the Lewis number is not small the Dufour effect has significant influence on the onset of convection if the Dufour number is of the order 10 or bigger. Then the range of oscillatory instabilities is drastically diminished, for not too large negative ψ there is first a destabilization of the conductive state below the stationary instability, for larger negative ψ a strong stabilization. The general trend is to deform and shift the stability curves r^c_{osc} and r^c_{stat} towards more negative ψ values. These effects can be understood qualitatively by using idealized free slip, permeable boundary conditions.

Finally we remark that in the conductive state of gases the concentration difference ΔC_{cond} between the plates is for the same r, ψ much bigger than in liquids. For the relative concentration difference one finds

$$\frac{\Delta\ C_{cond}}{C_m} = R\ \psi\ \frac{\kappa\nu}{\beta g d^3}\ \frac{1}{C_m} \tag{5}$$

where ΔC_{cond} and the mean concentration C_m are now and in the following *unscaled* concentrations, i. e. $0 \leq C_m \leq 1$. If one assumes concentration variations of, e. g., $|\Delta C_{cond}/C_m| = 0.1$ as an upper limit for the validity of the Oberbeck–Boussinesq approximation which demands that ψ does not vary significantly along the layer height one gets

$$|r\ \psi| < 0.1\ \gamma\, C_m \qquad \text{with} \quad \gamma = \frac{\beta g d^3}{1708\kappa^2\sigma}, \tag{6}$$

as a restriction[12] on $r\psi$. In a typical gaseous mixture with d=0.5 cm and σ=1 γ is about 1.5 whereas in ethanol–water mixtures γ is of the order 10^4. These changes are mainly due to κ being about 10^{-1} in gases, but 10^{-3} in liquids. To avoid difficulties with the validity of the Oberbeck–Boussinesq approximation it seems to be advisable to use convection cells with relatively large heights d, since the bound entering on the r.h.s. of (6) grows proportional to d^3.

This work was supported by Deutsche Forschungsgemeinschaft.

REFERENCES

1. See e. g. G. Ahlers and I. Rehberg, Phys. Rev. Lett. **56**, 1373 (1985); H. Gao and R.P. Behringer, Phys. Rev. A **35**, 3993 (1987); T. S. Sullivan and G. Ahlers, Phys. Rev. Lett. **61**, 78 (1988); P. Kolodner, C. M. Surko, and H. Williams: Physica D **37**, 319 (1989); J. Fineberg, E. Moses, and V. Steinberg, Phys. Rev. A **38**, 4939 (1988).
2. S. J. Linz and M. Lücke, Phys. Rev A **35**, 3997 (1987); Springer Series in Synergetics **41**, 292 (1988).
3. E. Knobloch and D. R. Moore, Phys. Rev. A **37**, 860 (1988).
4. M. C. Cross and K. Kim, Phys. Rev. A **37**, 3909 (1988); Phys. Rev. A **38**, 529 (1988).
5. S. J. Linz, Phys. Rev. A **40**, (1989).
6. W. Hort, S. J. Linz, and M. Lücke (unpublished).
7. See e. g. G. Z. Gershuni and E. M. Zhukhovitskii, *Convective Stability of Incompressible Fluids* (Keter, Jerusalem, 1976).
8. G. W. T. Lee, P. Lucas, and A. Tyler, J. Fluid Mech. **135**, 235 (1983). They considered the influence of the Dufour effect in liquid ^3He–^4He mixtures.
9. An early study of the Dufour effect in gaseous mixtures was done by P. L. G. Ybarra and M. G. Velarde, Geophys. Astrophys. Fluid Dyn. **13**, 83 (1979) by using a Galerkin truncation for no slip boundary conditions.
10. J. de Bruyn (private communication).
11. W. Hort, Diploma thesis (Universität Saarbrücken, 1989).
12. S. J. Linz, Ph. D. thesis (Universität Saarbrücken, 1989).

EXPERIMENTAL STUDY OF BINARY FLUID CONVECTION IN A QUASI 1-DIMENSIONAL CELL WITHOUT REFLECTIONS FROM THE SIDEWALLS

Wolfgang Schöpf and Ingo Rehberg

Physikalisches Institut
Universität Bayreuth
8580-Bayreuth (FRG)

With properly chosen parameters, convection in binary fluid mixtures sets in via a Hopf bifurcation leading to travelling waves. In this case one has to distinguish between a convectively unstable and an absolutely unstable situation. We present an observation in a cell where the fluid becomes unstable at the absolute instability point, not at the convective one as in most other experiments. This is done by preventing the sides of the cell from reflecting the travelling wave.

The experimental setup has been described elsewhere (Rehberg et al., 1987, Rehberg et al., 1988). The convection channel is cut out of a copper plate (Fig. 1). The bulk part of the channel is 3 mm high, 1.5 mm thick, and 18 mm long. The ramps adjacent to this central part decrease the height of the channel from 3 mm to 1 mm over a length of 26 mm by means of a parabolic curvature of the top and bottom. The bottom of the copper plate is heated by an electric heater and the top of the cell is kept at a constant temperature by a water circuit controlled better than ±0.01 °C. We have chosen such a thin cell in order to prevent three dimensional

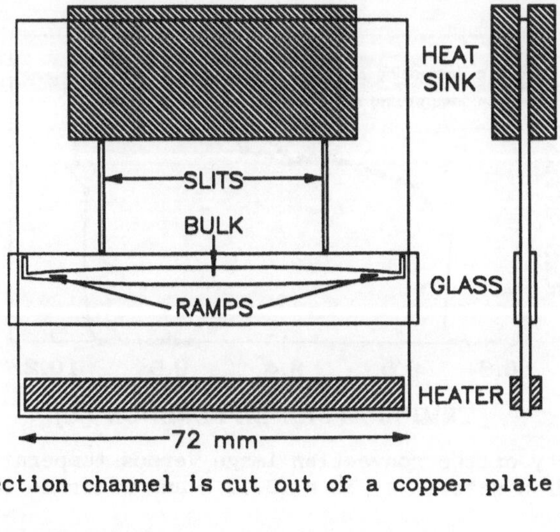

Fig. 1. The convection channel is cut out of a copper plate of $72 \times 72 \times 1.5$ mm^3.

Nonlinear Evolution of Spatio-Temporal Structures in Dissipative Continuous Systems
Edited by F.H. Busse and L. Kramer
Plenum Press, New York, 1990

effects. The purpose of the parabolic subcritical ramps in this experiment is to decrease the amplitude of the waves travelling into this ramp smoothly to zero because the systems becomes subcritical near the right and left sidewalls. So the convection dies out and we got no hint for any reflection from these walls. Our working fluid is a mixture of 16.98 wt.% of ethanol in water at a mean temperature of about 30 °C. To visualize the flow field we used the shadowgraph technique (Rasenat et al., 1989).

When heating from below convection sets in via a backward Hopf bifurcation as is well understood (Schöpf 1988, Schöpf et al., 1989). This leads to a hysteresis in the bifurcation diagram as shown in Fig. 2. Here the amplitude of the convection is shown as a function of the temperature difference between the top and the bottom of the convection channel. When increasing the applied temperature difference, stability is lost with respect to overturning convection at $\Delta T = 10.12$ K, and cooling down again the convection vanishes at the saddle node at $\Delta T = 8.84$ K. The important feature of this picture is the fact that this convection onset is well above the convective instability point, which is the generic one from linear stability analysis of the fundamental hydrodynamic equations (Knobloch et al., 1988, Cross et al., 1988, Schöpf 1988). This convective instability point is observed in most of the experiments, where small disturbances are able to grow and finally the entire convection cell is filled with rolls (Kolodner et al., 1986, Moses et al., 1986, Heinrichs et al., 1987, Kolodner et al., 1988). We are able to determine it by measuring the growth rate of pulses of travelling waves which are the answer of the system to applied heat pulses, similar to the procedure described by Kolodner et al. (1987). The system gets convectively unstable at a zero growth rate, and this point is approximately at 9.6 K. In our experimental situation all small disturbances are carried away and die out due to the nonreflecting sidewalls until the absolute instability point is reached. This point is characterized by a local increase of small disturbances, so that the onset of overturning convection can no longer be prevented. This feature can be understood theoretically by analysing the suitable amplitude equation and the absolute instability point is determined by the linear coefficients of this equation (Huerre, 1987).

The novel feature that we like to present here occurs in the range between the convective instability and the absolute instability point

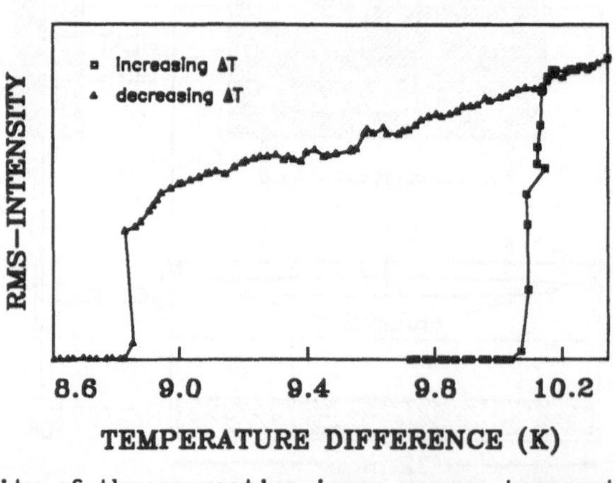

Fig. 2. Intensity of the convection image versus temperature difference. Convection sets in at $\Delta T = 10.12$ K and vanishes at $\Delta T = 8.84$ K.

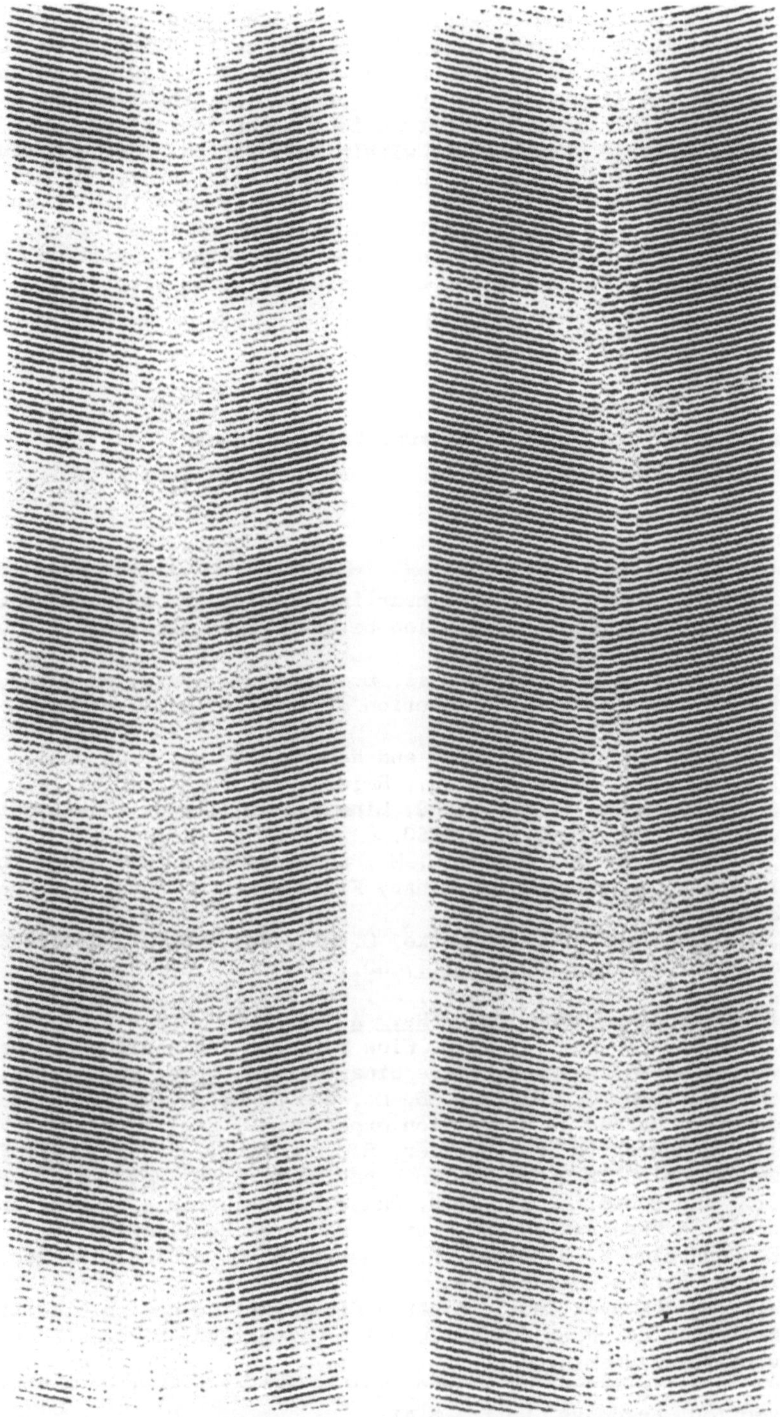

a) ΔT = 10.02 K. b) ΔT = 10.09 K.

Fig. 3. Time series of the small amplitude 'Blinking Convection'. Time goes upwards (4096 seconds) and the x-direction shows about 50 mm of the long dimension of the cell.

close to the second one. Here we observe the blinking behaviour illustrated by Fig. 3a) and b) which is a 4096 sec. time series of intensity lines along the long dimension of the cell. Time goes upwards and about 50 mm of the cell are shown. Randomly occuring islands of convection with the linear frequency can be seen in the left and right part of the cell. This state is no transient - our longest run was about ten days and no change was observed within this period. The amplitude of this kind of convection is less than 1% of the fully developed convection on the upper branch of Fig. 2. Fig. 3a) is taken at ΔT = 10.02 K and Fig. 3b) at ΔT = 10.09 K. It seems that the convection periods become longer with increasing temperature difference and also the amplitude of the structure increases. Further study of this state is in progress.

Acknowledgement

This work is supported by Stiftung Volkswagenwerk.

References

Cross, M. C., and Kim, K., 1988, Linear instability and the codimension-2 region in binary fluid convection between rigid impermeable boundaries, Phys. Rev. A, 37:3909.

Heinrichs R., Ahlers, G., and Cannell, D. S., 1987, Traveling waves and spatial variation in the convection of a binary mixture, Phys. Rev. A, 35:2761.

Huerre, P., 1987, in "Instabilities and Nonequilibrium Structures," E. Tirapequi, D. Villaroel, ed., Reidel, Dordrecht.

Knobloch, E., and Moore, D. R., 1988, Linear stability of experimental Soret convection, Phys. Rev. A, 37:860.

Kolodner, P., Passner, A., Surko, C. M., and Walden, R. W., 1986, Onset of Oscillatory Convection in a Binary Fluid Mixture, Phys. Rev. Lett, 56:2621.

Kolodner, P., Bensimon, D., and Surko, C. M., 1988, Traveling-Wave Convection in an Annulus, Phys. Rev. Lett., 60:1723.

Kolodner, P., Surko, C. M., Passner, A., and Williams, H. L., 1987, Pulses of oscillatory convection, Phys. Rev. A, 36:2499.

Moses, E., and Steinberg, V., 1986, Flow patterns and nonlinear behavior of traveling waves in a convective binary fluid, Phys. Rev. A, 34:693.

Rasenat, S., Hartung, G., Winkler, B. L., and Rehberg, I., 1989, The shadowgraph method in convection experiments, Exp. in Fluids, 7:412.

Rehberg, I., Bodenschatz, E., Winkler, B., and Busse, F. H., 1987, Forced Phase Diffusion in a Convection Experiment, Phys. Rev. Lett., 59:282.

Rehberg, I., and Busse, F. H., 1988, Phase Diffusion in a Ramped Convection Channel, in "Propagation in Systems Far from Equilibrium," J. E. Wesfreid, H. R. Brand, P. Manneville, G. Albinet, N. Boccara, ed., Springer, Berlin, Heidelberg.

Schöpf, W., 1988 "Konvektion in binären Flüssigkeiten und multikritisches Verhalten in der Nähe des Kodimension-2 Punktes", Diploma Thesis, Universität Bayreuth.

Schöpf, W., and Zimmermann, W., 1989, Multicritical Behavior in Binary Fluid Convection, Europhys. Lett., 8:41.

INTERFACIAL OSCILLATIONS AND SOLITONS DRIVEN BY THE MARANGONI EFFECT

Manuel G. Velarde and Xiao-Lin Chu

Facultad de Ciencias, U.N.E.D.
Apartado 60.141
Madrid 28.080, Spain

1. INTRODUCTION : A LIQUID OPEN TO AIR AND THE ROLE OF SURFACTANTS OR OTHER THERMAL CONSTRAINTS

Generally, interfacial waves or interfacial oscillations are damped by viscosity (1-3). However, if a non equilibrium distribution of surfactant, temperature, chemical potential or an electric field,... is imposed in the liquid, with mass and/or energy transfer across the surface, the Marangoni effect, i.e., stresses induced by the variation of the surface tension from point to point along the surface may transform the "thermal" or "chemical",...energy into convective motion, overtaking the viscous dissipation and thus sustaining the wave motions (4-18).

Let us study the evolution of the open surface of a liquid layer subjected to some thermal gradients. The motionless undisturbed surface is located at z=0 and, for simplicity, consider the bottom at an infinite depth. The linearized equations that disturbances upon the quiescent state obey are:

$$\nabla * \mathbf{v} = 0 \tag{1.1}$$

$$\rho \frac{\partial \mathbf{v}}{\partial \tau} = - \nabla p + \eta \nabla^2 \mathbf{v} \tag{1.2}$$

$$\frac{\partial c}{\partial t} - \beta^c w = D \nabla^2 c \tag{1.3}$$

where $\mathbf{v}$ =(u,w), p and c are velocity pressure and concentration of the surfactant, respectively. ρ is the density of the liquid, $\eta=\rho\nu$ the dynamic viscosity and ν the kinematic viscosity, D is the mass diffusivity and $\beta^c = (\partial c/\partial z)_0$ the gradient of surfactant at the quiescent state. Note that our analysis carries over almost *verbatim* to the heat transfer problem and we shall mark the difference only when strictly necessary.The disturbances obey the following (linearized) boundary conditions

$$\frac{\partial \zeta}{\partial t} = w \tag{1.4}$$

$$- T_0 \nabla_\Sigma^2 \zeta + g\rho\zeta - p - 2\eta \frac{\partial w}{\partial z} = 0 \tag{1.5}$$

Nonlinear Evolution of Spatio-Temporal Structures in Dissipative Continuous Systems
Edited by F.H. Busse and L. Kramer
Plenum Press, New York, 1990

$$\left(\frac{\partial T}{\partial \Gamma}\right)_0 \nabla_\Sigma \gamma - \eta\left(\nabla_\Sigma w + \frac{\partial u}{\partial z}\right) = 0 \qquad (1.6)$$

$$\frac{\partial \gamma}{\partial t} + \Gamma_0 \nabla_\Sigma * u_\Sigma - D_\Sigma \nabla_\Sigma^2 \gamma + D\frac{\partial c}{\partial z} = 0 \qquad (1.7)$$

$$\gamma = k^1 (c - \beta^c \zeta)_\Sigma \qquad (1.8)$$

where ζ is the surface deviation from the z=0 level and Γ is the excess surfactant concentration at the surface. The subscript "0" indicates a value in a reference state, γ is the disturbance upon Γ_0, g is the gravity acceleration and T is the surface tension. Subscript "Σ" accounts for either a value taken on the surface or a derivative along the surface.

For universality in presentation, we introduce new units to rescale the quantities in the equations. The capillary length:

$$l = \sqrt{\frac{T_0}{g\rho}} \qquad (1.9)$$

is chosen as our length scale; ν/l, l^2/ν, $\nu^2\rho/l^2$, $\beta^c l$ and Γ_0 are used as units for velocity, time, pressure, surfactant concentration and excess surface concentration, respectively. Thus eqs. (1.1)-(1.8) become:

$$\nabla * \mathbf{v} = 0 \qquad (1.10)$$

$$\frac{\partial \mathbf{v}}{\partial t} = -\nabla p + \nabla^2 \mathbf{v} \qquad (1.11)$$

$$\frac{\partial c}{\partial t} - w = S^{-1}\nabla^2 c \qquad (1.12)$$

with boundary conditions at z=ζ

$$\frac{\partial \zeta}{\partial t} = w \qquad (1.13)$$

$$-\frac{1}{SC}\nabla_\Sigma^2 \zeta + \frac{Bo}{SC}\zeta - p - 2\frac{\partial w}{\partial z} = 0 \qquad (1.14)$$

$$\frac{HE}{SH_z}\nabla_\Sigma v + \left(\nabla_\Sigma v + \frac{\partial u}{\partial z}\right) = 0 \qquad (1.15)$$

$$HS\left(\frac{\partial \gamma}{\partial t} + \nabla_\Sigma * u_\Sigma - S_\Sigma^{-1}\nabla_\Sigma^{2}\gamma\right) + \frac{\partial c}{\partial z} = 0 \qquad (1.16)$$

$$\gamma = \frac{H_z}{H}(c - \zeta)_\Sigma \qquad (1.17)$$

Note that although we have used the same notation for **v**,p,c,γ and ζ in these equations as in the dimensional case, all the units have changed and they are dimensionless now. The following dimensionless parameters have also been used : S= ν/D,Schmidt number;S_Σ= ν/D_Σ, surface Schmidt number; $C = \eta D/lT_0$, capillary number; $Bo=\rho g l^2/T_0$, Bond number (for simplicity is taken equal to unity); $E= -(\partial c/\partial\Gamma)_0 k^l\beta l^2/(\eta D)$, surfactant (elasticity) Marangoni number (when considering heat transfer just replace the concentration gradient by the temperature gradient, D by κ-thermometric conductivity; then E is called M); $H = \Gamma/(\beta^c l^2)$, surface excess surfactant number; and $H_z = k^l/l$, Langmuir adsorption number.

2. TRANSVERSE OSCILLATIONS : THRESHOLD VALUES

Assuming for simplicity that surface adsorption and accumulation have negligible effect for transverse surface waves eqs. (1.15)-(1.17) reduce to

$$\frac{Ea^2}{S}(c-\zeta)+\left(\frac{\partial^2}{\partial z^2}+a^2\right)w=0 \tag{2.1}$$

$$\frac{\partial c}{\partial z}=0 \tag{2.2}$$

Equation (1.11) gives the dynamic evolution equation of the liquid layer. It is valid in the volume as well as at the surface - a part of the liquid. On the other hand, we have a kinematic relation (1.13) on the surface. Then a straightforward stability analysis yields (8,9) that when

$$4a^2+\frac{Ea^3}{s\omega\sqrt{2S\omega}}>0\quad\left(\text{or } -E<\frac{4S\omega\sqrt{2S\omega}}{a}\right) \tag{2.3}$$

the oscillatory convection will be damped out. In contrast, when

$$4a^2+\frac{Ea^3}{s\omega\sqrt{2S\omega}}<0\quad\left(\text{or } -E>\frac{4S\omega\sqrt{2S\omega}}{a}\right) \tag{2.4}$$

the oscillation grows exponentially. At the neutral state

$$-E=\frac{4S\omega\sqrt{2S\omega}}{a} \tag{2.5}$$

The kinetic energy dissipated by viscosity and that produced by surface tension work just compensate each other, giving a sustained oscillator (8). The oscillation frequency is given by the dispersion relation

$$\frac{B_0+a^2}{SC}a-\omega^2=0 \tag{2.6}$$

By taking $dE(a,w(a))/da = 0$, the necessary condition for minimum yields the neutral curve, i.e., the threshold values for sustained transverse waves :

$$E_c^T=4\sqrt{10}\left(\frac{6S}{5\sqrt{5}C}\right)^{3/4}\approx -7.931\left(\frac{S}{C}\right)^{3/4} \tag{2.7}$$

$$\omega_c^T=\sqrt{\frac{6}{5^{3/2}}}\frac{1}{\sqrt{SC}}\approx\frac{0.7326}{\sqrt{SC}} \tag{2.8}$$

$$a_c^T=\frac{1}{\sqrt{5}}\approx 0.4472 \tag{2.9}$$

3. LONGITUDINAL OSCILLATIONS : THRESHOLD VALUES

For simplicity we assume that the deformability of the surface has negligible influence on the longitudinally oscillatory motion of the surfactant concentration which is indeed nonuniformly distributed along the open surface. Thus we set the capillary number, C, to zero. Also, it is known that surfactant accumulation on the surface affects mainly high fequency oscillatory convection and the frequency of longitudinal waves normally is small, so that the surfactant accumulation number H can be neglected. With these assumptions eqs. (1.13)-(1.17) reduce to

$$w = 0 \tag{3.1}$$

$$\frac{Ea^2}{S} c + \frac{\partial^2 w}{\partial z^2} = 0 \tag{3.2}$$

$$SH_z \frac{\partial c}{\partial t} = - \frac{\partial c}{\partial z} \tag{3.3}$$

Then a straightforward stability analysis (8,9,12) yields that when energy dissipation and Marangoni work compensate each other

$$1 + \frac{Ea^2}{S^{3/2} \omega^2} = 0 \tag{3.4}$$

Then

$$-E = \begin{cases} > \dfrac{S^{3/2} \omega^2}{a^2} & \text{explosion} \\ < \dfrac{S^{3/2} \omega^2}{a^2} & \text{damped motion} \end{cases}$$

In the neutral case we also have

$$SH_z \omega^2 + \frac{Ea^3}{S\omega} = 0 \tag{3.5}$$

Thus one expects that sustained longitudinal oscillations occur at the surface with dispersion relation

$$a_c = H_z \omega S^{1/2} \tag{3.6}$$

which generalizes earlier findings (5).

Comparison of (2.7) and (3.5) shows that both thresholds depend on the Schmidt (Prandtl) number. However, transverse waves are directly related to the interfacial deformation (measured by C) whereas longitudinal waves rather depend on Langmuir's adsorption (measured by H_z).Cross-over from one to another mode of instabilty is given by the condition: when S is greater (smaller) than $C^3/(7.931 H_z^2)^4$ we have longitudinal (transverse) waves first.

4. LIMIT CYCLE OSCILLATIONS

For illustration we shall concentrate on the case of transverse motions only. If we now consider the nonlinear extension of the problem posed in section 1 we have in dimensionless form

$$\nabla * v = 0 \tag{4.1}$$

$$\frac{\partial w}{\partial t} + u\frac{\partial w}{\partial x} + w\frac{\partial w}{\partial z} = -\frac{\partial p}{\partial z} + \nabla^2 w \tag{4.2a}$$

$$\frac{\partial u}{\partial t} + u\frac{\partial u}{\partial x} + w\frac{\partial u}{\partial z} = -\frac{\partial p}{\partial x} + \nabla^2 u \tag{4.2b}$$

and

$$\frac{\partial \theta}{\partial t} + u\frac{\partial \theta}{\partial x} + w\frac{\partial \theta}{\partial z} = w + P^{-1}\nabla^2\theta \tag{4.3}$$

where we have explicitely indicated in a two-dimensional problem the full nonlinear disturbance system. Besides, in order to rely our analysis to the standard Bènard problem we take here ß as temperature gradient rather than surfactant gradient. ß is positive when heating the layer from below. Thus eq. (4.3) is Fourier's heat equation . P is the Prandtl number, P= ν/κ, with κ the thermometric conductivity of the liquid. Its counterpart in mass transfer is the Schmidt number, S. In reciprocity with the coment given in Sec. 1 note that the analysis can be transposed almost *verbatim* to the case of surfactants.

The b.c (1.4)-(1.8) at the open surface need to be extended to the nonlinear case. Thus we now have

$$\frac{\partial \xi}{\partial t} = w - u\frac{\partial \xi}{\partial x} \tag{4.4}$$

$$p - \frac{B_0}{CP}\xi + \frac{1}{N^3}\left[\frac{1}{CP} - \frac{M}{P}(\theta - \xi)\right]\frac{\partial^2 \xi}{\partial x^2}$$

$$= \frac{2}{N^2}\left[\frac{\partial w}{\partial z} - \left(\frac{\partial u}{\partial z} + \frac{\partial w}{\partial x}\right)\frac{\partial \xi}{\partial x} + \frac{\partial u}{\partial x}\left(\frac{\partial \xi}{\partial x}\right)^2\right] \tag{4.5}$$

$$-\frac{M}{P}\left[\frac{\partial(\theta - \xi)}{\partial x} + \frac{\partial \theta}{\partial z}\frac{\partial \xi}{\partial x}\right]$$

$$= \frac{1}{N}\left\{\left(\frac{\partial u}{\partial z} + \frac{\partial w}{\partial x}\right)\left[1 - \left(\frac{\partial \xi}{\partial x}\right)^2\right] + 4\frac{\partial w}{\partial z}\frac{\partial \xi}{\partial x}\right\} \tag{4.6}$$

and

$$\frac{\partial \theta}{dz} = 0 \tag{4.7}$$

We see nonlinear contributions like the second term in the r.h.s. of the kinematic b.c. (4.4). Eq.(4.7) prescribes the heat flux at the open surface. $N = (1 + |\partial\xi/\partial x|^2)^{1/2}$. Again the air is assumed to be passive and weigthless with respect to the liquid.

The simplest approach to the above posed nonlinear problem is the single-mode analysis which is expected to be a useful description in a small enough neighborhood of the onset of overstability. Moreover, the more we move into low gravity the larger the capillary length becomes thus providing greater relevance to the single-mode approximation. On the other hand, transverse interfacial disturbances are expected to penetrate little in the liquid; the penetration depth depends indeed on the wavelength and frequency excited and on the viscosity of the liquid. The latter assumption gives relevance to the "potential" flow approximation to the time-dependent convection or in other terms to the limitation of the study to the high-frequency motions only. Thus for an arbitrary disturbance f(x,z,t) we set $f(x,z,t) \approx f(z,t)\exp(iax)$ and $\omega_0^2 = (B_0 + a^2)a/CP$ (Laplace's law). Using them, eq. (4.4) becomes

$$\frac{\partial \xi}{\partial t} = w + \xi \frac{\partial w}{\partial z} \tag{4.8}$$

On the other hand eq. (4.1) at z=ξ is

$$\frac{\partial w}{\partial t} = - \frac{B_0 + a^2/N^3}{PC} a\xi + \frac{M}{PN^3}(\theta - \xi)a^3\xi$$

$$- 2a\frac{1 + a^2\xi^2}{N^2}\frac{\partial w}{\partial z} - a^2\left(1 + \frac{2a\xi}{N^2}\right)w$$

$$+ \left(1 - \frac{2a\xi}{N^2}\right)\frac{\partial^2 w}{\partial z^2} \tag{4.9}$$

The zeroth-order (linear) disturbances are ξ_0,

$$w_0 = \frac{\partial \xi_0}{\partial t} e^{az} \tag{4.10}$$

and

$$\theta_0 \approx \xi_0 e^{az} + \frac{a}{\sqrt{2P\omega_0}}\left(\frac{1}{\omega_0}\frac{\partial \xi_0}{\partial t} - \xi_0\right) \exp\left(\sqrt{2P\omega_0}\; z\right) \tag{4.11}$$

where ω_0 denotes the harmonic frequency (2.6).

Consideration of the nonlinear terms in eqs. (4.8) and (4.9) up to cubic terms leads to

$$\frac{d^2\xi}{dt^2} + \delta\frac{d\xi}{dt} + [\omega_0^2 - \omega_0(\delta - 4a^2)]\xi$$

$$= - \omega_0^2 a\xi^2 + a\left(\frac{d\xi}{dt}\right)^2$$

$$+ (\delta - 4a^2)\left(\frac{d\xi}{dt} - \omega_0\xi\right) 2a\xi - 8a^3\xi\frac{d\xi}{dt}$$

$$- a^2\xi\left(\frac{d\xi}{dt}\right)^2 + \frac{11}{2}a^2\xi^2(\delta - 4a^2)\left(\frac{d\xi}{dt} - \omega_0\xi\right) + 16a^4\xi^2\frac{d\xi}{dt} + \frac{3a^5}{2CP}\xi^3 \tag{4.12}$$

where

$$\delta = \frac{Ma^3}{\sqrt{2}\,(P\omega_0)^{3/2}} + 4a^2 \tag{4.13}$$

At δ=0 we recover the results of Sec. 2. Positive (respectively, negative) values of δ account for subcritical (respectively, supercritical) motions. We have checked that indeed eq. (4.12) possesses limit cycle solution. This has been done both with the computer and using the time-derivative expansion procedure (16).

Note that for an effective gravitational acceleration of, say, 10^{-4}g , with g the standard value on earth, the predicted period of oscillation at onset is of the order of one to two minutes for most liquids. Threshold temperature gradients at overstability are of the order of a few K/cm for mercury and other liquids, including water.Thus aboard a spacecraft, in a low/microgravity environment the crucial test of our predictions can be obtained by making an experiment with a water-alcohol solution or some other liquids with a minimum of surface tension versus temperature (19-22). The suggested experiment is Bénard convection. Before the minimum is reached one expects steady polygonal cells (Bénard cells) whereas past the minimum, i.e., in the region where the surface tension of the liquid increases with increasing temperature, oscillations are predicted. In Sc. 6 we come back to the same experiment with, however, different initial conditions in order to trigger a soliton excitation (23-25).

5. SPACE MODULATED OSCILLATIONS

When due consideration is given to the full space dependence in Eqs. (4.1)-(4.7) a straightforward analysis yields a set of nonlinear *partial* differential equations to describe the evolution of the *transverse* wave motion. We have

$$\frac{\partial \zeta}{\partial t} = w - \frac{1}{a_c} w \frac{\partial^2 \zeta}{\partial x^2} \tag{5.1}$$

and

$$\frac{\partial w}{\partial t} = -\frac{a_c}{SC}\left\{Bo - \frac{\partial^2}{\partial x^2}\right\}\zeta - \frac{3 a_c}{2 S C}\left(\frac{\partial^2 \zeta}{\partial x^2}\right)\left(\frac{\partial \zeta}{\partial x}\right)^2 - 2 a_c^2 w + 2\frac{\partial^2 w}{\partial x^2}$$
$$+ 4 a_c^2 w \left(\frac{\partial \zeta}{\partial x}\right)^2 + 4 a_c (1 - 2 a_c \zeta)\frac{\partial}{\partial x}\left\{w \frac{\partial \zeta}{\partial x}\right\} + \frac{E_0 a_0}{S \omega_0 \sqrt{2 S \omega_0}}\left\{- a_c \frac{\partial^2 \zeta}{\partial x^2}\right.$$
$$\left. + \frac{3}{2}\left[\frac{\partial}{\partial x}\left(\frac{\partial \zeta}{\partial x}\right)^2\right]\frac{\partial}{\partial x} + \left[1 - 2 a_c \zeta + \frac{3}{2}\left(\frac{\partial \zeta}{\partial x}\right)^2\right]\frac{\partial^2}{\partial x^2}\right\}\left(\frac{\partial \zeta}{\partial t} - \omega_c \zeta\right) \tag{5.2}$$

which constitute our extension of the Stokes-Boussinesq wave theory (24) to a dissipative liquid layer subjected to Marangoni stresses. Work is at present underway in order to obtain solutions of these nonlinear partial equations, i.e., space modulated limit cycle oscillations.

6. SOLITONS EXCITED BY THE MARANGONI EFFECT

In the preceding sections we have discussed the onset and eventual nonlinear sustainment of some interfacial oscillations. These oscillations were either transverse or longitudinal waves. The former may be capillary-gravity waves that in a closed container are expected to develop as standing transverse periodic motions at the open surface of a liquid or at the interface between two liquids when the Marangoni effect is operating. In all the cases considered with the small viscous penetration depth or the high frequency limit the approximations used permitted assuming the "infinite" depth of the layer. Now we turn to purely traveling motions in the form of solitary waves with restriction to *shallow* layers. Work is now in progress towards a derivation of results for arbitrary depth (nonlinear Schroedinger equation).

We consider a shallow horizontal liquid layer of thickness "h" initially at rest and, as in the preceding sections, subjected to a transverse thermal gradient. Disturbances upon the quiescent state obey the continuity and Navier-Stokes equations to which we now add, to a first approximation, either Fourier's heat equation or Fick's mass diffusion equation. These equations are suplemented with the corresponding nonlinear boundary conditions at the bottom and at the open surface. Once more, for simplicity we shall restrict consideration to a two-dimensional geometry with x and z denoting the horizontal and vertical coordinates, respectively. β is the thermal gradient that again corresponds either to heat or mass diffusion (always positive when the heating is from the liquid side). Again as before θ denotes either temperature disturbance or surfactant concentration disturbance, so that D is heat or mass diffusivity. η and ν still account for dynamic and kinematic viscosity, respectively. All other quantities have already the meaning assigned in the preceding sections.

The "shallow layer approximation" demands that $h \ll 1$ (depth much smaller than wavelength of motions) and thus we can disregard buoyancy effects in the Navier-Stokes equations. Here "l" is the "wavelength" or maximum horizontal extent of the interfacial deformation ζ. For self-consistency in this section we recall the equations to be used. They are Eqs. (4.1)-(4.3)

$$\nabla^* v = 0 \tag{6.1}$$

$$\frac{\partial}{\partial t} v + v^* \nabla v = -\frac{1}{\rho} \nabla p + \nu \nabla^2 v \tag{6.2}$$

$$\frac{\partial}{\partial t} \theta = \beta w + k \nabla^2 \theta \tag{6.3}$$

The b.c. at the rigid bottom z=0 are

$$w = u = 0 \tag{6.4}$$

and

$$\theta = 0 \tag{6.5}$$

while at the surface $z=h+\zeta$

$$\frac{\partial}{\partial t} \zeta = w - u \frac{\partial}{\partial x} \zeta \tag{6.6}$$

$$p = \rho g \zeta + 2\mu \frac{\partial}{\partial z} w - \sigma_0 \nabla^2_\Sigma \zeta \tag{6.7}$$

$$\left[\frac{\partial}{\partial T} \sigma\right] \nabla_\Sigma (\theta - \beta \zeta) - \eta \left[\frac{\partial}{\partial z} u + \frac{\partial}{\partial x} w\right] = 0 \tag{6.8}$$

and

$$\frac{\partial}{\partial z} \theta = 0 \tag{6.9}$$

Note that we have considered the heat equation as a linear disturbance upon the nonlinear Navier-Stokes problem. Then a straightforward extension of the derivation presented by Lamb (23, Sec.6.1) for the standard KdV equation yields here (18)

$$\xi_t + \xi_y + \frac{5}{2} \varepsilon \xi \xi_y - \frac{30 - B_0}{60 B_0} \delta^2 \xi_{yyy} = \frac{\gamma}{2} \xi_{yy} \tag{6.10}$$

where $\gamma = 2(6)^{1/2}(4 + M)/3\ Re$, with $Re = C_o l/\nu$ and $C_o^2 = gh$. As before B_o is the Bond number. $\varepsilon = a/h$ and $\xi = \zeta/a$, with "a" the maximum value attainable by ζ. We see that setting $\gamma = 0$, eq. (6.10) provides the Korteweg-de Vries equation (23-25). $\gamma = 0$ demands either $\nu=0$, i.e., no viscosity, which is an irrelevant case in our analysis, or $M = -4$. The latter result indicates that for such negative value of the Marangoni number the open surface of the liquid layer heated from the air side is excitable in the form of a KdV soliton. Whether or not the soliton is stable can only be decided by studying the role of the nonlinear part in Eq. (6.3) left out here. However, what we can safely say is that, due to the Marangoni effect, $M = -4$ defines the threshold for soliton excitation in a quiescent shallow liquid layer subjected to a transverse thermal gradient. If, however, M is positive we are in the case of Bénard convection and we know that at M= 80 there is the onset of steady polygonal (mostly hexagonal) cell patterns. Thus here again an experiment with a liquid layer heated from below could provide a clear-cut test of our prediction. Once more it suffices to operate with a liquid having a minimum in the surface tension versus temperature curve. Before the minimum is reached we expect steady patterned convection and past the minimum here the prediction is the solitary travelling wave. Note that as the KdV equation corresponds to a nonlinear excitation the experimenter must strongly excite the liquid with, say, a sudden jump in the temperature gradient or with strong evaporation in the case of a volatile surfactant. Another possibility is to excite the interface with mechanic means, say, and once the soliton has been created, then the heat and/or mass transfer with the Marangoni effect is expected to help sustaining this excitation.

ACKNOWLEDGMENTS

This chapter is based upon research sponsored by CICYT (Spain) Grant PB 86-651 and by an EEC Grant. Both authors acknowledge fruitful discussions with Proff. Ph Drazin, J.K. Koster, H. Linde, A. Sanfeld and R. Sani and Dr. M. Hennenberg. The first author wishes to acknowledge the hospitality at the CNLS, Los Alamos Laboratory, and at the Center for Low-Gravity Fluid Mechanics, University of Colorado at Boulder, Colorado, where parts of this work were carried out. The second author wishes to acknowledge the hospitality at the Service de Chimie Physique, Université Libre de Bruxelles, where also parts of this work were carried out.

REFERENCES

1. Lamb, H., "Hydrodynamics", Dover, New York, 1932
2. Landau, L.D., and Lifshitz, E.M., "Fluid Mechanics", Pergamon, Oxford, 1959
3. Levich, B.G., "Physicochemical Hydrodynamics", Prentice-Hall, Inc., Englewood Cliffs, N.J., 1962
4. Velarde, M.G. and Normand,C., Sci. American **243**, 78 (1980)
5a. Legros, J.C., Sanfeld, A. and Velarde, M.G., in "Fluid Sciences and Materials Science in Space" (H.U. Walter, Ed.), pp. 83-140, Springer-Verlag, New York., 1987
5b. Velarde, M.G.(Ed.), "Phsysicochemical Hydrodynamics.Interfacial Phenomena", Plenum Press, New York., 1988
6. Linde,H., in "Dynamics and Instability of Fluid Interfaces" (T.S. Sørensen, Ed.), pp. 75-119. Springer-Verlag., Berlin, 1978
7. Linde, H., in "Convective Transport and Instability Phenomena" (J. Zierep and H. Oertel, Eds.), pp. 256-296. Braun-Verlag, Karlsruhe, 1982
8. Velarde, M.G. and Chu, X.-L., Phys. Scripta **T25**, 231 (1989)
9. Chu, X.-L. and Velarde, M.G., Physicochem. Hydrodyn. **10**, 727 (1988)
10. Chu, X.-L. and Velarde, M.G., J. Colloid Interface Sci. **131**, 471 (1989)
11. Velarde, M.G. and Chu, X.-L., Phys. Lett. **A 131**, 403 (1988)
12. Velarde, M.G. and Chu, X.-L., Il Nuovo Cimento **D 11**, 709 (1989)
13. Chu, X.-L., Velarde, M.G. and Castellanos, A., Il Nuovo Cimento **D 11**, 726 (1989)
14. Chu, X.-L. and Velarde, M.G., Il Nuovo Cimento **D 11**, 1615 (1989)
15. Chu, X.-L. and Velarde, M.G., Il Nuovo Cimento **D 11**, 1631 (1989)
16. Chu, X.-L. and Velarde, M.G., Phys. Lett. **A 136**, 126 (1989)
17. Velarde, M.G.and Chu, X.-L., in "Phase Transitions in Soft Condensed Matter" (T. Riste and D. Sherrington, Eds.), pp.139-143, Plenum Press, N.Y.,1989
18. Velarde, M.G. and Chu, X.-L., "Interfacial Instabilities", World Scientific, London (in preparation)
19. Vochten, R. and Petré, G., J. Colloid Interface Sci. **42**, 320 (1973)
20. Motomura, K., Iwanage,S.- I, Hayami,Y., Uryu, S. and Matuura, K.,J. Colloid Interface Sci. **80**, 32 (1981)
21. Desré, P.J. and Joud, J.C., Acta Astronaut. **8**, 407 (1981)
22. Legros, J.C., Limbourg-Fontaine, M.C. and Petré, G., Acta Astronaut. **11**, 143 (1984)
23. Lamb, G.L.," Elements of Soliton Theory", John Wiley, New York, 1980
24. Whitham, G.B., "Linear and Non-linear Waves", John Wiley, New York, 1974
25. Drazin, P.G. and Johnson, R.S., "Solitons : An Introduction", Cambridge Univ. Press, Cambridge, 1989

COUETTE FLOWS, ROLLERS, EMULSIONS, TALL TAYLOR CELLS, PHASE SEPARATION AND INVERSION, AND A CHAOTIC BUBBLE IN TAYLOR-COUETTE FLOW OF TWO IMMISCIBLE LIQUIDS

D.D. Joseph, P. Singh, and K. Chen

Department of Aerospace Engineering and Mechanics
University of Minnesota
107 Akerman Hall, Minneapolis, MN 55455

ACKNOWLEDGEMENTS

This research was supported by the National Science Foundation, the Army Research Office, and the Department of Energy.

ABSTRACT

Oil and water in equal proportion are set into motion between horizontal concentric cylinders when the inner one rotates. Many different flows are realized and described. In one regime many large bubbles of oil are formed. In a range of speeds where the water is Taylor unstable and the oil Taylor stable, we get strange Taylor cells of emulsified fluids whose length may be three or even four times larger than normal. The length of cells appears to be associated with effective properties of a non-uniform emulsion, so the cell sizes vary along the cylinder. At much higher speeds we get a fine grained emulsion which behaves like a pure fluid with normal Taylor cells. A second focus of the paper is on the mathematical description of the apparently chaotic trajectory of a small oil bubble moving between an eddy pair in a single Taylor cells trapped between the oil bands of a banded Couette flow. We defined a discrete autocorrelation sequence on a binary sequence associated with left and right transitions in the cell to show that the motion of the bubble is chaotic. A formula for a macroscopic Lyapunov exponent for chaos on binary sequences is derived and applied to the experiment and to the Lorenz equation to show how binary sequences can be used to discuss chaos in continuous systems. We use our results and recent results of Feeny and Moon (1989) to argue that Lyapunov exponents for switching sequences are not convenient measures for distinguishing between chaos (short range predictability) and white noise (no predictability).

The flows which develop between our rotating cylinders depend strongly on the material properties of the two liquids. A third focus of the paper is on dynamically maintained emulsions of two immiscible liquids with nearly matched density. The two fluids are 20 cp silicone oil and soybean oil with a very small density difference and small interfacial tension. The two fluids are vertically stratified by weight when the angular velocity is small. Then one fluid fingers into another. The fingers break into small bubbles driven by capillary instability. The bubbles may give rise to uniform emulsions which are unstable and break up into bands

Nonlinear Evolution of Spatio-Temporal Structures in Dissipative Continuous Systems
Edited by F.H. Busse and L. Kramer
Plenum Press, New York, 1990

of pure liquid separated by bands of emulsified liquid. We suggest that the mechanics of band formation is associated with the pressure deficit in the wake behind each microbubble.

INTRODUCTION

Some of the phenomena which are now to be described extend results first given in the paper of Joseph, Nguyen, and Beavers (1984). This paper is the only one we know to report results of experiments on the flow of two immiscible liquids in a Taylor-Couette apparatus. We shall refer to this paper as JNB. The papers by Y. Renardy and D. Joseph (1985) and by Guillopé, Joseph, Nguyen, and Rosso (1987) are the only theoretical studies of the special form of two-fluid flow between rotating cylinders called Couette flow. The paper by Joseph and Preziosi (1987) gives a theoretical explanation of rollers, which is another configuration which appears in our experiments between cylinders. In this paper we are going to describe some of the flows which can be observed when two immiscible liquids are set into motion between horizontal cylinders when the inner one rotates. Many different flows are realized: fingering flows, coarse and fine emulsions, phase-separated emulsions, lubricated flow, and banded Couette flows.

The nature of fluid-solid interactions takes a more important place in two-fluid dynamics than in single fluid dynamics. There is a competition between the two fluids as to which one will wet a solid boundary. The factors that enter into this competition are not understood. We get some kind of interaction between the physical chemistry of adsorption of fluid at the boundary of a solid with two-fluid dynamics. Wetting is not determined entirely by energy considerations, by contact angles, static or dynamic. The history of the motion also plays an important role in determining the places on the solid which are wet by one liquid or another (see Figure 3e).

Taylor-Couette flows of a single fluid are among the best understood of all fluid phenomenon. One reason for this is that the geometry is relatively simple for analysis and perfectly marvelous for experiments. The Taylor-Couette apparatus may also evolve as an apparatus of choice in the study of two-fluid dynamics, fluid-solid interaction, and in the study of dynamical properties of emulsions.

EXPERIMENTS

All of the experiments were carried out between two concentric cylinders with axis horizontal, perpendicular to gravity. The outer cylinder and the end plates are plexiglass. The inside diameter of the outer cylinder is 2.495 inches; the outside diameter is 2.986 inches. The inner cylinder is of aluminum with a diameter of 1.985 inches. This is a convenient reference for normal Taylor cells which combine two counter-rotating eddies approximately 0.515 inches in length. The length of the cylinder is 11.985 inches. The outer cylinder is fixed and the inner one rotates with angular velocity Ω.

Our Taylor apparatus uses two neoprene lip seals to prevent leakage. The shaft driving the inner cylinder is connected to a torque meter which has a provision for counting rpm. The torque meter is connected to a mechanical-digital converter which displays the value of the torque and rate of rotation. The digital signal is transferred to a Hewlett-Packard 87 and then sample averaged. Usually there is a fluctuation in the torque before the rotation reaches a new steady state. By monitoring the values displayed on the digital converter, we can determine when the motion is in steady state. We took torque data for some of the experiments which is reported in Figure 1. For uniformity each data point was taken one-half hour after establishing a changed condition. Obvious transients were well delayed after this time. Sometimes there is a slow emulsification which will eventually change the dynamics.

The torque values shown in Figures 1, 5, and 9 are not guaranteed because there is an unknown frictional torque due to the neoprene seals which varies with the speed. The decrease in the torque at small values of Ω is almost certainly an effect of seals and not of flow. At higher speed the fraction of the total torque due to the bearings is smaller. Perhaps torques measured at speeds in excess of 80 rpm are reasonably accurate.

The laboratory is temperature controlled at 25° Celsius. We used two different oils in the experiments with water and oil: Mobil heavy duty oil with density of 0.97 g/cm^3 and viscosity 0.95 poise, and SAE 30 motor oil with density of 0.886 g/cm^3 and viscosity of 0.98 poise. The major effect of the density difference occur in slow or lubricated flow in which the oil floats up. This effect is greater in SAE 30 oil-water systems than in the heavy oil-water system which is more nearly density matched.

A different set of experiments in the same apparatus were carried out with silicone oil and soybean oil (under the brand name of Crisco). The density and viscosity of the silicone oil is ρ=0.949 g/cm^3, μ=0.2 p and of the soybean oil is ρ=0.922 g/cm^3, μ = 0.46 p.

The interfacial tension between Mobil heavy-duty motor oil and tap water is 30.00 dyne/cm. The interfacial tension between SAE 30 motor oil and tap water is 9.2 dyne/cm. The interfacial tension between 0.2 p silicone oil and Crisco is 1.4 dyne/cm.

PARAMETERS

Six dimensionless parameters govern these flows: the viscosity ratio, the density ratio, the volume ratio, a capillary number, a Froude number based on $\Delta\rho g$ where $\Delta\rho$ is the density difference, and a Taylor (or Reynolds) number. We are going to collect data for systematic variations of the parameters later. It is useful to note that we usually get some form of emulsions when the Taylor number for the high viscosity constituent is larger than the critical one for Taylor instability. The critical angular velocities in our apparatus are calculated from the instability theory as [4.35, 409, 402, 91.7, 217] rpm for [water, SAE 30 motor oil, Mobil heavy-duty motor oil, 0.2 p silicone oil, Crisco] when one fluid rather than two fluids fills the gap. We say that two-phase flow between cylinders is bistable when one phase is stable, the other unstable. Bistable flow of water and motor oil occurs when $4.35<\Omega<409$ (or 402) rpm. Bistable flow of silicone and Crisco oil occurs when $91.7<\Omega<217$ rpm.

COUETTE FLOWS

Couette flows of two liquids are here defined as steady axisymmetric flow of two immiscible liquids between infinitely long rotating cylinders of radius a<b and angular velocities Ω_1 and Ω_2. These flows satisfy the Navier-Stokes equations, no-slip boundary conditions, and classical interface conditions at liquid-liquid interfaces. The Couette flows are a small class of steady and possibly nonaxisymmetric flows which could develop between rotating cylinders, but in contrast to the one-fluid case, Couette flows of two fluids are not unique, there is a continuum of solutions of at least two different types: layered and banded Couette flows.

Layered Couette flows

Layered Couette flows are one class of steady solutions. For two layers, one is on the inside, two outside, and the interface between them is at r=d In fact the steady solutions are not unique, there could be any number of layers of any thickness subject to the specification of the volume of each fluid and geometrical constraints. JNB showed that the layered flow with just two layers, the low viscosity liquid on the inner cylinder, uniquely minimizes the torque when the angular velocity difference is prescribed. They noted that the minimizing torque can be determined by minimization for two layers since we may always consider the problem for adjacent layers. For two layers, one inside, two outside, the torque M_L per length L is

$$M_L = \frac{a^2b^2(\Omega_2 - \Omega_1)}{(b^2 - a^2)k} \frac{\mu_1\mu_2(b^2 + ka^2)L}{(b^2\mu_1 + ka^2\mu_2)} \tag{1}$$

where

$$k = \frac{v_2}{v_1} \tag{2}$$

is the ratio of volumes. Suppose that we have oil and water, fixing the water volume v_w and oil volume. First we compute M when $\mu_1=U_w$, $v_1=v_w$ then compute M when $m_2=\mu_w$, $v_2=v_w$. The case with $\mu_1=\mu_w$ gives a smaller torque, so the water is on the inner cylinder r=a, oil outside.

The stability of layered Couette flow was studied by Renardy and Joseph (1985) and Guillopé et al. (1988). The linear theory with gravity neglected studied in their paper showed that a thin layer of the less-viscous fluid next to either cylinder is linearly stable and that it is possible to have stability with the less dense fluid lying outside. The stable configuration with less-viscous fluid next to the outer cylinder, layered Couette flows, have never been observed because of effects neglected. However various lubricated flows with less viscous fluid on the wall in the presence of gravity could be regarded as realizations of the torque-minimizing layered Couette flow. For examples of this type we refer the reader to Figures 7, 8, and 9, and especially Figure 10 of JNB and to the description of the flows shown in Figures 2e, f, and g of this paper.

Banded Couette flows

These flows are such that the two fluids are arranged in alternating bands rather than layers. This is an exact solution, the same solution as if there were no bands.

$$\mathbf{u} = \mathbf{e}_\theta V(r) = \mathbf{e}_\theta\left(Ar + \frac{B}{r}\right),$$

$$A = \frac{b^2\Omega_2 - a^2\Omega_1}{b^2 - a^2}, \tag{3}$$

$$B = (\Omega_1 - \Omega_2)\frac{a^2b^2}{b^2 - a^2}.$$

The interface between bands are set on the annular areas in the intersection of the gap and planes perpendicular to axis z of the cylinders. The velocity is automatically continuous; the shear stresses $\tau_{z\theta}$ and τ_{zr} are zero and the pressure is continuous across these planes.

Banded Couette flows are not unique, the width of the bands and their number is not determined by stated conditions. The torque on the cylinders is the sum of the torques of each band. The torque M_B on a band of length L, with viscosity μ, is given by

$$\begin{aligned} M_B &= \frac{L\mu}{2}\int_a^b r^3\left[\frac{d\left(\frac{v}{r}\right)}{dr}\right]^2 dr \\ &= L\mu a^2 b^2 \frac{(\Omega_2-\Omega_1)^2}{b^2-a^2}. \end{aligned} \tag{4}$$

Suppose L_1 is the total length of bands of fluid with viscosity μ_1 and L_2 with viscosity μ_2. Then

$$\begin{aligned} &L_1 + L_2 = L, \\ &\frac{L_2}{L_1} = \frac{V_2}{V_1} = k, \text{ and} \\ &M_B = \frac{(\mu_2 + k\mu_1)}{1+k} La^2b^2\frac{(\Omega_2-\Omega_1)^2}{b^2-a^2}. \end{aligned} \tag{5}$$

To compare the torques on the cylinders in banded and layered Couette flow we put

$$\begin{aligned} \frac{M_B}{M_L} &= \frac{\left(b^2\mu_1 + ka^2\mu_2\right)}{\mu_1\mu_2\left(b^2+ka^2\right)}\frac{(\mu_2+k\mu_1)}{(1+k)} \\ &= \frac{\left(m + k\frac{a^2}{b^2}\right)(1+mk)}{m\left(1+k\frac{a^2}{b^2}\right)(1+k)} \end{aligned} \tag{6}$$

where

$$m = \frac{\mu_1}{\mu_2}.$$

The minimum torque in layered Couette flow is when the low viscosity constituent is on the wall, m<1 The function (6) of m decreases from infinity at m=0 to one at m=1. Hence $M_B>M_L$; layered Couette flow has a smaller torque for the same angular velocity and volume fraction as banded Couette flow for which the oil must necessarily attach itself to both inner and outer layer.

Banded Couette flows are not lubricated. We have never seen a banded Couette flow without secondary motions, but the flows shown in Figures 3a and 10 are banded Couette flows in the water cells which have already become unstable to Taylor vortices. The active water cells in Figure 10 are the sites for the strange attractor which is described in the section, "Chaotic trajectories of oil bubbles in an unstable water cell."

ROLLERS

There can be a superficial resemblance between banded Couette flows and rollers. The rollers shown in Figures 26 and 28 of JNB look exactly like the banded Couette flows shown in Figure 10 of this paper, but the rollers are not attached to the outer cylinder; they rotate nearly as a rigid solid sheared only by water. Truncated rollers are shown in Figures 2e, f, and g.

EMULSIONS, TALL TAYLOR CELLS, CELL NUCLEATION

The generation of emulsion can occur in several ways. The fluids at rest are stratified by gravity. At slow speeds a stable interface of one fluid advances into the other. The advancing interface develops scallops at its leading edge as in Figure 2.1. These scallops become unstable and finger into the host fluid (Figures 4 and 6a). Bubbles form from capillary instability leading directly or eventually to emulsions. The average size of the bubbles in an emulsion decreases as the speed (shear) increases. A coarse emulsion has large oil bubbles (Figures 2c, d). All these emulsions are maintained by shearing. They collapse to stratification when the motion is stopped. Strictly speaking, emulsions are unsteady because the oil water interfaces are moving. We can realize transient uniform emulsions (Figures 4 and 6b), but they appear to be unstable because they cannot be maintained when the angular velocity of the inner cylinder is fixed (Figure 6c).

The forms taken by rotating flows of emulsions depend strongly on the fluids used. In the oil-water systems we get very long cells and the cell sizes are variable along the axis of the cylinder (Figures 2d, 3b, 3c). This nonuniformity may be due to a nonuniformity of the degree of emulsification, the bubble size, along the axis. In any event the long nonuniform cells are robustly stable to changes in the angular velocity. At much higher speeds the degree of emulsification increases and new cell boundaries nucleate, producing more cells. Eventually we get a very fine stable homogeneous emulsion which has square Taylor cells of the usual type (Figures 3d, 6e). These cells are robust; if the angular velocity is reduced the normal cells persist. This shows that fine emulsions have a different response than coarse emulsions even when the gap size and angular velocity are the same.

Typical sequences of flow types are exhibited in the photographs of heavy motor oil in water shown in Figure 2 and in the photographs of SAE 30 motor oil shown in Figure 3. Many of the transitions are evident also in the torque graph, Figure 1, associated with transitions in the heavy oil of Figure 2.

PHASE INVERSION

There is a critical $\phi = \frac{V_1}{V_1 + V_2} = \tilde{\phi}$ for *phase inversion*, which we shall now define. In our experiments $0.6<\tilde{\phi}<0.7$; when $\phi<0.6$, the liquid with the lower viscosity (silicone oil) fingers into the more viscous liquid (Crisco oil) and eventually we get silicone drops in Crisco oil. When $\phi>0.7$, we can sketch the same Figure 4; but the words Crisco and silicone are interchanged because Crisco, rather than silicone oil, fingers. In either case, the bubbly emulsions which exist before phase separation are uniform without structure. The bubbly emulsions all have smaller torques than the pure low viscosity constant (silicone oil) when Ω is small (see Figure 4). The torque is an increasing function of volume fraction up to phase inversion; after inversion ($\phi>0.7$) the torque appears to decrease with increasing ϕ. The "effective viscosity" of bubbly emulsions is smaller than the viscosity of its lowest viscosity constituent in pure form (silicone oil). There is a preference for low viscosity fingers to penetrate into the high viscosity fluid, $\phi<0.6$ Phase inversion shows that it is also possible to get a more viscous liquid to finger into a less viscous one.

The torque curves can be used to back out the effective viscosity of an emulsion. The effective viscosity of emulsified Crisco oil is smaller than the effective viscosity of emulsified Crisco oil after phase inversion.

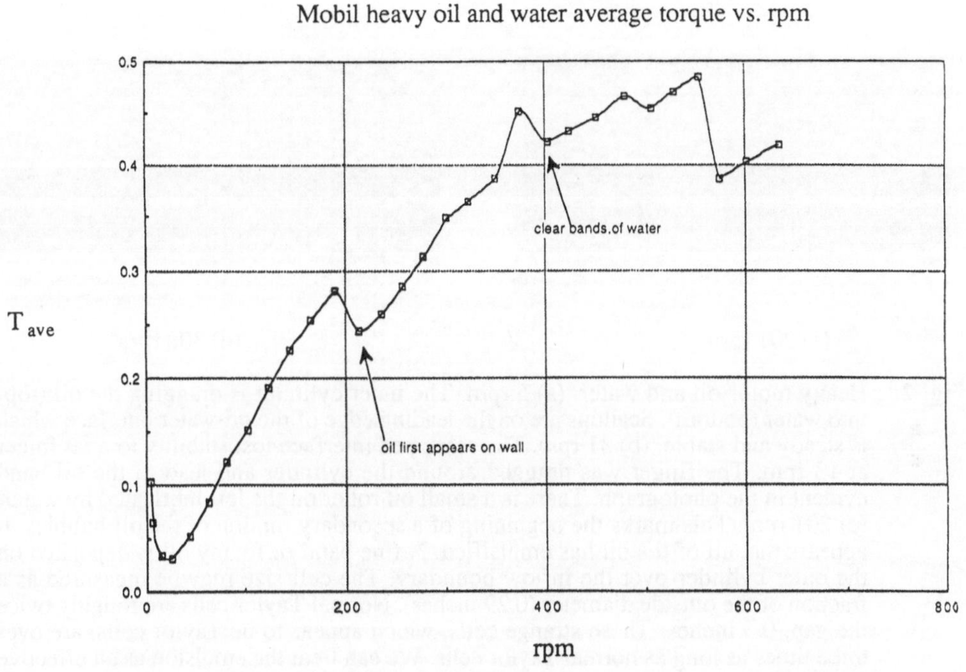

Fig. 1. Torque versus angular velocity for heavy motor oil and water. The decrease of the torque at the origin is due to the friction of the neoprene seal. We think that the oil foam which appears at $\Omega>200$ marks the inflow boundary of a Taylor cell for a coarse grained emulsion. The appearance of clear bands is associated with spontaneous clearing of oil from the inner cylinder, increasing lubrication.

PHASE SEPARATION

Apparently the state of uniform emulsification is unstable. At least, we have never been able to maintain a state of uniform emulsification, even after one is created (see Figure

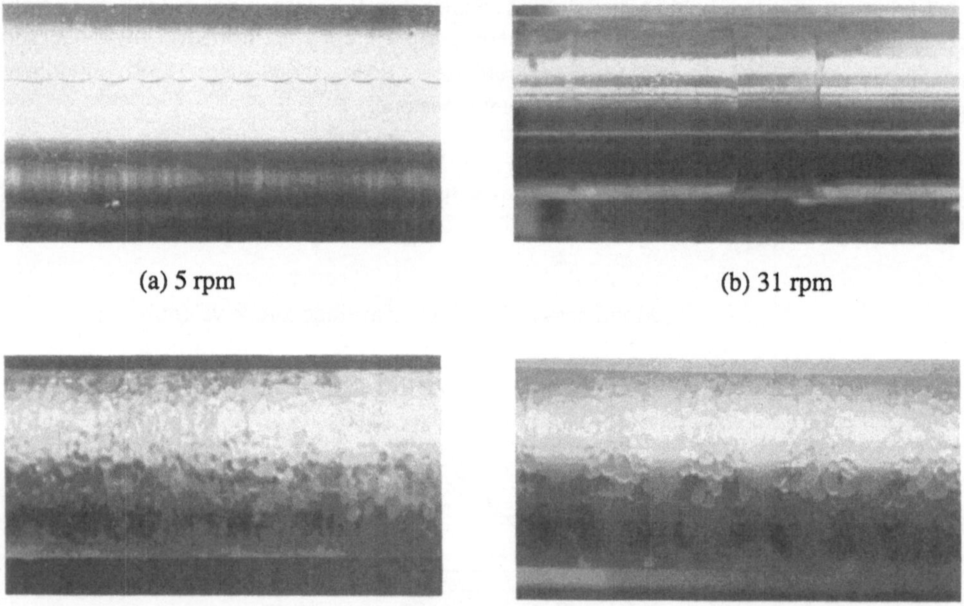

(a) 5 rpm (b) 31 rpm

(c) 201 rpm (d) 308 rpm

Fig. 2. Heavy motor oil and water. (a) 5 rpm. The inner cylinder is dragging the oil (top) into water (bottom). Scallops are on the leading edge of the oil-water interface which is steady and stable. (b) 31 rpm. The scalloped interface lost stability to a fat finger at 13 rpm. The finger was dragged around the cylinder and lead to the oil band evident in the photograph. There is a small oil roller on the left lubricated by water. (c) 201 rpm. This marks the beginning of a secondary motion of the oil bubbles. It appears that all of the oil has emulsified. A fine band of foamy oil is deposited on the outer cylinder over the inflow boundary. The cell size may be measured as a fraction of the outside diameter (0.29 inches). Normal Taylor cells are roughly twice the gap, 0.5 inches. These strange cells, which appear to be Taylor cells, are over three times as long as normal Taylor cells. We can treat the emulsion as an effective fluid with an effective viscosity. (d) 308 rpm. The oil bubbles are fine markers of the fluid motion elongating themselves in the direction of motion. The foamy oil bands are inflow boundaries. The outflow boundaries are between the inflow boundaries and define a center of orientational symmetry for the fluidized oil bubbles. The length of the cells is not uniform. The cells are three times longer than usual. *(Fig. 2 continued on next page.)*

(e) 500 rpm

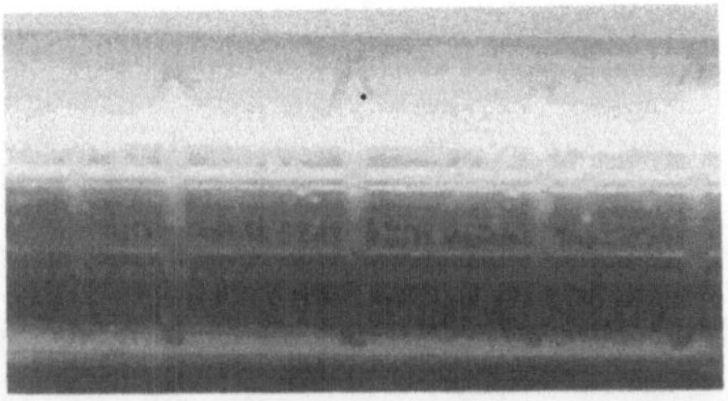

(f) 550 rpm

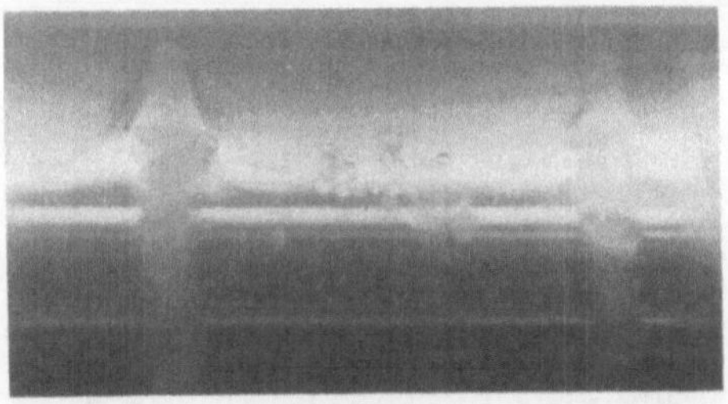

(g) 563 rpm

Fig. 2. *(continued)* Heavy motor oil and water. (e) 500 rpm. The emulsion has almost vanished. The oil on the outside cylinder is lubricated by water on the inside. Truncated rollers are on the inner cylinder. (f) 550 rpm. The oil at the top is shielded from the inner cylinder by a layer of water. It is a lubricated flow, like a layered Couette flow. There are well-lubricated oil rollers on the inner cylinder. (g) 563 rpm. There are bigger rollers and fewer drops. Everything is lubricated by water.

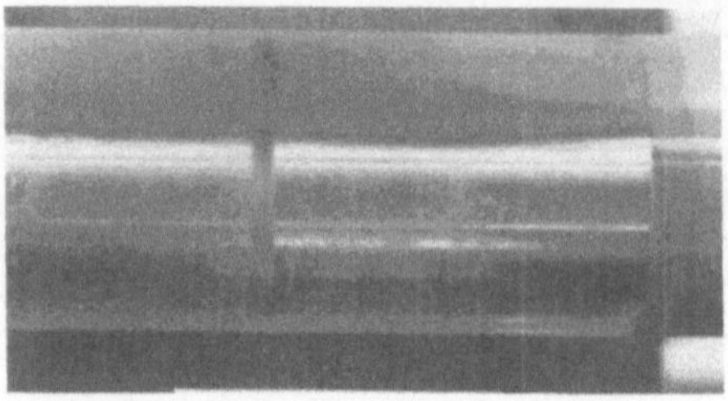

(a) 99 rpm

(b) 360 rpm

(c) 785 rpm

(d) 360 rpm

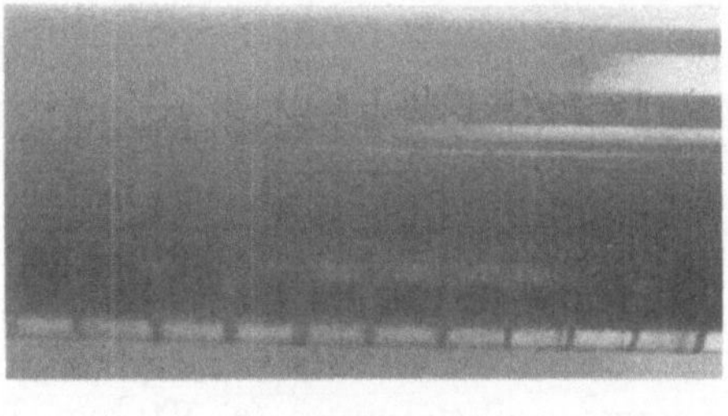

(e) 0 rpm

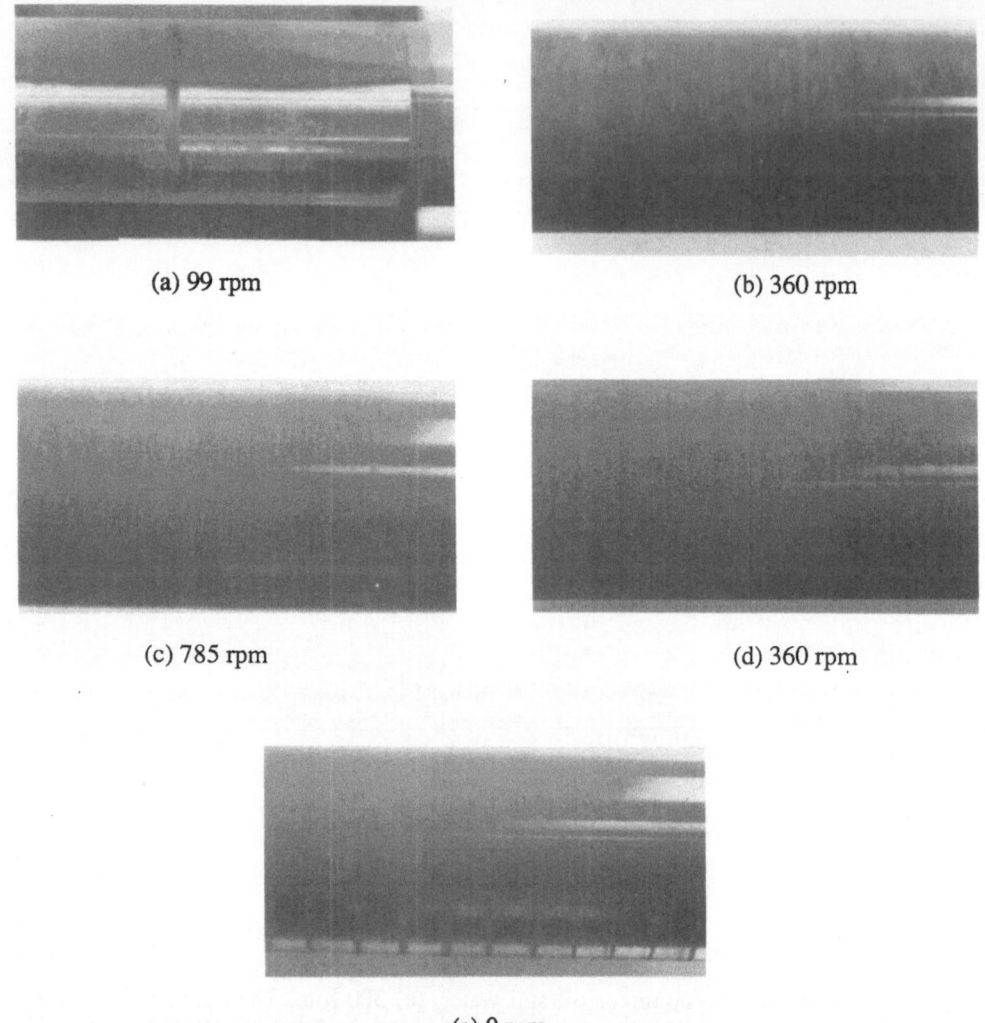

Fig. 3. SAE 30 motor oil and water, series 1. This series undergoes development somewhat like the one in Figure 2. (a) 99 rpm. A banded Couette is on the right and a water lubricated stratified flow on the left. The oil broke up at about 300 rpm and coarse grained emulsion is formed. (b) 360 rpm, (c) 785 rpm. The secondary motions are beautifully marked by the emulsions. The cell sizes are not uniform but do not change with speed. At higher speeds new cells nucleate and after several adjustments the long cells shorten to the normal length, twice the gap. This adjustment is complete at 2550 rpm. Then the speed is reduced to 360 rpm (d) and the normal cells are retained. The difference between (b) and (d) is due possibly to the fact that the emulsion in (d) is finer. (e) The speed was reduced from 360 rpm to zero. The oil foam at the outside of the inflow boundary is stuck to the wall and it won't come off. This shows clearly that the fluid which wets the wall in different places depends on the history of the motion.

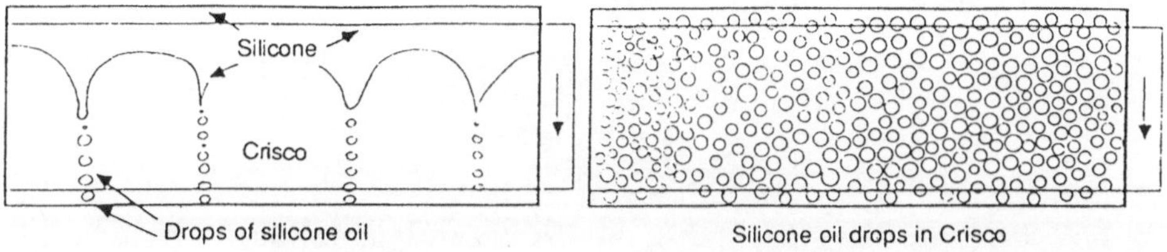

Fig. 4. Fingering instability leading to emulsion. The inner cylinder rotates.

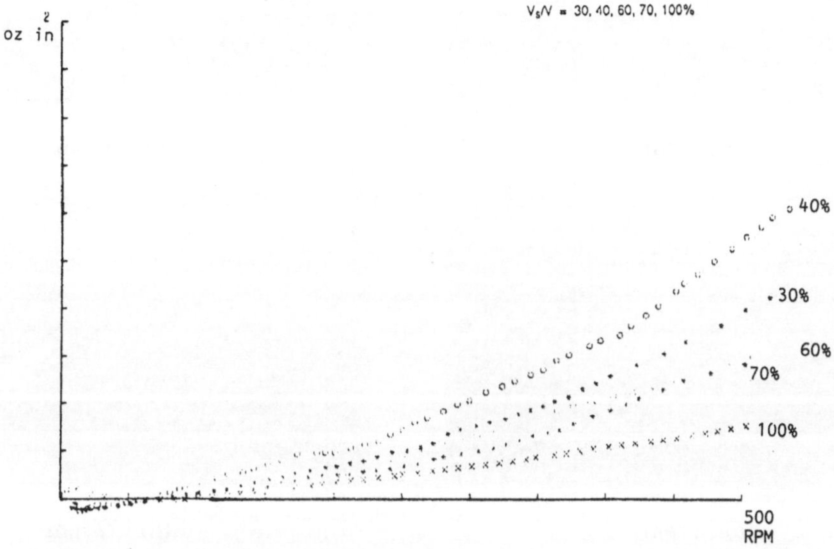

Fig. 5. Torque versus angular velocity with the volume fraction of silicone oil as a parameter.

6). The instability of uniform emulsions in the silicone oil and Crisco oil systems leads to a phase separation. The cause of this phase separation is not understood. It could not be a form of Taylor instability since it occurs at angular velocities well below the critical Taylor number, 91.7, for the silicone oil alone. In fact the phase separation may occur at all finite values of the speed or rotation of the inner cylinder. We think that the state of uniform emulsification is unstable because of wakes which tend to align bubbles in rows. This effect is very clear in beds of spherical particles fluidized by water discussed in the paper by Fortes et al. (1987). In that case there is a scenario called drafting, kissing, and tumbling. Drafting is the mechanism by which one sphere is sucked into the wake of another, as debris is pulled from the side of a road behind a fast-moving truck. The rear sphere accelerates in the wake of the forward sphere and they kiss. The kissing spheres are aligned with stream. The kissing spheres form a long body which is unstable to the same kind of turning couples that cause and aircraft to stall so that spheres tumble into more stable cross stream pairs. Falling drops and rising bubbles also draft but they don't tumble. Instead, they appear to align as we have seen in drop experiments in which heavy liquids are dropped into a long tube filled with an immiscible lighter liquid.

The drops always align and drafting is obvious. Side bubbles cannot be pulled uniformly in alignment, so there is a tendency to segregate. Clearly this explanation is tentative and incomplete.

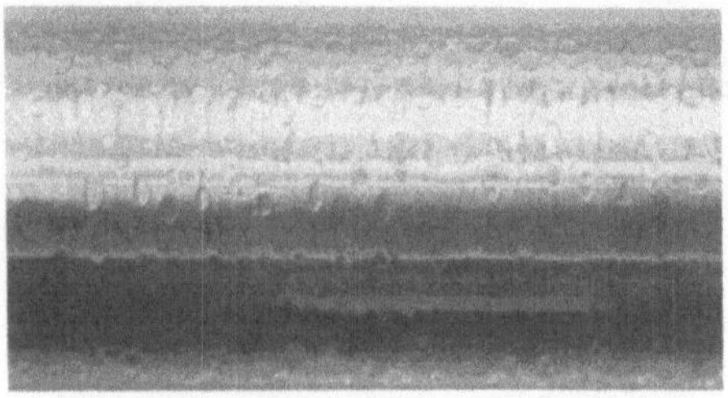

(a) 16 rpm

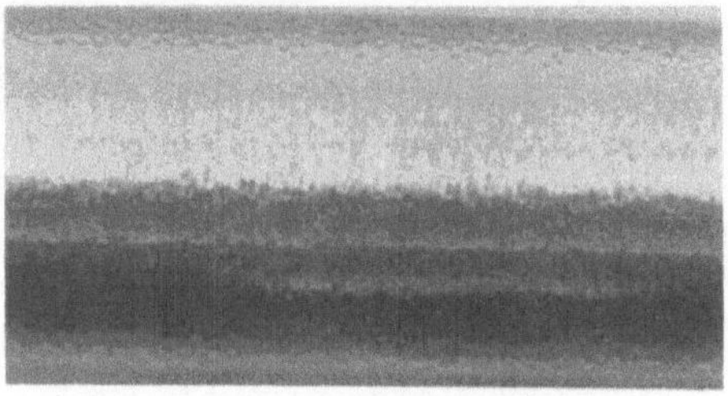

(b) 24 rpm

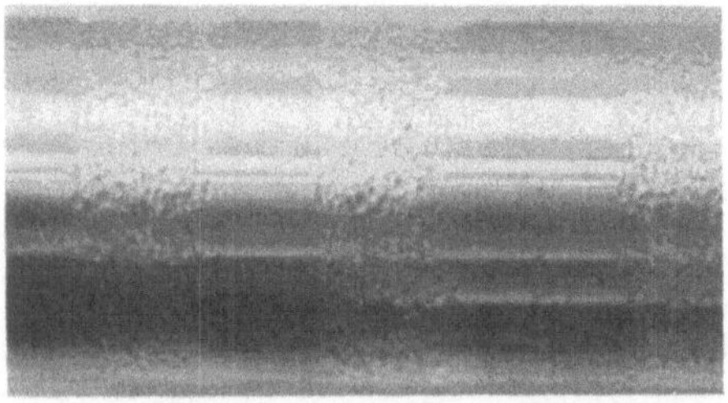

(c) 24 rpm

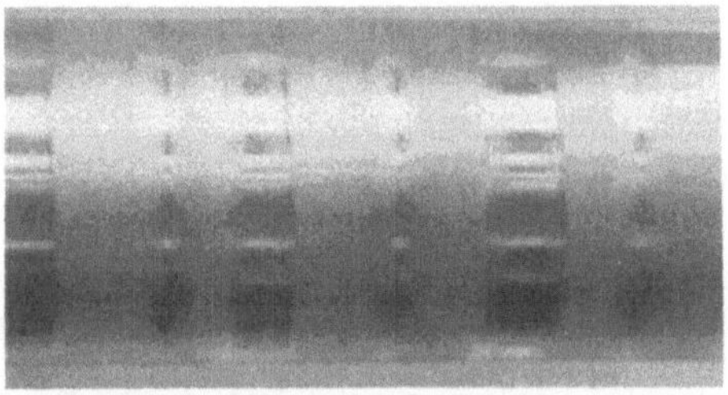

(d) 122 rpm

(e) 410 rpm

Fig. 6. Soybean (Crisco) oil and silicone oil in equal proportions. (a) 16 rpm. Silicone drops form as capillary instability after fingering. (b) "Uniform" emulsion at 24 rpm. (c) Phase separation at 24 rpm. (d) A finer emulsion of silicone oil at 122 rpm. Some of the phase boundaries are very distinct. (e) Stable emulsion with normal Taylor cells at 410 rpm.

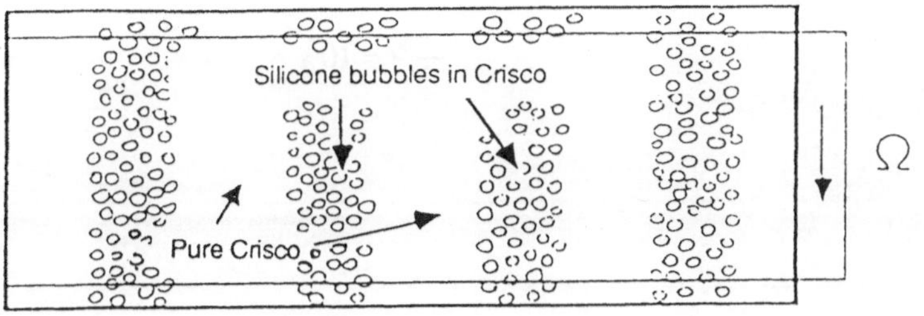

Fig. 7. Phase separation. Bands of emulsified silicone oil are separated by bands of pure Crisco oil, $\phi \leq 0.6$.

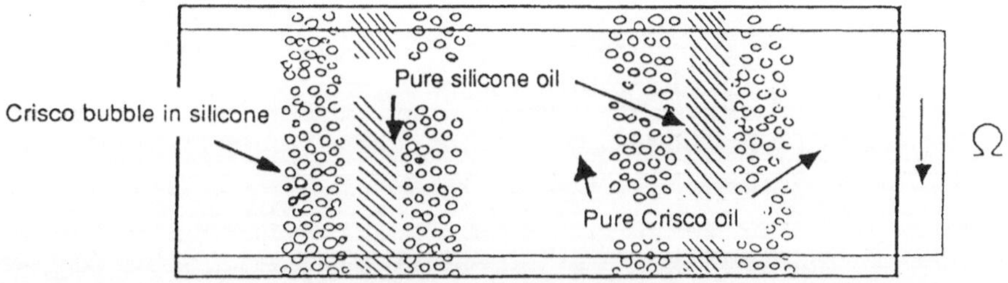

Fig. 8. Phase separation when Crisco oil fingers, $\phi = 0.7$.

PHASE INVERSION AND PHASE SEPARATION

When $\phi \leq 0.6$, symmetric equally spaced bands of an emulsion of silicone oil in Crisco are separated by bands of pure Crisco oil, as in Figure 7.

For $\phi = 0.7$, instead of silicone bubbles, we get Crisco bubbles, again with more or less symmetric equally spaced bands. In this case, a narrow band of pure Crisco oil is in the center of each band of emulsified Crisco oil, as in Figure 8.

The next event, as Ω increases, is the disappearance of the phase boundaries with uniform mixing, leading to a foamy emulsion that is unlike the grainy emulsion which develops after fingering. The foamy emulsions are more stable and take longer (5 to 10 minutes) to collapse when the rotation stops.

The appearance of foamy emulsions appears to coincide with the appearance of regular Taylor vortices. The secondary motion, which is generated from the instability of the hitherto stable Crisco oil in the bistable phase separated regime, may be the cause of the mixing leading to foamy emulsions. The critical angular velocity for the onset of Taylor vortices in the composite fluid, which we have called a foamy emulsion, can be determined visually; and, when the volume ration is small, ($\phi < 0.5$) forms a sharp break in the torque curve (see Figure 9).

If we assume that this emulsion acts basically like a Newtonian fluid when in the Taylor apparatus, we can use the formula for the critical Taylor number to back out the value of the effective viscosity from the critical angular velocity, which is evidenced either by visual observation or by the break point in the angular velocity-torque curve.

The next critical Taylor number signals the formation of uniform wavy vortices in the foamy emulsion. We could also try to identify material properties from this transition.

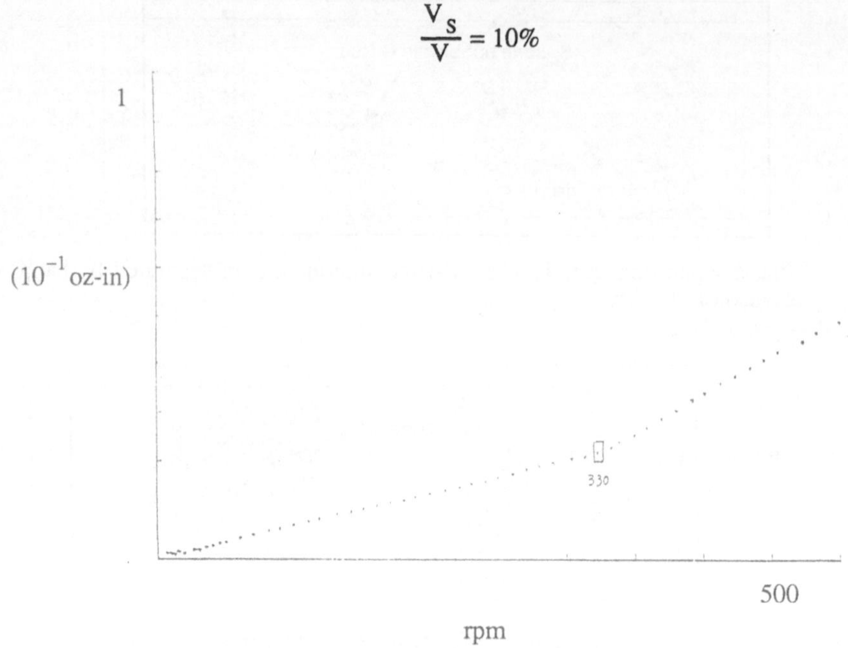

Fig. 9. Torque versus angular velocity for an emulsion of 10% silicone oil in Crisco. Taylor cells appear at 330 rpm.

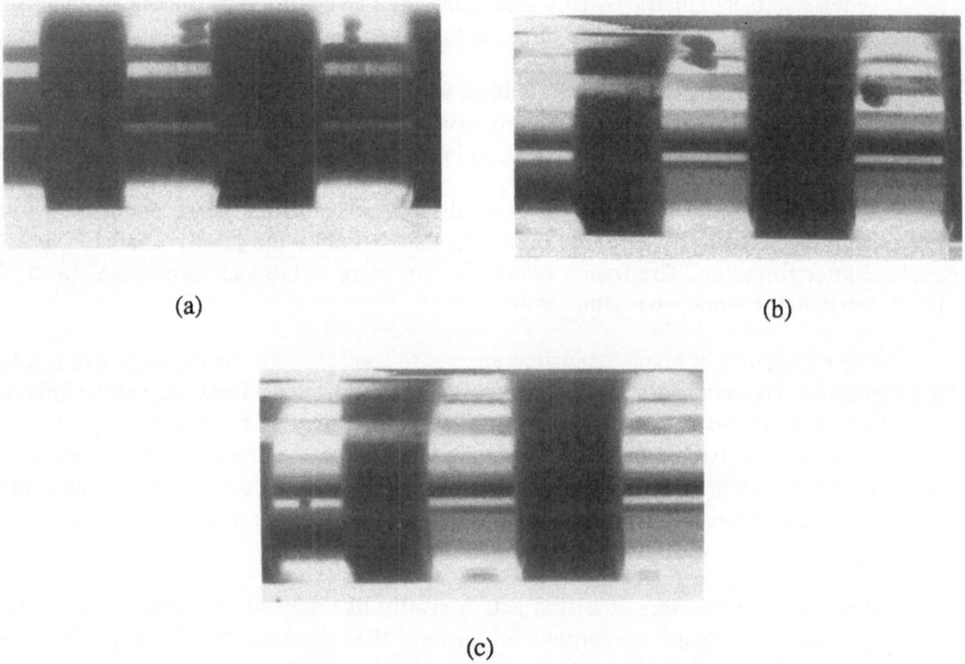

Fig. 10. Banded Couette flows of SAE 30 motor oil and water at 309 rpm. There is a small bubble in the unstable water cell on the left which is visible in (c) but not in (a) or (b). This small bubble undergoes an apparently chaotic motion. The large bubble appears to undergo chaotic switching from right to left.

Finally we note that at yet higher rotation rates the waves on the vortices seem to disappear, and, as far as the eye can see, the vortices are steady. This is unusual because it does not happen in the dynamics of Taylor vortices in single constituent liquids.

CHAOTIC TRAJECTORIES OF OIL BUBBLES IN AN UNSTABLE WATER CELL

In the course of experiments described in the section, "Emulsions, tall Taylor cells, cell nucleation," we found a motion which appears to us to be chaotic. At sufficiently high values of the angular velocity prior to emulsification of motor oil, some bubbles of oil are torn away from the oil bands. In some situations we were able to get one oil bubble into a Taylor cell. This oil bubble is carried round and round by water and is dragged around in the secondary motion due to Taylor instability. We made a video tape of this and some still photographs are shown in Figure 10. The small oil bubble on the left is the one for which the binary sequence is studied. Each time the oil drop goes around it is either in the left eddy or in the right eddy. We monitored about 3000 terms in the sequence LRLL... and assigned number minus one to left and one to right. It is difficult to get revealing still photographs of the motion of the small bubble in the leftmost water cell shown in Figure 10. However, the large bubble in the center water cell also executes a chaotic motion of a slightly different type as can be seen in the photographs.

Binary sequences

We are going to apply methods of estimation theory (see Singh and Joseph, 1989) to characterize the chaos in the binary number sequence generated by the bubble in our experiments. Consider a sequence $u(n)=\pm 1$ of binary numbers. We assume that the sequence is ergodic so that time averages are the same as ensemble averages. In our experiment the average

$$E[u(n)] = \frac{1}{N}\sum_{n=1}^{N} u(n) \to 0$$

when N is large, left and right or ± 1 are equally probable.

Singh and Joseph (1989) showed how to generate a binary sequence for chaotic trajectories of the Lorenz system $[\dot{x}, \dot{y}, \dot{z}]=[\sigma(y-x), rx-y-xz, xy-bx]$ for $(\sigma, b, r) =(10, \frac{8}{3}, 28)$. The binary sequence is generated by projecting the trajectories into the xz plane,as shown in Figure 11, and monitoring the crossing points of trajectories on the segments AB and CD of the line AD. The crossing times are put into correspondence with the sequence n of integers, left crossings on AB are recorded as $u(n)=1$, and the right crossings of CD, as $u(n)=-1$. The time averages of these sequences vanish for large N, independent of initial condition, so that left and right crossings are equally probably and we may assume that the sequences are ergodic.

Autocorrelations

An estimate of the autocorrelation function on an ergodic binary sequence can be obtained as follows:

$$r(n) = \frac{1}{N}\sum_{k=1}^{N} u(k+n)u(k), \quad n = 1,2,\ldots \quad N \gg n. \tag{7}$$

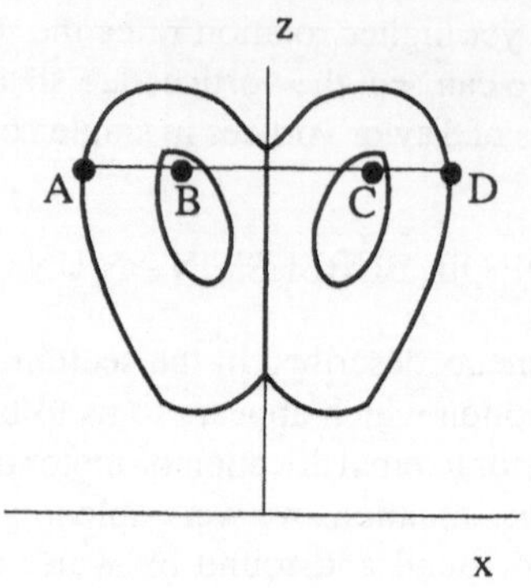

Fig. 11. The projected trajectories of the Lorenz attractor remain inside the butterfly region and outside the ovals around the fixed points.

The value r(1) represents the correlation between immediate neighbors (1, 2), (2, 3), (3, 4), etc. Value r(2) gives the correlation between separated pairs (1, 3), (2, 4), etc. A chaotic response is one for which $r(1)\neq 0$ and $r(n)\to 0$ for large n.

For the oil bubble autocorrelation values r(n)'s, for large n, are not uniformly close to zero because of the relatively small length of the sequence, N=3000 (Figure 12). We tried sequences of different length and found that r(n)'s, for large n's, approached zero uniformly as the length of the sequence was increased.

The Lorenz equations were integrated numerically using the NAG library. Subroutine DO2BBF was used for different tolerance levels in the range 10^{-4} to 10^{-10}. We projected into the xz plane and formed a binary number symbol sequence with 76,000 entries. The autocorrelation function is shown in Figure 13. The tolerance level in the numerical scheme had absolutely no effect on the nature of autocorrelation sequence, even though sequences generated were quite different for different tolerance levels. For large n, r(n) approached zero uniformly with the increase in length of the sequence, N.

In both cases the decay in the autocorrelations value is very rapid. For large n, autocorrelation values decrease monotonically with the length of the sequence. The decay of autocorrelation for the bubble is essentially complete after n=2, a substantial correlation exists only for r(1). The decay of correlation is slower for the Lorenz system with nonzero r(n) for n<6. We could say that the Lorenz system is less random.

Lyapunov exponents

Singh and Joseph (1989) derived a macroscopic Lyapunov exponent for binary sequence. Lyapunov exponents for continuous times are locally defined quantities which measure the tendency for chaotic trajectories to diverge exponentially for small time, on the average. One can define the first exponent by

$$\lambda = \frac{1}{t_{N+1} - t_1} \sum_{k=1}^{N} \log_2 \frac{d(t_{k+1})}{d_0(t_k)} \tag{8}$$

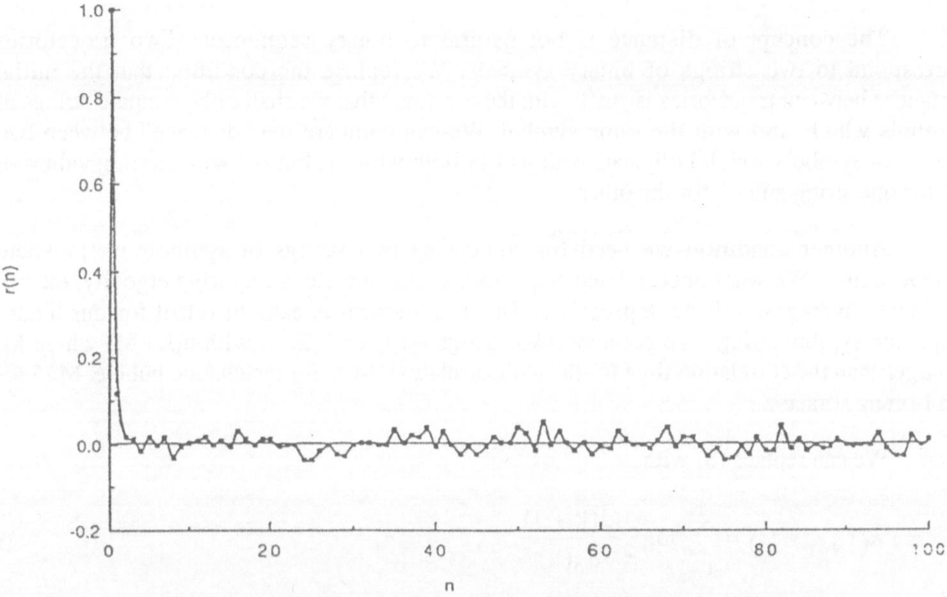

Fig. 12. The autocorrelation sequence for the oil bubble, N=3000.

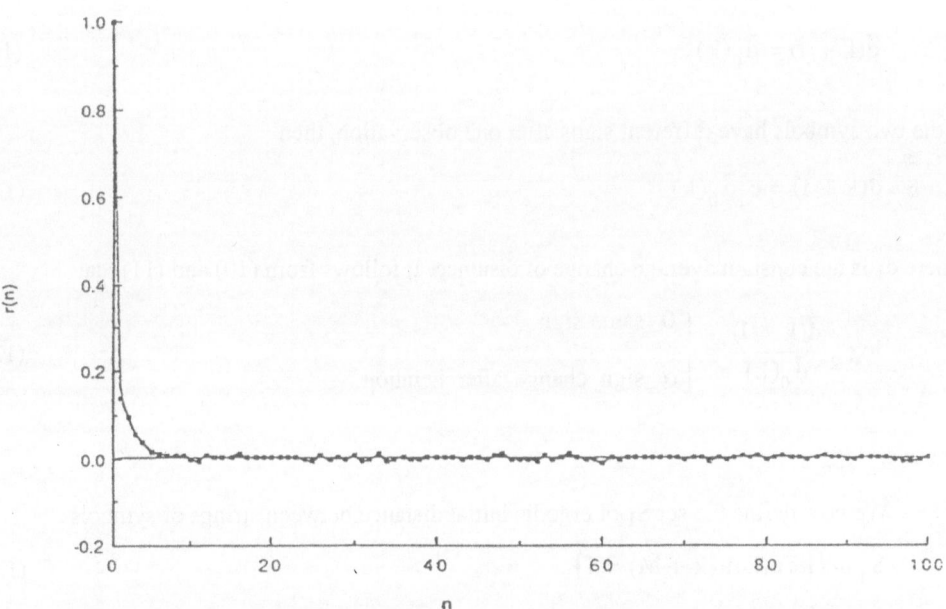

Fig. 13. The autocorrelation sequence for the Lorenz attractor, N=76,000.

where $d_0(t_k)$ is the initial distance between two trajectories at time t_k and $d(t_{k+1})$ is the distance between these two trajectories at time $t_{k+1}>t_k$. In the continuous case $d_0(t_k)$ and $d(t_{k+1})$ are infinitesimal and $N\rightarrow\infty$.

The concept of distance is not natural to binary sequences. Two trajectories correspond to two strings of binary symbols. We replace the condition that the initial distances between trajectories is small with the condition that we shall only compare strings of symbols which start with the same symbol. We can compare the "distance" between two strings of symbols which both start with u=1 or both with −1, but not with starting values of +1 for one string and −1 for the other.

Another condition we need for comparing two strings of symbols is statistical independence. We want uncorrelated sequences so that theorems requiring ergocity, the use of "time" averages, will be appropriate. This requirement is easy to fulfill for our binary sequence symbol string. We compare two strings u(k), k = 1, 2, ... with u(k + M) where M is larger than the correlation time for the autocorrelation, M>2 for the chaotic bubble, M>5 for the Lorenz attractor.

We can replace (8) with

$$\lambda(t_{N+1} - t_1) = \sum_{k=1}^{N} \log_2 \frac{\bar{d}(k+1)}{\bar{d}_0(k)} \tag{9}$$

where $\bar{d}_0(k)$ is the average "distance" between two statistically independent strings at the kth observation. If the two symbols at the k+1st observation have the same sign we say that the "distance" is unchanged, on the average

$$\bar{d}(k+1) = \bar{d}_0(k). \tag{10}$$

If the two symbols have different signs after one observation, then

$$\bar{d}(k+1) = c_1 \bar{d}_0(k) \tag{11}$$

where c_1 is the constant average change of distance. It follows from (10) and (11) that

$$\log_2 \frac{\bar{d}(k+1)}{\bar{d}_0(k)} = \begin{cases} 0 & \text{same sign} \\ \alpha & \text{sign change after iteration} \end{cases} \tag{12}$$

where $\alpha=\log_2 c_1$.

We now define the set S_1 of ergodic initial distance between strings of symbols

$$S_1 = \{k : u(k)u(k+M) = 1\}. \tag{13}$$

The complementary set is

$$S_2 = \{k : u(k)u(k+M) = -1\}. \tag{14}$$

Hence, we may write

$$\log_2 \frac{\bar{d}(k+1)}{\bar{d}_0(k)} = \frac{\alpha}{2}\{1 - u(k+1)u(k+1+M)\}$$

for all symbol sequences which have the same sign at the time k for all $k \in S_1$. Hence,

$$\sum_{k=1}^{N} \log_2 \frac{\bar{d}(k+1)}{\bar{d}_0(k)} = \frac{\alpha}{2} \sum_{k\in S_1} \{1 - u(k+1)u(k+1+M)\}. \tag{15}$$

The total number of k is N. Let N_1, N_2 be the number of k's in the sets S_1, S_2, and $N_1+N_2=N$. We have also that

$$\begin{aligned} N\,r(M) &= \sum_{k=1}^{N} u(k)u(k+M) = \sum_{k\in S_1} u(k)u(k+M) \\ &+ \sum_{k\in S_2} u(k)u(k+M) = N_1 - N_2 = 0. \end{aligned} \tag{16}$$

Since r(M)=0 when M is larger than the correlation "time." Hence $N_1=N_2=\frac{N}{2}$.

We next define the macroscopic Lyapunov exponent as the average value

$$\begin{aligned} \lambda_m &= \frac{1}{N_1} \sum_{k\in S_1} \log_2 \frac{\bar{d}(k+1)}{\bar{d}_0(k)} \\ &= \frac{\alpha}{N} \sum_{k\in S_1} \{1 - u(k+1)u(k+1+M)\}. \end{aligned} \tag{17}$$

This is related to the average Lyapunov exponents by

$$\frac{\lambda(t_{N+1} - t_1)}{N} = \lambda_M. \tag{18}$$

Singh and Joseph (1989) showed that

$$\lambda_m = \frac{\alpha}{2}[1 - r^2(1)]. \tag{19}$$

Lyapunov exponents and white noise

Singh and Joseph (1989) calculated the macroscopic Lyapunov exponent for the Lorenz system described in the section, "Binary sequences." They calculate α as follows. The average distance between starting trajectories on the line AB(=CD) of Figure 11 is

$$\frac{|AB|}{3} = \bar{d}_0(k).$$

The switching distance is $|AD|-|AB|+\bar{d}(k+1)$. Hence

$$\frac{\bar{d}(k+1)}{\bar{d}_0(k)} = 3\left[\frac{|AD|}{|AB|} - 1\right].$$

They found that $|AD|=4.31|AB|$. Then from (12) we calculate $\alpha=3.3$. The relation (18) between the average Lyapunov exponent λ and the macroscopic exponent λ_M may be simplified by putting $t_{N+1}-t_1=N\Delta T$ where ΔT is the average period. Then

$$\bar{\lambda} = \frac{\lambda_m}{\Delta T} = \frac{\alpha}{2\Delta T}\left(1 - r(1)^2\right)$$

where $\Delta T=0.7519$ sec. We get

$$\lambda_m=1.618 \text{ bits/period} .$$

The largest Lyapunov exponents computed directly for the Lorenz attractor is

$$\lambda=1.30 \text{ bits/period} .$$

Feeny and Moon (1989) have studied a chaotic dry friction oscillator using the method of binary sequences of Singh and Joseph (1989). They did an experiment with sliding friction in which an imposed change of the normal force caused the slider to stick. They also modeled their experiment with a second order forced ODE involving friction coefficient and normal load functions. They did Poincaré sections for the experiments with 2,048 symbols and for the differential equation with 10,000 symbols. The symbols form a string of binary numbers ± 1 corresponding to whether the motion is sticking or slipping at each pass through the Poincaré section. They measure distance on the Poincaré plot:

$$\bar{d}_0(k) = \frac{1}{3},\ \bar{d}(k+1) = 1.$$

Hence, using (19), they get $\alpha=\log_2 3=1.585$.

Feeny and Moon studied the tent map and logistic map using the formula (19) with $\alpha=1.585$. They calculated r(1) for $N=10^5$ and $N=2048$. The theoretical value of the largest Lyapunov exponent is $\lambda=1$ for both the tent map and the logistic map. They compute

$$\lambda_m = \left.\begin{matrix} 0.787515 \ (10^5 \text{ symbols}) \\ 0.787705 \ (2048 \text{ symbols}) \end{matrix}\right\} \text{tent map}$$

$$\lambda_m = \left.\begin{matrix} 0.791578 \ (10^5 \text{ symbols}) \\ 0.791116 \ (2048 \text{ symbols}) \end{matrix}\right\} \text{logistic map}$$

A binary autocorrelation was obtained for their experiments and numerically from the differential equation for a symbol string with $N=2048$. In both cases the autocorrelation r(1) is very small, less than ± 0.05. They calculate

$$\lambda_m = \begin{cases} 0.79055 \text{ experiment} \\ 0.79219 \text{ numerical integration} \end{cases}$$

The calculation of the exponent for the Poincaré map from the equations of motion gives

$$\lambda = 0.77 .$$

We draw the reader's attention to the fact that for all the calculations done by Feeny and Moon, they get

$$\lambda_m = \frac{\alpha}{2}\left[1 - r(1)^2\right] = 0.7925\left[1 - r(1)^2\right].$$

This shows that $r(1)^2$ is very small in the examples of the tent map, logistic map and experiments.

Short range predictability requires that r(1), r(2), ..., r(M) $\neq$ 0 for small M, r(n)$\to$0 for large n. For white noise, we have r(1)=0. The autocorrelation is good for distinguishing short range predictability and white noise. The macroscopic Lyapunov exponent is not useful for making this important distinction. In fact, the macroscopic Lyapunov exponent depends on distance through α, but λ_μ/α is universal, does not depend on distance and may be a more intrinsic measure of chaos. Certainly r(1) has a lot less information than the graph of r(n).

REFERENCES

Feeny, B.F. and Moon, F.C., 1990, Autocorrelation on symbol dynamics for a chaotic dry friction oscillation, *Phys. Letters A* (to appear).

Fortes, A., Joseph, D.D., and Lundgren, T.S., 1987, Nonlinear mechanics of fluidization of beds of spherical particles, *J. Fluid Mech.*, 177:467-483.

Guillopé, C., Joseph, D.D., Nguyen, K., and Rosso, F., 1987, Nonlinear stability of rotating flow of two fluids, *J. Theoretical & Applied Mech.*, 6:619-645.

Joseph, D.D., Nguyen, K., and Beavers, G.S., 1984, Nonuniqueness and stability of the configuration of flow of immiscible fluids with different viscosities, *J. Fluid Mech.*, 141:319-345.

Joseph, D.D., Preziosi, L., 1987, Stability of rigid motions and coating films in bicomponent flows of immiscible liquids, *J. Fluid Mech.*, 185:323-351.

Renardy, Y. and Joseph D.D., 1985, Couette flow of two fluids between concentric cylinders. *J. Fluid Mech.*, 150:381-394.

Singh, P. and Joseph, D.D., 1989, Autoregressive methods for chaos on binary sequences for the Lorenz attractor, *Phys. Letters A*, 135:247-253.

TRAVELING WAVES IN A PARTLY SOLIDIFIED MIXTURE

G. Zimmermann and U. Müller

Kernforschungszentrum Karlsruhe
Institut für Reaktorbauelemente
Postfach 3640, 7500 Karlsruhe 1
Federal Republic of Germany

1. INTRODUCTION

In the classical Bénard problem a horizontal, single component liquid layer is heated at the bottom and cooled at the top. If a critical temperature difference across the layer is exceeded, the heat conduction state becomes unstable and steady convection is observed.

However, in binary liquid mixtures with a negative thermal diffusion coefficient, time-dependent convection is found. A theoretical analysis shows that the critical temperature difference for the onset of time-periodic convection is lower than that for the steady state convection. Experimentally, linear traveling waves are observed at the onset of convection, that is, the whole convection pattern travels in a horizontal direction through the test cell at a constant velocity.

This paper reports investigations concerning the influence of a thin ice-layer in the upper part of the Bénard cell on the onset of convection. The deformable ice-layer is realized by lowering the temperature of the upper boundary below the melting temperature of the mixture. In chapter 2 a linear theory is outlined[1]. In chapter 3 experimental results[2] are presented. The theoretical and experimental results are discussed in chapter 4.

2. THEORY

The configuration is sketched in figure 1. The two horizontal, parallel plates of infinite extent are separated by a distance h. The lower plate at $z = 0$ is at the fixed temperature T_u, the upper plate at $z = h$ at the fixed temperature T_o. For a melting temperature T_s of the binary mixture higher than T_o we get a thin ice-layer in the upper part of the Bénard layer. In case of pure heat conduction, the height of the liquid and solid layer are h_L and $h\text{-}h_L$ respectively.

The material properties are the density ρ_0, the thermal diffusivity κ, the kinematic viscosity ν, the molecular diffusivity D, the Sorét coefficient S_0 and the expansion coefficients α and α' for thermal and solutal effects. The concentration C is related to the lean component of the mixture. Superscripts s are used to designate the solid properties.

Nonlinear Evolution of Spatio-Temporal Structures in Dissipative Continuous Systems
Edited by F.H. Busse and L. Kramer
Plenum Press, New York, 1990

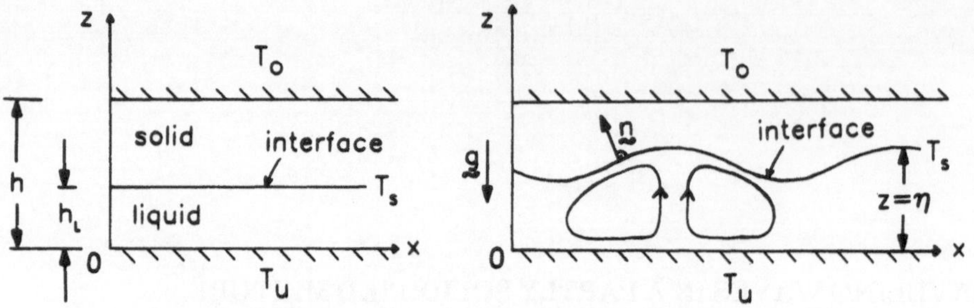

Figure 1. Schematic drawing of a partially solidified liquid layer in the conduction and convection state.

For small concentrations C, it is assumed that only one component is solidified. The heat in the solid phase is transported by conduction.

$$C^{(s)} = 0 \ , \tag{1}$$

$$\frac{\partial T^{(s)}}{\partial t} = \kappa^{(s)} \nabla^2 T^{(s)} \ . \tag{2}$$

In the liquid the Boussinesq equations are used.

$$\operatorname{div} \underset{\sim}{v} = 0 \ , \tag{3}$$

$$\frac{\partial v}{\partial t} + (\underset{\sim}{v}\nabla)\underset{\sim}{v} = -\frac{1}{\rho_0} p + \nu\nabla^2 \underset{\sim}{v} - [\alpha(T - T_s) - \alpha' C] g \underset{\sim}{e} \ , \tag{4}$$

$$\frac{\partial T}{\partial t} + (\underset{\sim}{v}\nabla)T = \kappa \nabla^2 T \ , \tag{5}$$

$$\frac{\partial C}{\partial t} + (\underset{\sim}{v}\nabla)C = \nabla \underset{\sim}{j} \ , \tag{6}$$

$$\underset{\sim}{j} = D[\nabla C - S_0 C(1 - C)\nabla T] \ . \tag{7}$$

A linear equation of state for the density in the buoyancy term of equation (4) is assumed. The unit vector $\underset{\sim}{e}$ is in the direction of the z-axis and $-g\underset{\sim}{e}$ is the gravity vector. The thermal diffusion, which couples the concentration field to the external temperature gradient, is given by equation (7).

The upper boundary at $z = h$ is at a fixed temperature $T_0 < T_s$:

$$T^{(s)} = T_0 \ . \tag{8}$$

At the solid-liquid interface, undercooling due to the surface energy is neglected:

$$T^{(s)} = T = T_s \quad . \tag{9}$$

In the first approximation, the melting temperature T_s of the mixture depends linearly on the concentration at the interface:

$$T_s = T_{s1} + mC \quad . \tag{10}$$

T_{s1} is the melting temperature in the case of $C = 0$ and m is the slope of the melting curve. For small deformations η of the interface which are dependent on the horizontal coordinates x and y the heat and mass balances at the interface are given by

$$\rho^{(s)} L \frac{\partial \eta}{\partial t} = [\lambda^{(s)} \nabla T^{(s)} - \lambda \nabla T] \underset{\sim}{n} \quad , \tag{11}$$

$$\rho^{(s)} C \frac{\partial \eta}{\partial t} = \rho_0 \underset{\sim}{j} \underset{\sim}{n} \quad . \tag{12}$$

L is the heat of fusion and $\underset{\sim}{n}$ denotes the unit vector normal to the interface. The kinematic and the no-slip conditions at the interface are given by

$$(\rho_o - \rho^{(s)}) \frac{\partial \eta}{\partial t} = \rho \underset{\sim}{v} \underset{\sim}{n} \tag{13}$$

and

$$\underset{\sim}{v} \underset{\sim}{t}_1 = \underset{\sim}{v} \underset{\sim}{t}_2 = 0 \quad , \tag{14}$$

where $\underset{\sim}{t}_1, \underset{\sim}{t}_2$ are the unit tangent vectors normal to $\underset{\sim}{n}$. At the lower boundary ($z \equiv 0$) a rigid, impermeable wall at a fixed temperature is assumed:

$$T = T_u, \underset{\sim}{v} = 0, \underset{\sim}{j} e = 0 \quad . \tag{15}$$

The set of differential equations, boundary and interfacial conditions are scaled and an equivalent set of dimensionless equations for the liquid and the solid phase is obtained. This set contains 12 independent dimensionless groups. The most important ones for this problem are listed:

$$\text{Rayleigh number} \quad Ra = \frac{\alpha g (T_u - T_s) h_L^3}{\nu \kappa} \quad , \tag{16}$$

$$\text{Biot number} \quad B = \frac{\lambda^{(s)}}{\lambda} \frac{(T_s - T_0)}{(T_u - T_s)} \quad , \tag{17}$$

$$\text{Separation coefficient} \quad \psi = C_0 (1 - C_0) \frac{\alpha'}{\alpha} S_0 \quad . \tag{18}$$

The Rayleigh number characterizes the stability of the system; the Biot number measures the thickness of the ice-layer. The separation coefficient is a measure for the mass diffusion due to temperature gradient and is here assumed to be negative. This particular group causes time-dependent behaviour at the onset of convection.

A linear perturbation analysis is conducted for the governing system about the static basic state. Assuming spatially periodic solutions the z-dependence of the perturbation quantities is separated. The key issue for this is an explicit form of the solution of the energy equation in the solid phase. The perturbation expansions result in a system of differential equations for the z-dependent perturbation quantities in the fluid and expressions for the corresponding boundary conditions at the lower plate and the interface.

The resulting complex boundary-eigenvalue problem is solved numerically by employing a shooting method. The evaluation was done with the aid of a computer code[3]. The calculated critical Rayleigh numbers and corresponding periods as a function of the Biot number are given in figures 2 and 3.

3. EXPERIMENT

The experiments are done in a test cell with length $l=200.0$ mm, depth $d=20.0$ mm and height $h=3.12$ mm. The horizontal boundaries consist of copper plates fixed at the temperatures T_u and T_o. The large sidewalls are made out of glass to allow the optical observation of the convection pattern from the side and to measure the height of the ice-layer with a stereo-microscope. The endwalls of the test cell are made out of teflon and are optimized with a computer code to avoid horizontal temperature gradients in the horizontal copper plates.

In the experiment a mixture of 15 % by weight of ethanol in water is used with a negative thermal diffusion coefficient $\psi=-0.43$ and a melting temperature $T_s \approx 7.6$ °C. The structure of the convection pattern is visualized by a differential interferometer with horizontal beamsplitting. The interferograms of the periodic pattern of convection rolls correspond qualitatively to the streamlines. In the traveling wave state the whole pattern moves continuously with a constant velocity through the cell and can be detected easily. The temperature fluctuations in the liquid layer are measured by thermocouples in the fluid. A traveling wave produces a time-periodic temperature signal belonging to the alternating passing by of cold downstreams and warm upstreams in the convection patterns.

During the experiments the thickness of the ice-layer is varied and the critical temperature difference as well as the periods at the onset of convection are measured. The horizontal ice-layer is generated by lowering the temperature of the upper boundary below the melting temperature in a quasi-steadily process. In the heat conducting state a plane interface is obtained. Then, at a fixed temperature T_o, the temperature of the lower bondary T_u is increased very slowly close to the critical temperature difference across the liquid layer. An additional increase in ΔT of less than 1 % above the critical value triggers the onset of convection.

A state described by linear traveling waves is found with a very low amplitude of the temperature fluctuations ($T' < 0.003\ \Delta T$). The amplitudes are slightly modulated because of local differences in the height of the liquid layer of about 0.02 mm. This induces local small variations of the Rayleigh number.

Typically after a few hours the amplitudes of the temperature fluctuations increase exponentially. The final state is a steady three-dimensional convection pattern of polygons. The interface is deformed and shows the structure of the underlying convection pattern. The experimental results are shown in figures 2 and 3.

4. DISCUSSION

Figure 2 shows the values of the measured and calculated critical Ralyeigh numbers as a function of the Biot number. With increasing B the value of $Ra_{c,osc}$ decreases as in the case of a pure liquid[5]. The reason for this behaviour is the increasing resistance for the heat flux at the upper boundary with increasing thickness of the ice-layer. The experimental values are found to be higher than the calculated ones for all B.

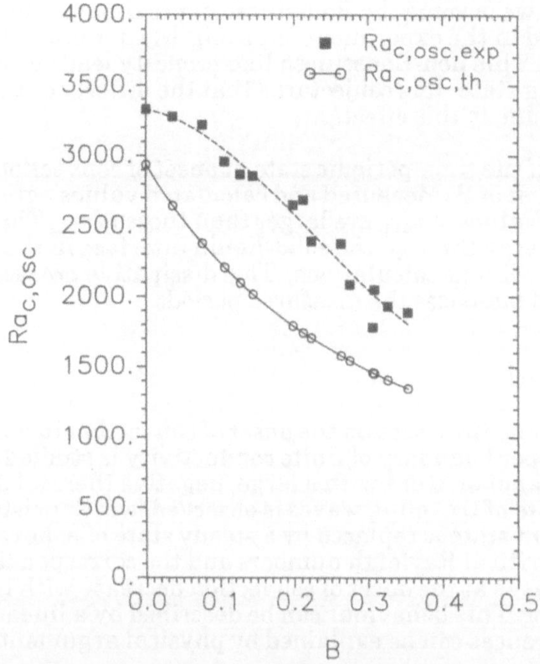

Figure 2. Measured and calculated values of the critical, oscillatory Rayleigh number $Ra_{c,osc}$ for different Biot number B.

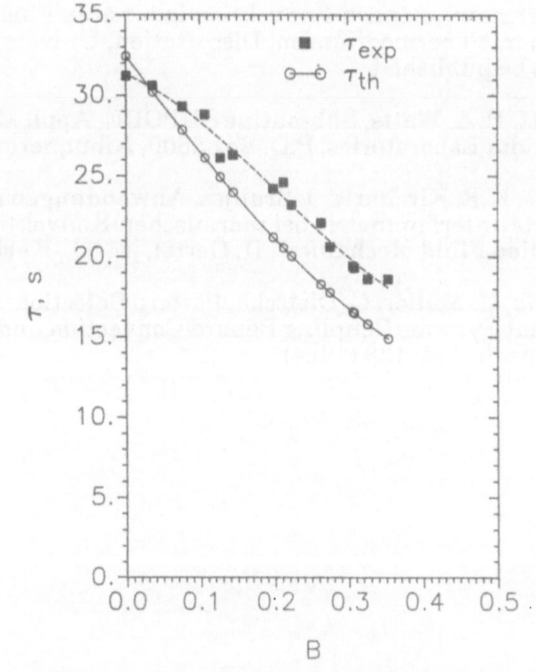

Figure 3. Measured and calculated values of the period τ at the onset of oscillatory convection for different Biot numbers B.

In the theory we assume the Boussinesq approximation to be valid. But the liquid mixture used in the experiment has a sligthly nonlinear temperature-dependent density. This non-Boussinesq like property leads to a stabilization of the heat conducting state. It is conjectured that the difference between theory and experiment is due to this effect.

The periods of the time-periodic state at onset of convection are shown in figure 3 as a function of B. Measured and calculated values agree reasonably well. For $B > 0$ the values of τ_{exp} are larger than those of τ_{th}. The reason for this is a transport of energy through the solid-liquid interface in the experiment not accounted for in the linear calculations. This dissipative process delays the traveling velocity and increases the measured periods.

5. CONCLUSION

In this paper the influence on the onset of convection in a Bénard problem of a deformable upper boundary of finite conductivity is studied. In the experiments a water-ethanol mixture with a large, negative thermal diffusion coefficient is used. A state of traveling waves is observed, which exists only as a transient. This transient state is replaced by a steady state of a hexagonal convection pattern. The critical Rayleigh numbers and the corresponding time periods of the traveling waves at the onset of convection decrease with increasing thickness of the ice-layer. This behaviour can be described by a linear theory. The quantitative differences can be explained by physical arguments.

6. REFERENCES

1. G. Zimmermann, U. Müller, S.H. Davis, Bénard Convection in a Partly Solidified Two-Component System, KfK report 4122, Karlsruhe (1986).

2. G. Zimmermann, Bénard-Konvektion in binären Flüssigkeitsmischungen mit Thermodiffusion, Dissertation, Universität Karlsruhe (1990), to be published.

3. M.R. Scott, H.A. Watts, Subroutine SUPORE; Appl. Math. Division 2623, Sandia Laboratories, P.O. Box 5800, Albuquerque, NM (1979).

4. K. Bühler, K.R. Kirchartz, J. Srulijes, Anwendungen der Differentialinterferometrie bei thermischen Konvektionsströmungen, in: "Applied Fluid Mechanics", H. Oertel, jr., ed., Karlsruhe (1978).

5. S.H. Davis, U. Müller, C. Dietsche, Pattern Selection in Single-Component Systems Coupling Bénard Convection and Solidification; J. Fluid Mech. 144: 133 (1984).

PHASE-MEAN DRIFT EQUATION FOR CONVECTION PATTERNS IN LARGE ASPECT RATIO CONTAINERS

A. C. Newell, T. Passot and M. Souli

Arizona Center for Mathematical Sciences
The University of Arizona
Tucson, AZ 85721

ABSTRACT: We present the phase diffusion and mean drift equation which describe the convection pattern in large aspect ratio containers for arbitrary large Rayleigh numbers. An exact agreement is found with the borders of the Busse balloon, concerning the long wavelength instabilities. We propose a calculation of the selected wavenumber which agrees closely with experiments and we predict a new instability which appears to be important in initiating time dependence. We predict also the Rayleigh numbers at which loss of spatial correlation due to global defect nucleation will occur.

I. INTRODUCTION

Many extended physical systems are correctly described by non-linear partial differential equations, of which very few can be exactly solved. Nevertheless, many interesting behaviours can be captured by standard perturbative techniques, when the stress parameter is small and/or there exists a spectral gap between the microscopic and the macroscopic scales. In this latter case falls the important subject of pattern forming systems which is our interest here, more particularly in the case of Rayleigh-Bénard convection. In order to replace our work among others, we will first briefly mention some of the analytical approaches which have been used in this context to extract informations from these non-linear PDE's.

When the stress parameter (here the deviation from the critical Rayleigh number) is small, it can serve as an expansion parameter for deriving perturbatively around a trivial basic state (here the conductive state) a weakly non-linear solution of the PDE (here the Oberbeck-Boussinesq (hereafter O.B.) equations). These solutions are found after solving "Stuart-Watson-Landau-type" amplitude equations [1], and they capture the time evolution and eventually the non-linear saturation of the order parameters (coefficients of the critical modes). This reduction to a dynamical system of finite order is particularly suitable when the physical phenomena do not involve many spatial scales, as this is the case for convection in order 1 aspect ratio boxes. It was particularly successful in studying the stability and competition of different planform

Nonlinear Evolution of Spatio-Temporal Structures in Dissipative Continuous Systems
Edited by F.H. Busse and L. Kramer
Plenum Press, New York, 1990

configurations (rolls, hexagons, ...) [2], and, in the slightly different context of truncated Galerkin expansions [3], it served also to illustrate transition to chaos.

When, in contrast, one deals with large aspect ratio containers, this approach will be usually inadequate. Even though periodic solutions exist and are stable against small scale perturbations (of the order of the depth of the container), such as rolls in the vertically symmetric case, they can still undergo instabilities to long wavelength spatial disturbances not described by the previous ODE's. This led Newell and Whitehead [4], and Segel [5], to use a multiple scale technique in order to derive a PDE (envelope equation) for the slowly varying complex amplitude A of almost parallel rolls; (A multimode approach is also possible). It reads:

$$\frac{\partial A}{\partial T} - \left(\frac{\partial}{\partial X} - \frac{i}{2k_c}\frac{\partial^2}{\partial Y^2}\right)^2 A = A - A^2A^* \tag{1.1}$$

This equation contains two long wavelength instabilities of simple roll solutions, found earlier by Busse [6]: the zig-zag and, Eckhaus [7], the Eckhaus instabilities. In order to capture the skewed-varicose instability, it is necessary to include an important effect missed in the previous equation, namely the coupling to the mean-drift velocity field found by Siggia and Zippelius [8].

When there is no privileged direction, as in natural experiments, this latter equation however fails to correctly describe the pattern which then consists of rolls of slowly varying orientation, possibly connected by defects such as grain Boundaries, disclinations, ... This reason led Cross and Newell [9] to use a different strategy, that they have illustrated on various models and that we carry-out for the O.B. equations in this paper. One wishes here to describe these complicated patterns after averaging over the small scales of the rolls, namely the period and the vertical direction. Let us here briefly mention the idea of the method.

As in the homogeneization technique [10] or in the theory of modulated train waves of Whitham [11], use is made of the inverse aspect ratio as an expansion parameter, and not of the stress parameter as previously. This relies on the above mentioned observation that the wavevector $\vec{k}$ changes slowly, that is over distances comparable to the box size L (except at defects) and renders possible the study at large Rayleigh numbers. The calculation is made locally, thus allowing one to preserve the original rotational degeneracy, around the stable fully non-linear straight parallel roll solution of the O.B. equations. The existence of such solution is assured by the previous work of Busse and colleagues [6], [12-17]. The full solution can be written

$$\vec{v}(\vec{x}, z, t) = \vec{f}(\theta, z, A) + \epsilon\, \vec{v}_1 + \epsilon^2\, \vec{v}_2 + \dots$$

where

$$\vec{v} = (u, v, w, T, p)$$

is the vector consisting of the velocity $\vec{u} = (u, v, w)$, the temperature T and the pressure p. The basic solution f is 2π-periodic in the phase θ and also depends on the vertical coordinate z, and the amplitude of the roll A. The assumption that the wavenumber $\vec{k} = \nabla_{\vec{x}}\theta$ is slowly varying can be written:

$$\theta = \frac{1}{\epsilon}\,\Theta(X, Y, T) = \frac{1}{\epsilon}\int \vec{k}(X, Y, T)\, d\vec{x}. \tag{1.2}$$

where $X = \epsilon x$; $Y = \epsilon y$; $T = \epsilon^2 t$ are the scales of the long wavelenglth disturbances. Note at this point that in contrast to the case of the Newell-Whitehead-Segel (hereafter N.W.S) equations, direction Y is here scaled in the same way as the direction X, since we want to preserve the rotational invariance. The iterates $\vec{v}_1$, $\vec{v}_2$, ... are calculated by linearizing around $\vec{f}(\theta, z, A)$. This leads to inhomogeneous singular linear systems of equations whose solvability condition provide the equations for the slowly varying wavevector $\vec{k}$ and a depth averaged mean drift velocity field $\vec{V}$. An important remark has to be made concerning the amplitude A. The constraint at leading order, when looking for the basic solution, is that $\vec{f}(\theta, z, A)$ be periodic in θ. This gives an algebraic relation between A and the wavenumber k. Thus in this approach, far from onset, the pattern turns out to be described (aside from the mean drift) by a simple scalar, the phase $\Theta(X, Y, T)$ because the amplitude is slaved to the wavenumber $k = |\nabla\Theta|$. This is however only valid when A is of order one. Near onset, the amplitude becomes small and for consistency the basic solution has to be calculated perturbatively. The equations have then to be linearized around the null solution and the linear operator obtained has an additional null vector. The consequence of this mathematical fact is that A becomes an active order parameter satisfying a PDE, rather than being passively slaved to k. In addition, when the wavenumber of the parallel rolls becomes close to the left boundary of the Busse balloon (see Figure 2) which it does both at high Prandtl numbers for finite Rayleigh numbers and at all Prandtl numbers for Rayleigh numbers near critical, the rolls lose their resistance to lateral bending and as a consequence another length scale of order $\sqrt{\epsilon}$ has to be introduced in this direction. Other terms proportional to Θ_{YYYY} become relevant in the phase equation. Recombining the equations for the phase Θ and the amplitude A, in terms of a single complex field $w = A\, e^{i\frac{\Theta}{\epsilon}}$ one recovers the N.W.S. equation for low amplitude convection.

In Section 2 we give the mathematical formulation and derivation of the phase-mean-drift equations. Section 3 discusses the results and shows the agreement with experimental observations, and predicts a new instability for circular target patterns which is important in initiating time dependence. It also predicts the Rayleigh numbers at which loss of spatial correlation due to global defect nucleation will occur. Finally, Section 4 is a brief conclusion which particularly discusses the possibility of extending these equations in order to describe the formation of defects.

II. DERIVATION OF THE EQUATIONS

The numerical calculation of the coefficients of the phase equation has been performed on the equation for the vorticity, thus eliminating the pressure for numerical conveniences. Nevertheless, we will present here the derivation based on the momentum formulation since it appears simpler. The consistency between both formulations has also been checked numerically.

We start from the O.B. equations with rigid-rigid boundary conditions, which read:

$$\sigma(\partial_t \vec{u} + (\vec{u}\nabla)\vec{u}) = -\nabla p + T\,\vec{z} + \nabla^2 \vec{u}. \tag{2.1}$$

$$\partial_t T + \vec{u}\,\nabla T = Rw + \nabla^2 T \tag{2.2}$$

$$\nabla \cdot \vec{u} = 0. \tag{2.3}$$

The velocity $\vec{u}$ has components (u, v, w), R is the Rayleigh number and σ the inverse Prandtl number. We have $\vec{u} = T = 0$ at $z = \pm\frac{1}{2}$. We first construct the periodic solution $\vec{v} = \vec{f}(\theta, z, A)$ by a Galerkin scheme, following Busse, and write:

$$u = \sum_{m,n} u_{mn}\, e^{im\theta}\, g_n'(z)$$

along with similar expressions for w, T, p replacing $g_n'(z)$ respectively by $g_n(z)$, $f_n(z)$ and $h_n(z)$. The summations are over $1 \le n \le N$; $-M \le m \le M$. The $g_n(z)$ are given in [16], $f_n(z)$ and $h_n(z)$ are simply sines and cosines. The pressure p is determined here up to a constant p_s. Reality and symmetry conditions require u_{mn} to be pure imaginary and w_{mn}, T_{mn}, p_{mn}, real; $u_{mn} = i\,\hat{U}_{mn}$ with $U_{-m,n} = U_{m,n}$ so that continuity equation simply reads:

$$w_{mn} = k_m\, U_{mn}. \tag{2.4}$$

Substituting these expansions into equations (2.1-2.3) gives us a system of non-linear algebraic equations for the Galerkin coefficients, obtained by projecting the horizontal momentum, the vertical momentum and heat equation onto the bases functions $e^{im\theta} g_n'(z)$; $e^{im\theta} g_n(z)$ and $e^{im\theta} f_n(z)$ respectively. The solution of these equations, for which all coefficients whose indices m and n sum to an odd number are zero, are the finite amplitude roll solutions of interest. For each R and σ, we solve for a set of values of k spanning the interval contained inside the marginal stability curve, using a Newton's method. We start at the right border, initializing the first iterate by an approximated solution, calculated perturbatively in power of the amplitude. Then, for other values of k, the preceeding non-linear solution is used as the guess. Convergence is checked by looking at the value of the Nusselt number, and we compared also with values found in [16]. Since derivatives with respect to k will be needed, we interpolate the field using cubic spline polynomials in k.

Now, as outlined in the introduction, we look for modulated solutions and solve the linear equations obtained by inserting in (2.1-2.3) the following expansion in power of the inverse aspect ratio ϵ:

$$\partial_z \to \partial_z$$

$$\nabla_{\vec{x}} \to \vec{k}\,\frac{\partial}{\partial\theta} + \epsilon\,\nabla_{\vec{X}}$$

$$\frac{\partial}{\partial t} \to \epsilon\,\Theta_T\,\frac{\partial}{\partial\theta} + \epsilon^2\,\frac{\partial}{\partial T}$$

$$\nabla^2 \to k^2\,\frac{\partial}{\partial\theta^2} + \frac{\partial^2}{\partial z^2} + \epsilon\,D\,\frac{\partial}{\partial\theta} + \epsilon^2\nabla_X^2$$

$$D = 2\vec{k}\,\nabla_{\vec{X}} + \nabla_{\vec{X}}\cdot\vec{k}.$$

We also introduce cross roll and along the roll horizontal velocities:

$$\begin{cases}\tilde{u} = \hat{k}\cdot\vec{u}\\ \tilde{v} = (\hat{k}\times\vec{u})\vec{z}\end{cases}$$

where

$$\hat{k} = \frac{\vec{k}}{k}.$$

We end up, at order ϵ and ϵ^2, with linear systems of the form M $\vec{\tilde{v}} = \vec{g}_j$ (j = 1, 2), where M is the operator obtained by linearizing (2.1-2.3) about $\vec{\tilde{v}} = \vec{\tilde{f}}$. This operator M is singular, since due to symmetries, $\frac{\partial\vec{\tilde{f}}}{\partial\theta}$ and $\frac{\partial\vec{\tilde{f}}}{\partial p}$ are elements of its kernel. When solving for $\vec{\tilde{v}}_1$ and $\vec{\tilde{v}}_2$, solvability conditions will then arise which will provide the phase diffusion and mean-drift equation respectively. Since this system is non self-adjoint, we must also solve for the homogeneous adjoint boundary value problem. One of its solution, associated with the mass conservation is $\vec{\tilde{v}}_A = (0, 0, 0, 0, 1)$. The corresponding solvability condition is satisfied for j = 1 but becomes non-trivial for j = 2, because the perturbed field u_1 and v_1 pick up non-periodic components due to slow horizontal Reynolds stresses:

$$\sigma\,\partial_X\,u_0^2 + \sigma\,\partial_Y(u_0v_0) \quad\text{and}\quad \sigma\,\partial_X(u_0v_0) + \sigma\,\partial_Y(v_0^2).$$

The inclusion of the slowly varying pressure p_S whose gradients enter when we solve for the velocity fields at order ϵ is necessary in order that the mean (over θ) components of the induced fields satisfy continuity. It turns out to be convenient to eliminate p_S by writing the final equations in a form which directly includes the mean drift velocity field represented by a stream function ψ. The system is thus closed but its resolution requires the exact calculation of the fluctuating part of $\tilde{u}_1$, $\tilde{v}_1$. No approximations obtained by averaging the momentum equation over θ and z will suffice because the vertical (z) structure of the induced mean-drift field is not trivial, and especially for low Prandtl numbers, is not well approximated by a Poiseuille-like profile. We have to use an extremely robust inversion method to solve these singular equations and we found out that the computation of the generalized inverse obtained through applying a singular value decomposition to the singular matrix M is ideally suited for this purpose. It is indeed not as sensitive to small errors as an usual computation of eigenvalues and moreover it saves computing time. For some cases a poor conditioning of the matrix M has to be taken care of by eliminating the incompressibility equation using (2.4). In this case, removing secular terms is first necessary before eliminating this equation.

Let us now write down the equations obtained after eliminating p_S between the two solvability conditions at j = 1 and j = 2, using the second one which expresses the incompressibility condition on $\vec{u}_1$ to introduce a stream function ψ. On thus write:

$$\langle \vec{u} \rangle = (\hat{k} \times \nabla \psi) . \vec{z}$$

where $\langle\ \rangle$ means averaging over θ and $\vec{z}$. The equations reads:

$$\Omega(A, k, R) = 0, \tag{2.5}$$

$$\Theta_T + \rho(k)\, \vec{V} \cdot \nabla \Theta + \frac{1}{\tau(k)} \nabla \cdot \vec{k}\, B(k) + 0(\epsilon) = 0, \tag{2.6}$$

$$\hat{z} \cdot \nabla \times \hat{k}\, \alpha(k)\, (\hat{k} \times \nabla \psi) \cdot \hat{z} - \nabla \cdot \hat{k}\, \beta(k)\, (\hat{k} \cdot \nabla \psi)$$

$$= \hat{z} \cdot \nabla \times \left(\sigma\, \hat{k} \nabla \cdot \vec{k} A^2 - \frac{\hat{k}}{\tau_\alpha(k)} \nabla \cdot \vec{k} B_\alpha(k) \right)$$

$$- \nabla \cdot \hat{k} \left(\nabla \times \vec{k} B_\beta(k) \right) \cdot \hat{z} + 0(\epsilon), \tag{2.7}$$

where $\vec{k} = \nabla \Theta$, $\vec{V} = \nabla \times \psi\, \hat{z}$, the quantities $\rho(k)$, $B(k)$, $\tau(k)$, $\alpha(k)$, $\beta(k)$, $B_\alpha(k)$, $\tau_\alpha(k)$ $B_\beta(k)$ are all functions of k which are explicitly calculated. Their graphs are shown in Figure 1.

The first equation expresses the periodicity condition of the basic roll solution and gives the amplitude A as a function of the wavenumber k and the Rayleigh number R. In the mean-drift equation, the first term on the right-hand side comes from the horizontal Reynolds stress whereas the second and third terms essentially come from the vertical Reynolds stresses and the complicated vertical structure of the mean-drift field. The across the roll and along the roll components have different vertical profiles. The Cross-Newell formula is recovered by taking $\alpha = \beta$ and ignoring B_α and B_β. The appropriate boundary conditions are that ψ is a streamline and that the roll axis is perpendicular to the boundary: $\hat{k} \cdot \hat{n} = 0$. However, for the case of circular target pattern, where a lateral thermal forcing is present, the last condition is replaced by the fact that the boundary is also a phase contour i.e. $\vec{k} \times \hat{n} = 0$. It is interesting to note that these questions are translationally and rotationally invariant and also Galilean invariant, even though the original rigid-rigid boundary value problem is not.

In the next section, we will give the first interesting comparisons which have been made between experimental results and stability analysis of this equation.

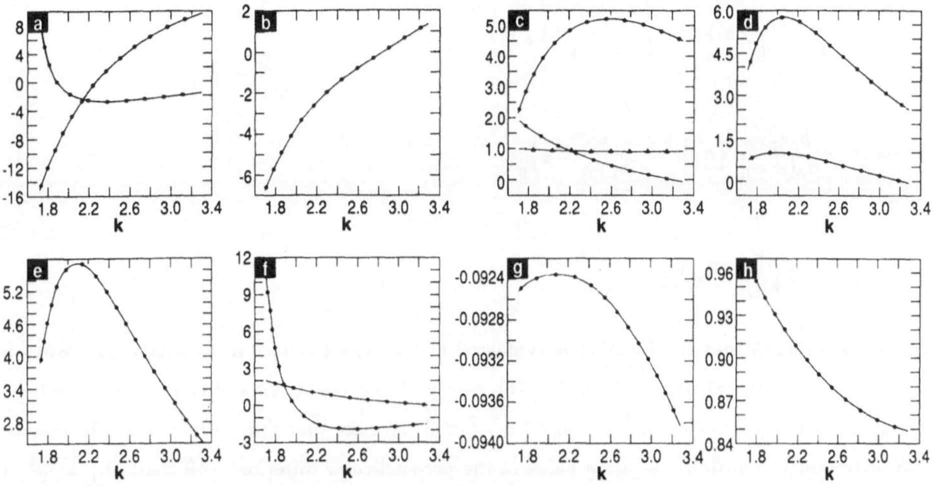

FIGURE 1. Graphs of (a) $D_{||}(b_1 + b_3)$ (marked with circles), a-1 (marked with squares) (b) $D_{||}\, b_3$, (c) $D_{||}\, b_2$ (circles), $-$ a (squares), c (triangles), (d) kB(k) circles, $A^2 = \frac{\langle \tilde{u}_0^2 \rangle}{k^2}$ (squares), (e) $A^2 = \frac{\langle \tilde{u}_0^2 \rangle}{k^2}$ on an expanded scale, (f) $D_{||}$ (circles), $D_{\perp}$ (squares), (g) ρ, (h) c, and (i) for R = 3000, P = 0.71.

III. SOME RESULTS

The first analysis we carried out on these equations is the study of the linear stability of straight parallel rolls:

$$\Theta_0 = k_0 X; \; \psi_0 = 0.$$

Setting $\Theta = \Theta_0 + \Phi\,(X, Y, T)$; $\psi = \Psi(X, Y, T)$ in (2.5-2.7), linearizing the resulting equation and substituting:

$$(\Phi, \Psi) = (\hat{\Phi}, \hat{\Psi}) \; \exp(i\, K_X X + i\, K_Y Y + \gamma T),$$

we obtain the dispersion relation

$$\frac{\gamma}{K^2}\left(\frac{-\tau_0}{(kB)_0}\right) = -1 - (a\text{-}1)S + \frac{S}{1+(c\text{-}1)S}\left(b_1 + b_3 - (b_1+b_2+b_3)S\right) \tag{3.1}$$

where $K^2 = K_X^2 + K_Y^2$ $S = K_Y^2/K^2$, and:

$$a = \frac{B_0}{(kB)'_0} = \frac{D_{\perp}}{D_{||}}$$

$$c = \frac{\alpha_0}{\beta_0}$$

$$b_1 = \frac{\rho_0 k_0^2 \tau_0}{\beta_0(-kB)_0'} \left\{ \sigma(kA^2)_0' - \frac{1}{k_0 \tau_\alpha}(kB_\alpha)_0' \right\}$$

$$b_2 = -\frac{\rho_0 k_0^2 \tau_0}{\beta_0(-kB)_0'} \left\{ \sigma\, A^2 - \frac{1}{k_0 \tau_\alpha}\, B_\alpha \right\}_0$$

$$b_3 = -\frac{\rho_0 k_0^2 \tau_0}{\beta_0(-kB)_0'} (B'_\beta)_0.$$

The index 0 means that the function is evaluated at k_0. The Eckhaus instability occurs when S = 0 (no Y dependence), for $\frac{(kB_0')}{\tau_0} > 0$. The purely Y dependent instability (S = 1) called the zig-zag instability occurs when $\frac{B_0}{\tau_0} + \rho_0\, k_0^2\, \sigma\, A_0^2 > 0$. We see that finite Prandtl number mean-drift effects require a finite negative value of the perpendicular diffusion coefficient $D_\perp = \frac{-B}{\tau}$ in order for the instability to be triggered. The skewed-varicose instability occurs for a value of S, $0 < S < 1$. Approximating c by 1 (which is a very good at moderate Rayleigh and Prandtl numbers), a necessary condition for the existence of skewed varicose instability is: $b_1 + b_3 > a - 1$. A summary of these results are plotted in Figure 2 where we can see that the borders of the Busse balloon corresponding to long-wavelength instabilities are reproduced to within an accuracy of less than one percent. This agreement serves as a non-trivial check of our theory.

Because of the influence of sidewalls boundaries, patches of circular rolls tend to be the dominant pattern convection in large aspect ratio boxes, and since we think that curvature of the rolls is predominant in both selecting the wavenumber and triggering new kind of instabilities, we investigate here the stability of circular target patterns. This is a simple model but which has the advantage to have been reproduced in laboratory by Steinberg, Ahlers, and Cannell [18]. There is therefore a very good possibility to check our calculation and conjecture with a real experiment.

For an exactly circular pattern (2.2) becomes

$$\frac{\partial \Theta}{\partial T} + \frac{1}{\tau r} \frac{\partial}{\partial r} (rkB) = 0 \tag{3.2}$$

when r is the radius in polar coordinates. A stationary assumption, together with the fact that (3.2) has to hold for every r, leads to $B(k) = 0$. The wavenumber k_B such that $B(k_B) = 0$ is then the one selected by the pattern. If the circular pattern is allowed to be time dependent, we can see that if $k < k_B$ then the focus is acting as a source of new rolls. On the other hand, for $k > k_B$ the umbilicus acts as a sink. Carrying out a linear stability analysis around such a circular patch by setting $\Theta(r, T) = k_0 r + k_0\, D(r, t) \sin m\theta$ and $\frac{-\rho_0 \tau_0}{-B_0'}\, \psi(r, T) = \phi(r, t) \cos m\theta$, where $r = \sqrt{X^2 + Y^2}$ and $\tan\theta = \frac{Y}{X}$, we find after rescaling the time and choosing $k_0 = k_B$:

$$D_t' - D_{rr}' - \frac{1}{r} D_r' + \frac{1}{r^2} D' = m\left(\frac{\phi}{r}\right)_r \tag{3.3}$$

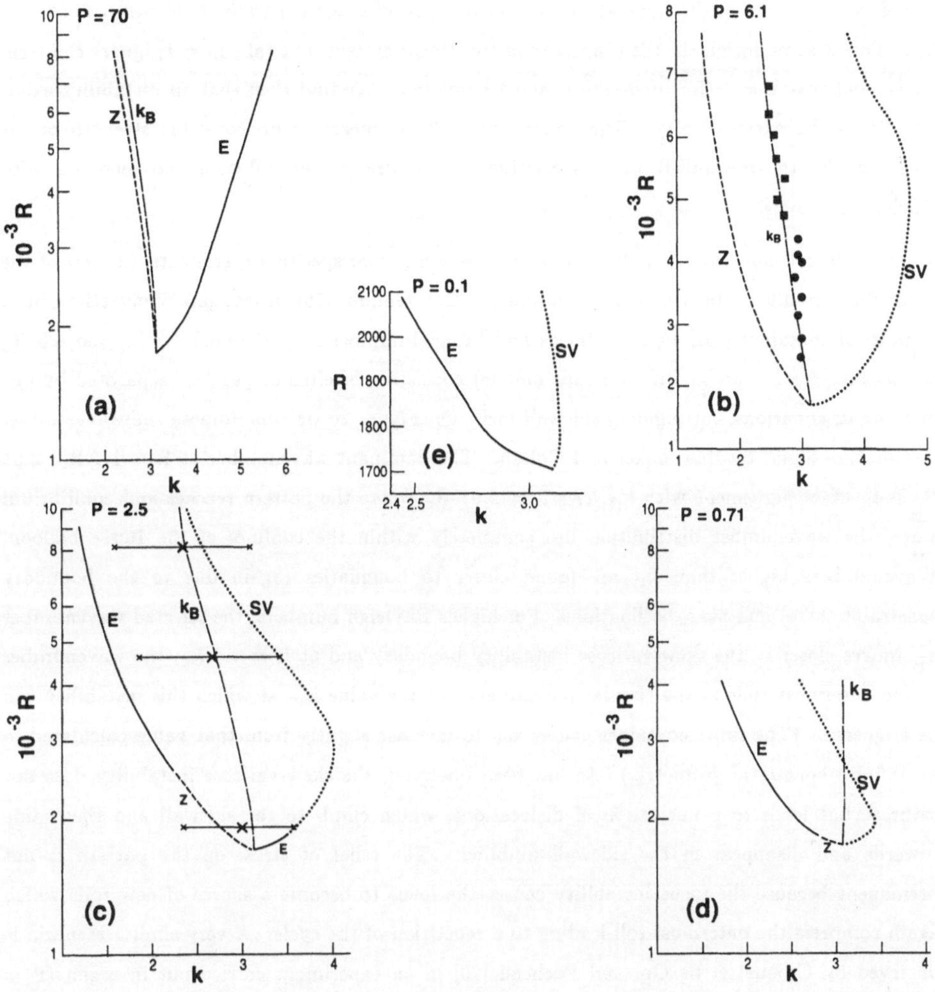

FIGURE 2. The long wave borders of the Busse balloon marked E (Eckhaus), and Z(zig-zag) and SV (skew-varicose) and the zeros k_B of B(k) as calculated from (2.5-2.7) for Prandtl number 70(2a), 6.1(2b), 2.5(2c), 0.71(2d) and 0.1(2e) respectively. The agreement with the calculations of Busse for the E, Z and SV borders is so good that the two sets of curves superimpose exactly. In Figure (2b) are shown the wavenumbers (as measured by $2\pi/\lambda$ where λ is the width of the second pair of rolls from the wall) of the $3\frac{1}{2}$(•), 3(■), roll equilibrium states as functions of Rayleigh number as the Rayleigh number is increased (taken from Steinberg, Ahlers and Cannell [18]). In Figure 1c the maxima (marked with ×) and the support of the wavenumber distribution are superimposed (taken from Heutmaker and Gollub [19]).

$$\phi_{rr} + \frac{1}{r}\,\phi_r - \frac{m^2 c}{r} = m(b_1+b_3)\,\frac{(rD')r}{r^2} - m\,b_3\,\frac{D'}{r^2} + m\,(m^2-1)b_2\,\frac{D}{r^3} \tag{3.4}$$

where the constant are the same as in the previous case of straight parallel rolls but evaluated at k_B. To get some insight in the properties of this linear system let's take $m = 1$, ignore the term $b_3 \frac{D'}{r^2}$ and take $c = 1$ (for moderate Prandtl numbers). We find then that an instability exists when $b_1 + b_3$ exceeds unity. The parameter $b_1 + b_3$ measures precisely the strength of the advection by the mean-drift field in relation to the strength of diffusion and increases with Rayleigh numbers.

We will now interpret the observations made in two specific experiments, in view of the preceeding results. In the first, Heutmaker and Gollub [19] investigate convection in a cylindrical container with aspect ratio 14 and Prandtl number 2.5. Between 1.2 R_c and 4.5 R_c the pattern reaches an equilibrium (not unique) consisting of circular patches, separated by one or more disclinations, surrounding sidewall foci. On Figure 2c we superimpose the wavenumber distribution found by Heutmaker and Gollub. The dominant wavenumber at $R = 1.1\ R_c$, 2.61 R_c is in exact agreement with k_B. At $R = 2.61\ R_c$ where the pattern reaches and equilibrium state, the wavenumber distribution lies completely within the confines of the Busse balloon. Wavenumbers higher than k_B are found closer to boundaries (again due to the boundary constraint there) and near disclinations. For higher Rayleigh numbers, the selected wavenumber k_B moves closer to the skew-varicose instability boundary and at $R = 4.5\ R_c$, the wavenumber of the outermost rolls in the circular patches exceeds the value k_{sv} at which this instability can be triggered. (The finite container causes k_{sv} to increase slightly from that value calculated in an infinite horizontal geometry.) As has been observed, the skew-varicose instability does not saturate but leads to a nucleation of dislocations, which climb to the sidewall and then glide towards and disappear in the sidewall umbilici. The relief of stress on the pattern is not permanent because the focus instability causes the focus to become a source of new rolls which again compress the outermost roll leading to a repetition of the cycle. A very similar scenario is observed by Croquette, Le Gal and Pocheau [20] in an experiment carried out in argon ($P = 0.7$) in a cylindrical container with aspect ratios 7 and 20. Note that our theory shows that the onset of time dependence is not due to k_B crossing the skew varicose boundary as had been conjectured. This crossing takes place until 7 R_c. At this point, we would suggest that dislocations would be nucleated all over the container and that the spatial coherence will be lost.

Steinberg, Ahlers and Cannell [18] investigated circular target patterns in a cylindrical container with aspect ratio of about 3 at a Prandtl number of 6.1. They initiated the target pattern by thermally forcing the sidewalls but found that, once initiated, the target patterns did not require the sidewall forcing to be sustained. They found that as the Rayleigh number increased, the pattern wavenumber increased from $3\frac{1}{2}$ to 3 to $2\frac{1}{2}$. The selected wavenumber is superposed on Figure 2b and agrees very closely with our calculated value. Moreover, we would predice that at about $R_{TP} = 3.5\ R_c$, the 3 roll solution destabilizes to a perturbation which breaks the circular symmetry and the circular pattern shown in Figure 3a deforms towards that in Figure 3b.

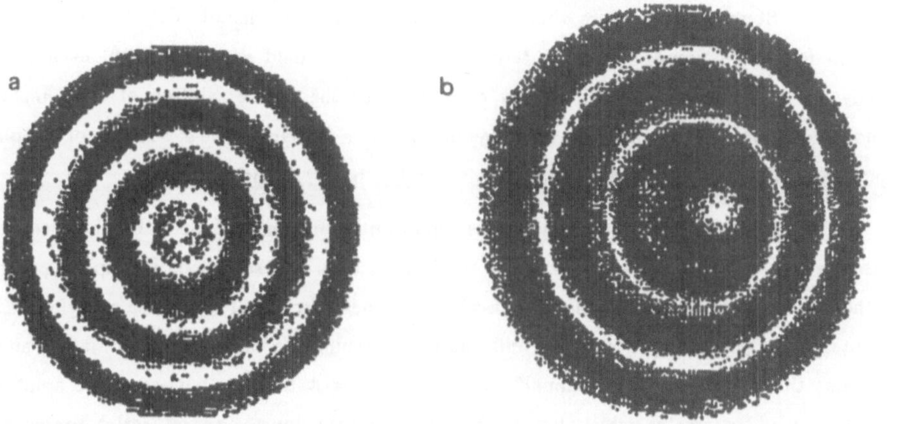

FIGURE 3. Equilibrium patterns taken from Steinberg, Ahlers and Cannell [18] at P = 6.1, Γ = 6 at two several Rayleigh numbers (a), a stable three roll state at $R = 3\ R_c$, (b) its distortion at $R = 4\ R_c$.

The amplitude of the distortion is a monotonically increasing function $R\text{-}R_{TP}$ and when the roll compression is sufficient, a dislocation is nucleated, which glides to the umbilicus thereby removing a roll and then the pattern relaxes to a circular target pattern again, this time with $2\frac{1}{2}$ rolls. The transitions are hysteretic, again consistent with our theory.

IV. CONCLUSION

In this paper, as a first step, we have derived the phase diffusion and mean-drift equation, assuming that the pattern is locally periodic, tested their prediction against known theoretical results, made some new predictions and compared them with experimental evidence. We want in this section briefly describe the next challenges and obstacles to be overcame in order to describe more precisely these complicated convection patterns.

The present theory appears at first sight much more general than the N.W.S. equation since it preserves rotational invariance and is not restricted to be close to onset. However, as already mentioned in the introduction, the equation obtained governs the evolution of the single phase Θ. The order parameter A is slaved to the wavenumber k through an algebraic equation. This has some important consequences, one of them being that solutions to these equations can lose regularity. When the wavenumber k goes outside the non-linear stability region, these equations take the form of a non-linear reverse heat equation. Even when introducing ϵ^2 order correction terms as in [9], which basically take the form of a bilaplacian of the phase, the phase equation blows up (for example, during the non-linear development of a skewed varicose instability, in regions where a necking of the phase contours, induced by the mean-drift velocity field, gives rise to the formation of a pair of dislocions). This point has been tested with direct numerical formulations of (2.5-2.7) using a Fourier-Collocation method. It is important here to stress that neither the phase equation nor the N.W.S. equation are meant to be valid in such situations. However, it appears that the N.W.S. equation can handle defect formation, the solution found remaining smooth in terms of the complex field $w = A \exp(i\theta)$ even though locally the wavenumber goes to infinity exactly as for the vortex solution of the Ginzburg-Landau equation. In contrast here, it is even not possible to define a weak solution of the phase equation as it stands. What has to be done is sketched briefly in the following.

In regions where the amplitude gets significantly smaller than one, the phase equation ceases to be valid, as already mentioned in the introduction. These regions correspond precisely to the locations of defect nucleation. A weakly non-linear version of the phase equation has to be obtained by perturbing around the null solution. Keeping the Cross-Newell formalism and at the same time expanding the "eikonal" equation (relating A to k) in powers of the amplitude (another small parameter which has to be related to the inverse aspect ratio) provides an intermediate formalism between the N.W.S. and the phase equation, which combines both the advantage of having a separate equation for the amplitude and allowing the wavevector $\vec{k}$ to point in any direction. The equations obtained can be, in simple cases, put in a form which resembles the "liquid crystal" equation:

$$\partial_T w = \Delta w + w - w^* w^2 \tag{4.1}$$

once the amplitude A and the phase Θ have been recombined into a single complex field $w = A \exp(i\Theta/\epsilon)$. This latter equation describe also defect solutions [21] the field w remaining smooth.

The project which is in progress consists in matching the two approaches in order to describe the pattern by the non-linear phase equation in regions of large amplitude and slowly varying wavevector, and by an equation of the latter type where the amplitude gets smaller and particularly where defects are found.

ACKNOWLEDGEMENTS

The authors wish to thank the Arizona Center for Mathematical Sciences (ACMS) for support. ACMS is sponsored by contracts AFS0SR-90 00021 and FQ 8671-9000589 with the University Research Initiative Program at the University of Arizona. Computations were made at the NSF Pittsburgh Supercomputing Center on a CRAY YMP.

REFERENCES

[1] Stuart, J. T., "On the nonlinear mechanics of wave disturbances in stable and unstable parallel flows. I. The basic behavior in plane Poiseuille flow," J. Fluid Mech., **9**, 353 (1960).

[2] Malkus, W. V. R. and Veronis, G., "Finite amplitude convection," J. Fluid Mech., 4, 225 (1958).

Schulter, A., Lortz, D. and Busse, F. H., "On the stability of steady finite amplitude convection," J. Fluid Mech., **23**, 129 (1965).

[3] Lorentz, E. N., "Deterministic non-periodic flow," Journal of Atmospheric Sciences, 20, 130 (1963).

[4] Newell, A. C. and Whitehead, J. A.,"Finite Bandwidth, Finite Amplitude Convection," J. Fluid Mech., **38**, 179 (1969).

[5] Segel, L. A., 1969, "Distant side-walls cause slow amplitude modulation of cellular convection," J. Fluid Mech., **38**, 203 (1969).

[6] Busse, F. H., "On the stability of two-dimensional convection in a layer heated from below," J. Math. Phys., **46**, 149-150 (1967).

[7] Eckhaus, A. W., Studies in Nonlinear Stability Theory, (New York: Springer-Verlag) (1965).

[8] Siggia, E. D. and Zippelius, A., "Pattern selection in Rayleigh-Bénard convection near threshold,:" Phys. Rev. Lett., **47**, 835 (1981b).

[9] Cross, M. C. and Newell, A. C., "Convection Patterns in Large Aspect Ratio Systems," Physica, 10D, 299-328 (1984).

[10] Bensoussan A., Lions, J. L. and Papanicolaou, G., "Asymptotic Analysis for Periodic Structures," Studies in Mathematics and its Applications, Vol. 5, North-Holland (1978).

[11] Whitham, G. B., "Linear and Nonlinear Waves," Wiley-Interscience (1974).

[12] Busse, F. H. and Whitehead, J. A., "Instabilities of convection rolls in a high Prandtl number fluid," J. Fluid Mech., 47, 305-320 (1971).

[13] Busse, F. H. and Whitehead, J. A. "Oscillatory and collective instabilities in large Prandtl number convection," J. Fluid Mech., **66**, 67-79 (1974).

[14] Busse, F. H., "Nonlinear properties of convection," Rep. Prog. Phys., 41, 1929-1967 (1978).

[15] Busse, F. H., "Transition to turbulence in Rayleigh-Bénard convection," Hydrodynamic Instabilities and the Transition to Turbulence, edited by H. L. Swinney and J. P. Gollub (Berlin: Springer-Verlag), 97 (1981).

[16] Clever, R. M. and Busse, F. H., "Transition to time dependent convection," J. Fluid Mech., **65**, 625-645 (1974).

Clever, R. M. and Busse, F. H., "Instabilities of convection rolls in a fluid of moderate Prandtl number," J. Fluid Mech., **91**, 319-335 (1979).

[17] Clever, R. M. and Busse, F. H., "Large wavelength convection rolls in low Prandtl number fluid," J. Appl. Math. Phys. Z. angew. math. Phys., **29**, 711-714 (1978).

[18] Steinberg, V., Ahlers, G., Cannell, D. S., "Pattern formation and wavenumber selection by Rayleigh-Bénard convection in a cylindrical container," Physica Scripta, **T13**, 135 (1985).

[19] Heutmaker, M. S. and Gollub J. P., "Wave-vector field of convective flow patterns," Phys. Rev. A, **35**, 242 (1987).

[20] Croquette, V., Le Gal, P. and Pocheau A., Physica Scripta, **T13**, 135 (1986).

[21] Bodebschatz, E., Pesch, W., and Kramer, L., "Structure and dynamics of dislocations in anisotropic pattern forming systems," Physica D, **32**, 135 (1988).

MODULATED TRAVELING WAVES IN NONEQUILIBRIUM SYSTEMS

M. Bestehorn, R. Friedrich, and H. Haken

Institut für Theoretische Physik und Synergetik
Universität Stuttgart
Pfaffenwaldring 57 / IV
D - 7000 Stuttgart 80

I. INTRODUCTION

Time periodic behaviour may arise in systems far from equilibrium due to an instability of a stationary state. If such systems are additionally able to produce spatial patterns a rich variety of phenomena occurs which have recently attracted experimental as well as theoretical interest. Experimental systems under consideration are the Taylor-Couette experiment with counter rotating cylinders [1], convection in binary fluid mixtures [2], as well as higher instabilities arising in the Bénard experiment [3]. Since interesting spatio-temporal behaviour occurs already close to instability the description of the system can be formulated in terms of the synergetic concepts of order parameters and their dynamics [4] because other degrees of freedom of the systems are enslaved. The present paper gives an overview over theoretical results which have been obtained by an examination of the generalized Ginzburg-Landau equation which describes the behaviour of the system close to onset in terms of a suitably defined order parameter [5]. Section II presents this generalized Ginzburg-Landau equation. Section III deals with one-dimensional traveling wave patterns and indicates a mechanism by which modulated traveling waves are generated. Section IV is devoted to two-dimensional traveling wave patterns.

II. THE ORDER PARAMETER EQUATION

The system under consideration is mathematically described by a state vector $\underline{q}(\underline{r},t)$ which obeys an evolution equation of the general form

$$\partial_t \underline{q}(\underline{r},t) = \underline{N}[\underline{q}(\underline{r},t),\underline{\sigma}]$$
$$= L[\underline{\sigma}]\ \underline{q}(\underline{r},t) + \Gamma : \underline{q}(\underline{r},t) : \underline{q}(\underline{r},) + \dots \qquad \text{(II.1)}$$

For the following we consider the geometry of large aspect ratio systems, i.e. systems which are widely extended in either one dimension (quasi-one-dimensional large aspect ratio systems) or two dimensions.

Nonlinear Evolution of Spatio-Temporal Structures in Dissipative Continuous Systems
Edited by F.H. Busse and L. Kramer
Plenum Press, New York, 1990

It is assumed that the system undergoes an oscillatory instability for a certain set of control parameter values $\underline{\sigma}_c$. In an infinitely extended system the linear eigenvalue problem determining the onset of instability takes the form

$$\lambda_j(\underline{k})\ \Phi_{j\underline{k}}(\underline{r}) = L[\underline{\sigma}]\ \Phi_{j\underline{k}}(\underline{r}) \tag{II.2}$$

Due to the translational invariance in x_1, x_2- direction the modes $\Phi_{j\underline{k}}(\underline{r})$ have the form of plane waves

$$\Phi_{j\underline{k}}(\underline{r}) = \phi_{j\underline{k}}(z)\ \exp[i\underline{k}\underline{x}] \tag{II.3}$$

with wave vector $\underline{k}$. The index j specifies the spatial behaviour of the modes in z-direction. Due to the rotational invariance $\underline{x} \to R\underline{x}$ the eigenvalues $\lambda_{j\underline{k}}$ depend only on $\underline{k}^2$.

In order to deal with the nonlinear problem the state vector $\underline{q}(\underline{r},t)$ is expanded into the set of normal modes

$$\underline{q}(\underline{r},.t) = \sum_j \xi_j(\underline{k},t)\ \phi_{j\underline{k}}(z)\ \exp i[\underline{k}\underline{x}] + c.c. \tag{II.4}$$

and one obtains a set of evolution equations for the amplitudes $\xi_j(\underline{k},t)$. By adiabatically eliminating the amplitudes $\xi_s(\underline{k},t)$ belonging to the bands of stable wave numbers one arrives at a closed set of amplitude equations for the amplitudes $\xi(\underline{k},t)$ of the modes belonging to the unstable band of wave numbers (here and in the following we suppress the index j=u). Introducing the order parameter field $\xi(\underline{x},t)$ as the Fourier transform of the amplitude $\xi(\underline{k},t)$ one arrives at the generalized Ginzburg-Landau equation for the oscillatory instability in large aspect ratio systems:

$$\begin{aligned} d_t\ \xi(\underline{x},t) = [\ \varepsilon + i\ \omega_c - (k_c^2+\Delta)^2 + i\ a\ (k_c^2+\Delta)\ ]\ \xi(\underline{x},t)\ + \\ \int d\underline{x}_1\ d\underline{x}_2\ d\underline{x}_3\ \Gamma(\underline{x},\underline{x}_1,\underline{x}_2,\underline{x}_3)\ \xi(\underline{x}_1,t)\ \xi(\underline{x}_2,t)\ \xi(\underline{x}_3,t)^* \end{aligned} \tag{II.5}$$

Here, only terms up to the third order in $\xi(\underline{k},t)$ have been included assuming a supercritical instability.

Due to the nonlocal nonlinear interaction the numerical treatment of the generalized Ginzburg-Landau equation is rather involved. However, it turns out that in many cases the nonlocal terms can be well-approximated by local ones (cf. the discussion in [5]):

$$\begin{aligned} d_t\ \xi(\underline{x},t) = [\ \varepsilon + i\ \omega_c - (k_c^2+\Delta)^2 + i\ a\ (k_c^2+\Delta)\ ]\ \xi(\underline{x},t)\ + \\ \alpha\ \xi(\underline{x},t)\ |\xi(\underline{x},t)|^2 + \beta\ \xi(\underline{x},t)\ |\nabla\xi(\underline{x},t)|^2 \end{aligned} \tag{II.6}$$

This is due to the fact that the approximation of the kernel Γ has to represent the interactions of modes accurately only for $|k|<2k_c$ if the nonlinearities in the basic equations are local, since then only stable modes with $|\underline{k}|<2k_c$ have to be taken into account in the present order of approximation.

III. ONE-DIMENSIONAL PATTERNS

III.1 Conventional theory: Standing and traveling waves

The conventional treatment of the oscillatory instability in quasi-one-dimensional systems with large aspect ratio consists in introducing periodic boundary conditions with a period corresponding to the wavelength of the most unstable mode. The investigation of the problem reduces then to the problem of Hopf-bifurcation in the presence of O(2)-symmetry. It is well-known that if the bifurcation is supercritical the amplitude equations have the form

$$d_t \xi^+ = \xi^+ [\varepsilon+i\omega - A |\xi^+|^2 - B |\xi^-|^2]$$
$$d_t \xi^- = \xi^- [\varepsilon+i\omega - A |\xi^-|^2 - B |\xi^+|^2] \qquad \text{(III.1)}$$

and either stable traveling wave solutions (ReA/ReB<1) or stable standing wave solutions (ReA/ReB>1) emerge.

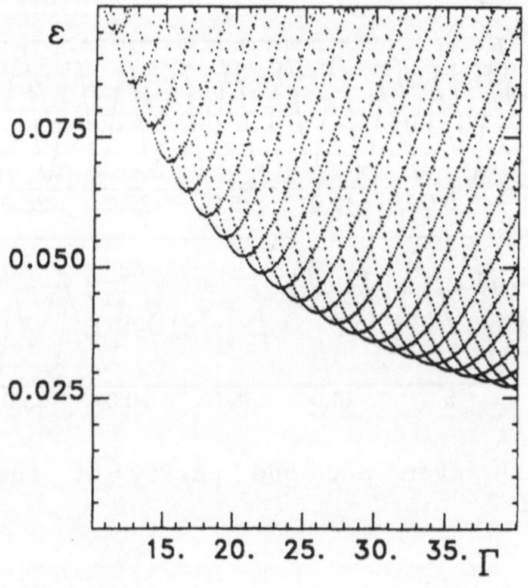

Fig. 1. The neutral curves for the onset of instability as a function of the aspect ratio.

III.2 Realistic boundary conditions

Other types of boundary conditions than periodic ones can change the bifurcation problem qualitatively as well as quantitatively. This fact is demonstrated by considering the order parameter equation (eq. II.6). Let us first consider the linear eigenvalue problem

$$\lambda\, \psi(x) = [\, \varepsilon + i\, \omega_c - (k_c^2+\Delta)^2 + i\, a\, (k_c^2+\Delta)\,]\, \psi(x) \tag{III.2}$$

$$\psi(-L/2) = \psi(L/2) = d_x\, \psi(-L/2) = d_x\, \psi(L/2) = 0$$

The modes can be classified with respect to their transformation property under inflection $x \to -x$. There are modes with odd and even parity:

$$\psi_u(x) = -\, \psi_u(-x) \quad ; \; \psi_g(x) = \psi_g(-x)$$

Additionally, the modes differ in the number of nodes, i. e. in the number of cells or rolls. Figure 1 shows typical neutral curves for the onset of instability of modes with n cells as a function of the aspect ratio Γ obtained by a numerical solution. For small values of Γ the modes with few cells become unstable first while with increasing aspect ratio modes with increasing numbers of cells become critical. There are specific values of the aspect ratio where a mode with n cells (even parity) and n+1 cells (odd parity) become unstable simultaneously. The spatial patterns described by the modes of the linear eigenvalue problem have the following form: They consist of two wave trains traveling with a certain frequency towards the boundaries (see fig. 2). The modes in

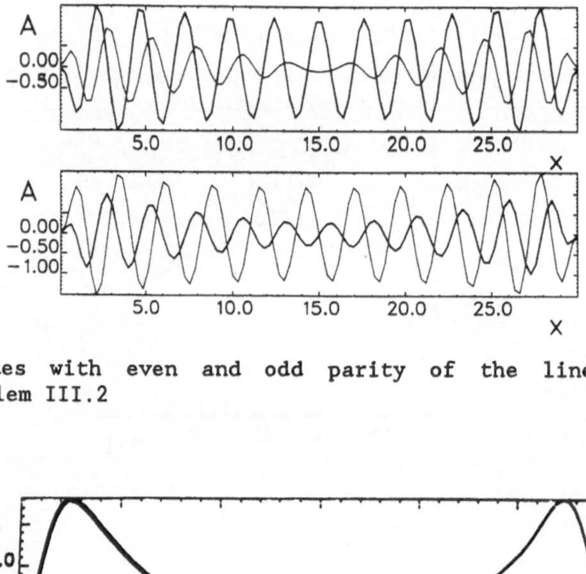

Fig. 2a. Modes with even and odd parity of the linear stability problem III.2

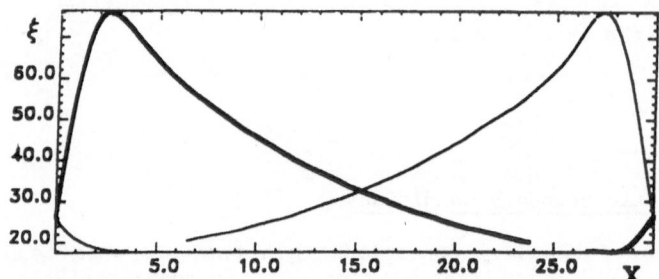

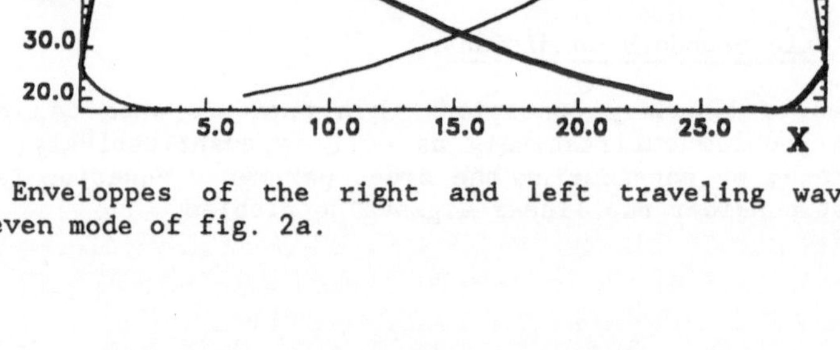

Fig. 2b. Enveloppes of the right and left traveling waves of the even mode of fig. 2a.

the case of realistic boundary conditions are thus topologically different from the ones in the case of periodic boundary conditions. The patterns contain a defect of the traveling wave structure. We note that the existence of this defect is essentially a linear phenomenon.

Since the modes of the linear eigenvalue problem for realistic boundary conditions can be topologically different from the ones for periodic boundary conditions the actually emerging patterns can not entirely be predicted by conventional (O(2) symmetric) bifurcation theory. In fact results based on this approach can be rather misleading. Instead of finding either traveling waves or standing waves the first bifurcation leads, in the supercritical case, for large values of a to a structure with two wave trains (see fig. 3). A first transition leading to standing waves is expected if the absolute value of a is small. Furthermore, it may happen that in cases where the O(2)-symmetric bifurcation theory gives subcritical bifurcations, nevertheless the transition is supercritical (and possibly vice versa). Thus the discrepancies between experiment and theoretical calculations raised in reference [1] for the case of the Taylor-Couette experiment are understandable.

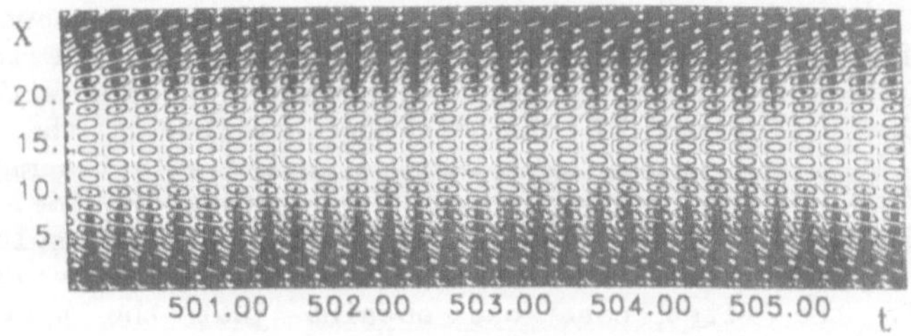

Fig. 3. Time development of the order parameter (symmetric state).

III.3 Modulated traveling waves: The blinking state

Modulated traveling waves are observed close to oscillatory instabilities in various nonequilibrium systems. We mention convection in binary mixtures ('blinking state') [6], the spiral vortex flow in the Taylor-Couette system with counterrotating cylinders [7] as well as at the oscillatory instability of finite convective rolls in the Bénard experiment [3]. Modulated traveling waves consist of two oppositely traveling wave trains whose intensities alternate periodically in time. By varying control parameters the modulated traveling waves may again become time periodic forming a so called 'confined state', a state for which the cellular waves are observed only in a localized region.

Modulated traveling waves have been numerically observed by M. Cross [8] using two coupled Ginzburg-Landau equations and M. Bestehorn, R. Friedrich, and H. Haken [5] using the order parameter equation eq. (II.6). In the following we shall consider the generalized Ginzburg-Landau equation (eq. II.6) in order to investigate the underlying mechanisms. To this end we represent the order parameter field $\xi(x,t)$ as a superposition of the modes of the linear eigenvalue problem (III.2). If we restrict ourselves to the close vicinity of a point in parameter space $[\varepsilon,\Gamma]$ where two modes with cell numbers n and n+1 become unstable simultaneously all other modes are enslaved and can be eliminated. The dynamics of the amplitudes of the unstable modes is governed by the following set of order parameter equations:

$$d_t A = [\lambda_A + i\omega_A]\, A + a_1\, |A|^2 A + a_2\, |B|^2 A + a_3\, B^2 A^*$$
$$d_t B = [\lambda_B + i\omega_B]\, B + b_1\, |B|^2 B + b_2\, |A|^2 B + b_3\, A^2 B^* \qquad \text{(III.3)}$$

For the sake of simplicity we assume that the bifurcation is supercritical. Otherwise fifth order terms have to be included, which, however, do not the change the basic ideas. The structure of the amplitude equations is fixed by the fact that they are invariant with respect to the transformations

$$A \to -A \quad , \quad B \to B \qquad \text{(III.4)}$$

This accounts for the different parity of the two modes. The coefficients a_i, b_i can be determined from the evolution equation (1) and, in turn, from the governing equations of the considered system.

A close inspection of the amplitude equations shows that two kinds of temporal behaviour are possible. In one case the ratio of the two amplitudes A and B is a periodic function of time with frequency close to the difference of the eigenfrequencies of the two modes. Second, for different parameter values, a locked solution may exist for which the ratio of the two amplitudes is constant in time. Let us now look at the spatial behaviour related with the two solutions. The spatial structure of two modes with n and n+1 cells is exhibited in fig. 2a. In the first case one obtains just the spatio-temporal behaviour of the blinking state (see fig. 4). A pattern corresponding to the locked solution is shown in fig. 5. Depending on the ratio of the amplitudes A/B the patterns are more or less confined. Nonsymmetric traveling wave patterns, as in the experiments of Croquette et al. [3], are obtained if the phase Φ oscillates around a certain mean value.

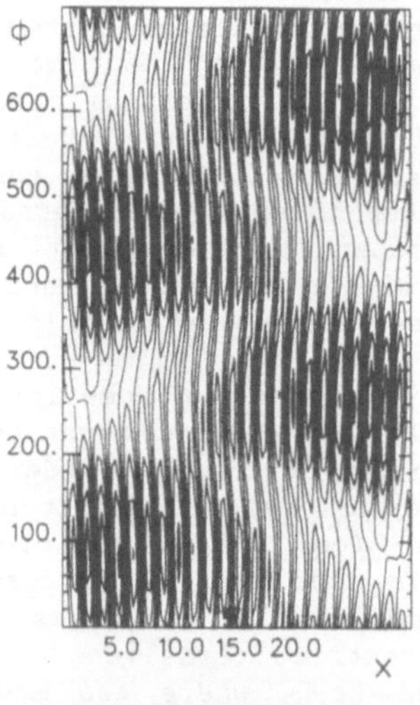

Fig. 4. Blinking state formed by the superposition Re(A $\phi_u(x)$ + B exp[iΦ] $\phi_g(x)$) of the modes of fig. 2a.

IV. TWO-DIMENSIONAL PATTERNS

Pattern formation in large aspect ratio systems close to a nonoscillatory instability like for instance the Bénard instability usually exhibits relaxational character. Frequently, a Ljapunov-functional exists which is minimized as for instance in the case of the Swift-Hohenberg equation. In case of an oscillatory instability such a functional does not exist and further phenomena are expected to occur already close to instability.

We have performed a numerical integration of the order parameter equation (II.6) using the values for the coefficients α, β as well as the coefficients of the linear terms obtained for the case of convection in binary fluid mixtures (free / free horizontal boundary conditions) for the limits of high Prandtl and small Lewis number (cf. the discussion in [5]).

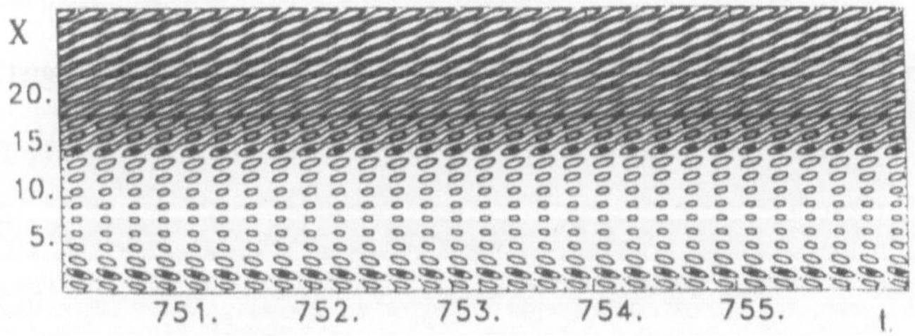

Fig. 5a. Temporal evolution of confined state.

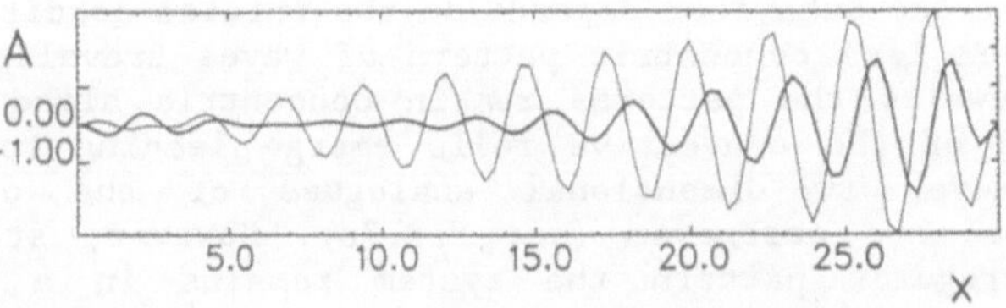

Fig. 5b. Confined state formed by the phase-locked superposition of the modes of fig.2a (real and imaginary part of the order parameter).

IV.1 <u>Traveling waves in rectangular regions</u>

Typical wave patterns arising in rectangular geometries are exhibited in figs. 6. Waves traveling to the right and waves traveling to the left are located at the right and left sides of the container, respectively. Close to onset the structure along the rolls is only slightly disturbed. For larger values of ε a transverse instability occurs. Types of 'blinking states', which however show a spatial modulation in transverse direction, are observed. Far above threshold the traveling waves show the tendency to be confined to localized regions. The obtained patterns are qualitatively similar to patterns observed experimentally (cf. [2] and references therein).

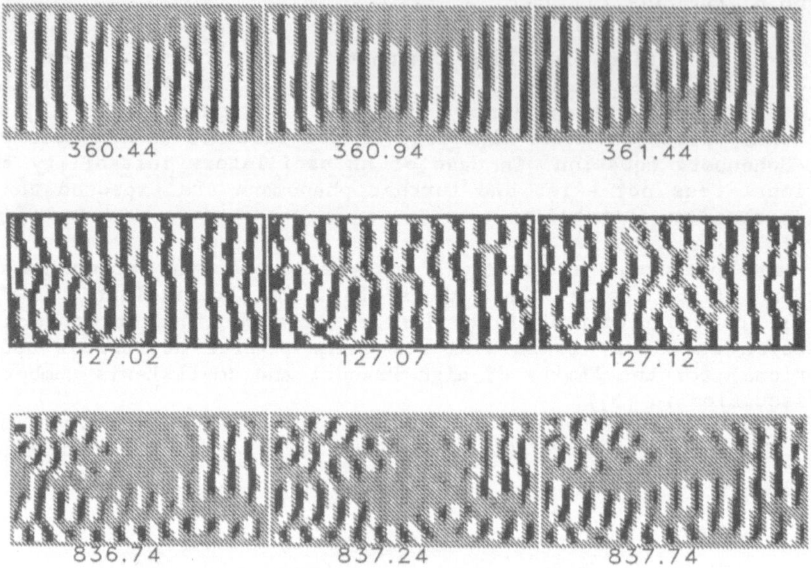

Fig. 6. Two-dimensional traveling wave patterns in rectangular geometry for increasing control parameter ε.

IV.2 Traveling waves in circular regions

Results for a circular fluid container are exhibited in figs. 7, 8. Close to onset the behaviour depends on the initial conditions. Starting from a more or less concentric pattern of waves traveling towards the circular sidewalls the patterns remain concentric although transverse instabilities of the convective rolls emerge leading to complex time evolution. Here, two-dimensional analogues of the one-dimensional blinking state are obsereved (see fig.7a). However, starting from a spatially irregular pattern the system remains in a more or less spatially disordered state, at least for very long times (see fig.7b). This underlines the ability of defects to generate spatio-temporal complexity in systems without relaxational character. Far from onset the tendency of the system to form confined states can be observed (see fig. 8). Thereby the convective rolls travel in azimuthal direction whereas the center of the container remains nearly free from activity. Intermittently, spiral-shaped wave trains emerge traveling from the center towards the boundaries.

Acknowledgements

We gratefully acknowledge useful discussions with K. Scheller, B. Hölle, and A. Greiner. We wish to thank the Volkswagenwerk foundation, Hannover, for financial support within its project on Synergetics.

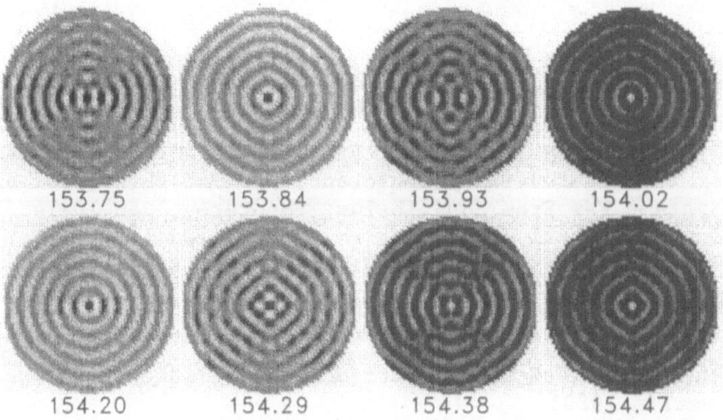

Fig. 7a. Two-dimensional blinking state in a circular geometry.

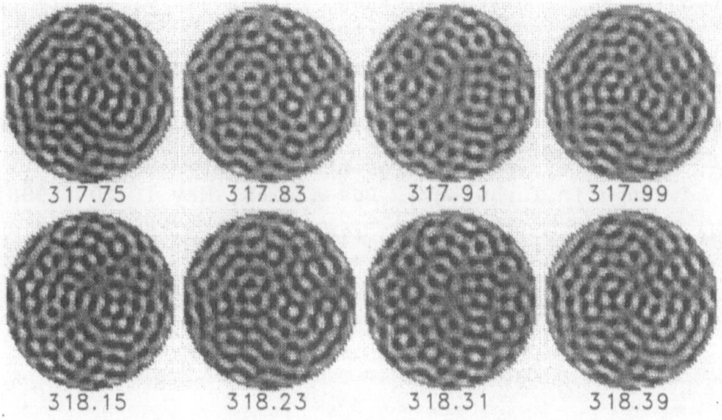

Fig. 7b. Same conditions as in fig. 7a starting form a pattern with defects.

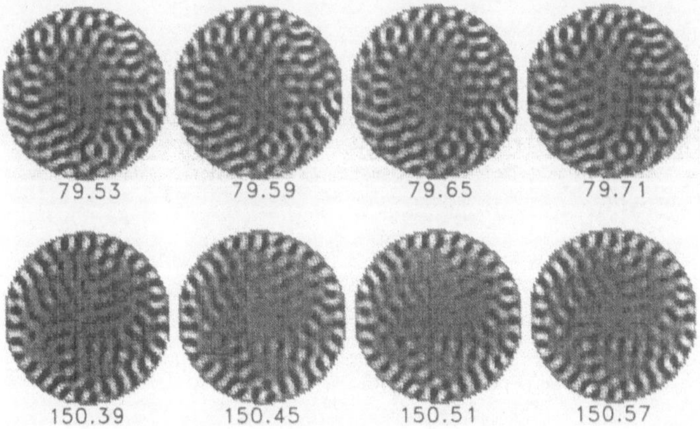

Fig. 8. Time evolution of traveling waves in circular geometry well above threshold.

References

[1] R. Tagg, W. S. Edwards, H. Swinney, and P. S. Marcus, Phys. Rev. A 39, 3734 (1989)

[2] V. Steinberg, E. Moses, and J. Fineberg in the proceedings of the International Conference on 'The Physics of Chaos and Systems Far From Equilibrium', Monterey, Jan 10-14, 1986, in the Journal of Nuclear Phys. B 2, 109 (1987);
P. Kolodner, A. Passner, H.L. Williams, and C. M. Surko, ibid., 97

[3] V.Croquette, H.Williams, Phys. Rev. A 39, 2765 (1989)
A.Chiffaudel, B.Perrin, and S.Fauve, Phys. Rev. A 39, 2761 (1989)

[4] H.Haken, 'Synergetics, An Introduction', (3rd ed., Springer, Berlin 1983)
H.Haken, 'Advanced Synergetics', (2.print., Springer, Berlin 1987)

[5] M.Bestehorn, R.Friedrich, H.Haken, Z.Phys. B 72, 265 (1988)
M.Bestehorn, R.Friedrich, H.Haken, Z.Phys. B 75, 265 (1989)
M.Bestehorn, R.Friedrich, H.Haken, Z.Phys. B 77, 151 (1989)
M.Bestehorn, R.Friedrich, H.Haken, Physica D 37, 295 (1989)

[6] J.Fineberg, E.Moses, and V.Steinberg, Phys. Rev. Lett. 61, 838 (1988)
P. Kolodner and C. M. Surko, Phys. Rev. Lett. 61, 842 (1988)

[7] R.Tagg, H. Swinney, poster presented conference on Advances in Turbulence, Los Alamos, May 16-20 (1988)

[8] M.Cross, Phys. Rev. A 38, 3593 (1988)

CONFINED STATES IN PHASE DYNAMICS

Helmut R. Brand

FB 7, Physik
Universität Essen
D 43 Essen 1
West Germany

and

Theoretische Physik
Universität Bayreuth
D 8580 Bayreuth
West Germany

INTRODUCTION

Recently two groups[1-3] have described the observation of confined states in an annulus near the onset of convection in binary fluid mixtures. These confined states are characterized by the fact that for part of the annulus a convective pattern is visible - that is the envelope of the convective pattern assumes a finite value - whereas the rest of the container is in the heat conduction state. These observations are now under intense investigation theoretically[4-6] and experimentally[2,3].

Here we discuss the analogous phenomenon in phase dynamics, namely the coexistence of two patterns showing different wavelengths in different parts of the cell, a concept we have introduced recently.[7,8] The concept of phase dynamics[9], the analogue of hydrodynamics for large aspect ratio pattern-forming nonequilibrium systems[10,11], has not only found experimental verification early on[12], but its usefulness far above onset of the instability has also been demonstrated recently experimentally[13].

The analogue of hydrodynamic variables are the phase variables, whose slow spatial and temporal variations characterize the changes of the wavelength of the pattern as a function of space and time. We elucidate that confined states in phase dynamics are an intrinsically nonlinear phenomenon[7] and we discuss the physical properties of these novel states[8].

Nonlinear Evolution of Spatio-Temporal Structures in Dissipative Continuous Systems
Edited by F.H. Busse and L. Kramer
Plenum Press, New York, 1990

We critically compare our predictions with recent experimental results on slot convection (the height of the cell is larger than the width) in a simple fluid[14-16] and on the Taylor instability for the flow in the gap between co-rotating cylinders[17,18] and we suggest further experiments.

We note that the present review makes use of our recent work on the subject[7,8,19]. In the next section we outline the concept and in the third section we show that for a stationary pattern the phase variation and the local wavevector satisfy a conservation law. In this section we also discuss similarities and differences with the corresponding equations for a propagating pattern. In the fourth section we remark that a generalized thermodynamic potential[20-22] can be found for the equations associated with the phase variations and the local wavevector of a stationary 1D pattern. We conclude with a brief summary and a perspective.

CONFINED STATES

Making use of general symmetry and invariance arguments[10,11], we find that for a one-dimensional (1D) stationary pattern one has only one phase variable ψ describing the spatial and temporal variations of the wavelength, which obeys the equation

$$\dot{\psi} = \left[D + E\psi_x + F(\psi_x)^2\right]\psi_{xx} - G\psi_{xxxx} \tag{2.1}$$

where we have kept the cubic nonlinearity in eq.(2.1) to guarantee that the solutions are bounded. Rewriting eq.(2.1) in terms of the local wavevector $q = \psi_x$:

$$\dot{q} = \left(D + Eq + Fq^2\right) q_{xx} + (E + 2Fq)(q_x)^2 - Gq_{xxxx} \tag{2.2}$$

we note that the linear terms are the same as in the phase equation.

Inspection of eq.(2.1) shows[7] that it can be derived from a Liapunov functional

$$\dot{\psi} = -\frac{\delta V}{\delta \psi} \tag{2.3}$$

with

$$V(\{\psi\}) = \int dx \left[\frac{D}{2}(\psi_x)^2 + \frac{E}{6}(\psi_x)^3 + \frac{F}{12}(\psi_x)^4 + \frac{G}{2}(\psi_{xx})^2\right] \tag{2.4}$$

and one finds that V is strictly non-increasing in time provided surface terms vanish. Global stability of $V(\{\psi\})$ is guaranteed assuming F and G to be positive. Since it is a Liapunov functional and due to its structure $V(\{\psi\})$ is purely diffusive and does not show oscillatory behavior. We note in passing that such a functional cannot be found straightforwardly for the KS equation.

We rewrite eq.(2.4) in terms of the wavevector to make contact with a generalized Ginzburg-Landau energy

$$V(\{q\}) = \int dx \left[\frac{D}{2}q^2 + \frac{E}{6}q^3 + \frac{F}{12}q^4 + \frac{G}{2}(q_x)^2\right] \tag{2.5}$$

Eq.(2.5) has the form of the Ginzburg-Landau energy for a weakly first order phase transition with the term proportional to G being the analogue of the gradient energy and interpreting the local wavevector q as the order parameter and where the gradients of the local wavevector characterize the length scale over which the local wavevector changes. Therefore the term proportional to G gives us the analogue of the square of the coherence length for the changes in the local wavelength. Depending on the signs and magnitude of D and E different scenarios for the nonlinear behavior of a stationary pattern emerge.

For a negative value of the phase diffusion coefficient D the fourth derivative in eq.(2.1) must be kept to stabilize the system for large wavevectors linearly. Here we focus on the case of positive phase diffusion coefficient, which is more easily accessible experimentally. To investigate the influence of the quadratic nonlinearity we examine the stationary solutions of eq.(2.1) with constant ψ_x. We find[7] that $V(\{q\})$ has two local minima and one maximum provided $E^2 > (16DF/3)$, that is $|E|$ must be sufficiently large. From the existence of two local minima it follows that two different wavelengths are locally stable and that one can thus have a stationary pattern, where one observes different wavelengths in different parts of the cell. For example a confined region of larger wavelength can be surrounded by a bulk region of smaller wavelength in a stationary situation. The fourth order derivative term in eq.(2.1) serves to smoothly connect the two regimes of different wavelengths as a function of space. For a detailed discussion of the shape of the potential we refer to ref.7.

We emphasize that the existence of the Liapunov functional guarantees that q is bounded, regardless of whether the "gradient energy" (proportional to G) is incorporated or not. To guarantee, however, that q_x is bounded the incorporation of the gradient energy is essential for the case that one deals with two wells.

There also exist experiments showing stationary localized states with a wavevector which is different from that in the bulk of the sample. For the case of slot convection studied by Dubois et al.[15,16] (compare also ref.14 for a detailed description of the set-up) the results presented here and the experimental observations are qualitatively similar. For both, the experiment and the model, one finds a region for which the wavelength in part of the cell is different from that in the bulk of the container and both regimes are completely stationary. To further check the applicability of our approach to the coexistence of states with different wavelengths in slot convection experimentally, it might be worthwhile to fit the envelope of the experimental data to the analytic solutions given below. Another interesting point is that the range over which stationary confined states exist in experiment is fairly wide and we will discuss in the next section a mechanism which can stabilize them over such a large range of the bifurcation parameter.

For the case of the dynamic domains observed for the flow between concentric co-rotating cylinders[17,18] the analogy is not quite as immediate as for the case of slot convection, since for the Taylor instability the wavelength variation is not strictly one-dimensional, but has spatial and temporal variations in the azimuthal direction as well. Focussing on the direction parallel to the cylinder axis, however, the same global picture as that outlined above for slot convection emerges: one has two locally stable states existing in different parts of the gap along the cylinders. More detailed work, however, will be needed to investigate the coupling to the motion in the azimuthal direction.

THE CONSERVATION LAW FOR THE PHASE AND FOR THE WAVEVECTOR

Inspection of the nonlinear phase equation (eq.(2.1)) shows it can be rewritten as

$$\dot{\psi} + \partial_x j^{\psi} = 0 \tag{3.1}$$

with

$$j^{\psi} = -[D\psi_x + \frac{E}{2}\psi_x^2 + \frac{F}{3}\psi_x^3 - G\psi_{xxx}] \tag{3.2}$$

that is as a conservation law for the phase with a phase current j^{ψ}. This conservation law takes the same form as conservation laws in hydrodynamics such as for example the conservation laws for density of linear momentum and energy density.

Furthermore we find that also the nonlinear equation for the local wavevector (eq.(2.2)) can be rewritten as a conservation law

$$\dot{q} + \partial_x j^{q} = 0 \tag{3.3}$$

with

$$j^{q} = -[Dq_x + Eqq_x + Fq_xq^2 - Gq_{xxx}] \tag{3.4}$$

These results have a few important implications (compare also ref.8 for a detailed discussion). Firstly we note that the existence of these conservation laws opens the possibility to pin the phase at the boundaries, while maintaining a constant envelope. This would not be possible if there would be sources and sinks for the phase in the bulk (compare also the discussion for the Kuramoto-Sivashinsky equation further below). For a pinned phase at the boundaries the number of unit cells (rolls, vortices etc.) will be conserved as long as there are no amplitude variations. This follows from the observation that for a pinned phase at the boundaries the integral over q is constant. This is the situation which occurs frequently in the experiments as for example in Benard convection. Therefore confined states in phase dynamics can exist over the whole range of parameter values for which the associated Liapunov functional given in ref.7 has two locally stable states. This observation sheds light on the question why the situation discussed here is qualitatively different from that arising close to the liquid-gas transition, which is also of first order. In the latter case any droplet nucleated either shrinks to zero or grows and thus droplets of finite size are never locally stable. Even at the coexistence point, at which the two minima are equal in depth, they are only neutrally stable and any perturbation will make the droplet grow or shrink.

From the considerations reviewed here, it also becomes clear how to implement boundary conditions such that a spatially homogeneous wavelength for the unit cells results.

This can be achieved for example by making use of the concept of ramps[23]. If the geometry of the set-up is changed in such a way that one goes smoothly from a value of the control parameter for which confined states are stable for suitably chosen boundary conditions to a value for which convection is unstable, the number of unit cells can adjust freely and we predict that in such a situation no confined states exist stably.

For arbitrary boundary conditions one might still get confined states as long transients, but there is in general no reason for such a state to be stable.

This analysis applies only to 1D stationary patterns. For the nonlinear phase equation applicable to a one dimensional pattern associated with propagation[10] - the Kuramoto-Sivashinsky equation[24-28]

$$\dot{\phi} = D\phi_{xx} - \tilde{D}\phi_{xxxx} + \frac{E}{2}\phi_x^2 \tag{3.5}$$

or, rewritten as an equation for the local wavevector,

$$\dot{q} = Dq_{xx} - \tilde{D}q_{xxxx} + Eqq_x \tag{3.6}$$

this is different. Eq.(3.5) cannot be written in the form of a conservation law, whereas for eq.(3.6) we find

$$\dot{q} + \partial_x j^q = 0 \tag{3.7}$$

where

$$j^q = -[Dq_x - \tilde{D}q_{xxx} + \frac{E}{2}q^2] \tag{3.8}$$

This does not mean, however, that one can find confined states of the type discussed above in this case for suitably chosen boundary conditions. This will depend for example on the sign of the diffusion coefficient D; for $D < 0$ eq.(3.8) is known to lead to weak turbulence[25]. But this inequality also corresponds to the regime, where one expects the generation of defects in the pattern[29] since it is Benjamin-Feir unstable[30]. We conclude that a consideration of the equation for the local wavevector alone is insufficient to guarantee the existence of confined states.

For a stationary pattern, the nonlinear equation for the local wavevector can be rewritten in another revealing form

$$\begin{aligned}\dot{q} &= \partial_{xx}[\frac{\delta V}{\delta q}] \\ &= \partial_{xx}[Dq + \frac{E}{2}q^2 + \frac{F}{3}q^3 - Gq_{xx}]\end{aligned} \tag{3.9}$$

and we have shown[19] that eq.(3.9) can be used to demonstrate that $V(q)$ is also a Liapunov functional for the local wavevector for vanishing boundary terms.

In addition, close inspection shows that eq.(3.9) takes the form of the Ginzburg-Landau equation for a conserved order parameter known e.g from spinodal decomposition[31,32]. Thus the local wavevector can serve as a conserved order parameter for one-dimensional stationary patterns.

The analogy with spinodal decomposition can be made even more explicit by introducing a shift of q: $q = q_0 - E/(2F)$:

$$\dot{q}_0 = \partial_{xx}[\{D - \frac{E^2}{(4F)}\}q_0 + \frac{F}{3}q_0^3 - Gq_{0xx}] \tag{3.10}$$

A similar observation has been made recently for modulated systems[33], for which a Ginzburg-Landau equation for a conserved order parameter has been derived.

THE GENERALIZED THERMODYNAMIC POTENTIAL

Our considerations in the last two sections have been purely deterministic. If we incorporate additve noise into eq.(2.1) we have

$$\dot{\psi} = \left[D + E\psi_x + F(\psi_x)^2\right]\psi_{xx} - G\psi_{xxxx} + \xi \tag{4.1}$$

Assuming the random force ξ to be Gaussian white and also delta-correlated spatially, eq.(4.1) can be converted into the associated Fokker-Planck equation, in this case a functional differential equation for the probability density $W(\{\psi\}, t)$

$$\dot{W} = \int dx\, \{\delta_\psi(x)[-(D + E\psi_x + F\psi_x^2)\psi_{xx} + G\psi_{xxxx} + Q\delta_\psi(x)]W\} \tag{4.2}$$

It is straightforward to check, that the conditional probability density obtained from eq.(4.2) satisfies the condition of detailed balance[20-22] and that the deterministic Liapunov functional discussed above can be used to rewrite eq.(4.2) in the form

$$\dot{W} = Q\int dx\, (\{\delta_\psi(x)\Phi(\{\psi\})\} + \delta_\psi(x)\}W) \tag{4.3}$$

with $\Phi(\{\psi\}) = Q^{-1}V\{\psi\})$

Thus one has for the stationary probability density $P \sim exp(-\Phi)$, which is known explicitly in the present case. Since it is a functional integral, however, the averaging required to evaluate e.g. time-independent correlation functions involves the evaluation of a functional integral.

We mention in passing that one cannot find easily such a generalized thermodynamic potential for the other prototype equation of nonlinear phase dynamics, the Kuramoto-Sivashinsky equation, as it arises for phases associated with traveling and standing waves.

For the equation for the local wavevector we draw again on the analogy of q with the concentration in spinodal decomposition. For the local wavevector the resulting Langevin equation reads

$$\dot{q} = [D + Eq + Fq^2]q_{xx} + [E + 2Fq]q_x^2 - Gq_{xxxx} + \xi \tag{4.4}$$

where the fluctuating forces are again assumed to have the properties $< \xi >= 0$ and $< \xi(x',t')\xi_\tau(x,t) >= Q\delta(x-x')\delta(t-t')$. Making use of the work on spinodal decomposition[31,32], one can show that the functional differential Fokker-Planck equation associated with eq.(4.4) for noise which is delta-correlated in space and time

$$\dot{W} = \int dx \{\delta_q(x)[-\partial_{xx}\frac{\delta V}{\delta q} + Q\delta_q(x)]W\} \tag{4.5}$$

satisfies detailed balance and has the stationary solution $P \sim exp(-\Phi)$ with $\Phi(\{q\}) = Q^{-1}V\{q\})$ with $V(q)$ from section 2.

The analysis of this section shows that for one-dimensional stationary patterns of the type investigated here the concept of a generalized thermodynamic potential is a useful one.

SOME EXACT SOLUTIONS OF THE NONLINEAR EQUATION FOR THE LOCAL WAVEVECTOR

To facilitate comparison of some of the predictions made in the second and third section with experimental results and to check the accuracy of numerical solutions at least asymptotically, it is always useful to have some analytic solutions of nonlinear evolution equations at one's disposal. We have found so far two classes of stationary solutions of the nonlinear equation for the local wavevector.

We start our discussion with eq.(3.10), which is already in the standard form known for conserved order parameters. Then we are looking for solutions of the equation

$$-Hq_0 + \frac{F}{3}q_0^3 - Gq_{0xx} = 0 \tag{5.1}$$

where $H = -D + E^2/(4F)$. In writing down eq.(5.1) we keep in mind that aside from the solutions of this equation, eq.(3.10) also allows for other solutions, for example those which correspond to a fixed value of the local wavevector q.

To analyze eq.(5.1) we proceed similar as Newell and Whitehead[34] for their analysis of the stationary solutions for the envelope equation as it arises for example for stationary Bénard convection.

Firstly there is a class of wall-like solutions

$$q_0 = \pm\sqrt{\frac{3H}{F}}tanh(\sqrt{\frac{H}{2G}}(x-x_0)) \tag{5.2}$$

where the presence of both signs in eq.(5.2) reflects the fact that the replacement of q_0 by $-q_0$ leaves eq.(3.10) invariant. This means that walls, which lead to an increase as well as those which lead to a decrease of the wavevector q are possible.

The second class of solutions, which might actually be related to the experimental results in slot convection[14-16], corresponds to localized solutions of the form

$$q_0^2 = \frac{3H}{F}(1 - sech^2[\sqrt{\frac{H}{2G}}(x - x_0)]) \tag{5.3}$$

For $x = x_0$, $q_0 = 0$ and for large distances of x from x_0, q_0^2 reaches asymptotically the value $3H/F$. Depending on the sign of q_0 chosen from eq.(5.3) one obtains a dip or a hump in the background wavevector. Experimentally one observes a hump in the wavelength, that is a dip in the wavevector and it might therefore be worthwhile to fit the experimental data corresponding to the variation of the wavevector to the expression given in eq.(5.3).

Naturally the question arises under what circumstances the two classes of stationary solutions given in eqs.(5.2) and (5.3) are stable and we hope to present our results concerning this question in the near future.

CONCLUSIONS AND PERSPECTIVE

We have discussed how one can obtain in the framework of phase dynamics, stationary states for which different wavelengths coexist in different parts of a cell and we have shown that this is an intrinsically nonlinear phenomenon. For a phase variable associated with a stationary pattern a Liapunov functional can be given. We have shown that for one-dimensional stationary patterns the nonlinear equations for both, the local wavevector and the phase changes, assume the form of a conservation law. Depending on the boundary conditions, one can have either confined states or a spatially homogeneous wavelength. We have examined recent experiments on slot convection and the Taylor instability for co-rotating cylinders in the light of the approach presented and we have suggested experiments to further test the concept introduced. Especially it seems highly desirable to implement the idea of a ramp geometry experimentally to test further the idea that the localized states observed e.g. in slot convection are confined states of the type discussed.

Acknowledgements

It is a pleasure to thank Guenter Ahlers, David Andereck, Bob Deissler, and Monique Dubois for stimulating discussions.
Support of this work by the Deutsche Forschungsgemeinschaft is gratefully acknowledged.

References

1) P. Kolodner, D. Bensimon and C.M. Surko, *Phys.Rev.Lett.* **60**, 1723 (1988)
2) J. Niemela, G. Ahlers and D.S. Cannell, *Bull.Am.Phys.Soc.* **33**, 2261 (1988), and to be published
3) D. Bensimon, P. Kolodner, C.M. Surko, H. Williams and V. Croquette, preprint, March 1989

4) O. Thual and S. Fauve, *J.Phys.(Paris)* **49**, 1829 (1988)
5) H.R. Brand and R.J. Deissler, *Phys.Rev.Lett.* **63**, xxxx (1989)
6) W. van Saarloos and P.C. Hohenberg, preprint, Dec. 1989
7) H.R. Brand and R.J. Deissler, *Phys.Rev.Lett.* **63**, 508 (1989)
8) H.R. Brand and R.J. Deissler, submitted for publication
9) Y. Pomeau and P. Manneville, *J.Phys.(Paris) Lett.* **40**, 609 (1979)
10) H.R. Brand, *Prog.Theor.Phys.* **71**, 1096 (1984)
11) H.R. Brand, p. 206, in *Propagation in Systems far from Equilibrium*, J.E. Wesfreid et al., Eds., Springer Series in Synergetics, Springer, Heidelberg, 1988
12) J.E. Wesfreid and V. Croquette, *Phys.Rev.Lett.* **45**, 634 (1980)
13) L. Ning, G. Ahlers, and D.S. Cannell, preprint, October 1989
14) M. Dubois, R. DaSilva, F. Daviaud, P. Bergé, and A. Petrov, *Europhys.Lett.* **8**, 135 (1989)
15) M. Dubois et al. in *The Geometry of Nonequilibrium* , eds. P. Huerre and P. Coullet, Plenum, New York, 1990, to appear and private communication
16) M. Dubois et al., to be published
17) G.W. Baxter and C.D. Andereck, *Phys.Rev.Lett.* **57**, 3046 (1986)
18) C.D. Andereck and G.W. Baxter, p.315 ff, in the proceedings quoted in ref.11
19) R.J. Deissler, Y.C. Lee, and H.R. Brand, to be published
20) R. Graham, *Phys.Rev.Lett.* **31**, 1479 (1973)
21) R. Graham, *Phys.Rev.* **A10**, 1762 (1974)
22) H. Haken, *Rev.Mod.Phys.* **47**, 67 (1975)
23) L. Kramer, E. Ben-Jacob, H. Brand, and M.C. Cross, *Phys.Rev.Lett.* **49**, 1891 (1982)
24) Y. Kuramoto and T. Tsuzuki, *Prog.Theor.Phys.* **54**, 687 (1976)
25) Y. Kuramoto, *Prog.Theor.Phys.Suppl.* **64**, 346 (1980)
26) H. Chaté and P. Manneville, *Phys.Rev.Lett.* **58**, 112 (1987)
27) P. Manneville, p.265 ff in ref.4
28) H.R. Brand and R.J. Deissler, *Phys.Rev.* **A39**, 462 (1989)
29) P. Coullet, L. Gil, and J. Lega *Phys.Rev.Lett.* **62**, 1619 (1989)
30) A.C. Newell, p. 122 in *Propagation in Systems far from Equilibrium*, J.E. Wesfreid et al., Eds., Springer Series in Synergetics, Springer, Heidelberg, 1988
31) J.S. Langer, *Ann.Phys.(N.Y.)* **65**, 53 (1971)
32) J.S. Langer, *Physica* **73**, 61 (1974)
33) H. Riecke, preprint
34) A.C. Newell and J.A. Whitehead, *J. Fluid Mech.* **38**, 279 (1969)

SMALL-SCALE EXCITATIONS IN LARGE SYSTEMS

P. Cessi,[1] E. A. Spiegel,[2] and W. R. Young[1]

[1]Scripps Institute of Oceanography
La Jolla, CA, 92093, USA

[2]Department of Astronomy
Columbia University
New York, NY, 10027, USA

1. AIMS

In the development of instability in extended systems, the prevalence of fluctuations is a disquieting reminder of our failure to understand completely the nonlinear aspects of instability theory. Experimental studies, such as those of Ahlers (1975) in thermal convection, show fluctuations that are plausibly interpreted in terms of the wandering of defects through the underlying patterns engendered by the instability. These defects are small-scale structures not easily accessible to the slowly varying wave theory that has been used in the study of unstable extended systems. However, they are not filtered out by the gauge field approach that Procaccia advocates in these proceedings. Perhaps they may relate to modes already known in the standard theory, as in Pocheau's (1989) interpretation of the vertical vorticity modes of convection theory as a gauge field.

Still, whatever the mathematical nature of the defect fields may be, it remains true that we shall need to isolate their exciting mechanism if the dynamics of small-scale structures in extended systems is fully to be understood. It is our aim, in this note, to propose a mechanism for generating such defects in unstable extended systems.

2. BACKGROUND

Suppose we have a system governed by a nonlinear partial differential equation of the form

$$\partial_t \mathbf{U}(\mathbf{x}, z, t) = \mathcal{L}\mathbf{U}(\mathbf{x}, z, t) + \mathcal{N}(\mathbf{U}(\mathbf{x}, z, t)), \tag{1}$$

where $\mathcal{L}$ is a linear operator depending on spatial derivatives but not on t or ∂_t, and $\mathcal{N}$ is a nonlinear operator. We assume that $\mathbf{U} = 0$ is a solution of (1). We are concerned with systems that are extended in the two directions of $\mathbf{x}$ and take z to be in the direction transverse to the channel.

Nonlinear Evolution of Spatio-Temporal Structures in Dissipative Continuous Systems
Edited by F.H. Busse and L. Kramer
Plenum Press, New York, 1990

If we start the system off close to the trivial solution, we can ascertain the initial behavior by a study of the associated linear problem

$$\partial_t \mathbf{U}(\mathbf{x}, z, t) = \mathcal{L}\mathbf{U}(\mathbf{x}, z, , t). \tag{2}$$

In the geometry we have adopted, (2) has solutions of the form $exp(st + i\mathbf{k}\cdot\mathbf{x})\phi_n(z;k)$, up to a multiplicative constant. When substituted into (2), these lead to an ODE for ϕ_n with s as eigenvalue and $n = 1, 2, \cdots$. Imposition of boundary conditions permits us, by solving this ODE, to find s as a function of n and k. We assume that this gives the standard onset situation: for $n = 1$ we have $s \approx \epsilon^2 - \xi(k^2 - k_c^2)^2$ when $k = |\mathbf{k}|$ is close to k_c, where ϵ, k_c and ξ are positive constants of the theory, and we are in the regime where ϵ^2 is small. For $n > 1$, we have $s < 0$ for all $\mathbf{k}$ and we assume for this discussion that there are no other modes in the problem that can be unstable or even nearly neutral.

In these circumstances, one normally seeks an approximation of the form

$$\mathbf{U}(\mathbf{x}, z, t) = \int A_{\mathbf{k}}(t) e^{i\mathbf{k}\cdot\mathbf{x}} \phi_1(z; k), \tag{3}$$

in which the modulation factor $A_{\mathbf{k}}$ satisfies the nonlinear evolution equation

$$\partial_t A_{\mathbf{k}} = s A_{\mathbf{k}} + \mathcal{F}([A_{\mathbf{k}}]), \tag{4}$$

where $\mathcal{F}$ is a strictly nonlinear functional. The issue we want to discuss turns on some subtleties in the existing derivations of such equations for extended systems.

3. ADIABATIC APPROXIMATION

Expression (3) represents the leading term in a modal expansion of the field $\mathbf{U}(\mathbf{x}, z, t)$. Corrections involve not only the unstable modes of amplitude $A_{\mathbf{k}}$, but also all the stable modes whose amplitudes we denote by $B_{\mathbf{k},n}$. So

$$\mathbf{U}(\mathbf{x}, z, t) = \int d\mathbf{k}\, e^{i\mathbf{k}\cdot\mathbf{x}} [A_{\mathbf{k}}(t)\phi_1(z;k) + \sum_{n=2}^{\infty} B_{\mathbf{k},n}(t)\phi_n(z;k)]. \tag{5}$$

Introducing the expression (5) in the equations of motion (1) and performing the appropriate projection in z, we obtain coupled equations for the amplitude of the modes $A_{\mathbf{k}}$ and $B_{\mathbf{k},n}$:

$$\begin{aligned} \partial_t A_{\mathbf{k}} &= \sigma_k A_{\mathbf{k}} + \int d\mathbf{p} \int d\mathbf{q} \sum_n^{\infty} F(\mathbf{p}, \mathbf{q}, \mathbf{k}) A_{\mathbf{q}} B_{\mathbf{q},n} + \cdots \\ \partial_t B_{\mathbf{k},n} &= \gamma_{k,n} B_{\mathbf{k},n} + \int d\mathbf{p} \int d\mathbf{q}\, G_n(\mathbf{p}, \mathbf{q}, \mathbf{k}) A_{\mathbf{p}} A_{\mathbf{q}} + \cdots. \end{aligned} \tag{6}$$

Here $\sigma_k = s(k, n = 1)$ denotes the (possibly) unstable eigenvalue while the eigenvalues associated with the stable modes are $\gamma_{k,n} = s(k, n > 1)$; the ellipses represent other nonlinear terms in the amplitudes $A_{\mathbf{k}}$ and $B_{\mathbf{k},n}$. In the regime where ϵ is small, only σ_k can be slightly positive, while $\gamma_{\mathbf{k},n}$ is always negative.

Because the amplitudes $B_{\mathbf{k},n}$ are rapidly damped, it is usually assumed that they are asymptotically controlled through the nonlinear terms by the slowly varying $A_{\mathbf{k}}$, for small ϵ. Specifically, the adiabatic approximation (Haken, 1978) is commonly used:

$$B_{\mathbf{k},n} \approx -(\gamma_{k,n})^{-1} \int d\mathbf{p} \int d\mathbf{q}\, G_n(\mathbf{p}, \mathbf{q}, \mathbf{k}) A_{\mathbf{p}} A_{\mathbf{q}}. \tag{7}$$

The approximate expression (7) can be improved by iteratively adding the neglected time derivative, calculated using the evolution equation for $A_{\mathbf{k}}$:

$$B_{\mathbf{k},n} \approx -(\gamma_{k,n})^{-1}\int d\mathbf{p}\int d\mathbf{q}\, G_n(\mathbf{p},\mathbf{q},\mathbf{k})A_{\mathbf{p}}A_{\mathbf{q}}[1+(\gamma_{k,n})^{-1}(\sigma_p+\sigma_q) + (\gamma_{k,n})^{-2}(\sigma_p+\sigma_q)^2+\cdots] + O(A_{\mathbf{k}}^3). \tag{8}$$

The series in the integrand can be summed using the property that $\sum_{n=0}^{\infty} z^n = (1-z)^{-1}$ to get

$$B_{\mathbf{k},n} = \int d\mathbf{p}\int d\mathbf{q}\, G_n(\mathbf{p},\mathbf{q},\mathbf{k})A_{\mathbf{p}}A_{\mathbf{q}}(\sigma_p+\sigma_q-\gamma_{k,n})^{-1} + O(A_{\mathbf{k}}^3). \tag{9}$$

Therefore, the final expression for the damped modal amplitude $B_{\mathbf{k},n}$ contains a denominator that may vanish. While the eigenvalue $\gamma_{k,n}$ is always negative, the eigenvalue σ_k is positive only when $|\mathbf{k}| \approx k_c$ and it is negative for all the other wavenumbers. Unless the amplitude, $A_{\mathbf{k}}$, is exactly zero for $|\mathbf{k}|$ lying outside a thin ring in the $\mathbf{k}$-plane of mean radius k_c, the adiabatic elimination leads to singularities resulting from resonances with and among modes that are, according to linear theory, strongly damped. The generation of resonances among stable modes can be seen also by using the normal form approach (Coullet and Spiegel, 1988).

The difficulty associated with the elimination of resonant stable modes has been commonly circumvented by assuming that the spectrum of the gravest vertical eigenfunction has compact support in $\mathbf{k}$, either explicitly or implicitly (as in slowly varying wave theory). That this assumption may have serious physical consequences is suggested by an analogous situation arising in the study of plasma waves with the Vlasov equation. There, a similar assumption of compact support of the one-particle distribution function, f_0, has sometimes been invoked to remove singularities. But this assumption cuts out Landau damping (Van Kampen, 1955). The same terms can also give rise to instabilities, depending on the shape of the distribution function. As Van Kampen (1955) has remarked in the plasma context, ''the assumption of a rigorously cut off f_0 is rather artificial, and it is unsatisfactory that the calculation should fail for an f_0 that decreases rapidly without actually vanishing."

In his treatment of Vlasov equation, Van Kampen allowed for the resonant modes by including delta functions of the vanishing denominators, with suitable constant factors. These factors represent choices about how to integrate around the singularities. Such a procedure is called for in the derivation of amplitude evolution equations in the style of (4). We do not go into such technical questions here. Rather, we wish to bring out the possible significance of the extra terms that arise in the amplitude-evolution equations with a simple model.

4. A RESONANT CHAIN

Consider first a chain of N coupled modes, which are linearly damped and nonlinearly excited:

$$\dot{y}_n = -2^n y_n + y_{n-1}^2 \tag{10}$$

for $n = 0,1,2,...,N$, where $y_{-1} \equiv 0$ in the first equation. This model is reminiscent of one recently discussed by Pikovsky (1989), but here we have focused on a chain of resonant stable modes. Because all the modes are damped in (10), the trivial solution $y_n = 0$, for $n = 0,1,2,...,N$, is linearly stable. Asymptotically, all the modes approach this solution exponentially and the higher the cutoff ''wavenumber'', N, the larger the decay rate. However, in this strongly resonant example, interesting things may happen on the way to the asymptotic trivial state. Indeed, for infinite N, the system can develop a singularity in finite time.

With the initial conditions $y_n = Y\delta_{n0}$, (10) has the solution $y_0 = Y\,exp(-t)$ and, for $n = 1, 2, ..., N$, we get

$$\begin{aligned} y_n &= Y t^{p-1} e^{-pt}/a_n \\ \text{with} \quad a_n &= a_{n-1}^2(p-1), \quad a_1 = 1 \quad \text{and} \quad p \equiv 2^n. \end{aligned} \tag{11}$$

The maximum value of y_n, is achieved at $t = t_n = 1 - 2^{-n}$. For $n \to \infty$, this maximum approaches

$$y_n(t_n) \approx 4p\left(\frac{Y}{Y_0}\right)^p, \tag{12}$$

where Y_0 is a constant approximately equal to 6.9. Thus, for suitable initial conditions, we get large excitation of the highest modes in finite time, even though the trivial state is eventually reached for finite N.

The situation may be entirely different if the resonant chain is coupled to a slightly unstable mode. Denote the amplitude of the unstable mode by u and its growth rate by ϵ. Its equation of motion is

$$\dot{u} = \epsilon u + u(\lambda_0 y_0 + \lambda_1 y_1 + \ldots + \lambda_N y_N). \tag{13}$$

We have included nonlinear terms, which may saturate the instability if the constants λ_n are chosen appropriately. To couple (10) back to (13) we make the identification $u = y_{-1}$ in (10), which still applies for $n = 0, 1, 2, ..., N$.

To illustrate the new situation, we show a numerical solution of the combined equations (10) and (13) in the accompanying figure for $N = 5$ and for the parameter values indicated. For this solution we used the initial conditions $u = 2\epsilon, y_n = 0, n = 0, 1, ..., N$. In contrast to the previous situation, where all the modes decay after their initial outburst, we have a series of such bursts because of the unstable mode. Here again, the most rapidly damped mode goes to the largest amplitude. There is an asymptotic tendency toward center-manifold behavior, in which the stable modes decay to steady values, but that consummation is achieved only very slowly. We can always choose the parameters to make the bursting continue for times long compared to any normal observation period.

5. CONCLUSIONS

When we try to simplify the dynamics of unstable finite systems with *discrete* spectra, the center manifold theorem tells us that, given conditions, dimensional reduction is possible at marginality. When we carry out dimensional reduction with the system off the critical condition, we may hit on a resonance, which typically can be avoided by changing the control parameters a bit. When we try the same approach in extended systems with *continuous* spectra, we get irremovable resonances. Altering the control parameter just transfers the trouble to a neighboring wave number. It is then no wonder that there is no center manifold theorem for systems with continuous spectra.

For many purposes, the resonances may not be of great significance and will merely change the final approach of the stable modes to their slavish equilibrium amplitudes from exponential to algebraic. But if these stable modes have initial conditions far from their final amplitudes, the approach to the asymptotic state may be complicated, as we have tried to illustrate with our simple model. The behavior seen in the model, provides a basis for understanding the generation of small-scale structures beyond the threshhold of instability, through resonances among highly damped modes.

Generally, if one tries to derive the amplitude-evolution equation (4), there will appear additional terms from the poles in (9). These terms will be com-

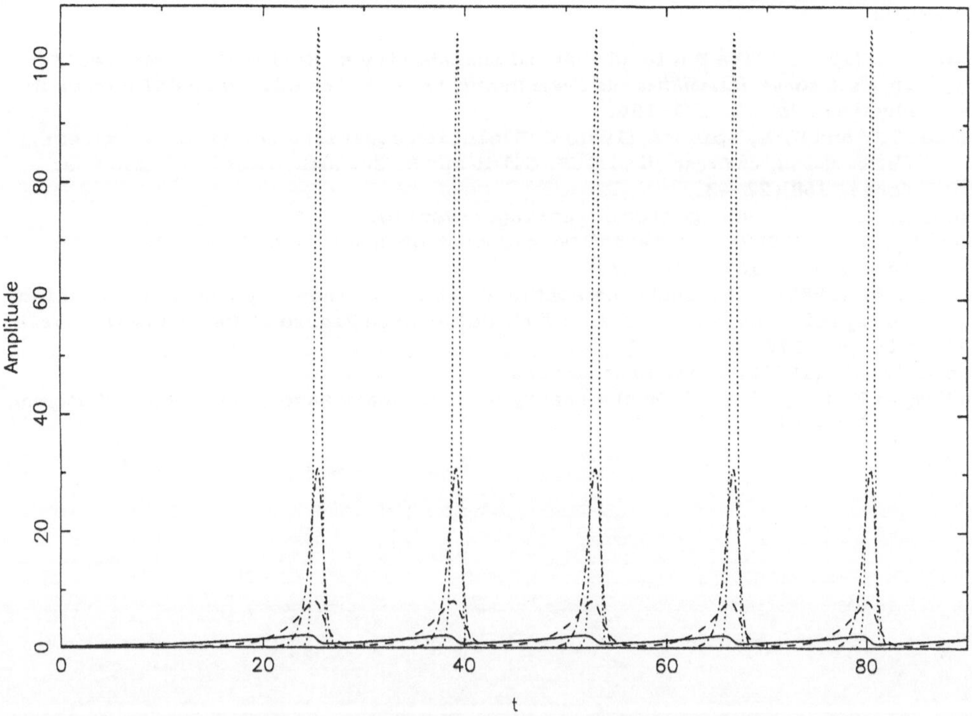

Figure1. Amplitudes of the modes, u (solid line), y_1 (dashed line), y_3 (dash-dotted line), and y_5 (dotted line), as functions of time for $\epsilon = 0.1$, $\lambda_n = \epsilon 2^{n-5}/10$ and $u(0) = 2\epsilon, y_n(0) = 0$.

plicated for they typically involve multiple integrals, even though one of the integrals may be done by going around the singularity. We do not have a general method for discussing such terms, beyond what is known in plasma physics, though we hope that a normal form theory may eventually be developed for them. Our aim here is to suggest that these resonant terms may have significant physical consequences in the resonant shuddering of extended systems.

We have tried to indicate how it may be possible to excite modes that have very high wavenumber by transferring excitation to them through a resonant chain, as in the cracking of a whip. The number of such cracks will clearly depend on details, such as the spatial and temporal spectrum of the stable modes and on the nature of the instability. Our model has allowed only a simple instability with a countable number of stable modes, but even that seems enough to produce interesting effects.

ACKNOWLEDGEMENTS

We are grateful to E.T. Scharlemann for very helpful discussions of the issues of §§2 and 3. A part of this work was done during the GFD Summer Program of W.H.O.I. in 1989. We acknowledge financial support from the N.S.F. under grant PHY87-04250, from the Air Force under grant AFOSR89-0012, and from the Navy under contract ONR N00014-87-K-0005.

REFERENCES

Ahlers, G. (1975) ''The Rayleigh-Bénard instability at Helium temperatures'', in *Fluctuations, Instabilities and Phase transitions*, T. Riste ed., Nato ASI Series B: Physics, Vol 11, 181-194.

Coullet, P. and E. A. Spiegel (1988) ''Evolution equations for extended sytems,' *Energy stability and convection*, G.P. Galdi and B. Straughan eds., Pitman Res. Notes, **168**, 22-43.

Haken, H. (1978) ''Synergetics,'' Springer-Verlag.

Pikovsky, A. S. (1989). ''Spatial Development of Chaos in Nonlinear Media,'' *Phys. Lett. A*, **137**, 121-127.

Pocheau, A. (1989). ''Structures spatiales et turbulence de phase en convection de Rayleigh-Bénard'', *Thése d'État*, Université Pierre et Marie Curie, Paris VII, pp. 176.

Procaccia, I. (1990), These proceedings.

Van Kampen N. G. (1955). ''On the theory of stationary waves in plasmas,'' *Physica*, **21**, 949-963.

BOUND STATES OF INTERACTING LOCALIZED STRUCTURES

Christian Elphick, G. R. Ierley, Oded Regev and E. A. Spiegel

Physics Department
Universidad Técnica F. Santa María
Valparaíso 110-V Chile

Department of Mathematical Sciences
Michigan Technological University
Houghton, MI 49931

Physics Department
Technion
Haifa 32000, Israel

Department of Astronomy
Columbia University
New York, NY 10027

Localized or solitary structures are frequently formed in extended systems under the combined effects of instability and dissipation.[1] The effective particle approach widely used for integrable systems[2] and in quantum field theory[3] can also be used for such systems. Many of the nonlinear PDEs encountered in macroscopic physics can be thus reduced to ODEs that give insight into the full problem, particularly when the original system is invariant under a continuous group. Each solitary structure is assigned a set of the group parameters and these become collective coordinates characterizing the state of the system.[4,5]

Work of this kind has been done in a variety of physical problems possessing translational invariance.[6,7] Remarkably, it appears that such methods have not yet been developed for systems having Galilean invariance as well, perhaps because there are some delicate points in this extension. We report here how this can be carried through and note some interesting issues that arise when the localized structures are not symmetric.

To be explicit, we consider nonlinear PDEs of the form:

$$\partial_t u + u\partial_x u + \nu\partial_x^2 u + \mu\partial_x^3 u + \lambda\partial_x^4 u = 0. \tag{1}$$

This equation appears in the study of long waves on a thin layer of viscous fluid flowing down an inclined plane,[8] the so-called Kapitza problem. Well-known special cases of (1) include the Burgers equation ($\mu = 0,\ \lambda = 0$), the Korteweg-de Vries equation ($\nu = 0,\ \lambda = 0$),[9] and the Kuramoto-Sivashinsky equation ($\mu = 0$).[10] In all cases, (1) may be derived from the complex time-dependent Ginzburg-Landau equation that describes the amplitude modulation of unstable waves in some extended systems.

Nonlinear Evolution of Spatio-Temporal Structures in Dissipative Continuous Systems
Edited by F.H. Busse and L. Kramer
Plenum Press, New York, 1990

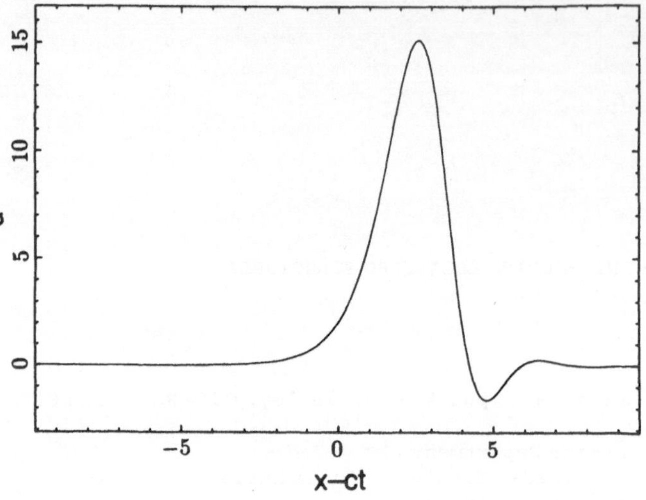

Fig. 1. A homoclinic orbit (pulse) from (2) for $\lambda = 1, \mu = 1, \nu = 2$.

The first two special cases are completely integrable systems which are useful in testing our approach, but they have scale invariances that would complicate our presentation. As to the K-S equation, we find it less interesting than the general case of (1) since it lacks dispersion. It may be seen that, as a result, its associated ODE has no attractor.

What we mean by an associated ODE is seen on substituting $u(x,t) = H(\xi)$ into (1) where $\xi = x - ct$. This leads to an ODE for H that we may integrate once to obtain

$$\lambda H''' + \mu H'' + \nu H' + \frac{1}{2}H^2 - cH = 0. \tag{2}$$

The choice of zero integration constant is equivalent to choosing boundary conditions that put the system into a selected inertial frame. This equation is known to have both periodic and chaotic solutions and this richness is behind the complexity we find here.

The solutions connecting the fixed points in the phase space of (2) to themselves (homoclinic orbits or pulses[11]) or to each other (heteroclinic or fronts[9]) describe localized structures in (1). In each, there is a core where the properties change rapidly and to which we assign a width, σ. When the ODE is autonomous, the pulse or front typically approaches its asymptotic values exponentially as $\xi \to \pm\infty$. Because of the invariance properties of (1), if we have a solution $H(\xi)$ then $H(\xi - Y)$ is also a solution as is $H(\xi - X) + V$, where Y and V are constants and $\dot{X} = V$.

We next mention some key features of our derivation of the general equations for the coordinates and velocities of localized structures. For definiteness, we use the example of the homoclinic solution of equation (1): the pulse shown in figure 1 was obtained for $c = c_0 = 5.51$ in the case $\lambda = \mu = 1$; $\nu = 2$. In studying N pulses, we assign values of X and V, the parameters of the Galilean symmetry group, as the individual positions and velocities of each pulse, calling these X_i and V_i, where $i = 1, 2, \ldots, N$. A particular value of c selects that inertial frame in which a single pulse is a steady solution. For several pulses in interaction, steady solutions may be found in other frames.

When the mutual separations of the structures are all larger than their width, σ, we suppose that they are elementary excitations. That is, we neglect the consequences of strong distortions that may create or destroy pulses. Indeed, when the separations of the structures exceeds a few σ, the interactions among them are exponentially weak and we need to take only nearest-neighbor interactions into account.

In the neighborhood of the i-th pulse we seek an approximation of the form

$$u(x,t) = H(\xi - Y - X_i) + V_i + H(\xi - Y - X_{i-1}) + H(\xi - Y - X_{i+1}) + R, \tag{3}$$

which represents a pulse, its nearest neighbors and a small correction, R. The correction includes the effects of other than nearest neighbors. Since the superposition (3) is not an exact solution, we allow the group parameters to depend on time in partial compensation.

In gauge field theory, the parameters are allowed to depend continuously on x as well as t. Here, by writing (3) only for the neighborhood of the i-th pulse, where $|\xi - Y - X_i| < \sigma$, we introduce a discrete dependence on the effective spatial variable i. As in phase dynamics, we obtain this dependence from a solvability condition that renders the asymptotic calculation of R possible, in the limit of large pulse separation. We do not have the space here to provide the derivation of these equations of motion for the group parameters, and will spell that out elsewhere. However, the general form of these equations is consistent with the symmetries of (1) and can be anticipated from them.

First, the equations of motion of the localized features depend only on their *relative* positions and velocities because of the two symmetry groups. Second, the interactions are linear in the velocities because (1) is quasi-linear. Finally, only linear terms in $\dot{Y}$ appear because (1) is translationally invariant and linear in first derivatives in time. Thus, the equations have the following form:

$$\dot{X}_i = V_i, \tag{4}$$

$$\begin{aligned}\dot{V}_i = F_L(X_i - X_{i-1}) + F_R(X_{i+1} - X_i) + \dot{Y}\,[D_L(X_i - X_{i-1}) + D_R(X_{i+1} - X_i)] \\ + (V_i - V_{i-1})\,G_L(X_i - X_{i-1}) - (V_{i+1} - V_i)\,G_R(X_{i+1} - X_i),\end{aligned} \tag{5}$$

where F_I, G_I and D_I with $I = L(eft), R(ight)$ depend on the arguments indicated. The full calculations provide the explicit dependences as linear functionals of the localized structures.

Equations (4) and (5) are not closed since they contain $2N + 1$ variables. But an equation for Y can be derived on applying the condition that $\mathcal{J} = \int u dx$ is a conserved quantity, which follows from (1). Application of this conservation law to (3) leads to the condition

$$2\sigma \sum_{i=1}^{N} \dot{V}_i = 0. \tag{6}$$

Like a Lagrange multiplier, Y is the enforcer of this constraint.

Far from its core, the pulse is exponentially small, and the functions F and G are easily evaluated. In that case, they are also exponentials, which decay to the appropriate asymptotic values:

$$F_I(Z) = f_I e^{-\beta_I Z} \cos(\omega_I Z + \phi_I), \qquad G_I(Z) = g_I e^{-\beta_I Z} \cos(\omega_I Z + \psi_I), \tag{7}$$

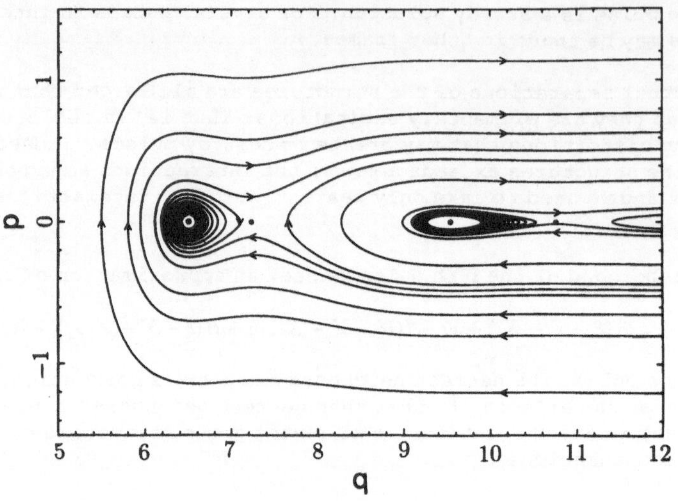

Fig. 2. A portion of the phase portrait for eqns. 7 showing the fixed points. The values of the constants are, in units with $\sigma = 1$: $f_R = -644.0, f_L = -214.6, g_R = 29.9, g_L = -38.7, \beta_R = 1.2, \beta_L = 1.0, \omega_R = 0, \omega_L = 1.9, \phi_R = 0, \phi_L = -0.38, \psi_R = 0, \psi_L = 0.92$.

where the index I may again take the values R or L and all the quantities except Z are constants derived from the pulse shape. To this degree of approximation, $D_L + D_R = [H(\sigma) - H(-\sigma)]/2\sigma \equiv \Delta H = const$. The numerical values of the constants for our example are given in the caption of figure 1.

The two-body problem illustrates the way Y enters into the Galilean dynamics of localized structures. For $N = 2$, we introduce the new coordinates and velocities

$$Q = \frac{1}{2}(X_1 + X_2), \quad q = X_2 - X_1, \quad P = \frac{1}{2}(V_1 + V_2), \quad p = V_2 - V_1. \tag{8}$$

Then the relative motion is governed by

$$\dot{q} = p, \tag{9a}$$
$$\dot{p} = f_L e^{-\beta_L q} cos(\omega_L q + \phi_L) - f_R e^{-\beta_R q} + p\,[g_L e^{-\beta_L q} cos(\omega_L q + \psi_L) + g_R e^{-\beta_R q}] \tag{9b}$$

while the motion of the entire system is given by:

$$\dot{Q} = P, \tag{10a}$$
$$\dot{P} = 0, \tag{10b}$$
$$2\,\Delta H\,\dot{Y} = -[f_L e^{-\beta_L q} cos(\omega_L q + \phi_L) + f_R e^{-\beta_R q}] - p\,[g_L e^{-\beta_L q} cos(\omega_L q + \psi_L) - g_R e^{-\beta_R q}] \tag{10c}$$

Eqs. (9) may be solved independently of (10). In Fig. 2, we show the positive q portion of the phase portrait for equations (9), where the unit of q is the pulse width, σ. We see a few of the countably many fixed points that arise in (9). These are alternately stable foci (or spiral points) and saddle points. The basins of attraction of the stable spirals decrease in size as the separation of the pair of structures increases. The ''ground'' state of the implied bound pair has a separation of about 6σ, so the approximations used here are safe for the description of these aspects of the dynamics.

A bound state of (9) corresponds to a solution of the associated ODE. We find numerical solutions of (2) in good agreement with such solutions in regard to pulse spacings and the velocities of the reference frames. These bound states have steady frame velocities other than that of the single pulse. When, in a two-body interaction, a bound state is formed, the pair is able to go to the correct frame because of the $\dot{Y}$ term in (5). Once it is there, the right side of (10c) vanishes identically. If we had not included Y, the equations of motion, derived in the usual way, would still provide the initial acceleration to the new frame in leading order. However, without higher order terms, this acceleration would lead to breakdown of the approximation. This issue does not arise in examples where the pulses are symmetric about their cores, but it is vital in a number of cases like ours.

Bound states can be expected to appear in any PDE whose associated ODE is (2). Indeed, the existence of such solutions has been proven for reaction-diffusion equations, which have translational invariance but *not* Galilean invariance.[12] The lack of Galilean invariance of course makes the pulse dynamics and the scattering problems quite different from the results shown in Fig. 2.

To get some idea of what may happen during hard collisions, when the pulses approach to within a distance σ from each other, we need to use the full functional form of the pulse shapes (instead of just the exponential tails). When we do this, we get the phase portrait shown in Fig. 3, on a larger scale than in Fig. 2. An additional saddle point has appeared at $q = 0$, $p = 0$, and the modified topology of the solution curves implies the existence of a critical relative velocity p_c (about 11.2 for the parameters of this example). Below p_c (in magnitude), particles incoming from infinite separation are reflected back, while above it they penetrate to the other side. However, we must emphasize that results for cases with strong overlap are outside of our strict region of validity and should be considered only suggestive. We shall report on the comparison with numerical solutions of (2) in a longer paper.

In the limit of large N, we can immediately learn something by setting $V_i = V_0 = const$ and $-\dot{Y}\Delta H \equiv A = const$. This corresponds to a constant velocity of the whole pattern of N pulses in an asymptotic steady state. Then (4) and (5) reduce to the pattern map:

$$F_R(\Delta_{n+1}) = A - F_L(\Delta_n), \qquad \Delta_n \equiv X_n - X_{n-1}, \tag{11}$$

or explicitly (when Δ_n are larger than σ):

$$f_R\, exp(-\beta_R\, \Delta_{n+1}) = A - f_L\, exp(-\beta_L\, \Delta_n)\, cos(\omega_L\, \Delta_n + \phi_L). \tag{12}$$

For an infinite train of pulses, there is always a fixed point of the map corresponding to uniformly spaced pulses, and (12) is a relation between the spacing and A.

The map (12) has the same form as the Poincaré map for flows arising from ODEs of the general form (2), when the parameters λ, μ, ν are close to the ones used by us. Such maps can be derived using the arguments of Shil'nikov[11] or of Melnikov[13]. In the present work, this map tells us how successive pulses are spaced. It is known that such spacings may be uniform, periodic or *spatially* chaotic (when $\beta_L/\beta_R < 1$, as in our case).

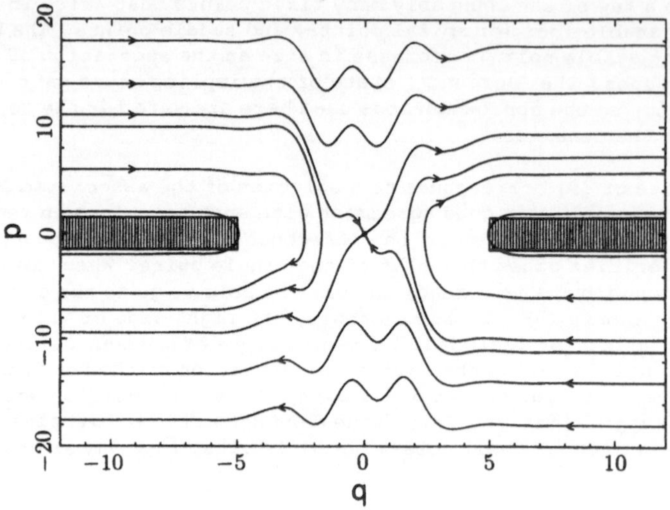

Fig. 3. The phase portrait for the two body problem suggested by eqns. (9), when the full functional form for the pulses is used. Within the shaded regions, this figure is indistinguishable from the more detailed view of Fig. 2.

For the map explicitly given by (12), we may solve $\Delta_{n+1} = \mathcal{F}(\Delta_n)$. Instability of the fixed point $\Delta_n \equiv \Delta_* = const$ depends on the parameter $\alpha = \mathcal{F}'(\Delta_*)$. The necessary and sufficient condition for instability of the map is $|\alpha| > 1$.[14] For pulse trains without Galilean invariance, the condition for instability is $1 + \alpha < 0$.[7] For the present case, with Galilean invariance, we have again the condition $1 + \alpha < 0$ for instability with, in addition, the condition that $G_R(\Delta_*) + G_L(\Delta_*) > 0$. The latter may be interpreted as a condition of negative effective friction. The instability of the uniformly spaced pattern leads to a pairing of pulses, which doubles the spatial period of the pattern.

Such insights lead us to think that the effective particle approach is of value in studying the dynamics of localized objects such as defects. Thus, they help us rationalize some of the salient features of the existing numerical experiments on (1),[15,16] which reveal a steady state consisting of ''a row of solitary pulses of ... equilibrium amplitudes".[15] Regular arrays of pulses have also been seen in numerical simulations of interacting localized structures in the Ginzburg-Landau equation,[17] and they correspond to solutions of the pattern map (11). But just how successful the method will be can be found only after more detailed quantitative comparison with numerical experiments and the laboratory experiments described elsewhere in this volume. In such comparisons, it proves useful to try to reformulate the effective particle approach in terms of the boundary conditions appropriate to those situations. Thus, we have performed extensive simulations on the motions of localized solutions of (1) with periodic boundary conditions and will report them elsewhere. For the present, we are quite content to have been able to sketch some of the remarkable collective behavior revealed by the equations of motion of localized structures.

The appearance of discrete bound states and the consequent possibility of forming large ''molecules'' of coherent structures is an intriguing phenomenon which has given us a new outlook on pattern theory. We shall elaborate on these issues elsewhere. We shall also return to the continuous limit of our dynamical equations (4)-(5), as if they were difference equations, to make contact with the theory of phase dynamics.[18] Above all, it seems worthwhile to have in hand a suitable description of the Galilean dynamics of interacting localized structures. Just as many ODEs have in common Poincaré maps of the same form, so too, it appears that whole classes of PDEs will share a common set of dynamical equations for their localized structures. This may point a way to a classification of such PDEs.

ACKNOWLEDGEMENTS

We are grateful to E. Meron for his valuable participation in the beginning of this work and N.H. Baker for making available his computer code for solving ODEs. Our collaboration on this project was made possible by support from the NSF under grant PHY87-04250 and from the Air Force under grant AFOSR89-0012.

REFERENCES

1. *Solitons and Coherent Structures* edited by D. K. Campbell, A. C. Newell, R. J. Schreiffer, H. Segur, Physica (Amsterdam) 18D (1986).
2. M. J. Ablowitz and H. Segur, *Solitons and the Inverse Scattering Transform* (SIAM, Philadelphia, 1981).
3. R. Rajaraman, Phys. Rev. D 15, 2806 (1977).
4. C. Elphick and E. Meron, preprint (1989).
5. C. Elphick and E. A. Spiegel, Proc. Summer Inst. in GFD, G. Flierl, ed. (Woods Hole Oceanographic Inst., 1989).
6. P. Coullet, C. Elphick and D. Repaux, Phys. Rev. Lett. 58, 431 (1987).
7. C. Elphick, E. Meron, and E. A. Spiegel, Phys. Rev. Lett. 61, 496 (1988); SIAM J. Appl. Math., in press.
8. J. Topper and T. Kawahara, J. Phys. Soc. Jpn. 44, 663 (1978).
9. G. B. Whitham, *Linear and Nonlinear Waves* (Wiley-Interscience, New York, 1974).
10. Y. Kuramoto, *Chemical Oscillations, Waves and Turbulence* (Springer, Berlin, 1984).
11. A. Arneodo, P. H. Coullet, E. A. Spiegel, and C. Tresser, Physica (Amsterdam) 14D, 327 (1985).
12. J. Evans, N. Fenichel, J. Feroe, SIAM J. Appl. Math. 42, 219 (1982), and J. Feroe *ibid.* 235.
13. P. Coullet and C. Elphick, Phys. Lett. A 121, 233 (1987).
14. P. Collet and J.-P. Eckmann, '*Iterated Maps on the Interval,* Progress in Physics Vol. 1 (Birkhäuser, Boston, 1980).
15. S. Toh and T. Kawahara, J. Phys. Soc. (Japan) 54, 1257 (1985).
16. T. Kawahara, Phys. Rev. Lett. 51, 381 (1984).
17. C. S. Bretherton and E. A. Spiegel, Phys. Lett. 96A, 152 (1983).
18. S. Fauve, Proc. Summer Inst. in GFD, G. Veronis, ed. (Woods Hole Oceanographic Inst., 1985).

NONLINEAR DYNAMICS OF PARTICLE-LIKE

STATES OF MULTIDIMENSIONAL FIELDS

I.S. Aranson, K.A. Gorshkov, A.S. Lomov, and M.I. Rabinovich

Institute of Applied Physics, USSR Academy of Sciences
46 Uljanov.Str., 603600, Gorky, USSR

1. INTRODUCTION

In recent five-ten years there have appeared quite a number of papers concerned with various multidimensional spatial patterns in nonlinear media: periodic and quasiperiodic lattices, localized (particle-like) solutions, spiral structures, vortices, and so on (Makhan'kov, 1983; Haken, 1983; Linde, 1984; Kuramoto, 1984). As a rule, unified methods are applied for the analysis of the structures of different origins and the results obtained can be used in different fields of physics and not only physics. This means that we witness the advent of a new field of nonlinear dynamics -the theory of structures in nonlinear media. Many problems of this theory were posed and solved before, in the theory of nonlinear waves, in the first place. However, they were, ordinarily , concerned (except, perhaps, hydrodynamics) with one- dimensional structures: solitons, shock waves, dissipative structures (combustion fronts, etc.). The investigation of multidimensional structures that was necessitated by astrophysics, the physics of atmosphere and ocean, the nonlinear field theory, microelectronics and biophysics, gave new formulations of the problems. For example, the study of structure bifurcations, i.e., the off-beat variation of their topology when the governing parameter passes the critical value; the spatial interaction of structures, that is the formation of bound spatial states, including "planet-like" states; their regular or chaotic spatio-temporal dynamics explaining, in particular, many manifestations of nonlinear field turbulence. The latter is also known as spatio-temporal chaos of structures or topological turbulence. The main problems of the multidimensional theory are the mechanisms of the spatial localization of structures, their stability and the interaction of such particle-like patterns. All these problems are considered in our paper.

1. THE FORMULATION OF THE PROBLEMS

The particle-like solutions, i.e., the localized solutions of a nonlinear field, are the solutions decreasing rather fast from the

Nonlinear Evolution of Spatio-Temporal Structures in Dissipative Continuous Systems
Edited by F.H. Busse and L. Kramer
Plenum Press, New York, 1990

localization maximum to the periphery. A more accurate and formal definition is based on the convergence of some integrals.

In order to work out the approach to the construction of a theory of localized structures we shall consider laboratory (Dieker, Pindak and Meyer, 1986; Olsen, 1985) and computer (Gorshkov, Lomov and Rabinovich, 1989) experiments.

In spite of the diversity of the media and experimental situations in which localized structures are observed, these structures are remarkably universal and the origin of their universality is, yet, to be elucidated. However, even now it is clear that the space dimensions, the form of nolinearity and the type of spatial dispersion are most essential for pattern formation. Note that identical particle-like solutions are possible even in the fields with different dynamics, in particular, in conservative and nonequilibrium dissipative fields, if their nonlinear characteristics and dispersion features are equal.

Let us put aside for the present the localization mechanisms and consider some peculiarities of the individual dynamics and interaction of structures. To a rough approximation based on various experiments one can arrive at a conclusion that there are "strong" and "weak" interactions of particle-like solutions. When the interaction is weak the localized structures (dislocations, solitons,etc.) are spaced rather widely from one an other and the field of one structure at the center of the localization of another one can be considered to be weak. This facilitates the construction of a consistent theory of such adiabatic interactions. Strong coupling is the interaction during which particle -like solutions change qualitatively, appear and die "colliding" with each other. The result of strong interaction depends on the particle prehistory and field characteristics. We believe that in the description of the nonlinear dynamics of ensembles of structures the laws of strong interactions can be formulated as the rules according to which mutual conversions occur with subsequent transition to weak interactions that can be described approximately taking into account the conservation of the localized object structure.

Apparently, not every nonlinear dynamics of localized structures can be described within the interaction scheme presented above("adiabatic" -weak interaction and "bifurcation" - strong interaction). However, experiment furnishes ample example. Therefore it is essential to formulate general models describing local experimental situations, on the one hand, and allowing for a fairly complete analytic and computer investigation, on the other hand. We shall take as such models generalized gradient systems, i.e., systems in the form

$$T(\partial/\partial t)U = -\delta F/\partial t + \varepsilon G(U,r,t), \quad \varepsilon \ll 1 \qquad (1)$$

Here U is a set of physical variables; F is a functional having the sense of the free energy of the system (medium, field); G is the nonlinear operator that takes into account the external field effect, the deviation of (1) from the potential system and other factors and $T(\partial/\partial t)$ is a linear differential operator.

For $\varepsilon=0$ the dynamics of the particle-like solutions of (1) is determined by the form of the operator $T(\partial/\partial t)$. If $T(\partial/\partial t)=\partial/\partial t$, (1) is a conventional gradient system where all solutions are statical, as $t\to\infty$, and correspond to the local minima of the functional F (in this case F is the Lyapunov functional). From here it follows that the localized structures in such media either become fixed or go to the infinity, or die, for example, as a result of merging.

When the perturbations are small ($\varepsilon\ll1$), particle-like structures will not be statical in dissipative media either, they will interact with external fields, walk randomly, change their shape gradually, etc.

It will be shown below that the identification and investigation of models of the form (1) contribute to a better understanding of the processes of localized pattern formation and interaction.

2. TOPOLOGY OF MULTIDIMENSIONAL STRUCTURES: EXAMPLES

It seems to be impossible to classify the topological properties of localized structures when the theory of such nonlinear patterns is only on the verge of its formulation. Therefore we shall restrict ourselves to some examples which, however, demonstrate vividly the topological diversity of possible localized structures.

Consider a class of gradient systems whose free energy functional can be expanded in field series and in field gradient series near the point where the homogeneous state loses its stability (the physical nature of the field isn't specified). It should be borne in mind that the scalar realfield U may lose stability both in a soft and in a hard fashion and instability is possible not only at maximal but also at finite scales. The resulting expansion ($F=\int F dr$) has a form

$$F_1(U,\nabla U)=\alpha U^2 + \beta U^3 + \delta U^4 + \xi \nabla U)^2 + \zeta(\nabla U)^4 \quad (2)$$

to which eq. (2) corresponds. A similar expansion for a complex scalar field has a form

$$F_2(U,\nabla U)=\alpha|U|^2 + \beta|U|^4 + \delta|U|^6 + \xi|\nabla U|^2 + \zeta|\nabla U|^4 \quad (3)$$

(here guage invariance, i.e. the phase independence of F_2 , is taken into account). The gradient equation corresponding to such a free energy has a form of a generalized Ginzburg-Landau equation (Haken,1983; Malomed, 1986)[+)].

The gradient models (2) and (3) under study are related directly to the potential models of Hamiltonian fields:

$$\partial^2 U/\partial^2 t = - \delta F_1/\delta U \quad (4)$$

and

$$\partial U/\partial t = - i\delta F_1/\delta U \quad (5)$$

Apparently, the stable localized structures revealed within the gradient models (2) or (3) turn the functionals in the right-hand sides of (4) and (5) to zero as well. From this it follows, in particular, that thelocalized structures found in this fashion correspond to the localizedstates of the appropriate Hamiltonian fields. The stability of thesestates,however, needs further investigation. The point is that aminimal potential energy at these states isn't, yet, a guarantee of thestability of the found localized solutions with respect to the resonanceperturbations, i.e. , to the excitation or intrinsic degrees of freedom of such structures-particles.

+) The fields, described by the functionals with simple Laplacian terms $|\nabla U|^2$, corresponding to the nonlinear diffusion equations, have no stable multidimensional localized solutions (Aranson and Rabinovich, 1983; Hobart, 1963). The necessary stability conditions of localized structures in such models are the hard excitation of the medium and the finite scale of initial instability.

Consider as an example of possible existence of stable stationary localized structures the models of convection with hard excitation. We shall distinguish between two situations, depending on the type of medium instability: a) aperiodic shortwave instability described by a generalized Swift-Hohenberg equation (Swift and Hohenberg, 1977; Walgraff, 1985) (an analog of this equation was used for the description of convection in an ordinary liquid):

$$U_t=-U + \beta U^2 - U^3 - (k_o^2 + \Delta)^2 U \tag{6}$$

and b) oscillatory shortwave instability described by a generalized Ginzburg-Landau equation (Malomed, 1986; Moses and Stainberg, 1986; Rehberg et al., 1988; Kolodner, Bensimon and Surko, 1988) (this situation occurs in binary liquid convection):

$$U_t=-U + \beta|U|^2U - |U|^4U - (k_o^2 + \Delta)^2U \tag{7}$$

The analyses of these models have much in common, therefore we shall consider them in parallel.

The Hamiltonian models, associated with these equations, are

$$U_{tt}=-U + \beta U^2 - U^3 - (k_o^2 + \Delta)^2 U \tag{8}$$

(the generalized Klein - Gordon equation) and

$$U_t=i(-U + \beta|U|^2U - |U|^4U - (k_o^2 + \Delta)^2)U \tag{9}$$

(the generalized nonlinear Schrodinger equation).

We shall investigate the solutions that are localized in the sense of integral boundedness : $\int|U|^2 d\vec{r} < \infty$.

Simple steady-state solutions of this type are axisymmetric ones, i.e.

$$U(x,y)=U_o(\rho), \text{ where } \rho^2=(x^2+y^2) \text{ or } \rho^2=(x^2+y^2+z^2) \tag{10}$$

The model (7) also has a solution containing the angular dependence $\theta=arctg(y/x)$, $\varphi=const$ +):

$$U(x,y)=\Phi(\rho)e^{\pm i(m\theta+\varphi)} \tag{11}$$

These solutions, when $m = 0$, will be called spiral solutions with a topological charge m (by analogy with Hagan (1982)). As $\rho\to\infty$, the localized solutions decrease exponentially, with the decrease being oscillatory, i.e.,

$$U \approx e^{-\alpha\rho}cos(\eta\rho) \tag{12}$$

where

$$\alpha=|Re(i-k_o^2)^{1/2}|, \quad \eta=|Im(i-k_o^2)^{1/2}| \tag{13}$$

The localized solutions are stable only in the range $\beta_o\le\beta\le\beta_1$. The boundaries of this range can be estimated approximately based on the following considerations.

The models (6) and (7) can be represented in a gradient form

$$\partial U/\partial t = - \delta F_3/\delta U \tag{14}$$

where F_3 is a "free energy" functional

+) A three-dimensional analog of this solution is a toroidal vortex.

$$F_3=\int(U^2/2 - \beta U^3/3 + U^4/4 + ((k_o^2+\Delta)U)^2)/2)d\vec{r}$$

for the model (6) and

$$\partial U/\partial t = - \delta F_4/\delta U^* \quad (15)$$

where

$$F_4=\int(|U|^2 - \beta|U|^4/2 + |U|^6/3 + |(k_o^2+\Delta)U)^2|)d\vec{r}$$

for the model (7).

We can rewrite the expression for F_3 in the following form

$$F_3=\int U^2(1/2 - \beta U/3 + U^2/4) + ((k_o^2+\Delta)U)^2)/2)d\vec{r} \quad (16)$$

It is apparent that when $0 < \beta \leq \beta_o^{(1)}=3/2^{1/2}=2.12....$ the expression is valid for all U: $1/2-\beta U/3+U^2/4 \geq 0$. Consequently, the functional is nonnegative when $\beta \leq 2.12...$ and reaches its global minimum when $U=0$. In view of its gradient form (14) may only decrease on the trajectories of eq. (6). Hence, any localized structures collapse when $\beta \leq \beta_o^{(1)}$.

Following the same considerations for the model (7) we obtain a nonnegative F_4 for $0 < \beta \leq \beta_o^{(2)}=4/3^{1/2}=2.3094...$

The right-hand boundary of the range of β_1 is estimated assuming that there is no spreading of structures in systems (6) and (7), i.e.,under the condition of "energetically unprofitable" propagation of transfer fronts resulting in the transformation of a localized state into a nonlocalized one. For sufficiently high β there exists stable homogeneous equilibrium $|U|^2=const$ in (6) and (7),which corresponds to a nonlocalized state. The corresponding critical value of β is

$$\bar{\beta}_1 = 2\sqrt{1 + k_o^4} \quad (17)$$

The amplitude of stable equilibrium for (6) is given by

$$U_m = \beta/2 +(\beta^2/2- 1 - k_o^4)^{1/2} \quad (18)$$

and for (7) by

$$|U_m|^2 = \beta/2 +(\beta^2/2- 1 - k_o^4)^{1/2} \quad (19)$$

Consider a solution of (6) in the form of a cylindric front

$$U(r) = \begin{cases} U_m , & \rho \leq \rho_o , \rho_o \gg 1 \\ U_m \exp(-\alpha(\rho-\rho_o) \cos(\eta\rho) , & \rho > \rho_o \end{cases} \quad (20)$$

Such a solution occurs with spreading, i.e., when the initial solution (10) transforms to a nonlocalized state. The solution (20) is a two-dimensional analog of the transfer front in one-dimensional active media. Let us substitute (20) into the functional F_3 and trace its variations with increasing ρ_o. Then the contribution from the exponentially small "tail" can be neglected and we obtain approximately

$$F_3 \approx \pi\rho_o^2(1 + k_o^4 - \beta U_m^2)/4 \tag{21}$$

When $\beta U_m/3 \geq 1 + k_o^4$, the increase of ρ_o leads to the decrease of F_3, i.e., the spreading is "energetically profitable" and the localized structures are likely to be unstable. Using (18) we readily obtain the condition of unspreading for the model (6):

$$\beta_1^{(1)} \leq 3\sqrt{(1+k_o^4)/2} = 2.12\sqrt{1 + k_o^4} \tag{22}$$

and, in a similar fashion, for the model (7):

$$\beta_1^{(2)} \leq 4\sqrt{(1+k_o^4)/3} = 2.31\sqrt{1 + k_o^4}$$

Thus, both the models have a range of β where localized solutions may be stable. These qualitative results were confirmed in a numerical simulation on the model (6) and (7) (Figs.1 a and b). We used the following parameters in our experiments: β=2.5, and k_o=1. The integration was performed by an implicit split-step method employing FFT. The size of the integration domain was 40×40 and the number of FFT harmonics was 64×64 (Aranson et al., 1989)

Localized structures of the form (10) were observed for the model (6) (see Fig. 2). The stability of solutions with m =0, 1, 2 ... was investigated for the model (7). Only simple spirals (m=±1) turned out to be stable. It is not excluded that for other values of the parameter β solutions with $|m|\neq 1$ may also be stable.

Thus, the upper and the lower estimates for stability regions of a localized structure relative to the parameter β is as follows:

$$2.12 \leq \beta \leq 2.12\sqrt{1 + k_o^4} \tag{23a}$$

for the model (6) and

$$2.31 \leq \beta \leq 2.31\sqrt{1 + k_o^4} \tag{23b}$$

for the model (7).

The described localized two- and three-dimensional particles are stable indeed. The common feature for these 'elementary particles' is the field behaviour in the structure periphery: the field decays exponentially, oscillating with the characteristic spatial scale $1/k_o$. The presence of these oscillations (typical of localized spirals as well) is caused by spatial dispersion in the models under study (6-9).

3. WEAK INTERACTION. ADIABATIC THEORY

The general concept of the adiabatic description of the dynamics of particle-like solutions is quite clear and close to that in the investigation of the interaction and evolution under the action of various perturbations of nonlinear solitary waves, solitons (Gorshkov and Ostrovsky, 1981; Lonngren and Scott, 1978). In this respect the approach presented below can be considered as the development of the basic idea set forth in those papers concerning the localized patterns of arbitrary physical origin.

With the perturbations (inhomogeneous or nonstationary medium, exposure to external forces, the effect of weak fields on other particle-like solutions, etc.) taken into account, the solution of interest near the given localized structure can be sought in the form of a series

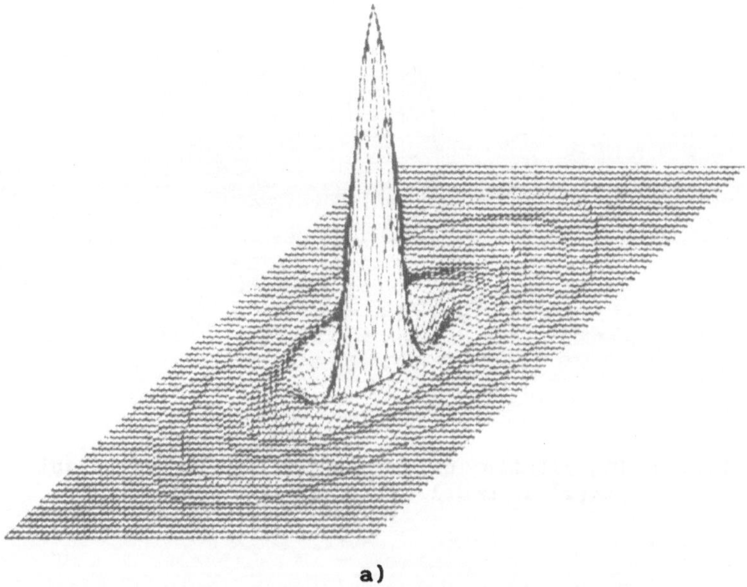

a)

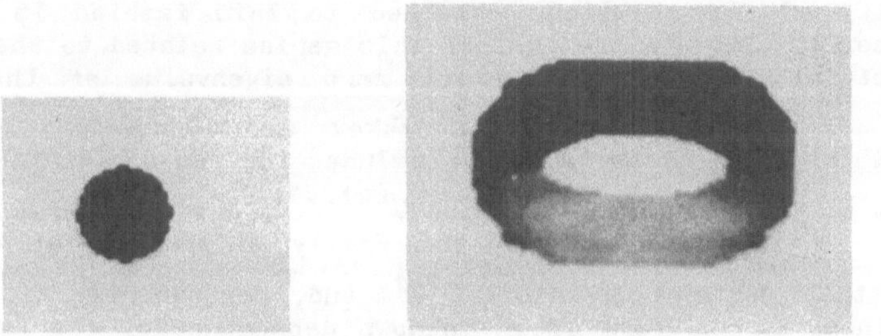

b)

Fig. 1. (a) Distribution of $U(x,y)$ for localized axi-symmetric stable solution in models (6) or (8); (b) "Elementary particles": a ball and a torus. The parameter values are: β=2.5 and k_o=1.

$$U(\vec{r},t) = U^{(o)}(\vec{r}) + \sum \varepsilon^n U^{(n)}(\vec{r},t) \tag{24}$$

where the dominant term of the series, $U^{(o)}(\vec{r})$, is a known elementary localized solution and ε is a small parameter of the problem. When we deal with the interaction of localized structures this parameter is equal to the ratio of the fields of foreign localized structures at the site of the given localized structure to the maximal value of the field of this structure. Equations (1) and (24) yield in a standard fashion the following successive approximations for $U^{(n)}(\vec{r})$:

$$\hat{L}\, U^{(n)}(\vec{r}) = H^{(n)} \tag{25}$$

Here $H^{(n)}$ contains only the functions of preceding approximations and the specified weak fields of foreign localized structures. The solution of the linear evolution problem (25) can be constructed by the expansion of

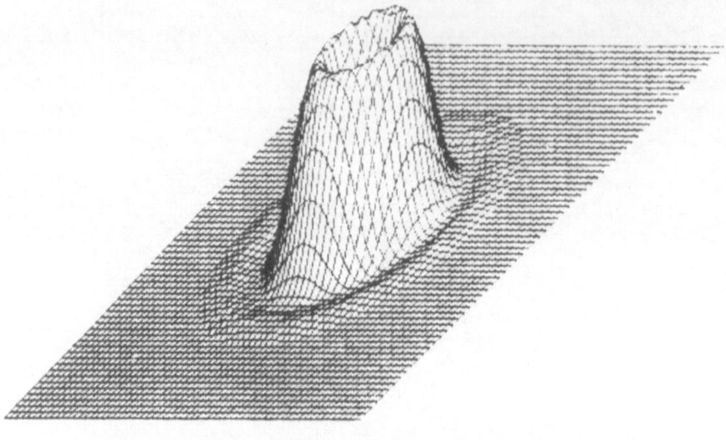

Fig. 2. Distribution of the $|U(x,y,)|^2$ for a stable spiral in model (7): β=2.5 and k_o=1.

$U^{(n)}(\vec{r})$ in a complete set of eigenfunctions of the operator $\hat{L}$. A typical peculiarity of the solutions obtained in this fashion is a secular divergence (in time) of the part of this series related to the localized eigenfunctions belonging to the zero eigenvalue of the discrete spectrum of the operator $\hat{L}$ (Gorshkov and Ostrovsky, 1981). These localized eigenfunctions always include the ones obtained by the variation of the generating solution $U^{(o)}(\vec{r},\vec{C})$ over the parameters $C(\vec{C} = \{C_1, C_2 \ldots C_n\})$ characterizing the family of such solutions. This suggests that introducing into $U^{(o)}(\vec{r})$ and, consequently, into $H^{(n)}$ an arbitrariness in the form of an unknown dependence of the parameters $\vec{C}$ on slow time ($\tau=\varepsilon t$), the secular divergence caused by these eigenfunctions can be suppressed, provided that the modified right-hand parts of H_* in (25) ($H_*=\vec{C}_t \nabla_C U + H$) are orthogonal to the corresponding eigenfunctions of the operator conjugate to $\hat{L}$:

$$\int U^+ H_*^{(n)} d\vec{r} = 0 \qquad (26)$$

If a subsystem of the localized functions corresponding to a zero eigenvalue consists only of $\partial U^{(o)}/\partial C_l$ $\{l=1,2\ldots\}$, then it may be argued, at least for the solutions to a first approximation (n=1), that with the use of the orthogonality conditions (26) the primary divergence is eliminated and the initial distributed system can be described approximately by a reduced finite-dimensional system:

$$\sum_{l=1}^{m} \int U_{C_k}^+ U_{C_l} \frac{dC_l}{dt} d\vec{r} = \int U_{C_k}^+ H^{(1)} d\vec{r} \qquad (27)$$

The most essential problem that remains uncertain when using (27) is the determination of the number of parameters characterizing the families of localized structures of interest. This problem, in turn, is closely connected with the symmetry of initial field equations and with the structure of localized solutions. Thus, the number of the "external" parameters describing the position and the orientation of static localized structures in space is maximal (six) in the case of an isotropic and a homogeneous model (1): three angular parameters $(\varphi_o,\theta_o,\eta_o)$ and center coordinates $(\vec{r}=\{x_o,y_o,z_o\})$. In a conservative model (1), besides these six "external" parameters, four more may appear: three velocity components (V_x,V_y,V_z) and energy that are, evidently, due to the inertial transformation invariance of the coordinate system and to the time shift invariance.

3.1. Interaction of Elementary Ball-Type Structures. First consider the problem in terms of the nonconservative model (6) . In this case the linear operator L has a form

$$\hat{L} = \partial/\partial t + (k_o^2+\Delta)^2 + 1 - 2\beta U^{(o)}(\rho) + 3(U^{(o)}(\rho))^2 \tag{28}$$

and is a self-conjugate one. Because the variable coefficients of the operator L depend only on ρ, all their eigenfunctions can be sought by the separation of variables $U^+(\vec{r})=Y_p^{(q)}(\theta,\varphi)U^+(\rho)$ where $Y_p^{(q)}(\theta,\varphi)$ are spherical functions. The eigenfunctions needed in (28) correspond to the translational mode and are proportional to Y_1, therefore they can be written as

$$\vec{r}U^+(\rho) = \nabla_r U^{(o)}(\rho) \tag{29}$$

The gradient structure of eigenfunctions and the spherical geometry of $\hat{L}$ $U^{(o)}$, contribute to an exquisite form of eq. (27). Thus, owing to the symmetry of $U^{(o)}$, we may obtain the following equations for the ball centers:

$$M\frac{d\vec{r}_{12}}{dt} = \int\nabla_r U^{(o)}(\rho)\, H_{12}^{(1)}\, d\vec{r} \tag{30}$$

where $M=\int|U^{(o)}|^2 d\vec{r}/3$. The gradient structure of the eigenfunctions of specifies the gradient form of the right-hand parts of (30). Indeed, the right-hand part of (25) written, for example, for the first particle, in the first approximation is equal to

$$H_1^{(1)} = 3(U_1^{(o)})^2 U_2^{(o)} - 2\beta U_1^{(o)} U_2^{(o)} \tag{31}$$

where $U_1^{(o)}$ depends only on ρ while $U_2^{(o)}$ depends on $|\vec{r} - \vec{R}|$ in the coordinate system originating from the first particle; and $|\vec{R}|=|\vec{r} - \vec{R}|$ is the distance between the particles. Substituting (31) into (30) and integrating it by parts, yields an expression in the form

$$M\frac{d\vec{r}_1}{dt} = \int\nabla_r U_1^{(o)} H_1^{(1)}\, d\vec{r} = \int\nabla_r U_1^{(o)}\,[3(U_1^{(o)})^2 U_2^{(o)} - 2\beta U_1^{(o)} U_2^{(o)}]\,d\vec{r} \tag{32}$$

Because the operations ∇ and $\int$ may exchange places, eq. (30) will take on a gradient form. Replacing $U_2^{(o)}(r)$ by the asymptotic forms (12) , we obtain (30) in an explicit form

$$M\frac{d\vec{r}_1}{dt}=I\nabla_{r_1}\frac{e^{-\alpha R}}{R}\cos\eta R \qquad I=const \tag{33}$$

An equation for $\vec{r}_2$ can be obtained in a similar fashion. Combining these equations (the right-hand parts depend only on R) we finally obtain an equation for R:

$$M\frac{d\vec{r}_1}{dt}=I\nabla_{R}\frac{e^{-\alpha R}}{R}\cos\eta R \tag{34}$$

In this case there is a countable (infinite) number of the equilibrium points that correspond to the bound states of the elementary structures of the sphere type; the motion starting with an arbitrary initial distance $R(t=0)$ between the particles ceases at the nearest stable equilibrium state. Equation (33) is easily generalized to the case of the interaction of an arbitrary number of localized structures by adding the terms $exp(-\alpha R_{1j})cos(\eta R_{1j})/R_{1j}$ and $R_{1j}=|\vec{r}_1 - \vec{r}_j|$ to the right-hand part of (33). It is already difficult to enumerate the configurations of the bound states of an arbitrary number of particles (regular and irregular polyhedrons and polygons, finite and infinite lattices, etc.). All these bound states may be multiplied by a similar extension of the figures until they reach the next stable equilibrium state (this, of course, affects the stability factor of the bound state). Examples of stable three-dimensional bound states, derived in our numerical simulations, are presented in paper by Gorshkov et. al. (1989).

To conclude this section we shall consider in brief the problem of the interaction of localized structures of the same type but in terms of the conservative model (8). In this case the family of statical solutions is part of the points of the common family that is described by three more parameters (velocity $\vec{V}$) and the derivation of the equation of interaction needs, generally speaking, to know the structural dependence of the generating solution on $\vec{V}$. However, the derivation of the sought equations in the limit $\vec{V} \to 0$ is analogous to the one described above. The only difference is that all transformations are performed in the second approximation (the first approximation gives an apparent relation (cf. Gorshkov and Ostrovsky (1982):

$$\frac{d\vec{V}}{dt}=\nabla_R\frac{e^{-\alpha R}}{R}\cos\eta R \qquad \frac{d\vec{R}}{dt}=\vec{V} \tag{35}$$

Equation (35) is an example of a classical two-body problem. It is well known that this problem may be reduced to squaring. Note that all statical bound states of systems (35) and (33) coincide completely. The essential difference is that in this case non- statical bound states that correspond to the mutual rotation of the particles around the common center of gravity are also possible. This stable rotating 'planetary' system of two balls (the 'particle' with spin) , obtained numerically, is shown in Fig. 3.

3.2 Spiral Wave Interaction. A particular form of the equation of spiralmotion can be obtained knowing the structure of the eigenfunctionsbelonging to the zero eigenvalue of the discrete spectrum and the structure of L . It is easily seen that L of the linearized problem (7) is a self-conjugate operator and the eigenfunctions obtained

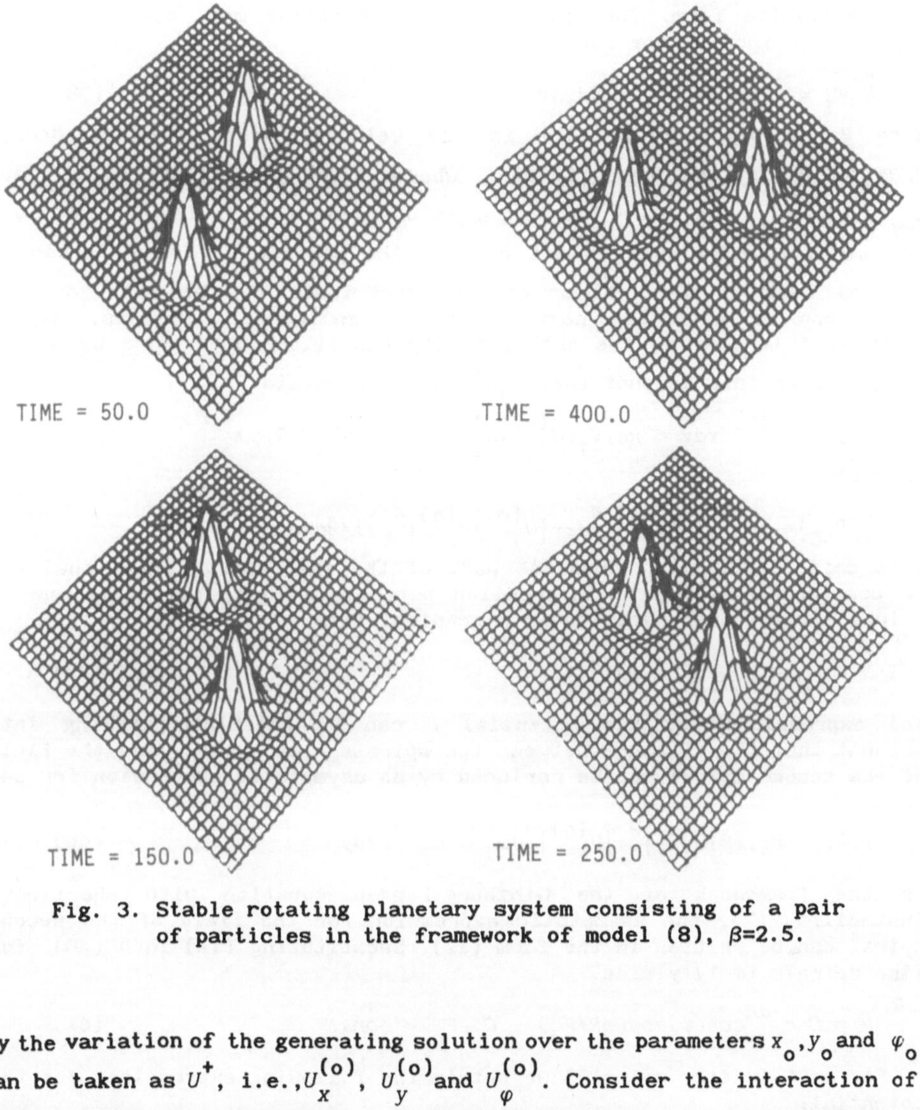

Fig. 3. Stable rotating planetary system consisting of a pair of particles in the framework of model (8), β=2.5.

by the variation of the generating solution over the parameters x_o, y_o and φ_o can be taken as U^+, i.e., $U_x^{(o)}$, $U_y^{(o)}$ and $U_\varphi^{(o)}$. Consider the interaction of widely spaced spirals. Then a two-spiral solution can be represented approximately as a superposition of individual spirals:

$$U = \Phi^{(o)}(\rho_1)e^{i(\theta_1-\varphi_1)} + \Phi^{(o)}(\rho_1)e^{i(m\theta_2-\varphi_2)} \qquad (36)$$

Like spirals (rotating in one direction) correspond to m=1 , while unlike spirals (rotating in opposite directions) correspond to m=-1. The correction $H_1^{(1)}$ to the first spiral is written in the form

$$H_1^{(1)}=Q-\partial U_1^{(o)}/\partial t;\; Q=(\beta|U_1^{(o)}+U_2^{(o)}|^2-|U_1^{(o)}+U_2^{(o)}|^4)(U_1^{(o)}+U_2^{(o)}) -$$
$$-\beta(|U_1^{(o)}|^2U_1^{(o)}+|U_2^{(o)}|^2U_2^{(o)})+|U_1^{(o)}|^4U_1^{(o)}+|U_2^{(o)}|^4U_2^{(o)} \qquad (37)$$

The transformation of (27) gives the equations of spiral motion in a more exquisite form. The orthogonality conditions may, apparently, be represented in a vector form

$$\hat{M}\vec{V}_1 = -Re\int(\nabla_1 U^{(o)})^* Q dxdy \qquad (38)$$

Here $V=(\partial x_1/\partial t, \partial y_1/\partial t, \partial\varphi_1/\partial t)$ is the velocity vector, $\nabla=(\partial/\partial x_1, \partial/\partial y_1, \partial/\partial\varphi_1)$, $\hat{M}$ is the mass tensor, where $M_{xx}=M_{yy}= |\nabla U|^2 dxdy/2$ and $M_{\varphi\varphi}=\int|U|^2 dxdy$, and Q are the perturbations generated in the interaction. The terms in (37) that do not contain the product of U_1 and U_2 can be omitted because they either do not contribute to the orthogonality conditions or have the next order of smallness. Besides, it is convenient to rewrite the orthogonality conditions in a more symmetric form. Taking into account that $\nabla_1 U_2^{(o)}= 0$, we obtain

$$Re\int\nabla_1 U_1^{(o)*} Q dxdy = Re\int\nabla_1(U_1^{(o)}+U_2^{(o)})^* Q dxdy=\nabla_1 P(R,\varphi)$$

The value

$$P(R,\varphi)=\int\{\beta|U_1^{(o)}+U_2^{(o)}|^4/2-|U_1^{(o)}+U_2^{(o)}|^6/3\}/2dxdy;\quad \varphi=\varphi_1-\varphi_2$$

that coincides with a nonsquare part of the free energy functional can be considered as a pair interaction potential. Then, the equations of spiral motion can be written in a gradient form

$$\hat{M}\vec{V}_{1,2} = -\nabla_{1,2}P(R,\varphi) \qquad (39)$$

The expressions for the potential P can be calculated taking into account that the distance between the spirals R is large and the field of the second spiral can be replaced by an asymptotic expression for $\rho\to\infty$ Then

$$P(R,\varphi)=Re\int\{\beta|U_1^{(o)}|^2-|U_1^{(o)}|^4\}U_1^{(o)*}U_2^{(o)}dxdy \qquad (40)$$

In the framework of the Ginzburg-Landau equation with short-wave instability (7), the asymptotic expression for the field of the second spiral can be written in the form (12). Substituting (12) into (40) for like spirals (m=1) yields

$$P \approx C e^{-\alpha R}\cos(\varphi)\cos(\eta R+\xi_o) \;;\; C,\ \xi_o = const \qquad (41)$$

For m=-1 (unlike spirals) we obtain the following expression for the potential:

$$P \approx C e^{-\alpha R}\cos(\varphi+2\Psi_{12})\cos(\eta R+\xi_o)$$

where $\Psi_{12} = arctg(y_1-y_2)/(x_1-x_2)$ is the viewing angle of the second spiral from the site of the first spiral.

In either case these potentials have a countable number of stable fixed points corresponding to bound states: spiral dipoles (when m=-1) and double spirals (when m=+1). These states were observed in a direct numerical experiment with eq.(7). The results are shown in the paper by Aranson and Rabinovich (1989).

3.3 Chaotic Drift of Localized Structures. In this section we shall consider a case demonstrating a nontrivial dynamics of localized structures in an external field taking as an example a weakly perturbed equation (7)

$$U_t=-U + \beta|U|^2U - |U|^4U - (k_o^2 + \Delta)^2U + i\beta'|U|^2U + \varepsilon|U|^2 f(x,y) \quad (42)$$

where $|U|^2$ describes the nonlinear response of the medium to the external effect whose spatial distribution is characterized by the function $f(x,y)$ that is smooth in comparison with the size of the spiral. Equations (38) describing the evolution of the spiral parameters (x_o, y_o and φ_o) are obtained using the eigenfunctions same as in the problem of interaction described in sect.3.2. Substituting $H^{(1)}= i\beta'|U^{(o)}|^2U^{(o)}+\varepsilon|U^{(o)}|^2 f(x,y)-U_t^{(o)}$ into (38) we obtain

$$\begin{aligned} dx_o/dt &= B\, Re\{fe^{i\varphi_o}\} \\ dy_o/dt &= B\, Im\{fe^{i\varphi_o}\} \\ d\vee_o/dt &= \Omega \end{aligned} \quad (43)$$

where $B=\varepsilon\int\rho\Phi(\rho)^3 d\rho/(3m_{xx})$ and $\Omega=\beta'\int\rho\Phi(\rho)^4 d\rho/m_{\varphi\varphi}$.

Let us represent the function $f(x,y)$ in the form$f(x,y)=cos(ky)+isin(kx)$ and analyse, first, the situation when the spiral does not rotate ($\beta'=\Omega=0$). Then (43) can be represented in the form of a Hamiltonian system with a Hamiltonian $H=B/k(sin(ky_o)+cos(kx_o))$

$$\begin{aligned} dx_o/dt &= B\, cos(ky_o) = \partial H/\partial y_o \\ dy_o/dt &= B\, sin(kx_o) = -\partial H/\partial x_o \end{aligned} \quad (44)$$

It is seen that system (44) has a separatrix network covering all the x,y -plane. It is apparent that the separatrix network will disintegrate when the nonstationary perturbations are weak and a stochastic spider-web similar to that considered by Chernikov et. al. (1987) will appear. The stochastic spider-web is indicative of possible spiral wave stochastic drift along the x, y -plane at arbitrary long distances. Taking into account the spiral rotation ($\Omega\neq0$) we shall get nonstationary (time periodic) perturbations. In this case we shall obtain the equations for the spiral center coordinates which depend on time explicitly thus guaranteeing the destruction of the separatrix network in the unperturbed system (44) (Aranson and Rabinovich, 1989).

We have no numerical verification of two-dimensional random walk of localized structures in the framework of eq.(42). We have confirmed a one-dimensional random walk of the soliton of a nonlinear Schrodinger equation (Aranson et. al, 1989), and in the two- dimensional eq. (8) with periodic inhomogeneity (Aranson, Lomov, and Rabinovich, to be published). We hope to prove the above results in our papers to come.

It is remarkable that the model (8) permits to understand the mechanisms of the birth of spatio-temporal disorder in purely dynamic models of the field. One of the primary mechanisms of this type is the formation of the localized states and their random walk in space as a result of interaction with other "particles" or regular fields. Thus, under certain initial conditions the dynamics of the system

$$U_{tt}=-U + \beta U^2 - U^3 - (k_o^2 + \Delta)^2U + U\, f(x)f(y), \quad f(x)=a_0+a_1cos(kx) + \ldots \quad (45)$$

can be considered as the interaction of one "particle" with a periodically inhomogeneous field. In two-dimensional space, in particular, eq. (27) will take on the form

$$d^2x_0/dt^2=\sin(kx_0)(a_0+a_1\cos(ky_0))$$

$$d^2y_0/dt^2=\sin(ky_0)(a_0+a_1\cos(kx_0)) \quad (46)$$

System (46) is known to describe random walk, which were also observed in direct simulations on eq. (45) (see Fig. 4).

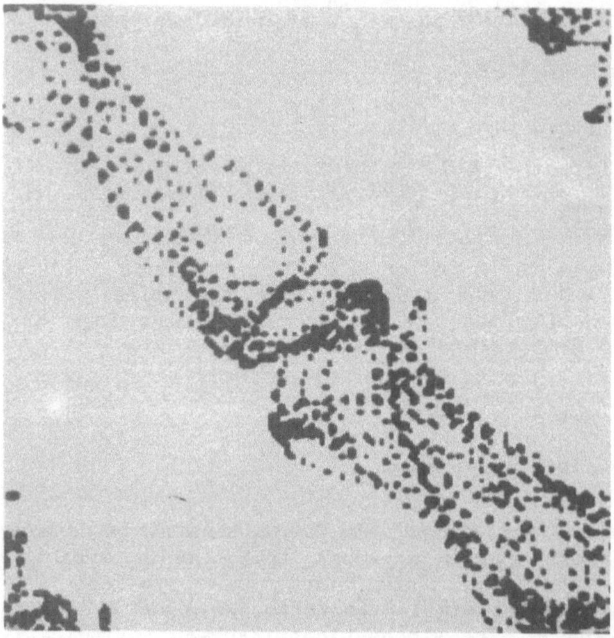

Fig. 4. Random walk of a localized structure within the model (45). The integration domain size is 64×64 and the number of harmonics is 128. The integration time is 6433.

REFERENCES

Aranson I. S., Gorshkov K. A. and Rabinovich M. I., 1989, Phys. Lett. A, 139:65.

Aranson I. S. and Rabinovich. M. I., 1989, J. Physics A , 22.

Chernikov A. A., Sagdeev R.Z., Usikov D. A. and Zaslavsky G. M., 1987, Phys. Lett. A, 125:101.

Dieker S. B., Pindak R. and Meyer R. B., Phys. Rev. Lett.,1986, 56:1819.

Gorshkov K. A., Lomov A. S. and Rabinovich M. I., 1989, Phys. Lett. A, 137:50.

Gorshkov K. A. and Ostrovsky L. A., 1981, Physica D, 3:428.

Hagan. P. S., 1982, SIAM J Appl. Math. , 42:726.

Haken. H., 1983, "Synergetics. Hierarchy of Instabilities in Self-Organized Systems and Devices". Springer, N. Y.

Hobart. R. H., 1963, Proc. Phys. Soc., 82:201.

Joets A. and Ribotta R., 1988, Phys. Rev. Lett., 60:2164.

Kolodner P., Bensimon P. and Surko C. H., 1988, Phys. Rev. Lett., 60:1723.

Kuramoto Y., 1984, "Chemical Oscillations. Waves and Turbulence", Springer, Berlin.

Linde H., 1984, Topological Similarities in Dissipative Structures, in: "Self-Organization Autowaves and Structures Far from Equilibrium," V.I.Krinski, ed, Springer, Berlin.

Lonngren K. and Scott E., 1978, "Solitons in Action," Academic Press, N.Y.

Malomed B. A., 1986, DAN SSSR, 291:327.

Makhan'kov. V. G. , 1983, Fiz. Elementarn. Chastits i Jadra, 14:123, (in Russian).

Moses E. and Steinberg V., 1986, Phys. Rev. A, 34:693.

Olsen. J. L., 1985, J. Low Temp. Phys., 61:167.

Rehberg I., Rasenat S., Fineberg J., Juarer M. and Steinberg V., 1988, Phys. Rev. Lett., 61:2443.

Swift J. and Hohenberg P. C., 1977, Phys. Rev. A, 15:319.

Walgraff D., 1989, Flow Field Effects on Dynamical Instabilities, in: "Instabilities and Nonequilibrium Structures, II," E. Torapegui and D. Villaroel ed., Kluner Academic Publisher.

CAUSTICS OF NONLINEAR WAVES AND RELATED QUESTIONS

Yves Pomeau
Laboratoire de Physique Statistique
24, rue Lhomond
75231, Paris Cedex 05, France

Abstract

Patterns of parallel and equidistant layers are rather common in physical systems, as smectic liquid crystals or Rayleigh-Bénard rolls in thermal convection. Although a minimisation principle would impose perfectly straight layers, boundary conditions may change this when they impose the layers to be parallel to a closed smooth curve. Then a Huygens-like construction allows to draw the full pattern and yields caustics in general for linear wave equations. I show that, in nonlinear systems those caustics are to be replaced by grain boundaries, and cusps by ends of those grain boundaries. I study too the equivalent of the diffraction dressing of those grain boundaries by using a phase equation approach.

The linear Helmholtz equation for the amplitude F of waves with wavenumber k in free space reads:

$$(\Delta + k^2)\ F = 0 \tag{1}$$

where Δ is the usual Laplacian, that will be considered here in two space dimensions: $\Delta = \frac{d^2}{dx^2} + \frac{d^2}{dy^2}$, x and y being the Cartesian coordinates. This equation can be solved in the limit of geometrical optics when the curvature of the wavefront, that is a line of Cartesian equation F = constant is much less than k. However the occurence of caustics leads to situations where this limit of geometrical optics cannot be taken everywhere. We shall consider the so called cusp singularity (this is a singularity in the limit of geometrical optics but not for the original Helmholtz equation), where two caustics merge (see figure 1).

A form of solution of the Helmholtz equation near this cusp follows from the Huygens principle and may be written as a Fresnel integral in the form[1]:

$$F = \frac{e^{ikx}}{(2\pi k)^{1/2}} \int ds\ \exp\ i\Phi\ (x,y;s) \tag{2}$$

Nonlinear Evolution of Spatio-Temporal Structures in Dissipative Continuous Systems
Edited by F.H. Busse and L. Kramer
Plenum Press, New York, 1990

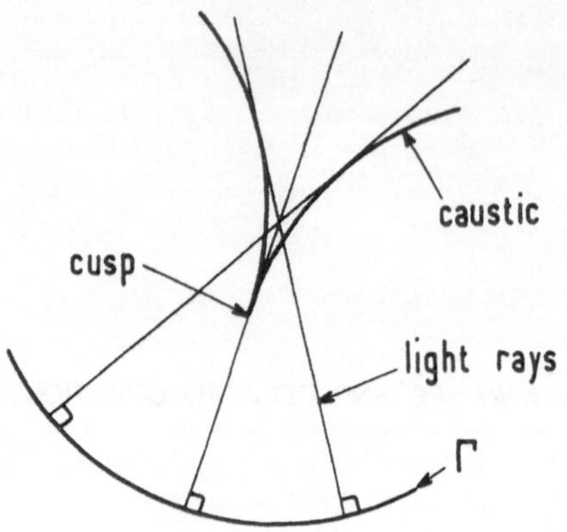

Fig. 1. This is the classical Huygens construction from a smooth wavefront Γ. The enveloppe of the normals to Γ is the locus of the center of curvature of Γ and it is generically made of caustics merging at cusps. Diffraction is important near the caustics and at the cusp.

where

$\Phi(x,y;s) = \dfrac{Ls^4 - s^2x/2 + sy}{\lambda}$ is a dimensionless phase, λ the wavelength $2\pi/k$, and the dummy variable s may be seen as the curvilinear distance along a smooth wavefront and near the ray tangent to the cusp. The length L is much larger than λ and is of order of the radius of curvature of the smooth wavesurface from which the Huygens construction is made.

Outside of a region near the cusp (here at x=y=0, with our choice for the origin of coordinates) and near the caustics, the integral on the r.h.s. of (2) is computed by steepest descent, equivalent here to the limit of geometrical optics. Each extremum of the phase Φ with respect to the s-variable yields a contribution to F. In a region of extent x ë $(\lambda R)^{1/2}$ and y ë $(\lambda^3 R)^{1/4}$, near x ë y ë 0 (x=y=0 being the tip of the cusp), the limit of geometrical optics breaks down and after proper rescaling one is left with the solution by Pearcey[1] for the diffraction "dressing" of the cups.

In the present note I consider how this picture is transformed for nonlinear wave equations. I will have in mind two physical situations:

i) Rayleigh-Bénard rolls contained inside a closed two dimensional curve (that would correspond to near threshold situations where the third dimension can be neglected). This might be thought mathematically, although in an oversimplified form as the solution of the nonlinear partial differential equation:

$$\left[\varepsilon - (\Delta + k^2)^2\right] G = G^3 \qquad (3.a)$$

$$-\Delta G/k^2 = G = K\varepsilon^{1/2}, \text{ on the external boundary with K large.} \qquad (3.b)$$

As this is an Euler-Lagrange equation for an energy functional it makes sense to seek the absolute minimum of this energy, that defines the optimal solution. The b.c. is such that rolls of wavelength $2\pi/k$ tend to be parallel to the outer boundary, at least in the limit of a small positive ε in which we shall deal from now on. Indeed it is known[2] that Rayleigh-Bénard rolls tend instead to be perpendicular to the outer boundaries if natural b.c. are retained. In the limit of an outer boundary with a radius of curvature much larger than $\varepsilon^{-1/2}$, and than $2\pi/k$, one expects the optimal structure to be made almost everywhere of locally parallel rolls with an amplitude and wavelength that would be optimal for an infinite system. Our problem now is to find how to bend on large scales this roll system in order to fit the b.c. Before, I describe another physical situation where a related problem appears.

ii) this is the situation of a smectic liquid crystal. Such a crystal, at least for the less complicated form, is made of equidistant layers[3] that may bend on large scale to comply with boundary conditions and/or defects. Suppose that the layers have to be parallel to an outer boundary, that is a closed convex and arbitrary (= not a circle) curve with a very large radius of curvature (again we restrict ourselves to 2D situations). If one starts to build the structure one will construct an inward going set of parallel layers, the first one being the boundary itself, and this will create a cusp defect at some point. If I understand well[3,4] what is predicted in the literature to happen in that situation, the structure of this crystal should be an Appolonian packing: one fits first into the outer boundary the largest possible circle (this one being then filled with concentric layers), then the largest one in the space not yet filled by this largest circle and so on. Indeed this leaves at the end defects (actually smoothed up or even discarded when reaching the small layer thickness). The Hausdorff dimension of the set of defects left by this Appolonian packing is believed to be universal and near 1.3058[4], bigger than the dimension 1 of the grain boundary generated by the construction described below. This would mean that this Appolonian packing is not optimal in terms of energy. It would be interesting to know experimentally if those structures with line defects (or surfaces in a 3D sample) may occur.

Coming back to the original problem, I will show first that the bending of the rolls does change the total energy by a finite amount only in the limit of a very large system (this neglects perhaps logarithms of the size of the system, but they are irrelevant at the present degree of approximation).

By expansion in the small curvature of the rolls, as done in ref. 5.a. for slightly bent rolls, one finds that the perturbation to the energy density brought by the curvature, given that straight "rolls" are optimal, is of order $(\frac{\lambda}{R})^2$ times the energy of the unperturbed state, R being the radius of curvature of the roll structure. Thus the perturbation brought to a 2D pattern of rolls by their bending is of order of this perturbation times the total area, of order R^2 if the bending originates from b.c. effect. Thus this bending energy tends to a constant (up to logarithms perhaps) as the size of the system increases indefinitely.

From those remarks the construction of the equiphase lines of the optimal structure is identical to the one for the solution of the Helmholtz equation in the geometrical limit, once the wavenumber is set to the optimal value for the non linear equation (3). Whence it makes sense to consider the problem of formation of cusps for the equation (3) whenever the optimal pattern is made of weakly bent rolls.

Indeed the occurence of caustics and of cusps for such a nonlinear equation is governed by the same geometrical phenomenon as for the Helmholtz equation: inside a cusp more than one ray goes through each point. This multiplicity of rays is not a problem with linear wave equations: one adds the contribution of all rays, as done for the Fresnel integral (2) in the limit of geometrical optics. But there is no obvious superposition principle for nonlinear equations as (3). Nevertheless, at least for small positive ε one can find by perturbation in ε steady solutions made by the superposition of finitely many rolls with different orientations. But those solutions are not optimal and if we would choose one of them to fill in the area inside a cusp (that is where three rays at least go through each point) one would lose a surface term in the energy, compared to the energy of a pattern with only one roll orientation. This implies that there is no caustics in the optimal state: those caustics would be at the border between a one-ray domain and a three-rays domain, this last one being too costly in energy for the ground state as just shown.

Thus in this ground state, the local solution inside the cusp is to be found by choosing by continuity the same ray (among the three possible ones) as the one outside of the cusp and discarding the other ray contributions. This keeps the continuity of the roll orientation.

But it is also true that one cannot go from one side of the cusp to the other by keeping the same choice among the three possible rays and then merge continuously with the outside rays, since precisely there is an exchange between two extrema of the phase function inside a cusp. Said in another way, a transition has to occur somewhere inside the cusp between the two different choices of rays, or orientations of the equiphase lines, each one being defined by continuity with the outside orientation.

In the x-y plane, this transition has to occur on a line, again in order not to increase too much the energy, because the structure is not optimal in this transition domain. Now we have to specify a way to draw this transition line. A familiar example of the effect of nonlinearities on cusp singularities is[6] the formation of shocks in the (position, time) space for Riemann simple waves in 1D, as described by the Burgers-Hopf equation for instance in the small dissipation limit. Then the trajectory of the shock wave follows from an equal area construction- or Maxwell rule. There on each side of the shock only one (instead of three) characteristics goes through each point and characteristics cannot cross the shock. In the problem under consideration the equal area construction is replaced by a condition of mechanical equilibrium of the grain boundary, where the two different roll orientations meet. As the "primitive" equations are isotropic, this equilibrium implies that the grain boundary is the bissectrix of the two rays (see figure 2). Then the cusp itself is the end point of the grain boundary.

In this framework, the pattern generated by the solution of (3) inside a smooth curve Γ may be seen as follows: one draws first the set of normals (or rays) to Γ and moves inward along those normals. If this is done at constant speed, one finds at some time a cusp where two caustics would merge in the case of the Helmholtz equation. In the present case on the contrary one forgets the caustics and replaces the cusp by the end of a grain boundary. The precise shape of this grain boundary is defined by two conditions, first no ray crosses through, then it is locally the bissectrix of the two rays (or equivalently of the two roll orientations) ending on each side.

This construction reduces the curve Γ to its "skeleton" of grain boundaries. It is an interesting problem to know if one can come back from this skeleton to Γ, up to an obvious choice of the optical length along the rays. I intend to study this problem in a future publication.

Another question of interest is the equivalent of the Pearcey integral for the non linear wave equation (3) in the weakly nonlinear regime. In the linear problem, the length scales for the diffraction near the cusp are fixed by combining the macroscopic length L entering into the definition of the phase Φ and the wavelength itself. Here another parameter enters into the game, that is ε and this introduces also large length scales, as $\varepsilon^{-1/2}$ or $\varepsilon^{-1/4}$. Those length scales being a priori independent of L, it makes sense to try to estimate the size of the domain near the cusp where the non linear "geometrical optics" cannot be applied.

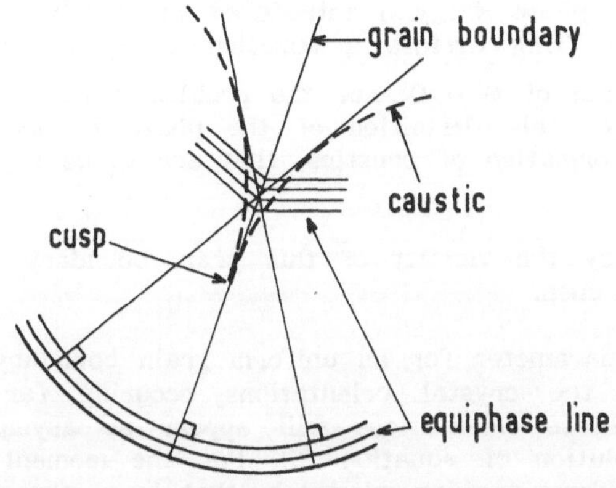

Fig. 2. Same as figure 1, but for a nonlinear wave equation. The caustics disappear and are replaced by a "grain boundary" where equiphase lines with different orientation merge. The classical cusp becomes here the end of a grain boundary.

Near the cusp the pattern that we consider is made of rolls almost perpendicular to the x-direction, and so one can use there the general amplitude equation[7]. Let χ be the complex amplitude of a solution of (3.a) with a phase factor as e^{ikx}. Then, with proper rescaling the slow spatial variations of this amplitude are described by the solution of:

$$\varepsilon\chi - \left(\frac{\partial}{\partial x} + \frac{i\lambda\partial^2}{2\partial y^2}\right)^2\chi - \chi^2\chi^* = 0 \qquad (4)$$

where χ^* is the complex conjugate of χ.

The solutions we are looking for are near optimal, so that the modulus of the amplitude is close to $\varepsilon^{1/2}$, and they can be seen at the dominant order (the meaning of this will be explained later on) as the product of this amplitude with a slowly varying phase factor, as $\varepsilon^{1/2}\exp[i\Psi(x,y)]$. This slowness is defined by the condition that the modulus of the amplitude χ adjusts itself quickly to changes of Ψ. The typical length scale for the modulus is $\varepsilon^{-1/2}$ along the x-direction and $\varepsilon^{-1/4}$ along the y-direction. Thus the phase approximation to be developped below will be valid if the variations of Ψ are on scales much longer than those typical length scales.

In this limit, the phase Ψ obeys the following equation:

$$- 2 \frac{\partial^2\Psi}{\partial y^2} \Theta + \frac{\lambda^2}{4} \frac{\partial^4\Psi}{\partial y^4} - 2 \frac{\partial\Psi}{\partial y}\frac{\partial\Theta}{\partial y} - \frac{\partial\Theta}{\partial x} = 0 \qquad (5)$$

where $\Theta = \frac{\partial\Psi}{\partial x} + \frac{\lambda}{2}\left(\frac{\partial\Psi}{\partial y}\right)^2$ is the covariant phase gradient[8]. The equation (5) is derived from (4) by plugging in $\chi = \varepsilon^{1/2}\exp(i\Psi)$ and setting to zero all terms proportional to i exp $(i\Psi)$. Indeed this yields[9] the solubility condition allowing to derive phase equations from amplitude equations. The limit of geometrical optics considered previously is recovered at once by noticing that if this phase is to vary slowly, the highest derivative on the left hand side of (5) is negligible so that (5) is satisfied if Θ is zero. Consider the phase $\Phi(x,y;s)$ introduced previously at a value of s such that $\frac{\partial\Phi}{\partial s} = 0$. This defines a function of x and y only and this function is a solution of $\Theta = 0$. For the problem under consideration, and as explained before, this definition of the phase is done by continuity, avoiding thus the formation of caustics, that are to be replaced here by a grain boundary.

Below we study the vicinity of this grain boundary and its merging with the tip of the cusp.

The relevant parameter for an uniform grain boundary is the angular mismatch between the crystal orientations occuring far away. In the present problem, those orientations will appear as asymptotic conditions imposed to the solution of equation (5). For the moment we shall limit ourselves to an uniform angular mismatch, that is to the class of solution of (5) of the form:

$\Psi(x,y) = -q^2x + \omega(y)$. At large distances from the grain boundary, the limit of geometrical optics applies and (5) reduces to $\Theta = 0$ that has here the solution:

$\omega(y) = \pm \left(\frac{2q^2}{\lambda}\right)^{1/2} y$, as y tends to $\pm\infty$. In the core of the grain boundary itself, ω is given by the solution of (5) when restricted to function Ψ of the above form:

$$- 2 \frac{d^2\omega}{dy^2}\left[- q^2 + \frac{\lambda}{2}\left(\frac{d\omega}{dy}\right)^2\right] + \frac{\lambda}{4}\frac{d^4\omega}{dy^4} - 2\lambda\left(\frac{d\omega}{dy}\right)^2 \frac{d^2\omega}{dy^2} = 0 \qquad (6)$$

A first integration gives the Newton like equation:

$\frac{d^2\Omega}{dy^2} = \frac{4}{\lambda}(\lambda\Omega^3 - 2q^2\Omega)$, where $\Omega = \frac{d\omega}{dy}$, and where the boundary condition is satisfied for the trajectory with the limit behavior $\Omega(y) = \pm\left(\frac{2q^2}{\lambda}\right)^{1/2}$ as y tends to $\pm\infty$. From this last equation the thickness of the grain boundary in the y-direction is of order $\frac{\lambda^{1/2}}{q}$. To relate that to the problem under consideration, we have to write this thickness in terms of the mismatch angle of the two crystal orientations merging at the grain boundary. Let α be this (small) angle, thus the quantity called q^2 is $-\alpha/2\lambda$, so that the thickness is of order $\lambda\alpha^{-1/2}$, that has to be much less than the "macroscopic" length (i.e. L in our notation) that enters into the formulation of the problem for the phase in the limit of geometrical optics. In particular the angle α varies on a length scale L, much larger than the thickness itself at non zero α, and the inner structure of the grain boundary may be found along this grain boundary by an adiabatic assumption: one considers the mismatch angle (and thus q) as almost independent on x, and then retains for each value of x the phase variation across the grain boundary as given by the solution of (6). Furthermore the phase approximation can be retained inside the grain boundary if its thickness is much larger than the typical distance for the variation of the modulus of the amplitude. This happens at small α, the precise condition being given below.

The adiabatic assumption made before cannot be true to the tip of the cusp, since there the angle α tends to 0 as $(x/L)^{1/2}$ (as it follows from the approximation of geometrical optics) and changes rapidly as x tends to zero. Therein diffraction dresses[1] the solution of the Helmholtz equation. Here instead we have to solve the full equation (5). This one can be put into a dimensionless form with the length scale $(\lambda L)^{1/2}$ for x, and $\lambda^{3/4} L^{1/4}$ for y, although F is of order 1. Those scalings are consistent, as they should, with the large distance behavior of the sought solution. This one is given by the grain boundary separating two domains where only one ray of the geometrical optics solution contributes. This in a sense completes the determination of the equivalent of the Pearcey integral for our problem.

However the inner length scales defined near the tip have to be much longer than the ones for the modulus of the amplitude, to keep the validity of the phase approach. This requires the inequalities: $\lambda\varepsilon^{-1/2} < (\lambda L)^{1/2}$, and $\lambda\varepsilon^{-1/4} < \lambda^{3/4} L^{1/4}$ to hold. To be consistent as far as the physical scales are concerned, we have used in those inequalities a dimensionless ε whence the form of the left hand sides. Both inequalities are equivalent to $\varepsilon > \lambda/L$, although the bifurcation toward a non trivial solution of (1) as ε increases occurs at ε ë $(\lambda/L)^2$. This is in agreement with a remark already done[5] (for axisymmetric patterns) that the phase of the solution becomes relevant at ε bigger than or equal to λ/L. A slight extension of those arguments allows to define the domain of validity of the phase equation (6) for the inner structure of the grain boundary. As already said, this limit requires that the angle α is small enough to keep the thickness $\lambda\alpha^{-1/2}$ much larger than the typical length of variation for the modulus of χ, that is $\lambda\varepsilon^{-1/4}$. Since α is of order $(x/L)^{1/2}$, the phase equation (6) can be retained in the range $(\lambda L)^{1/2} < x < \varepsilon L$, where x measures the distance along the grain boundary from its end.

Let us end this note with some comment on the more general problem of finding the large scale "phase structure" of a solution of equation (3.a) with b.c. imposing an arbitrary dependence of the phase on an external closed boundary. Yet we have considered the special case of a b.c. imposing a constant phase on this outer boundary and shown how to get a solution by drawing straight "rays" normal to this boundary and then replacing the locus of the center of curvatures of the boundary by its skeleton of grain boundaries, as described in the text. It is not obvious that other external b.c. could be handled the same way. The opposite to the situation studied here is the one where the b.c. are such that the rolls are orthogonal to the outer boundary, as it occurs in real Rayleigh-Bénard convection without forcing on this boundary. Indeed, since this boundary is curved in general, it cannot be taken as a straight ray perpendicular everywhere to the local roll orientation. This brings us back to the mathematical origin of the straight "light" rays in our construction. This comes from the eikonal solution of the Helmholtz equation that reads

$$(\text{grad}\ \Psi)^2 = k^2 \qquad (7)$$

where Ψ is the phase of the solution of the Helmholtz equation (the same equation appears in the theory of magnetic domains[9]). The general solution of this equation is:

$\Psi/k = Z(\alpha) + x \cos\alpha + y \sin\alpha$, with the condition that Ψ is stationary with respect to α, and where Z(.) is a 2π periodic function depending on the shape of one equiphase line for instance. The equation $\Theta = 0$ that was introduced before is a simple consequence of this by expansion near $\alpha = 0$, that is for nearly parallel rolls. Henceforth, in order to represent non straight "light rays", one has to change equation (7) in a drastic way. The solution to this kind of problem has been known for some time, and turns out to be quite simple in 2D geometries, as considered here. It implies basically that rays may be curved because of a small density of dislocations. The mathematics of this is as follows:

Let **n** be the local normal to the rolls. In the eikonal limit described by equation (7), this is a pure gradient: **n** = **grad**Ψ. To represent curved light rays, it is enough to add a non gradient term to **n**. This is done in general in 2D by introducing a non harmonic function $\zeta(x,y)$ such that, in coordinates:

$$n_x = \frac{\partial\Psi}{\partial x} - \frac{\partial\zeta}{\partial y}$$

$$n_y = \frac{\partial\Psi}{\partial y} + \frac{\partial\zeta}{\partial x}$$

so that the number density of dislocations (or the number of dislocations per unit area), which is proportional to **rot n** is equal to $-\Delta\zeta$. Should such dislocations be added to the phase field, one can only say that the form of the covariant gradient Θ should be changed into:

$\Theta' = \frac{\partial\Psi}{\partial x} - \frac{\partial\zeta}{\partial y} + \frac{\lambda}{2}\left(\frac{\partial\Psi}{\partial y} + \frac{\partial\zeta}{\partial x}\right)^2$. Assuming that the field ζ is known, one has not anymore a simple explicit solution of the equation $\Theta' = 0$, even though ray paths can still be drawn. The difficulty now is to take into account the extra energy introduced by the dislocation field, and to add it to the total energy and to minimize it. We intend to come back to this in a future publication.

Acknowledgement

I have benefited from discussions with Jacques Prost, Harry Suhl and Alan Newell. Part of this work has been done when I was at the Department of Mathematics of the University of Arizona where I was supported in part by the Center for Complex systems.

References

1. T. Pearcey, Phil. Mag. 37, 311 (1946).

2. S. Zaleski, Y. Pomeau and A. Pumir, Phys. Rev. A29, 366 (1984).

3. P. G. de Gennes "The Physics of liquid crystals", Clarendon, Oxford (1974).

4. R. Bidaux, N. Boccara, G. Sarma, L. de Sèze, P. G. de Gennes and O. Parodi, J. de Phys. 34 661 (1973); B. Mandelbrot "Fractals, form, chance and dimension", Freeman and Co, San Francisco (1977).

5.a. Y. Pomeau and P. Manneville, J. de Physique 42, 1067 (1981).

5.b. Y. Pomeau, S. Zaleski and P. Manneville ZAMP 36, 367 (1985).

6. See for instance: T. Poston, I. Stewart "Catastrophe theory and its applications", Pitman, London (1978).

7. L. A. Segel, J. of Fluid Mech. 38, 203 (1969); A. C. Newell, J. A. Whitehead, J. of Fluid Mech. 38, 279 (1969).

8. J. Prost, Y. Pomeau and E. Guyon, preprint (April 1989); E. Guyon, communication at the 19th Statphys. Conf., Rio de Janeiro (Brazil), August 1989.

9. Y. Pomeau and P. Manneville, J. de Phys. Lettres L40, 609 (1979).

10. P. Bryant, H. Suhl, Appl. Phys. Lett. 54, 78 (1989).

PHASE DYNAMICS AND PROPAGATION OF DISLOCATIONS

L.M. Pismen and Daniel Zinemanas

Department of Chemical Engineering
Technion – Israel Institute of Technology
Haifa 32000, Israel

1. Introduction

Much attention has been attracted in recent years to the dynamics of distorted two-dimensional patterns in non-equilibrium systems, in particular, to the dynamics of defects that are necessarily present in realistic striped patterns, such as Rayleigh-Bénard convection rolls, observed either experimentally or computationally, and play the crucial role in the overall organization of patterns in extended systems and their long-time evolution. We refer to Newell [1] for a comprehensive overview of the field.

Quantitative observations of pattern dynamics and motion of defects have been largely provided by numerical experiments with the two-dimensional Swift-Hohenberg model [2,3] and, more recently, with the real and complex Landau-Ginzburg (LG) equation [4,5]. Experimental data with best spatial resolution were obtained for convective patterns in liquid crystals [6], including recent quantitative measurements of the climb velocity of dislocations [7].

If defects were absent, computations based on universal equations of phase dynamics [1,8] would be sufficient for studying dynamics of weakly distorted patterns in large aspect ratio systems. The phase diffusion approximation, however, breaks down in the immediate vicinity of defects, where the macroscopic parameters (wave number and orientation) have to change over distances comparable with the basic wavelength of the pattern. Failing to incorporate the most important feature determining in a qualitative way the geometry of the pattern, phase equations could not be used beyond such primitive tasks as the linear stability analysis, and had to give way in realistic computations to full underlying "microscopic" equations or to their simplified models.

Nonlinear Evolution of Spatio-Temporal Structures in Dissipative Continuous Systems
Edited by F.H. Busse and L. Kramer
Plenum Press, New York, 1990

One can argue that the failure of phase equations to describe the structure of defects locally still cannot be seen as a decisive disqualifying factor, inasmuch as defects are stable topological features that leave their signature in the far field where equations of phase dynamics are expected to hold. It seems feasible therefore that dynamics of defects, to the extent it is determined by the far field, might also follow laws of a universal nature (independent on a particular underlying system), and be understood within the framework of phase dynamics.

This paper presents an attempt to compute realistic distorted patterns, incorporating formation and motion of defects, using phase equations alone, without resorting to the solution of the full underlying problem. We shall work with the simplest "generic" equation of phase dynamics, due to Cross and Newell [8], that describes evolution of weakly distorted patterns far from the onset. The numerical algorithm, described in Section 4, uses a computational device imitating the action of local Eckhaus instability to graft the nucleation and propagation of dislocations upon the large-scale dynamics obeying the phase equation. This allows to compute, using a rough grid with a single point per roll, realistic patterns evolving in about the same way as patterns constructed via a much more laborous fine-grid solution of model two-dimensional equations.

Looking beyond this visual semblance, we shall investigate in detail the motion of an isolated dislocation in a striped pattern, and compute its stationary velocity as a function of the prevailing wave number and the size of the box. This problem has been extensively studied theoretically, but is still not fully understood. Siggia and Zippelius [9], Pomeau *et al.* [10] and Tesauro and Cross [3] studied the dislocation dynamics either close to a primary symmetry-breaking bifurcation, or in the vicinity of an "optimal" wave number corresponding to vanishing transverse phase diffusivity. The deviation from the onset or from the optimal wave number appear to be natural small parameters of the problem. It is stated in Ref.3 that calculating the climb velocity must depend otherwise on full numerical computation, and as a result, "the velocity will presumably be O(1) in the basic time units of the problem". This, if true, would be rather unfortunate since, under conditions when the transverse diffusivity vanishes, one has to scale the coordinates directed along and across the wave vector anisotropically, and to use a far more complicated phase equation. Moreover, the relevant phase equation should be nonlinear, though linear equations were used in Refs. 3,9.

Bodenschatz *et al.* [4] extended the theory to non-isotropic (liquid crystal) systems and found that under these conditions the problem actually simplifies, inasmuch as the amplitude equation has a simple LG form. They computed the propagation velocity, under conditions when it is small and, respectively, the phase correction due to the moving defect is nearly isotropic, by combining the far field asymptotics obeying a linear phase diffusion equation with the numerical solution near the defect core. The nearly isotropic situation was also considered by Neu [10] who used, however, the wave number itself, rather than the deviation from criticality, as the small parameter of the problem.

The above analytical studies were mostly aimed at elucidating the dependence of the defect propagation velocity on the wave number. The influence of the size of the box entered only through a weak logarithmic dependence due to a short-scale divergency in the vicinity of the defect. A remarkable exception was the inverse proportionality of the dislocation velocity to the squared number of rolls verified computationally by Pomeau *et al.* [10]. This strong dependence was apparently caused by a diffusional spreading of a phase perturbation introduced by a moving defect, that also manifested itself in relaxation of the propagation velocity to a lower stationary value. The slow-down of an isolated defect in a large box indicates that the ratio of the basic wavelength to the size of the box, *i.e.* the inverse number of rolls $\epsilon = N^{-1}$, should appear as a natural small parameter of the problem. This justifies the application of the phase equation, and even of its linearized version, for computing the far field in extended systems where the wave number increment due to adding a single roll is $O(\epsilon)$.

The analysis of the motion of a single defect in Sections 2, 3 will follow the same general line as in Refs. [10,3,4], and rely on the equivalence (in the leading order) of an arbitrary pattern-forming system described by the generic phase equation to a potential system with an effective Lagrangian yielding identical phase diffusivities. We are lead to the conclusion that the defect velocity is proportional to $\epsilon \ln(1/\epsilon)$. This correlation well fits computational data obtained using the numerical algorithm of Section 4.

2. Solution of linearized phase equation

We assume, as usually, that an underlying "microscopic" nonlinear system is invariant to planar rotations and translations, and is verified by a family of 2π-periodic stationary solutions (one-dimensional structures) that depend on planar coordinates $\mathbf{x}$ through a single variable — eikonal $\theta(\mathbf{x})$, with $\mathbf{k} = \nabla\theta = \text{const}$. The solutions are parametrized by the wavenumber $k = |\mathbf{k}|$, that can vary within certain limits $k_{min} \le k \le k_{max}$. Structures distorted on a long scale are characterized by a wave vector $\mathbf{k}$ slowly varying throughout an extended region; their evolution unfolds on an extended time scale, and obeys equations of phase dynamics that can be constructed by setting $\nabla\mathbf{k} = O(\epsilon)$ and expanding the underlying system in powers of the scale ratio ϵ.

If microscopic equations contain scalar state variables only (as in reaction-diffusion problems or in convective problems with a vanishing vertical component of the vorticity field), the phase equation can contain only scalars formed with the help of k, $\mathbf{n}=\mathbf{k}/k$ and ∇, which are, in the leading order, $\mathbf{n}\cdot\nabla k$ and $\nabla\cdot\mathbf{n}$. This immediately leads to the generic phase diffusion equation

$$-\omega = D_{\|}\mathbf{n}\cdot\nabla k + kD_{\perp}\nabla\cdot\mathbf{n} \,, \tag{2.1}$$

where $\omega=-\partial\theta/\partial t$ is an $O(\epsilon)$ "frequency", and $D_{\|}$, $D_{\perp}$ are, respectively, longitudinal (along the wave vector $\mathbf{k}$) and transverse (along a line θ=const) phase diffusivities, that both depend on a particular underlying problem and are, generally, functions of the wave number k.

The phase diffusion equation can be linearized when deviations of the wave number are small. We shall consider weak distortions of a regular striped stationary pattern with the wave number k_0 and the wave vector oriented along the x axis, and look for a solution corresponding to a single defect propagating along the y axis with a constant velocity v (in the following, it is assumed, without loss of generality, $v>0$). The small parameter ϵ implied in the derivation of the phase diffusion equation will be identified with the ratio of the basic wavelength to the size of the box along the x axis. The total phase can be presented as $\theta=\epsilon^{-1}k_0x+\tilde{\theta}(x,y)$. In the frame moving with the velocity v, the solution is stationary, and the correction $\tilde{\theta}$ verifies

$$v\frac{\partial\tilde{\theta}}{\partial y} + D_{\|}\frac{\partial^2\tilde{\theta}}{\partial x^2} + D_{\perp}\frac{\partial^2\tilde{\theta}}{\partial y^2} = 0, \tag{2.2}$$

where the values of phase diffusivities $D_{\|}$, $D_{\perp}$ correspond to the prevalent wave number k_0. The scaling of eq. (2.2) implies that time is measured on the scale of the characteristic macroscopic diffusional time. The propagation velocity, which has been rescaled in (2.2) to macroscopic units, must be $O(\epsilon)$ when measured on the basic short scale in order to balance the convective and diffusional terms.

In the presence of a dislocation, the phase is not defined globally as a continuous univalued function. The correction to the wave vector $\tilde{\mathbf{k}}=\nabla\tilde{\theta}$ is subject to the integral condition $\oint\tilde{\mathbf{k}}\cdot d\mathbf{l}=2\pi$, where the integration is carried out around an arbitrary contour enclosing the defect. Supposing that the defect is located at the origin, this can be accomodated by making $\tilde{\theta}$ to jump by 2π across the negative y axis.

All parameters of eq. (2.2) can be eliminated by rescaling $x\to 2\xi(D_{\perp}D_{\|})^{1/2}/v$, $y\to 2\eta D_{\perp}/v$, bringing it thereby to the form

$$2\frac{\partial\tilde{\theta}}{\partial\eta} + \frac{\partial^2\tilde{\theta}}{\partial\xi^2} + \frac{\partial^2\tilde{\theta}}{\partial\eta^2} = 0 \,. \tag{2.3}$$

As in Refs. [12,13], eq. (2.3) subject to the jump condition is solved via introducing an adjoint function $\phi(\xi,\eta)$ that satisfies

$$2\frac{\partial\phi}{\partial\eta} + \frac{\partial^2\phi}{\partial\xi^2} + \frac{\partial^2\phi}{\partial\eta^2} = 2\pi\,\delta(\xi)\delta(\eta) \,, \tag{2.4}$$

$$\frac{\partial\tilde{\theta}}{\partial\xi} = -\left[\frac{\partial\phi}{\partial\eta} + 2\phi\right], \qquad \frac{\partial\tilde{\theta}}{\partial\eta} = \frac{\partial\phi}{\partial\xi} . \tag{2.5}$$

Using eq. (2.5) in the contour integral $\oint\nabla\tilde{\theta}\cdot dl$ and applying the Gauss theorem we see that this integral equals 2π as required, by virtue of eq. (2.4). It is easily checked that eq. (2.3) is satisfied automatically due to (2.5), while the integrability of eq. (2.5) is insured by eq. (2.4). The solution of eq. (2.4) in the infinite region is

$$\phi = -e^{-\varsigma\cos\alpha}K_0(\varsigma) , \tag{2.6}$$

where $\varsigma = (\xi^2+\eta^2)^{1/2}$, $\alpha = \arctan(\xi/\eta)$, and K_0 is a modified Bessel function. The components of the wave vector read from (2.5), (2.6) are

$$k_x - \hat{k}_0 = e^{-\eta}[K_0(\varsigma) - \cos\alpha\, K_1(\varsigma)], \qquad k_y = e^{-\eta}\sin\alpha\, K_1(\varsigma), \tag{2.7}$$

with $\hat{k}_0 = 2k_0(D_\perp D_\|)^{1/2}/v$. In the original units,

$$k_x = k_0 - \tfrac{1}{2}\epsilon v(D_\| D_\perp)^{-1/2}\exp(-vy/2D_\perp)[K_0(\varsigma) - \cos\alpha\, K_1(\varsigma)],$$

$$k_y = (\epsilon v/2D_\perp)\exp(-vy/2D_\perp)\sin\alpha\, K_1(\varsigma), \tag{2.8}$$

Using the asymptotics of Bessel functions at $\varsigma\to 0$, $K_0 = \ln(2/\varsigma)$, $K_1 = 1/\varsigma$, we see that a correction to the wave vector becomes $O(1)$ (yielding an increase of the wave number) when $\varsigma = O(\epsilon)$. This corresponds to a vicinity of the defect comparable to the basic short scale of the pattern. Thus, the outer solution (2.7) has to be matched in the inner region with a full solution of the underlying system, rather than with a solution of the nonlinear phase equation. As a consequence, the propagation velocity v, that would be determined by matching conditions, must depend on detailed short-scale dynamics, even though it is $O(\epsilon)$ when measured on the short time scale.

Matching with the full solution is an undesirable operation that, in any case, cannot yield universal results, inasmuch as one would have to work with a specific underlying system. It is preferrable therefore to use integral conditions derived for an equivalent potential system, as described in the next Section.

A correction to the above solution accounting for a finite size along the x axis can be obtained in the following way. Let the width of the box be $L = 2\pi N k_0^{-1} = 2\pi/\epsilon k_0$, and assume periodic or no-flux boundary conditions at $x = \pm L/2$. Then the solution can be expressed using the Green's function for the infinite region and introducing an array of images of the source spaced by L along the axis $y=0$; the dimensionless spacing is $\Lambda = vL/2(D_\perp D_\|)^{1/2}$. The resulting expression for dimensionless components of the wave vector is

$$k_x - \hat{k}_0 = e^{-\eta}\sum_{n=-\infty}^{\infty}[K_0(\varsigma_n) - \cos\alpha_n\, K_1(\varsigma_n)], \qquad k_y = e^{-\eta}\sum_{n=-\infty}^{\infty}\sin\alpha_n\, K_1(\varsigma_n), \tag{2.9}$$

where $\varsigma_n = [(\xi - n\Lambda)^2 + \eta^2]^{1/2}$, $\alpha_n = \arctan[(\xi - n\Lambda)/\eta]$. The corresponding limiting forms at $\varsigma_n\to\infty$, $\eta<0$ are

$$k_x-\hat{k}_0=\sqrt{\frac{\pi}{2}}\sum_{n=-\infty}^{\infty} e^{-(\varsigma_n-|\eta|)}(\varsigma_n+|\eta|)\varsigma_n^{-3/2},$$

$$k_y=\sqrt{\frac{\pi}{2}}\sum_{n=-\infty}^{\infty} e^{-(\varsigma_n-|\eta|)}(\xi-nL)\varsigma_n^{-3/2}. \tag{2.10}$$

When $|\eta|$ is very large, the principal contribution into the sum comes from a large number of small terms with $n\Lambda\ll|\eta|$. The sum can be therefore approximated in this limit by an integral. This gives, neglecting $\xi=O(1)$,

$$k_x-\hat{k}_0=\sqrt{\frac{2\pi}{|\eta|}}\int_{-\infty}^{\infty} e^{-\Lambda^2 n^2/2\eta}\,dn = \frac{2\pi}{\Lambda} \tag{2.11}$$

Thus, the summation over images removes the weak decay of the wave number correction in the far field behind the moving dislocation, and gives the correct expression for the change of the wave number. The respective asymptotic expression for k_y is

$$k_y=\sqrt{\frac{2\pi}{|\eta|^3}}\int_{-\infty}^{\infty}(\xi-\Lambda n)e^{-(\xi-\Lambda n)^2/2\eta}\,dn = \frac{2\pi\xi}{\Lambda|\eta|} \tag{2.12}$$

3. Variational principle for phase equation

An equivalent form of the phase equation can be obtained starting with a complex field that obeys evolution equations possessing a gradient structure

$$\dot{u} = -\delta\mathcal{L}/\delta\bar{u}, \qquad \dot{\bar{u}} = -\delta\mathcal{L}/\delta u \tag{3.1}$$

Here $\mathcal{L}(u,\bar{u},\nabla)$ is a real scalar Lagrangian density, the dot denotes the time derivative, and the overbar marks the complex conjugate. The integral form of eq. (3.1) is

$$\int [\dot{u}\delta\bar{u}+\dot{\bar{u}}\delta u+\delta\mathcal{L}(u,\bar{u},\nabla)]d^2\mathbf{x} = 0. \tag{3.2}$$

Presenting u in the polar form $u=ae^{i\theta}$, we rewrite this expression as

$$\int [\dot{a}\delta a-\omega a^2\delta\theta+\tfrac{1}{2}\delta\hat{\mathcal{L}}(a,i\mathbf{k})]d^2\mathbf{x} = 0. \tag{3.3}$$

The reduced Lagrangian density $\hat{\mathcal{L}}$ may contain (in case the Lagrangian $\mathcal{L}$ includes second or higher order derivatives) derivatives of $\mathbf{k}$ as well as the wave vector itself; for any given $\mathcal{L}$, it can be obtained by straightforward computation. It is clear that, provided the Lagrangian is analytical, the non-differential part of $\hat{\mathcal{L}}$ contains only even powers of a, which can be replaced by the "density" $\rho=|u|^2=a^2$. The wave vector can enter $\hat{\mathcal{L}}$ only through rotationally invariant combinations $q=k^2$, $\nabla\cdot\mathbf{k}$, $\mathbf{k}\cdot\nabla a$, etc. When the Lagrangian is varied in eq. (3.3), one has to keep in mind that variations of $\mathbf{k}$ are restricted by the condition $\delta\mathbf{k}=\delta\nabla\theta$. This leads to dynamic equations of the real amplitude a and eikonal θ

$$\dot{a} = -\tfrac{1}{2}\partial\hat{\mathcal{L}}/\partial a, \qquad \omega = -\tfrac{1}{2}a^{-2}\nabla\cdot\partial\hat{\mathcal{L}}/\partial\mathbf{k}\,. \tag{3.4}$$

If $\mathbf{k}$ changes on a long scale, so that its derivatives are small, the amplitude follows the phase adiabatically, and is defined, to the leading order, by the condition $\partial\hat{\mathcal{L}}/\partial a=0$. Under the same conditions, the leading term in $\hat{\mathcal{L}}$ depends on k only through the norm $q=k^2$. Therefore the phase equation above can be presented, in the leading order, as

$$\omega = -\rho^{-1}\nabla\cdot(\mathbf{k}\,\partial\mathcal{L}/\partial q) , \tag{3.5}$$

Eq. (3.5) is equivalent to eq. (2.1) with the phase diffusivities

$$D_{\perp}=\rho^{-1}\frac{\partial\mathcal{L}}{\partial q},\quad D_{\parallel}=\rho^{-1}\Big(1+2q\frac{d}{dq}\Big)\frac{\partial\mathcal{L}}{\partial q}. \tag{3.6}$$

For example, for the real LG equation

$$\dot{u} = \nabla^2 u+u-|u|^2 u , \tag{3.7}$$

the Lagrangian density and the reduced Lagrangian are

$$\mathcal{L} = |\nabla u|^2-|u|^2+\tfrac{1}{2}|u|^4 , \quad \hat{\mathcal{L}} = |\nabla a|^2-(1-k^2)a^2+\tfrac{1}{2}a^4 . \tag{3.8}$$

The corresponding quasistationary density ρ and phase diffusivities are

$$\rho=1-q,\quad D_{\parallel}=\rho^{-1}(1-3q),\quad D_{\perp}=1, \tag{3.9}$$

It is clear that the form of the phase equation is not dependent on the presence of a gradient structure, since the same form, with ρ, $\partial\hat{\mathcal{L}}/\partial q$ replaced by arbitrary functions of k, can be obtained starting from any rotationally invariant system by either scaling and symmetry arguments or formal expansion. Moreover, given phase diffusivities corresponding to a certain microscopic system, the equivalent Lagrangian $\hat{\mathcal{L}}(q)$ can be obtained by "solving" eqs. (3.6). Based on this approach, Cross and Newell [8] showed the way to construct an effective potential of a real field. Construction of the effective Lagrangian of a complex field, which can be seen as an analog of the transformation to action-angle variables in classical mechanics, is carried out in a similar manner.

Combining the two expressions (3.6) gives a differential equation for the effective density ρ:

$$\frac{d\ln\rho}{dk} = D_{\parallel} - \frac{d(kD_{\perp})}{dk} . \tag{3.10}$$

This equation can be integratied using as an "initial" condition $\rho(k_{max})=0$; then the solution defines $\partial\hat{\mathcal{L}}/\partial q=\rho D_{\perp}$. The latter can be interpreted as a full, as well as partial derivative, since $\partial\hat{\mathcal{L}}/\partial\rho=0$ is the condition defining $\rho(k)$; thus the function $\hat{\mathcal{L}}(q)$ at the extremum can be obtained simply by integration. One would not try, of course, to reconstruct the entire Lagrangian, since the system behaves in a quasipotential fashion only as long as deviations from the one-dimensional structure $\rho(k)$ remain small.

Following Pomeau *et al.* [10], one can compute the propagation velocity using two equivalent integral expressions that determine the rate of change of the Lagrangian with time. It follows from eq. (3.4) that, in the leading order,

$$\frac{d}{dt}\int\hat{\mathcal{L}}d^2\mathbf{x}=\int\frac{\partial\hat{\mathcal{L}}}{\partial\mathbf{k}}\cdot\dot{\mathbf{k}}\,d^2\mathbf{x}=-\int\frac{\partial\hat{\mathcal{L}}}{\partial\mathbf{k}}\cdot\nabla\omega\,d^2\mathbf{x}=\int\omega\nabla\cdot\frac{\partial\hat{\mathcal{L}}}{\partial\mathbf{k}}d^2\mathbf{x}=-2\int\rho\omega^2d^2\mathbf{x}=-2v^2\int\rho k_y^2d^2\mathbf{x}.$$

The last expression is valid for an isolated dislocation steadily propagating along the y axis (normal to $\mathbf{k}$). On the other hand, the change of the Lagrangian is caused in this case just by an $O(\epsilon)$ change of the wave number in a strip of the width $v dt$ parallel to $\mathbf{k}$, and its rate equals to $-vF$, where $F = -\partial \mathcal{L}/\partial k$ is the Peach-Köhler force. Thus,

$$v = F/2\int \rho k_y^2 d^2\mathbf{x}. \tag{3.11}$$

Using eq. (2.8), we express the integral in eq. (3.11) as

$$\rho(k_0)\int k_y^2 d^2\mathbf{x} = \rho(D_{\parallel}/D_{\perp})^{1/2}\int_0^{\infty} z K_1^2(z)dz \int_0^{2\pi} e^{-2z\cos\alpha}\sin^2\alpha d\alpha =$$
$$= \pi\rho(D_{\parallel}/D_{\perp})^{1/2}\int_0^{\infty} I_1(2z)K_1^2(z)dz . \tag{3.12}$$

The integrand in the last expression decays as $z^{-3/2}$ at $z\to\infty$, and diverges as z^{-1} at $z\to 0$. The computed value should depend therefore logarithmically on a short-scale cutoff at $z=O(\epsilon)$. Finite-size corrections to k_y are not essential, since it decays sufficiently fast to avoid large-distance divergencies in the viscosity integral, which accumulates largely at small distances.

We conclude, that, in basic short units, v scales as

$$v \propto \epsilon/\ln\tfrac{1}{\epsilon} . \tag{3.13}$$

An apparent disagreement with results of Bodenschatz *et al.* [4], who found no stong dependence on the box size (beyond the logarithmic correction), is due to a different role played by the LG equation. In Ref. [4] it is an envelope equation defining the amplitude of a short-scale pattern containing a dislocation. Here, however, the LG equation is just a model underlying system that possesses a family of stationary inhomogeneous solutions; the defects pertain to these solutions, and not to any substructure existing on a still shorter scale. On the other hand the above scaling law has nothing in common with a similar result of Neu [11] that involves a small parameter of a different nature.

Note that the defect motion in the present case is not influenced by coupling with any field other than $\mathbf{k}$ itself, as in Ref.7. A long-range field with a non-vanishing nonsingular curl compatible with the symmetry of amplitude equations, like the vertical vorticity field, can be, indeed, an important factor in the pattern evolution in convective systems; this field, however, cannot appear in long-scale equations deriving from scalar problems, and cannot be solely responsible for the motion of defects.

4. Numerical solution of the phase equation

The numerical solution of the phase equation described below is based directly on eq. (2.1) interpreted as a Lagrangian equation defining the displacement of lines of constant phase. The normal velocity of a line θ=const is defined as $v = \omega/k$, and can be computed using local values of the gradient of the wave number and of the curvature of lines of constant phase $\nabla\cdot\mathbf{n}$. When both phase diffusivities are positive, the lines of constant phase move in such a manner that inhomogeneities in their spacing and orientation are attenuated. The longitudinal diffusivity plays the role of "metric elasticity" tending to keep lines differing by a fixed phase increment evenly spaced, while the transverse diffusivity acts as a "line tension". On the other hand, if k increases at some location beyond the limit of Eckhaus instability, so that the longitudinal diffusivity becomes negative, adjacent lines of constant phase are attracted to each other, and are apt to collapse, thus leading to the formation of a phase singularity.

The appearance of negative diffusivities is, as a rule, catastrophic for any computational procedure. This, however, does not need to be true in the case of phase diffusion, since only $\theta(\text{mod } 2\pi)$ is physically relevant, and the "runaway" due to negative diffusivity is arrested after a dislocation has been formed. This suggests the following modification of the Lagrangian computational algorithm extending it to unstable situations and allowing to incorporate formation and motion of defects. Suppose, for convenience, that the phase increment between adjacent lines $\theta=$const is set at 2π. At each time step, the position of each line is updated using eq. (2.1), and local values of the wave number k, its longitudinal derivatives, and curvature $\nabla\cdot\mathbf{n}$ are recomputed. If k exceeds locally the value k_{max}, a defect is inserted by joining two lines coming close one to another at this location. The value k_{max} provides the natural cutoff limit, inasmuch as the zero-order pattern is not defined on shorter wavelengths; this limit is achieved promptly after the longitudinal diffusivity changes sign at the limit of Eckhaus instability, unless transverse diffusion acts to dissipate a local perturbation, thus preventing the defect nucleation. Further convergence of adjacent lines $\theta=$const leads to the growth of the joint line segment, which is equivalent to the defect propagation in the transverse (normal to $\mathbf{k}$) direction. The defect propagation is, thus, imposed kinematically by convergent motion of adjacent lines of constant phase caused by local Eckhaus instability. Similar dynamics might be observed at the alternative limit $k=k_{max}$ where, conversely, a new phase line would have to be inserted; long waves are, however, usually eliminated through a different mechanism of zigzag instability.

The phase equation in the form (2.1) is particularly suitable for implementation of the above algorithm, since neither the normal derivative of the wavenumber nor the tangential derivative of the orientation angle do diverge anywhere along the line normal to $\mathbf{k}$ and passing through a defect – not even at the point of the phase singularity itself (it is the tangential derivative of the wavenumber that diverges there).

In most computations, we have used expressions for phase diffusivities

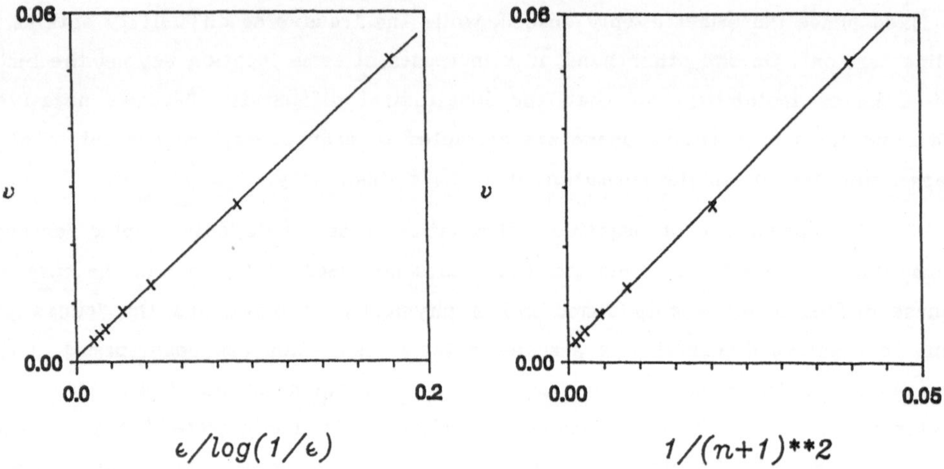

Fig.1. Propagation velocity of a single dislocation as a function of the number of rolls n or its inverse ϵ for the Landau-Ginzburg equation. Crosses mark computed data.

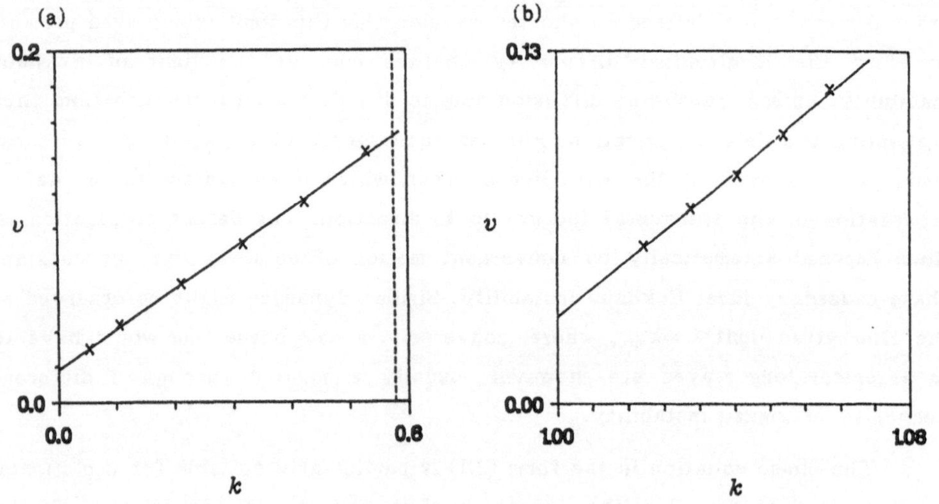

Fig.2. Propagation velocity of a single dislocation as a function of the wave number for (a) Landau-Ginzburg equation rolls and (b) complexified Swift-Hohenberg model. Crosses mark computed data. The dashed line shows the limit of Eckhaus instability.

(3.9) corresponding to the LG equation (3.7). The boundary conditions are periodicity in the longitudinal (x) direction and perpendicularity of rolls to the walls in the transverse (y) direction. The numerical algorithm based on the phase equation, requiring just one grid point per roll, is far more efficient than solving the underlying LG equation directly. It reflects the basic mechanism of propagation, though is not expected to be quantitatively correct in the $O(\epsilon)$ core region. This, fortunately, is not essential for verifying the scaling law (3.13) up to a proportionality constant. Fig. 1a shows the velocity of a single defect as a function of the inverse number of rolls ϵ at a fixed value of the prevalent wave number k. The data are perfectly correlated by eq. (3.13). One can observe, however, that the correlation $v \propto (N+1)^{-2}$ suggested by Pomeau *et al.* [10] fits the data nearly as well (Fig. 1b).

Computations of the dependence of v on the prevailing wave number, either analytical or numerical, with the help of the phase equation are not totally reliable, due to an uncertain contribution of the core region. The data for the LG equation shown in Fig. 2a were obtained in a box containing initially 15 rolls. Since the spatial intervals are inversely proportional to k, the time increment was changed accordingly in order to avoid slow-down and yet keep numerical stability. The data are well fitted by the correlation $v \propto k^2$, up to the limit of Eckhaus instability at $k=3^{-1/2}$ (indicated by the dashed line in Fig. 2a).

For comparison, Fig. 2b shows results of computations using the values of phase diffusivities corresponding to the complexified Swift-Hohenberg model

$$\dot{u} = -(1+\nabla^2)^2 u + \mu^2 u - |u|^2 u \ . \tag{4.1}$$

This model possesses an optimal wave number $k_c=1$, and, strictly speaking, one has to use a more complicated phase equation, employing anisotropic scaling in the longitudinal and transverse direction, in the vicinity of this point. Nevertheless, the data obtained solving the simple generic equation (2.1) satisfactory reproduce the well-known velocity correlation $v \propto (k-k_c)^{3/2}$ [9,3]. Taking note that near the optimal wave number $F \propto D_{\perp} \propto k-k_c$, one can see that the same dependence directly follows from eqs. (3.11), (3.12).

The series of snapshots shown in Fig. 3 depicts the spontaneous formation of defects in the course of the long-time relaxation of a pattern described by the LG equation in a circular cell with rolls normal to the boundary. This cell, of the radius 130.1 (measured in dimensionless short-scale units), contained initially 29 rolls which were more compressed at the center. The higher compression of rolls induces nucleation of defects in this region. Subsequently, the defects move toward the cell boundary, and new ones are formed, as the average wavelength gradually increases.

The nucleation and motion of defects was also studied in a circular cell

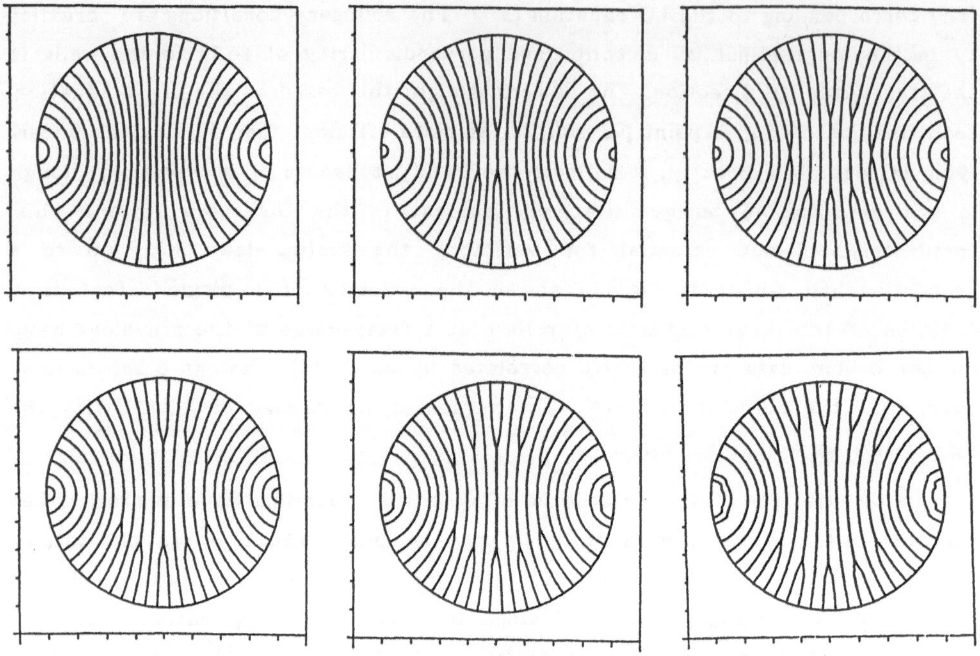

Fig.3. Evolution of a pattern obeying the Landau-Ginzburg equation in a circular cell under conditions when rolls are normal to the containing walls. The shown snapshots correspond to time moments 0, 20, 60, 150, 200, and 300 (measured in dimensionless short-scale units).

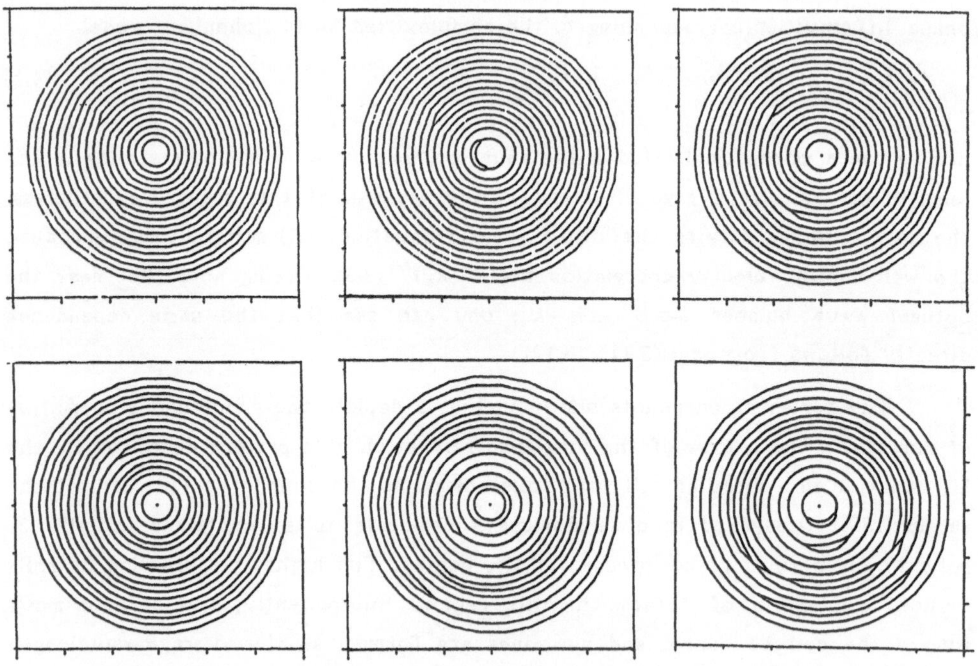

Fig.4. The same as in Fig.3, but under conditions when rolls are parallel to the containing walls. The shown snapshots correspond to time moments 0, 33, 70, 140, 200, and 230.

with rolls oriented paralel to the border, as shown in Fig. 4. Initially, the cell contained 17 rolls, and a pair of defects was included in order to break the circular symmetry. As the pattern evolves, the number of rolls decreases mainly via nucleation of pairs of dislocations in the high-curvature inner region. The nucleation is induced by defects already present at outer locations.

Both above examples demonstrate that realistic evolution of patterns involving formation and motion of defects can be indeed simulated using the algorithm based on the phase equation. Computations can be extended in a straightforward way to stable wave patterns, where similar climbing motion of dislocations is observed in a frame moving with the group velocity corresponding to the prevailing wave number. A corrected phase equation incorporating stabilizing fourth order terms is, however, required in situations involving the zigzag or Benjamin-Feir instabilities.

Acknowledgement

This work has been supported by the US – Israel Binational Science Foundation.

References

[1] A.C.Newell, in *Propagation in Systems Far from Equilibrium*, J.E.Wesfreid *et al.*, eds., Springer, Berlin, 1988, p. 122.

[2] H.S.Greenside and W.M.Coughran, Phys. Rev. **A30** (1984) 398.

[3] G.Tesauro and M.C.Cross, Phys. Rev. **A34** (1986) 1363.

[4] E.Bodenschatz, W.Pesch and L.Kramer, Physica **32D** (1988) 135.

[5] P.Coullet, L.Gil, and J.Lega, Phys. Rev. Lett., **62** (1989) 1619.

[6] M.Lowe and J.P.Gollub, Phys. Rev. Lett., **55** (1985) 2575; E.Bodenschatz, W.Zimmermann, and L.Kramer, J. Physique **49** (1988) 1875; A.Joets and R.Ribotta, J. Physique **50** (1989) C3-171.

[7] G.Goren, I.Procaccia, S.Rasenat and V.Steinberg, Phys.Rev.Lett., **63** (1989) 1237.

[8] M.C.Cross and A.C.Newell, Physica **10D** (1984) 299.

[9] E.Siggia and A.Zippelius, Phys. Rev. **A24** (1981) 1036.

[10] Y.Pomeau, S.Zaleski, and P.Manneville, Phys. Rev. **A27** (1983) 2710.

[11] J.Neu (1989), unpublished.

[12] E.Dubois-Violette, E.Guazelli, and J.Prost, Phil. Mag. **A48** (1983) 727.

SPIRALS IN EXCITABLE CHEMICAL MEDIA: FROM ARCHIMEDIAN TO NON-ARCHIMEDIAN GEOMETRY

Stefan C. Müller and Theo Plesser

Max-Planck-Institut für Ernährungsphysiologie
Rheinlanddamm 201, D-4600 Dortmund 1, FRG

CHEMICAL WAVE PROPAGATION

Propagation phenomena under far from equilibrium conditions are observed in many nonlinear physical and biological systems (for an overview see [1]). In particular, travelling waves are a characteristic spatio-temporal pattern occurring in excitable media. They have been studied in much detail in chemical model systems, mostly in the Belousov-Zhabotinskii (BZ) reaction prepared with excitable reaction kinetics [2,3]. In this solution malonic acid is oxidized and decarboxylized by bromous compounds to form organic products. The reaction takes place in the presence of a metal ion redox catalyst, usually ferroin, which changes its colour during the excitation,involving a transition from the reduced to the oxidized state. Thus, in a thin solution layer chemical waves are detectable by their blue fronts indicating the presence of the oxidizing state ferriin, which travel through a quiescent red solution layer where ferroin, the reduced form of the catalyst, prevails. The spatial distribution of the catalyst concentration in such waves can be determined quantitatively since devices of 1D and 2D spectrophotometry have become available [4-6]. Improved evaluation techniques allow for the detailed analysis of wave features such as shapes and profiles under systematic variation of system parameters, in particular of the initial chemical concentrations [7-10].

SPIRAL WAVES AS A FUNCTION OF CHEMICAL COMPOSITION

Since the discovery of travelling chemical waves there has been particular interest in studying the phenomenon of rotating spirals. Much of the early work was devoted to spirals evolving in a "standard" recipe [11] where their geometry is very close to (or practically identical with) that of an Archimedian spiral and the spiral tip performs a steady circular motion around the rotation center [6,12]. There had been early indications, however, that the regular geometry of a spiral changes and the apparent stability of spiral rotation is lost, when the initial concentrations are appropriately changed [13].

Nonlinear Evolution of Spatio-Temporal Structures in Dissipative Continuous Systems
Edited by F.H. Busse and L. Kramer
Plenum Press, New York, 1990

The crucial property resulting from the choice of the initial system parameters is the level of excitability of the solution. In a highly excitable solution waves are easily generated by various kinds of perturbations, e.g. a hot platinum wire, dust particles and other inhomogeneities, or they emerge spontaneously at the boundary of the container. In solutions of low excitabilty perturbations for triggering a wave have to be large, e.g by a sufficient heat or electrochemical pulse or a drop of chemical reagent.

If the change in chemical composition results in a sufficiently low excitability of the medium, there appear distorted spirals and the tip starts to perform a looping motion (often referred to as "meandering", although the authors think that this term is not appropriate). This has been first pointed out by Winfree [13] and by Agladze [14]. Only recently more attention was given to a systematic approach in terms of concentration dependence of characteristic parameters describing the complex dynamics of the distorted spirals [15-17].

In order to further quantify the properties of spiral waves in media of low excitability we have conducted a detailed study concerning the dependence of spiral rotation on the initial concentration of sulfuric acid, i.e. its initial pH value. We report in the following on some aspects of our findings regarding the geometry and velocity of the spiral waves, and the temporal behaviour of their rotation center.

INVESTIGATION OF WAVE PHENOMENA BY 2D SPECTROPHOTOMETRY

Our investigations are based on an apparatus for space resolved spectrophotometry in two dimensions which consists of an optical precision set-up mounted on an optical table, a sensitive video camera with a broad spectral response and a video frame buffer linked to a computer system. The schematic in Fig. 1 shows the main components [6]. A homogeneous parallel light beam is transmitted through the solution layer at a specific wavelength, imaged on the target of the camera, transferred to the video frame buffer and stored in the memory of a computer. One image consists of 512 x 512 picture elements (pixel) each having theoretically one out of 256 possible grey levels. With this raster and intensity resolution the information content of one full frame amounts to a quarter of a MByte. Image acquisition and storage is feasible at a rate of 30 frames per minute.

The spatial resolution depends on the photolense system and can be increased up to about 0.5 μm per pixel if an inverted microscope is used. For our purpose the maximum spatial resolution was chosen to be 15.6 μm/pixel. When recording a movie on a Umatic video recorder the frames are stored on tape at a rate of 25 frames per second, but images have significantly less spatial and intensity resolution and the finite lag time of the camera has to be taken into account [5,18]. In this work we combined the recording of single images with optimum intensity resolution and of video movies having reduced intensity resolution.

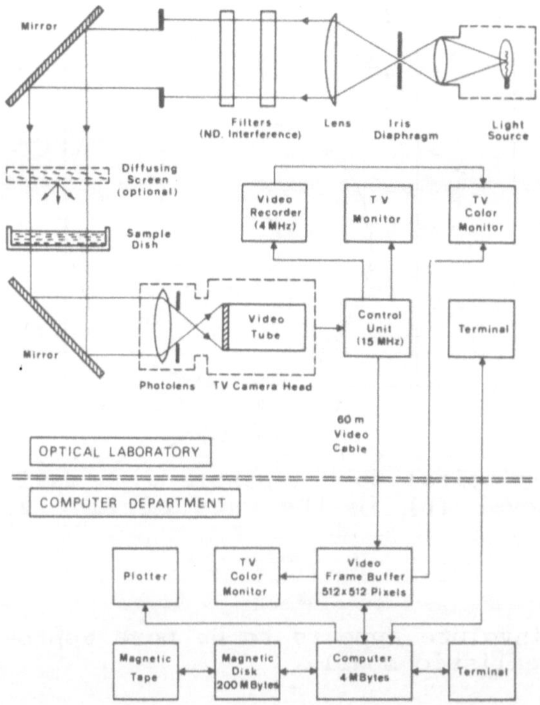

Fig. 1. Schematic of the experimental arrangement for computerized 2D spectrophotometry. (Fig. 1 from [6])

ARCHIMEDIAN SPIRALS WITH STATIONARY CENTERS

From previous spectrophotometric measurements of regular spirals in a standard recipe we know that their shape is practically identical with that of an Archimedian spiral [8]. In the snapshot of a digital image shown in Fig.2 the pixel sites of maximum and minimum intensity (that is highest and lowest concentration of ferriin are enhanced in black and white. These isoconcentration levels are very well fitted by Archimedian spirals as convincingly demonstrated in Fig.3. The geometry of the pattern can be described by 4 parameters only: the two coordinates of the spiral center, the pitch, and the angu-lar position. We also chose involutes of a circle which exclude a circular region around the center and found that they fit these data either well. In view of the properties of

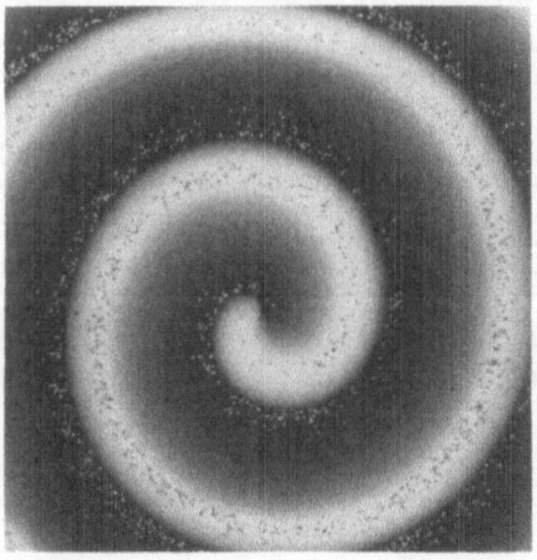

Fig. 2. Regular spiral wave obtained in a standard solution, the maximum grey level (crest) is enhanced in black, the minimum level in white (from [8])

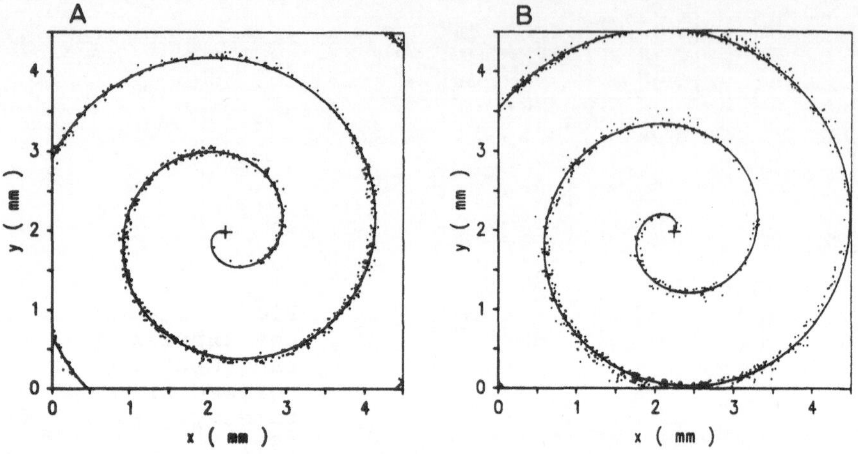

Fig. 3. Fit of an Archimedian spiral to the pixels with maximum (A) and minimum grey level (B) in the wave of Fig. 2. (from [8])

the core of the spiral the involute appears to be more appropriate for comparison with realistic models.

The structure of the core region in these regular spirals was determined by an overlay technique as described in detail in [8]. To summarize, its main properties are (compare Fig. 4 below): At the center of spiral rotation there is a singular site of diameter < 30 μm that remains quasi-stationary at a concentration slightly above the level of maximum reduction in the spiral pattern. For a large number of revolutions its position does not change in space. The spiral core is a region with a diameter of about 0.7 mm in which a smooth and gradual transition takes place: from the stationary site towards the outer area of the pattern the amplitude of the emanating wave grows to full height, that is, in the outer area each volume element becomes excited to a maximum degree once per revolution.

In solutions prepared with the standard or similar recipes only rare indications of irregularities were found. They are usually quite small and can be detected more easily during the aging stage of the solution. Typcially, we observe the formation of patch-like bright speckles [8] of higher transmitted intensity that cannot be detected with the naked eye. The overlay in Fig. 4 leads to a contrast enhancement and therefore to a clear presentation of the spatial extent of such a speckle pattern. At the same time it illustrates its temporal evolution during outward propagation. The onset of speckle formation may well be related to compositional changes of the solution during aging, but a close inspection of the regular spiral in its initial stage reveals a weak "precursor" modulation along the spiral crest that might be taken as an indication of inherent deviations from a mathematically perfect shape.

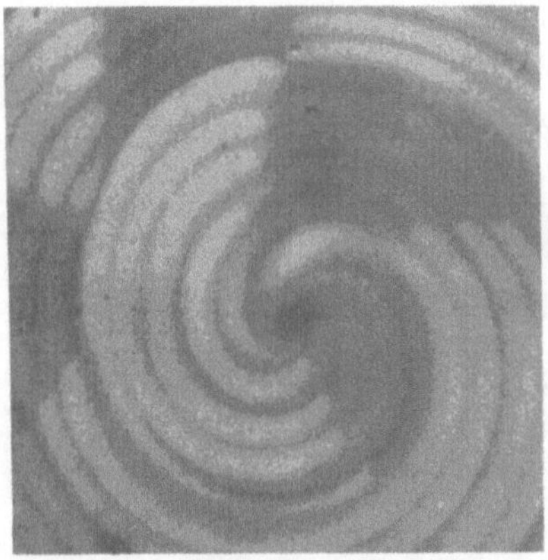

Fig. 4. Slight irregularities (speckles) detected in the spiral of the previous picture (Fig. 3) about 16 min after the start of the experiment and shown in an overlay of 6 consecutive images taken at 3s intervals (Fig. 2a from [19])

NON-ARCHIMEDIAN SPIRALS WITH TRANSIENT ROTATION CENTERS

We varied the initial chemical composition of excitable BZ reagent by preparing solutions with sulfuric acid concentrations both higher (up to 0.7M) and lower (down to 0.15M) than in the standard composition (0.37M). This choice leads to high (low pH) respectively low excitability (high pH). In a qualitative summary, the spiral pitch and the wave amplitude decrease with increased proton concentration while the propagation velocity increases. (For quantitative details see [17].)

As another remarkable feature, the geometry of the spirals changes from Archimedian to distorted non-Archimedian geometry. This becomes clearly visible below 0.26 M H_2SO_4, but is possibly detectable already at higher sulfuric acid concentration. The change in geometry, pitch, and amplitude is illustrated in Fig. 5 by direct comparison of spirals in solutions with initially 0.7 M and 0.19 M H_2SO_4.

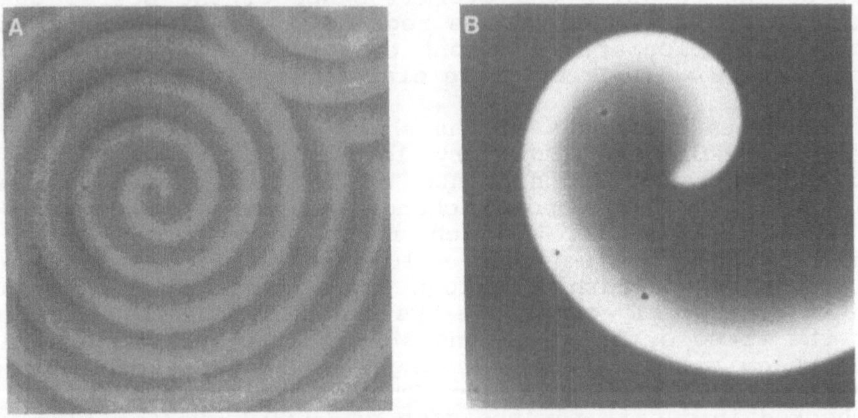

Fig. 5. (A) Image of an Archimedian ($[H_2SO_4]$= 0.7M) and (B) of a non-Archimedian spiral ($[H_2SO_4]$=0.19 M). Observation area in both pictures: 7.0 x 7.0 mm^2

In Fig. 6 the different shape and amplitude of the spiral tip for the case of high and low excitability is presented in three-dimensional perspective by plotting the measured concentration along the third axis of a 3D coordinate system.

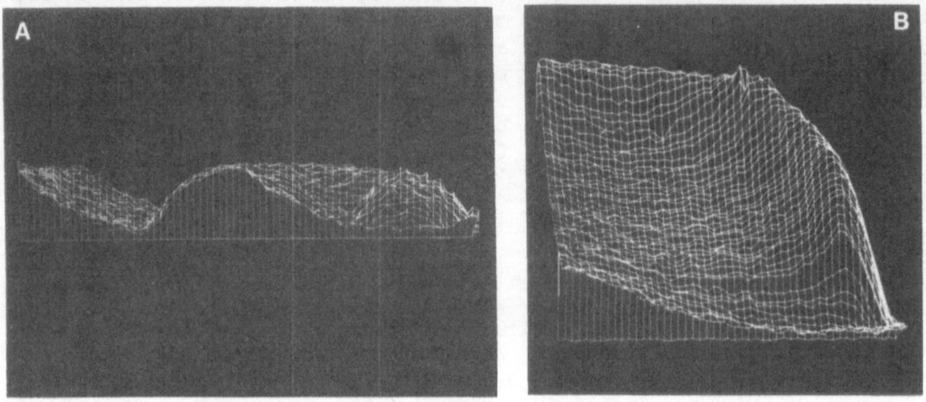

Fig. 6. Front view of the tip of the spiral wave of Fig. 5A (left) and B (right) in three-dimensional perspective. Scales are the same in both pictures

In this presentation the viewpoint of the observer is located in the x-y plane of the pattern and in front of the spiral tip. It clearly shows the large difference between the amplitudes of the two spiral waves. The shape of the concentration profiles at the tip is quite symmetric in the Archimedian spiral, whereas pronounced differences in the gradients on either side of the tip are observed in the distorted spiral.

In the distorted case the pitch of the spiral (the distance between consecutive whorls) is no longer constant but varies considerably as a function of arc length. This implies a complex variation of the local curvature along the front. In fact, during rotation the overall shape of the spiral changes in that a displacement of the regions of smallest pitch takes place. In particular, the front curvature in the area closest to the tip is subject to quite drastic variations.

The dynamic changes of the shape of the tip region in the case of a solution with very low excitabilty are shown in Fig. 7. The assembled sequence of 15 consecutive 2D images indicates both the temporal changes of the tip curvature and the trace of the tip which first moves along an almost straight path, then performs a loop and continues again along a slightly curved path. Previous investigations of the tip motion in such weakly excitable systems have already established that the trace of the tip consists of a number of consecutive loops which are frequently arranged on the circumference of a larger circle. Thus the trace resembles a prolate epicycloid, as checked in detail for the case of $[H_2SO_4]$ =0.19 M [17].

Here we show further evidence for the looping motion in a system with initially 0.15 M sulfuric acid by constructing an overlay image as derived from 29 consecutive snapshots taken at 6s intervals (Fig. 8). During this time the tip moves around a temporary rotation center the properties of which are

Fig. 7. Sequence of 15 images of a spiral travelling in a solution with $[H_2SO_4]$=0.15 M, taken at 12s intervals

comparable to those found for the Archimedian case. At the center of the dark spot in Fig. 8 the ferroin concentration remains stationary, until after completing one turn the tipmoves to an adjacent location where it will perform the subsequent loop. Then a wave will "sweep" across the previously unexcited spot and the temporary center will disappear.

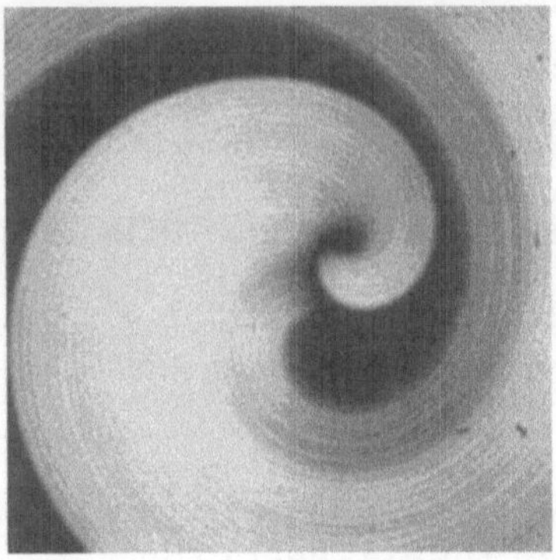

Fig. 8. Overlay of 29 consecutive images of a non-Archimedian spiral wave ($[H_2SO_4]$=0.15 M], taken at 6s intervals

REFERENCES

1. "Propagation in Systems Far from Equilibrium", J.F. Wesfreid, H.R. Brand, P. Manneville, G. Albinet, and N. Boccara, eds., Springer, Heidelberg, 1987.
2. "Oscillations and Traveling Waves in Chemical Systems", R.J., Field, and M. Burger, eds., Wiley, New York (1985).

3. J. Ross, S.C. Müller, and Ch. Vidal, Chemical Waves Science 240:460 (1988).
4. P. Wood, and J. Ross, A Quantitative Study of Chemical Waves in the Belousov-Zhabotinskii Reaction, J. Chem. Phys. 82:1924 (1985).
5. S.C. Müller, Th. Plesser, and B. Hess, Two-Dimensional Spectrophotometry with High Spatial and Temporal Resolution by Digital Video Techniques and Powerful Computers, Anal. Biochem. 146:125 (1985).
6. S.C. Müller, Th. Plesser, and B. Hess, Two-Dimensional Spectrophotometry and Pseudo-Color Representation of Chemical Reaction Patterns, Naturwissenschaften 73:165 (1986).
7. A. Pagola, and C. Vidal, Wave Profile and Speed near the Core of a Target Pattern in the Belousov-Zhabotinskii Reaction, J. Phys. Chem. 91:501 (1987).
8. S.C. Müller, Th. Plesser, and B. Hess, Two-Dimensional Spectrophotometry of Spiral Wave Propagation in the Belousov-Zhabotinsky Reaction. I. Experiments and Digital Data Representation, Physica D24:71 (1987).
S.C. Müller, Th. Plesser, and B.Hess, Two-Dimensional Spectrophotometry of Spiral Wave Propagation in the Belousov-Zhabotinskii Reaction. II. Geometric and Kinematic Parameters, Physica D24:87 (1987).
9. Zs. Nagy-Ungvarai, J.J. Tyson, and B. Hess, Experimental Study of the Chemical Waves in the Cerium Catalyzed BZ-Reaction. I. Velocity of Trigger Waves, J. Phys. Chem. 93:707 (1989).
10. Zs. Nagy-Ungvarai, J.J. Tyson, S.C. Müller, and B. Hess, Experimental Study of the Chemical Waves in the Ce-Catalyzed BZ-Reaction. II. Concentration Profiles J. Phys. Chem. 93:2760 (1989).
11. A.T. Winfree, Spiral Waves of Chemical Activity, Science 175:634 (1972).
12. S.C. Müller, Th. Plesser, and B. Hess, The Structure of the Core of the Spiral Wave in the Belousov-Zhabotinskii Reaction, Science 230:661 (1985).
13. A.T. Winfree, Scroll-Shaped Waves of Chemical Activity in Three Dimenions, Science 181:937 (1973).
14. K.I. Agladze, Proc. of the Biological Center of Academy of Science, Pushchino, USSR (1983).
15. K.I. Agladze, A.V. Panfilov, and A.N. Rudenko, Nonstationary Rotation of Spiral Waves: Three-Dimensional Effect, Physica D29:409 (1988).
16. W. Jahnke, W.E. Skaggs, and A.T. Winfree, Chemical Vortex Dynamics in the Belousov-Zhabotinskii Reaction and in the Two-Variable Oregonator Model, J. Phys. Chem. 93 740 (1989).
17. Th. Plesser, S.C. Müller, and B. Hess, Spiral Wave Dynamics as a Function of Proton Concentrationin the Ferroin-Catalyzed Belousvo-Zhabotinskii Reaction, J. Phys. Chem., submitted.
18. H. Miike, S.C. Müller, and B. Hess, Oscillatory Hydrodynamic Flow Induced by Chemical Waves, Chem. Phys. Lett. 144: 515 (1988).
19. Th. Plesser, H. Miike, S.C. Müller, and K.H. Winters, Some Pecularities in Propagating Chemical Waves and their Relation to Hydrodynamic Flow, Harwell Report TP. 1267, Harwell Laboratory, England (1987).

CHEMICAL PATTERN DYNAMICS BY CHEMICALLY INDUCED HYDRODYNAMIC FLOW

Hidetoshi Miike and Stefan C. Müller

Faculty of Engineering
Yamaguchi University
755 Ube, Japan

Max-Planck-Institut für
Ernährungsphysiologie
D-4600 Dortmund, FRG

INTRODUCTION

Pattern formation due to the coupling between chemical reaction and hydrodynamic flow has recently attracted increasing attention.[1,2] Examples are "mosaic" patterns and the deformation and irregular decomposition of chemical waves observed in an uncovered layer of the unstirred Belousov-Zhabotinskii (BZ) solution.[3-6] Various attempts have been made to explain the origin of the structures by the effects of convective flow caused by evaporative cooling and/or the exothermicity of the reaction.

We carried out direct and quantitative measurements of hydrodynamic flow occurring in an excitable BZ-solution layer, using space-resolved microscope video imaging techniques. Thus we detected the following hydrodynamic instabilities generated by chemical wave propagation, even in layers with a covered liquid/gas interface:[7,8]

a) A global flow structure traveling with a circular wave is induced spontaneously.

b) An oscillatory hydrodynamic flow accompanied with periodically varying deformations and irregular decomposition of chemical wave fronts is excited by the periodic passage of spiral wave trains.

MATERIALS AND METHODS

Sample Preparation. A quiescent, excitable solution of the BZ-reaction was obtained by preparing a mixture of 48 mM NaBr, 340 mM $NaBrO_3$, 95 mM $CH_2(COOH)_2$, and 378 mM H_2SO_4. About 5 min after mixing, the catalyst and indicator ferroin (3.5 mM) was added. All solutions were filtered with Millipore filter (0.22 µm). A volume of the mixture was poured into a dust free Petri dish of 7.0 cm diameter at 25 ± 1°C. The depth of the solution layer was 0.85 ± 0.05 mm.

2D-Velocimetry and 2D-Spectrophotometry. In order to investigate the correlation between chemical pattern dynamics and hydrodynamic flow, 2D-velocimetry and 2D-

Nonlinear Evolution of Spatio-Temporal Structures in Dissipative Continuous Systems
Edited by F.H. Busse and L. Kramer
Plenum Press, New York, 1990

spectrophotometry based on microscope video imaging techniques were applied.[7] For measurement of flow, polystyrene particles (diameter 0.48 μm) serving as scattering centers were mixed into the BZ-solution and illuminated at a slightly tilted angle by He-Ne laser light (632.8 nm). Together with a homogeneous transmitted light beam (490 nm), hydrodynamic flow and propagation of chemical activity could be observed simultaneously.

HYDRODYNAMIC FLOWS WITH CIRCULAR WAVES

Traveling Hydrodynamic Flows. Experiments were carried out by triggering circular waves at the boundary of a layer with a covered liquid/gas interface. In Fig.1, a typical example of the temporal evolution of this pattern is shown. Wave fronts with almost constant period travel through the layer. Note that no CO_2 bubbles were nucleated and no additional uncontrolled waves emerged from the dish boundaries.

In Fig.2, time traces of flow velocity measured at the center of this system near the surface (a) and the bottom (b) are shown. Near the surface, the flow first proceeds opposite to wave propagation and increases as the wave front approaches the observation area (Fig.2 a). After wave passage, it suddenly reverses its direction and then moves with the wave. By contrast, the hydrodynamic flow near the bottom shows the inverse behavior (Fig.2 b).

Schematic profiles of waves and hydrodynamic flow as derived from these experiments are shown in Fig.3. These profiles suggest that a global cellular structure of the hydrodynamic flow travels with the propagating circular wave fronts.

Hydrodynamic Flow Induced by Curvature Effect. The enhanced flow shown in the shaded part of Fig.2 a was caused by a collision of waves close to the observation area resulting in the formation of very sharp cusps. Because of the large negative curvature (K), the normal velocity (N) of the cusp waves becomes very high. The relationship between N and K is :[9,10]

$$N = \mu - DK \qquad (1)$$

where μ is the velocity of plane waves, and D is the diffusion coefficient of the active chemical species. We find that this rapid propagation of the cusp waves causes a pronounced

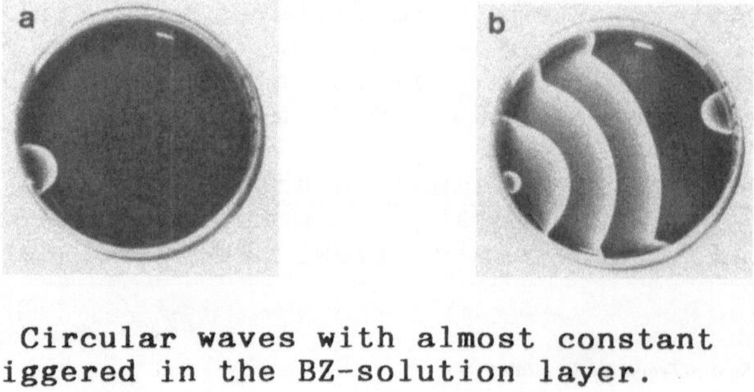

Fig.1. Circular waves with almost constant period are triggered in the BZ-solution layer.

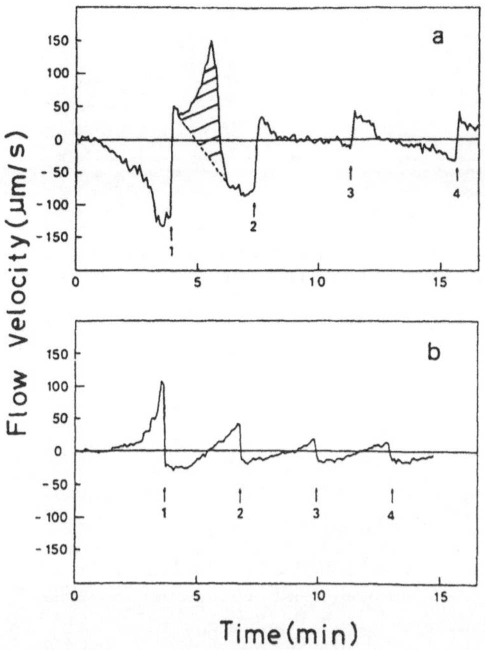

Fig.2. Time trace of flow velocity measured near the surface (a) and near the bottom (b) of the circular wave pattern in the BZ solution layer (see Fig.1 b).

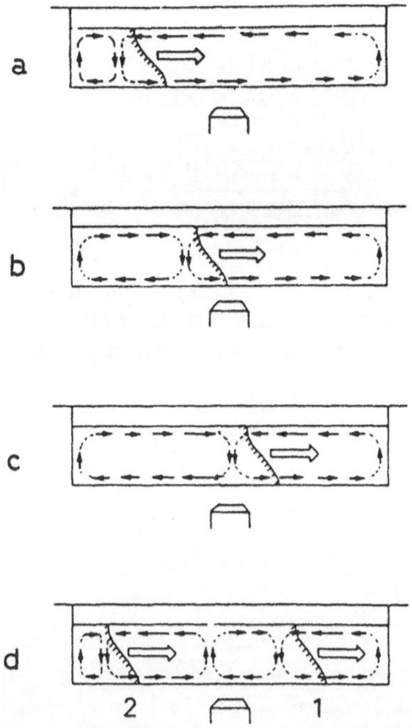

Fig.3. Schematic of hydrodynamic flow induced by chemical waves. A cellular structure of the hydrodynamic flow is traveling with the wave fronts.

increase of the flow velocity. This suggests that the hydrodynamic flow is strongly influenced by curvature effects on the chemical waves.

HYDRODYNAMIC FLOWS WITH SPIRAL WAVES

Oscillatory Hydrodynamic Flow. The initiation of a pair of spiral waves near the center of an excitable BZ solution layer is shown in Fig.4. The measurement of flow velocity is carried out at the center of the dish. Figure 5 shows a time trace of flow velocity measured near the surface of the spiral wave pattern. Arrows at the bottom indicate passage of wave fronts through the detection area at the dish center. There are three phases:

1) 0 - 2 min: alternating flow induced by a few discrete waves in a chemically well reduced medium (compare Fig.2),
2) 3 - 9 min: unidirectional, mean flow with a velocity up to 20 µm/s,
3) 10-19 min: oscillatory hydrodynamic flow with an amplitude up to 150 µm/s correlated with the periodic passage of wave trains.

In the oscillatory region (phase 3), we measured the velocity variation of the hydrodynamic flow along the vertical direction. We found that the velocity changes in an oscillatory manner as schematically shown in Fig.6.

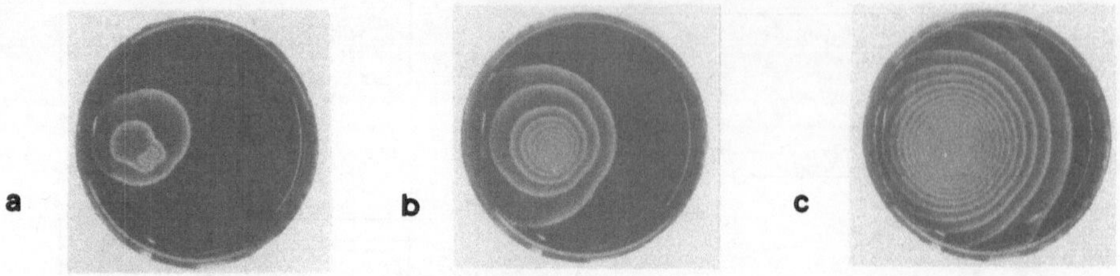

Fig.4. Temporal development of spiral wave trains produced in a dish in which wave initiation at the boundary was suppressed.

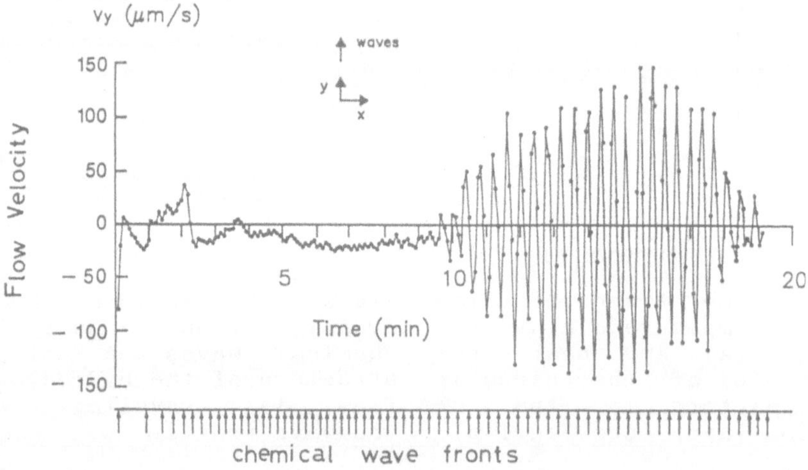

Fig.5. Time trace of flow velocity measured near the surface of the spiral wave pattern in the BZ solution layer. An oscillatory flow is induced in the later stage of the reaction.

Deformation and Decomposition. The dynamic behavior of the propagating chemical waves under the influence of oscillatory hydrodynamic flow was observed by 2D-spectrophotometry in a relatively large area (13 mm x 13 mm). Periodically varying deformations (see a-b) and onset of irregular decomposition (see c) of chemical waves in a layer with a covered liquid/gas interface are shown in Fig.7. Apparently, these structural changes are caused by the global flow oscillation, as drawn in Fig.6, and thus provide indirect evidence for its existence. The mechanism of the oscillatory flow remains an open question. For an explanation, not only the dependence of surface tension on chemical composition and possibly reaction heat (Marangoni effect) may be relevant, but also effects of curvature and chemical entrainment have to be taken into account.[8]

CONCLUSIONS

In summary our results show:

1) Propagation of a circular wave induces a hydrodynamic flow traveling with the wave front.
2) A chemical wave front having high curvature strongly influences the hydrodynamic flow.

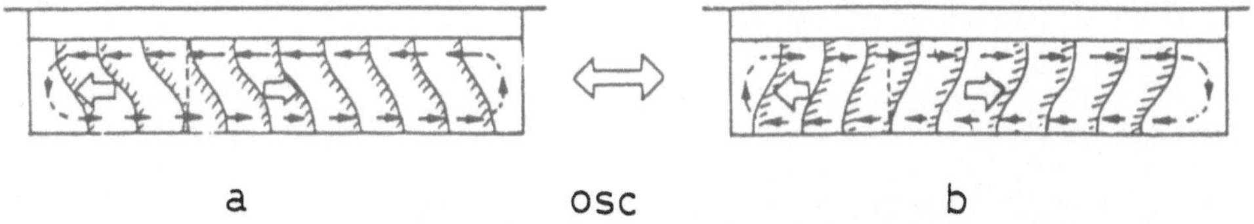

Fig.6. Schematic profile of hydrodynamic flow oscillation induced in the BZ solution layer.

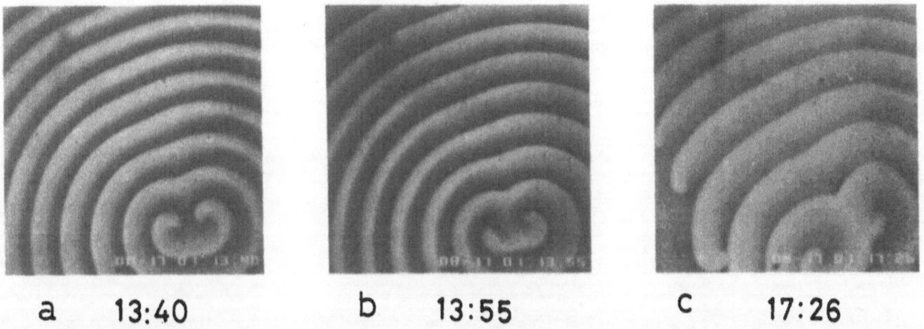

Fig.7. Periodically varying deformation (a,b) and onset of irregular decomposition (c) of chemical wave fronts caused by the oscillatory hydrodynamic flow.

3) A successive train of chemical waves induces an oscillating hydrodynamic flow. As a consequence, periodically varying deformations of chemical wave profile occur.
4) A feed-back mechanism due to curvature effects on the flow velocity occurring during wave deformation is suggested to play a major role for the establishment of flow oscillations.

ACKNOWLEDGMENTS

The authors wish to thank Prof.B.Hess, Dr.Th.Plesser and Dr.H.Hashimoto for helpful discussions; Mr.U.Heidecke, Mr.K.Dreher, Ms.E.Schlüter and Ms.G.Schulte for technical assistance. This work was supported by the Stiftung Volkswagenwerk, Hannover.

REFERENCES

1. C.Vidal and P.Hanusse, Int.Rev.Phys.Chem. 5(1986)1.
2. P.Borckmans and G.Dewel, in: From Chemical to Biological Organization, eds. M.Markus, S.C.Müller and G.Nicolis (Springer Ser. in Synergetics, Berlin, 1988) p.114.
3. A.M.Zhabotinskii and A.N.Zaikin, J.theor.Biol. 40(1973)45.
4. K.Showalter, J.Chem. Phys. 73(1980)3735.
5. K.I.Agladze, V.I.Krinski and A.M.Pertsov, Nature 308(1984) 834.
6. S.C.Müller, Th.Plesser and B.Hess, Physica 24D(1987)125.
7. H.Miike,S.C.Müller and B.Hess, Chem.Phys.Lett.114(1988)515.
8. H.Miike,S.C.Müller and B.Hess, Phys.Rev.Lett. 61(1988)2109.
9. J.P.Keener and J.J.Tyson, Science 239(1988)1284.
10. Y.Kuramoto, Prog. theor. Phys. 63(1980)1885.

STABILITY OF TRAVELLING WAVES IN THE BELOUSOV-ZHABOTINSKII REACTION

David A. Kessler* and Herbert Levine+

* Physics Department
University of Michigan
Ann Arbor, MI 48109 USA

+ Department of Physics
and
Institute for Nonlinear Science
University of California, San Diego
La Jolla, CA 92093

INTRODUCTION

As we have seen in Dr. Müller's talk, the Belousov-Zhabotinskii (BZ) reaction leads to a rich variety of non-equilibrium spatial structures and serves as a paradigm for pattern formation in excitable media[1]. In this talk, we will focus on the simplest such pattern, the planar travelling wave, in which regions of excited and quiescent reagents move uniformly through space (Fig. 1). Understanding this structure is a necessary first step towards a complete picture of more complex structures such as the target[2], the rotating spiral[3] or, in three dimensions, the scroll[4]. This is particularly true as these patterns far from their centers asymptotically approximate planar travelling waves.

Starting from the work of Keener and Tyson[5,6], there has recently been considerable progress in finding travelling wave solutions. The most tractable model for this phenomenon is the piece-wise linear Oregonator[6,7]. In this simplification of the Oregonator model discussed by Dr. Müller, the disparity in time-scales for the two reaction kinetics is exploited to reduce the system to an effective dynamics for the slow reaction in each of the "excited" and "quiescent" regions. The resulting dynamics is further approximated to be linear in each of the two "phases"[8]. This set of approximations allow for an semi-analytic treatment of the dispersion relation, giving the wavelengths of the excited and quiescent regions λ_+, λ_- in terms of the velocity.[6,7]

Nonlinear Evolution of Spatio-Temporal Structures in Dissipative Continuous Systems
Edited by F.H. Busse and L. Kramer
Plenum Press, New York, 1990

Experimentally, planar travelling waves are seen[9,10], albeit in only the upper range of the velocity band. Also, it is usually assumed[11] that target patterns are generated simply by a pacemaker acting as a local source for spherically spreading waves. This means that these wave solution are stable, at least in some parameter range. As we lower the velocity, though, there may be instabilities which could in principle give rise to more complex spatial structure or temporal dynamics. These instabilities, beyond their own intrinsic interest, are important for further testing the current models beyond the presently calculated dispersion relation. They also clearly will have implications for the dynamics of the more complex patterns.

With the above in mind, the purpose of this paper is to compute the spectrum of linear oscillations around the aforementioned planar solutions. We first review the formulation of the PLO along with the dispersion relation calculation and the various scaling limits. In section III, we study the linear stability within the context of the quasistatic approximation. This approximation, which assumes the time-scale for interfacial motion is slow compared to that for diffusion, greatly simplifies the problem and thereby allows some insight into the instability mechanism. Our major result here is the identification of a set of unstable modes corresponding to real eigenvalues crossing zero, representing transverse steady-state deformations of the wave within some wavevector band. This instability occurs below some critical velocity.

In sec. IV, we turn to the full calculation, relaxing the above-mentioned quasistatic approximation. We show that as the velocity is lowered, there occurs both the aforementioned transverse instability as well as a Hopf bifurcation at zero wavevector. The value of the diffusion constant ratio governs which of these instabilities occurs first. We also discuss briefly the effect of non-zero Bloch wavevector in the longitudinal direction. Section V presents the weakly non-linear analysis of the steady-state bifurcation and section VI presents our conclusions.

PIECE-WISE LINEAR OREGONATOR

In this section, we briefly review the theory of travelling waves in the piece-wise linear Oregonator (PLO)[6,7]. The basic Oregonator model is that of two chemical species, u and v which react and diffuse. The concentrations obey

$$\frac{\partial u}{\partial t} = \epsilon_0 \nabla^2 u + \frac{f(u,v)}{\epsilon_0} \; ; \tag{1a}$$

$$\frac{\partial v}{\partial t} = \epsilon_0 \frac{D_v}{D_u} \nabla^2 v + g(u,v) \; . \tag{1b}$$

Here $g(u,v)$ is a linear function of u, v while f is non-linear with a "cubic" shape, so that for every v in a some range it possesses 3 roots, $u_- < u_0 < u_+$. The nullclines of f and g are presented in Fig. 1.

For small ϵ_0, we can find thin (width $\sim \epsilon_0$) interfacial reaction zones where u changes from one of the stable roots, u_-, u_+ to the other, with v approximately constant across the zone. Then, the concentration v obeys the two separate equations

$$\begin{aligned} \frac{\partial v_+}{\partial t} &= \epsilon_0 \frac{D_v}{D_u} \nabla^2 v_+ + g(u_+(v),v) \; , \\ \frac{\partial v_-}{\partial t} &= \epsilon_0 \frac{D_v}{D_u} \nabla^2 v_- + g(u_-(v),v) \end{aligned} \tag{2}$$

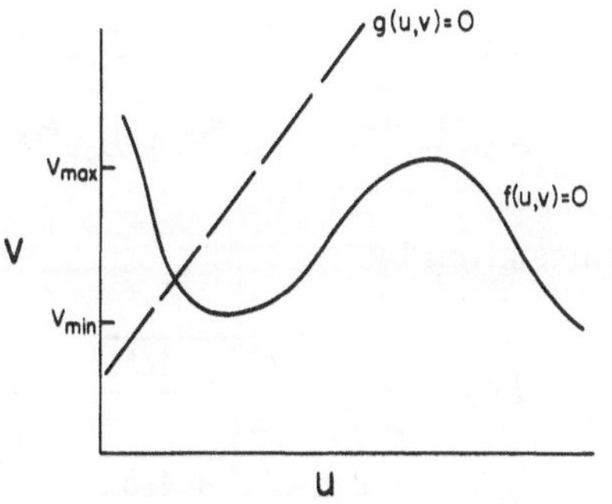

Fig. 1. Nullclines for the reaction-rate functions f, g.

in the excited and quiescent regions respectively. These equations are then supplemented by the boundary conditions

$$v_+ = v_- = \nu(c_n + \epsilon_0 \kappa) \; ; \tag{3a}$$
$$v'_+ = v'_- \; . \tag{3b}$$

imposed at the boundary between + and − regions. Here ν is a function derived by solving the u-field equation (1a) in the thin interface and κ is the curvature of the interface, whose normal velocity is c_n.

It is convenient at this point to change notation slightly. As the curvature term does not enter into the unperturbed planar problem, the unperturbed solutions are parametrized by the single parameter $\epsilon \equiv \epsilon_0 \frac{D_v}{D_u}$. We identify then the strength of the curvature term in the boundary condition Eq. (3a) by the "surface tension" $\gamma \equiv \epsilon_0$ in analogy with the similar Gibbs-Thomson condition which arises in first order phase transitions[13].

The above field equation (2) for v, while a major improvement over the original coupled (u, v) system, is nonlinear and so still analytically intractable. To make progress it is useful[7] to linearize (2) in each phase about the stall concentration v_s for which the interface between the phases is stationary. We thus take

$$g(u_\pm(v), v) = \pm a_\pm - b_\pm(v - v_s) \tag{4}$$
$$\nu(c_n + \gamma\kappa) = (c_n + \gamma\kappa)\nu'_0 + v_s \tag{5}$$

where $a_\pm$, $b_\pm$ and ν'_0 are all constants. This is quite accurate as long as v is not close to v_{max}/ν'_0, where v_{max} is the location of the local maximum on the f nullcline. Approximate values of the parameters for this linearized version of the Oregonator (which we will denote as the piece-wise linear Oregonator - PLO) are

$$a_+ = \frac{11}{16}, \quad a_- = \frac{1}{16}, \quad b_+ = 7, \quad b_- = 1, \quad \nu'_0 = \frac{\sqrt{2}}{30} \; . \tag{6}$$

The problem of travelling waves in the PLO model can be set up as follows. We assume a solution periodic in the y-direction moving with velocity c. The width of the excited (quiescent) region is λ_+ (λ_-). The general solution of the v-field equation (2) in each region is

$$v_+ = v_s + \frac{a_+}{b_+} + B_1 e^{P_1 y} + B_2 e^{P_2 y}$$
$$v_- = v_s - \frac{a_-}{b_-} + B_3 e^{P_3 y} + B_4 e^{P_4 y} \tag{7}$$

with the spatial growth rates given by

$$P_{1,2} = \frac{-c \pm \sqrt{c^2 + 4\epsilon b_+}}{2\epsilon}$$
$$P_{3,4} = \frac{-c \pm \sqrt{c^2 + 4\epsilon b_-}}{2\epsilon} . \tag{8}$$

For a given c, the four coefficients B_i and the two widths $\lambda_\pm$ are determined by the six boundary conditions derived from the interface conditions (3):

$$v_+(0) = v_-(0) = v_s + \nu_0' c; \qquad v_+(\lambda_+) = v_-(-\lambda_-) = v_s - \nu_0' c;$$
$$v_+'(0) = v_-'(0); \qquad v_+'(\lambda_+) = v_-'(-\lambda_-) . \tag{9}$$

A study of these equations reveals the following structure. At very small ϵ, there exist travelling waves of all velocities. One can distinguish three distinct regimes. If the velocity c is $O(1)$, diffusion in v can be completely neglected. At c such that $c \sim \epsilon^{1/3}$, a regime originally identified by Fife[6] emerges. Here, diffusion cannot be neglected; instead, the $b_\pm(v - v_s)$ term can be dropped. One can then show that $\lambda_- = \frac{a_+}{a_-}\lambda_+$ and hence the ratio $\frac{\lambda_-}{\lambda_+}$ is independent of c. Also, it is interesting to note that in this regime the linearization on which the PLO rests can be rigorously justified. This Fife regime eventually breaks down as $c \to 0$. In fact, there exists a critical $\epsilon = \epsilon_c$ such that the solution branch itself ends before $c = 0$ for all $\epsilon > \epsilon_c$. For the parameters given above, $\epsilon_c \simeq 3.5 \times 10^{-3}$.

LINEAR STABILITY: QUASISTATIC

Having derived the band of travelling wave solutions, we can now proceed to study the stability of these solutions. Clearly, a perturbation can be labelled by a transverse Fourier wavevector q. Also, since the original pattern is periodic in y, we can choose perturbations which satisfy Bloch's theorem $\delta \sim e^{i\alpha y}\phi$ where ϕ is periodic.

Let us assume that the interfaces at $y = -\lambda_-$, 0 and $y = \lambda_+$ are deformed respectively to

$$y = -\lambda_- + \delta_-(t) \cos qx ;$$
$$y = \delta_+(t) \cos qx ;$$
$$y = \lambda_+ + \delta_-(t) e^{i\alpha} \cos qx . \tag{10}$$

This will induce new terms in the v field:

$$v_+ = v_+^0 + A_1 \cos qx \, e^{k_1 y + \omega t} + A_2 \cos qx \, e^{k_2 y + \omega t} ;$$
$$v_- = v_-^0 + A_3 \cos qx \, e^{k_3 y + \omega t} + A_4 \cos qx \, e^{k_4 y + \omega t} . \tag{11}$$

The spatial growth rates of the perturbation are given by

$$k_{1,2} = \frac{-c \pm \sqrt{c^2 + 4\epsilon(\epsilon q^2 + \omega + b_+)}}{2\epsilon},$$
$$k_{3,4} = \frac{-c \pm \sqrt{c^2 + 4\epsilon(\epsilon q^2 + \omega + b_-)}}{2\epsilon} \tag{12}$$

and ω is the growth rate: $\delta_+(t) = e^{\omega t}\delta_+$. Again the interfacial conditions (3) yield a system of six equations, linear in the six unknowns A_i, $\delta_\pm$:

$$\begin{aligned}
A_1 + A_2 + (P_1B_1 + P_2B_2)\delta_+ &= \nu_0'\delta_+(\omega + \gamma q^2);\\
(A_1e^{k_1\lambda_+} + A_2e^{k_2\lambda_+})e^{-i\alpha} + (P_1B_1e^{P_1\lambda_+} + P_2B_2e^{P_2\lambda_+})\delta_- &= -\nu_0'\delta_-(\omega + \gamma q^2);\\
A_1 + A_2 &= A_3 + A_4;\\
(A_1e^{k_1\lambda_+} + A_2e^{k_2\lambda_+})e^{-i\alpha} &= A_3e^{-k_3\lambda_-} + A_4e^{-k_4\lambda_-};\\
k_1A_1 + k_2A_2 + (P_1^2B_1 + P_2^2B_2)\delta_+ &= k_3A_3 + k_4A_4 + (P_3^2B_3 + P_4^2B_4)\delta_+;\\
k_1A_1e^{k_1\lambda_+} + k_2A_2e^{k_2\lambda_+} + (P_1^2B_1e^{P_1\lambda_+} + P_2^2B_2e^{P_2\lambda_+})\delta_-e^{i\alpha} &\\
= k_3A_3e^{-k_3\lambda_-} + k_4A_4e^{-k_4\lambda_-} &+ (P_3^2B_3e^{-P_3\lambda_-} + P_4^2B_4e^{-P_4\lambda_-})\delta_-.
\end{aligned} \tag{13}$$

We treat these equations as follows. The last four equations can be used to eliminate the field coefficients A_i. Substituting these into the first two equations yields a 2x2 matrix equation

$$\begin{pmatrix} L_{11} & L_{12} \\ L_{21} & L_{22} \end{pmatrix}\begin{pmatrix} \delta_+ \\ \delta_- \end{pmatrix} = (\omega + \gamma q^2)\nu_0'\begin{pmatrix} \delta_+ \\ \delta_- \end{pmatrix} \tag{14}$$

where

$$\begin{aligned}
L_{11} &= \frac{\partial A_1}{\partial \delta_+} + \frac{\partial A_2}{\partial \delta_+} + P_1B_1 + P_2B_2;\\
L_{12} &= \frac{\partial A_1}{\partial \delta_-} + \frac{\partial A_2}{\partial \delta_-};\\
L_{21} &= -(\frac{\partial A_1}{\partial \delta_+}e^{k_1\lambda_+} + \frac{\partial A_2}{\partial \delta_+}e^{k_2\lambda_+})e^{-i\alpha};\\
L_{22} &= -(\frac{\partial A_1}{\partial \delta_-}e^{k_1\lambda_+} + \frac{\partial A_2}{\partial \delta_-}e^{k_2\lambda_+})e^{-i\alpha} - P_1B_1e^{P_1\lambda_+} - P_2B_2e^{P_2\lambda_+}.
\end{aligned} \tag{15}$$

This immediately yields the eigenvalue equations

$$\omega = -\gamma q^2 + \frac{L_{11} + L_{22} \pm \sqrt{(L_{11} - L_{22})^2 + 4L_{12}L_{21}}}{2\nu_0'}. \tag{16}$$

Since the L's depend on ω (through the k_i), this is actually a non-linear set of equations for the two modes at fixed q.

As a first pass, we will ignore the possibility of a Hopf bifurcation so that the onset of the instability (if it exists) corresponds to ω=0. If we are interested in solutions of (16) for $\omega = 0$, however, we are free to set ω to zero in the L's. This corresponds to the well-known quasistatic approximation, where one assumes that the motion of the interface is slow compared to the relaxation rate of the concentration field. We will see in the next section that this approach is appropriate if the ratio γ/ϵ is small.

Within the quasistatic approximation, it is now a straightforward exercise to compute ω_{qs} as a function of q and c. As we are interested in possible instabilities it suffices to consider the branch of ω_{qs} corresponding to the positive sign in Eq. 16. The results for the case $\alpha = 0$ (periodic perturbations) with the PLO parameters (6) and $\epsilon = .01$, $\gamma = .008$ are displayed in Fig. 2, which presents $\omega_{qs}(q)$ for the positive branch for three values of c. We see that an instability sets in at finite q as c is lowered below $c \approx .58$.

The source of this instability is easy to see. Consider a localized (large q) perturbation on the interface. If a point on the interface moves forward a distance δ, the value of v at that point increases by an amount $(\frac{dv}{dx})^0\delta$. In the absence of surface tension, then, the velocity of that point would increase by $(\frac{dv}{dx})^0\delta/\nu_0'$. Since the quantity $(\frac{dv}{dx})^0$ is always positive, this gives rise to an instability. At large q, moreover, the interfaces are decoupled and due to the requirement of continuity of the first derivative of v on the interface, no shift in the field is induced and so the above argument is exact. The effect of surface tension is simply to shift ω by $-\gamma q^2$. The large-q dispersion relation is then (in the quasistatic limit)

$$\omega_{qs} \rightarrow (\frac{dv}{dx})^0/\nu_0' - \gamma q^2 \tag{17}$$

Now, since $(\frac{dv}{dx})^0$ increases as c decreases, the instability becomes stronger for decreasing c and is eventually able to overcome the stabilizing effect of surface tension. At small q, the dominant effect is the stabilizing effect of the coupling of the interfaces. Of course, the vanishing of ω_{qs} on the positive branch at $q = 0$ is due to the overall translation symmetry of the problem. The stability of the negative (optical) branch can be seen most easily in the Fife limit. Here, $\frac{\lambda_+}{\lambda_-}$ is a constant independent of c, so a longitudinal perturbation which alters this ratio is disallowed. Thus ω_{qs} in the Fife limit actually approaches $-\infty$ as q goes to zero on the optical branch. We now briefly investigate the effect of breaking the longitudinal periodicity of the perturbation. In Fig. 3 we set $\gamma = .01$, and $c = .5$, and plot the variation of the growth rate of the real mode with Bloch phase factor α, for q =10, 15, 20. Note that for the larger values of q, there is almost no variation with α; this represents an effective decoupling of different planes. Clearly, the effect of a transverse oscillation of large q decays sufficiently rapidly that the longitudinal structure is largely irrelevant. In all cases, however, finite α stabilizes the pattern. Hence, the most unstable perturbation is a transverse deformation which is exactly periodic.

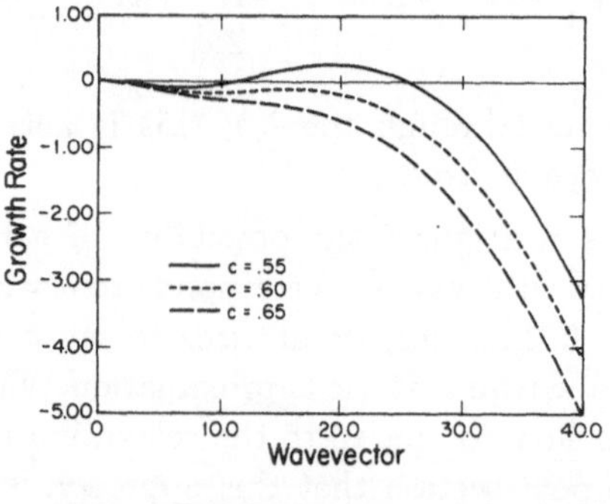

Fig. 2. Quasistatic growth rate, ω_{qs} vs. wavevector,q, for $c = .65, .60, .55$ in the PLO with $\epsilon = .01, \gamma = .008$.

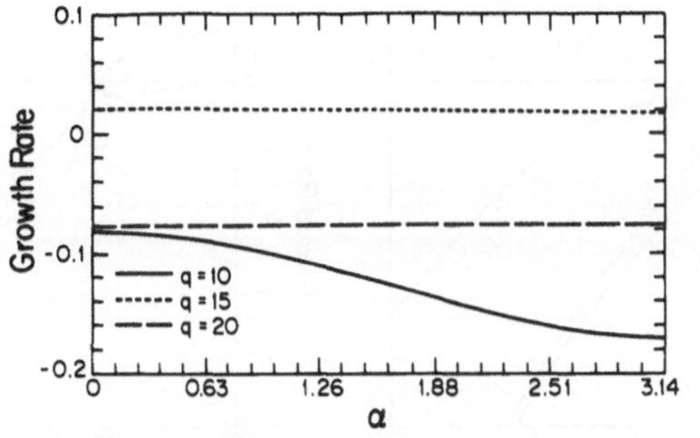

Fig. 3. Quasistatic growth rate vs. Bloch index α for several wavevectors at $\gamma = \epsilon = .01$,c= .5.

LINEAR STABILITY: EXACT

We now return to the problem of solving the original stability equations, restoring the nonlocality in time dropped in the quasistatic treatment. In general, we expect to find an *infinite* number for *each* wavevector q. This is because a set of initial conditions for the (linearized) time dependent problem would necessitate giving values for the field everywhere in space, not just on the interface.

Before proceeding to the calculation, we can address the question of the quality of the quasistatic answers for modes with ω nonzero. (As we have already discussed, the quasistatic answers are exact for any modes with ω exactly zero.) To begin, we note that the dispersion relation (16) takes the schematic form

$$\omega = -\gamma q^2 + F(\epsilon q^2 + \omega; c) \tag{18}$$

where the function F is known implicitly, but depends on ω and q only through the combination $\omega + \epsilon q^2$; this follows from the form of the spatial growth rates (12). Now, the quasistatic limit is defined by dropping the ω dependence in F, yielding

$$\omega_{qs} = -\gamma q^2 + F(\epsilon q^2; c) \tag{19}$$

The onset of the spatial instability is signaled by finding values of q and c such that $\omega_{qs} = 0$, and $\frac{\partial \omega_{qs}}{\partial q^2} = 0$. The latter implies $\gamma = \epsilon F'(\epsilon q^2; c)$ where the $'$ denotes derivative with respect to the first argument.

What does the above condition imply for the actual eigenvalue ω? Obviously, $\omega_{qs} = 0$ guarantees $\omega = 0$. Around this point, we thus have

$$\omega = \omega_{qs} + F(\epsilon q^2 + \omega; c) - F(\epsilon q^2; c) \approx \omega_{qs} + \omega\, F'(\epsilon q^2; c) \tag{20}$$

Substituting the value of F', we arrive at the final relation

$$\omega \approx \frac{\omega_{qs}}{1 - \gamma/\epsilon} \tag{21}$$

The above result shows that at small γ/ϵ, the true growth rate near the instability is close to that predicted by the quasistatic approximation. As $\gamma \rightarrow \epsilon$

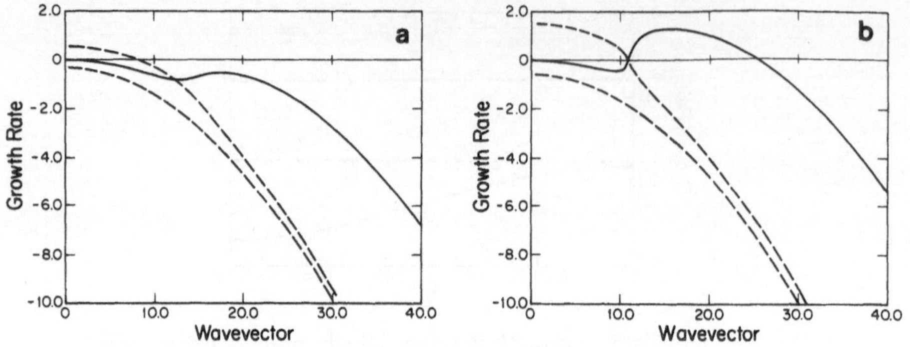

Fig. 4. Growth rate vs. wavevector of least stable modes for a) $c = .55$, b) c=.6, with $\gamma = .008, \epsilon = .01$. Solid lines: real modes; Dashed lines: complex modes.

however, the deviation becomes alarmingly large. At exactly $\gamma = \epsilon$, this equation implies that the first (time) derivative term in the true equation actually vanishes, and the bifurcation becomes degenerate. In fact, this degeneracy signals the collision of a real instability with a Hopf bifurcation; this is completely missed by the quasistatic approach. Finally, as γ becomes larger, the actual spectrum is not simply related to the quasistatic one and the computations of the last section are no longer useful.

Let us now turn to the numerical results. In the following, we focus on the case $\alpha = 0$, where the instability is strongest. Let us again consider $\gamma = .008$, $\epsilon = .01$. In Fig. 4a, we have plotted the growth rate (i.e., Re ω) of the three least stable branches for $c = .55$. Just as we saw within the quasistatic calculation, there is an unstable real mode as c is below .58. But this is not the entire story! Note that there is also an unstable complex branch, with a maximum growth rate at $q = 0$. To track this branch, we present in Fig. 4b the analogous results for $c = .6$. Here, the real branch has stabilized, as indicated by the quasistatic calculation, but the complex branch is still unstable at small q. So, for these parameters, the travelling wave would presumably start exhibiting spatially uniform amplitude oscillations, before it could undergo the bifurcation to transverse structure.

In Table 1 we show the real and imaginary parts of ω at zero wavevector as a function of the velocity. This indicates that the Hopf bifurcation occurs at approximately $c = .671$. Note that since this instability occurs at $q = 0$, it does not depend on the value of γ.

As we lower the value of γ, the onset of the spatial instability occurs at larger velocity. At $\gamma = .00621$, this instability occurs at $c = .671$ which is precisely where

Table 1. ω vs. c for the Unstable Mode at $q = 0$

c	Re ω	Im ω
.6	.560	2.425
.625	.283	2.337
.65	.105	2.239
.675	−.020	2.150
.70	−.112	2.067

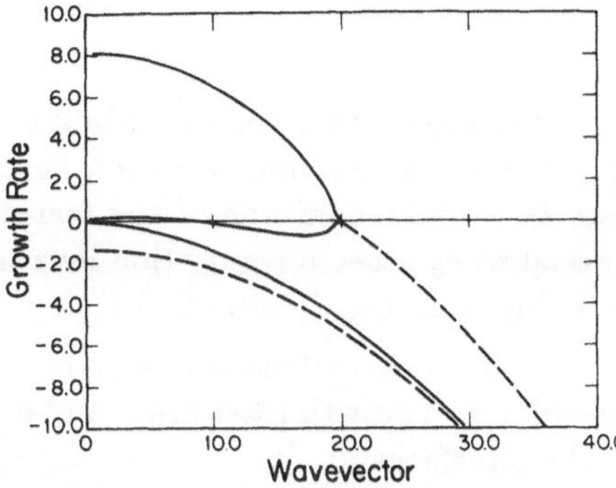

Fig. 5. Growth rate vs. wavevector of least stable modes for c=.4, $\gamma = .02, \epsilon = .0$. Solid: real modes, Dashed: complex modes.

the Hopf bifurcation occurs. At still smaller γ the spatial instability dominates and there should be a chance of observing corrugated travelling waves. This will be explored further in the following section using the standard tools of bifurcation theory.

Let us now turn to the case $\gamma = .012$, at the same value of ϵ. In Fig. 5, we present the growth rate of several of the lowest modes at $c = .4$. This point is of course well below the critical c for the Hopf bifurcation and so the system clearly exhibits unstable modes. The branch which connects to the translation zero mode is purely real. Unlike the previous case, the growth rate of this mode never turns up to give rise to a spatial instability.

It is interesting to understand the connection between the true modes and what would have occurred if we had used an obviously inaccurate quasistatic approximation. We would then have found a branch similar to that for $c = .55$ in Fig. 2, with zero modes at $q \approx 7.5$ and $q \approx 20$ and positive growth rates in between. As already mentioned, any point for which $\omega = 0$ solves the quasistatic equation is also a point where it solves the full equation; hence the actual spectrum must also have zero eigenvalues at these wavevectors. Here this occurs because the two relevant branches are purely *real*. So, as they cross zero to become stable (at $q \approx 20$ and $q \approx 7.5$), they correspond to the roots seen in the misleading quasistatic approach.[14] These modes then meet and become a complex conjugate pair at large wavevector. The last mode shown is always complex with an imaginary part of around 3.2.

We have here considered the case of fixed ϵ, letting γ vary. The effects of varying ϵ are rather minor as long as we maintain the ratio of γ to ϵ.

NON-LINEAR STABILITY

We have just seen that for small γ there is a possible instability at finite transverse wavevector q which occurs as c is lowered below a critical value. At any such bifurcation, the next question concerns the form of the non-linear terms close to onset. The general symmetry of the problem guarantees that there exist a coupled pair of equations of the form

$$\frac{d\delta_q}{dt} = \omega\delta_q + \rho\delta_q^3 \tag{22}$$

$$\frac{d\delta_0}{dt} = \sigma\delta_q^2 \tag{23}$$

Here δ_q is the amplitude of the eigenvector corresponding to the (almost) marginal mode with wavevector q; we will use the normalization convention that this mode is given by $\delta_q(1, \Phi)$. Also, δ_0 is the magnitude of the translation zero mode. The absence of terms proportional to δ_0 reflects translation invariance; the presence of σ reflects a velocity change induced by the deformation δ_q.

The basic method[15] is to do perturbation theory by expanding in the amplitudes δ_q and δ_0. To linear order, the growth rates are ω and 0 respectively. To second order, we set the shift of the interfaces to

$$y_+^{(2)} = \Psi_2^{(+)} \cos 2qx e^{2\omega t} + \Psi_0^{(+)} e^{2\omega t} \tag{24}$$

$$y_-^{(2)} = \Psi_2^{(-)} \cos 2qx e^{2\omega t} + \Psi_0^{(-)} e^{2\omega t} \tag{25}$$

The Ψ_2 terms are both unknown; the Ψ_0 coefficients are related to each other by the fact that there is only one additional mode at $q = 0$, necessarily orthogonal to the pure translation. Hence, there is one fewer unknown for the zero wavevector sector.

We assume a similar form for the x and t dependencies of the concentrations v_+, v_-. The one subtlety in the calculation is the inclusion of the inhomogenous terms, proportional to σ) in v induced by the time-dependence of the coefficients of the first-order term for v.

As in the first order calculation, substituting v into the boundary conditions gives rise to six coupled linear equations for each transverse wavevector mode. For the $\cos 2qx$ system, the six unknowns are $\Psi_2^{(+)}$, $\Psi_2^{(-)}$ and the 4 coefficients of the $\cos 2qx$ modes in $v_\pm$. This is non-resonant behavior. For the $q = 0$ mode, there are again six unknowns: $\Psi_0^{(+)}$, the 4 field coefficients and σ. So, σ is determined by this procedure. Finally, the $\cos qx$ system which naively results from the product $\delta_0\delta_q$, is satisfied identically due to translation invariance. We leave the detailed form of the equations to a longer paper.[20]

The more interesting coefficient is, of course, ρ. This determines whether the bifurcation is subcritical or supercritical and hence whether we can expect to experimentally find small amplitude deformed travelling waves. To do this, we must extend the above computation to third order. To go to third order, everything is similar but even more tedious. The only Fourier mode that we need to study is $\cos qx$. Again the appropriate inhomogeneous terms, now including one proportional to ρ, must be added. Substituting this into the boundary conditions gives, as usual, a 6x6 linear system. The unknowns are the 4 homogeneous field coefficients, the magnitude of the orthogonal shift at wavevector q, $\{\Psi_1^{(+)}, \Psi_1^{(-)}\}$ and the value of ρ.

We have carried out the above calculation for the standard set of parameters at $\epsilon = .01$, with γ varying from .004 to .006. As shown in the last section, the steady-state bifurcation at non-zero q is the dominant instability in this parameter range. For each value of γ, we choose c and q to exactly correspond to the instability threshold. The results are given in Table 2.

The most important result is that ρ is negative, signaling a supercritical bifurcation. This means that over this parameter range, there should exist small amplitude deformed waves. These states should be observable experimentally.

Table 2. Bifurcation Parameters ρ, σ as a Function of γ for $\epsilon = .01$

γ	c	q	$\rho(\times 10^4)$	$\sigma(\times 10^2)$
.004	.843	20.49	−7.72	1.66
.00425	.816	20.35	−6.26	1.44
.0045	.793	20.16	−5.20	1.25
.00475	.771	19.97	−4.23	1.09
.005	.751	19.76	−3.51	0.95
.00525	.733	19.54	−2.94	0.83
.0055	.715	19.33	−2.48	0.72
.00575	.699	19.11	−2.10	0.63
.006	.683	18.90	−1.78	0.55

Recall that we have shown that the bifurcation becomes degenerate at $\gamma = \epsilon$. This feature of the spectrum also shows up in the nonlinear calculation; specifically, if we naively continue the above table until $\gamma = .01$, we find that $\rho \to \infty$. This occurs precisely because the linear term in time derivatives is vanishing and the amplitude equation (22) need to be replaced by the more general form

$$\frac{d^2\delta_q}{dt^2} + \alpha\frac{d\delta_q}{dt} = \tilde{\omega}\delta_q + \tilde{\rho}\delta_q^2 \tag{26}$$

Here α is a second unfolding parameter which goes to zero linearly as γ approaches ϵ. The divergence of ρ predicted in this manner ($\rho \approx \tilde{\rho}/\alpha$) provides a useful check on our numerical procedures.

CONCLUSIONS

In summary, we have carried out the linear and nonlinear stability analysis around travelling wave solutions in the piecewise linear Oregonator model. Our results indicate

1) There exists an instability to transverse corrugation which sets in below a critical velocity. The instability occurs at finite wavevector. For small enough ϵ, the instability disappears as the velocity is decreased below a second threshold.

2) There also exists a Hopf bifurcation which is strongest at $q = 0$. Roughly, at large γ the Hopf bifurcation governs the breakdown of the band of travelling waves, whereas for small γ the spatial instability occurs at higher c.

3) The nonlinear analysis predicts that the steady bifurcation will be supercritical in a wide range of parameters. It might therefore be worthwhile to search for nonlinear states which would exist close to onset.

Results similar to our linear stability calculations have been reported in recent paper by Ohta et al[16] dealing with pulses in excitable systems. Also, qualitatively similar behavior occurs during the process of explosive crystallization[17].

How do these predictions compare to existing experiments? In the paper by Dockery, Keener and Tyson[6], there is a report of unpublished data by Winfree regarding planar travelling waves. There is also some work on planar waves by Ross et al[10]. In both of these experiments, there appears to be reasonable agreement[18] between the dispersion relation and the measured waves periods. Experimentally, the waves seem to disappear if the velocity is lowered. We now can hypothesize that this is due to the instability discovered here. The location of the instability ($c \approx .6$) is roughly coincident with the onset of the various unstable modes discussed here.

Of course, there is a need for a much more detailed study of these waves before any conclusions can be reached. There do not seem to be any experimental results concerning exactly what happens if one tries to set up a wave with too low a velocity. The simplest control parameter is the frequency (or equivalently, the period) of the travelling wave. Therefore, the best chance of observing an interesting transition is to reduce the period and look for some instability. Since the two species Oregonator is probably not all that accurate quantitatively, it makes sense to heuristically investigate several differing reaction "recipes" to try to reach a regime such that the spatial instability occurs *before* the Hopf bifurcation.

Finally, we offer one conjecture. Recently, there have been several observations[19] of an oscillatory instability of the tip region in the rotating spiral pattern. The amplitude of this oscillation seems to vanish as one moves outward from the tip. Recall[5,7] that the usual spiral solution must asymptotically approach a planar travelling wave. Now assume that the asymptotic solution is in the stable regime. As we get closer to the tip the normal velocity decreases, eventually approaching zero as we reach the tip which moves purely tangentially in a circle. It might be possible that if the rate of decrease of velocity is fast enough, the Hopf bifurcation found here gives rise to an unstable tip region and therefore to the observed oscillation. Needless to say, this extremely crude guess needs to be checked using a real calculation which takes into account the curvature of the wavefronts in the spiral structure.

The work of D.A.K. was supported in part by Dept. of Energy Grant. No. DE-FG-02-85ER54189; H.L. was supported in part by U.S. Defense Research Projects Agency Grant no. AFOSR-F49620-87-C-0117.

REFERENCES

1. For a general review see R.J. Field and M. Burger, "Oscillations and Travelling Waves in Chemical Systems," Wiley (1985).
2. A.N. Zaikin and A. M. Zkabotinskii, Nature **225**, 535 (1970); A. T. Winfree, Science **175** 634 (1972).
3. See e.g. S. C. Muller, T. Plesser and B. Hess, Physica **24D**, 87 (1987).
4. A. T. Winfree and S. H. Strogatz, Nature **311**, 611 (1984).
5. J.P. Keener and J.J. Tyson, Physica **21D**, 307 (1986); J.J. Tyson, Siam J App Math, **46**, 1039 (1986).
6. J.D. Dockery, J.P. Keener and J.J. Tyson, Physica **30D**, 177 (1988).
7. D. Kessler and H. Levine, Physica **D37**, 1 (1989).
8. This idea seems to be a fairly common one in the reaction-diffusion literature.
9. A. Winfree, unpublished data quoted in ref. 6.
10. J. Ross, S. C. Muller and C. Vidal, J. Phys. Chem **92**, 163 (1988).
11. P. Fife, J. Stat. Phys. **39**, 687 (1985).

12. P. Fife, "Non-Equilibrium Dynamics in Chemical Systems," C. Vidal and A. Pacault ed., Springer-Verlag (1984).

13. For a discussion of the physics of interfaces in first order phase transitions, see D. Kessler, J. Koplik and H. Levine, Adv. in Phys. **37**, 255 (1988); J. S. Langer in "Chance and Matter", J. Souletie ed., North Holland (Amsterdam, 1987).

14. Note that the lower of these two modes must go through zero at $q = 0$ at precisely the point where the velocity versus wavelength curve turns around.[6,7] This is because such a saddle node bifurcation always involves an exchange of stabilities.

15. The basic methodology being used is the so-called direct method, given e.g. in D. Wollkind, D.B. Oulton and R. Sriraganathan, J. de Physique **45**, 505 (1984).

16. T. Ohta, M. Mimura, and R. Kobayashi, Physica **34D**, 115 (1989).

17. W. VanSaarloos and J. D. Weeks, Phys. Rev. Lett **51**, 1046 (1983); Physica **12D**, 115 (1984).

18. It should be noted that the Tyson-Fife Oregonator does not possess travelling wave solutions of the type studied here for $c \geq .8$, as can be seen from Fig. 5 of ref. 6. The experimental data, however, are only for $c \geq 1.$. This discrepancy bears further investigation.

19. H. Swinney and G.S. Skinner, private communication; W. Jahnke, W. F. Skaggs and A. T. Winfree, J. Phys. Chem **93**, 740 (1989).

20. D.A. Kessler and H. Levine, to appear in Physical Review A.

THE FORMATION AND PROPAGATION OF FRONTS AT THE ELECTRIC FIELD INDUCED BEND FREEDERICKSZ TRANSITION

P. Palffy-Muhoray[†], H.J. Yuan[†] and B.J. Frisken*

Liquid Crystal Institute and Department of Physics[†]
Kent State University
Kent, OH 44242, U.S.A.

and

W. van Saarloos

AT&T Bell Laboratories
Murray Hill, NJ 07974, U.S.A.

ABSTRACT

The electric field induced bend Freedericksz transition in 5CB is first order. At the threshold value of the applied voltage, the deformed and undeformed states can coexist in equilibrium. The width of the front separating the two states gives a measure of the abruptness of the transition. If the applied voltage differs from its threshold value, the stable state invades the metastable one, and the front separating the two states moves. We present data for the velocity of front propagation in 5CB, and compare results with the predictions of theory.

INTRODUCTION

A nematic liquid crystal sample aligned between parallel glass plates can undergo a transition from a uniform state to an elastically deformed one under the influence of external fields. If the transition is first order, then, at the transition, the undeformed and deformed states can coexist in equilibrium. These two states are separated by an interface, or a front. At the transition, this interface does not move, but if the electric field is increased above the threshold value, the interface starts to move: the region where the director field is deformed expands into the undeformed domain. Upon decreasing the field below the threshold, the interface moves in the other direction. The problem of front propagation into unstable and metastable states has received considerable theoretical[1] and experimental[2] attention recently.

We have recently studied the electric field induced bend Freedericksz transition[3] and shown that this transition in the nematic liquid crystal 5CB (4-cyano-4'-n-pentylbiphenyl) is first order. Furthermore, we found that the phase behaviour of this system is well described by an approximate Landau free energy

*Current address: Department of Physics, University of California at Santa Barbara, Santa Barbara, CA 39106, U.S.A.

Nonlinear Evolution of Spatio-Temporal Structures in Dissipative Continuous Systems
Edited by F.H. Busse and L. Kramer
Plenum Press, New York, 1990

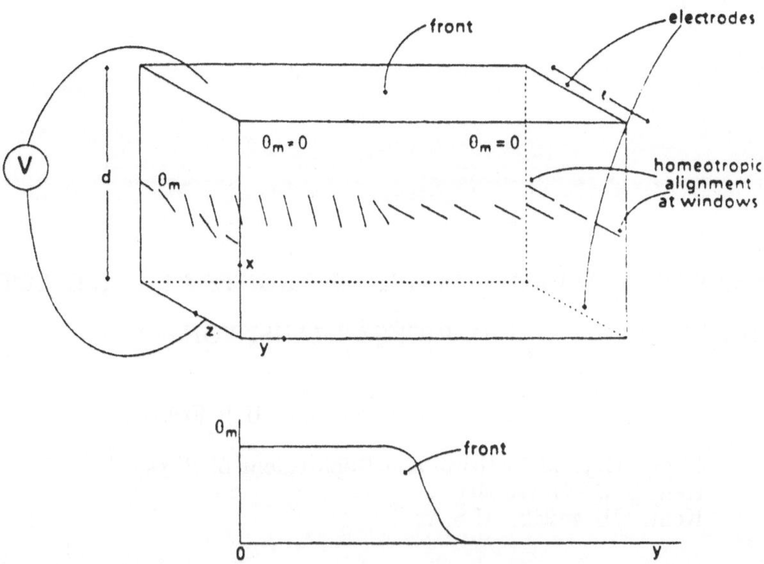

Figure 1. Schematic of the cell and director configuration.

expansion. This system therefore appeared well suited to the study of front propagation, since the material properties of 5CB are well known.

The cells used are sketched in Fig. 1 and consist of a rectangular slab of 5CB between parallel glass windows with initial uniform alignment of the nematic director perpendicular to the windows. A voltage applied to a pair of planar electrodes perpendicular to the windows gives rise to an elastic deformation of the liquid crystal such that the director away from the windows makes a nonzero angle with the window normal. Since this transition is first order, distorted and undistorted domains can coexist in equilibrium. The transition may be made more strongly first order by applying a magnetic field perpendicular to the windows. Because of the large refractive index anisotropy of the material, the two states are readily distinquishable, and the interface is easily seen by illuminating with polarized light. The free energy density difference between the distorted and undistorted states is determined by the voltage applied to the electrodes. Changing this voltage causes one state to invade the other, giving rise to front propagation.

THEORY

In the electric field induced bend Freedericksz transition, the orientation dependent part of the energy density associated with the electric field is[4]

$\mathcal{S}_{EL} = -\frac{1}{2}\,\frac{\epsilon_\perp (\frac{V}{d})^2}{(1-u\,n_x^2)}$, where $u = 1-\frac{\epsilon_\perp}{\epsilon_{//}}$, $\epsilon_\perp$ and $\epsilon_{||}$ are the principal values of the dielectric tensor, V is the voltage applied to the electrodes, d is the separation between two electrodes, and n_x is the component of the director $\hat{n}$ in the x direction (see Fig. 1). Because of the form of $\mathcal{S}_{EL}$, this transition can be first order[3,4]. Assuming that the director field has the form $\hat{n}=(\sin\theta,\ 0,\ \cos\theta)$, where $\theta=\theta_m(y)\sin(\pi z/l)$ is the angle between director and the z axis, and expanding the free energy density in powers of θ_m, we get[5] up to terms of order $\theta_m{}^6$

$$\mathfrak{F} = \frac{a}{2}\theta_m{}^2 + \frac{b}{4}\theta_m{}^4 + \frac{c}{6}\,\theta_m{}^6 + \frac{d}{2}\,\left(\frac{\partial\theta_m}{\partial y}\right)^2 , \qquad (1)$$

where the Frank free energy is measured in units of π^2 A $K_3/(4\ l)$, A is the area of the sample, l is the thickness between the glass plates, K_1 , K_2 , and K_3 are the splay, twist, and bend elastic constants. The constants a,b,c, and d depend on material parameters and the applied fields.

We note that as the transition turns out to be relatively strongly first order, there is no priori reason to expect the power series expansion in Eq.(1) to be accurate. Fortunately, both numerical studies of $\mathfrak{F}$ and the experiments[3] show that (1) is surprisingly accurate, and this allows us to compare our dynamical measurements directly with theoretical results derived from (1). For 5CB we have b<0, and a first order transition takes place at $16ac=3b^2$. At this value, θ_m jumps from the $\theta_m=0$ to the $\theta_m=\sqrt{-\frac{3b}{4c}}$ state.

From Eq.(1), the torque balance condition gives

$$\gamma_{eff}\frac{\partial\theta_m}{\partial t} = d\,\frac{\partial^2\theta_m}{\partial y^2} - a\,\theta_m - b\,\theta_m{}^3 - c\,\theta_m{}^5, \tag{2}$$

the time dependent Ginzburg-Landau equation. Here $\gamma_{eff}=\gamma/2$ with γ the twist viscosity. In the two phase region a>0, b<0, the uniformly propagating solution describing the motion of an interface with boundary conditions $\theta_m(y=-\infty)=\theta_0$ and $\theta_m(y=+\infty)=0$ is[1]

$$\theta_m = \left[\sqrt{\frac{c}{a}}\ e^{2\,(y-vt)\,\theta_o{}^2\sqrt{\frac{c}{3d}}} + \frac{1}{\theta_o{}^2}\right]^{-\frac{1}{2}}, \tag{3}$$

and it propagates with velocity v given by

$$v = \frac{b}{\gamma_{eff}}\sqrt{\frac{d}{3c}}\left[-1 + 2\sqrt{1-\frac{4ac}{b^2}}\ \right]. \tag{4}$$

The width w of the front, from Eq.(3), is given by $w=(1/2\theta_o{}^2)(3d/c)^{1/2}$. At the transition, $16ac=3b^2$ and v = 0, we get $w=-(2/b)(dc/3)^{1/2}$. Indeed, the width of the front is finite and inversely proportional to |b|, which measures the strength of the first order transition. If the voltage differs from its threshold value, the stable phase displaces the metastable one and the front moves with a velocity given by Eq.(4). This behaviour is analogous to that described by Cladis et al.[2] in their 'dynamical test of phase transition order'.

EXPERIMENT

We studied front propagation in a homeotropically aligned 5CB cell. In this alignment, the nematic director at the windows is perpendicular to the glass. The aligment was achieved by surface treatment of the glass by a silane compound prior to assembing the cell. The cell dimensions were 30mm x 4mm x 0.5mm. The cell was slightly wedge shaped in the y direction, with a wedge angle of 0.05 radians to allow localization of the front. The electrodes were stainless steel plates, 1cm in width to minimize fringe effects. Because the physical properties of liquid crystals sensitively depend on temperature, the cell was placed in a temperature controlled housing and the sample temperature was thermostatted at $24.847 \pm 0.001\ ^{0}C$ in this experiment. A 0.17 T magnetic field in the z direction was applied used to make the transition more strongly first order. [The magnetic field was taken into account in evaluating the constants a, b and c in Eq.(1)]. In our experiment, the electrode separation (and hence a, b and c) varied linearly with y, and on varying the applied voltage, the front moved to its equilibrium position with an exponentially decaying velocity.

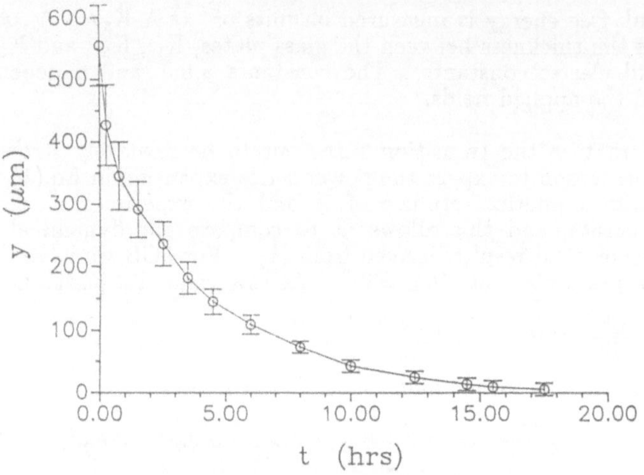

Figure 2. Front position y versus time t.

The cell was observed through holes in the magnet pole pieces via a specially constructed microscope. The resulting images were recorded on vidotape and photographed. The direction of motion of the front was observed to change depending on whether the applied voltage was increased or decreased.

From Eq.(4), we obtain that in our slightly wedge shaped cell the velocity should decay as $v = v_i\, e^{-t/\tau}$ with time constant

$$\tau = -\frac{2\,\gamma_{eff}}{3\,\alpha\, b}\sqrt{\frac{3\,c}{d}}\quad . \tag{5}$$

Here α is defined by $4ac/b^2=3(1+\alpha\, y)/4$ and, as stated, $\gamma_{eff} = \gamma/2$ with γ the usual twist viscosity of 5CB.

Figure 3. Photograph of the cell with transmitted light polarized along the x direction. The dark domains are the deformed regions and the white domains are the undeformed regions.

Figure 4. Photograph of the cell showing the spacing of domains near the equilibrium position.

The front position as a function of time was measured and is shown in Fig.2. From the plot of ln y vs. time, we estimate τ=3.6 hrs. The time constant τ calculated from Eq.(5) shows a strong dependence on the values of the dielectric and diamagnetic susceptibility anisotropies. Using values for the material constants of 5CB which are within the uncertainties published in the literature, we get values in excellent agreement with experiment. The details of these comparisons will be published elsewhere[5].

As can be seen from the photograph in Fig. 3, instead of observing two homogeneous states, we observed a periodic modulation in the deformed region of the cell. This periodic modulation has been reported elsewhere[3,6]. We view each domain as being bounded by two fronts, one on either side. As the applied voltage is changed, the two fronts on the two sides of each domain move in opposite directions. We assume that the domain walls of adjacent domains are prevented from merging by the opposite sense of director deformation in the domains. As expected, the width of the undeformed region (white region in the pictures) between the black domains decreases away from the equilibrium position of the front. This effect can be seen in Fig. 4. When the applied voltage was increased so that only the deformed state was stable, then we observed nucleation and growth of new domains. When the applied voltage was decreased so that only the undeformed state was stable, we observed melting of domains.

CONCLUSIONS

We have observed front propagation in a nematic cell in the vicinity of the electric field induced bend Freedericksz transition. The existence of a stable front separating undeformed and deformed domains is additional evidence that the transition is first order. The time constant of the propagating front calculated from theory using the material parameters of 5CB was found to be in good agreement with the experimentally observed value.

ACKNOWLEDGEMENTS

We are grateful to P.E. Cladis and V.G. Kamensky for illuminating discussions, and to H. Lin and J.Y. Kim for their help with the photographs.

REFERENCES

1. See W. van Saarloos, Phys. Rev. Lett. 58, 2571 (1987); Phys. Rev. A37, 211 (1988); Phys. Rev. A 39, 6367 (1989), and references therein.
2. P.E. Cladis, W. van Saarloos, D.A. Huse, J.S. Patel, J.W. Goodby and P.L. Finn, Phys. Rev. Lett. 62, 1764 (1989).
3. B.J. Frisken and P. Palffy-Muhoray, Phys. Rev. A 39, 1513 (1989).
4. S.M. Arakelyan, A.S. Karayan and Yu. S. Chilingaryan, Dokl. Akad. Nauk. SSSR 275, 52 (1984) [Sov. Phys.-Dokl. 29, 202 (1984)].
5. P. Palffy-Muhoray, H.J. Yuan, B.J. Frisken, and W. van Saarloos, (to be published).
6. D.A. Allender, B.J. Frisken and P. Palffy-Muhoray, Liq. Cryst. 5, 735 (1989).

THE DYNAMICS OF TRANSIENT STRUCTURES IN THE FREDERICKSZ TRANSITION

B.L. Winkler, A. Buka*, L. Kramer,
I. Rehberg, and M. de la Torre Juarez

Universität Bayreuth;
8580 Bayreuth; Postfach 101251, West Germany

*Permanent Address:
Central Research Inst. for Physics
H-1525 Budapest 114 P.O.B.49. Hungary

A new pattern occuring during the Fredericksz transition has been described recently (Buka et al, 1989). We report on detailed experimental observations, clarifying the relation between the dynamics of the pattern and the dynamics of the director reorientation.

EXPERIMENTAL SETUP

A nematic liquid crystal is homogeneously aligned between two parallel transparent electrodes as seen in figure 1. The distance d between these electrodes is 100μm and the sample is 2cm by 1cm wide. It is mounted on a polarizing microscope equipped with a hot stage for temperature control with a accuracy of ±0.05K by means of a water circuit. The sample is illuminated from below with a light beam polarized parallel to the director, which represents the average orientation of the liquid crystal molecules and the optical axis of the system. Thus spatial modulations in the director angle become visible due to birefringence. In this experiment the liquid crystal 5CB (Merck K15, 4-pentyl-4'cyanobiphenyl) is used.

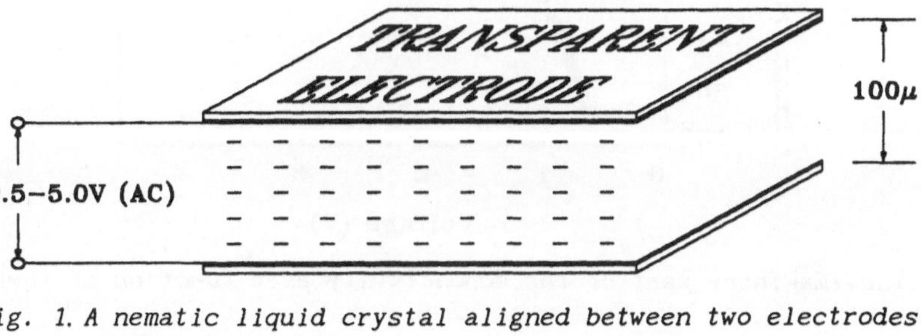

Fig. 1. A nematic liquid crystal aligned between two electrodes.

Nonlinear Evolution of Spatio-Temporal Structures in Dissipative Continuous Systems
Edited by F.H. Busse and L. Kramer
Plenum Press, New York, 1990

STATIC MEASUREMENTS

Under the influence of an electric or magnetic field applied across the layer the liquid crystal undergoes a Fredericksz Transition to an elastically deformed state. This Fredericksz transition (P.G. de Gennes, 1975; L.M. Blinov, 1983) has a mechanical analog in the Euler buckling problem and occurs in a similar fashion in the anisotropic A phase of superfluid ^3He. In the presence of an electric or magnetic field there is a competition between two torques:

1) Due to the elastic interactions between the molecules the director prefers to be parallel to the glass plates.
2) Due to the positive anisotropy of the dielectric tensor of the substance ($\varepsilon_{\parallel}=18$, $\varepsilon_{\perp}=6.25$ at 20°C) the director tends to be parallel to the electric field.

Both torgues increase with the deviation angle θ from the initial alignment. If the voltage applied across the layer exceeds a threshold value the electrical torgue increases faster than the restoring elastic torgue and therefore θ grows until, there is a satisfaction due to nonlinear effects. For known boundary condition this deviation $\theta(z)$ may be calculated as a function of the vertical position z in the cell.

The two electrodes of our sample may be considered as a capacitor. The capacity is sensitive to changes in the dielectric permittivity ε of the medium between the electrodes. As the dielectric permittivity is anisotropic for liquid crystals, it depends on the orientation of the director: $\varepsilon_{zz}(\theta) = \varepsilon_{\perp}\cdot\cos^2\theta+\varepsilon_{\parallel}\cdot\sin^2\theta$. For small θ the sample's capacity is: $C(\theta) = C_0 \cdot \left(1 + \frac{\varepsilon_a}{\varepsilon_{\perp}} \cdot \theta^2(0) \right)$ where $\varepsilon_a = \varepsilon_{\parallel} - \varepsilon_{\perp}$.

In figure 2 the imaginary part of the conductivity of the sample, which is directly proportional to the cell's capacity, is plotted versus the voltage applied across the cell. The measurement was carried out at 30°C. The voltage was varied step by step and after every change the director configuration in the liquid crystal was allowed to reach the equilibrium by waiting 625 seconds. The points marked by the open symbols were obtained while increasing the voltage, the closed symbols while decreasing the voltage. Only every 6th measured point has been marked by a symbol. In our case the threshold voltage was 0.65V.

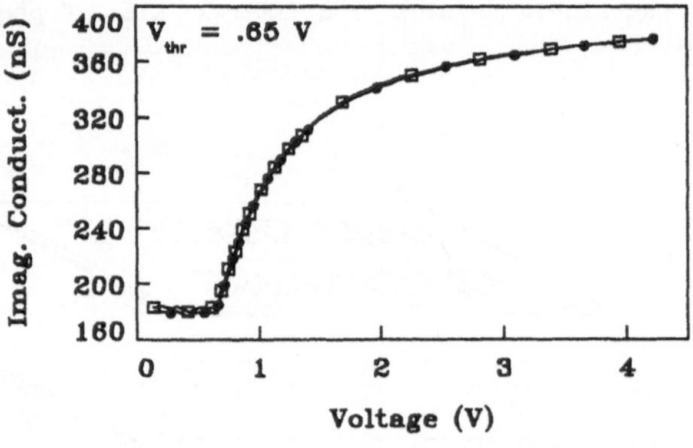

Fig. 2. The imaginary part of the conductivity as a function of the applied voltage.

DYNAMIC MEASUREMENTS

We now let the voltage jump from 0V to a supercritical value (above Fredericksz threshold) and measure the time dependent conductivity as well as the light intensity transmitted through the sample. Figure 3 shows a photo of the transient pattern we observe after jumping to the supercritical voltage.(Buka et al, 1989) The lines have a preferred orientation parallel to the director alignment. This pattern has nothing in common with the well known domain walls. Domain walls appear when in one region of the sample the director turns clockwise and in the neighbouring region the director turns counterclockwise. This domain wall between two symmetric solutions can not simply fade away. The only possible way to disappear is that a closed curve of lines which surrounds no other lines shrinks to a point. This is a slow process. We see these domain walls in our samples, too. They appear about every 5 - 10mm and seem to have no preferred orientation.

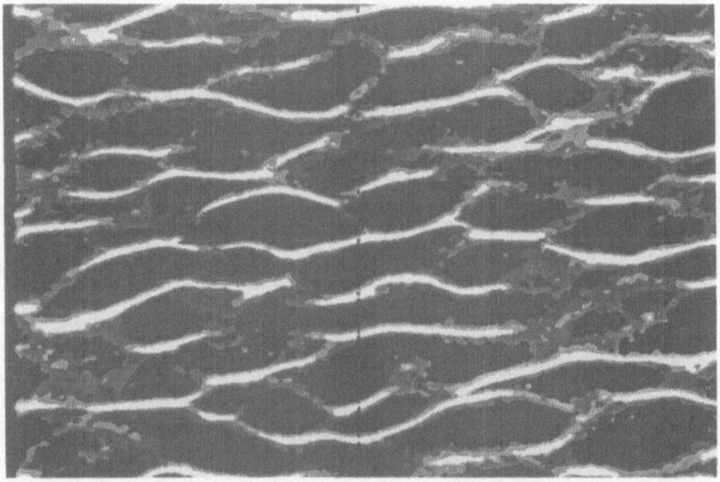

Fig. 3. Photograph of the observed pattern. The lines are parallel to the initial director orientation.

The patterns we observe appear after turning on the field and then simply fade away approximately as fast as they come up. They are seen within regions where the director turns in one direction. The structure has a wavelength of the order of the thickness of the sample. Thus there is sufficient space to observe our pattern between the domain walls.A speculation about the nature of this line pattern has been presented elsewhere (Buka et al, 1989)

At the position marked by the dashed line we record the light intensity by means of a line camera mounted on the microscope as a function of time after switching on the field. The camera has a digitizer with 6 bit intensity resolution (64 grey levels) and a spatial resolution of 1728 pixels and is connected to a microcomputer. In figure 4 intensity profiles are shown for three different values of $\varepsilon = (V^2 - V_{thr}^2)/ V_{thr}^2$ where V_{thr} is the threshold voltage of the Fredericksz transition. Consecutive intensity profiles are plotted on top of each other. For small ε one observes only fluctuations of the director. For high ε a clear pattern forms and then vanishes again.

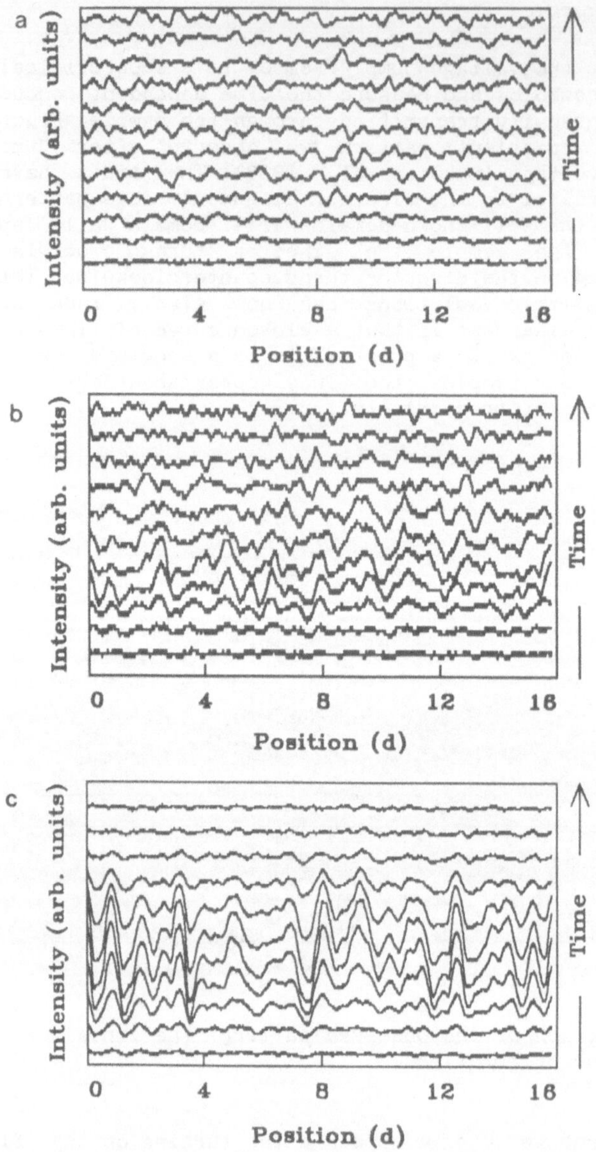

Fig. 4. Light intensity profiles as a function of time after turning on the field.
a) ε=0.66, 14.3 sec between two consecutive profiles, first profile 9.3 sec after turning on the field.
b) ε=4.25, 1.8 sec between two consecutive profiles, first profile 2.2 sec after turning on the field.
c) ε=17.43, 0.4 sec between two consecutive profiles, first profile 0.7 sec after turning on the field

Simultaneously with the light intensity profiles we measure the conductivity of the sample. In figure 5 the time dependence of the conductivity is plotted in the lower frames. To obtain some information of the strength of the modulation of the director angle we calculate the contrast from the intensity lines by taking the root mean square deviation of the intensity profiles. The upper frames in figure 5 show how the contrast changes with time.

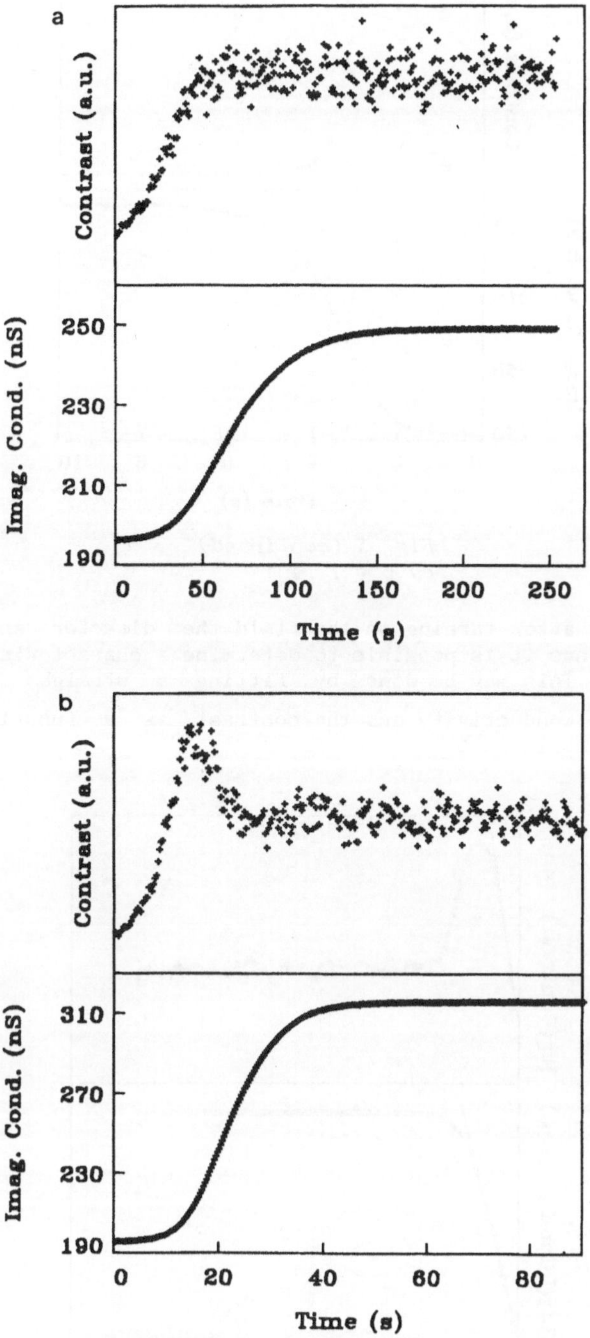

Fig. 5. Dependence of the contrast (upper frames) and of the conductivity (lower frames) on the time after switching on the field.
a) ε = 0.66
b) ε = 4.25

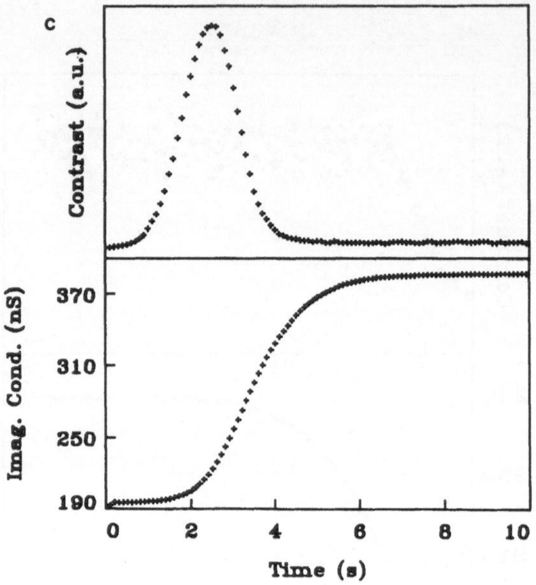

Fig. 5 (continued)
c) ε = 17.43

Immediately after turning on the field the director angle θ grows exponentially. Thus it is possible to determine a characteristic time τ or a growth rate $\frac{1}{\tau}$. This may be done by fitting a straight line to the logarithm of the conductivity and the contrast as a function of time, which is shown in figure 6.

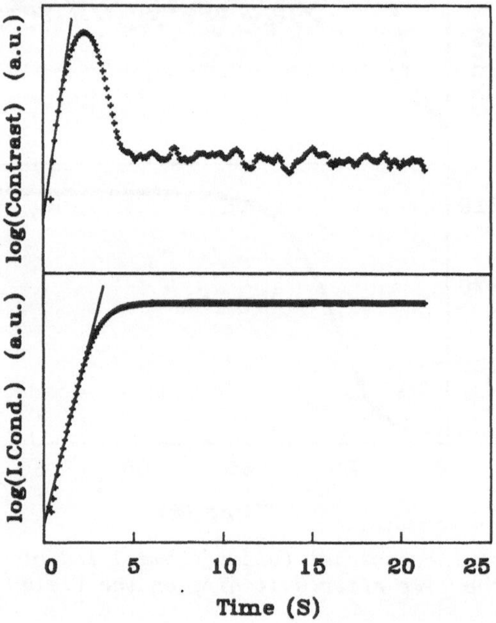

Fig. 6. Determination of the growth rate of contrast and conductivity after switching on the field.

This procedure may be done for various ε. The results are shown in figure 7. In the upper frame the growth rate $\frac{1}{\tau}$ for the conductivity measuremets is shown as a function of ε. For the contrast the growth rate $\frac{1}{\tau}$ is shown in the lower frame. As seen already from figures 4 and 5 the formation of the transient pattern only sets in above a second threshold voltage, higher than the threshold for the Fredericksz transition. We found this second threshold to be $\varepsilon = \varepsilon_p = 1.1 \pm 0.1$ for 30°C. A theoretical search for this instability is in progress.

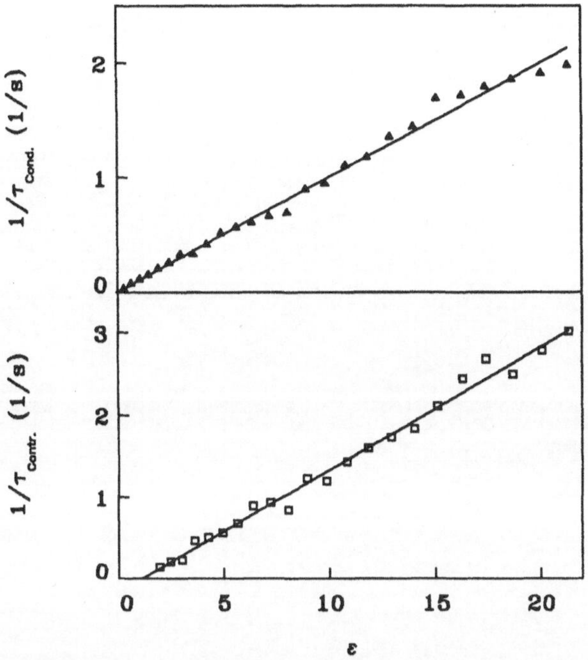

Fig. 7. Dependence of the growth rates on the driving force ε.
Upper frame: Growth rate of the conductivity
Lower frame: Growth rate of the contrast.

This work was supported by the Deutsche Forschungsgemeinschaft. A.B. acknowledges the A. von Humboldt Foundation. M.T.J. would like to thank the Ministerio de Educacion y Ciencia for financial support.

REFERENCES

de Gennes, P.G., 1975, "The Physics of Liquid Crystals", Clarendon Press, Oxford

Blinov, L.M., 1983,"Electroptical and Magnetoptical Properties of Liquid Crystals", Wiley, New York

Buka, A.,de la Torre Juarez, M., Kramer, L., Rehberg, I., 1989, Transient structures in the Fredericksz transition, Phys Rev A 40, N 11

SIDEBAND INSTABILITY OF MODULATED TRAVELING WAVE CONVECTION

M. de la Torre Juárez, W. Zimmermann, and I.Rehberg

Physikalisches Institut, Universität Bayreuth, (FRG)

INTRODUCTION

It has been shown theoretically (Riecke et al. 1988, Walgraef 1988), that in a system exhibiting a supercritical Hopf bifurcation a temporal modulation of the driving force with a modulation frequency ω_m of about double of the Hopf frequency can stabilize standing waves (SW). An experimental verification was presented (Rehberg et al. 1988) for the electro-hydrodynamic convection of liquid crystals, where the driving ac voltage is modulated as $V(t)= V_c\cdot\cos(\omega\cdot t)\cdot[1+\varepsilon+b\cdot\cos(\omega_m\cdot t)]$, with b being the modulation and ε the reduced driving amplitude. The theoretical model is very similar to the one describing parametrically excited waves which are known to exhibit Benjamin-Feir turbulence that is characterized by a transfer of energy from the fundamental Fourier mode to the side bands (Craik 1985). When increasing ε for a constant modulation amplitude b the simplified theoretical model predicts a supercritical bifurcation from SW to modulated traveling waves (TW). In the experiment SW become unstable via a different mechanism which we clarify here. The scenario includes the appearance of the sideband instability, defects and stable undulated rolls of a very short wavelength.

TRANSITION FROM STANDING TO TRAVELING WAVES

The solution describing a system of traveling waves can be written as

$$u(x,t)= A_1(X,T)\cdot\exp(i\cdot(q_c\cdot x+\omega_c\cdot t)) + A_2(X,T)\cdot\exp(i\cdot(q_c\cdot x-\omega_c\cdot t)) + c.c.$$

where $u(x,t)$ describes one of the observable quantities and A_1 and A_2 are the slowly varying amplitudes of the left and right TW, respectively. The fast variable $u(x,t)$ is translational invariant ($x \to x + d$) and has reflectional symmetry ($x \to -x$). Using these symmetries the normal form valid for small amplitudes near a supercritical Hopf-bifurcation is (Ioss 1987):

$$\partial_t A_1 = [(\mu+i\nu)-(1+i\beta)|A_1|^2-(\delta+i\gamma)|A_2|^2]A_1 + O(A_1^5)$$
$$\partial_t A_2 = [(\mu-i\nu)-(1-i\beta)|A_2|^2-(\delta-i\gamma)|A_1|^2]A_2 + O(A_1^5)$$

As long as $\delta > 0$ left TW or right TW are stable with respect to SW.

When the driving frequency is time modulated with the frequency $\omega_m \simeq 2\cdot\omega_c$ the system acquires a new invariance under the transformation $t \to t + 2\pi/\omega_m$ and leads to a linear coupling between A_1 and A_2 proportional to the modulation amplitude b:

Nonlinear Evolution of Spatio-Temporal Structures in Dissipative Continuous Systems
Edited by F.H. Busse and L. Kramer
Plenum Press, New York, 1990

$$\partial_t A_1 = [(\mu+i\nu)-(1+i\beta)|A_1|^2-(\delta+i\gamma)|A_2|^2]A_1 + \lambda A_2 + O(A_1^5)$$
$$\partial_t A_2 = [(\mu-i\nu)-(1-i\beta)|A_2|^2-(\delta-i\gamma)|A_1|^2]A_2 + \lambda A_1 + O(A_1^5) \qquad \text{(I)}$$

where the phases have been chosen adequately to make $\lambda \propto b$ real. For $\lambda \neq 0$ the simple solutions of this equation are SW or modulated TW instead the TW solutions.

In large aspect ratio systems long wavelength perturbations can destabilize the coherent pattern. Spatial derivatives have to be added to (I) in order to describe this (Fauve 1987):

$$(\partial_t+s\partial_x)A_1 = [(\mu+i\nu)+(\varphi+i\eta)\partial_{xx}-(1+i\beta)|A_1|^2-(\delta+i\gamma)|A_2|^2]A_1+\lambda A_2+O(A_1^5)$$
$$(\partial_t-s\partial_x)A_2 = [(\mu-i\nu)+(\varphi-i\eta)\partial_{xx}-(1-i\beta)|A_2|^2-(\delta-i\gamma)|A_1|^2]A_2+\lambda A_1+O(A_1^5) \qquad \text{(II)}$$

The linear stability analysis of (I) around the trivial state $A_1=A_2=0$ shows a Hopf bifurcation at $\mu=0$, $\lambda^2<\nu^2$ leading to modulated TW; and a steady one at $\mu^2+\nu^2=\lambda^2$, $\lambda^2>\nu^2$ that gives rise to SW. One can also see that at an $\varepsilon>0$ for $\lambda^2>\nu^2$ the SW become unstable against modulated TW along a threshold line of cubic order in ε. This transition could not be seen in the experiment. This is understandable when a stability analysis based on the more realistic model (II) is done for long wavelength perturbations. It turns out that the SW become unstable against spatial modulations before the transition to temporally modulated TW occurs. This kind of instability is called sideband instability since it destabilizes the Fourier modes close to the critical one. One example of sideband instability for a steady bifurcation is the Eckhaus instability in a dissipative system like liquid crystals (Lowe and Gollub 1985). Experimentally it was first described in the case of an oscillatory bifurcation in a conservative system, namely trains of water waves propagating downstream from a wavemaker (Benjamin and Feir 1966), thus the name Benjamin-Feir instability (BF) is also used for this instability.

An example for this situation is shown in the phase diagram (Fig. 1, from Rehberg et al. 1988). Here along the closed circles the SW become unstable against a solution with alternating SW and TW even for $\varepsilon<0$ (Fig. 2). This line can be explained by taking into account the effect of spatial modulations. A good fit of equation (II) to this line was possible with values of the coefficients of the order of magnitude of the ones measured by de la Torre Juárez and Rehberg (1989) and in addition the line obtained for the saddle node bifurcation predicted for (I) fitted very well with the threshold shift measured for $\varepsilon<0$ and $\lambda^2<\nu^2$.

EXPERIMENTAL SETUP

The sample studied consisted of one cell of 15μm thickness filled with Merck Phase V. This is a nematic liquid crystal mixture whose parameter values are mostly unknown. However most of the ones appearing in the context of the simplified model (Riecke et al. 1988) could be estimated and reported elsewhere (de la Torre Juárez and Rehberg 1989). The measurements presented here were done in a different cell than those by Rehberg et al. (1988) and de la Torre and Rehberg (1989), but all three cells were filled with Phase V and had 15μm thickness, so their behaviour is expected to be very similar.

The cell has been treated to impose an initial homogeneously planar orientation on the director field. It was kept inside a thermalized box to achieve a temperature stabilization of about 0.005 K. It was also hermetically sealed in order to avoid impurities to come from outside into the fluid. This stabilized the values of the bifurcation diagrams over month.

The measurements were done using a shadowgraph technique (Rasenat et al. 1989) by illuminating the sample along the direction of the electric field. The data were taken with a CCD camera mounted on a microscope and digitized by a computer to be stored and treated. The basic treatment consisted in taking a whole image with a resolution of 512x512 pixels or measuring along a line parallel to the initial orientation of the director of the sample (x-direction).

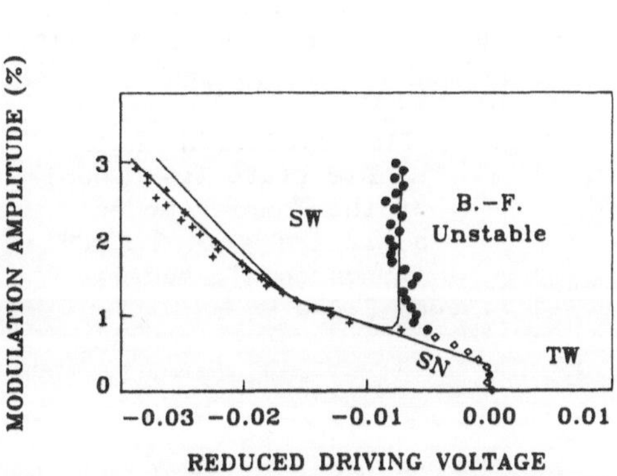

Fig. 1. Phase diagram and lines obtained by fitting the theory to the stability threshold. Open dimonds: transition to TW. Crosses: transition to SW. Solid circles: transition to the unstable regime.

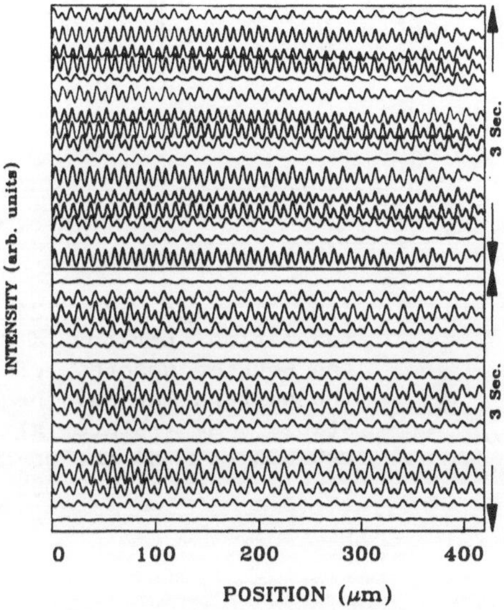

Fig. 2. Time evolution of stable SW (below) and the unstable solution.

EXPERIMENTAL RESULTS

The SW state on the left of the instability line shows a well defined wavelength stable in time as shown in Fig. 3. Here we have averaged the Fourier spectra of the structure along the x direction over 24 periods of the modulation frequency after having waited for the transient response to pass. The SW spectrum (down) was made for $\varepsilon=-0.005$. The narrow wave number peak indicates that the structure stays stable in time. At a higher value of $\varepsilon=0.055$ the peak becomes wider and shifted to another wave number. This broadening appears because now there is no stable wave number and defects appear continuously in the structure.

To give proof that this instability is the predicted one, we imposed a periodic state of the type $u(t)\cdot\exp(iq_c x)$ as shown in Fig. 3a for $\varepsilon=-0.005$ and jumped into the unstable region, where the spatial perturbations are supposed to grow. In Fig. 4 the temporal evolution of the fundamental mode

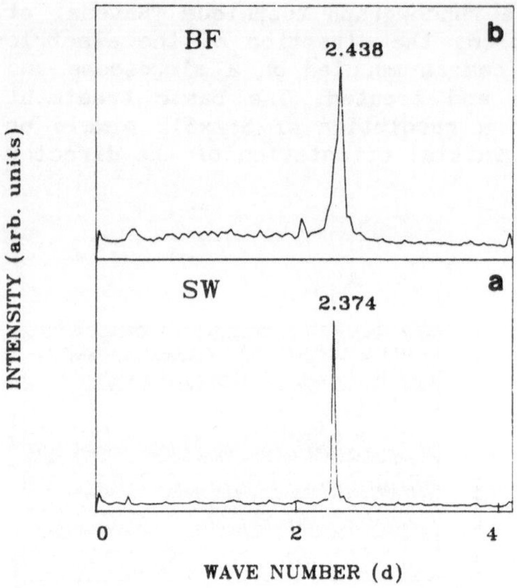

Fig. 3. Time average over 24 modulation periods of the Fourier spectra of the pattern in a) the stable SW state at ε=0.005; b) the unstable state at ε=0.055

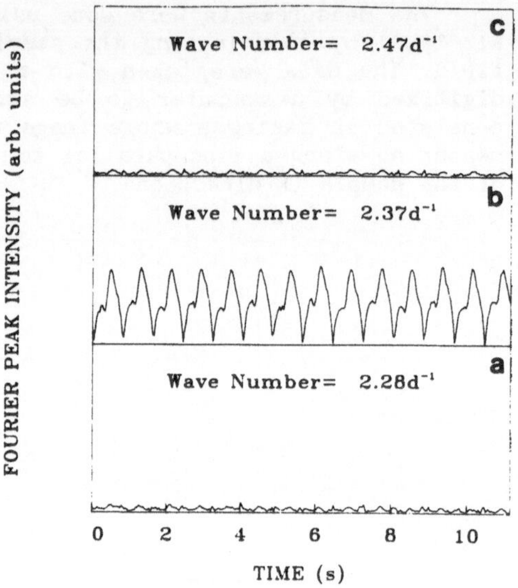

Fig. 4. Time evolution in the stable state (ε=-0.005) of the Fourier modes: a) $2.28d^{-1}$; b) $2.37d^{-1}$ (fundamental mode); c) $2.47d^{-1}$.

is plotted in the middle and compared to two neighboring modes. The resolution of our spectra is $\Delta q=0.03d^{-1}$. We are showing as an example the third next mode in our discrete serie. The time evolution of the same three modes q_n during the transient decay from the periodic structure is shown in Fig. 5. One can see that while the fundamental mode decreases its intensity, the modes on the right and on the left of it show a positive growth factor. This experimental behaviour corresponds clearly to the definition of a sideband instability.

We measured the structure function (Rehberg et al. 1989) of the final state resulting after this transients to quantify the properties of the unstable regime. We made a time average over 512 modulation periods for constant modulation b=5% and slowly increasing values of ε. It shows a decay in both directions (Fig. 6) and a very interesting additional property: In the x-direction (perpendicular to the roll axis) above ε=0.20 a slight wavelength change. This wavelength shift in the x-direction is correlated with the appearance of finite periodicity in the y-direction. For ε=0.02 and 0.03 the wavelength parallel to the roll axis changes from about 36d to 30d, and above ε=0.04 makes a big jump into values of the order of magnitude of 8d. This corresponds to a spatial structure of Zig-Zag rolls of a very short wavelength. The spatial structure of this solution is shown in Fig. 7 where 7a corresponds to SW with normal straight rolls, 7b corresponds to the instability where the defects start to appear, and a short wavelength sets in in the y-direction. In Fig. 7c one can see the stable structure corresponding to the high wave number Zig-Zags. Further measurements demonstrating the relationship between these two instabilities are in progress.

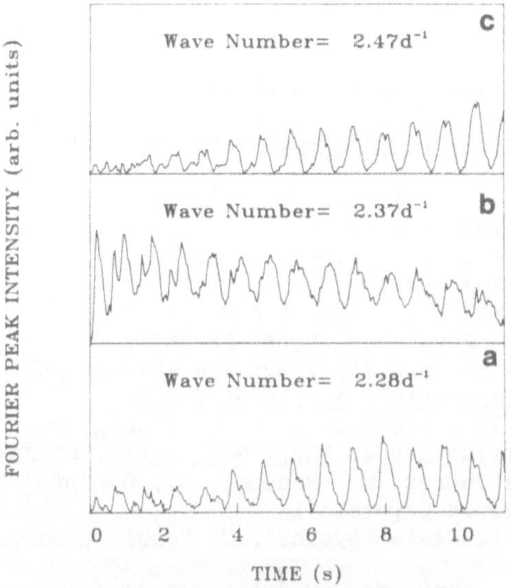

Fig. 5. Time evolution during the jump from SW into the unstable state at ε=0.04 of the Fourier modes: a) $2.28d^{-1}$; b) $2.37d^{-1}$ (the fundamental mode); c) $2.47d^{-1}$.

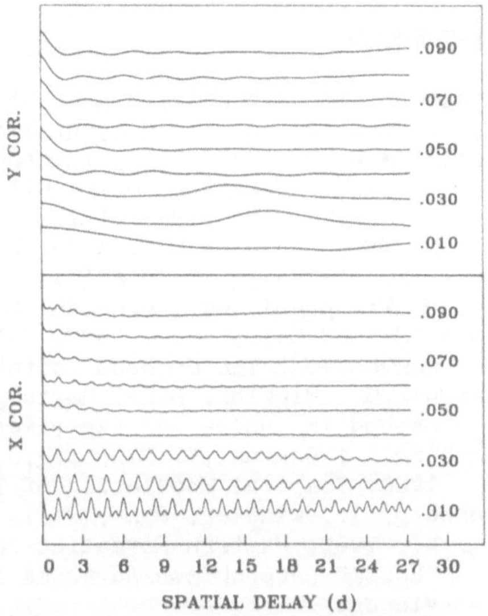

Fig. 6. Spatial correlations of the structure in x- and y-direction averaged over 512 modulation periods for a constant modulation frequency of 1.3 Hz as a function of ε for b=7%.

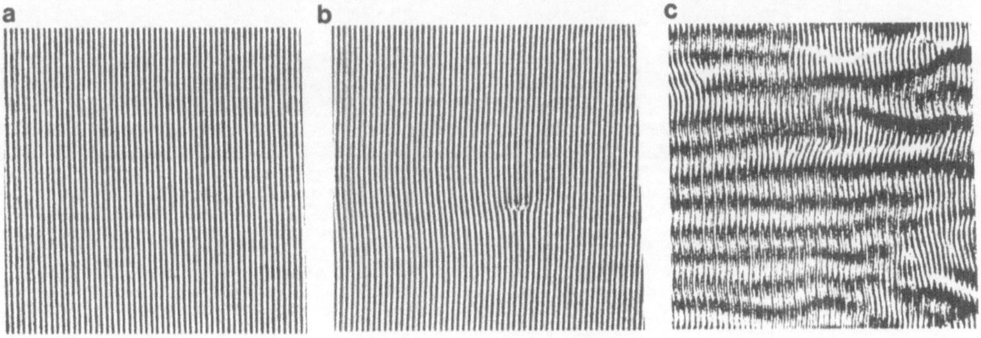

Fig. 7. Spatial structures observed at a) ε=-0.005 b) ε=0.04 c) ε=0.08).

ACKNOWLEDGMENT

The measurements were supported by Deutsche Forschungsgemeinschaft. M.T.J. acknowledges financial support by the Spanish Ministerio de Educación y Ciencia. Its a pleasure to thank F.H. Busse for very useful comments and suggestions.

REFERENCES

Benjamin, T.B and Feir, J.E., 1967; The disintegration of wave trains on deep water, J. Fluid Mech. 27:417.

Craik, A.D.D., 1985; "Wave interactions and fluid flows" (and references therein), Cambridge University Press.

Fauve, S. 1987; in "Instabilities and Nonequilibrium Structures", D. Tirapegui, ed., D.Reidel Publishing Company.

Ioss, E., 1987; in "Instabilities and Nonequilibrium Structures", D. Tirapegui, ed., D.Reidel Publishing Company.

Lowe, M. and Gollub,J.P. 1985, Pattern Selection near the Onset of Convection: The Eckhaus Instability, Phys. Rev. Lett. 55:2575.

Rasenat,S., Winkler, B.L., Hartung, G., Rehberg,I., 1989; The shadograph method in convection experiments, Experiments in fluids 7:412.

Rehberg,I., Rasenat,S. Fineberg,J., de la Torre Juárez, M. Steinberg, V., 1988, Temporal Modulation of Travelling Waves, Phys. Rev. Lett., 61:2449.

Rehberg, I., Winkler, B.L., de la Torre Juárez, M., Rasenat, S., Schöpf, W., 1989, Pattern Formation in a Liquid Crystal, in "Festkörperprobleme/Advances in Solid State Physics", U. Rössler, ed., Vieweg.

Riecke, H., Crawford, J.D., Knobloch,E., 1988, Time-Modulated Oscillatory Convection , Phys. Rev. Lett., 61:1942;

de la Torre Juárez, M. ,Rehberg, I., 1989, Experiments with travelling waves in electrohydrodynamic convection, in "New Trends in Nonlinear Dynamics: the Geometry of Nonequilibrium", P.Huerre and P.Coullet, ed., NATO-ASI Series, Plenum Press.

Walgraef, D., 1988, External Forcing of Spatio-Temporal Patterns, Europhysics Lett. 7, 485.

SPATIAL INSTABILITIES AND DEFECT ORDERING IN SOLIDS

D. Walgraef †
Service de Chimie-Physique, Université Libre de Bruxelles
B-1050, Brussels, Belgium
and
N.M.Ghoniem
Mechanical, Aerospace and Nuclear Engineering Department
University of California, Los Angeles, CA 90024, USA

ABSTRACT

Spatial instabilities leading to the formation of defect patterns and microstructures seem to appear generically in solids driven away from thermal equilibrium by physico-chemical constraints. Some of these instabilities have been recently studied within the framework of dynamical models for the defect densities. The basic properties of these models which take into account the motion and interaction of defects are reviewed. It is shown on a specific example, namely the ordering of vacancy loops in irradiated materials, how the diffusion and nonlinear interactions may trigger the formation of defect microstructures .

INTRODUCTION

When crystalline materials are driven away from thermal equilibrium by external constraints (monotonous or cyclic loadings, corrosion, laser or particle irradiation,...), they display several types of instabilities associated with the ordering of defect populations and which modify their macroscopic properties [**1-3**]. Typically, the defect densities increase significantly and homogeneous defect distributions become unstable. The resulting microstructures are associated with the inhomogeneity of deformation, with strain localization, inhomogeneous swelling, etc. They also act as initiating centers for micro-crack nucleation, influence crack propagation, void lattice formation ,.... and have thus a very practical importance.

Furthermore these phenomena are related to the strong nonequilibrium conditions imposed to the material and they can usually not be interpreted by classical thermodynamical or mechanical concepts. Hence, new theoretical tools are needed to describe and understand these and many other aspects of today's materials science and solid state physics. Effectively, during the last years the whole field of materials science and related technologies has experienced a complete renewal. For example, by using techniques corresponding to strong non equilibrium conditions, it is now possible to escape from the constraints of equilibrium thermodynamics and to process

† Senior Research Associate, National Fund for Scientific Research (Belgium).

Nonlinear Evolution of Spatio-Temporal Structures in Dissipative Continuous Systems
Edited by F.H. Busse and L. Kramer
Plenum Press, New York, 1990

totally new materials structures including different types of glasses, nano- and quasicrystals, superlattices, ... These techniques include ion implantation, laser beam surface melting as well as electron beam heating. For example in laser annealing, after the extremely rapid melting of a shallow layer of material followed by a comparably fast recrystallization, microstructures with superior resistance to friction, corrosion, ... may be frozen into place. Ultra-rapid solidification of alloys may trigger the formation of quasicrystalline structures. Point defect and void patterning in irradiated systems are related to the precipitation of solid solutions. Finally the self-organisation of dislocation populations in stressed or irradiated materials is associated with plastic instabilities and deformation localization. It is obvious that traditional concepts are not adapted to describe such far from equilibrium phenomena and progress in their understanding should be related to the explicit use of genuine nonequilibrium techniques, nonlinear dynamics and instability theory [**3**].

These methods are becoming more and more popular in the study of the behavior of materials subjected to agressive mechanical, thermal or chemical environments. For example, since it appears that defect microstructures result from a dynamical equilibrium between different processes, kinetic models were proposed to describe this collective behavior by taking into account the motion (diffusion, transport, ...) and nonlinear interactions (annihilation, pinning, clustering,...) between defects [**3-4**] and I will discuss here the basics of a dynamical description of the collective behavior of defect populations in driven solids.

Such a description is usually based on reaction-diffusion schemes and we know that, when several populations with different mobilities interact via sufficiently high nonlinear processes, pattern forming instabilities may be expected [**5**]. Such instabilities have been, and are still, investigated in various fields including hydrodynamics, chemistry, biology and nonlinear optics. Besides the determination of the critical wavelength , it is also essential to investigate the geometry, the symmetries, and the stability ranges of the selected structures. As extensively discussed elsewhere , the difficulties of the post-bifurcation analysis lie in the fact that the complexity of the dynamics does not allow, in general, the attainement of analytic solutions for the different variables. However near the instability or bifurcation points, the dynamics may be reduced to much simpler forms by taking advantage of the time and space scale separation between stable and unstable modes and by projecting the dynamics on its unstable manifold. The resulting slow mode dynamics which governs the system evolution on its longest time scale becomes similar to the time dependent Landau-Ginzburg dynamics describing phase transitions in equilibrium systems. This description leads then to amplitude equations for the patterns and allows the derivation of their phase dynamics. Pattern selection and stability may then be discussed in this framework as shown in several contributions of this volume.

In driven materials, it is only recently that the problem of pattern selection, stability, symmetry, or of the influence of the underlying lattice , has been adressed along these lines. Preliminary results are promising, however fundamental questions raised by the anisotropies and the three-dimensional character of the samples remain open.

REACTION-TRANSPORT DYNAMICS FOR DEFECT POPULATIONS

Strained or irradiated metals and alloys present several types of spatial structures including dislocation microstructures (slip , kink or shear bands, Lüders bands, dislocation cells,...), vacancy loops, bubbles, cavity and void lattices. These structures which originate in the spatial organization of point or line defect populations have a strong influence on the macroscopic properties of the materials and on their resistance to external constraints. Hence, the understanding of the formation, selection and stability properties of these defect patterns is of primary importance both from fundamental and from technological point of views.

Several attempts have recently been made to describe these phenomena in the framework of kinetic models for the defect populations [3-6]. These models are based on the fundamental elements of the collective behavior of each defect population which are :
- their motion (diffusion for point defects such as interstitials and vacancies, glide, climb and cross-slip for dislocations),
- their interactions which correpond for example to recombination of point defects, capture or emission of point defects by microstructures and defect creation mechanisms induced by the external constraint.

Defect Motion

Different types of motion may be observed, according to the nature of the defects considered. Point defects such as interstitials and vacancies diffuse in the crystal and the corresponding diffusion coefficient are computed by using energetic consideration. It turns out that their relative value depends on temperature; in the case of 316 steel, for example, we have [10]:

$$\frac{D_v}{D_i} = 6.10^2 exp(-1eV/k_B T)m^2 s^{-1}$$

Hence, according to the temperature of the material, they may differ by orders of magnitude.

The motion of dislocations corresponds to glide, climb or cross-slip. Glide is the fastest one and is typical of the plastic regime when dislocations move almost freely on well-defined crystalline planes. Dislocations of opposite Burgers vectors move in opposite directions. In real materials, there are also obstacles to dislocation motion : impurities, stacking faults, grain boundaries, dislocation clusters forming the so-called forest of immobile dislocations. Line defects may also have very different mobilities, which is a perequisite for pattern forming instabilities in reaction-diffusion systems.

Defect Creation and Multiplication

Point defects are always present in crystalline solids but much larger amounts of such defects are created under external constraints. Under irradiation, ballistic and cascade effects provide creation mechanisms for interstitials and vacancies. The clustering or collapse of point defects leads to the formation of interstitial or vacancy loops. Dislocation creation, on the other hand, is a direct consequence of the deformation process, but the increase of the dislocation density is mainly due to multiplication effects such as the Frank-Read or Bardeen-Herring mechanisms [7].

Defect Interactions

The process mentioned above are essentially linear. The nonlinearities of the dynamics are mainly due to defect interactions. In fact, a few of them dominate the defect dynamics. A particularly important phenomenon is the pair annihilation of defects of opposite nature (annihilation of dislocations of opposite Burgers vectors, recombination of interstitials and vacancies) which leads to quadratic nonlinearities. Other quadratic nonlinearities are related to the formation of dislocation dipoles or to the capture of point defects by dislocations and loops. Higher order nonlinearities may be associated with the pinning of free dislocations by clusters of the nearly immobile dislocations of the forest [4].

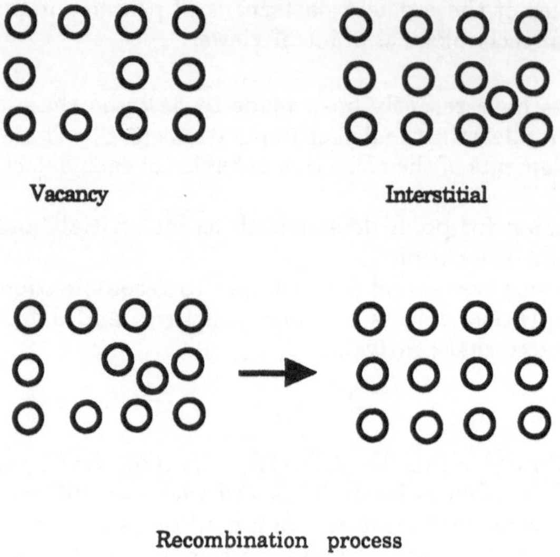

Figure 1 . Schematic representation of point defects and of their recombination process.

The next section will illustrate how these effects may be incorporated in a dynamical analysis of the collective behavior of point defects and vacancy loops in irradiated metals and alloys and describe their spontaneous ordering.

VACANCY LOOP ORDERING IN IRRADIATED MATERIALS

The ordering of vacancy loops occurs frequently in metals and alloys irradiated at moderate doses and high temperatures [8]. The uniform distribution of loops which are created by cascade collapse becomes unstable beyond some threshold related to various parameters such as the irradiation dose, the kinetic damage rate, the bias in the migration of point defects to loops and network dislocations, etc... A minimal model has been proposed by Murphy to describe the dynamics of defect populations in metals and alloys under particle irradiation[9-10]. It is based on the rate theory of radiation damage originally developed by Bullough, Eyre and Krishan [10], and expanded further by Ghoniem amd Kulcinski [12] to include the dynamics of point defects in the fully dynamic rate theory. The network dislocations which are also present in the material are assumed to have a constant uniform distribution and interstitial loops are neglected. By taking into account the basic mechanisms cited above, the kinetic equation for the defect concentrations are written as follows, where c_v corresponds to vacancies, c_i to interstitials, ρ_L and ρ_N to vacancy loops and network dislocation densities :

$$\partial_t c_i = K - \alpha c_i c_v + D_i \nabla^2 c_i - D_i c_i (Z_{iN} \rho_N + Z_{iL} \rho_L)$$
$$\partial_t c_v = K(1-\nu) - \alpha c_i c_v + D_v \nabla^2 c_v - D_v (Z_{vN}(c_v - \bar{c}_{vN})\rho_N + Z_{vL}(c_v - \bar{c}_{vL})\rho_L)$$
$$\partial_t \rho_L = \frac{1}{|\vec{b}| r_L^0}[(\nu K - \rho_L (D_i Z_{iL} c_i - D_v Z_{vL}(c_v - \bar{c}_{vL}))] \qquad (1)$$

where K is the displacement damage rate and ν the cascade collapse efficiency; D_i

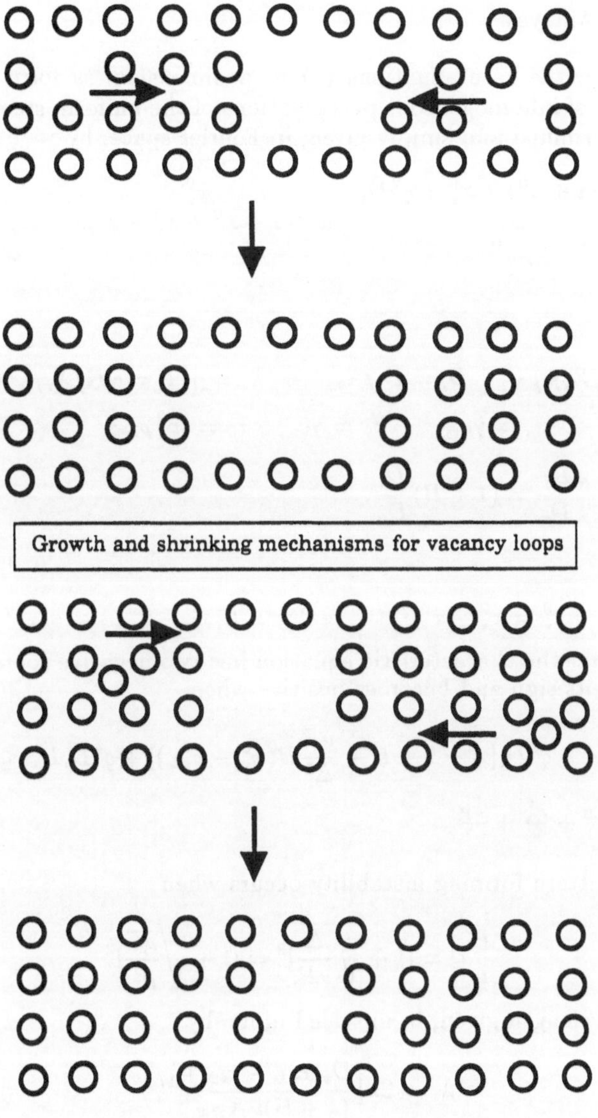

Figure 2 . Schematic representation of the interactions between point defects and vacancy loops which lead to the growth or shrinking of these loops.

and D_v are the diffusion coefficients, α is the recombination coefficient, $\vec{b}$ the Burgers vector, r_L^0 the mean vacancy loop radius and $Z_{..}$ the bias factors which will be approximated by $Z_{iL} = Z_{iN} = 1 + B$ and $Z_{vL} = Z_{vN} = 1$. $\bar{c}_{vN}$ and $\bar{c}_{vL}$ are the thermally emitted vacancies from network dislocations and vacancy loops. The various coefficients appearing in these rate equations may be computed theoretically or related to measurable quantities [**12**].

Linear Stability Analysis

On writing the rate equations (1) in a dimensionless form [**12**], the linear evolution matrix of inhomogeneous perturbations of the uniform steady state (x_i^0, x_v^0, x^0) is easily determined and simply given, in Fourier space, by :

$$\begin{bmatrix} \omega + \mu(1 + x^0) + x_v^0 + q^2\bar{D}_i & x_i^0 & \mu x_i^0 \\ x_v^0 & \omega + 1 + x^0 + x_i^0 + q^2\bar{D}_v & x_v^0 - \bar{x}_{vL} \\ \mu x^0 & -x^0 & \omega + \Delta \end{bmatrix} \tag{2}$$

where

$$\lambda_v = D_v Z_{vN}\rho_N, \quad \bar{D}_. = D_./\lambda_v, \quad \alpha/\lambda_v = \gamma, P = \gamma K/\lambda_v, \quad \tau = \lambda_v t,$$
$$x = \frac{\rho_L}{\rho_N}, \quad x_i = \gamma c_i, \quad x_v = \gamma c_v, \quad \tau_0 = b r_L^0 \rho_N \gamma$$
$$\mu = \frac{Z_{iN} D_i}{Z_{vN} D_v} = (1 + B)\frac{D_i}{D_v}. \tag{3}$$

and

$$\mu x_i^0 = x_v^0 - \bar{x}_{vN}, \qquad \nu P = x^0(\bar{x}_{vL} - \bar{x}_{vN}) = x^0\Delta \tag{4}$$

It turns out that the characteristic equation has two negative roots while the third one may change its sign and becomes positive when

$$\mu A + q^2\bar{D}_i[A + 1 + B - \frac{\nu P}{\Delta^2}B(x_v^0 - \bar{x}_{vL})] + q^4\bar{D}_i\bar{D}_v \leq 0 \tag{5}$$

where $A = 1 + x_i^0 + \frac{x_v^0}{\mu} + \frac{\nu P}{\Delta}$

Hence, a pattern forming instability occurs when

$$\frac{B}{\nu}|_c = [1 + \sqrt{\frac{\Delta}{\nu P}}]^2 = [1 + \sqrt{\frac{\rho_N}{\rho_L^0}}]^2 \tag{6}$$

for a critical wavelength given, in unscaled units, by:

$$\lambda_c = 2\pi[\frac{D_v(\bar{c}_{vL} - \bar{c}_{vN})}{(1 + B)\nu K \rho_N}]^{1/4} \tag{7}$$

We see that it decreases with increasing network dislocation density, cascade collapse efficiency and damage rate. On the other hand, its temperature dependence is more difficult to asses since D_v is an increasing function of the temperature while $(\bar{c}_{vL} - \bar{c}_{vN})$ is a decreasing function of the temperature and its global behavior may vary from material to material. As an example, consider 316 SS steel irradiated at 500^0 C with a displacement damage rate of 10^{-6} dpa s^{-1} [**12**]. The critical wavelength is nearly 1.24 μm for solution-annealed material with a typical dislocation density of 10^{13} m^{-2}. The wavelength is smaller for cold-worked material, on the order of 0.39 μm for a dislocation density of 10^{15} m^{-2}.

Since the orientation and the number of wavevectors underlying the possible structures are not fixed by this linear analysis, a nonlinear analysis in the post bifurcation regime is needed to discuss the emergence of stable patterns.

The various coefficients which represent defect creation, annihilation, and migration to dislocations and vacancy loops may be obtained from experimental data or theoretical analysis. K is the displacement damage rate and ϵ the cascade collapse efficiency. D_i and D_v are the diffusion coefficients, α the recombination coefficient, b the length of the Burgers vector. $Z_{.,.}$ are the bias factors and $\bar{c}_{v,.}$ the thermally emitted vacancies from the microstructures.

The Amplitude Equations for the Microstructures

In the systems considered here, the fluctuations of vacancy and interstitial concentrations evolve much more rapidly than vacancy loop density fluctuations. Hence the dynamics may be reduced through a multiple scale analysis [13] or the adiabatic elimination of the fast variables [15] and the slow mode or order parameter-like variable will be associated with the vacancy loop density. On expanding the point defect concentrations as a power series in the vacancy loop density one obtains [13] :

$$\partial_t \sigma(\vec{x},t) = [\epsilon - \xi_0^2(q_c^2 + \nabla^2)^2]\sigma(\vec{x},t) + v\sigma^2(\vec{x},t) - u\sigma^3(\vec{x},t) \tag{8}$$

where $\sigma(\vec{x},t) = x(\vec{x},t) - x^0$, $b = B/\nu$, $\frac{b-b_c}{b_c} = \epsilon$, $\xi_0^2 \propto (x_0)^{-1/2}$, $v = 2/(x_0)^{3/2}$ and $u = 2/(x_0)^{5/2}$.

For $b > b_c$, the stable solutions of this Landau-Ginzburg type of dynamics correspond to [5] :

(1) roll or wall structures associated with spatial modulations of the order parameter (here the vacancy loop density) in one direction. They appear via a second orderlike transition, or supercritical bifurcation.

(2) rodlike hexagonal or triangular structures appearing via a first orderlike transition (subcritical bifurcations) defined by the following amplitude equations:

$$\tau \dot{A}_i = [\epsilon + \frac{4\xi_0^2}{q_c^2}(\vec{q}_i.\vec{\nabla})^2]A_i + vA_{i-1}^*A_{i+1}^* - 3u[|A_i|^2 + 2\sum_{j\neq i}|A_j|^2]A_i \tag{9}$$

where $\sigma(\vec{x},t) = \sum_{i=1}^3 A_i(\vec{x},t)e^{i\vec{q}_i.\vec{x}} + c.c.$, with $|\vec{q}_i| = q_c$, and $\vec{q}_1 + \vec{q}_2 + \vec{q}_3 = 0$

The stable steady state is given by $|A_i| = A = \frac{1}{30u}[v + \sqrt{v^2 + 60u\epsilon}]$.

(3) bcc lattices or filamental structures of cubic symmetry, also associated with a subcritical bifurcation and defined similarly to hexagonal structures but with six pairs of wavevectors. In this case, the corresponding steady state is then given by:

$$\sigma(\vec{x}) = A[\cos\frac{q_c}{\sqrt{2}}x\cos\frac{q_c}{\sqrt{2}}y + \cos\frac{q_c}{\sqrt{2}}y\cos\frac{q_c}{\sqrt{2}}z + \cos\frac{q_c}{\sqrt{2}}z\cos\frac{q_c}{\sqrt{2}}x] \tag{10}$$

with $A = \frac{1}{33u}[v + \sqrt{v^2 + 33u\epsilon}]$

When the bifurcation parameter is increased, the 2d and 3d structures may in turn become unstable (the hexagonal structure for $\epsilon > \frac{4v^2}{3u}$ and the bcc structure for $\epsilon > 3\frac{v^2}{u}$). Hence, between threshold and $3v^2/u$, bcc dislocation structures should be expected while above this limit the structure should consist of regularly spaced planes of maximum density.

The bifurcation diagram for bcc and planar wall structures is sketched in Fig.3.

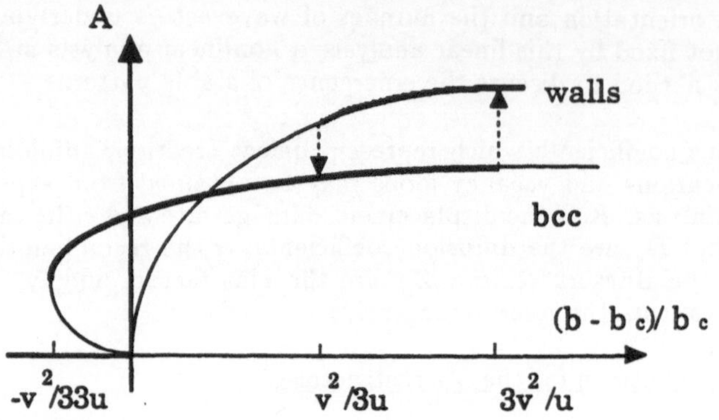

Figure 3 . Schematic bifurcation diagram associated with the amplitude equation for the microstructures showing the transition and the hysteresis loop between bcc and wall structures (heavy lines represent stable states, and thin lines represent unstable states).

In the case of an anisotropic diffusion of interstitials as in hcp materials where the mobility of interstitials is much larger in the basal planes than between these planes, it is easy to show with the same method, that the stable patterns for vacancy loops correspond to planar arrays with planes of maximum density parallel to the planes of high interstitial mobility in agreement with experimental observations [16].

The result of the present analysis obtained in the weakly nonlinear regime beyond the pattern forming instability should of course be confronted to detailed experimental investigations and to at least 2-D numerical simulations. It is interesting to note that recent experimental observations by Jäger et al.[8] indicate that spatial microstructure modulation is a general phenomenon under ion-irradiation conditions where cascade are produced. Over a limited range of conditions, Jäger observed that the wavelength was insensitive to temperature, dose rate, and type of primary knock-on atom. Dislocation loops and tangles were found to be arranged in planar arrangements (walls), with a wavelength of 0.03 to 0.06 μm. While the model discussed here is in general agreement with these experimental observations, it should be viewed as a step towards a generalized theoretical explanation of the nature of microstructure ordering under irradiation.

CONCLUSION

It has been shown that the coupling between reaction and transport may induce pattern forming instabilities in defect distributions in driven crystalline materials. For example, uniform distributions of point defects such as interstitials and vacancies in irradiated materials may become unstable and lead to the precipitation of solid solutions, to the nucleation of voids and of void lattices. In this framework, it was shown that vacancy loop ordering occur under very general conditions in irradiated metals and alloys. It mainly results fom the different mobilities and bias in the migration of point defects to line defects such as vacancy loops or network dislocations. Structures with different symetries may be simultaneously stable beyond the primary bifurcation. For example, when the diffusion and interactions of point defects are isotropic the maxima of the vacancy loop density may either correspond to bcc lattices or planar arrays. Hence, these structures could be in non parallel orientation, i.e. with a structure different from the structure of the host lattice.On increasing further the displacement damage rate bcc lattices become unstable and a first-orderlike transition should occur to planar structures. In the case of anisotropic interstitial diffusion, planar structures should be the rule. Hence, since the symmetry of the defect structures is a crucial issue in irradiated materials [17], the present discussion shows that a careful study of the post-bifurcation regime is needed to test the relevance of particular kinetic models to the interpretation of experimental observations.

Furthermore a coherent description of materials instabilities associated with the spatio-temporal organisation of defects will hopefully lead to a deeper understanding of these phenomena. Due to the strong nonequilibrium conditions under which they occur, classical mechanical or thermodynamical considerations are not sufficient and we need an important input from nonlinear dynamics and instability theory. Effectively, despite the huge complexity of the defect dynamics, even in the case of phenomenological models, valuable information may be obtained via the reduced dynamics near instability points leading to a possible description of the pattern selection and stability properties in the post-bifurcation regime. Hence, by the combination of the results of bifurcation analysis, amplitude equation formalism and numerical simulations we hence expect significant breakthroughs in the understanding and prediction of the effects of materials instabilities on the macroscopic behavior of driven or degrading solids.

Acknowledgements. Financial assistance through a grant of the U.S. Department of Energy, Office of Fusion Energy (DOE grant No. DE-FG03-84ER52110 with UCLA) is gratefully acknowledged.

References

1. G.Nicolis and F.Baras, "Chemical Instabilities, Applications in Chemistry, Engineering, Geology and Materials Science," Reidel, Dordrecht, 1983.
2. D.Walgraef, "Patterns, Defects and Microstructures in Nonequilibrium Systems," Martinus Nijhoff, Dordrecht, 1987.
3. G.Martin and L.P.Kubin, "Nonlinear Phenomena in Materials Science," Trans tech, Aedermannsdorf (Switzerland), 1988.
4. D.Walgraef and E.C.Aifantis, Res Mechanica **23** (1988), p. 161.
5. D.Walgraef, in "Nonlinear Phenomena in Materials Science," G. Martin and L.P. Kubin eds., Transtech, Aedermannsdorf (Switzerland), 1988, p. 77.
6. a)G.Martin, Phys.Rev. **B30** (1984), p. 1424.
6. b)K.Krishan, Radiat.Eff. **66** (1982), p. 121.
7. D.Hull and D.J.Bacon, "Introduction to Dislocations," 3rd edition, Pergamon, Oxford, 1984.
8. W.Jäger, P.Ehrhart and W.Schilling, in "Nonlinear Phenomena in Materials Science," G.Martin and L.P.Kubin eds., Transtech, Aedermannsdorf (Switzerland), 1988, p. 279.
9. S.M.Murphy, Europhys.Lett **3** (1987), p. 1267.
10. S.Murphy, in "Nonlinear Phenomena in Materials Science," G. Martin and L.P. Kubin eds., Transtech, Aedermannsdorf (Switzerland), 1988, p. 295.
11. R.Bullough, B.L.Eyre and K.Krishan, J.Nucl.Mat. **44** (1975), p. 121.
12. N.M.Ghoniem and G.L.Kulcinski, Radiation Effects **39** (1978), p. 47.
13. D.Walgraef and N.M.Ghoniem, Phys.Rev. **B39** (1989), p. 8867.

14. A.C.Newell, in "Lectures in Applied Mathematics," vol.15, M.Kac, ed., American Mathematical Society, 1974, p. 157.
15. H.Haken, "Synergetics: an Introduction," 3rd ed., Springer, Berlin.
16. J.H.Evans, Mater.Sci.Forum **15-18** (1987), p. 869.
17. R.W.Cahn, Nature **329** (1987), p. 284.

MUTUAL INTERPLAY BETWEEN THE BREAK-UP OF SPATIAL ORDER AND THE ONSET OF LOW-DIMENSIONAL TEMPORAL CHAOS IN AN EXEMPLARY SEMICONDUCTOR SYSTEM

Jürgen Parisi

Physical Institute, University of Tübingen
D-7400 Tübingen, Fed. Rep. Germany

INTRODUCTION

Partially all branches of modern science ranging from physics through chemistry and biology to economics and sociology deal with complex nonlinear systems the dynamics of which may acquire a macroscopic spatial, temporal, or functional structure without specific interference from the outside. Such ubiquitous processes of spontaneous self-organization can in general be formulated as nonequilibrium order-disorder phase transitions. The basic idea for the underlying unifying approach stems from that of synergetics[1] and information thermodynamics.[2] It implies that we consider open systems capable of decomposing into a potentially large number of competing individual subparts. In our quest to understand how structures are generated by nature, the mutual interaction among these subsystems, or say variables, is of fundamental importance in the neighborhood of critical instability points where only a few collective degrees of freedom, often called order parameters, dominate the global system behavior. Those coherent variables force the subsystems to join an organized motion, just giving the total system its specific structure or order. Most characteristically, it turns out that the detailed nature of any particular subsystem becomes unessential, ensuing the universal character of pattern-forming processes.

As physics copes with ever more ambitious problems in structure formation, especially those appearing in the space and time domain, we focus on nonlinear current transport phenomena in semiconductors providing a prototype situation with dynamical behavior ranging from orderly to chaotic.[3] In particular, recent experiments on impurity impact ionization avalanche breakdown in extrinsic germanium and gallium arsenide at low temperatures have demonstrated the self-sustained development of both filamentary spatial and oscillatory temporal dissipative structures in the formerly homogeneous semiconductor.[4-8] This kind of nonequilibrium phase transition between different conducting states results from the autocatalytic nature of impurity impact ionization generating mobile charge carriers.[9] The simple and direct experimental accessibility via advanced measurement techniques favors semiconductors as a nearly ideal study object for complex nonlinear dynamics compared to other physical systems.[10] Further representing a convenient reaction-diffusion system[11,12] that exhibits distinct universal features, the present semiconductor system may receive general significance for many synergetic systems in nature. Finally, in view of the rapidly growing application of semiconductor technologies, the understanding, control, and

Nonlinear Evolution of Spatio-Temporal Structures in Dissipative Continuous Systems
Edited by F.H. Busse and L. Kramer
Plenum Press, New York, 1990

possible exploitation of sources of instability in these systems have considerable practical importance.

This paper provides a brief survey of our recent experimental investigations on the spatio-temporal nonlinear current flow in the post-breakdown region of p-germanium at liquid-helium temperatures. Spatially, we report on filamentary flow patterns developing during avalanche breakdown. Temporally, we outline the typical universal scenarios demonstrated by means of the self-generated oscillatory behavior of distinct state variables. The central theme, however, concerns the relationship between the onset of low-dimensional chaotic dynamics and the break-up of spatial order during current filamentation. Such cooperative phenomena induced by the avalanche breakdown kinetics in semiconductors can be interpreted in terms of a phenomenological reaction-diffusion model known from chemical systems theory.

EXPERIMENT

Our experimental system consists of single-crystalline p-doped germanium, electrically driven into low-temperature avalanche breakdown via impurity impact ionization. Having the typical dimensions of about $0.2 \times 2 \times 5\ mm^3$ and an indium acceptor concentration of about $3 \times 10^{14}\ cm^{-3}$, the extrinsic germanium crystal carries properly arranged ohmic aluminum contacts evaporated on one of the two largest surfaces. To provide the outer ohmic contacts with an electric field, a d.c. bias voltage was applied to the series combination of the sample and a load resistor. A d.c. magnetic field parallel or perpendicular to the broad sample surfaces could also be applied by a superconducting solenoid surrounding the semiconductor sample. The resulting current was found from the voltage drop at the load resistor. Various inner probe contacts served for monitoring — independent of the total sample voltage — the corresponding partial voltages along the sample.[13] Utilizing low-temperature scanning electron microscopy, two-dimensional images of both stationary and dynamical current structures were obtained by scanning the specimen surface with an electron beam chopped properly and by recording the beam-induced current change in the voltage-biased sample as a function of the beam coordinate. As described elsewhere[7,8,14] in detail, the present imaging method combines a low-temperature stage with a commercial scanning electron microscope such that both spatial and temporal measurements can be performed simultaneously. During the experiments, the semiconductor sample was always kept at liquid-helium temperatures (less than 10 K) and carefully protected against external electromagnetic irradiation (visible, far infrared).

Analogous to the corresponding processes of structure formation in gaseous plasma discharges and atmospheric lightning, impact ionization of the shallow impurity acceptors can be achieved in the bulk of the homogeneously doped semiconductor at low temperatures. In the temperature regime of liquid helium, most of the charge carriers are frozen out at the impurities. Since the ionization energy is only about 10 meV and electron-phonon scattering is strongly reduced, avalanche breakdown due to an abrupt multiplication of mobile charge carriers already takes place at electric fields of a few V/cm and persists until nearly all impurities are ionized.[15] The underlying nonequilibrium phase transition from a low conducting state to a high conducting state is directly reflected in strongly nonlinear regions of negative differential resistivity in the microscopic current-density versus electric-field characteristic.[9,16] Accordingly, the autocatalytic process of impurity impact ionization also leads to a strongly nonlinear curvature of the macroscopic (measured) current-voltage characteristic (sometimes with S-shaped negative differential resistance), the nonlinearity occurring just beyond the voltage corresponding to the critical electric field where the current increases by many orders of magnitude (typically, from a few nA in the pre-breakdown up to a few mA in the post-breakdown region). Under slight variation of distinct control parameters (electric field, magnetic field, and temperature in the range of some 10^{-6} V/cm, 10^{-5} T, and 10^{-3} K, respectively) the resulting electric current flow displays a wide variety of spatial and temporal dissipative structures.

RESULTS

The complex spatial behavior of our semiconductor system can be globally visualized by means of two-dimensional imaging of the current filament structures via low-temperature scanning electron microscopy. Current filaments represent high conducting channels embedded within a much lower conducting semiconductor medium. As reported elsewhere[5,17,18] in detail, nucleation and growth of filamentary current patterns in the nonlinear post-breakdown regime are often accompanied by abrupt changes between different stable filament configurations via noisy current instabilities. Here a common phenomenon is that there exists a minimum as well as a maximum value of the filament current. On the one hand, with decreasing applied electric field, the filament size can not decrease continuously to zero, but abruptly decays to zero at a distinct minimum diameter. On the other hand, with increasing applied electric field, a continuous growth of existing filaments takes place only up to a distinct maximum size beyond which an additional filament is nucleated. The ensuing multifilamentary current flow becomes more and more homogeneous if the semiconductor system further approaches its linear post-breakdown region at higher electric fields.

The strongly nonlinear post-breakdown regime of the measured current-voltage characteristic is further associated with the appearance of self-generated current and voltage oscillations.[19,20] Both the integral current and the partial voltages along the sample display — superimposed upon the steady d.c. signals of typically a few mA and some hundred mV, respectively — temporal oscillations with a relative amplitude of about 10^{-3} in the frequency range 0.1 - 100 kHz. By means of slightly varying the distinct control parameters, the temporal behavior of the system variables (current, partial voltages) changes dramatically, exhibiting the typical universal scenarios of chaotic nonlinear systems.[21-25] The observed dynamical phenomena include the intermittent switching between different oscillation modes (Pomeau-Manneville scenario), the period-doubling cascade to chaos (Feigenbaum-Großmann scenario), and the quasiperiodic route to chaos (Ruelle-Takens-Newhouse scenario), as well as the suppression of quasiperiodicity through frequency locking. Moreover, we found a transition from low-dimensional ordinary chaos to higher-dimensional hyperchaos (Rössler scenario). On the ladder towards more fully developed turbulence, we discovered a chaotic hierarchy of strange attractors.[26,27] In accordance with their gradually increasing fractal dimensionality and entropy, the different chaotic states are further discriminated by distinct numbers of positive Lyapunov characteristic exponents, determining mutually independent directions of stretching and folding-over of near-by trajectories in phase space and, thus, reflecting the order of chaos.

Experimental evidence of the abrupt structural change between different dynamical states associated with an apparent loss of spatial coherence among different sample parts indicates a break-up of the semiconductor system from strongly coupled into more independent subsystems. In this way, new actively participating degrees of freedom are gained reflecting increasing complexity of the system. Turbulent dynamics may thus be ascribed to nonlinear coupling between competing localized oscillation centers intrinsic to the present multicomponent semiconductor system.

So far, we have demonstrated experimentally the simultaneous existence of spatially separated oscillatory subsystems as well as their long-range interaction.[23,28-31] Due to the extremely high lattice heat conductivity of germanium at low temperatures, the only relevant spatial coupling mechanism results from energy diffusion via acoustical phonons emitted by the hot charge carriers. There was an additional experimental finding that the underlying nonlinear physics reveals critical phase transition behavior by varying the temperature at constant electric and magnetic field.[32] Most importantly, we disclosed an upper bound for the onset of structure formation (first-order phase transition) and avalanche breakdown (second-order phase transition) at critical temperatures of about 5.5 K and 7.2 K, respectively.

Summarizing the results discussed hitherto, our experiments deal with a challenging example of a macroscopic synergetic system, consisting of spatially localized and diffusively coupled subsystems which by themselves show oscillatory behavior. Depending sensitively upon distinct control parameters, the resulting spatio-temporal current flow undergoes various symmetry-breaking phase transitions typical for nonlinear dynamical systems. In the following, we intend to stress the crucial role of the mutual interplay between the self-organized formation of filamentary current structures in space, at one side, and of oscillatory current and/or voltage structures in time, at the other side.

Figure 1 illustrates two-dimensional images of both stationary and dynamical current structures recorded by the help of low-temperature scanning electron microscopy. Here we have applied the voltage bias to the triangular ohmic contacts the tips of which are indicated by the full markers (the open triangle means a nonactive contact). The surface of the intervening sample part was scanned by the electron beam (horizontal traces), and the beam-induced current change was plotted vertically (y-modulation). For obtaining the necessary signal detection sensitivity, the images were taken at different beam-modulation frequencies via conventional lock-in techniques. The well-established stationary current filament pattern (Fig. 1a) oriented vertically can be recognized through pronounced response signals originating from its enveloping boundaries where the local conductivity of the sample is most sensitive to the beam-injected perturbation.[5] Note that the electron beam power employed (typically, less than 1 μW) was always small compared to the Joule heat dissipated in the sample during avalanche breakdown (typically, in the mW range). Moreover, for the least undisturbed recording of the intrinsic filamentary current flow, the modulation frequency used (5 kHz) had to be chosen lower than the inverse signal relaxation times (typically, about 10 μs). The spatial resolution limit of approximately 10 μm attainable for stationary current images has been determined from individual line scans (not shown here) where the amplitude of the beam-induced response signal is directly measured as a function of the beam position perpendicular to the direction of the current flow.

Simultaneously, the dynamical part of the spatially inhomogeneous current distribution can be visualized by resonant interaction with the electron beam applied. One only has to modulate the beam just at the intrinsic frequency of the spontaneous oscillation observed in the unirradiated sample and to scan over the specimen surface, in order to pick up the regions oscillating spontaneously. Note that a distinct resonance signal arises in the temporal current response if the focus of the properly modulated electron beam approaches the location of the oscillation center. This novel resonance imaging method thus provides spatially resolved analysis of the nonlinear system dynamics.[8] The pronounced dynamical current structure (Fig. 1b) taken at resonant electron beam excitation clearly identifies spontaneously oscillating sample regions on the r.h.s. boundary of the current filament extending in vertical direction (cf. Fig. 1a). Of course, the modulation frequency

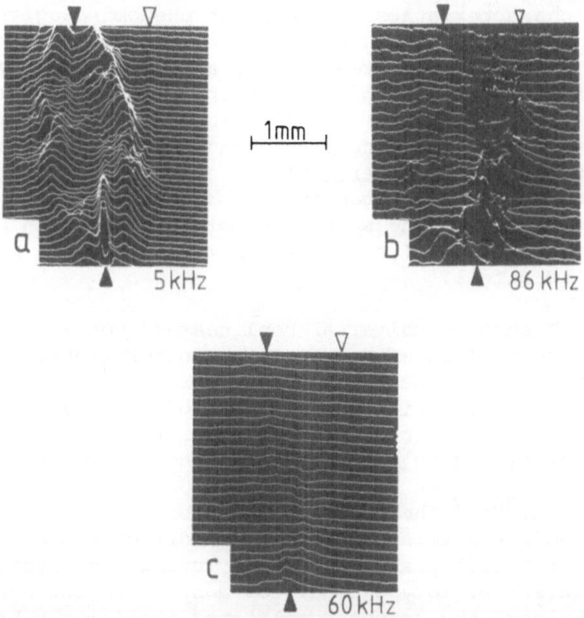

Fig. 1. Imaging of spatio-temporal current flow patterns during semiconductor avalanche breakdown via low-temperature scanning electron microscopy: (a) standard image of a stationary current filament obtained at low electron-beam modulation (5 kHz); (b) resonance image of spontaneously oscillating current filament regions obtained at resonant electron-beam modulation (86 kHz); (c) off-resonance image obtained at nonresonant electron-beam modulation (60 kHz). The following control parameters were kept constant: bias voltage 1.4450 V, load resistance 1 Ω, magnetic field 0 T, bath temperature 4.2 K.

used (86 kHz) is far too high in view of the typical sample response times necessary to obtain the standard images of stationary current patterns. In comparison, the off-resonance case (Fig. 1c) due to nonresonant electron beam excitation neither displays a stationary filament nor an oscillatory resonance structure, as expected. Apparently, the modulation frequency used (60 kHz) is too far from the intrinsic frequency of the spontaneous current oscillation (86 kHz) for receiving any resonant response signal.

From our experiments (including results not shown in Fig. 1), we conclude that the origin of different oscillatory modes derives from the simultaneous existence of spatially separated oscillation centers located close to distinct boundary regions of the current filaments. More generally, the spatial discrimination of temporal current flow dynamics in the vicinity of different conducting phases yields a powerful tool for gaining deeper insight into the mutual interplay between spatial and temporal current structures developing during semiconductor breakdown. Hereto, the electron beam was demonstrated to act as an exemplary control parameter which can be manipulated both spatially and temporally.

MODEL APPROACH

The above cooperative transport phenomena display a striking similarity to the behavioral characteristics of a generic phenomenological model based on the well-known Rashevsky-Turing theory of morphogenesis.[33,34] The simplest version of Turing's model can be realized by a two-cellular symmetrical reaction-diffusion system consisting of cross-inhibitorily coupled, potentially oscillating two-variable subsystems.[11,12] This 4-D flow is capable of generating spontaneous symmetry-breaking phase transitions in the symmetrical homogeneous reaction system. Moreover, the differentiated system flow may further bifurcate from stable morphogenesis to boiling-type turbulence. Numerical evidence and analysis of the spatio-temporal correlation between the two compartments under variation of distinct control parameters have been presented elsewhere[35] in detail.

On the basis of such an universal nonlinear ad-hoc model, first attempts towards the development of an adequate semiconductor transport theory involving the concrete physical mechanisms of hot-carrier dynamics[30,31] look highly promising. Here energy relaxation oscillations of two electronic subsystems are driven by the autocatalytic process of impurity impact ionization and coupled through heat diffusion via long-range phonon propagation. Therefrom, analytical conditions for both filamentary and oscillatory instabilities can be explicitly derived.[36,37] The dynamical possibilities of avalanche breakdown kinetics in semiconductors turn out to be directly projected onto the main behavioral characteristics of a generic Turing-type system, representing the most convenient prototype model for many different synergetic systems in nature. Nevertheless, the rather complicated nonlinear behavior of the underlying physics can be only roughly approximated with the simple kinetics of the activator and inhibitor model variables considered. Clearly, the development of a microscopic theory for the detailed understanding of the carrier transport phenomena in an extrinsic semiconductor at low temperatures still represents a challenging task.

FINAL REMARKS

Our experimental semiconductor system has been shown to display both filamentary spatial and oscillatory temporal current structures during low-temperature avalanche breakdown. We have demonstrated a close connection between the stationary filament pattern and the spontaneous oscillations in the current flow. Beyond its imaging capability, the electron beam served as a highly effective localized control parameter.

It is appreciated, however, that the underlying complex nonlinear dynamics will — in a sense — generally be affected by the imaging method employed. Due to the extreme sensitivity of any nonlinear dynamical system against smallest variations of the control parameters involved, we have to keep in mind that difficulties on principle may arise in the interpretation of our experimental results. Even though the inherent system dynamics of the present semiconductor experiment has been unfolded to some extent, the crucial role of the interwoven relationship with the unwanted interference resulting from the measuring environment still remains an unsolved question. There is little hope that a continuation of the promising advances attained in semiconductor physics via certain fundamental interdisciplinary concepts may open up the possibility of a thorough understanding embracing the total multicomponent system, in order to cope with self-organizing phenomena.

ACKNOWLEDGMENTS

The author is indebted to Wilfried Clauß, Minh Duong-van, Rudolf P. Huebener, Klaus M. Mayer, Joachim Peinke, Uwe Rau, Brigitte Röhricht, Otto E. Rössler, Eckehard Schöll, and Ruedi Stoop for continuous fruitful collaboration. Part of this work was supported financially by the Deutsche Forschungsgemeinschaft.

REFERENCES

1. H. Haken, "Advanced Synergetics", Springer, Berlin (1989).
2. H. Haken, "Information and Self-Organization", Springer, Berlin (1988).
3. Y. Abe, ed., "Nonlinear and Chaotic Transport Phenomena in Semiconductors", special issue of Appl. Phys. A, vol. 48, pp. 93 - 191, Springer, Berlin (1989), and references therein.
4. R.P. Huebener, K.M. Mayer, J. Parisi, J. Peinke, and B. Röhricht, Chaos in semiconductors, Nucl. Phys. B (Proc. Suppl.) 2 : 3 (1987).
5. K.M. Mayer, J. Peinke, B. Röhricht, J. Parisi, and R.P. Huebener, Spatial and temporal current instabilities in germanium, Physica Scripta T 19 : 505 (1987).
6. J. Peinke, J. Parisi, B. Röhricht, K.M. Mayer, U. Rau, and R.P. Huebener, Spatio-temporal instabilities in the electric breakdown of p-germanium, Solid State Electron. 31 : 817 (1988).
7. K.M. Mayer, J. Parisi, and R.P. Huebener, Imaging of self-generated multifilamentary current patterns in GaAs, Z. Phys. B - Condensed Matter 71 : 171 (1988).
8. K.M. Mayer, J. Parisi, J. Peinke, and R.P. Huebener, Resonance imaging of dynamical filamentary current structures in a semiconductor, Physica D 32 : 306 (1988).
9. E. Schöll, "Nonequilibrium Phase Transitions in Semiconductors", Springer, Berlin (1987).
10. R.P. Huebener, J. Peinke, and J. Parisi, Experimental progress in the nonlinear behavior of semiconductors, Appl. Phys. A 48 : 107 (1989).
11. B. Röhricht, J. Parisi, J. Peinke, and O.E. Rössler, A simple morphogenetic reaction-diffusion model describing nonlinear transport phenomena in semiconductors, Z. Phys. B - Condensed Matter 65 : 259 (1986).
12. J. Parisi, J. Peinke, B. Röhricht, U. Rau, M. Klein, and O.E. Rössler, Comparison between a generic reaction-diffusion model and a synergetic semiconductor system, Z. Naturforsch. 42 a : 655 (1987).
13. J. Peinke, J. Parisi, B. Röhricht, K.M. Mayer, U. Rau, W. Clauß, R.P. Huebener, G. Jungwirt, and W. Prettl, Classification of current instabilities during low-temperature breakdown in germanium, Appl. Phys. A 48 : 155 (1989).
14. R.P. Huebener, Scanning electron microscopy at very low temperatures, in: "Advances in Electronics and Electron Physics", vol. 70, p. 1, P.W. Hawkes, ed., Academic Press, New York (1988).
15. J. Parisi, U. Rau, J. Peinke, and K.M. Mayer, Determination of electric transport properties in the pre- and post-breakdown regime of p-germanium, Z. Phys. B - Condensed Matter 72 : 225 (1988).
16. J. Peinke, D.B. Schmid, B. Röhricht, and J. Parisi, Positive and negative differential resistance in electrical conductors, Z. Phys. B - Condensed Matter 66 : 65 (1987).
17. K.M. Mayer, R. Gross, J. Parisi, J. Peinke, and R.P. Huebener, Spatially resolved observation of current filament dynamics in semiconductors, Solid State Commun. 63 : 55 (1987).
18. U. Rau, J. Peinke, J. Parisi, and R.P. Huebener, Switching behavior of current filaments in p-germanium connected in parallel, Z. Phys. B - Condensed Matter 71 : 305 (1988).
19. J. Peinke, J. Parisi, A. Mühlbach, and R.P. Huebener, Different types of current instabilities during low-temperature avalanche breakdown in p-germanium, Z. Naturforsch. 42 a : 441 (1987).
20. J. Peinke, U. Rau, W. Clauß, R. Richter, and J. Parisi, Critical dynamics near the onset of spontaneous oscillations in p-germanium, Europhys. Lett. 9 : 743 (1989).

21. J. Peinke, A. Mühlbach, R.P. Huebener, and J. Parisi, Spontaneous oscillations and chaos in p-germanium, Phys. Lett. 108 A : 407 (1985).
22. J. Peinke, B. Röhricht, A. Mühlbach, J. Parisi, Ch. Nöldeke, R.P. Huebener, and O.E. Rössler, Hyperchaos in the post-breakdown regime of p-germanium, Z. Naturforsch. 40 a : 562 (1985).
23. J. Peinke, J. Parisi, B. Röhricht, B. Wessely, and K.M. Mayer, Quasiperiodicity and mode locking of undriven spontaneous oscillations in germanium crystals, Z. Naturforsch. 42 a : 841 (1987).
24. U. Rau, J. Peinke, J. Parisi, R.P. Huebener, and E. Schöll, Exemplary locking sequence during self-generated quasiperiodicity of extrinsic germanium, Phys. Lett. 124 A : 335 (1987).
25. J. Peinke, J. Parisi, R.P. Huebener, M. Duong-van, and P. Keller, Quasiperiodic behavior of d.c. biased semiconductor electronic breakdown, Europhys. Lett. (to be published).
26. R. Stoop, J. Peinke, J. Parisi, B. Röhricht, and R.P. Huebener, A p-Ge semiconductor experiment showing chaos and hyperchaos, Physica D 35 : 425 (1989).
27. J. Parisi, J. Peinke, R.P. Huebener, R. Stoop, and M. Duong-van, Evidence of chaotic hierarchy in a semiconductor experiment, Z. Natúrforsch. a (to be published).
28. B. Röhricht, B. Wessely, J. Parisi, and J. Peinke, Crosstalk of the dynamical dissipative behavior between different parts in a current-carrying semiconductor, Appl. Phys. Lett. 48 : 233 (1986).
29. B. Röhricht, J. Parisi, J. Peinke, and R.P. Huebener, Spontaneous resistance oscillations in p-germanium at low temperatures and their spatial correlation, Z. Phys. B - Condensed Matter 66 : 515 (1987).
30. E. Schöll, J. Parisi, B. Röhricht, J. Peinke, and R.P. Huebener, Spatial correlations of chaotic oscillations in the post-breakdown regime of p-Ge, Phys. Lett. 119 A : 419 (1987).
31. E. Schöll, H. Naber, J. Parisi, B. Röhricht, J. Peinke, and S. Uba, Resonance transition of the spatial correlation factor of self-generated oscillations in the post-breakdown regime of p-Ge, Z. Naturforsch. a (to be published).
32. B. Röhricht, R.P. Huebener, J. Parisi, and M. Weise, Nonequilibrium critical and multicritical phase transitions in low-temperature electronic transport of p-germanium, Phys. Rev. Lett. 61 : 2600 (1988).
33. N. Rashevsky, An approach to the mathematical biophysics of biological self-regulation and of cell polarity, Bull. Math. Biophys. 2 : 15 (1940); Further contributions to the theory of cell polarity and self-regulation, Bull. Math. Biophys. 2 : 65 (1940); Physicomathematical aspects of some problems of organic form, Bull. Math. Biophys. 2 : 109 (1940).
34. A.M. Turing, The chemical basis of morphogenesis, Phil. Trans. Roy. Soc. London B 237 : 37 (1952).
35. B. Röhricht, J. Parisi, J. Peinke, and O.E. Rössler, Breakdown of symmetry in an exemplary Turing system, Dynamics and Stability of Systems (to be published).
36. E. Schöll, Nonequilibrium phase transitions and chaos in semiconductors, J. Phys. Chem. Solids 49 : 651 (1988).
37. E. Schöll, Instabilities in semiconductors including chaotic phenomena, Physica Scripta T 29 : 152 (1989).

ELECTROCONVECTION IN A FREELY SUSPENDED FILM OF SMECTIC A LIQUID CRYSTAL

Stephen W. Morris[a], John R. de Bruyn[b], and A.D. May[a,c]

(a) Department of Physics, University of Toronto
60 St. George St. Toronto, Ontario, Canada M5S 1A7
(b) Department of Physics, Memorial University of Newfoundland
St. John's, Newfoundland, Canada A1B 3X7
(c) Ontario Laser and Lightwave Research Centre
60 St. George St. Toronto, Ontario, Canada M5S 1A7

INTRODUCTION

Liquid crystals in the smectic A phase have a layered structure [1]. They are liquid–like in the plane of the layers, but act like a soft solid in the perpendicular direction. A film of smectic A, supported only at its edges, forms a nearly two–dimensional liquid [2]. We observe that such a film can be driven into convection by an electric field applied in the plane of the film [3]. Two types of convection are observed; one which requires the injection of charge at the electrodes [4,5] and one which is due to charged impurities already present in the material [6,7]. The director orientation is unaffected by the motion of the liquid. In contrast to the behaviour of nematic or isotropic films [8,9], we find that the free surfaces of the smectic are not deformed by the flow. No significant hysteresis is observed at the bifurcation to the convecting state. The flow patterns, shown if Fig. 1, resemble two–dimensional versions of conventional convection rolls. This instability is an interesting new example of a low dimensional pattern–forming system [10].

EXPERIMENT

The liquid crystal we studied was octylcyanobiphenyl (8CB), which is smectic A at room temperature and has a positive dielectric anisotropy. In order to control the species and concentration of impurities, we doped the material with 1 mMol/litre of tetracyanoquinodimethane (TCNQ) which has been shown to give strong unipolar injection of negative ions when used as a dopant in nematics [11].

Nonlinear Evolution of Spatio-Temporal Structures in Dissipative Continuous Systems
Edited by F.H. Busse and L. Kramer
Plenum Press, New York, 1990

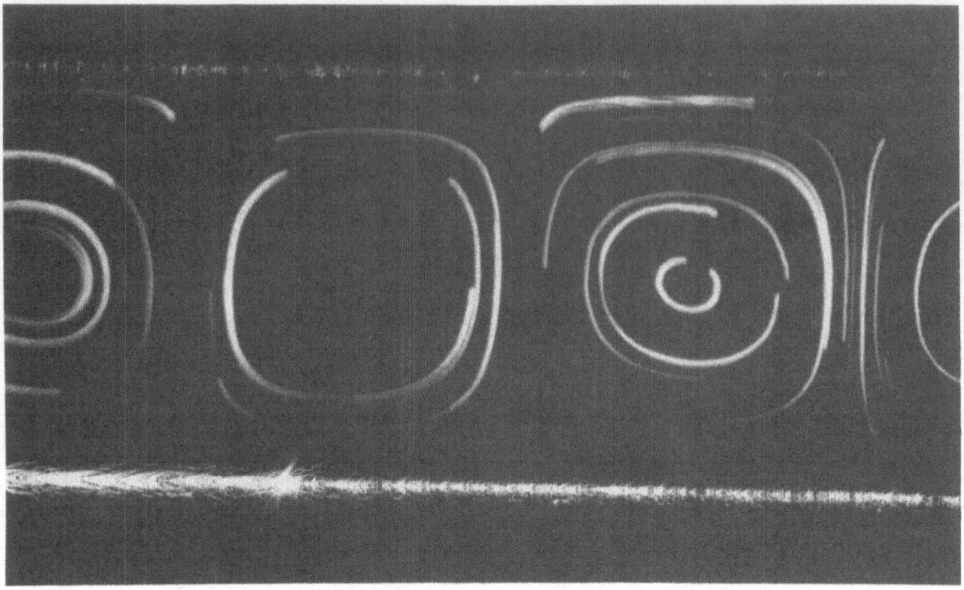

FIG. 1. Photograph showing the roll pattern observed in a contact holder above the critical voltage. Only a part of the holder is shown.

We used two film holders with different electrode configurations to separate the effects of injected charge. The two types of holder are shown in the insets of Fig. 2. In one configuration the long edges of the rectangular film were directly supported by 15μm tungsten wire electrodes. We will refer to this as the *contact* holder. In the other configuration the long edges of the film were supported by nylon threads and the electrodes were placed parallel to but separated from them by a 125μm air gap. We will refer to this as the *non–contact* holder. In both cases the short edges of the film were supported by a glass wiper on one end and by a nylon thread on the other. The film was formed by drawing the wiper across the holder at about 10 μm/s, stopping at the desired length. The width of the film was 1.5 mm and the length was greater than 15mm.

To visualize the flow, we directed a 10mW Argon ion laser beam through the film and observed the light scattered by dust particles with a low power microscope. The field of view covered the whole width of the film. The laser was also used to measure the reflectivity of the film as a function of wavelength, which we fit to theory to find the thickness [3,12].

RESULTS

When a constant voltage is applied to the electrodes in the contact holder, there is a well defined critical voltage above which the film convects steadily in a roll

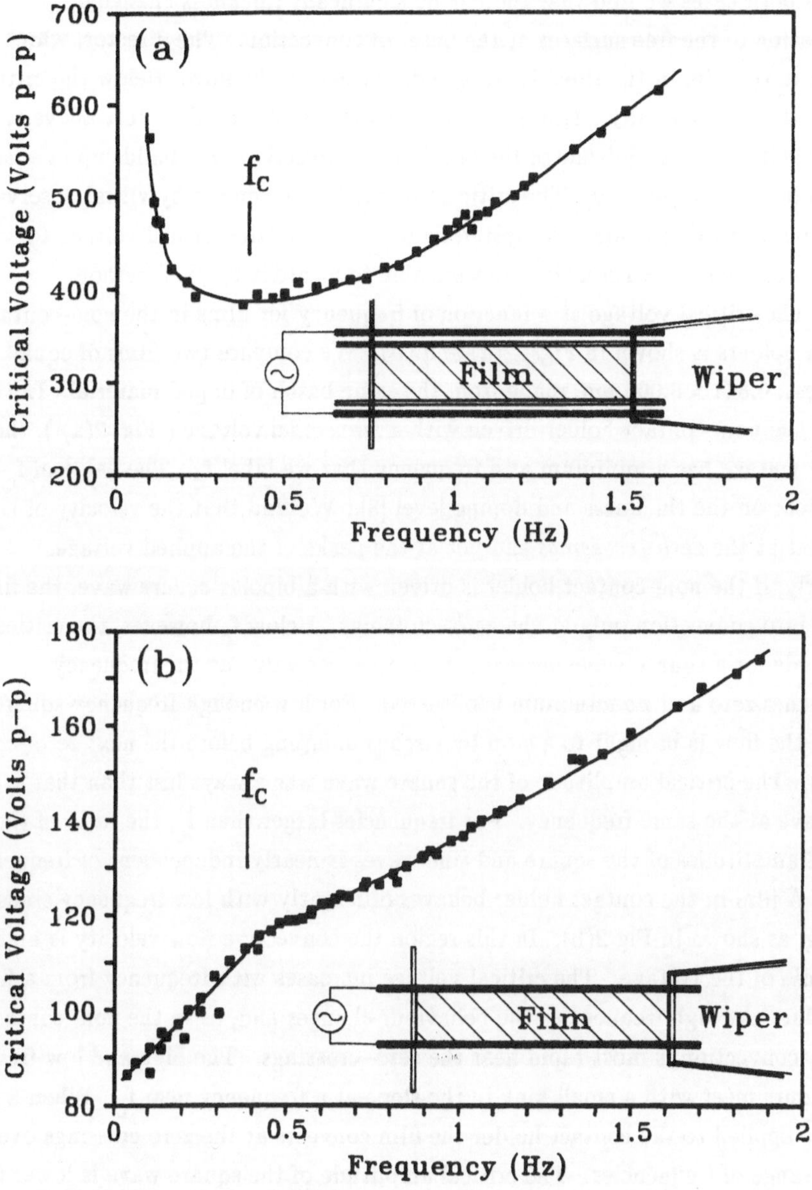

FIG. 2. The frequency dependence of the critical voltage for films in the non–contact (a) and contact (b) film holders (insets). The solid lines are guides to the eye. The transition is less sharp at low frequencies in (b) resulting in increased scatter in the critical voltage values.

pattern, as shown in Fig. 1. With an AC applied voltage, the rolls rock back and forth, reversing their senses of rotation at the frequency of the applied voltage. In the non–contact holder the behaviour is similar, except that no convection is observed at DC. In both cases we find that there is no significant thickness change or deformation of the free surfaces at the onset of convection. The director, which is normal to the film in the smectic A, is undisturbed by the flow. Below the critical voltage there is generally a slow disorganized stirring of the film. Just above the critical voltage the amplitude of the oscillating convective flow builds up over several periods of the applied field. The critical voltage is determined by visual observation of the onset of this motion. No hysteresis was found in the critical voltage to within the level of accuracy with which we were able to locate it by this method.

The critical voltage as a function of frequency for films in the non–contact and contact holders is shown in Figs. 2(a) and (b). We compare two films of equal thickness, 0.841 ± 0.006 μm, made from the same batch of doped material. In the case of the non–contact holder driven with a *sinusoidal* voltage (Fig. 2(a)), the critical voltage has a minimum at a frequency that we label f_c. The value of f_c is dependent on the thickness and doping level [3]. We find that the velocity of the flow is largest at the *zero–crossings* and not at the peaks of the applied voltage. Similarly, if the non–contact holder is driven with a bipolar *square* wave, the film breaks into convection only at the zero–crossings. Below f_c however, the critical amplitude for a square wave decreases to a constant value as the frequency approaches zero and *no* minimum is observed. For low enough frequency square waves, the flow is brought to a stop by viscous damping before the next zero–crossing arrives. The critical amplitude of the square wave was always less than that of the sine wave at the same frequency. For frequencies larger than f_c, the ratio of the critical amplitudes of the square and sine waves is nearly independent of frequency.

A film in the contact holder behaves differently with low frequency sinusoidal driving, as shown in Fig 2(b). In this region the convective flow velocity is highest at the *peaks* of the voltage. The critical voltage increases with frequency from a finite DC value. At high frequencies the behaviour changes and, as in the non–contact holder, convection is most rapid near the zero–crossings. The high and low frequency behaviours meet with a small kink in the slope at a frequency near f_c. When a square wave is applied to the contact holder the film convects at the zero crossings over the whole range of frequencies. The critical amplitude of the square wave is lower than that of a sine wave at all frequencies and above f_c they are proportional, just as in the non–contact holder. No kink in the slope occurs in the critical amplitude curve in the square wave case.

DISCUSSION

The above observations are consistent with a model in which there are two distinct driving mechanisms for convection. One mechanism operates in the non–contact holder and in the contact holder above f_c. In these cases we observed that the convection rolls oscillate at the same frequency as the applied voltage, but out of phase with it. This clue suggests the following mechanism. The applied field causes the charged impurities already present in the film to separate into screening layers. If the screening layers take a time τ to form, they cannot follow the field if it oscillates at a frequency above $1/\tau$. A phase delay between the screening charges and the field gives rise to an oscillating force [6,7]. At frequencies much higher than $1/\tau$, the screening layers do not have time to form and the oscillating force is small. At frequencies much less than $1/\tau$ the screening layers follow the field so that the phase delay, and hence the force, is again small. The force is maximized at a frequency near $1/\tau$. The critical voltage at any given frequency is the voltage required to produce a force which is sufficiently large to overcome the viscosity and destabilize the film. Thus the critical voltage has a minimum at a frequency $f_c \approx 1/\tau$ in the non–contact holder. The square wave observations are similarly explained: the force is largest when the field changes most rapidly, hence the film convects at the zero–crossings. We call this the *intrinsic* mechanism, since it does not depend on injected charge.

In the contact holder below f_c, the observations suggest that a second mechanism operates in which charge *injection* is important [4,5,14]. Charges in the form of ions injected from the electrodes are superimposed on the intrinsic charges, forming a distribution which can become unstable at high fields. Hence, convection occurs at the peaks of the applied voltage. At f_c there is a crossover of the two mechanisms. Below f_c, the film may still be destabilized by the intrinsic mechanism if the field changes rapidly enough, as it does at the zero–crossings of a square wave of sufficient amplitude.

We have also made some preliminary measurements of the current through the film. Fig. 3 shows the IV curve when the voltage is swept through the onset of convection in a contact holder. At the critical voltage the current shows a small upward kink. There is a large hysteresis in the shape of the IV curve, with higher currents being produced during increasing voltage and lower ones on decreasing voltage. The area of this hysteresis loop over successive voltage sweeps decreases exponentially with a time constant of approximately 3 hours. A similar hysteresis has been observed in DC convection experiments in nematics [13], where it has been attributed to the contribution of the diffusion of charge to the current.

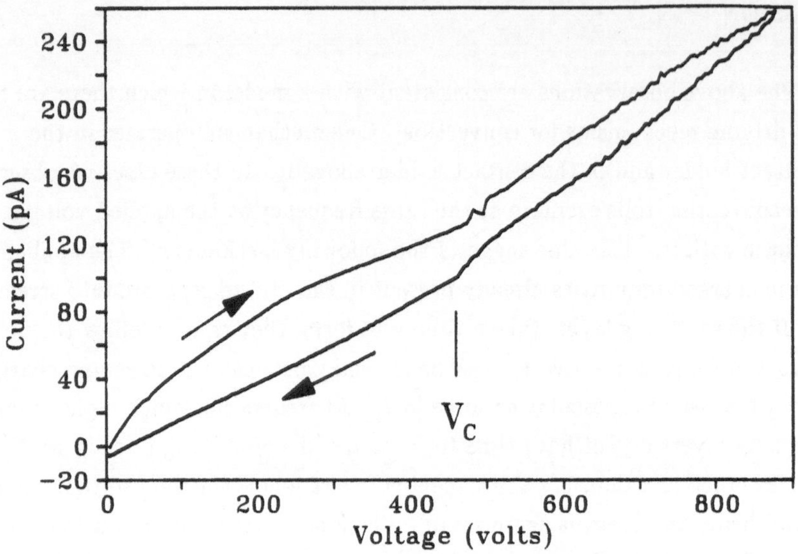

FIG 3. The current through the film, showing the hysteresis as the voltage is swept up and down at 2.25 V/s. An upward kink occurs at the critical voltage, reflecting the increased transport of charge due to convection.

CONCLUSION

The freely suspended smectic A film offers a uniquely two–dimensional convecting system. The bifurcation at the onset of convection appears to be forward and is unaccompanied by thickness changes or molecular reorientation. The data suggest that there are two distinct driving mechanisms which operate in different regimes of frequency. In future experiments we plan to study in more detail the mechanism of the convection as well as stability and wavenumber selection [10] in the roll pattern.

This work was supported by the Natural Sciences and Engineering Research Council of Canada and by the Province of Ontario through the Ontario Laser and Lightwave Research Centre.

REFERENCES

[1] L. M. Blinov, *Electro–optical and Magneto–optical Properties of Liquid Crystals*, Wiley, NY, (1983).

[2] R. Pindak and D. Moncton, *Phys. Today*, **35**(5), 57 (1982).

[3] S. W. Morris, J. R. de Bruyn and A. D. May, submitted to Physical Review.

[4] N. J. Felici, *J. Phys.* (Paris), **37**, C1–117, (1979).

[5] J. C. LaCroix, P. Atten and E. J. Hopfinger, *J. Fluid Mech.*, **69**, 539, (1975).
[6] M. I. Barnik, L. M. Blinov, S. A. Pikin and A. N. Trufanov, *Sov. Phys. JETP*, **45**, 396, (1977).
[7] P. Atten, B. Malraison and S. Ali Kani, *J. Electrost.*, **12**, 477, (1982).
[8] S. Faetti, L. Fronzoni and P. A. Rolla, *J. Chem. Phys.*, **79**, 5054, (1983).
[9] S. Faetti, L. Fronzoni and P. A. Rolla, *J. Phys.* (Paris), **40**, C3–497, (1979).
[10] G. Ahlers, *Complex Systems, Vol. 7, Santa Fe Institute Studies in the Sciences of Complexity*, ed. Dan Stein, Addison, Reading Mass., (1989).
[11] R. Cocco, F. Gaspard and R. Herino, *J. de Chim. Phys.*, **76**, 383, (1979).
[12] C. Rosenblatt and N. M. Amer, *Appl. Phys. Lett.*, **36**, 432, (1980).
[13] M. Nakagawa and T. Akahane, *J. Phys. Soc. Jpn.*, **52**, 3773, (1983) and *J. Phys. Soc. Jpn.*, **52**, 3782, (1983).
[14] R. J. Turnbull, *J. Phys. D*, **6**, 1745, (1973).

SYMMETRY BREAKING IN NONLINEAR CONVECTION

N. O. Weiss

Department of Applied Mathematics and Theoretical Physics
University of Cambridge, Cambridge, UK

ABSTRACT

In idealized two-dimensional convection the system possesses D_2 symmetry and the fundamental solution (a single roll) has point symmetry about its axis. This symmetry can be broken in a pitchfork bifurcation giving rise to mixed-mode solutions connecting branches of pure single-roll and two-roll solutions. Numerical experiments provide examples of symmetry-breaking in the nonlinear regime, where the bifurcation structure can be related to physical properties of the flow. In thermosolutal convection oscillations lose first spatial and then temporal symmetry. In compressible magnetoconvection there are transitions from steady single-roll solutions to mixed-mode quasiperiodic and periodic solutions, and also from single-roll standing wave to two-roll travelling wave solutions. Three-dimensional convection allows a richer variety of transitions.

1. INTRODUCTION

Interest in temporal chaos has been superseded by the need to understand changing spatial structures. This is especially true of three-dimensional convection, where complicated structures appear as a result of subsidiary bifurcations linking different solution branches. We can begin by considering interactions between two pure modes which lead to mixed-mode solutions, taking simple two-dimensional behaviour as a model where symmetries and symmetry-breaking bifurcations can be classified. The examples discussed here are drawn from numerical experiments carried out in collaboration with Derek Brownjohn, Neal Hurlburt, Dan McKenzie, Dan Moore, Michael Proctor and Janet Wilkins. I shall focus on thermosolutal convection and then go on to discuss compressible magnetoconvection and to comment briefly on three-dimensional convection. Our concern here is with mathematics rather than with physics but the most interesting feature of these numerical experiments is the interplay between physical behaviour and the constraints imposed by bifurcation theory.

Nonlinear Evolution of Spatio-Temporal Structures in Dissipative Continuous Systems
Edited by F.H. Busse and L. Kramer
Plenum Press, New York, 1990

2. SYMMETRIES OF TWO-DIMENSIONAL CONVECTION

Let us first consider idealized Boussinesq convection in the region $\{0 \leq x \leq \lambda;\ 0 < z < 1\}$ with stress-free boundaries at $z = 0, 1$. At the lateral boundaries we may either impose reflection symmetry or require that solutions be periodic with period λ; these choices correspond to Z_2 and $O(2)$ symmetry respectively. For the rest of this section and in section 3 we restrict our attention to systems with Z_2 symmetry. Then the normalized temperature $T(x, z, t) = 1 - z + \theta(x, z, t)$ and the velocity is described by a stream function

$$\psi(x, z, t) = \sum_m \sum_n \psi_{mn}(t) \sin \frac{m\pi x}{\lambda} \sin n\pi z \tag{1}$$

with a corresponding expression for θ. Now the system has both left-right and up-down symmetry. Thus it has the symmetry of a rectangle, described by the group D_2, with symmetries

$$m_x : \quad (x, z) \to (\lambda - x, z)\ ,\ (\psi, \theta) \to (-\psi, \theta)\ , \tag{2}$$

$$m_z : \quad (x, z,) \to (x, 1 - z)\ ,\ (\psi, \theta) \to (-\psi, -\theta)\ , \tag{3}$$

$$i = m_x m_z : \quad (x, z) \to (\lambda - x, 1 - z)\ ,\ (\psi, \theta) \to (\psi, -\theta)\ . \tag{4}$$

Solutions may have one or other of these symmetries. The point symmetry i implies that $(m + n)$ is even in (1), corresponding e.g. to a single roll in the domain. Reflection symmetry m_x about the plane $x = \frac{1}{2}\lambda$ implies that m is even in (1), as for a solution with two rolls. Similarly, m_z (reflection about $z = \frac{1}{2}$), requires n even, as for two stacked rolls, while the full group symmetry D_2 requires that m, n both be even, as for a four roll solution. Now there is a trivial solution $\psi = \theta = 0$ with D_2 symmetry and for any choice of λ there is a fundamental eigenfunction of the linear problem with symmetry i, corresponding to a single-roll solution, which becomes unstable when the thermal Rayleigh number $R_T = R_0(\lambda)$. Since the system preserves this symmetry (Veronis 1966) two branches of point-symmetric single-roll solutions, related by m_x or m_z, emerge from the pitchfork bifurcation at R_0. We can choose λ so that the fundamental mode is the first to become unstable, giving rise to stable single-roll solutions. Then any change of symmetry is associated with a secondary bifurcation leading to the formation of mixed-mode solutions. We shall be concerned with interactions between single-roll and two-roll solutions (with symmetries i and m_x respectively) and the appearance of mixed-mode solutions in which both symmetries are broken.

Stable, steady mixed-mode solutions of this type appear, for example, in Boussinesq magnetoconvection (Nagata, Proctor & Weiss 1989) and the interacting solution branches can be described by evolution equations of the form

$$\left.\begin{aligned} \dot{A}_1 &= \mu_1 A_1 - A_1^3 - A_1 A_2^2\ , \\ \dot{A}_2 &= \mu_2 A_2 - A_2^3 - c A_1^2 A_2 \end{aligned}\right\} \tag{5}$$

(Stuart 1962; Segel 1962; Knobloch & Guckenheimer 1983) which are simplified versions of the normal form equation for a double pitchfork bifurcation (Guckenheimer & Holmes 1986). Here A_1, A_2 represent the amplitudes of pure single-roll and two-roll solutions, while μ_1, μ_2 are control parameters. Stable mixed-mode solutions can exist if $0 < c < 1$.

3. TIME-DEPENDENT THERMOSOLUTAL CONVECTION

In a layer with an imposed vertical gradient in solute concentration, measured by a solutal Rayleigh number R_S which is large, instability sets in at a Hopf bifurcation provided the ratio τ of the solutal to the thermal diffusivity is sufficiently small. Time-dependent convection provides a richer variety of symmetry-breaking bifurcations. The scaled solute concentration $S(x,z,t) = \frac{1}{2} - z + \Sigma(x,z,t)$ and the scaled density $\rho = R_S S - R_T T$. Dynamical behaviour is particularly interesting in the regime where $R_S >> R_T - R_S > 0$: then the static layer is almost neutrally stratified but any motion immediately leads to strong horizontal density gradients owing to the different diffusion rates.

Nonlinear behaviour has been investigated in a series of numerical experiments (Veronis 1968; Huppert & Moore 1976; Knobloch et al. 1986b) which focussed on complicated time-dependence in a system with imposed point-symmetry. Symmetry-breaking has recently been investigated by Moore, Weiss & Wilkins (1989c). It is convenient to describe nonlinear behaviour by two global quantities, the thermal Nusselt number

$$N_T = 1 - \lambda^{-1} \int_0^\lambda (\partial\theta/\partial z) dx \ , \tag{6}$$

measured at $z = 0$ or at $z = 1$, and the rms velocity V such that

$$V^2 = (2\lambda)^{-1} \int_0^1 \int_0^\lambda |\nabla\psi|^2 dx dz \ . \tag{7}$$

These computations used a centred finite-difference scheme with second order accuracy in space and time and the number of mesh intervals $N_z = 32$. Bifurcations are affected by discretization but careful checks in which N_z was varied in the range $16 \leq N_z \leq 128$ confirm that the bifurcation structure obtained for $N_z = 32$ is robust, though further refinements of the mesh can shift individual bifurcation values of R_T by about 0.2% (Moore, Weiss & Wilkins 1989a, b).

The computations covered behaviour with $R_S = 10^4$, $\tau = 10^{-\frac{1}{2}}$, $\lambda = 1.5$ and a Prandtl number $\sigma = 1$. Stable steady solutions existed for $R_T \geq 10700$ and always retained point symmetry. A branch of point symmetric periodic solutions bifurcates from the trivial solution at $R_T \approx 77200$. Solutions on this branch possess symmetry in time as well as space: a solution of period P has the symmetry

$$t_z : \quad (x,z,t) \to (x, 1-z, t+\tfrac{1}{2}P) \ , \ (\psi,\theta,\Sigma) \to (-\psi,-\theta,-\Sigma) \ , \tag{8}$$

so that advancing time by half a period is equivalent to the reflection symmetry m_z. This temporal symmetry is broken in a pitchfork bifurcation at $R_T = 9150$, where the $S1$ solutions give way to $P1$ solutions which subsequently undergo period-doubling bifurcations which eventually lead to (spatially symmetric) chaos associated with the Shil'nikov mechanism (Knobloch et al. 1986b; Moore et al. 1989b). Now it follows from (7) and (8) that the rms velocity V has a period $\frac{1}{2}P$ and that

$$N_T(z=0,t) = N_T(z=1, t+\tfrac{1}{2}P) \ . \tag{9}$$

Since $N_T(z=0,t) = N_T(z=1,t)$ for any solution with the point symmetry i, from (4), the Nusselt number N_T has period $\frac{1}{2}P$ for oscillatory solutions possessing the symmetries t_z and i. In addition, from (4) and (8) we obtain the symmetry

$$t_x : \quad (x,z,t) \to (\lambda - x, z, t+\tfrac{1}{2}P), \quad (\psi,\theta,\Sigma) \to (-\psi,\theta,\Sigma) \tag{10}$$

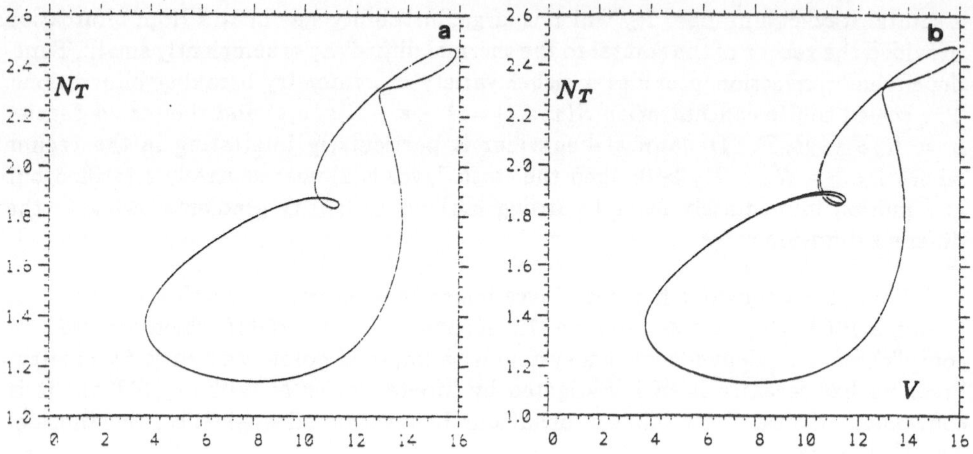

Figure 1. Point-symmetric solutions in thermosolutal convection. Orbits in the $(V N_T)$-plane with $N_z = 32$ for (a) $R_T = 11080$ $(S1)$, (b) $R_T = 11090$ $(P1)$.

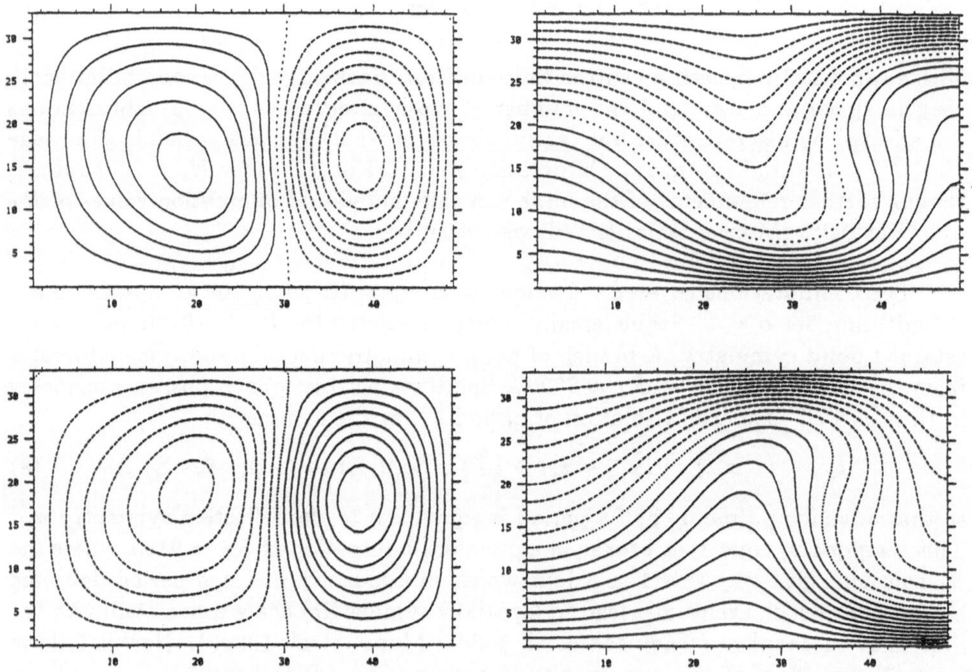

Figure 2. Spatial asymmetry. Streamlines and isotherms for an $S1$ solution with $R_T = 11100$ at two times separated by an interval $\frac{1}{2}P$. Note the loss of point-symmetry and retention of the symmetry t_z.

which relates a shift by half a period in time to the reflection symmetry m_x.

When point symmetry is explicitly imposed, with $N_z = 32$, there is a periodic $S1$ solution at $R_T = 11080$. Both V and N_T have period $\frac{1}{2}P$ and Figure 1(a) shows the limit cycle projected onto the VN_T-plane. The symmetry t_z is broken by $R_T = 11090$, where there is a $P1$ solution for which both V and N_T have period P as shown in Figure 1(b). At $R_T = 11100$ the period has doubled but the bifurcation points are shifted as the mesh is refined so that there is an $S1$ solution at $R_T = 11100$ for $N_z \geq 64$ (Moore *et al.* 1989a).

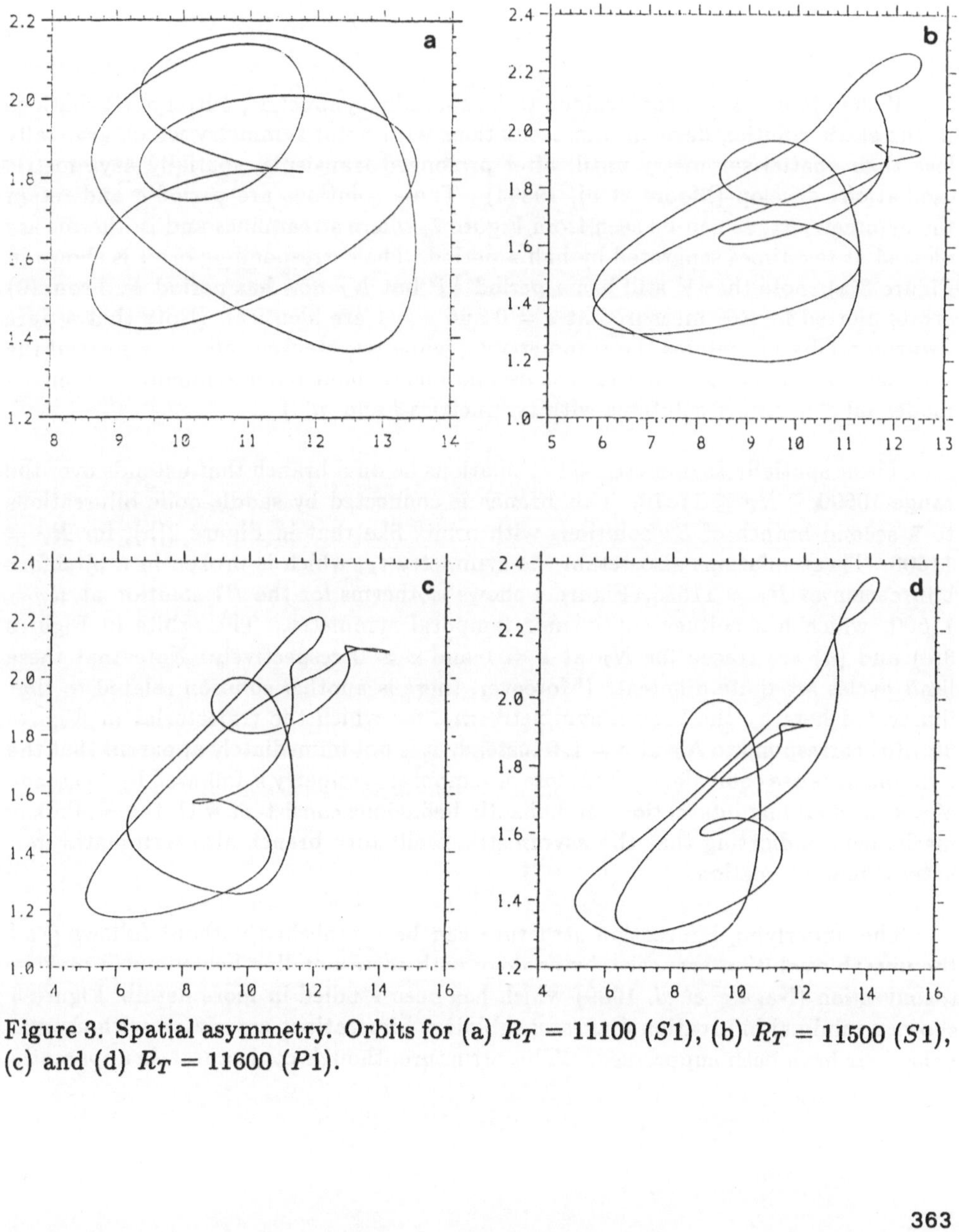

Figure 3. Spatial asymmetry. Orbits for (a) $R_T = 11100$ ($S1$), (b) $R_T = 11500$ ($S1$), (c) and (d) $R_T = 11600$ ($P1$).

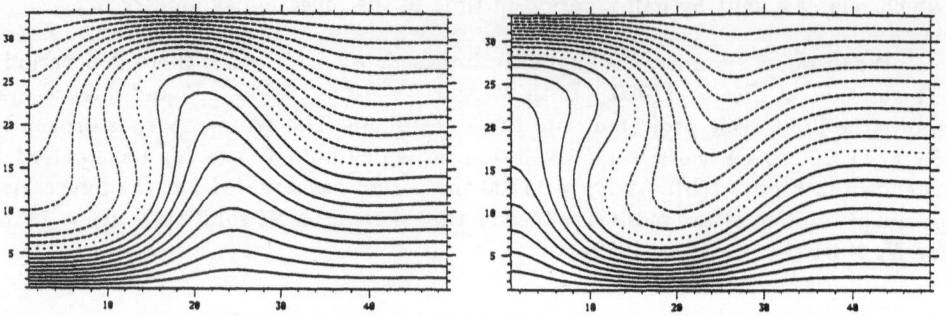

Figure 4. Loss of spatial and temporal symmetry. Isotherms for $P1$ solution at two times separated by an interval $\frac{1}{2}P$.

If the system is not constrained to be spatially symmetric, initial perturbations to the static solution develop into oscillations with point symmetry which gradually lose their spatial symmetry until, after prolonged transients, spatially asymmetric oscillations develop (Moore et al. 1989a). These solutions are periodic and retain the symmetry t_z, as can be seen from Figure 2, where streamlines and isotherms are plotted at two times separated by half a period. The corresponding orbit is shown in Figure 3(a): note that V still has a period $\frac{1}{2}P$ but N_T now has period P. From (9) orbits plotted for N_T measured at $z = 0$ and $z = 1$ are identical. (Note that a pure two-roll oscillation retains the symmetry t_z, while two stacked rolls have instead the symmetry t_x. This suggests that the oscillations without point symmetry are mixed modes involving pure solutions with symmetries i and m_x.)

These spatially asymmetric $S1$ oscillations lie on a branch that extends over the range $10690 \leq R_T \leq 11570$. This branch is connected by saddle-node bifurcations to a second branch of $S1$ solutions with orbits like that in Figure 3(b), for $R_T = 11500$. These solutions also retain the symmetry t_z, which is broken in a pitchfork bifurcation at $R_T \approx 11535$. Figure 4 shows isotherms for the $P1$ solution at $R_T = 11600$, which has neither spatial nor temporal symmetry. The orbits in Figures 3(c) and (d) are traced for N_T at $z = 1$ and $z = 0$ respectively. Note that these limit cycles are quite different. (Moreover, there is another solution related to that illustrated here by the broken symmetry m_z, for which the trajectories in Figures 3(c),(d) correspond to N_T at $z = 1, 0$ instead. It is not immediately apparent that the two solutions are equivalent.) This loss of temporal symmetry is followed by a cascade of period-doubling bifurcations and chaotic behaviour consistent with the Shil'nikov mechanism, indicating that the asymmetric oscillatory branch also terminates in a heteroclinic bifurcation.

The underlying bifurcation structure can be established without following all the unstable solution branches, by analogy with corresponding behaviour in magnetoconvection (Nagata *et al.* 1989) which has been studied in more detail. Figure 5 shows a minimal bifurcation diagram; subidiary bifurcations associated with chaotic behaviour have been suppressed. This structure, though conjectural, is apparently

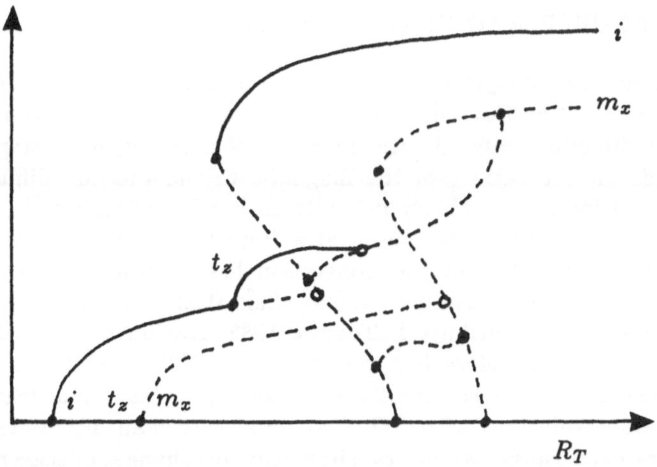

Figure 5. Minimal conjectural bifurcation diagram for mixed-mode thermosolutal convection. Full (broken) lines indicate stable (unstable) periodic or steady solutions, with symmetries indicated.

unique. The structure of the pure single-roll and two-roll branches can be derived from the Bogdanov normal form equations (Knobloch & Proctor 1981; Coullet & Spiegel 1983; Guckenheimer & Holmes 1986) by adding higher order terms (Dangelmayr, Armbruster & Neveling 1985). The mixed-mode oscillatory branch ends in a heteroclinic bifurcation which requires the existence of a mixed-mode steady branch and consideration of the relevant eigenvalues shows that there has to be a second mixed-mode steady branch. The full structure for thermosolutal convection corresponds to the unfolding of a codimension-three bifurcation described by a simplified version of the normal form equations for a double Bogdanov bifurcation of codimension four with higher-order terms added to describe the stable steady branches:

$$\left.\begin{aligned}\ddot{A}_1 + (\nu - C_1A_1^2 + F_1A_2^2)\dot{A}_1 + (\mu_1 + D_1A_1^2 + G_1A_2^2 - E_1A_1^4)A_1 = 0\ ,\\ \ddot{A}_2 + (\nu - C_2A_2^2 + F_2A_1^2)\dot{A}_2 + (\mu_2 + D_2A_2^2 + G_2A_1^2 - E_2A_2^4)A_2 = 0\ .\end{aligned}\right\} \tag{11}$$

This system is clearly more complicated than (5) and it has not yet been investigated.

A further variety of behaviour appears if the lateral boundary conditions are varied. For instance, doubling the aspect ratio, so that $\lambda = 3$, introduces pure multiroll solutions corresponding to normal modes with $m = 1, 2, 3, 4$ in equation (1). The nonlinear solution obtained at $R_T = 11100$ by reflecting the asymmetric rolls in Figure 2 about the plane $x = 1.5$ is no longer attracting and asymmetric, aperiodic behaviour apparently persists indefinitely (Moore et al. 1989a). This is a simple example of spatiotemporal chaos. Furthermore, if the fixed boundaries at $x = 0$, λ are replaced by periodic boundary conditions, so that the system has $O(2)$ rather than Z_2 symmetry then travelling wave solutions are preferred (Bretherton & Spiegel 1983; Knobloch et al. 1986a). These waves are steady in the comoving frame: they do not possess the point symmetry (4) but instead have the symmetry

$$(x, z) \to (x + \tfrac{1}{2}\lambda, 1 - z)\ ,\ (\psi, \theta, \Sigma) \to (-\psi, -\theta, -\Sigma) \tag{12}$$

(Barten et al. 1989).

4. COMPRESSIBLE MAGNETOCONVECTION

In a compressible layer the pressure and density stratification destroys the up-down symmetry of the Boussinesq approximation. Any spatial symmetries therefore affect the x-direction only. In the presence of a strong magnetic field behaviour now depends on the ratio ς of the magnetic to the thermal diffusivity; if $\varsigma > 1$ the trivial solution loses stability in a pitchfork bifurcation but if $\varsigma < 1$ throughout the layer convection typically sets in at a Hopf bifurcation. Time-dependent two-dimensional compressible magnetoconvection has been systematically investigated in numerical experiments using a two-step Lax–Wendroff code with periodic lateral boundary conditions (Hurlburt & Toomre 1988; Hurlburt et al. 1989; Weiss et al. 1990). The system therefore has the symmetry $O(2) = SO(2) \times Z_2$. There exist standing wave and steady solutions with Z_2 symmetry as well as travelling wave and modulated wave solutions with $SO(2)$ symmetry. I shall discuss examples of both types in parameter ranges where transitions involve changes of scale and mixed-mode solutions.

(a) Secondary Oscillations in a Stratified Layer

In a stratified layer with uniform electrical and thermal conductivities the ratio ς is proportional to the density and increases with depth. Convection in a stratified layer with $\varsigma_0 \le \varsigma \le 11\varsigma_0$ and $\lambda = \frac{4}{3}$ has been studied in a parameter range where all stable solutions have Z_2 symmetry (Weiss et al. 1990); we therefore assume reflection symmetry about $x = 0$, without loss of generality. The magnetic field is described by a flux function $A = \lambda - x + \chi(x, z, t)$ and pure two-roll solutions have the reflection symmetry

$$m_x : \quad (x, z) \to (\lambda - x, z) \ , \ (\psi, \theta, \chi) \to (-\psi, \theta, -\chi) \ , \tag{13}$$

while pure four-roll solutions have in addition the symmetry

$$m'_x : \quad (x, z) \to (\lambda' - x, z) \ , \ (\psi, \theta, \chi) \to (-\psi, \theta, -\chi) \ , \tag{14}$$

where $\lambda' = \frac{1}{2}\lambda$. In the simplest case with $\varsigma_0 = 1$ two branches of steady two-roll solutions with the symmetry m_x emerge from a pitchfork bifurcation; these branches are stable and related by the symmetry m'_x. When $\varsigma_0 = 0.1$ there is a Hopf bifurcation leading to pure two-roll oscillations of period P with the symmetry m_x and the temporal symmetry

$$t'_x : \quad (x, z, t) \to (\lambda' - x, z, t + \tfrac{1}{2}P) \ , \ (\psi, \theta, \chi) \to (-\psi, \theta, -\chi) \ . \tag{15}$$

A layer with $\varsigma < 1$ at the top and $\varsigma > 1$ at the bottom (which models behaviour in a sunspot) is more interesting. Interactions between two-roll and four-roll solutions have been followed for a layer with $\varsigma_0 = 0.2$, when the primary bifurcations from the trivial solution are both pitchforks. Four-roll solutions were tracked by restricting the calculation to a box of width λ' and there is a branch of steady solutions with the symmetries m_x, m'_x as indicated schematically in Figure 6. The corresponding branch of two-roll solutions with symmetry m_x is stable over the range $3100 \le R_T \le 45500$ but solutions change in form as subsidiary rolls develop in the flux sheets. We

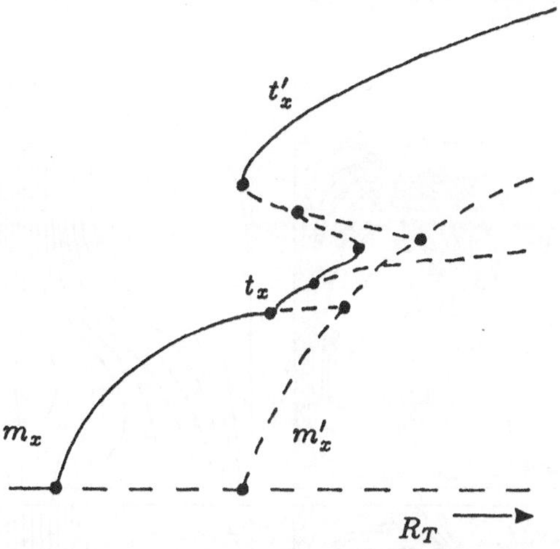

Figure 6. Schematic bifurcation diagram for magnetoconvection in a stratified compressible layer, with symmetries indicated.

conjecture that it terminates in a pitchfork bifurcation on the four-roll branch at $R_T \approx 48000$. (Since there is no up-down symmetry to break this transition does not require an intermediate bifurcation leading to the formation of a mixed-mode branch.)

The steady two-roll solutions lose stability in a Hopf bifurcation at which the symmetry m_x is broken giving rise to a branch of periodic solutions with the temporal symmetry

$$t_x: \quad (x,z,t) \to (\lambda - x, z, t + \tfrac{1}{2}P) \ , \ (\psi,\theta,\chi) \to (-\psi,\theta,-\chi) \ . \tag{16}$$

This branch rapidly undergoes a second Hopf bifurcation at which the symmetry t_x is broken, leading to quasiperiodic oscillations which lose stability in a saddle-node bifurcation at $R_T \approx 47000$. Trajectories are then attracted to a second branch of periodic oscillations which is stable for $R_T \geq 44500$. Solutions on this branch have four rolls with the symmetry m_x which are modulated in time without reversing their sense of motion, as illustrated in Figure 7, and possess the temporal symmetry t'_x. Figure 6 shows a minimal bifurcation structure for this problem. Note that this conjectural pattern would have been much simpler if the symmetry m_x had been explicitly imposed, as in the previous section.

(b) Travelling Waves in a Strong Magnetic Field

In a weakly stratified layer with a strong magnetic field (such that the ratio of gas pressure to magnetic pressure $\beta = 6$ for the static solution) there is a transition from stable two-roll standing wave solutions to stable four-roll travelling wave solutions for $\varsigma_0 = 0.071$ and $\lambda = 2$ (Hurlburt et al. 1989). The first bifurcation from the trivial solution is a Hopf bifurcation giving rise to two branches of two-roll solutions (e.g. Stewart 1988). One corresponds to standing waves (SW) with Z_2 symmetry plus the

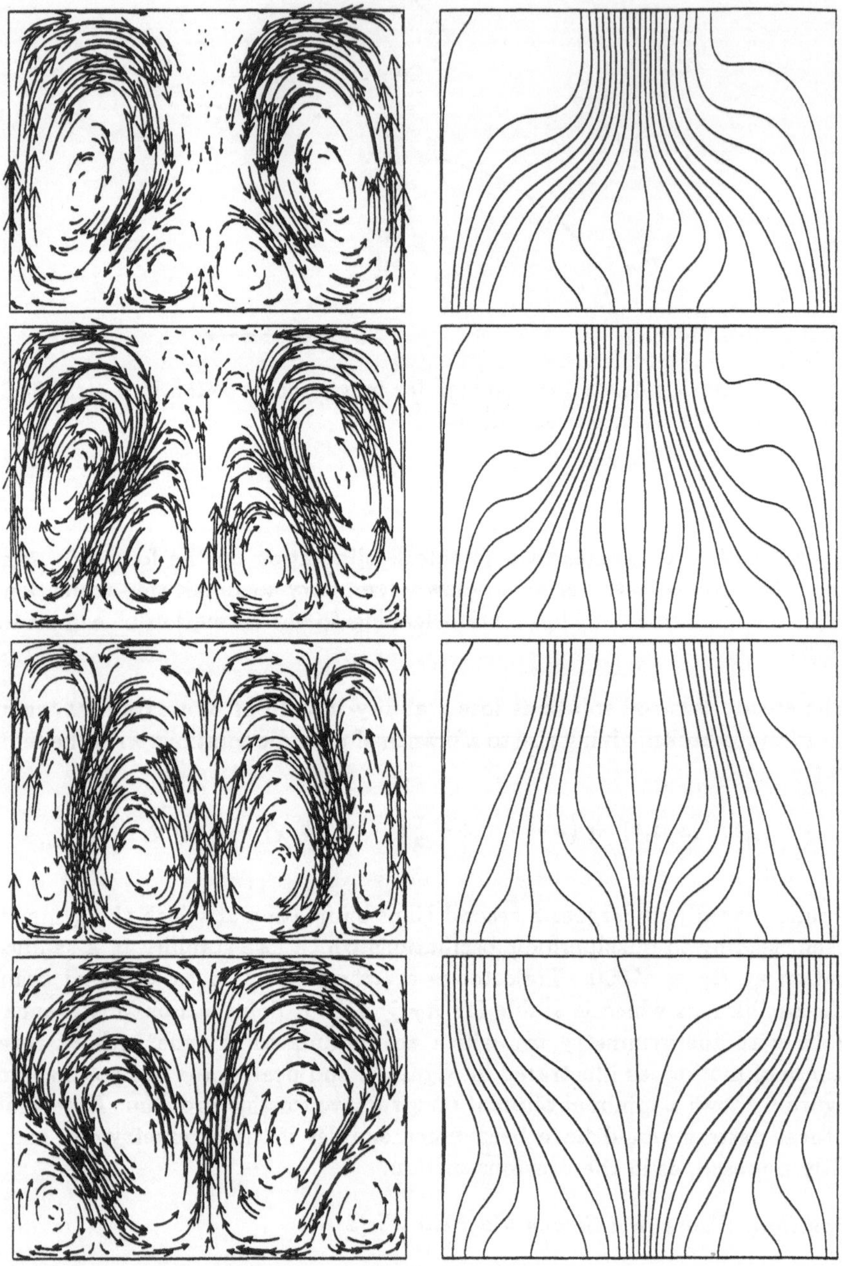

Figure 7. Modulated oscillatory convection: streaklines and fieldlines at equally spaced intervals in time for $R_T = 100000$.

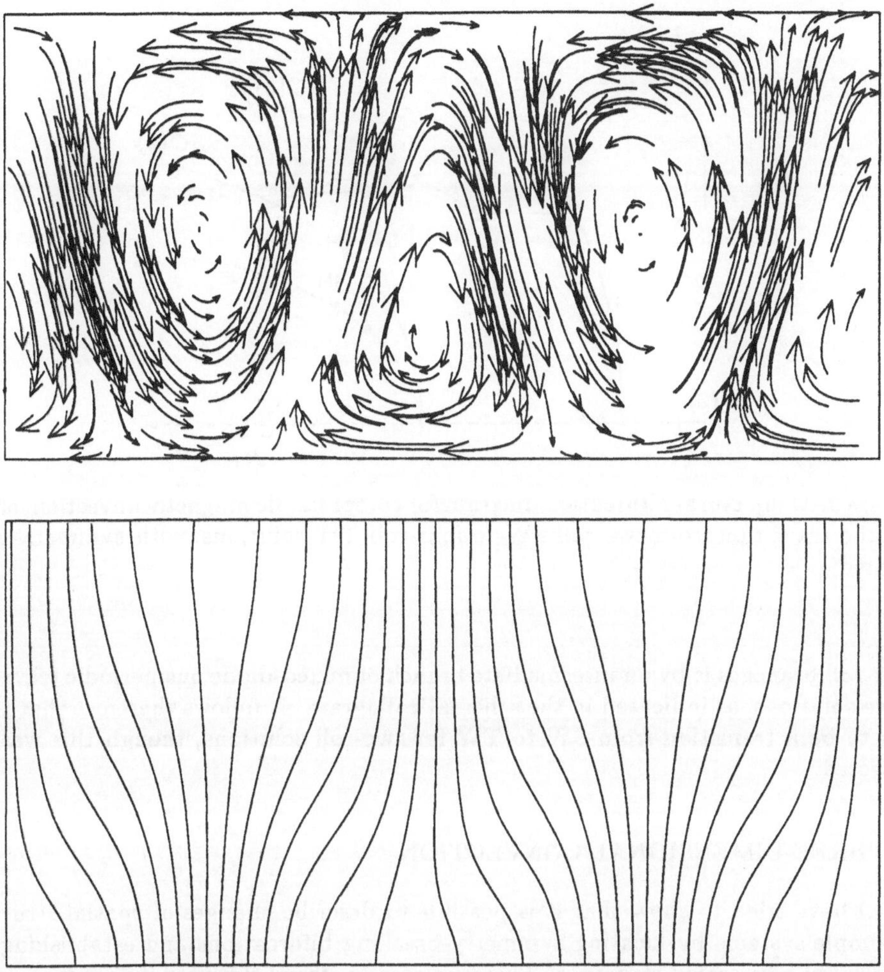

Figure 8. Streaklines and fieldlines for a rightward travelling wave solution with $R_T = 56000$ and $\beta = 6$.

spatial symmetry m_x and the temporal symmetry t'_x. The second corresponds to left and rightward travelling waves (TW) with $SO(2)$ symmetry. (Owing to lack of up-down symmetry there is no symmetry that corresponds to (12).) In the immediate neighbourhood of the Hopf bifurcation the SW solution is stable and the TW are unstable. Once again the four-roll solutions can be investigated by restricting the calculation to a box of width λ' with $0(2)$ symmetry. The SW are again stable near the Hopf bifurcation but as R_T is increased stability is transferred to TW with the form illustrated in Figure 8. This requires an intermediate branch of modulated wave (MW) solutions, which may be unstable and has not been detected.

The transition from two-roll SW solutions to four-roll TW solutions with the symmetry $x \to \lambda' + x$ (cf. Figure 8) in the comoving frame can be described by the conjectural bifurcation structure in Figure 9. The transition from the two-roll to the

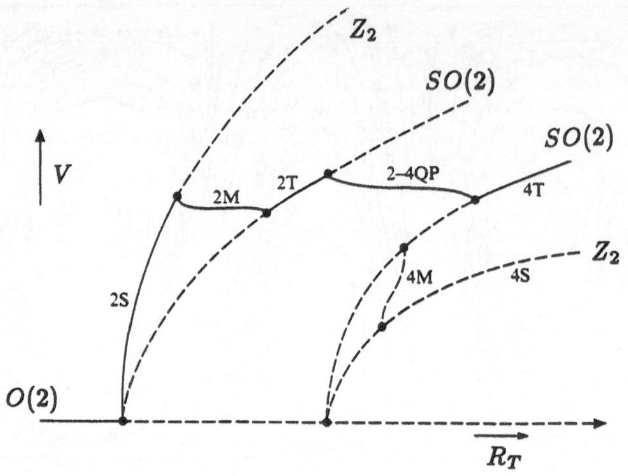

Figure 9. Conjectural bifurcation diagram for compressible magnetoconvection, showing the transition from two-roll SW to four-roll TW solutions, with symmetries indicated.

four-roll branches is by an intermediate branch of mixed-mode qusiperiodic travelling wave solutions, as indicated in the schematic diagram. It follows therefore that there has to be a transition from SW to TW for two-roll solutions, though this was not detected.

5. THREE-DIMENSIONAL CONVECTION

I have tried to show that it is possible to describe changes of spatial structure in simple systems by locating symmetry-breaking bifurcations and establishing the associated bifurcation structure. This is relatively straightforward if only two modes are involved (as here) but rapidly becomes more complicated as more modes become unstable in primary bifurcations, leading to the formation of more and more mixed modes in subsidiary bifurcations (Nagata et al. 1989). Nevertheless, it is possible, at least in principle, to describe the development of complicated spatial structures and of spatiotemporal chaos by following this procedure.

The two-dimensional systems analysed here are only a curtain-raiser to more realistic but much more difficult three-dimensional problems. For example, experiments on convection in a fluid with variable viscosity (golden syrup) show transitions involving a great variety of planforms (White 1988). A general description of symmetries and symmetry-breaking can be developed by taking over the systematic classification established for three-dimensional crystal structures (McKenzie 1988). This makes it possible to identify allowable symmetry-breaking bifurcations and to describe behaviour in their neighbourhoods.

Some forms of three-dimensional convection are amenable to a simpler approach. Convection driven by internal heating in a fluid with infinite Prandtl number is a promising example. Two-dimensional numerical experiments show that motion is dominated by sinking plumes and that the preferred horizontal scale decreases as

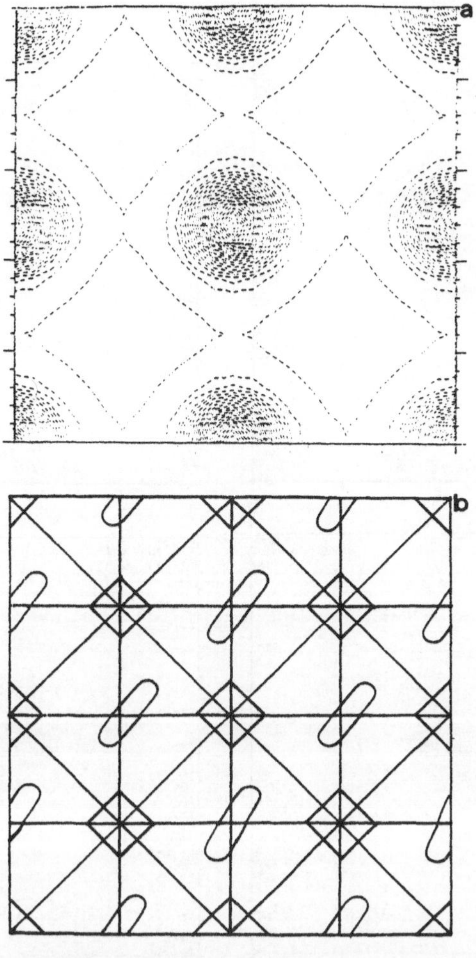

Figure 10. Steady three-dimensional convection driven by internal heating. (a) Isotherms in the mid-plane ($z = \frac{1}{2}$) for $R_T = 2 \times 10^5$. (b) Symmetries of this solution, showing dyad and tetrad axes and mirror planes.

R_T is increased. These transitions again involve one-roll and two-roll interactions (Lennie et al. 1988). For three-dimensional convection in the region $\{0 \leq x \leq \lambda;\ 0 \leq y \leq \lambda;\ 0 \leq z \leq 1\}$ the planform can conveniently be represented by the temperature field at $z = \frac{1}{2}$, where the pattern is defined by the location of sinking plumes. Thus the full three-dimensional problem can be reduced to two dimensions and becomes simpler to describe. In a series of experiments with Z_2 symmetry in x and y and $\lambda = 2$ more plumes appeared as R_T was increased (McKenzie et al. 1990). At $R_T = 2 \times 10^5$ there was a steady solution with plumes disposed on a regular square lattice of side $\frac{1}{2}\lambda$, illustrated in Figure 10(a), and D_4 symmetry about each plume, as indicated in Figure 10(b). By $R_T = 3.6 \times 10^5$ this had undergone a Hopf bifurcation in which the tetrad symmetry about the central plume was broken, leading to a periodic solution with D_2 symmetry about $x = y = \frac{1}{2}\lambda$, as shown in Figure 11. A further increase in Rayleigh number, to $R_T = 5 \times 10^5$ leads to a progressive loss of symmetry and spatiotemporal chaos (Weiss 1987). This system offers a rich variety of behaviour which has yet to be fully analysed.

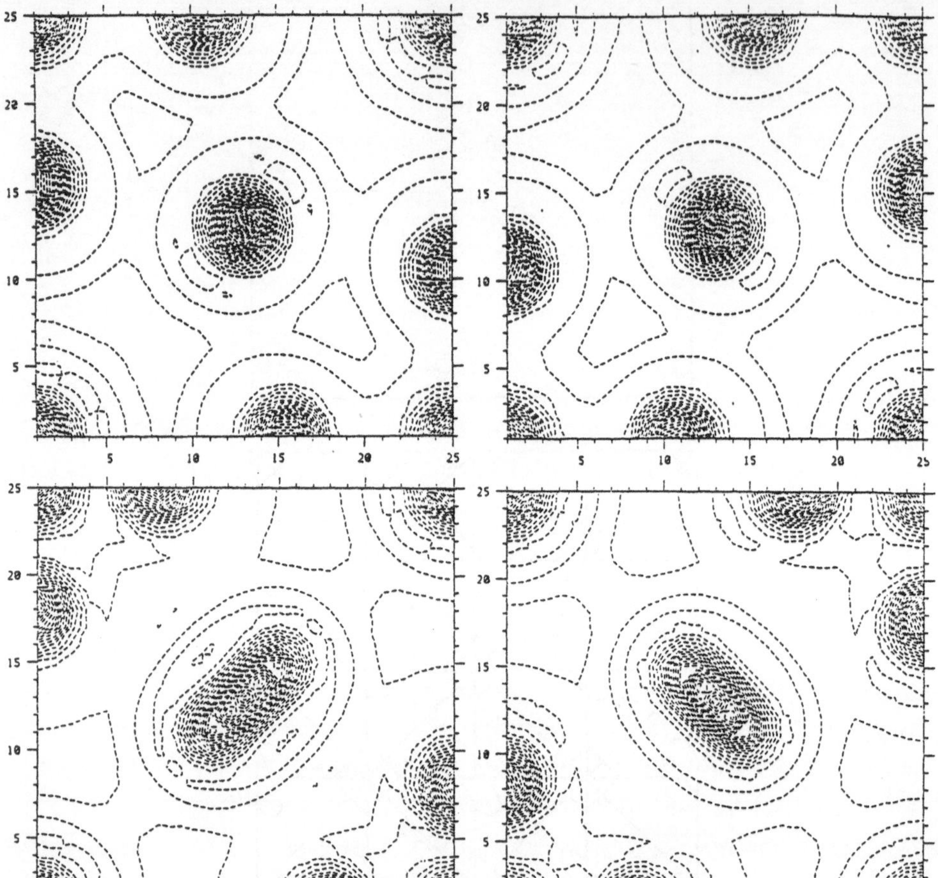

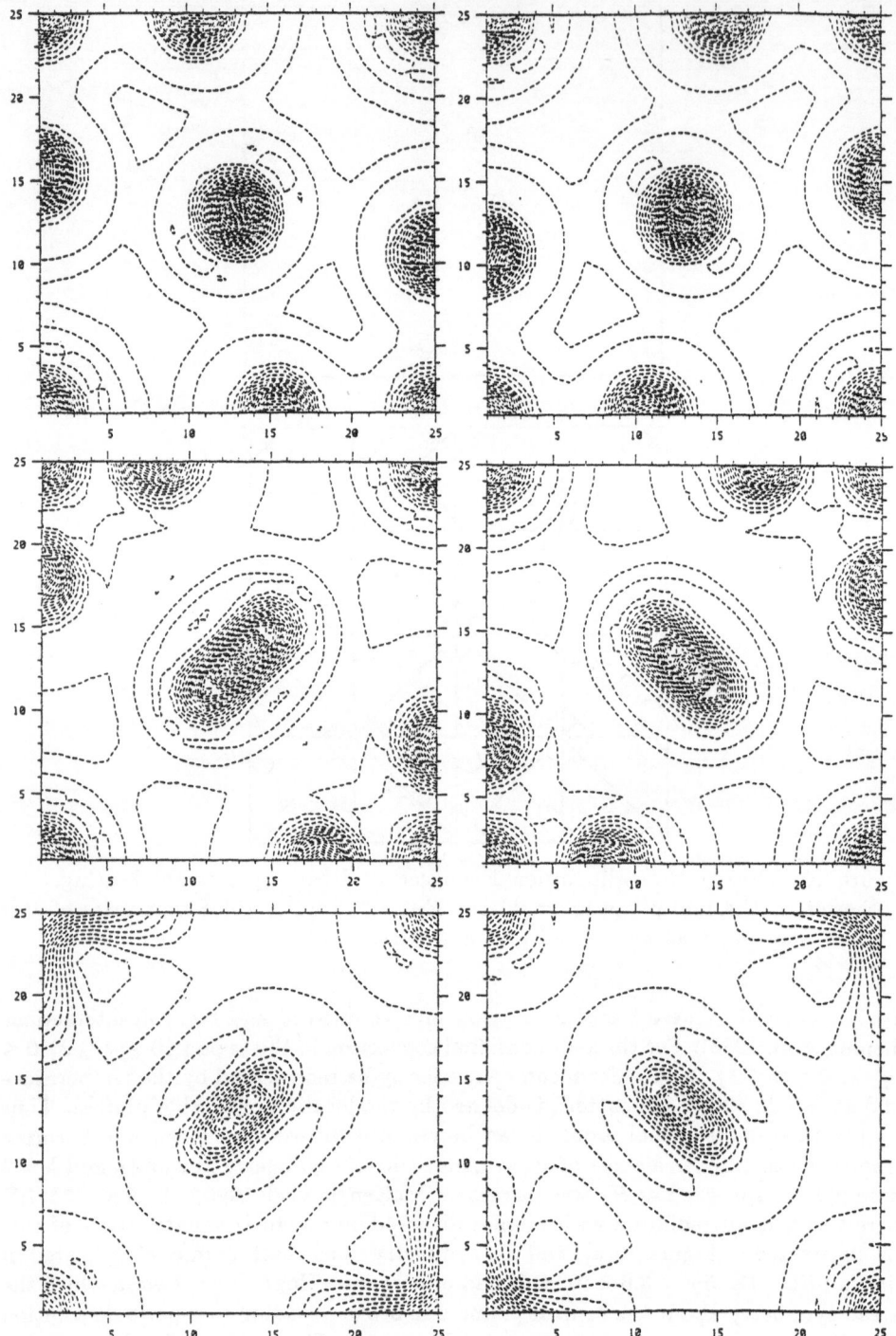

Figure 11. Isotherms in the mid-plane at equally spaced intervals in time for a periodic solution with $R_T = 5 \times 10^5$.

REFERENCES

Barten, W., Lücke, M., Hort, W. and Kamps, M., 1989, Fully developed travelling-wave convection in binary fluid mixtures, *Phys. Rev. Lett.*, 63:376.

Bretherton, C. S. and Spiegel, E. A., 1983, Intermittency through modulational instability, *Phys. Lett.*, 96A:152.

Coullet, P. H. and Spiegel, E. A., 1983, Amplitude equations for systems with competing instabilities, *SIAM J. Appl. Math.* 43:776.

Dangelmayr, G., Armbruster, D. and Neveling, M., 1985, A codimension three bifurcation for the laser with saturable absorber, *Z. Phys. B*, 59:365.

Guckenheimer, J. and Holmes, P., 1986, "Nonlinear oscillations, dynamical systems and bifurcations of vector fields," 2nd ed., Springer, New York.

Huppert, H. E. and Moore, D. R. 1976, Nonlinear double-diffusive convection, *J. Fluid Mech.*, 166:409.

Hurlburt, N. E., Proctor, M. R. E., Weiss, N. O. and Brownjohn, D. P., 1989, Nonlinear compressible magnetoconvection Part 1. Travelling waves and oscillations, *J. Fluid Mech.*, 207:587.

Hurlburt, N. E. and Toomre, J., 1988, Magnetic fields interacting with nonlinear compressible convection, *Astrophys. J.*, 327:920.

Knobloch, E., Deane, A. E., Toomre, J. and Moore, D. R., 1986a, Doubly diffusive waves, *Contemp. Maths*, 56:203.

Knobloch, E. and Guckenheimer, J., 1983, Convective transitions induced by a varying aspect ratio, *Phys. Rev. A*, 27:408.

Knobloch, E., Moore, D. R., Toomre, J. and Weiss, N. O., 1986b, Transitions to chaos in two-dimensional double-diffusive convection, *J. Fluid Mech.*, 166:409.

Knobloch, E. and Proctor, M. R. E., 1981, Nonlinear periodic convection in double-diffusive systems, *J. Fluid Mech.*, 108:291.

Lennie, T. B., McKenzie, D. P., Moore, D. R. and Weiss, N. O., 1988, The breakdown of steady convection, *J. Fluid Mech.*, 188:47.

McKenzie, D. P., 1988, The symmetry of convective transitions in space and time, *J. Fluid Mech.*, 191:287.

McKenzie, D. P., Moore, D. R., Weiss, N. O. and Wilkins, J. M., 1990, in preparation.

Moore, D. R., Weiss, N. O. and Wilkins, J. M., 1989a, Symmetry-breaking in thermosolutal convection, *Phys. Lett. A*, submitted.

Moore, D. R., Weiss, N. O. and Wilkins, J. M., 1989b, The reliability of numerical experiments: transitions to chaos in thermosolutal convection, *Nonlinearity*, submitted.

Moore, D. R., Weiss, N. O. and Wilkins, J. M., 1989c, Asymmetric oscillations in thermosolutal convection, *J. Fluid Mech.*, to be submitted.

Nagata, M., Proctor, M. R. E. and Weiss, N. O. 1989, Transitions to asymmetry in magnetoconvection, *Geophys. Astrophys. Fluid Dyn.*, in press.

Segel, L. A., 1962, The non-linear interaction of two disturbances in the thermal convection problem, *J. Fluid Mech.*, 14:97.

Stewart, I. N., 1988, Bifurcations with symmetry, *in* "New directions in dynamical systems," T. Bedford and J. W. Swift, eds, p. 233, Cambridge University Press, Cambridge.

Stuart, J. T., 1962, Non-linear effects in hydrodynamic stability, *in* "Applied Mechanics," F. Rolla and W. T. Koiter, eds, p. 63, Elsevier, Amsterdam.

Veronis, G., 1966, Large-amplitude Bénard convection, *J. Fluid Mech.*, 26:49.

Veronis, G., 1968, Effect of a stabilising gradient of solute on thermal convection, *J. Fluid Mech.*, 34:315.

Weiss, N. O., 1987, Dynamics of convection, *Proc. Roy. Soc. Lond. A*, 413:71.

Weiss, N. O., Brownjohn, D. P., Hurlburt, N. E. and Proctor, M. R. E., 1990, *Mon. Not. Roy. Astr. Soc.*, submitted.

White, D. B., 1988, The planforms and onset of convection with a temperature-dependent viscosity, *J. Fluid Mech.*, 191:247.

CHAOS AND THE EFFECT OF NOISE FOR THE DOUBLE HOPF BIFURCATION WITH 2:1 RESONANCE

Michael R.E. Proctor and David W. Hughes

Department of Applied Mathematics and Theoretical Physics
University of Cambridge
Silver Street, Cambridge CB3 9EW, England

INTRODUCTION

A major goal of bifurcation theory is the identification and classification of important degenerate bifurcations of low codimension. Although many have been extensively studied (for a non-exhaustive list see Guckenheimer & Holmes 1983) key exceptions include degenerate Hopf or oscillatory bifurcations with strong resonance. In this paper we consider the problem of 2:1 resonance, so that at the bifurcation point there are considered to be two neutrally stable infinitesimal oscillatory solutions with frequencies in the ratio 2:1. The normal form for this problem was discussed by Knobloch & Proctor (1988) who noted that it contains quadratic as well as the usual non-resonant cubic terms, leading to a system of three coupled real equations, unlike the non-resonant case where the phase equation decouples from the other two. Knobloch & Proctor investigated the normal form including all terms up to cubic order — they found many of the principal transitions, but did not discover any régime in which chaos occurred, and did not identify regions in the unfolding diagram where the dynamics was controlled by the linear and quadratic terms alone. In the present paper we rectify this deficiency by showing that when the linear growth rates $\gamma_{1,2}$ of the modes of normalised frequency $1, 2$ satisfy

$$(-\gamma_1) \gg \gamma_2 > 0$$

there are indeed bounded solutions when the normal form is truncated at quadratic order. Furthermore, in this régime the governing o.d.e.'s may be replaced by a one-dimensional map, which can be calculated analytically in certain special cases. We show that the usual period-doubling sequences etc. exist for this map, and that in addition there can be multiple stable solutions.

Finally, we show that this normal form can be extremely sensitive to the effects of additive noise, when the latter destroys an important invariant line in the phase space. Using methods developed for an allied problem by Hughes & Proctor (1990a) we give a simple characterisation of the effects of the noise in terms of the map, and thus show when it may be expected to have a significant effect on the dynamics.

REDUCTION OF THE NORMAL FORM TO A ONE-DIMENSIONAL MAP

Using the standard techniques of equivariant bifurcation theory, Knobloch & Proctor (1988) showed that the normal form for the problem at hand can be expressed as two coupled complex equations for the amplitudes A_1, A_2 of the modes proportional to

Nonlinear Evolution of Spatio-Temporal Structures in Dissipative Continuous Systems
Edited by F.H. Busse and L. Kramer
Plenum Press, New York, 1990

$e^{i\omega t}, e^{2i\omega t}$, respectively, which take the form

$$\left.\begin{aligned} \dot{A}_1 &= \gamma_1 A_1 + A_1^* A_2 + O(3), \\ \dot{A}_2 &= (\gamma_2 + i\delta)A_2 - A_1^2 e^{i\Phi} + O(3), \end{aligned}\right\} \tag{1}$$

where $O(3)$ represents terms of cubic and higher order, δ is the detuning parameter and Φ is the single non-linear coefficient that determines the nature of the dynamics near the polycritical point. Because of the time invariance of the problem equations (1) may be reduced to a three-dimensional real system. Introducing the real variables X, Y, Z according to the relations

$$X = |A_2| \cos\phi, \qquad Y = |A_2| \sin\phi, \qquad Z = |A_1|^2,$$

$$\text{where } \phi = \arg(A_2) - 2\arg(A_1) \tag{2}$$

(see, for example, Vyshkind & Rabinovich 1976), we obtain the system

$$\left.\begin{aligned} \dot{X} &= \gamma X + \delta Y - 2Y^2 + Z\cos\Phi, \\ \dot{Y} &= \gamma Y - \delta X + 2XY + Z\sin\Phi, \\ \dot{Z} &= -2Z(1+X), \end{aligned}\right\} \tag{3}$$

where we have scaled time so that $\gamma_1 = 1$ and have dropped the suffix on γ_2. These equations are to be analysed for small γ, with δ, Φ of order unity. Now when $\Phi = 0$ equations (3) are the same as those arising in a problem of three-wave resonance (see, for example, Vyshkind & Rabinovich 1976, Wersinger *et al.* 1980) and this special case has been extensively investigated in the case of small γ by Hughes & Proctor (1990b) using methods developed by Hughes & Proctor (1990a). We therefore merely summarise the important steps, which are unaltered in essence when Φ is non-zero.

When $\gamma \ll 1$ the evolution of the system (after initial transients) can be divided into two different phases which follow each other according to well-defined rules:

Slow phase. Here Z is negligible while Y is close to $\delta/2$. If on input $X = X_0$ with $0 > X_0 > -1$, X decreases according to the approximate equation $\dot{X} = \gamma(X + \delta^2/4X)$. This equation and (3c) can be solved to find Z in terms of X, and if on output $X = X_1 < -1$ then X_0 and X_1 are defined as those values where $Z \gg O(\exp(-1/\gamma))$. They are then related by

$$f(X_0) = f(X_1), \text{ where } f(X) = \int_0^X \frac{x(1+x)}{x^2 + \delta^2/4} dx. \tag{4}$$

For small γ the system spends almost all of its time in the slow phase. On exit from this phase Z becomes of order unity and X will begin to increase, at least if $|\Phi|$ is not too large. We then enter the

Fast Phase. Here the non-linear terms in the X and Y equations dominate, and the terms in γ can be ignored. Now if $\gamma = 0$ the line $Y = \delta/2$, $Z = 0$ is composed of degenerate fixed points and, for a range of values of Φ in the neighbourhood of zero, trajectories leaving the vicinity of this line for $X = X_1 < -1$ will eventually return to it for X equal to $X_2 > -1$. We may then write $X_2 = g(X_1; \delta, \Phi)$. In the special case $\Phi = 0$ the form of g is easy to calculate since the trajectory splits into two simple parts. First Z becomes significant while $(Y - \delta/2)$ remains small. Then it is easily seen that $(X+1)^2 + Z \approx \text{const.} \approx (X_1 + 1)^2$, so that on return to the invariant line $X = X_* = -(2 + X_1)$. If $-1 < X_* < 0$, X_* may be identified with X_2, while if $X_* > 0$ there is a further phase in which Z is small and X and Y change so that $X^2 + Y^2 \approx \text{const.}$ Thus at the next return to the invariant line $X_* = -|2 + X_1|$. If this value lies between 0 and -1 it may be identified with X_2, while otherwise the sequence repeats as often as necessary until X *is* in the right range. Then Z becomes small and the slow phase supervenes. For this picture to work, it is of course necessary that the operation of the fast phase always brings X closer to the origin; when $\Phi \neq 0$ this only occurs for X sufficiently small. We summarise the relevant theory for small Φ below but the main result is simply stated; the fast phase is attracting if $|X| < \text{const.}\Phi^{-1}$.

The above considerations allow us to construct a map relating maxima in $|X|$ (for negative X) attained at the end of successive slow phases. Noting that these can occur only for $X < -1$, we find that for successive such maxima X_0, X_1

$$f(X_1) = f(g(X_0)) \tag{5}$$

$$\text{where for } \Phi = 0,\ g(X) = -|2 + \tilde{X}|$$

$$\text{and } \tilde{X} = X + 2N \text{ for } -3 - 2N < X < -3 - 2(N-1).$$

When $\Phi \neq 0$, the function g is not so straightforward and has to be calculated numerically.

Notice that this first approximation to the map is in fact independent of γ! Not only is this unsatisfactory for the construction of bifurcation diagrams etc., but it leaves the map with various unrealistic topological properties (see Hughes & Proctor 1990b). In Figures 1 and 2 we show graphs of the map as calculated in equation (5) for $\delta = 0.2$, $\gamma = 5 \times 10^{-3}$ and $\Phi = 0$, $\Phi = 0.1$ respectively, compared with the results of numerical simulations of the governing o.d.e.'s. It can be seen that the agreement is generally very good — however, there is a systematic error for $X_0 \leq -3$ and a more pronounced error in the immediate neighbourhood of $X_0 = -3$. The map is not so accurate for larger $|X_0|$ since, as described above, the fast phase involves more oscillations, and small systematic errors in each step are thus compounded. Following a maximum in the vicinity of $X_0 = -3 + 2N(N \geq 0)$ the errors are more significant since the fast phase takes X_2 (defined above) close to -1 where the distinction between fast and slow phases disappears. We have shown in Hughes & Proctor (1990b) that a detailed treatment near the fixed point at $X = -1$ yields corrections to the map given by (5) which are of order $|\gamma \ln \gamma|^{1/2}$ for $\Phi = 0$, and there is no reason to suppose that the correction for non-zero Φ will be qualitatively different, though we have not investigated this case in detail. The corrected map is shown in Figure 1 also.

BEHAVIOUR FOR SMALL NON-ZERO Φ

Since the remainder of the analysis in the paper will be for the special case $\Phi = 0$, it is important to establish that this value of Φ does not lead to non-generic behaviour. The forms of the maps indicate that there is little change when Φ is different from zero, with the important exception that the fast phase does not result in $|X|$ being reduced if it is sufficiently large initially. We can carry the analysis through a considerable distance when $\Phi \ll 1$ and X, Y and $Z^{1/2}$ are $O(\Phi^{-1})$ on input to the fast phase. Writing $\tilde{X} = X\Phi$, $\tilde{Y} = Y\Phi$, $\tilde{Z} = Z\Phi^2$, $X' = \dot{X}\Phi$, etc. and discarding terms of order Φ^2 and smaller, we arrive at the following system (recall that we can set $\gamma = 0$ in the fast phase; the tildes have been dropped):

$$\left.\begin{aligned} X' &= -2Y^2 + Z + \Phi\delta Y, \\ Y' &= 2XY + \Phi(Z - \delta X), \\ Z' &= -2XZ - 2\Phi Z. \end{aligned}\right\} \tag{6}$$

At leading order the system is now Hamiltonian, with the two invariants

$$E = X^2 + Y^2 + Z, \qquad C = YZ. \tag{7}$$

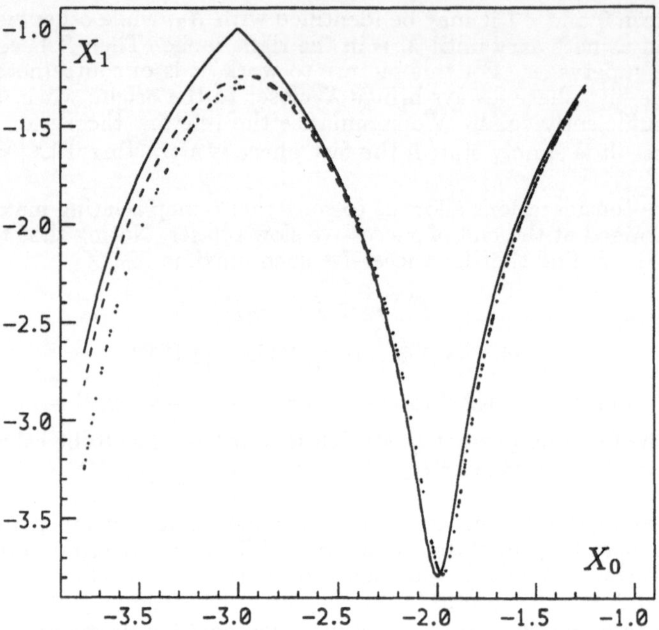

FIGURE 1. Map relating maxima of $|X|$, X_0 and X_1, attained at the end of successive slow phases. The dots are obtained from numerical integration of equations (3); the full curve is the analytic map (5); the dashed curve is the improved map, described by Hughes & Proctor (1990b). $\delta = 0.2$, $\gamma = 5 \times 10^{-3}$, $\Phi = 0$.

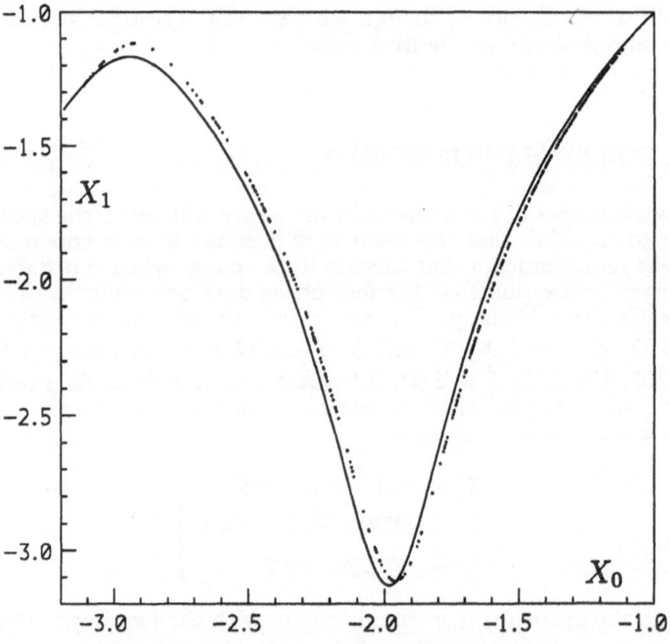

FIGURE 2. As Figure 1 but with $\Phi = 0.1$.

The slow evolution of these quantities due to the $O(\Phi)$ terms can be shown to occur according to the equations

$$E_T = 2C - 2\langle Z\rangle, \qquad C_T = -2C + 2E\langle Z\rangle/3, \tag{8}$$

where T is a slow timescale and $\langle Z\rangle$ is the average of Z over an orbit of the Hamiltonian system, and is a function of E and C (interestingly, δ does not enter at this level). Numerical integration of equations (8) shows that when Y and Z are initially very small, as they are at the start of the fast phase, then if E is initially greater than some value $E_0 \approx 4.41$ the trajectories escape to infinity, while for $E < E_0$ the orbits decay until the effects of non-zero Φ are insignificant. Full details of the analysis may be found in Hughes & Proctor (1990c).

Can the fast phase ever begin for $X^2 > E_0$? The maximum possible value of $|X|$ at the end of the slow phase is given, according to equation (4), by the solution X_c of $f(X_c) = 0$. For small δ, $X_c \approx \ln\delta$, so that the fast phase will be well behaved if Φ is small, and $\Phi|\ln\delta| < E_0^{1/2}$.

RESULTS AND DISCUSSION

Since the map takes such a simple form when $\Phi = 0$ we have been able to conduct an extensive survey of the parameter space. Figure 3 shows a bifurcation diagram constructed by varying δ for fixed γ and iterating the map to find an attractor. The non-trivial fixed point at $X = -1$ loses stability when $\delta = 2$, and a periodic orbit appears (here represented by a fixed point of the map). This eventually becomes unstable to a period-doubling bifurcation and the usual period-doubling sequence ensues, culminating in chaotic solutions (of the map, at least). For smaller values of δ further periodic windows appear associated with orbits that go through several oscillations in the fast phase. These results establish conclusively that chaos can occur in this normal form, and that the neglect of the cubic terms does not lead to unbounded solutions, at least when Φ is sufficiently small.

An important feature of the map is its multimodal nature. Because all the maxima occur at more or less the same height, as do the minima, only the maximum and minimum closest to the origin are significant. Bimodal maps have been investigated in some detail by Mackay & Tresser (1987) who showed (i) that there are superstable orbits associated with both extrema, and (ii) that there are regions of parameter space where there is more than one stable orbit. In Figure 4 we show plots (in $\delta-\gamma$ space) of the location of several superstable orbits. Where curves corresponding to orbits of different character cross there is a codimension two singularity associated with hysteresis. Figure 5 exhibits bifurcation diagrams near one of these points which clearly show that the realised solution can depend on the starting point. Although these crossings all occur for finite values of γ, as the periods of the orbits increase the associated γ's decrease, so that it is fairly easy to get into a region where the o.d.e.'s are accurately represented by the map. (For larger values of γ there is still a map, but the theory does not give the extrema correctly.)

Finally, we remark briefly on the effect of noise on the solutions. During the slow phase the system spends much time very close to the invariant plane $Z = 0$. Since the maximum of $|X|$ actually occurs when $Z = O(\gamma)$ or less, examination of equation (3c) reveals that the minimum value of Z during the fast phase can be as small as $\gamma e^{-1/\gamma}$. This is extremely small for small vaues of γ, and in consequence rounding error and underflows in the computer program become significant in determining the actual minimum of Z during a numerical integration of equations (3) as they stand (this problem may be alleviated however by exponentiation of the variable Z). For most problems the errors thus introduced would be insignificant, but here the eventual value of X at the start of the fast phase depends crucially on the value that Z is allowed to attain during the slow phase. If the equations are thought of as mimicing

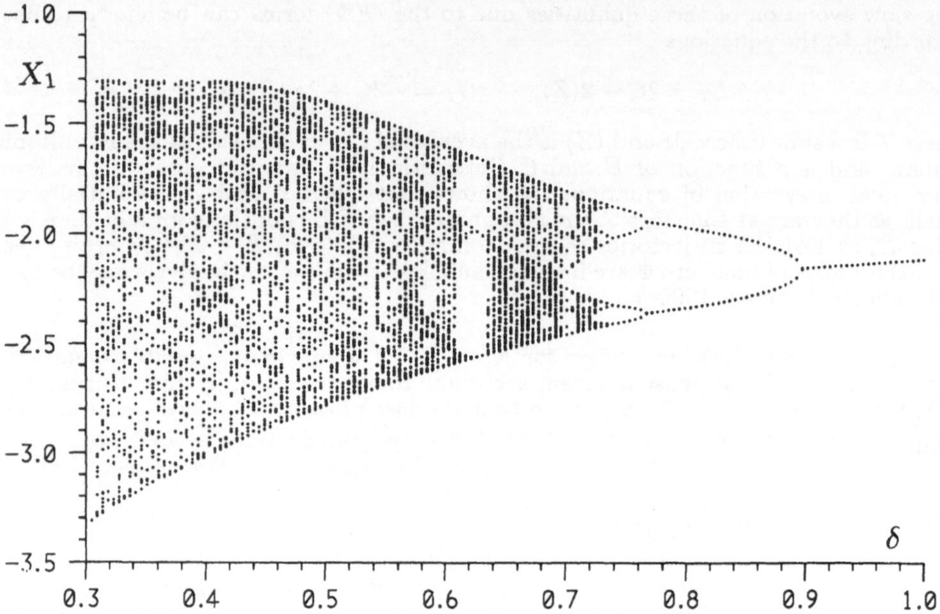

FIGURE 3. Bifurcation diagram for $\gamma = 10^{-2}$ with δ as the abscissa and iterates of the map as the ordinates. All iterations are started at $X = -2$.

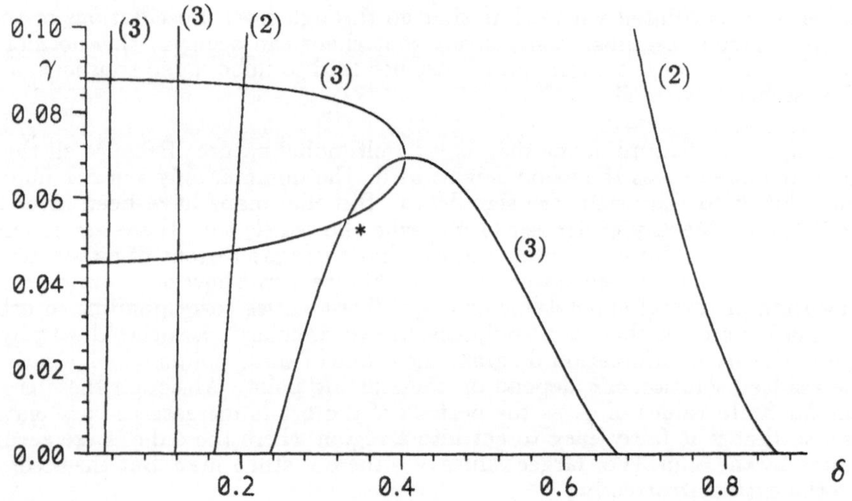

FIGURE 4. Location in $\delta - \gamma$ space of the superstable orbits of periods 2 and 3.

a physical process, then the effect of a small amount of noise will be the same as that of the numerical noise introduced by rounding error. A similar sensitive dependence on noise was discussed in detail by Hughes & Proctor (1990a) for a related problem and an extensive discussion of the present system is given in Hughes & Proctor (1990b). Here we note merely that additive noise is likely to become important in (3c) when its r.m.s. value, ϵ say, is such that $\alpha = |\gamma \ln \epsilon|$ is of order unity. This point is illustrated by Figure 6 which shows two numerical integrations of equations (3), one with no noise,

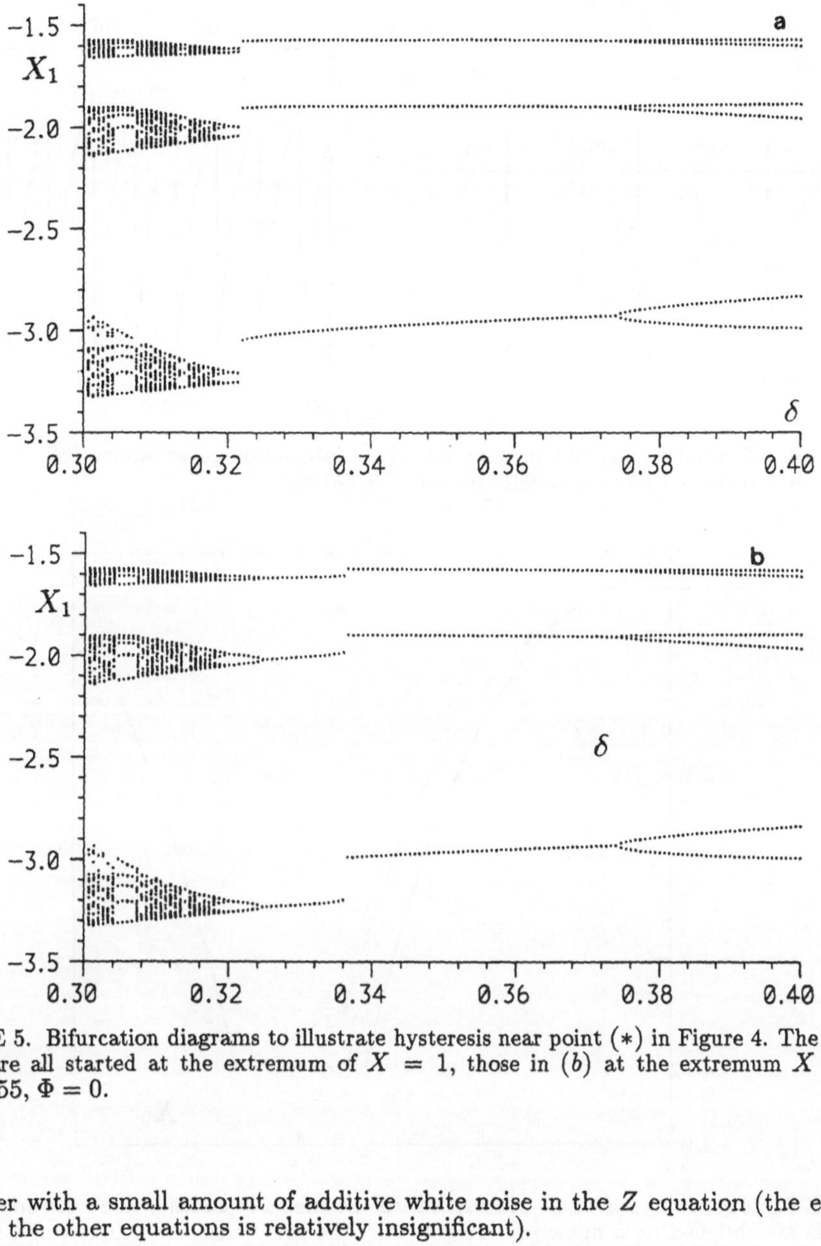

FIGURE 5. Bifurcation diagrams to illustrate hysteresis near point (*) in Figure 4. The iterates in (a) are all started at the extremum of $X = 1$, those in (b) at the extremum $X = -2$. $\gamma = 0.055$, $\Phi = 0$.

the other with a small amount of additive white noise in the Z equation (the effect of noise in the other equations is relatively insignificant).

For a quantitative investigation of the effects of noise we can use our map. In the presence of noise the value of $|X|$ at the end of the slow phase will be the lesser of the value in the absence of noise and that corresponding to $|Z|$ being $O(\epsilon)$ at $X = -1$. (This heuristic description can be confirmed by a full analysis of the associated Fokker-Planck equation (c.f. Hughes & Proctor 1990a).) Thus we modify the maps of Figure 1 by cutting them off at the point where the noise takes over. Figure 7 shows such a map, constructed as above, compared with numerical simulations.

We end the discussion of noise by noting that even when it is significant in the map, orbits of the map may never intersect the region which is altered by the noise.

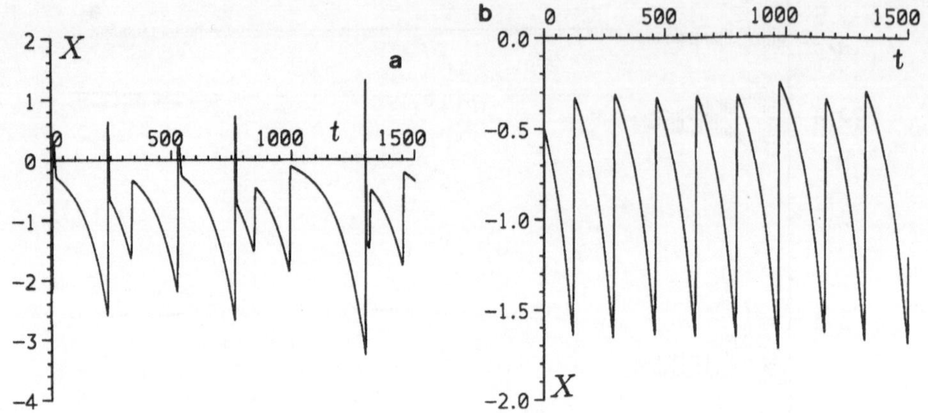

FIGURE 6. X versus time, obtained by numerical integration of equations (3). $\gamma = 0.01$, $\delta = 0.2$, $\Phi = 0$. In (a) there is no additive noise, in (b) $\epsilon = 10^{-8}$.

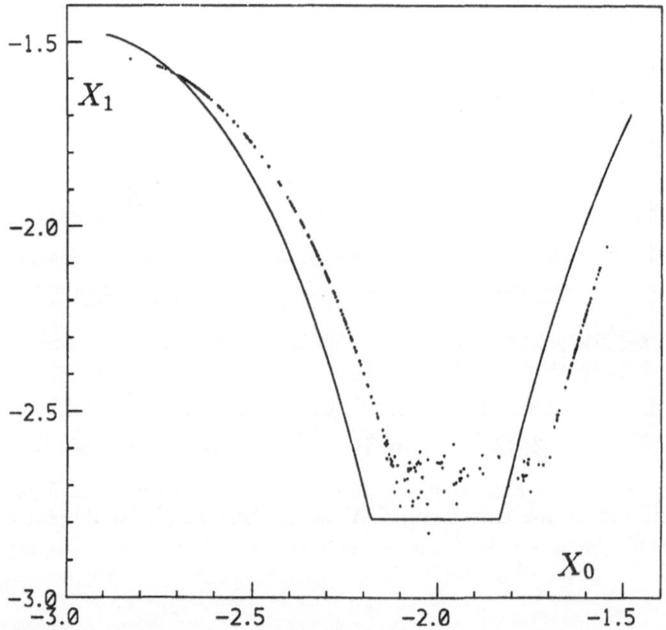

FIGURE 7. Map of X_1 versus X_0, incorporating the effects of additive noise in equation (3c). The dots are obtained from numerical integration of the equations. $\delta = 0.3$, $\gamma = 0.03$, $\Phi = 0$, $\epsilon = 10^{-11}$.

Figure 8 shows a bifurcation diagram for the map altered by the effect of noise, with the same value of γ as a previous case. It can be seen that there are regions of anomalous periodic orbits (which in a proper simulation would be 'noisily periodic' due to the effect of the noise), together with regions that are unchanged from the noiseless case.

In conclusion, we have shown that the normal form for the Hopf bifurcation with 2:1 resonance, when truncated at quadratic order, has bounded orbits in certain parameter ranges, and exhibits the kind of chaos (through period-doubling sequences) associated with a one-dimensional map. Although the special case $\Phi = 0$ is the only

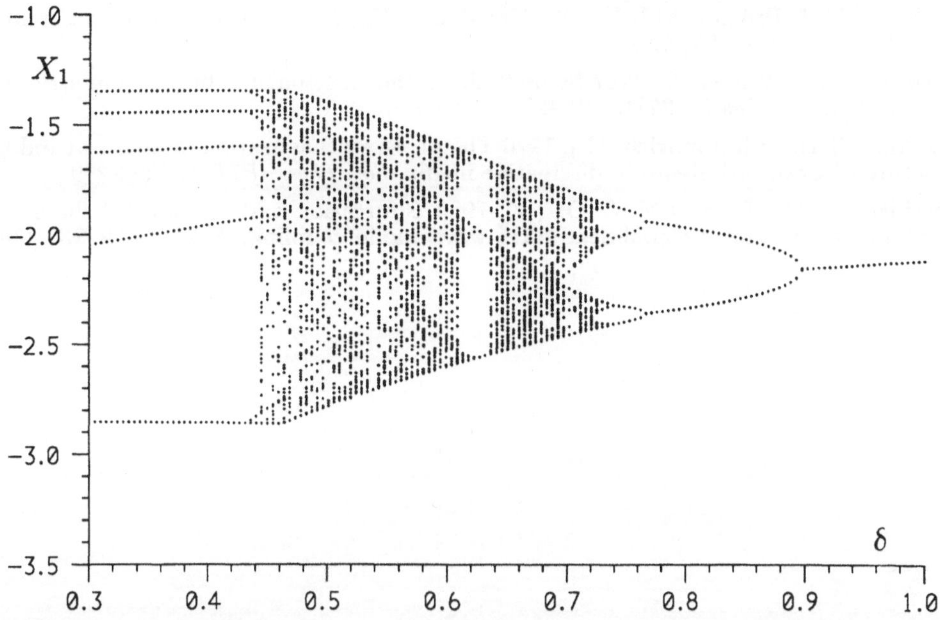

FIGURE 8. As Figure 3 but with additive noise of r.m.s. ϵ. $\gamma = 10^{-2}$ and $\gamma|\ln\epsilon| = 0.8$.

one where a fully analytical description of the map is available, and the only one extensively investigated, the dynamics for small non-zero Φ is shown to be almost the same provided that the starting values are not too far from the origin. The technique allows also the investigation of the effects of small amplitude additive noise. We emphasise that the methods of analysis are non-standard; conventional small-amplitude or centre manifold theory would not have been able to reveal the period-doubling bifurcations. Similar methods are likely to prove of considerable value in the analysis of other high codimension normal forms.

ACKNOWLEDGEMENTS

We are grateful to P. Glendinning for helpful discussions. This research was supported by the Science and Engineering Research Council of Great Britain and by Trinity College, Cambridge.

REFERENCES

Guckenheimer, J. & Holmes, P. 1983 *Nonlinear Oscillations, Dynamical Systems, and Bifurcations of Vector Fields.* Springer-Verlag.

Hughes, D.W. & Proctor, M.R.E. 1990a A low-order model for the shear instability of convection: chaos and the effect of noise. *Nonlinearity* **3**.

Hughes, D.W. & Proctor, M.R.E. 1990b Chaos and the effect of noise in a model of three-wave mode coupling *(in preparation).*

Hughes, D.W. & Proctor, M.R.E. 1990c Chaos and the effect of noise in the double Hopf bifurcation with 2:1 resonance *(in preparation).*

Knobloch, E. & Proctor, M.R.E. 1988 The double Hopf bifurcation with 2:1 resonance. *Proc. R. Soc. Lond.* **A415**, 61-90.

Mackay, R.S. & Tresser, C. 1987 Some flesh on the skeleton: the bifurcation structure of bimodal maps. *Physica* **27**D, 412-422.

Vyshkind, S.Ya. & Rabinovich, M.I. 1976 The phase stochastization mechanism and the structure of wave turbulence in dissipative media. *Sov. Phys. JETP* **44**, 292-299.

Wersinger, J.-M., Finn, J.M. & Ott, E. 1980 Bifurcation and "strange" behavior in instability saturation by nonlinear three-wave mode coupling. *Phys. Fluids* **23**, 1142-1154.

CODIMENSION 2 BIFURCATION IN BINARY CONVECTION WITH SQUARE SYMMETRY

Dieter Armbruster

Institut für Informationsverarbeitung
Universität Tübingen, W. Germany

1 Introduction

Codimension two bifurcations with double zero eigenvalues (Takens-Bogdanov bifurcations) and square symmetries have recently been shown to exhibit chaotic behaviour [AGK]. Unlike chaotic solutions in other Codimension 2 situations [GH] the chaotic behaviour here is found in a "large" region in parameter space - a wedge with positive angle originating at the singularity. Hence Codimension 2 singularities with square symmetry should be particularly attractive from an experimental point of view. The intuitive reason for the appearance of these "large" chaotic regions is the existence of a nonintegrable Hamiltonian with two degrees of freedom in a scaling limit of the D_4 - symmetric Takens-Bogdanov normal form. This nonintegrable Hamiltonian system creates stochastic regions in phase space which, for a certain range of unfolding parameters are not quenched to the fixed points when one takes the dissipation into account. Section 2 gives a short overview on the D_4-symmetric Takens-Bogdanov normal form and its unfolding. We analyse the phase space for typical parameters and discuss chaotic and nonchaotic solutions. Section 3 deduces the physical meaning of these solutions in real space as opposed to phase space. In Section 4 we calculate the nonlinear terms in the normal form for the specific system of two sets of orthogonal convection rolls in a mixture of He_3/He_4 and analyse the resulting normal form. Section 5 deals with the implication of this theory for some experiments by Moses and Steinberg [MS] and Le Gal et al. [LeG].

2 Nilpotent bifurcations with D_4-symmetry

We study vector fields which are symmetric with respect to D_4, the eight element symmetry group of the square, acting on two-dimensional Euclidean space $R^4 = C \times C$ with coordinates u=x+iz, v=y+iw via two reflections

$$\begin{array}{lccc} \tau: & (x,y,z,w) & \rightarrow & (z,w,x,y) \\ \rho: & (x,y,z,w) & \rightarrow & (-x,-y,z,w). \end{array} \quad (1)$$

Nonlinear Evolution of Spatio-Temporal Structures in Dissipative Continuous Systems
Edited by F.H. Busse and L. Kramer
Plenum Press, New York, 1990

Within this class of vector fields we analyse the equivalent to a Takens-Bogdanov bifurcation [GH] which is characterized by vector fields whose linear part have a nilpotent Jordan normal form $\begin{bmatrix} 0 & 1 \\ 0 & 0 \end{bmatrix}$. We show in [AGK] that the most general representation of this bifurcation problem is given, up to cubic order, by the D_4-equivariant normal form:

$$\begin{aligned} x' &= y + \epsilon^2 fz(zy - wx) \\ y' &= \mu x + x(a(x^2+z^2) + bz^2) + \\ &\quad \epsilon[\nu y + y(c(x^2+z^2) + ez^2) + dx(xy+zw)] + \epsilon^2 fw(zy - wx) \\ z' &= w - \epsilon^2 fx(zy - wx) \\ w' &= \mu z + z(a(x^2+z^2) + bx^2) + \\ &\quad \epsilon[\nu w + w(c(x^2+z^2) + ex^2) + dz(xy+zw)] - \epsilon^2 fy(zy - wx) \end{aligned} \tag{2}$$

Here ϵ is a scaling parameter, μ and ν are unfolding parameters and a,b,c,d,e and f are order 1 parameters. When $\epsilon = 0$ the system (2) becomes Hamiltonian with Hamiltonian function

$$H = \frac{1}{2}(y^2 + w^2) - \frac{\mu}{2}(x^2+z^2) - \frac{a}{4}(x^2+z^2)^2 - \frac{b}{2}x^2z^2 \tag{3}$$

We note, that for two values of b, H possesses a higher symmetry and becomes integrable: For b = 0 we have the Hamiltonian corresponding to an 0(2)-symmetric Takens-Bogdanov bifurcation [DK] and for b = -a the system decouples into two independent Duffing oscillators. One can prove via perturbation theory that, for values of b close to these integable values, the Hamiltonian H is nonintegrable and that transversal intersections of homoclinic orbits exist, leading to Smale's horseshoe maps and chaotic trajectories. In addition to these chaotic trajectories we find regular solutions living on 2-tori which manifest itself as doubly periodic behaviour as in Figure 1a-c. Typical chaotic solutions for the near integrable situations (b = 0.1) and (b = 0.9) are shown in Figure 2b-c. Note that intermittency plays a decisive role for these chaotic solutions: Take for example Figure 2b. This solution is basically doubly periodic and moves through all four quadrants of the (x,z) projection of the phase space. However, it stays unpredictably long in one of these quadrants and changes its direction of rotation in (x,z) space stochastically. A Poincare section shows a closed orbit with a small but finite thickness, thus revealing a very weakly chaotic, modulated travelling wave. A similar situation is shown in the simulation in Figure 2c. There appear to be quasiattractors on which a trajectory stays for a very long time until eventually it leaves these "attractors" and goes to another quasiattractor. Other simulations show the same behaviour but with a final approach to a fixed point, i.e. we have extremely long chaotic transients which were actually found experimentally in a similar situation for Faraday resonance [Gollub, personal communication]. In addition to these solutions one finds the typical resonant trajectories of KAM theory which produce resonant islands in Poincare sections. Allowing now $\epsilon \neq 0$ which amounts to switching on the dissipation, suppresses most of these chaotic and resonant solutions and leads to fixed points. There is, however, one region in parameter space where the chaotic dynamics persists which can be understood in the following way: The D_4-symmetry forces invariant subspaces in (2) given by $x = y = 0, (z,w) \neq 0$, and $z = w = 0, (x,y) \neq 0$ and $x = \pm z, y = \pm w$, respectively. Equilibria in the former two subspaces are called pure mode solutions while equilibria in the latter subspaces are mixed mode solutions. Within these subspaces the system is equivalent to a $Z(2)$-equivariant Takens-Bogdanov bifurcation whose dynamics is well understood [GH].

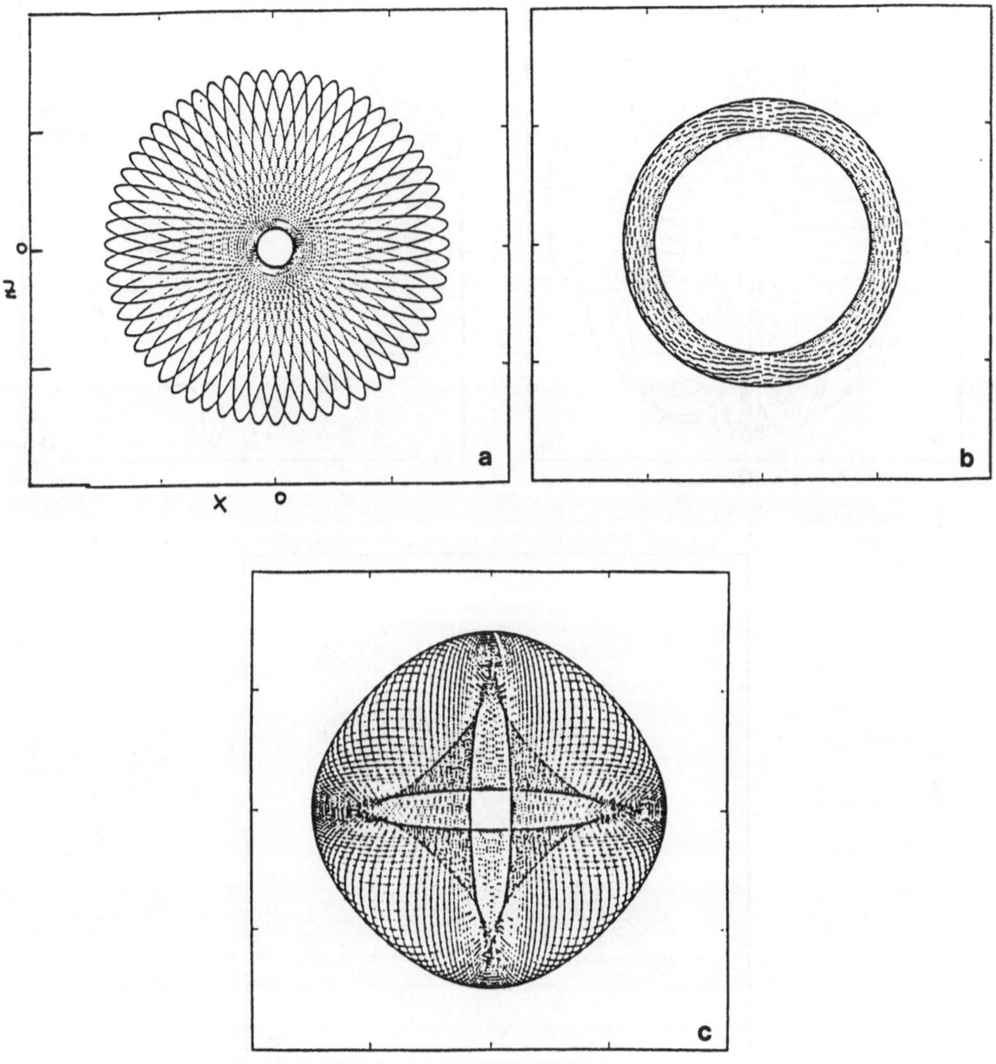

Figure 1. Projection onto (x,z)-space of typical quasiperiodic solutions for the Hamiltonian system (3) for a = 1 and $\mu < 0$: a.) O(2)-symmetric, b = 0, b.) b = 0.1 and c.) b = 0.5.

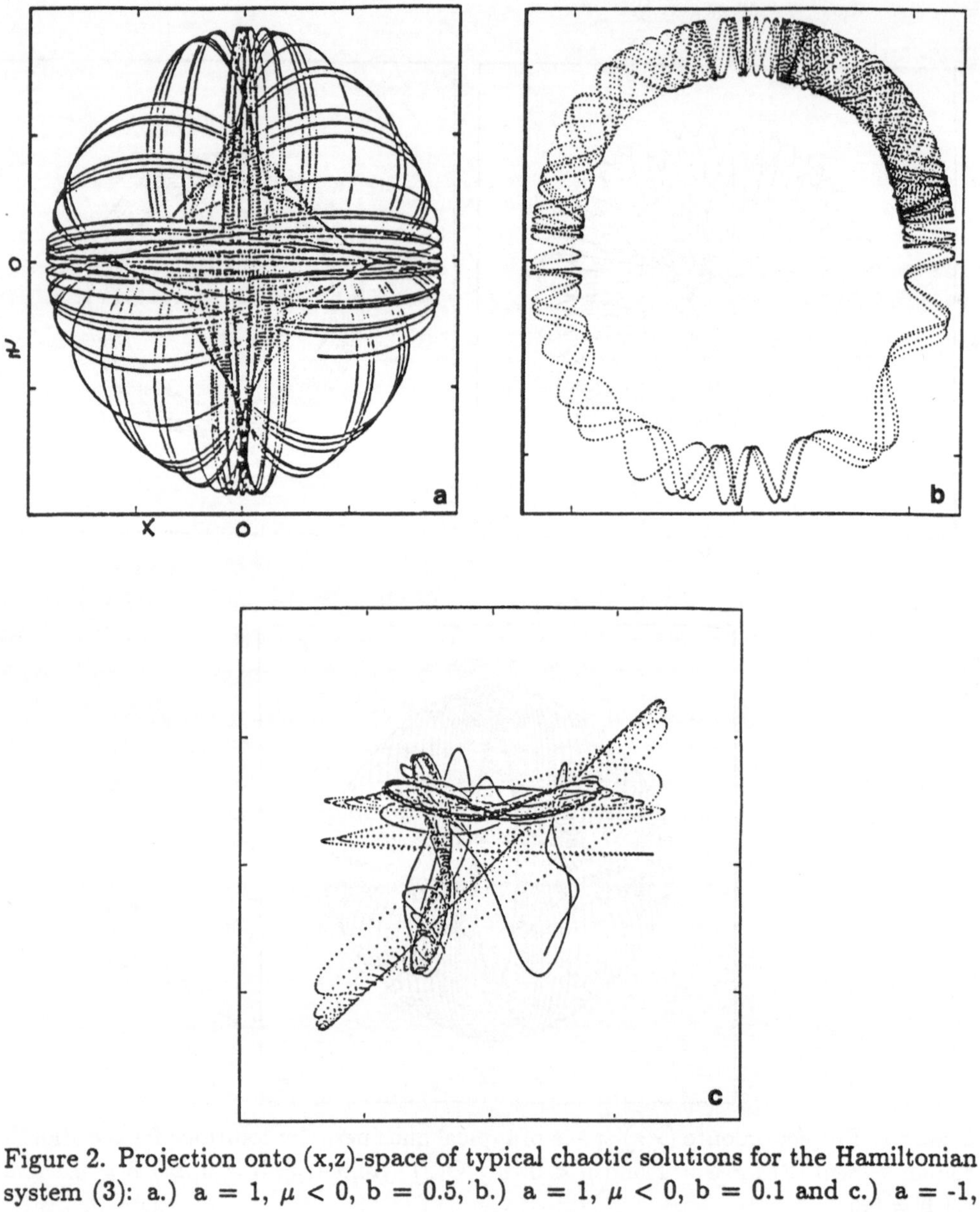

Figure 2. Projection onto (x,z)-space of typical chaotic solutions for the Hamiltonian system (3): a.) $a = 1$, $\mu < 0$, $b = 0.5$, b.) $a = 1$, $\mu < 0$, $b = 0.1$ and c.) $a = -1$, $\mu > 0$, $b = 0.9$.

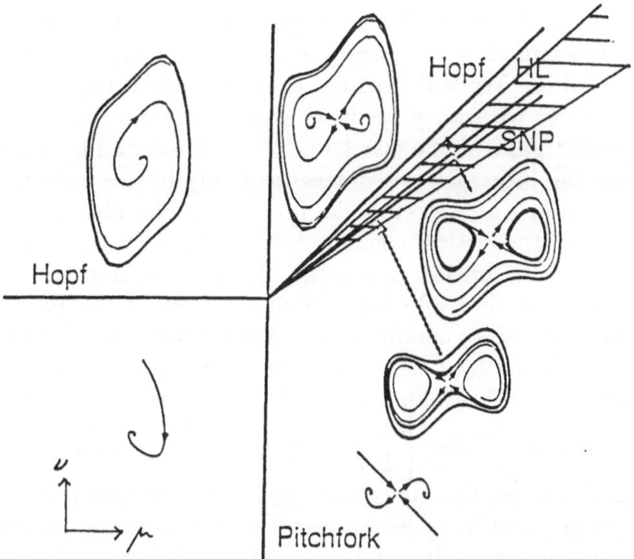

Figure 3. Bifurcation diagram and phase portraits for a Z(2)-symmetric Takens-Bogdanov bifurcation with $a < 0, c + d < 0$. The hatched region indicates the chaotic wedge for the full D_4-symmetric normal form (2).

In particular there is a homoclinic bifurcation where the two limit cycles around the nontrivial equilibria form a figure 8 homoclinic loop connecting the origin to itself and then coalesce to one large limit cycle (c.f. Figure 3). The homoclinic bifurcation curve in (μ, ν)-space can be calculated by balancing linear and nonlinear dissipation (which hence must be of opposite sign) around the figure 8 trajectory such that the homoclinic loop, which is a typical Hamiltonian structure, prevails. While this is true for equation (2) exactly in the invariant subspaces, trajectories near the homoclinic loop may, however, be unstable in the transverse direction for a parameter region near this bifurcation point. Hence if one allows initial conditions slightly off these invariant subspaces then for parameters near the homoclinic bifurcation one still expects an almost balancing of linear and nonlinear damping and therefore a dynamical behaviour which is determined by the underlying Hamiltonian, which here is chaotic. Figure 3 shows the bifurcation set and representative phase diagrams for the Z(2)-symmetric bifurcation diagram - augmented by the numerically determined chaotic wedge that one finds for initial conditions off the invariant subspaces.

3 D_4-symmetric convection

Binary mixtures have proved to be one of the most accessible physical systems exhibiting Codimension 2 bifurcations. Experiments for water/alcohol [Wa] or He_3/He_4 mixtures [AR] did not only confirm theoretical predictions but also challenged the theory and lead among others to the analysis of O(2)-symmetric Takens-Bogdanov bifurcations [DK] and to the development of defect dynamics. Hence it seems natural to analyse the implications of any new Codimension 2 theory on binary convection. A typical analysis of that kind starts with the Navier-Stokes equation plus two conservation laws for energy and mass and reduces these p.d.e.s onto a center manifold. This center manifold is spanned by the amplitudes of the marginally stable eigenmodes of the linearized p.d.e.. We will assume, that we have a typical binary convection problem in a container with a square base. Furthermore we assume that the system is small and that the gaps in the spectrum of the linearized operator are large thus guaranteeing that the bifurcation is determined only by the critical eigenmodes and that no continuum of unstable eigenmodes interferes. This also guarantees that, apart from the symmetry of the critical eigenmodes, there is no further symmetry involved. In particular, we do *not* consider a system which is translationally invariant in the plane, hence we will not find solutions which travel in space. In this case and with the simplest lateral boundary conditions the horizontal parts of these eigenmodes are of the form $trig(k_\chi \chi) trig(k_\eta \eta)$ where trig stands for sin or cos. This leads to two different types of marginally stable modes: If the critical wave number k_c is the same in χ and η direction, i.e. $k_c = \begin{bmatrix} k \\ k \end{bmatrix}$, then we have a trivial representation of the D_4-symmetry in the sense that only one type of convection roll becomes unstable. If this is the case the resulting center manifold will be one-dimensional and leads, for the Boussinesq approximation and permeable and free slip boundary conditions, to a pitchfork bifurcation with trivial dynamics. If, however, the initial wave vector leads to different wavelengths in χ and η direction $k_c = \begin{bmatrix} m \\ n \end{bmatrix}$, $m \neq n$, then, with every mode $\cos(n\chi)\cos(m\eta)$, we also excite the symmetrical mode $\cos(m\chi)\cos(n\eta)$. Therefore the D_4 symmetry acts as in Section 1 with a reflection and a permutation on the amplitudes x and z of these critical modes.

Typical solutions of (2) represent the following convection patterns: Pure mode equilibria represent stationary convection, governed totally by one type of rolls (m, n) or (n, m). Mixed mode equilibria describe stationary patterns resulting from an equal superposition of both types. The limit cycles in an invariant subspace (called standing waves in an O(2)-symmetric context) represent periodic oscillations of the stationary patterns whereas the limit cycles which are nearly circular around the origin in the (x,z)-projection (called travelling waves in an O(2)-symmetric context) are characterized by an alternating appearance and disappearance of the pure and mixed stationary convection patterns. We will continue to call the latter travelling waves with the understanding that there is nothing travelling in real space but only in phase space. Similarly there are doubly periodic solutions which, in phase space, look very much like the modulated travelling waves of an O(2) system. Finally the analysis of (2) predicts different limit cycles off the invariant planes, and subharmonic and chaotic solutions whose experimental signature is unclear but which can easily be discerned via a Fourier analysis of the experimental data.

4 Calculating the normal form

Recently Müller and Lücke [ML] have performed a Galerkin approximation to the problem discussed in the last section. They considered two orthogonal convection rolls (in χ and η direction respectively) and used the following critical modes for the velocity field.

$$u(\mathbf{x}, t) = \begin{bmatrix} u_1 \\ u_2 \\ u_3 \end{bmatrix} = A \begin{bmatrix} \sin k\chi \cos \pi z \\ 0 \\ \cos k\chi \sin \pi z \end{bmatrix} + B \begin{bmatrix} 0 \\ \sin k\eta \cos \pi z \\ \cos k\eta \sin \pi z \end{bmatrix}, \tag{4}$$

the temperature field

$$\Theta(\mathbf{x},t) = C\cos k\chi \sin \pi z + D\cos k\eta \sin \pi z, \tag{5}$$

and a combined field $\zeta(\mathbf{x},t) = c(\mathbf{x},t) - \psi\Theta(\mathbf{x},t)$, where c is the concentration field and ψ describes the separation ratio

$$\zeta(\mathbf{x},t) = E\cos k\chi + F\cos k\eta + G\cos \pi z. \tag{6}$$

Keeping all these modes real but using an infinitely extended system induces a representation of the D_4-symmetry on the mode amplitudes: Interchanging the space coordinates η and χ leads to a permutation of the amplitudes (A,B), (C,D) and (E,F). Translation of a convection pattern by half the wavelength $\frac{\lambda}{2}$ creates the reflection symmetry $(A,C,E) \rightarrow -(A,C,E)$ and $(B,D,F) \rightarrow -(B,D,F)$ for each of the rolls, respectively. Performing the Galerkin reduction on the Overbeck-Boussinesq equations we find after scaling:

$$\begin{aligned}
\tau\dot{X} &= -\sigma X + \sigma[(1+\psi)Y + \frac{8}{\pi^2}U] + \begin{bmatrix} X_2 \\ X_1 \end{bmatrix} S && (7)\\
\tau\dot{Y} &= -Y + (r-Z)X + \begin{bmatrix} X_2 \\ X_1 \end{bmatrix} T && (8)\\
\tau\dot{U} &= -\frac{L}{3}U + \psi Y + VX && (9)\\
\tau\dot{Z} &= -\frac{8}{3}(Z - X\cdot Y) && (10)\\
\tau\dot{V} &= -\frac{8}{3}(\frac{L}{4}V + \frac{2}{3}\psi Z + \frac{1}{2}X\cdot U) && (11)\\
\tau\dot{S} &= -\frac{10}{3}\sigma S + \frac{3}{5}\sigma(1+\psi)T - \frac{4}{5}X_1X_2 && (12)\\
\tau\dot{T} &= rS - \frac{10}{3}T - \frac{2}{3}(X_1Y_2 + X_2Y_1) && (13)
\end{aligned}$$

Where X,Y,U have two components and the scaling connection is given as

$$(A,B,C,D,E,F,G) \leftrightarrow (X_1,X_2,Y_1,Y_2,U_1,U_2,V) \tag{14}$$

The other mode amplitudes S,T and Z describe 3 noncritical eigenmodes of the temperature field and ζ, which are excited via the nonlinearity. τ here is an irrelevant constant , σ is the Prandtl number, $r = \frac{R}{R_0^c} = \frac{R}{27/4}$ is the reduced Rayleigh number and $L = \frac{mass\ diffusion}{thermal\ diffusion}$ is the Lewis number. It can be shown that the modes parameterized by S,T and Z are the only ones which contribute to the dynamics of the center manifold of the Takens-Bogdanov bifurcation up to third order. Hence we can calculate the coefficients of the normal form (2) by performing a center manifold reduction and a normal form transformation onto (2) from the system of o.d.e's (7 - 13) instead of dealing with the partial differential equation of the Overbeck-Boussinesq approximation. Even with this simplification the reduction in algebraically too complicated in general, hence we insert $\sigma = 0.6$ for the Prandtl number and $L = 0.003$ for the Lewis number which are typical values for He_3/He_4 mixtures. Therefore we consider the results of this reduction and of the following simulation not as something quantitatively predictive but rather as prototypical, which they are anyhow due to the unrealistic boundary conditions. For convection in He_3/He_4 mixtures with the above parameters the resulting Codimension 2 bifurcation point exists at

$r = r_c = 1.027, \psi = \psi_c = 3.2034 \cdot 10^{-4}$. Using a Computer Algebra program we can find the coefficients in the normal form (2).

$$
\begin{aligned}
\dot{x} &= y + \epsilon^2\, 3.3165 z(zy - wx) \\
\dot{y} &= (0.31405\tilde{\psi} + 0.0036303\tilde{r})x + x(x^2+z^2) - 0.007742xz^2 \\
&\quad +\epsilon[(0.18766\tilde{\psi} + 0.37029\tilde{r})y + \\
&\quad y(x^2+z^2) - 74.3573x(zw+xy) - 1.7052z^2y] + \epsilon^2\, 3.3165w(zy - wx) \\
\dot{z} &= \ldots \\
\dot{w} &= \ldots
\end{aligned} \tag{15}
$$

where $r = r_c + \tilde{r}, \psi = \psi_c + \tilde{\psi}$. Inspection of (15) leads to some interesting facts: Reduction onto the invariant subspaces $x = y = 0$ (or $z = w = 0$) and $x = \pm z, y = \pm w$, respectively, shows that near the Codimension two bifurcation point the stationary bifurcation given by $y = 0$ is subcritical while the Hopf- bifurcation is supercritical leading to stable standing waves in these subspaces. However, as expected (since this is true also for the O(2)-symmetric case), they are unstable with respect to perturbations orthogonal to these subspaces. Another interesting fact is the large discrepancy of the various nonlinear parameters. In particular the Hamiltonian system (i.e. $\epsilon = 0$) is nearly O(2)-symmetric and has only a very small symmetry breaking term. This is true also, albeit on a smaller scale, for the dissipative terms. Here the dominant term is the one associated to x(zw + xy), which when written in polar coordinates $x + iz = re^{i\phi}$, is of the form $r^2\dot{r}$. Hence we expect two very different time scales in the simulations of (15): Initial transients will decay very fast to an almost azimutal dynamics given by $\dot{r} \approx 0$. The second time scale will be very slow and is determined by the O(2)-symmetry breaking terms. For an O(2)-symmetric system there is a whole group orbit of conjugate fixed points and standing waves induced by rotating the subspace $\phi = constant, \dot{\phi} = 0$. Fixed points and standing waves are neutrally-stable in the direction of this group orbit. Once one breaks this symmetry one induces a motion in this direction whose velocity scales with the symmetry breaking parameters. Hence, for small O(2)-symmetry breaking terms this motion is extremly slow. Intuitively, the reason for the nearly 0(2)-symmetric normal form lies in the choice of eigenfunctions for the Galerkin projection and in the structure of the Boussinesq approximations: The only modes in [ML] which lead to terms in (7-13), which are not equivariant with respect to a rotation in the plane (X_1, X_2),and (Y_1, Y_2) and (U_1, U_2) are the eigenfunctions $cosk\chi\ cosk\eta\ sin2\pi z$ for the velocity field and the temperature field which are associated with the two amplitudes S and T. One expects that realistic lateral boundary conditions or eigenfunctions which are more typical for a square (i.e. no component of the critical wave vector becomes zero, $mn \neq 0$) lead to an increase of the D_4-typical terms in the normal form.

Simulating (2) with $\epsilon = 1, \mu = -0.003, \nu = 0.001$, a = 1, b = -0.008, c = 1, d = -75, e = -2, f = 3.31 leads to the travelling wave shown in Figure 4a. A projection of this solution into the (x, z)-plane shows an almost circular solution thus showing no influence of the D_4-symmetry. This solution is only weakly stable: Reducing e to -1.7 leads to a very slowly radially increasing wave which eventually diverges. Keeping the nonlinear parameters as above but increasing the unfolding parameters to $(\mu, \nu) = (-0.01, 0.00338)$ we find a modulated travelling wave whose Poincare section is shown in Figure 4b. Note the symmetry of the two parts of the Poincare section. It appears that the associated torus is topologically equivalent to a Klein bottle. Increasing the unfolding parameters further leads to instability - obviously the heteroclinic bifurcation with the subcritical stationary branch has occurred. The

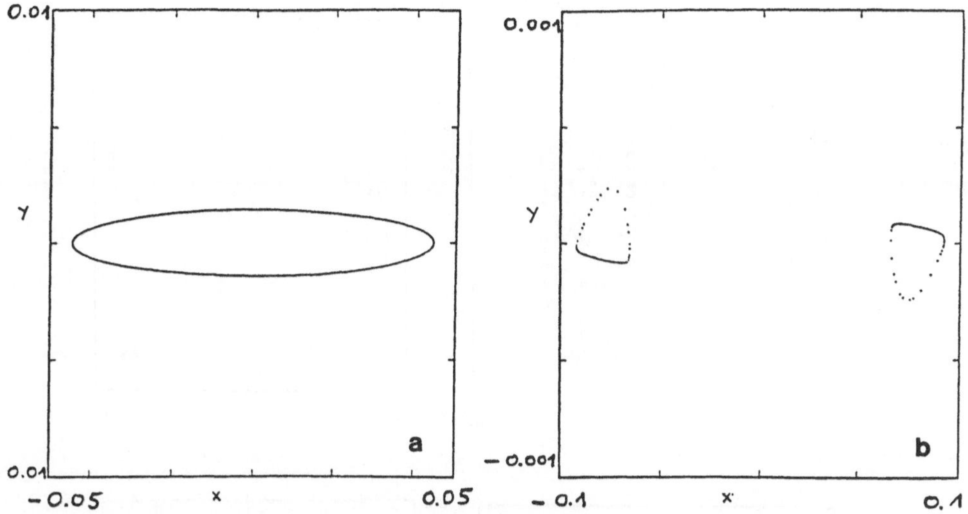

Figure 4. Simulation of (2) for parameter values as given in text. a.) Projection of a travelling wave solution into the (x,y)-plane. b.) Projection of a Poincare section at z = 0 into (x,y)-plane. The invariant curve indicates an incommensurable motion on a 2-torus. Note the different scaling in x and y direction.

above values for the unfolding parameters μ and ν lead to very practicable values for the separation ratio and the Rayleigh number: $(\mu, \nu) = (-0.003, 0.001)$ corresponds, according to (15) to $(\tilde{\psi}, \tilde{r}) = (-0.009, 0.007)$.

Keeping the typical features of (15) (i.e. subcritical steady state branch, supercritical periodic branch and strong nonlinear radial damping) but increasing the O(2)-symmetry breaking terms we perform simulations for a = 1, b = 0.2, c = 1, d = - 75, e = 10, f = 3.31. Also, in order to get nearer to the degenerate bifurcation point we choose $\epsilon = 0.1$. By fixing $\mu = -0.09$ and varying ν from 0.41 to 0.422 we find a new phenomenon: A period doubling scenario to chaos for an invariant curve in a Poincare section as shown in Figure 5. The associated trajectory in the full 4-dimensional phase space is a doubly periodic motion on a torus, one of whose frequencies does not change very much whereas the other shows the period doubling route to chaos. One can interpret the projection of such a trajectory in Figure 5c as a doubly periodic motion on a kind of knotted or doubled up torus. It is not surprising that in a parameter range which is in between those period doubling bifurcations there are other interesting things happening - e.g. a resonant phase locking of the two frequencies on a doubled torus as shown in Figure 6.

5 Implications for the interpretation of some real experiments

There exist two experiments by Moses and Steinberg [MS] and Le Gal et. al. [LeG], repectively, which are dealing with the preference of square convection patterns over rolls in the so called Soret regime of binary convection. Neither one of these experiments tries to measure near the Codimension 2 bifurcation point and hence in both cases the steady state bifurcation branch is supercritical. Therefore the normal form (15) does not apply directly to these experiments. [MS] find a very detailed bifurca-

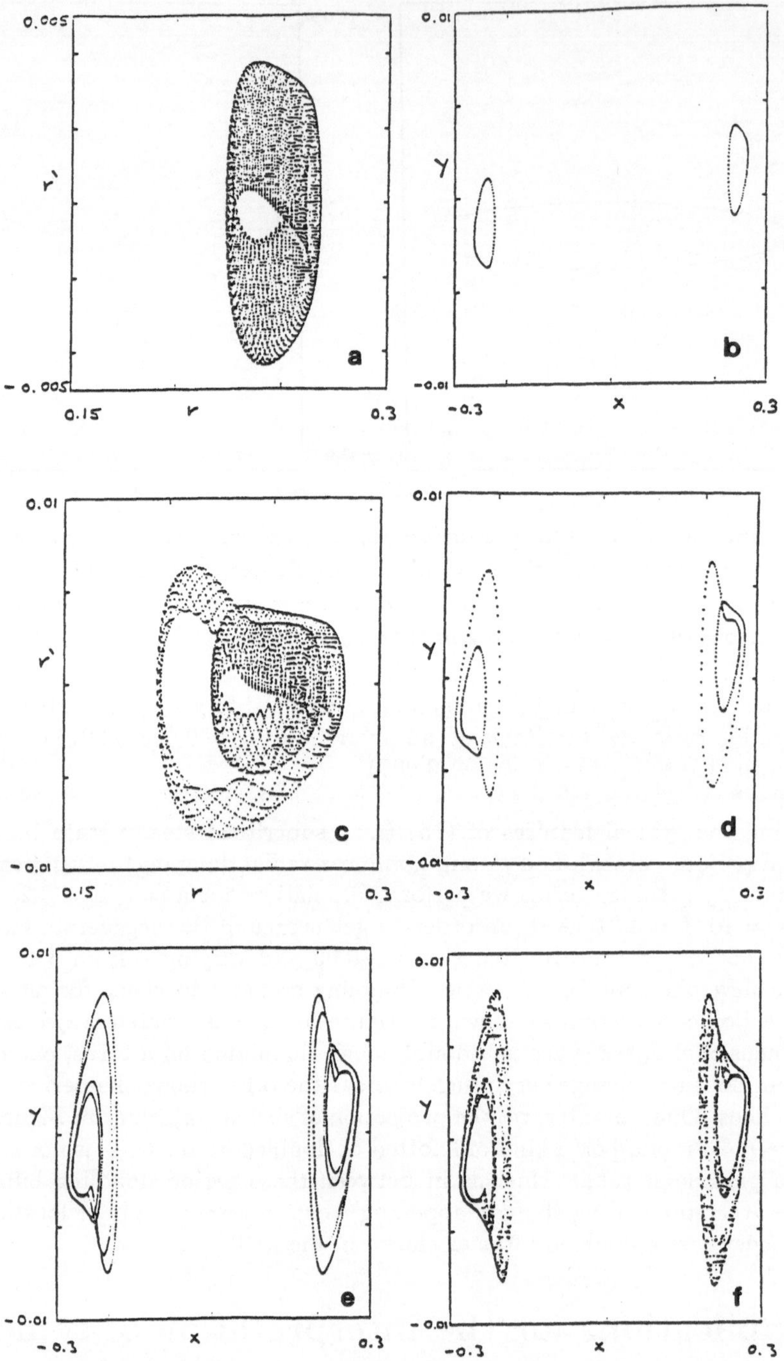

Figure 5. Simulation of (2) with parameters as given in text shows a period doubling scenario for an invariant curve: a.) $\nu = 0.41$. Projection of a modulated travelling wave into the polar coordinate plane (r,r'). Note the different scaling for r and r'. b.) The (x,y)-projection of the Poincare section of the trajectory in a.) at z = 0 shows the familiar closed invariant curves. c.) As in a.) with $\nu = 0.42$; d.) the associated Poincare section shows a period doubled invariant curve. e.) Period four can be found for $\nu = 0.4215$ and chaotic bands can be recognized in f.) for $\nu = 0.4218$.

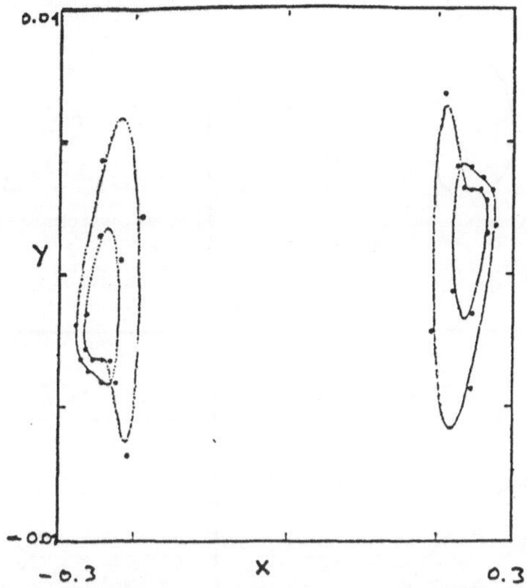

Figure 6. Poincare section of a simulation of (2) as in Figure 5c. The section consists of 30 isolated points. In order to indicate that this describes a phase locked motion on the period doubled torus we have copied the corresponding invariant curve from Figure 5c faintly into the picture.

tion sequence which so far not only cannot be explained for this specific experiment but which has not been seen before theoretically and seems to be ungeneric. The simulations of [ML] of the o.d.e's (7 - 13) could not explain this experiment without introducing the additional hypotheses of lateral heat fluxes. We will show here that this assumption is unnecessary and that the observed bifurcation scenario is a natural sequence for some typical parameter values of (2).

Let us first describe the experimental findings in terms of phase portraits: The relevant basic convection patterns are again two orthogonal sets of rolls whose amplitude we denote with x and z. We start in the Soret regime where one finds a stable square convection pattern which we represent as a mixed mode equilibrium (Figure 7 a). This mixed mode develops a Hopf bifurcation and leads to a small harmonic limit cycle around it (Figure 7 b). The limit cycle grows into a relaxation oscillation between the two types of orthogonal rolls. In terms of a phase portrait this implies that the limit cycle is almost heteroclinic and comes close to the invariant planes z = w = 0 and x = y = 0, respectively (Figure 7 c). The next measured structure are stable rolls (Figure 7 d). This last transition is not a regular generic bifurcation.

In order to reproduce such a bifurcation sequence from the D_4-symmetric Takens-Bogdanov normal form we start from (2) with a = -1 and c = -1. As mentioned in Section 1, Equation (2) becomes a Hamiltonian system for $\epsilon = 0$ with a potential

$$V(x,z) = \frac{1}{4}(x^2+z^2)^2 - \frac{b}{2}x^2z^2 - \frac{\mu}{2}(x^2+z^2) \tag{16}$$

For small b this is the Mexican hat potential postulated by [ML] in order to explain their simulations. In particular for $b > 0$ the mixed mode solutions belong to a lower energy level and for $b < 0$ the pure mode solutions are stable. Hence in order to reproduce Figure 7 we choose arbitraryly $\nu = 0.8$, b = 0.01, $\epsilon = 0.1$ d = -4 e = -2 and vary μ. At $\mu = 0.39$ the formerly stable mixed mode becomes unstable and induces

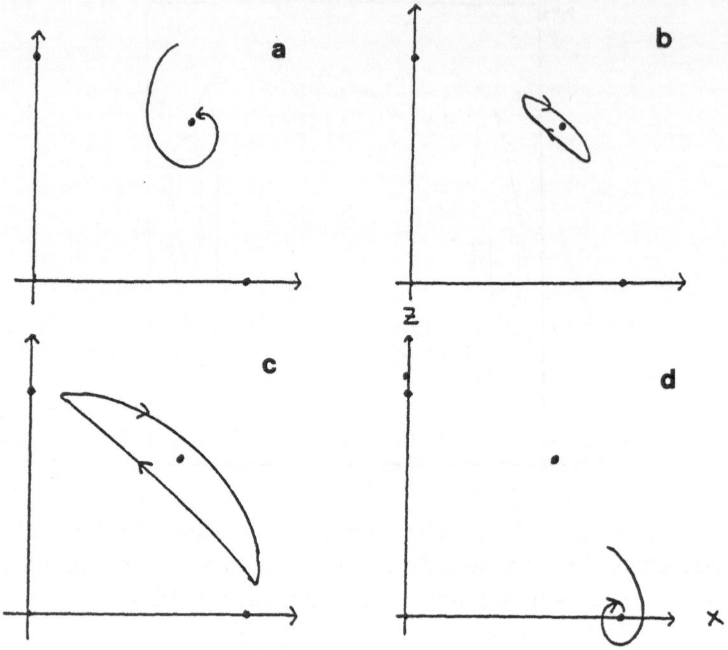

Figure 7. Schematic phase portraits describing the experimentally found bifurcation sequence [MS]

the harmonic oscillation in Figure 8a and b. Reducing μ to $\mu = 0.35$ increases the limit cycle nearly to a heteroclinic cycle and for $\mu = 0.33$ we find the travelling wave shown in Figure 8d. This travelling wave rotates in (x,z) space and slows down when crossing the x or z axes – an indication of its nearly heteroclinic nature. According to the interpretation of the experimental data by [ML] such a travelling wave has not been observed. Instead one finds stable rolls. Such a transition can however easily be arranged. Changing b from a small positive to a small negative value and keeping all other parameters fixed at the values for Figure 8d, leads to the extremely weakly attracting pure mode solution in Figure 9. A small change in this nonlinear parameter with a variation of r and ψ seems possible and since b is typically very small, as our model calculations in Section 4 showed, it is highly likely that b can change sign.

Our analysis suggests that a D_4-symmetric Takens-Bogdanov bifurcation may lead to novel periodic, quasiperiodic and chaotic solutions for convection in binary mixtures. We hope that we have convinced some experimentalists that it is worth while to perform more detailed experiments near the Codimension 2 bifurcation point in a square container. We have shown that even for bifurcations which exist in parameter ranges not in the neighbourhood of the Codimension 2 bifurcation point the normal form (2) gives an intuitive guideline for the interpretation of the experimental data. Work on the normal form for parameters typical for alcohol/water mixtures and for the interaction of critical eigenmodes which are already of square type is in progress. It is hoped that for these examples more of the chaotic solutions which are inherent in the normal form (2) can be found. Similar experiments for periodically driven surface waves [SG] show complex dynamics for nearly square containers, even when the the system showed only fixed points for an exactly square container. We hope to develop a general theory for the symmetry breaking from D_4 to $Z(2) \times Z(2)$ at Takens-Bogdanov bifurcations and apply it to the problem of binary convection.

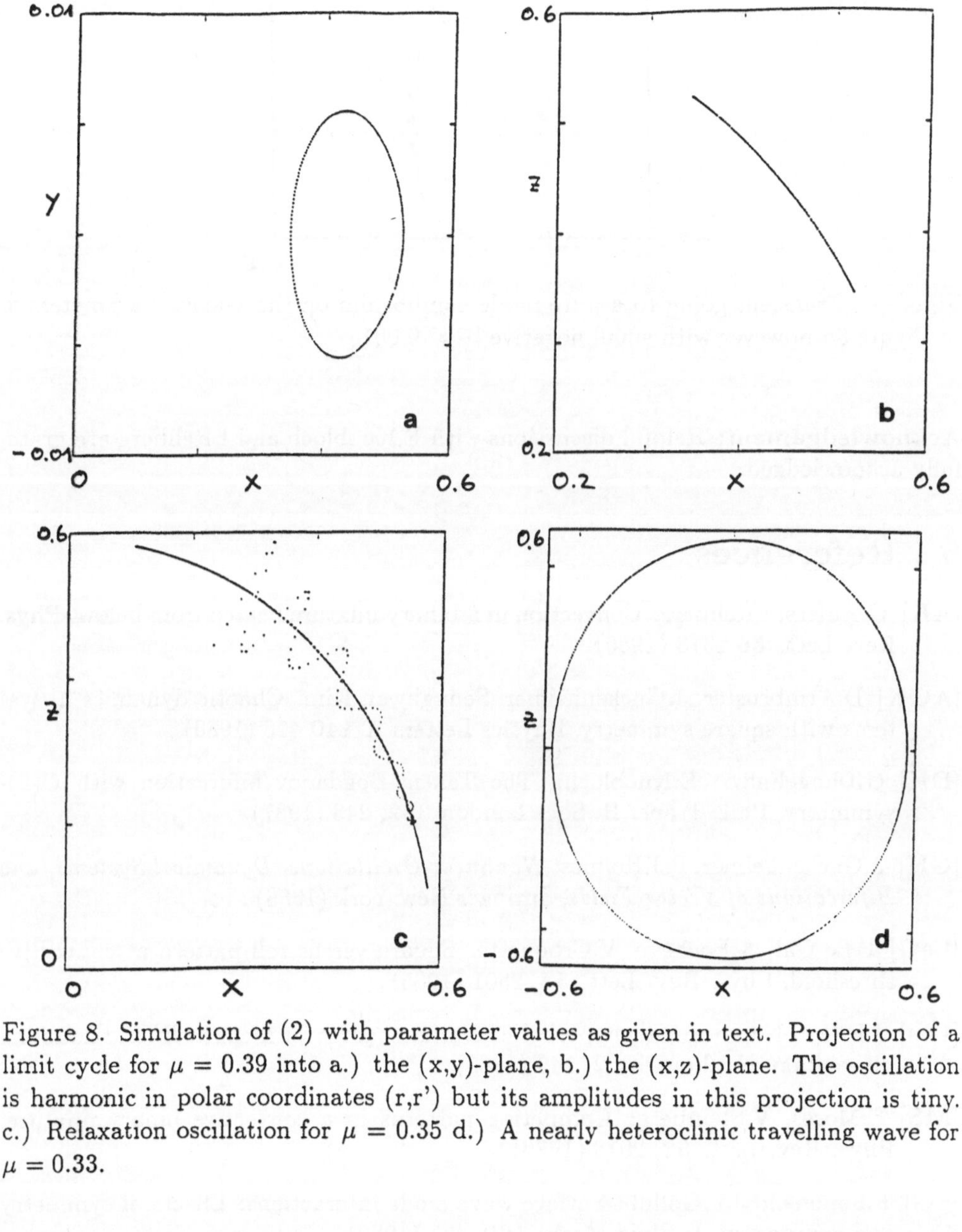

Figure 8. Simulation of (2) with parameter values as given in text. Projection of a limit cycle for $\mu = 0.39$ into a.) the (x,y)-plane, b.) the (x,z)-plane. The oscillation is harmonic in polar coordinates (r,r') but its amplitudes in this projection is tiny. c.) Relaxation oscillation for $\mu = 0.35$ d.) A nearly heteroclinic travelling wave for $\mu = 0.33$.

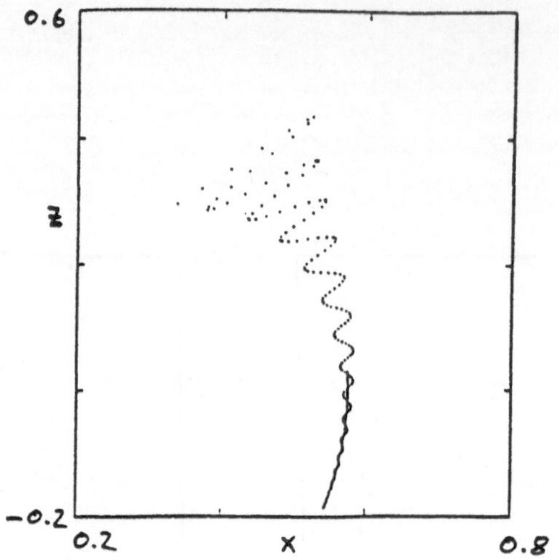

Figure 9. Transient going to a pure mode equilibrium on the x-axis. Parameters as in Figure 8d however with small negative b = -0.001.

Acknowledgement: Helpful discussions with E.Knobloch and I.Rehberg are gratefully acknowledged.

6 References

[**AR**] G.Ahlers, I.Rehberg: Convection in a binary mixture heated from below, Phys. Rev. Lett. **56** 1373 (1986)

[**AGK**] D.Armbruster, J.Guckenheimer, Seunghwan Kim: Chaotic dynamics in systems with square symmetry, Physics Letters A **140** 416 (1989)

[**DK**] G.Dangelmayr, E.Knobloch: The Takens-Bogdanov bifurcation with O(2)-symmetry, Phil. Trans. R. Soc. London **322**, 243 (1987)

[**GH**] J.Guckenheimer, P.J.Holmes: *Nonlinear Oscillations, Dynamical Systems, and Bifurcations of Vector Fields*, Springer New York (1986)

[**LeG**] P.Le Gal, A.Pocheau, V.Croquette: Square versus roll pattern at convective threshold, Phys. Rev. Lett. **54**, 2501 (1985)

[**Los**] J.E.Los: Non-normally hyperbolic invariant curves for maps in R^3 and doubling bifurcation, Nonlinearity **2**, 149 (1989)

[**MS**] E.Moses, V.Steinberg: Competing patterns in a convective binary mixture, Phys. Rev. Lett. **57**, 2018 (1986)

[**SG**] F.Simonelli, J.P.Gollub: Surface wave mode interactions: Effects of symmetry and degeneracy, J. Fluid Mech. **199**, 471 (1989)

[**Wa**] R.W.Walden, P.Kolodner, A.Passner, C.M.Surko: Travelling waves and chaos in convection in binary fluid mixtures, Phys. Rev. Lett. **55**, 496 (1985)

DYNAMICS OF SLOWLY VARYING WAVETRAINS IN FINITE GEOMETRY

Gerhard Dangelmayr
Institut für Informationsverarbeitung, Universität Tübingen
Köstlinstr. 6, D-7400 Tübingen, W. Germany

Edgar Knobloch
Department of Physics, University of California
Berkeley, California, CA 94720, USA

ABSTRACT. The effect of distant endwalls on a Hopf bifurcation from a translation-invariant state is considered. The walls break the translation symmetry with the result that the initial bifurcation is to standing waves with a fixed phase. Travelling waves (TW') appear in a secondary pitchfork bifurcation. A new two-frequency state (MW') is present at small amplitude only. The theory is applied to systems, such as binary fluid convection, that are described by coupled complex Ginzburg-Landau equations. As a result the TW' and MW' are tentatively identified with the confined travelling wave states and the blinking states, respectively, observed both experimentally and numerically in systems with sufficiently large group velocity.

1 Introduction

There has been much recent interest in the dynamics of slowly varying wave-trains. Much of this interest is motivated by the observation of travelling waves in binary fluid convection [1] and the subsequent discovery of the so-called "confined" [2] and "blinking" [3] states. In these experiments the term travelling wave (hereafter TW') is used to refer to a state of approximately parallel convection rolls drifting in a direction perpendicular to their axes. Such states typically arise from the nonlinear evolution of an oscillatory, or Hopf, bifurcation, and are observed in experiments in long narrow boxes. The TW' differs from the usual convection rolls in a moving reference frame in having well-developed phase differences between the velocity, thermal and concentration fields [4], which decrease with the phase velocity of the wave, as the applied Rayleigh number is raised. In a finite geometry a pure travelling wave (hereafter TW) cannot exist, however. In contrast to a pure TW there is no moving reference frame in which the TW' becomes a time-independent spatially periodic state. Indeed, the experiments reveal that when a roll reaches the endwall it shrinks and a new one is born at the far wall. It is evident, therefore, that the ratio of the length of the container to that of the roll wavelength is an important parameter in the problem. The so-called confined states are TW' that do not fill the whole container; in these states the TW' typically propagates towards the nearest endwall. Finally, the blinking states are those in which the confined state oscillates, generally irregularly, between a confined left-travelling wave near the left wall and a confined right-travelling wave near the right wall.

All of the above states were in fact first suggested theoretically. To study the TW periodic boundary conditions in the horizontal are natural [5], and have for this reason been frequently

Nonlinear Evolution of Spatio-Temporal Structures in Dissipative Continuous Systems
Edited by F.H. Busse and L. Kramer
Plenum Press, New York, 1990

adopted. Cross [6] was the first to study TW' in a finite geometry, emphasizing in particular the role played by reflection from the endwalls. At the same time numerical experiments on closely related doubly diffusive convection in finite geometry revealed the existence of the blinking states [7]. Both the confined and blinking states were studied in detail by Cross [8] using coupled Ginzburg-Landau equations for the (complex) amplitudes A and B of left- and right-travelling waves. These equations describe the evolution of small-amplitude slowly-varying wave trains produced through a Hopf bifurcation from a trivial (conduction) state in a continuous translation-invariant system. Such waves are described by an asymptotic expansion of the form

$$u(x,t;\epsilon) \sim \epsilon V_1 e^{i\omega_c t}\{A(X,T)\, e^{ik_c x} + \bar{B}(X,T)\, e^{-ik_c x}\} + c.c. + O(\epsilon^2)\,, \tag{1}$$

where $0 < \epsilon \ll 1$ is a small parameter proportional to $(R - R_c)^{1/2}$, with R the Rayleigh number and R_c its critical value for the Hopf bifurcation. The slow spatial and temporal scales X and T are defined by

$$X = \epsilon x, \qquad T = \epsilon^2 t, \tag{2}$$

while k_c denotes the critical wavenumber for which the bifurcation first appears, and ω_c the corresponding frequency.

The complex envelope functions A and B obey evolution equations of the form [9]

$$A_T = \{\partial_X^2 + s\partial_X + \Lambda + b|B|^2 + (a+b)|A|^2\}A \tag{3a}$$

$$B_T = \{\partial_X^2 - s\partial_X + \bar{\Lambda} + \bar{b}|A|^2 + (\bar{a}+\bar{b})|B|^2\}B, \tag{3b}$$

where s is related to the group velocity,

$$s = \frac{1}{\epsilon}\left(\frac{\partial \omega}{\partial k}\right)_c \tag{4}$$

and like the complex coefficients a, b is assumed to be $O(1)$. Note that this assumption formally restricts the applicability of equations (3) to systems with $(\partial\omega/\partial k)_c = O(\epsilon)$. The quantity Λ is also complex, its real part proportional to $R - R_c$ and its imaginary part to $(\partial\omega/\partial R)_c(R - R_c)$. In the absence of spatial variation these equations reduce to the normal form for the Hopf bifurcation with O(2) symmetry [10] [11]. Consequently, the coefficients a, b can be deduced from calculations that assume spatial periodicity with period $2\pi/k_c$. In addition the dispersive terms are linear, and hence follow directly from the dispersion relation. The coefficients in equations (3) have been calculated for realistic boundary conditions in [12] (linear terms) and in [13] (nonlinear terms).

The discussion that follows makes important use of the symmetry properties of equations (3). These are given by the operations

$$T: (A,B) \to e^{i\psi}(A,B), \qquad x \to x + \psi/k_c \tag{5a}$$

$$R: (A,B) \to (\bar{B},\bar{A}), \qquad x \to -x \tag{5b}$$

implied by translation and reflection symmetry of the system. Together these operations constitute the appropriate action of the group O(2) on the amplitudes A and B. An additional symmetry,

$$P: (A,B) \to (e^{i\phi}A, e^{-i\phi}B), \qquad t \to t + \phi/\omega_c\,, \tag{5c}$$

follows from a phase-shift invariance which is a consequence of the normal form transformations that typically have to be performed to put the evolution equations into the form (3) [11].

Depending on the boundary conditions the symmetry group of equations (3) may be increased (periodic boundary conditions), unchanged (Dirichlet or Neumann boundary conditions) or decreased (Robin boundary conditions). Some of the subtleties associated with

Dirichlet or Neumann boundary conditions are discussed in [14] [15]. In the present article it is the last case that is of interest. When equations (3) describe waves in a large container, the translation symmetry T is broken by the boundary conditions. These conditions are [8] [16]

$$A - \epsilon(\mu A_X + \nu \bar{B}_X) = 0 \quad , \quad B - \epsilon(\mu B_X + \nu \bar{A}_X) = 0 \quad \text{at } X = 1 \tag{6a}$$

$$A + \epsilon(\bar{\mu} A_X + \bar{\nu} \bar{B}_X) = 0 \quad , \quad B + \epsilon(\bar{\mu} B_X + \bar{\nu} \bar{A}_X) = 0 \quad \text{at } X = -1, \tag{6b}$$

where μ and ν are complex coefficients that can be explicitly computed from the underlying equations. Observe that the boundary conditions (6) break only the symmetry (5a), while preserving the symmetries (5b,c).

The purpose of the present article is to describe, from the point of view of bifurcation theory, the effect of breaking the translation symmetry. A preliminary account of this work has been given elsewhere[17], and a detailed analysis has now been completed [18]. The analysis treats the symmetry-breaking effects as small (cf. eq.(6)), and hence is able to discuss the effect of distant endwalls as a perturbation of a system equivariant under the symmetries (5). In the case where it is the reflection symmetry that is broken, such a procedure can be justified rigorously [19]. The analysis that follows focuses on the small-amplitude solutions of equations (3, 6). In the next section we derive the appropriate normal form near the initial instability. In section 3 we summarize the properties of the unperturbed problem, i.e., one equivariant with respect to the symmetries (5). In section 4 we specialize the existing analysis [18] to the parameter values used by Cross [8] in his numerical calculations, and draw a number of conclusions about the system (3, 6). The implications of these results are discussed in section 5.

2 Center manifold reduction

In the absence of spatial variation equations (3) are already in normal form [11]. With the spatial terms included they can be reduced to a corresponding normal form near the onset of instability by means of a center manifold reduction. Since ϵ in the boundary conditions (6) is a small parameter, we write

$$A = \epsilon^{1/2} A_0 + \epsilon^{3/2} A_1 + \ldots, \qquad B = \epsilon^{1/2} B_0 + \epsilon^{3/2} B_1 + \ldots, \tag{7}$$

and assume that the coefficients a, b satisfy the nondegeneracy conditions [11]

$$a_r \neq 0, \qquad b_r \neq 0, \qquad a_r + 2b_r \neq 0, \tag{8}$$

where the subscript r denotes the real part. Let $T' = \epsilon T$ be a super-slow time and $\Lambda - \Lambda_c = \epsilon\Delta$, where Λ_c is to be determined. At $O(\epsilon^{1/2})$ we have

$$[\partial_X^2 + s\partial_X + \Lambda_c]A_0 = 0, \qquad A_0(\pm 1) = 0 \tag{9a}$$

$$[\partial_X^2 - s\partial_X + \bar{\Lambda}_c]B_0 = 0, \qquad B_0(\pm 1) = 0. \tag{9b}$$

Consequently, the (real) eigenvalue Λ_c is given by

$$\Lambda_c = (\frac{\pi}{2})^2 + (\frac{s}{2})^2 \tag{10}$$

with the corresponding eigenfunctions

$$A_0 = v \exp(-\frac{s}{2}X)\cos(\frac{\pi}{2}X), \quad B_0 = w \exp(\frac{s}{2}X)\cos(\frac{\pi}{2}X), \tag{11}$$

where v and w are complex amplitudes. At $O(\epsilon^{3/2})$ we have

$$[\partial_X^2 + s\partial_X + \Lambda_c]A_1 = A_{0T'} - [\Delta + b|A_0|^2 + (a+b)|B_0|^2]A_0 \tag{12a}$$

$$[\partial_X^2 - s\partial_X + \Lambda_c]B_1 = B_{0T'} - [\bar{\Delta} + \bar{b}|B_0|^2 + (\bar{a}+\bar{b})|A_0|^2]B_0, \tag{12b}$$

subject to the boundary conditions

$$A_1 - (\mu A_{0X} + \nu \bar{B}_{0X}) = 0 \quad , \quad B_1 - (\mu B_{0X} + \nu \bar{A}_{0X}) = 0 \quad \text{at } X = 1 \tag{13a}$$

$$A_1 + (\bar{\mu} A_{0X} + \bar{\nu} \bar{B}_{0X}) = 0 \quad , \quad B_1 + (\bar{\mu} B_{0X} + \bar{\nu} \bar{A}_{0X}) = 0 \quad \text{at } X = -1. \tag{13b}$$

Note that (9b) is adjoint to (9a) and vice versa. Thus to solve (12a) we multiply by B_0 and integrate over the interval $[-1, 1]$, and similarly for (12b). The solvability conditions yield

$$\dot{v} = \{\lambda + \tilde{a}|w|^2 + \tilde{b}(|v|^2 + |w|^2)\}v + d\bar{w} \tag{14a}$$

$$\dot{w} = \{\bar{\lambda} + \bar{\tilde{a}}|v|^2 + \bar{\tilde{b}}(|v|^2 + |w|^2)\}w + \bar{d}\bar{v}, \tag{14b}$$

where

$$\lambda = \Delta - (\pi^2/2)\mu_r, \qquad d = -(\pi^2/4)(\nu e^s + \bar{\nu} e^{-s}) \tag{15}$$

and

$$(\tilde{a}, \tilde{b}) = K(a, b), \qquad K = \frac{3\pi^4 \sinh s}{4s\Lambda_c(s^2 + 4\pi^2)}. \tag{16}$$

Here the dot denotes differentiation with respect to T'. Note that equations (14) are independent of ϵ showing that (7) is the correct scaling for the center manifold reduction. The choice of scaling relies on the nondegeneracy conditions (8) holding at $\lambda = d = 0$; should these conditions fail a different scaling has to be employed in order to bring essential fifth and seventh order terms into equations (14). Identical equations are obtained by applying an ϵ-dependent linear transformation which to $O(\epsilon)$ leads to pure Dirichlet boundary conditions, but at the expense of introducing additional terms into the governing equations (3). Then the usual center manifold reduction of the transformed problem with the ϵ-independent boundary conditions also leads to equations (14)-(16).

Observe that equations (15) commute with the symmetries [17]

$$R : (v, w) \rightarrow (\bar{w}, \bar{v}) \tag{17b}$$

$$P : (v, w) \rightarrow (e^{i\phi}v, e^{-i\phi}w), \tag{17c}$$

but no longer with the symmetry

$$T : (v, w) \rightarrow (e^{i\psi}v, e^{i\psi}w). \tag{17a}$$

Note, however, that the dynamics of (14) are accesible perturbatively in the limit $nu \rightarrow 0$ when T is broken only weakly.

3 The unperturbed problem

When $\mu = \nu = 0$ the translation invariance (17a) is responsible for the decoupling of the amplitude and phase equations [10] [11]. In terms of the real variables x_j, θ_j, $j = 1, 2$, defined by

$$v = x_1 e^{i\theta_1}, \qquad w = x_2 e^{i\theta_2} \tag{18}$$

equations (15) become

$$\dot{x}_1 = \{\lambda_r + \tilde{a}_r x_2^2 + \tilde{b}_r S^2\}x_1 \tag{19a}$$

$$\dot{x}_2 = \{\lambda_r + \tilde{a}_r x_1^2 + \tilde{b}_r S^2\}x_2 \tag{19b}$$

$$\dot{\theta}_1 = \lambda_i + \tilde{a}_i x_2^2 + \tilde{b}_i S^2 \tag{19c}$$

$$\dot{\theta}_2 = -\lambda_i - \tilde{a}_i x_1^2 - \tilde{b}_i S^2, \tag{19d}$$

where $S^2 \equiv x_1^2 + x_2^2$. Consequently equations (19a,b) suffice to determine the dynamics. These equations have the following three types of solutions (x_1, x_2): the trivial solution $(0, 0)$, the pure travelling wave $(x, 0)$ or $(0, x)$, and the standing wave (x, x). All other solutions are transient. In figure 1 we summarize the results of analyzing the amplitude equations (19a,b) under the nondegeneracy conditions (8). The figure shows the $(\tilde{a}_r, \tilde{b}_r)$-plane together with the bifurcation diagrams characteristic of each of the six regions. Observe that as a consequence of the O(2) spatial symmetry both TW and SW bifurcate simultaneously, with stable solutions present only when both bifurcate supercritically.

It is important to note that equations (19) admit a whole circle of standing waves since the overall phase $\theta_1 + \theta_2$ is arbitrary. The standing waves are therefore neutrally stable with respect to translations. In contrast, the travelling waves have a spatio-temporal symmetry: they are unchanged by a translation followed by an appropriate evolution (forwards or backwards) in time. The reflection R also acts differently on the two types of solutions: it leaves SW unchanged, but interchanges left- and right-travelling waves. Consequently the TW solutions are isolated.

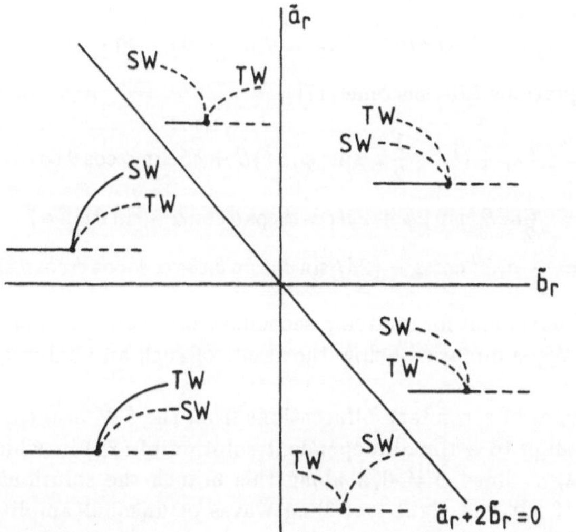

Figure 1. Division of the $(\tilde{a}_r, \tilde{b}_r)$- plane by the axes and the line $\tilde{a}_r + 2\tilde{b}_r = 0$ into six regions giving rise to different bifurcation diagrams for the unperturbed problem. Dashed branches are unstable.

4 The perturbed problem

When $\mu \neq 0$, $\nu \neq 0$ the translation symmetry is absent with profound consequences for the dynamics of equations (15). In this case the total phase $\theta = \theta_1 + \theta_2$ no longer decouples and equations (19) are replaced by

$$\dot{x}_1 = (\lambda_r + \tilde{a}_r x_2^2 + \tilde{b}_r S^2) x_1 + \delta x_2 \cos(\theta - \alpha) \quad (20a)$$

$$\dot{x}_2 = (\lambda_r + \tilde{a}_r x_1^2 + \tilde{b}_r S^2) x_2 + \delta x_1 \cos(\theta + \alpha) \quad (20b)$$

$$\dot{\theta} = \tilde{a}_i (x_2^2 - x_1^2) - \delta \frac{x_2}{x_1} \sin(\theta + \alpha) - \delta \frac{x_1}{x_2} \sin(\theta - \alpha), \quad (20c)$$

where

$$d = \delta e^{i\alpha}, \qquad 0 \le \alpha \le 2\pi \tag{21}$$

Consequently TW of the form $(x_1, x_2) = (x, 0)$ are no longer solutions. Solutions in the form of standing waves (x, x) continue to exist, but there is no longer a whole circle of them. Instead the endwalls select two values of θ, namely $\theta = 0, \pi$, by a process that may be called phase pinning. We refer to these as SW_0 and SW_π, respectively. From (20) their amplitudes and frequencies are given by

$$\lambda_r + (\frac{1}{2}\tilde{a}_r + \tilde{b}_r)S^2 \pm \delta \cos\alpha = 0 \tag{22a}$$

$$\Omega_{0,\pi} = \lambda_i + (\frac{1}{2}\tilde{a}_i + \tilde{b}_i)S^2 \pm \delta \sin\alpha. \tag{22b}$$

Thus $SW_{0,\pi}$ no longer bifurcate simultaneously from $S = 0$. Consequently, the effect at small amplitude of the presence of endwalls is indeed to set up standing waves with a fixed overall phase, as might intuitively be expected.

To study the counterpart of the TW solutions in the unperturbed problem we introduce the variable ϕ defined by

$$x_1 = S\cos(\phi/2), \quad x_2 = S\sin(\phi/2), \quad 0 \le \phi \le \pi, \tag{23}$$

in terms of which equations (20) become [17]

$$\dot{S} = \{\lambda_r + (\tilde{b}_r + \frac{1}{2}\tilde{a}_r \sin^2\phi)S^2\}S + \delta S \sin\phi\cos\theta\cos\alpha \tag{24a}$$

$$\dot{\phi} = \frac{1}{2}\tilde{a}_r S^2 \sin 2\phi + 2\delta(\cos\phi\cos\theta\cos\alpha - \sin\theta\sin\alpha) \tag{24b}$$

$$\dot{\theta} = -\tilde{a}_i S^2 \cos\phi - (2\delta/\sin\phi)(\sin\theta\cos\alpha + \cos\phi\cos\theta\sin\alpha). \tag{24c}$$

These variables are convenient for studying secondary bifurcations from the two SW branches ($\phi = \frac{\pi}{2}$, $\theta = 0, \pi$) . We summarize below the result of such an analysis, and refer to [18] for the details.

There are two types of secondary bifurcations from the SW branches. There is a steady-state bifurcation leading to a time-independent solution of (24) in which all three variables (S, ϕ, θ) vary with λ_r . Since $\phi \ne 0, \pi$ along this branch the solutions may be interpreted as a superposition of left-and right-travelling waves of unequal amplitudes ($\phi \ne \pi/2$). In contrast to the unperturbed problem in which such a superposition results in a modulated (i.e. quasiperiodic) travelling wave [20], in the present case the superposition is characterized by a single frequency $\dot{\theta}_1 = -\dot{\theta}_2 = \lambda_i + O(S^2)$, a fact that follows immediately from the condition for a fixed point in θ . Since, as S or equivalently λ_r increases, $\phi \to 0$ or π, this solution looks more and more like a pure travelling wave at large amplitudes. Consequently we refer to it as TW'. Since the bifurcation is in fact a pitchfork, there are two such solutions, distinguished by their direction of propagation.

The $SW_{0,\pi}$ branches may also lose stability in a secondary Hopf bifurcation. This bifurcation leads to an oscillation in both θ and ϕ, and so is an oscillation in the amplitudes x_1, x_2. We refer to this solution as MW' since it differs in character from the MW solution that occurs in the unperturbed problem when (x_1, x_2) with $x_1 \ne x_2$ becomes a possible solution [20]. In the (v, w)-space the MW' solution produces motion on a two-torus. Of the two frequencies that are present, the first is associated with the standing wave component, while the new frequency ω_2 results in a slow oscillation in ϕ and hence in x_1, x_2. Of particular interest is the fact that the MW' branch does not persist to large amplitudes. As λ_r increases it terminates on the TW' branch in either a global bifurcation or a tertiary Hopf bifurcation. The MW' branch is thus forced by the endwalls and has no counterpart in the unperturbed problem.

A nearly complete classification of the various secondary and tertiary bifurcations as a function of the perturbation phase α and the coefficients $\tilde{a}_r, \tilde{b}_r, \tilde{a}_i, \tilde{b}_i$ may be found in [18].

In the remainder of this section we relate the solutions determined above to the analysis of equations (3,6) carried out by Cross [8]. Cross argues, quite plausibly, that, in order for the group velocity $(\partial\omega/\partial k)_c$ to be $O(\epsilon)$, the propagation effects should be small, and consequently that one may take $a_i = b_i = 0$; his choice of the coefficients a_r, b_r implies

$$\tilde{a}_r = -3K, \qquad \tilde{b}_r = -K. \tag{25a}$$

With this choice both TW and SW bifurcate supercritically in the unperturbed problem, with TW stable and SW unstable. If we define $\tilde{a} = |\tilde{a}| \exp i\gamma$, then (25a) implies

$$\gamma = \pi. \tag{25b}$$

In addition Cross chooses μ, ν to be real and negative, implying that

$$\alpha = 0. \tag{25c}$$

In [18] we have classified the perturbed problem as a function of the parameters $k = 2\tilde{b}_r/|\tilde{a}|$ and $c = \cos\gamma$. In view of (25) the system studied by Cross is characterized by $k = -2/3$ and $c = -1$ and consequently falls into region III.5 of the (k, c)-plane in our classification. Since, in addition, $\beta \equiv \alpha - \gamma = -\pi$, the corresponding bifurcation diagram is as shown in fig. 2.

Observe that the first bifurcation is to SW_0. This bifurcation is supercritical, and the SW_0 branch is therefore stable near the bifurcation. As λ_r increases, the SW_0 branch loses stability to the TW' branch. This bifurcation is also supercritical, and the TW' branch produced is stable. There are no further local bifurcations. In particular for this choice of parameters the MW' branch is not present near the primary bifurcation.

In fig. 3 we show the TW' solution, reconstructed from (7,11,12), for $s = 1, 3, 6$. Both $A' = A_0 + \epsilon A_1$ and $B' = B_0 + \epsilon B_1$ are shown in the interval $-1 \leq X \leq 1$ for (a) λ_r near the secondary bifurcation, and (b) a significantly larger value of λ_r. It should be observed that the group velocity s enters only through the scale factor K of the coefficients a, b and the imperfection amplitude δ, and thus is absent from the present analysis, except in so far as it affects the structure of the eigenfunctions (11). The larger s is the more confined are the eigenfunctions to a neighborhood of the corresponding endwalls, the influence of the endwalls increasing with s through δ. These conclusions are in qualitative agreement with the numerical solutions to equations (3,6) [8]. Note that Cross scales equations (3) so that $\Lambda = 1$ while allowing the length l of the domain to vary. In contrast we have kept Λ as a bifurcation parameter, fixing the length l. These procedures are related by the scaling $\Lambda = l^2/4$, $s' = 2s/l$, where s' denotes Cross' group velocity. Furthermore, his amplitudes A_L, A_R are related to A, B according to $A_L = (2/l)\tilde{A}$ and $A_R = (2/l)B$. Since our analysis is valid when Λ is close to Λ_c we require Cross' parameters (s', l) to be near the curve $s' = 2[1 - (\pi/l)^2]^{1/2}$. Note that, since $0 \leq s' < 2$ for all s, we cannot describe the regime $s' > 2$. The values $s = 1, 3, 6$ chosen in fig. 3 correspond to Cross' parameters (s', l) being close to (0.6, 3.3), (1.4, 4.3), (1.8, 6.8), respectively.

In fig. 4 we show the effect of increasing $\beta \equiv \alpha - \gamma$ from $-\pi$ to $-\pi/2$ in region III.5; such a change can be accomplished by varying the parameter ν in the boundary conditions (6) without changing the parameters c and k. Note that the important quantity is the phase α of the endwall perturbation, and not its magnitude (provided the latter remains small). It should be noted in this context that the argument of ν depends on the length L of the container [6] [16]. Hence β can be varied simply by varying L, as already emphasized in the introduction. Fig. 4a, valid for $-\pi < \beta_1 < \beta < \delta_1 < -\pi/2$, shows that the bifurcation to TW' is now subcritical. Consequently, the TW' branch is initially unstable, although it acquires stability at a tertiary saddle-node bifurcation in order for the large amplitude behaviour to remain unchanged. For $\delta_1 < \beta < -\pi/2$ (fig. 4(b)), the SW_0 branch first loses stability at the secondary Hopf bifurcation producing a stable MW' branch. With increasing λ_r this branch terminates on the subcritical TW' branch in a global bifurcation in which the new

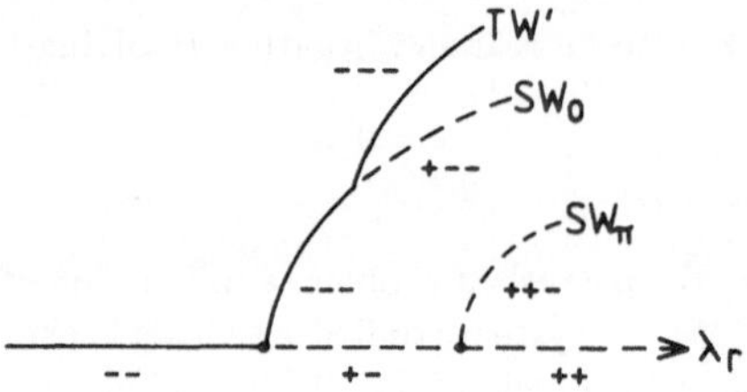

Figure 2. Bifurcation diagram corresponding to the system studied by Cross. The signs denote the stability assignments, i.e., the signs of the real parts of the eigenvalues determined by (24).

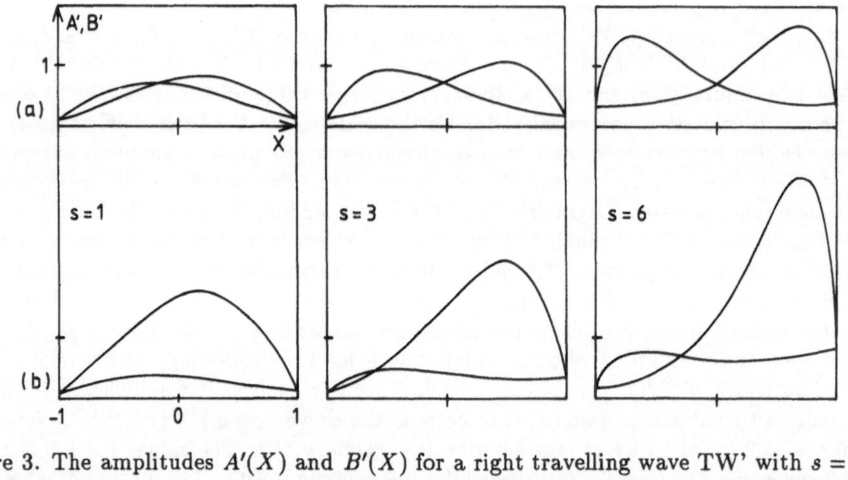

Figure 3. The amplitudes $A'(X)$ and $B'(X)$ for a right travelling wave TW' with $s = 1, 3, 6$ and $\epsilon\mu = \epsilon\nu = -0.03$. (a): λ_r close to the secondary bifurcation ($\lambda_r = 0.67$). (b): a significantly larger value of λ_r ($\lambda_r = 1.99$).

frequency ω_2 vanishes. As shown in fig. 5 the MW' solutions bear a strong resemblance to the blinking states, even though $s' < 2$. This is because ω_2 is the frequency of oscillations in the amplitudes x_1, x_2 . Thus for half the period $2\pi/\omega_2$ we have $x_1 > x_2$ and left-travelling waves dominate while for the other half $x_1 < x_2$ and right-travelling waves dominate, at least close to the secondary Hopf bifurcation. Note that since the Hopf frequency is small ($\omega_2 = O(S^2)$), the period between reversals of the direction of propagation of the wave pattern will be long compared with the period of the standing wave (SW_0) or the travelling wave (TW'). Note also that the period $2\pi/\omega_2$ lengthens as λ_r increases; after the global bifurcation the blinking state is replaced by TW'. Further transitions occur as β increases between $-\pi/2$ and 0 [18], and a similar analysis for the remaining regions in the (c, k) plane has also been carried out [18].

5 Discussion and conclusion

In this report we have studied the effects of breaking the translation symmetry on continuum systems undergoing a Hopf bifurcation from a translation-invariant equilibrium. Following our earlier work [17] we have shown that the primary effects of distant sidewalls are that (i) pure travelling waves (TW) are no longer possible; (ii) only standing waves (SW) bifurcate from the trivial state, with an overall phase that is no longer arbitrary; (iii) a counterpart of TW, denoted by TW', bifurcates in a secondary bifurcation from one of the two SW branches, and (iv) a two frequency mode MW' can appear via a secondary Hopf bifurcation from an SW branch. This mode is local in Rayleigh number and amplitude, and requires the broken translation symmetry for its presence[17].

We have applied these ideas to the coupled complex Ginzburg-Landau equations that describe the evolution of slowly varying wave trains generated by such a Hopf bifurcation. A reduction procedure, which we described in detail then yields a normal form (eq. (14)) for the evolution of the dominant modes near the onset of the instability. This normal form provides a generic description of the effects of broken translation symmetry, and can be derived on the basis of symmetry principles alone [17]. As a result we have made a tentative identification between the states denoted by TW' and the confined travelling waves, and between MW' and the blinking states, observed in experiments on binary fluid convection, and described by Cross [8] with the help of the coupled Ginzburg-Landau equations. This identification is incomplete because we cannot realize the regime $s' > 2$ near the primary bifurcation, although we can approach the crossover regime $s' \approx 2$ for $s \gg 1$. For the parameter values used by Cross in his numerical simulations we have found that the initial bifurcation is to stable standing waves (SW_0), followed by a secondary pitchfork to a pair of stable TW'. For slightly different boundary conditions we found that the standing waves first lose stability to the MW' state before a hysteretic transition to TW' takes place. Although the qualitative picture based on these results is similar to that determined by Cross, we emphasize that our results are valid for amplitudes of order $\epsilon\sqrt{\lambda_r}$ only, while Cross' calculations are carried out for amplitudes of order ϵ (i.e. $O(1)$ in the scaled variables). In addition, we assume in accordance with the scaling implied by the asymptotic analysis that the symmetry breaking terms in the boundary conditions (6) are $O(\epsilon)$ with $\epsilon \ll 1$. This scaling is not followed by Cross in his numerical study [8]. We emphasize that the solutions described here and in a forthcoming detailed paper [18] are asymptotic solutions of the coupled complex Ginsburg-Landau equations on a long but finite interval. The correspondence with both the experiments [2] [3] and Cross' numerical work [8] is therefore encouraging.

It should be observed that we have discussed the simplest possible symmetry-breaking effects. In particular we have used boundary conditions that are identical at both ends of the container. The interplay between perturbations that break the translation symmetry and those that break the reflection symmetry [19],[21] is beyond the scope of the present work.

We note finally that the generalization of the Swift-Hohenberg equation [22] employed recently by Bestehorn et al. [23] to simulate convection in binary fluid mixtures also reduces

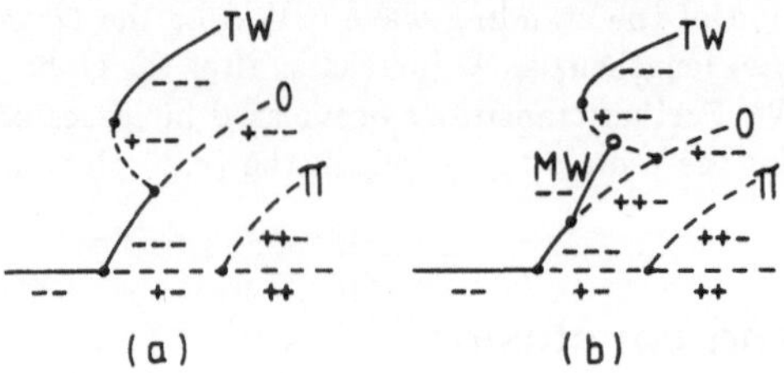

Figure 4. Bifurcation diagrams for increasing $\beta \equiv \alpha - \gamma$ from $-\pi$ to $-\pi/2$. (a) $\beta_1 < \beta < \delta_1$, (b) $\delta_1 < \beta < -\pi/2$. The angles β_1, δ_1 $(-\pi < \beta_1 < \delta_1 < -\pi/2)$ are determined in [18]. For $-\pi < \beta < \beta_1$ the diagram is as in figure 2.

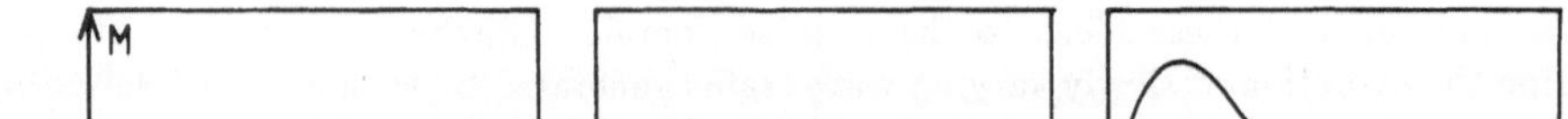

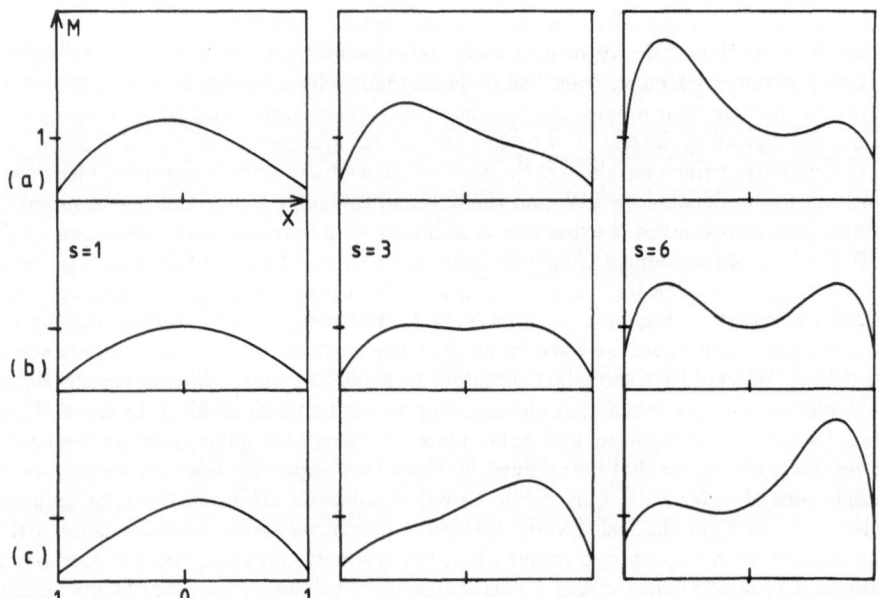

Figure 5. Modulation amplitudes $M = (|A'|^2 + |B'|^2)^{1/2}$ along a MW' solution as functions of X for three fixed times T during one period of oscillation; $\epsilon\mu = \epsilon\nu = -0.03$. (a): left-travelling waves dominate, (b): mixed state, (c): right-travelling waves dominate.

to the coupled complex Ginsburg-Landau equations for slowly varying wave-trains. This equation has, however, the advantage that it can also describe wavetrains that are not slowly varying. For distant endwalls these equations can also be reduced to the normal form (15), and confined and blinking states identified analytically.

Acknowledgement

We thank M. Wegelin for producing figures 3 and 5 and B. Dangelmayr for preparing the manuscript. The work of E. Knobloch was supported by NSF/DARPA under grant DMS-8814702.

References

[1] R.W. Walden, P. Kolodner, A. Passner and C.M. Surko, *Traveling waves and chaos in convection in binary fluid mixtures*, Phys. Rev. Lett. **55**: 496 (1985).

[2] E. Moses, J. Fineberg and V. Steinberg, *Multistability and confined traveling-wave patterns in a convecting binary mixture*, Phys. Rev. **A 35**: 2757 (1987); R. Heinrichs, G. Ahlers and D.S. Cannell, *Traveling waves and spatial variation in the convection of a binary mixture*, Phys. Rev. **A 35**: 2761 (1987).

[3] J. Fineberg, E. Moses and V. Steinberg, *Spatially and temporally modulated traveling-wave pattern in convecting binary mixtures*, Phys. Rev. Lett. **61**: 838 (1988); P. Kolodner and C.M. Surko, *Weakly nonlinear traveling-wave convection*, Phys. Rev. Lett. **61**: 842 (1988).

[4] A.E. Deane, E. Knobloch, J. Toomre, *Traveling waves and chaos in thermosolutal convection*, Phys. Rev. **A 36**: 2862 (1987); S.J. Linz, M. Lücke, H.W. Müller and J. Niederländer, *Convection in binary fluid mixtures: Traveling waves and lateral currents*, Phys. Rev. **A 38**: 5727 (1988).

[5] C. Bretherton and E.A. Spiegel, *Intermittency through modulational instability*, Phys. Lett. **96 A**: 152 (1983); E. Knobloch, *Doubly diffusive waves*, in *Doubly Diffusive Motions*, FED-Vol. **24**, N.E. Bixler and E.A. Spiegel, eds., ASME, New York (1985).

[6] M.C. Cross, *Traveling and standing waves in binary-fluid convection in finite geometries*, Phys. Rev. Lett. **57**: 2935 (1986).

[7] A.E. Deane, E. Knobloch and J. Toomre, *Doubly diffusive waves*, in *Proc. Int. Conf. on Fluid Mechanics (Beijing 1987)*, Peking University Press (1987); A.E. Deane, E. Knobloch and J. Toomre, *Traveling waves in large-aspect-ratio thermosolutal convection*, Phys. Rev. **A 37**: 1817 (1988).

[8] M.C. Cross, *Structure of nonlinear traveling-wave states in finite geometries*, Phys. Rev. **A 38**: 3593 (1988).

[9] P. Coullet, S. Fauve and E. Tirapegui, *Large scale instability of nonlinear standing waves*, J. Phys. Lettres **46**: L-787 (1985).

[10] M. Golubitsky and I. Stewart, *Hopf bifurcation in the presence of symmetry*, Arch. Rat. Mech. and Anal. **87**: 107 (1985).

[11] E. Knobloch, *Oscillatory convection in binary mixtures*, Phys. Rev. **A 34**: 1538 (1986).

[12] M.C. Cross and K. Kim, *Linear instability and the codimension-2 region in binary fluid convection between rigid impermeable boundaries*, Phys. Rev. **A 37**: 3909 (1988).

[13] W. Schöpf and W. Zimmermann, *Multicritical behaviour in binary fluid convection*, Europhys. Lett. **8**: 41 (1989).

[14] G. Dangelmayr and D. Armbruster, *Steady state mode interactions in the presence of O(2) symmetry and in non-flux boundary conditions*, Contemp. Math. **56**: 53 (1986).

[15] J.D. Crawford, M. Golubitsky, M.G.M. Gomas, E. Knobloch and I. Stewart, *Boundary conditions as symmetry constraints*, in *Proc. Warwick Symp. on Singularity Theory and its Applications*, to appear (1990).

[16] M.C. Cross, P.G. Daniels, P.C. Hohenberg and E.D. Siggia, *Phase-winding solutions in a finite container above the convective threshold*, J. Fluid Mech. **127**: 155 (1983); see also ref. [8].

[17] G. Dangelmayr and E. Knobloch, *On the Hopf bifurcation with broken O(2) symmetry*, in *The Physics of Structure Formation: Theory and Simulation*, W. Güttinger and G. Dangelmayr, eds., Springer-Verlag, Berlin (1987); G. Dangelmayr and E. Knobloch, *Hopf bifurcation in reaction-diffusion equations with broken translation symmetry*, in *Proc. of the Int. Conf. on Bifurcation Theory and its Numerical Anlaysis*, Li Kaitai, J. Marsden, M. Golubitsky, G. Iooss, eds., Xian Jiatong University Press, Xian (1989).

[18] G. Dangelmayr and E. Knobloch, *Hopf bifurcation with broken circular symmetry*, Nonlinearity (submitted).

[19] S. van Gils and J. Mallet-Paret, *Hopf bifurcation and symmetry: travelling and standing waves on the circle*, Proc. Roy. Soc. Edinburgh **104 A**: 279 (1986).

[20] E. Knobloch, *On the degenerate Hopf bifurcation with O(2) symmetry*, Contemp. Math. **56**: 193 (1986).

[21] J.D. Crawford and E. Knobloch, *On degenerate Hopf bifurcation with broken O(2) symmetry*, Nonlinearity **1**: 617 (1988).

[22] T. Ohta and K. Kawasaki, *Euclidean invariant phase dynamics for propagating pattern*, Physica **27 D**: 21 (1987); M. Bestehorn, R. Friedrich and H. Haken, *The oscillatory instability of a spatially homogeneous state in large aspect ratio systems of fluid dynamics*, Z. Phys. **B 72**: 265 (1988).

[23] M. Bestehorn, R. Friedrich and H. Haken, *Two-dimensional traveling wave patterns in nonequilibrium systems*, Z. Phys. **B 75**: 265 (1989).

KINKS AND SOLITONS IN THE GENERALIZED GINZBURG-LANDAU EQUATION

Boris A. Malomed* and Alexander A. Nepomnyashchy**

*P. P. Shirshov Institute for Oceanology of the USSR Academy of Sciences, 23 Krasikov St., Moscow 117218, USSR
**Institute for Continuous Media Mechanics of the Ural Branch of the USSR Academy of Sciences, 1 Academician Korolev St., Perm, 614061, USSR

INTRODUCTION

The present paper is devoted to the study of localized patterns in models in which a trivial homogeneous state is stable against infinitesimal disturbances, but can be triggered into a nontrivial oscillatory state by a finite disturbance. A well-known example of a physical medium that demonstrates this property is a layer of a binary liquid heated from below, where oscillatory convection sets in via a subcritical bifurcation.

A simplest model of such a system is based on the generalized Ginzburg-Landau (GL) equation put forward in Ref. 1 :

$$u_t = \gamma u + (1 + i\beta)u_{xx} + (1 + i\alpha)|u|^2 u - 3(1 + i\delta)|u|^4 u/16 \qquad (1)$$

where the coefficients γ, β, α, and δ are all real.

The trivial solution to Eq. (1), u = 0, ts stable in the subcritical range $\gamma < 0$. A nontrivial subcritical traveling-wave (TW) solution to Eq. (1),

$$u = a_0 \exp\left[i(kx - \omega t)\right], \qquad (2a)$$

$$a_0^2 = 4\left[2 \pm \sqrt{4 + 3(\gamma - k^2)}\right]/3, \qquad (2b)$$

$$\omega = (\alpha - \beta k^2)a_0^2 - 3\delta a_0^4/16,$$

Nonlinear Evolution of Spatio-Temporal Structures in Dissipative Continuous Systems
Edited by F.H. Busse and L. Kramer
Plenum Press, New York, 1990

exists in the wavenumber region

$$k^2 \lesssim k_0^2 = \gamma + 4/3, \qquad (3)$$

provided γ exceeds a threshold value:

$$\gamma \geq \gamma_{thr} = -4/3. \qquad (4)$$

Among the two TW solutions corresponding to different signs in Eq. (2b), only the one corresponding to the upper sign (i. e., a TW with the larger amplitude) is stable against infinitesimal disturbances.

The coexistence of the stable trivial solution with the stable TW suggests that there may also exist a kink-like configuration interpolating between the two stable states. Recently, an exact kink solution to Eq. (1) has been found in Ref. 2. The main idea of the present work is that, under certain conditions, a kink and an antikink may form a stable localized bound state (a "soliton"). The soliton can be described analytically in two opposite cases: when the dispersion coefficients α and β are large (while the coefficient δ remains small in comparison with them), and when all the coefficients α, β, and δ are small. In the former case, the stable soliton has been investigated by one of the present authors in the Appendix to Ref. 3. The objective of the present paper is to study stable soliton-like solutions in the opposite case, when the dispersion coefficients α, β, and δ in Eq. (1) are small.

THE KINK SOLUTION

The kink-like solution to Eq. (1) found in Ref. 2 has, in a general case, a rather cumbersome form. For our purpose, it is more convinient to find the kink as an approximate solution, starting from the particular case $\alpha = \beta = \delta = 0$, $\gamma = -1$, when Eq. (1) has an exact elementary solution

$$u_0(x) = 2\left[1 + \exp(-2\sigma(x - \xi))\right]^{-1/2}. \qquad (5)$$

Here $\sigma = \pm 1$ and ξ are the kink's polarity and "center-of-mass" coordinate. Now, assuming

$$\alpha^2 \sim \beta^2 \sim \delta^2 \sim (\gamma + 1) \qquad (6)$$

to be small, we will construct a kink solution based upon the zeroth approximation (5).

It is convinient to search for a solution in the form $u(x,t) = a(x,t)\exp\left[i\phi(x,t)\right]$, a and ϕ being the real amplitude and phase. The corresponding evolution equations ensuing from (1) are

$$a_t = \gamma a + a^3 - 3a^5/16 + a_{xx} - a\phi_x^2 -$$

$$\beta(a\phi_{xx} + 2a_x\phi_x), \tag{7a}$$

$$\phi_t = \alpha a^2 - 3\delta a^4/16 + \phi_{xx} + 2a^{-1}a_x\phi_x +$$

$$\beta(a^{-1}a_{xx} - \phi_x^2). \tag{7b}$$

Under the assumption (6), we look for a kink solution to Eqs. (7) in the form

$$a = u_0(z) + a_2(z) + \ldots, \qquad \phi = \omega t + \int k(z)dz,$$

$$k(z) = k_1(z) + \ldots, \; z \equiv x - ct, \tag{8}$$

where it is implied

$$\omega \sim \alpha, \; c \sim \alpha^2, \; k_1 \sim \alpha, \; a_2 \sim \alpha^2.$$

Let us set, for definiteness, $\sigma = +1$ in the expression (5). Straitforward calculation[4] yields

$$\omega = 4\alpha - 3\delta, \tag{9}$$

$$k_1(z) = A + Bu_0^2(z), \tag{10}$$

$$c = -2(\gamma + 1) + 2A^2 + 8AB + 16B^2 + 2\beta(8B/3 + A), \tag{11}$$

where

$$A \equiv 2\alpha - \beta/2 - 3\delta/2, \; B \equiv 3(\beta - \delta)/16. \tag{12}$$

According to Eqs. (10), (12), and (5),

$$k_\infty = k_1(z = +\infty) = 2\alpha + \beta/4 - 9\delta/4. \tag{13}$$

Equation (13) implies that the kink accomplishes the wavenumber selection. While the wavenumber k of the TW solution (2a) and (2b) is an arbitrary parameter (restricted by the inequality (3)), the only value of k at which the TW may be matched via the kink to the trivial solution is given by the expression (13).

The expression (11) implies that in the presence of the weak dispersion the equilibrium value $\gamma = -1$, at which the boundary between the zero and nonzero phases (the kink) is quiescent, changes to

$$\gamma_* = -1 + A^2 + 4AB + 8B^2 + \beta(8B/3 + A). \tag{14}$$

A KINK-ANTIKINK BOUND STATE

Let us proceed to consideration of localized states described by Eq. (1). First at all, in the dispersionless case $\alpha = \beta = \delta = 0$ this equation has an exact localized solution

$$u^2(x) = -4\gamma\left[1 + \sqrt{1+\gamma}\ \cosh(2\sqrt{-\gamma}\ x)\right]^{-1} \tag{15}$$

(provided $\gamma + 1 > 0$), which is surely unstable. Note that in the limit $0 < 1 + \gamma << 1$ the solution (1) describes a bound state of a kink and an antikink separated by the large distance

$$L_0 = \ln(1 + \gamma)^{-1}/2 \tag{16}$$

(Fig. 1). This bound state is unstable because the kink and antikink attract each other, while the factor $\gamma + 1 > 0$ gives rise to an effective "pressure" inside the localized state.

Let us demonstrate how a kink-antikink pair may form a stable bound state in the case of weak dispersion. To this end, let us consider the structure of the local wavenumber field k(x) in the configuration shown in Fig. 1. If the kink-antikink interaction is negligible, inside the nonzero phase k(x) must take the values $+ k_\infty$ and $- k_\infty$ in the regions adjacent to the left and right kinks, respectively (k_∞ is defined in Eq. (13)). To match the opposite values $\pm k_\infty$, we need to arrange a transient layer. That layer can be described in the so-called geometric optics approximation[5] based on the assumption that the local amplitude a(x, t) and wavenumber $k(x,t) \equiv \phi_x$ (see Eqs. (7)) vary at large spatial and time scales. Besides, we will assume $k^2 << 1$. Under this assumption, at the zeroth order of the GO technique Eq. (7) yields the quasistationary relation between a^2 and k^2:

$$a^2 \approx 2(2 - \phi_x^2). \tag{17}$$

Insertion of Eq. (17) into Eq. (7) yields, at the next order, the nonlinear phase diffusion equation

$$\phi_t = \phi_{xx} + (4\alpha - 3\delta) - (2\alpha + \beta + 3\delta)\phi_x^2. \tag{18}$$

Let us look for its solution in the form

$$\phi = \omega t + \int k(x)\,dx, \tag{19}$$

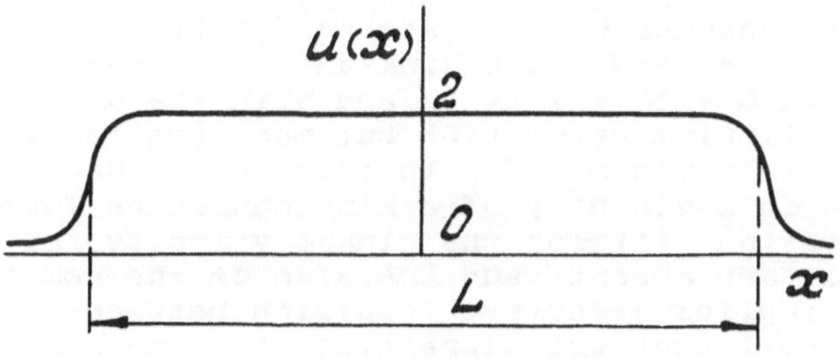

Fig. 1. A kink-antikink bound state.

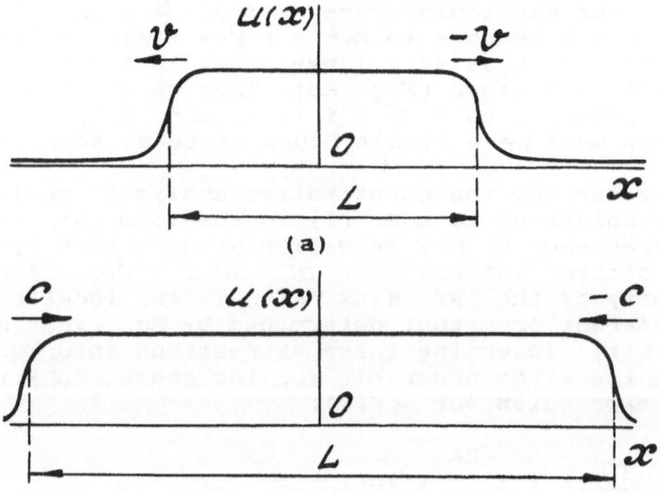

Fig. 2. Formation of the stable kink-antikink bound state: (a) $1 \ll L \ll l$; (b) $L \gg l$, l being the transient layer's width (23). The arrows indicate the signs of the kink's velocities: (a) The velocity v is given by the expression (11) without the dispersion term, i. e., $v = -2(\gamma + 1)$; (b) The velocity c is given by the full expression (11) which is assumed to be positive.

$$\omega = 4\alpha - 3\delta - (2\alpha + \beta + 3\delta)\bar{k}^2, \tag{20}$$

where $\pm\bar{k}$ are the values of $k(x)$ at $x = \mp\infty$. Substitution of Eqs. (19) and (20) into Eq. (18) brings us the eventual equation for the local wavenumber:

$$dk/dx = (2\alpha + \beta + 3\delta)(k^2 - \bar{k}^2). \tag{21}$$

Equation (21) has an evident solution

$$k(x) = -\bar{k}\tanh(k(2\alpha + \beta + 3\delta)x), \tag{22}$$

which describes the necessary transient layer between $k = +\bar{k}$ and $k = -\bar{k}$. As seen from Eq. (22), a characteristic width of the transient layer is

$$l \sim 1/\alpha\bar{k} \sim \alpha^{-2} \tag{23}$$

where we have assumed $\bar{k} \sim \alpha$ (see Eq. (13)).

Now let us consider the kink-antikink configuration (Fig. 2a) with $0 < \gamma + 1 \ll 1$, and with the distance L larger than the equilibrium value (16) but much smaller than the transient layer's width (23). In this case, the local wave-number field in a vicinity of a kink cannot be formed, i. e., in the expression (11) for the kink's velocity the dispersion-induced terms are absent, and the sign of the remaining term $-2(\gamma + 1)$ implies effective repulsion between the kink and antikink (recall that the expression (11) pertains to the left kink in Fig. 2a). Next, let us assume that the entire expression (11) is positive (according to Eq. (12), this is surely possible in the particular case $\alpha \neq 0$, $\beta = \delta = 0$, when the inequality $c > 0$ reduces to $\alpha^2 > (\gamma + 1)/4$). In the case when the distance L is very large, $L \gg 1$, the kinks will move to meet each other (Fig. 2b). Thus we conclude that in the case considered ($0 < \gamma + 1 \ll 1$, $\alpha^2, \beta^2, \delta^2 \ll 1$, $c > 0$) there must be a stable bound state at some intermediate value of L.

Let us develop the quantitative analysis. We look for a stationary solutions to Eqs. (7) in the form (8). We admit that the frequency ω may be different from that (9) due to the kink-antikink interaction: $\omega = \omega_1 + \delta\omega$. Accordingly, in a vicinity of the left kink (Fig. 1) the local wavenumber may be different from that determined by Eq. (10): $k(x) = k_1(x) + \delta k(x)$. Inserting these expressions into Eq. (7b) and taking the first order (i. e., the order $\sim \alpha$) yield the following expression for $\delta k(x)$:

$$\delta k(x) = \delta\omega (1 + e^{-2x}) \ln(1 + e^{2x})/2, \qquad (24)$$

where we have used the expression (5) for $u_0(x)$. At $x \to +\infty$ the solution (24) takes the asymptotic form

$$\delta k(x) = \delta\omega \cdot x \,. \qquad (25)$$

Similar analysis of the solution near the right kink (Fig. 1) yields the same asymptotic form (25) at $x \to -\infty$. To match the solutions, we must set

$$\delta\omega = 2k_\infty/L, \qquad (26)$$

L being the size of the bound state sought for.

At last, let us find a corresponding correction δc to the kink's velocity (11). Inserting the expressions (5), (13), (24), and (26) into the resolvability condition of the equation determining the value a_2, one can find a correction $\delta c \sim L^{-1}$ in an explicit form. We will write a full expression for the particular case $\beta = \delta = 0$: $c = -8\alpha^2 L^{-1}$. The corresponding full velocity

$$c + \delta c = -2(\gamma + 1) + 8\alpha^2 - 8\alpha^2 L^{-1}$$

(see Eq. (11)) vanishes at $L = L_0$, where

$$L_0 = 4\alpha^2 (\gamma_* - \gamma)^{-1}, \qquad (27)$$

γ_* being the shifted equilibrium value of γ, see Eq. (14).

CONCLUSION

We have demonstrated that Eq. (1) supports stable soliton-like solutions in two opposite limit cases. Recently, Eq. (1) with arbitrary values of parameters was a subjest of numerical investigations performed in Ref. 6. The numerical results of Ref. 6 provide a natural crossover between the two abovementioned analytically tractable limit cases.

In the same time, the existence of the unstable soliton-like solution which goes over into the one (15) in dispersionless case suggests a possibility of a bifurcation of merger of the two "solitons" (the stable and unstable ones). This issue requires further numerical analysis.

Stable localized states of a large size have been observed in experiments on traveling-wave subcritical convection in a narrow annular channel[7]. However, a size of the localized state observed in experiments could vary in wide limits, while the size of the stable "soliton" in the model (1) is uniquely determined. It seems more feasible that the stable "soliton" solution to (1) may be employed to interprete the quasi-one-dimensional (strongly stretched) localized states observed in the recent experiments on traveling-wave convection in a layer of a nematic liquid crystals[8].

An abridged version of the present paper has been published in Ref. 4. When submitting the full version of the paper, we have been kindly informed by L. Kramer about the preprint [9] in which some ideas close to ours are put forward.

We are indebted to A. Joets, A. V. Klyachkin, P. Kolodner, L. Kramer, and A. C. Newell for useful discussions and remarks.

REFERENCES

1. V. I. Petviashvili and A. M. Sergeev, Spiral solitons in active media with an excitation threshold, Dokl. AN SSSR (Sov. Phys. - Doklady) 276:1380 (1984).
2. A. V. Klyachkin, Modulational instability and autowaves in active media described by nonlinear equations of the Ginzburg-Landau type, Preprint (1989).
3. B. A. Malomed, Evolution of nonsoliton and "quasiclassical" wavetrains in nonlinear Schrödinger and Korteweg-de Vries equations with dissipative perturbations, Physica D, 29:155 (1987).
4. B. A. Malomed and A. A. Nepomnyashchy, Kinks and solitons in a generalized Ginzburg-Landau equation, Proc. of the IY Int. Workshop on Nonlinear and Turbulent Processes in Physics, Naukova Dumka, Kiev, 2:291 (1989).
5. B. A. Malomed, Nonsteady waves in distributed dynamical systems, Physica D 8:343 (1983).
6. O. Thual and S. Fauve, Localized structures generated by subcritical instabilities, J. Phys. 49:1829 (1988).
7. P. Kolodner, D. Bensimon, and C. M. Surko, Traveling-wave convection in an annulus, Phys. Rev. Lett. 60:1723 (1988).
8. A. Joets and R. Ribotta, Localized, time-dependent state in the convection of a nematic liquid crystal, Phys. Rev. Lett. 60:2164 (1988).
9. V. Hakim, P. Jakobsen, and Y. Pomeau, Fronts vs solitary waves in nonequilibrium systems, Preprint (1989).

ONSET OF CHAOS IN THE GENERALIZED GINZBURG-LANDAU EQUATION

Boris A. Malomed* and Alexander A. Nepomnyashchy**

*P. P. Shirshov Institute for Oceanology of the USSR Academy of Sciences, 23 Krasikov St., Moscow 117218, USSR

**Institute for Continuous Media Mechanics of the Ural Branch of the USSR Academy of Sciences, 1 Academician Korolev St., Perm, 614061, USSR

Study of chaos in the generalized Ginzburg-Landau equation (GLE)

$$iu_t + u_{xx} + 2|u|^2 u = i\epsilon_1 u - i\epsilon_3 |u|^2 u + i\epsilon_2 u_{xx} \qquad (1)$$

is a subject of great current interest, see, e.g., Refs. 1-6. We will consider Eq. (1) with the periodic boundary condition

$$u(x) = u(x + 2\pi/k). \qquad (2)$$

It has been proved[4] that in this case chaotic regimes (revealed previously in numerical simulations[1,3]) are accounted for by a finite-dimensional strange attractor in an infinite-dimensional phase space of the system (1),(2). One cannot explicitly describe that attractor by a system of a finite number of ordinary differential equations. However, we will demonstrate in the present work that, in the near-nonlinear-Schroedinger (NNS) regime, i.e., at $\epsilon_2, \epsilon_3 << 1$, the onset of chaos is governed by a three-dimensional dynamical system which, depending on values of parameters, either coinsides exactly with the famous Lorenz model[7] or gives rise to a new chaos-generating model allied to the Lorenz one.

Our starting point is the evident spatially homogeneous solution to Eq. (1):

$$u = a_0 \exp(2ia_0^2 t), \qquad (3)$$

where the amplitude a_0 is uniquely determined by the balance between gain ($\sim \epsilon_1$) and dissipation ($\sim \epsilon_3$):

Nonlinear Evolution of Spatio-Temporal Structures in Dissipative Continuous Systems
Edited by F.H. Busse and L. Kramer
Plenum Press, New York, 1990

$$a_0^2 = \epsilon_1 / \epsilon_3 . \tag{4}$$

As is well known, the solution (4) is subject to the modulational instability[8] against small perturbations with the wavenumbers $0 < q^2 < 4a_0^2$. Due to the periodic boundary condition (2), only the discrete wavenumbers $q = nk$ ($n = 0, 1, 2, \dots$) are admitted. It is natural to expect that the onset of chaos takes place when the minimum admitted wavenumber $q = k$ gets close to the modulational instability threshold $4a_0^2$:

$$0 < \varkappa^2 \equiv (4a_0^2)^{-1} (4a_0^2 - k^2) \ll 1. \tag{5}$$

Thus, in the NNS regime, the onset of chaos is governed by three small positive parameters ϵ_2, ϵ_3, and $\varkappa$. As we will see below, of basic interest is the range $\varkappa \sim \epsilon_2 \sim \epsilon_3$, i.e., as a matter of fact, we will deal with two independent parameters $A \equiv \epsilon_2/\epsilon_3$ and

$$K \equiv (\varkappa / \epsilon_3)^2 , \tag{5'}$$

which take values ~ 1.

In the case (5), when only the minimum admitted wavenumber lies inside the modulational instability range near its edge, it is natural to represent the wave field in the form $u(x,t) = a(x,t) \exp(i\phi(x,t))$,

$$a = a_0 + b_0(t) + b_1(t)\cos(kx) + b_2(t)\cos(2kx) + \dots, \tag{6a}$$

$$\phi = 2ia_0^2 t + \psi_0(t) + \psi_1(t)\cos(kx) + \psi_2(t)\cos(2kx) + \dots, \tag{6b}$$

where a_0 is given by Eq. (4), and it is implied $b_j \ll a_0$ ($j=0,1,2$). It is easy to see that, due to the underlying assumption $\varkappa^2 \ll 1$ (see Eq. (5)), the role of higher harmonics ($j > 3$) omitted in the expressions (6) is indeed negligible, while the zeroth and second harmonics must be retained.

Insertion of Eqs. (6) into Eq. (1) gives rise to the following three-dimensional dynamical system:

$$d^2 B_1/dT^2 + 2(1 + 4A)\, dB_1/dT - 8(2K - A - 2A^2)B_1 - 32 B_0 B_1 = 0, \tag{7a}$$

$$dB_0/dT + 2 B_0 = -(1/2)\, B_1\, dB_1/dT - (5 + 4A)\, B_1^2/2, \tag{7b}$$

where

$$B_1 \equiv (\epsilon_1 \epsilon_3)^{-1/2}, \quad B_0 \equiv (\epsilon_1 \epsilon_3^2)^{-1/2} b_0, \quad T \equiv \epsilon t.$$

Note that the system (7), without the first term in the right-hand side of Eq. (7b), coincides with the Shimizu-Morioka system that demonstrates chaotic dynamics[9].

The transformation

$$B_0 \equiv - 4A-3 \ (4A+5)z/4, \quad B_1 \equiv (4A-3)x/4,$$

$$dB_1/dT \equiv (4A-3)(4A+5)(y-x)/4, \quad T = |4A - 3|^{-1} \tau$$

brings the system (7) into the form

$$dx/d\tau = \sigma(y-x), \quad dy/d\tau = -sy + rx - xz, \quad dz/d\tau = -bz + xy, \tag{8}$$

where $s \equiv \mathrm{sgn}(4A - 3)$, $\sigma \equiv (4A + 5)|4A - 3|^{-1}$, $b \equiv 2|4A - 3|^{-1}$, and $r \equiv s + 8(2K - A - 2A^2)(4A + 5)^{-1}|4A-3|^{-1}$.

In the case $s = +1$, i.e., $A > 3/4$, the system (8) is nothing but the famous Lorenz system[7]. So, in this case the onset of chaos in the NNS regime goes by the well-known route studied in detail for the Lorenz model[10].

In the opposite case, $s = -1$, i.e., $A < 3/4$, dynamics of the system (7) must be studied anew.

Diagram of different dynamical regimes of the system (7) obtained by means of numerical simulations and qualitative analysis is depicted in Fig. 1. The trivial stationary point $B_0 = B_1 = 0$ is stable provided $2K - A - 2A^2 < 0$, and is a saddle in the opposite case. At the instability threshold (solid line in Fig. 1), a fork-like forward bifurcation takes place, giving rise to two stable stationary points with the coordinates

$$B_0 = - (2K - A - 2A^2), \tag{9a}$$

$$B_1 = \pm \{(2K - A - 2A^2)/(5 + 4A)\}^{1/2}. \tag{9b}$$

From the standpoint of the underlying GLE (1) (see Eqs. (6)), the stationary point (9) is a quasi-stationary pulsating wave. With the growth of K (at A fixed), the points (9) undergo a Hopf bifurcation at

$$K = K_0 = (5 + 40A + 76A^2 + 32A^3)/12 \tag{10}$$

(line 0 in Fig. 1). The Hopf bifurcation is, respectively, forward or backward in the ranges $A < A_*$ or $A > A_*$, where $A_* = 0.44$. A stable limit cycle generated by the forward bifurcation is depicted in Fig. 2a.

As is known[10], the onset and subsequent evolution of chaos are determined by homoclinic trajectories of the saddle. Lines of existence of several simplest trajectories are shown (on the parametric plane) in Fig. 3. Bufurcations of

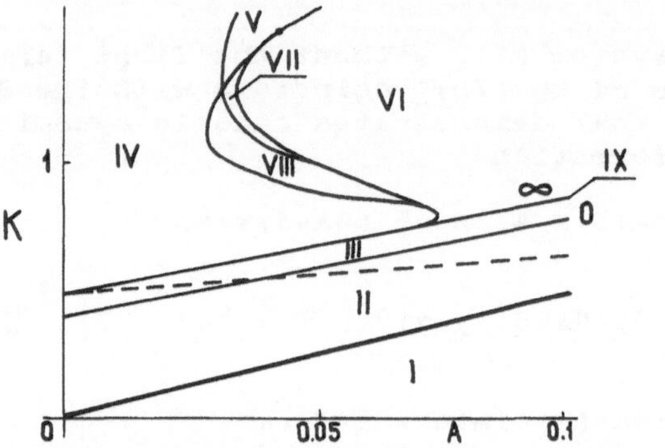

Fig. 1. The diagram of stable dynamical regimes of the system (7). The dashed line separates the regions where the saddle quantity (11) is, respectively, negative (beneath the dashed line) and positive (over it). The solid line and the ones marked 0 and ∞ depict, respectively, the instability threshold of the trivial stationary point, the Hopf bifurcation, and the lower boundary of the strange attractor. In the parametric regions marked by Roman figures the following dynamical regimes are stable: (I) The trivial stationary point; (II) The non-trivial stationary points (9); (III) The one-loop limit cycle (see Fig. 2a); (IV) The two-loop symmetric limit cycle (see Fig. 2b); (V) The two-loop asymmetric limit cycle (see Fig. 2c); (VI) The strange attractor with windows of multi-loop limit cycles; (VII) Coexistence (hysteresis) of the regimes V and VI; (VIII) Coexistence of the regimes IV and VI; (IX) Coexistence of the regimes III and VI (the region IX is very narrow).

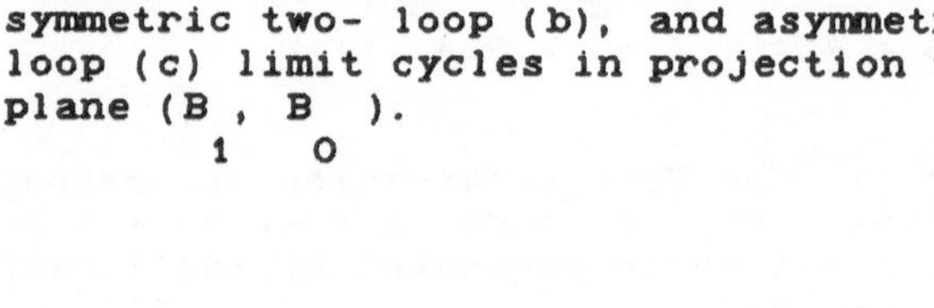

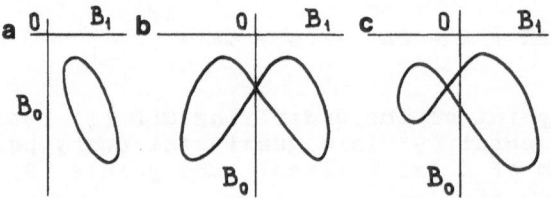

Fig. 2. Schematic description of the one-loop (a), symmetric two- loop (b), and asymmetric two-loop (c) limit cycles in projection to the plane (B_1, B_0).

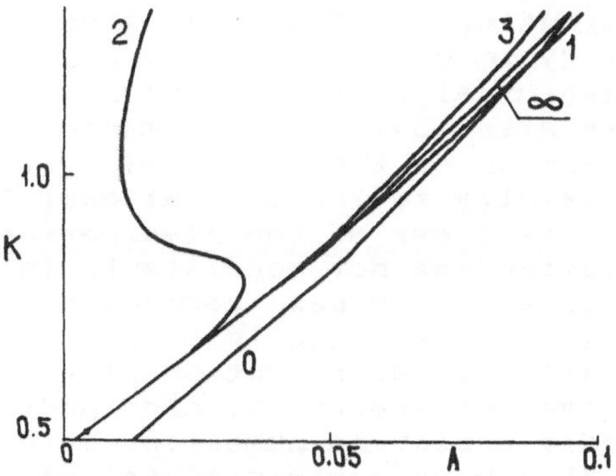

Fig. 3. The lines of existence of the homoclinic trajectories in the parametric plane (K, A). The lines 1, 2, and 3 pertain to, respectively, one, two, and three-loop trajectories. The lines 0 and ∞ are the same as in Fig. 1.

homoclinic trajectories strongly depend on the sign of the so-called saddle quantity λ equal to the sum of two largest exponents of the saddle point $B_0 = B_1 = 0$:

$$\lambda = \{2\sqrt{(\sigma - 1)^2 + 4\sigma(r + 1)} + 1 - 3\sigma\}/4. \tag{11}$$

In the case $\lambda < 0$, crossing the line 1 of existence of the one-loop homoclinic trajectory (in the direction of larger K, see Fig. 3) gives rise to a two-loop limit cycle shown in Fig. 2b. At $\lambda > 0$, the sequence of bifurcations is similar to that in the Lorenz system (in which λ is always positive). The lower boundary of existence of a Lorenz-type strange attractor is the line "∞" in Figs. 1 and 3. In contrast with Lorenz system, at $A < 0.2$ this boundary lies over the Hopf bifurcation line 0 (Figs. 1 and 3). From the viewpoint of the underlying GLE (1), the appearance of the strange attractor in the dynamical system (7) with the growth of the parameter K (see Eq. (5′)) means a transition of the regular wave regimes to chaotic waves when going into the range of the modulational instability. This seems to be a fundamental mechanism of onset of chaos in the NNS regime due to interplay of the modulational instability[8] with the gain and dissipation.

As well as in Lorenz system, at sufficiently large K the only remaining attractor is the symmetric two-loop limit cycle (Fig. 2b), which can be explicitly described in terms of elliptic functions at $K \to \infty$. With the decrease of K at A not too small, the symmetric two-loop cycle loses stability via a fork-like bifurcation that gives rise to two stable asymmetric two-loop cycles (Fig. 2c). Next, the asymmetric cycles undergo a Feigenbaum's sequence of period doublings similarly to the Lorenz system[10].

In conclusion, it is relevant to note that another

attempt of approximating the Ginzburg-Landau equation by a three-dimensional dynamical system was undertaken in earlier Ref. 2. For that dynamical system, which is different from the one (7), chaos was also observed in numerical simulations. However, the derivation of the dynamical system in Ref. 2 did not employ the proximity to the modulational instability threshold and the smallness of the dissipation parameters, so that the derivation was not consistent. In this connection, let us emphasize that in our case corrections from higher modes omitted in Eqs. (6) are negligible.

Detailed results of investigation of the dynamical system (7) will be given elsewhere. We are indebted to M. A. Zaks (who has kindly performed numerical calculations necessary to plot Fig. 3) for valuable discussions and suggestions.

REFERENCES

1. Y. Kuramoto and T. Tsuzuki, Persistent propagation of concentration waves in dissipative media far from equilibrium, Progr. Theor. Phys. 55:356 (1976).
2. M. I. Rabinovich and A. L. Fabrikant, Stochastic self-modulation of waves in nonequilibrium media, Zh. Exp. Teor. Fiz. (Sov. Phys. - JETP) 77:617 (1979).
3. H. T. Moon, P. Huerre, and L. G. Redekopp, Transition to chaos in the Ginzburg-Landau equation, Physica D 7:135 (1983).
4. C. R. Doering, J. D. Gibbon, D. D. Holm, and B. Nicolaenko, Low-dimensional behaviour in the complex Ginzburg-Landau equation, Nonlinearity 1:279 (1988).
5. A. S. Pikovsky, Spatial development of chaos in nonlinear media, Phys. Lett. A 137:121 (1989).
6. G. M. Dudko and A. N. Slavin, Transition from modulational instability to chaos in films of iron-yttrium garnet (IYG), Fiz. Tv. Tela (Sov. Phys. - Sol. St. Phys.) 31:114 (1989).
7. E. N. Lorenz, Deterministic nonperiodic flow, J. Atmos. Sci. 20:130 (1963).
8. T. B. Benjamin and J. E. Feir, The disintegration of wave trains on deep water, J. Fluid Mech. 27:417 (1967).
9. T. Shimizu and N. Morioka, On the bifurcation of a symmetric limit cycle to an asymmetric one in a simple model, Phys. Lett. A 76:201 (1980).
10. C. Sparrow, "The Lorenz Equations: Bifurcations, Chaos and Strange Attractors", Springer, New York (1982).

TURBULENCE AND LINEAR STABILITY IN A DISCRETE GINZBURG-LANDAU MODEL

Tomas Bohr and Anders W. Pedersen

The Niels Bohr Institute
Blegdamsvej 17
DK-2100 Copenhagen

Mogens H. Jensen

NORDITA
Blegdamsvej 17
DK-2100 Copenhagen

David A. Rand

Mathematics Institute
University of Warwick
Coventry CV4 7AL
England

INTRODUCTION

The Complex Ginzburg-Landau partial differential equation appears in many interesting none-quilibrium dynamical systems. It describes an extended system close to a global Hopf bifurcation [1] such as occurs e.g. in oscillatory chemical reactions like the Belousov-Zhabotinsky reaction [2]. In two recent papers [3-4] we have discussed a discrete, "map lattice" version of this equation, analysed the dynamics of vortices and the onset of turbulence. The main results were that vortices can get bound together in "entangled" states where their cores do not move and that the system has a well-defined transition to turbulence *below* the linear instability threshold for the uniform state. It remains to be seen which of our results will be valid for the continuum Ginzburg-Landau equation; but recently an analytic treatment of the motion of a pair of vortices leads to bound states analogously to our entangled states [5].

It was recently conjectured by Newell [6] that the transition to turbulence might be connected to linear instability; not, however, of the uniform state, but of a nonuniform plane wave state. The reason is that the vortices create spiral waves which, far away from the cores, approach plane waves with a well defined wavenumber set by the parameters [7], and thus a dilute system of vortices is dominated by plane waves of a specific selected wavelength. In the following we shall compare the transition points obtained by simulations of the coupled map lattice [3] with those of linear stability theory.

Nonlinear Evolution of Spatio-Temporal Structures in Dissipative Continuous Systems
Edited by F.H. Busse and L. Kramer
Plenum Press, New York, 1990

THE COMPLEX GINZBURG-LANDAU EQUATION

The Complex Ginzburg-Landau equation is derived by looking at slow motion close to a Hopf bifurcation. Assume that there is some parameter μ which makes the chemical reaction in a *stirred reactor* undergo a Hopf bifurcation. In other words, μ changes the local dynamics from a stable fixed point for $\mu<0$ to a stable limit (+ an unstable fixed point) for $\mu>0$, and for small, positive μ the limit cycle has a small radius which scales like $\sqrt{\mu}$. The frequency of the limit cycle does not approach zero at the transition but some constant ω_0 (for $\mu<0$ this is approximately the frequency of the transients leading to the fixed point). Close to the transition we can approximate the chemical concentrations by their lowest harmonics

$$\rho(\vec{x}) = \rho_0 + A(\vec{x},t)e^{i\omega_0 t} + \overline{A(\vec{x},t)}e^{-i\omega_0 t} \tag{1}$$

where the order parameter, A is assumed to be slowly varying in space and time and the bar denotes complex conjugation. It is important to note that A is necessarily complex due to the time dependence of ρ and it can be shown [8,9] that the time evolution of A, to lowest order in $|A|$, $|\dot{A}|$ and $|\nabla A|$, is described by the so-called complex Ginzburg-Landau equation

$$\dot{A} = \mu A - (1+i\alpha)|A|^2 A + (1+i\beta)\nabla^2 A \tag{2}$$

where μ, α and β are real numbers. The parameter μ is the usual Landau coefficient: Negative μ implies a quiescent state (A=0) whereas positive μ gives nonzero values to the order parameter. In fact there is a homogeneous solution

$$A = \sqrt{\mu}e^{-i\alpha\mu t} \tag{3}$$

for positive μ and it is seen that the frequency of this periodic state is $\omega = \alpha\mu$ (so the frequency of the chemical state is $\omega_0 - \alpha\mu$). The parameter β is related to differences between the diffusivities of the different species [9]. It is important to note that both α and β introduce a preferred sense of rotation in the complex A-plane so we anticipate that the relation between their <u>signs</u> will play a significant role. On the other hand, complex conjugation of (2) leads to an equation for $\bar{A}$ of the same form, but with opposite signs of α and β, and therefore the simultaneous change of sign of these two parameters does not affect the dynamics. In the following we can thus, without loss of generality, restrict our attention to the case $\beta \le 0$. We further restrict ourselves to two-dimensional systems i.e. chemical reactions in a shallow dish.

The linear stability of the homogeneous state (3) can be investigated by standard techniques [1]. If we look at weak perturbations of (3) in the form

$$A = (\sqrt{\mu} + \rho(x,y,t))e^{i(-\alpha\mu t + \phi(x,y,t))} \tag{4}$$

and treat (2) to linear order in ρ and ϕ we find that ρ and ϕ will decay exponentially as long as $1+\alpha\beta > 0$, whereas long wavelength modes ($|\vec{k}| < k_c \propto |1+\alpha\beta|^{1/2}$) will be exponentially enhanced if $1+\alpha\beta < 0$. In particular, if α and β have the same sign, (3) is always linearly stable.

The turbulent states seem closely connected to the appearance and dynamics of *topological defects* in the form of *vortices,* singularities of the angle field. The total variation of the angle over a closed loop, i.e. $\Delta\phi = \oint d\phi$, doesn't necessarily vanish if the loop encloses vortex centers. Instead $\Delta\phi = 2\pi n$, where the integer n is the total vorticity of the region enclosed. In the vortex center the angle is not defined so for the order parameter itself ($A = re^{i\phi}$) to remain well-defined requires that r vanishes in the center. Typical vortex ($\Delta\phi = 2\pi$) and antivortex ($\Delta\phi = -2\pi$) configurations are shown in Fig.1.

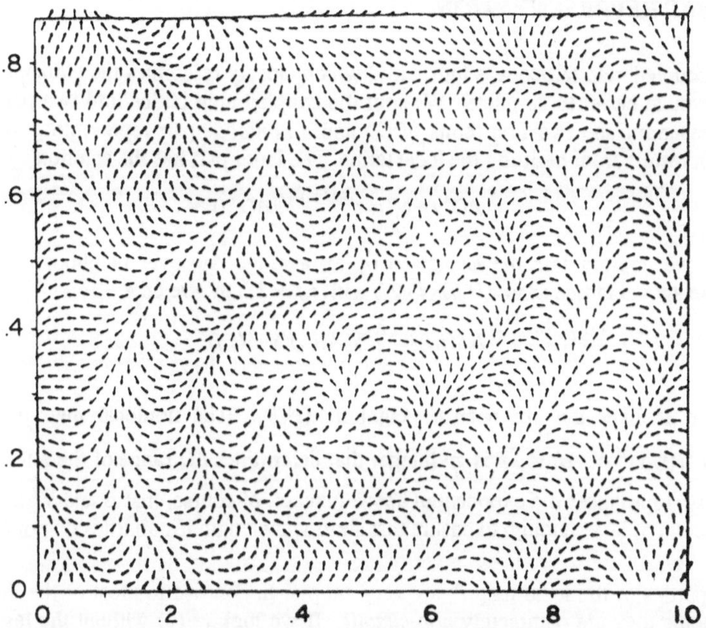

Fig.1. A vortex-antivortex configuration. Note the spiral arms and the characteristic wavelength.

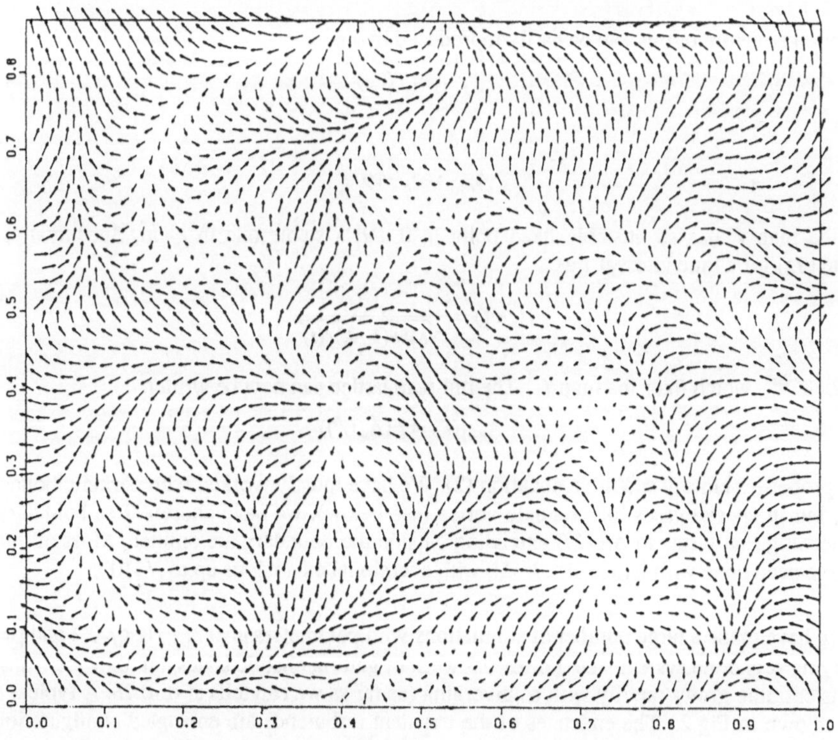

Fig.2. A turbulent state with several moving vortices and antivortices after 500 iterations. The parameters are: α=0.75, β=–1.0, τ=1.0, τ_0=0.2, μ=0.2.

COUPLED MAP REPRESENTATION

In [3] a coupled map lattice was introduced which closely resembles the Ginzburg-Landau equation (2) but is much easier to simulate. We split the coupled map lattice into two parts: a *local* map $A' = F(A)$ representing the two first terms of (2) and a *nonlocal* part representing the complex heat equation which results from omitting the local terms. The latter part has the solution

$$A(\vec{x},t+\tau_0) = e^{\tau_0(1+i\beta)\nabla^2} A(\vec{x},t) \tag{5}$$

On the lattice we approximate the Laplacian by an average ΔA over neighbors. Thus on a two-dimensional hexagonal lattice $(x,y)=(i+j/2,\sqrt{3}/2j)$ with $i,j = 1,\ldots,N$ we take

$$\Delta A(i,j) = \frac{2}{3}\sum_{i'j'} A(i',j') - A(i,j) \tag{6}$$

where the sum is over the six nearest neighbors (i',j'). The nonlocal map is then given by $\tilde{A} = (1 + \frac{\tau_0}{M}(1+i\beta)\Delta)^M A$. Here M is an integer that determines the range of the effective interaction. The limit $M \rightarrow \infty$ reproduces the exponential above (except, of course, that Δ and ∇^2 are not the same). We take M around 5, large enough to ensure that short wavelength instabilities do not occur.

The properties of the local map F are very simple. In contrast to most of the literature on coupled map systems they are completely *non-chaotic*. If we look at (2) without the last term it can be written as

$$\dot{r} = \mu r - r^3 \tag{7a}$$

$$\dot{\phi} = -\alpha r^2 \tag{7b}$$

The general structure of this can be easily reproduced by maps

$$r_{n+1} = f(r_n) \tag{8a}$$

and

$$\phi_{n+1} = \phi_n - \tau\alpha r_n^2 \tag{8b}$$

where the map f has an unstable fixed point in 0 and a stable one in $r=\sqrt{\mu}$. Specifically, one can integrate (7a) for a time interval τ as

$$r(t+\tau) = \frac{\sqrt{\mu}r(t)}{\sqrt{\lambda\mu+(1-\lambda)r(t)^2}} \tag{9}$$

where $\lambda = e^{-2\mu\tau}$, which fixes the map f. The full map lattice can now be written

$$A_{n+1}(\vec{r}) = F(\tilde{A}_n(\vec{r})) \tag{10}$$

and its properties closely resemble equation (2). We have mostly worked with periodic boundary conditions, but in special cases like a single vortex (see below), we have chosen "free boundary" conditions meaning that terms in $\Delta A\,(i,j)$ extending outside the boundary are omitted. Most of our simulations have been done with lattice size N=50 with occasional excursions up to N=100.

We have made a large number of simulations with parameters μ=0.2, τ=1, τ_0=0.2 and β=−1 fixed and varying α and the system size. For $\alpha \approx 0.5$ we start getting chaotic states, but they are *transient* and the transient time grows rapidly with α, diverging (as far as we can tell) at $\alpha \approx 0.75$. A typical turbulent state is shown in Fig.2. The endstates of the transient turbulence are entangled configurations where only few vortices are present and where one or two of have outgrown the others. One such state is shown in Fig.3. In the transient period we can determine a "finite time" Lyapunov exponent with reasonable accuracy and we find a well defined transition point at which it vanishes. Fig.4 shows this exponent as function of α for the above parameters and the transition point is close to α=0.5. In [4] the dependence of the transient times on the lattice size was investigated and it was conjectured that they would actually diverge with system size and thus that the transition point found above would be the true transition to turbulence for an infinite system. In the following we shall compare this value (and one for a different β) to linear instability threshold values calculated for our model.

PLANE WAVE SOLUTIONS

The map (10) can be written

$$A_{n+1} = F(e^{\tau_0(1+i\beta)\Delta} A_n) \tag{11}$$

with the complex function

$$F(z) = f(|z|^2)e^{-i\tau\alpha|z|^2}z \tag{12}$$

where

$$f(x) = \frac{\sqrt{\mu}}{\sqrt{\mu\lambda+(1-\lambda)x}} \tag{13}$$

and Δ is the discrete Laplacian defined in (6). The discrete model allows plane wave solutions (as does eqn.(2))

$$A_n(\vec{x}) = R_{\vec{k}} e^{i(\vec{k}\cdot\vec{x}-\omega n)}. \tag{14}$$

Substituting (15) into (11-14) gives the dispersion relation

$$\omega_{\vec{k}} = \tau\alpha p - \beta\tau_0\Delta_{\vec{k}} \tag{15}$$

where $p = R_{\vec{k}}^2 e^{2\tau_0\Delta_{\vec{k}}}$ must satisfy

$$f(p) = e^{-\tau_0\Delta_{\vec{k}}} \tag{16}$$

and $\Delta_{\vec{k}}$ is the diagonal Fourier representation of Δ defined through $\Delta e^{i\vec{k}\cdot\vec{x}} = e^{i\vec{k}\cdot\vec{x}}\Delta_{\vec{k}}$. For the hexagonal lattice $\Delta_{\vec{k}} = \frac{4}{3}\sum_i(\cos(\vec{k}\cdot\vec{e_i})-1)$ where the sum extends over the three basic vectors of the lattice. At small k $\Delta_{\vec{k}}$ always approaches $-k^2$. Using (13) we can solve (16) as

$$p = f^{-1}(e^{-\tau_0\Delta_{\vec{k}}}) = \frac{\mu}{1-\lambda}(e^{2\tau_0\Delta_{\vec{k}}} - \lambda) \tag{17}$$

and

$$\omega_{\vec{k}} = \tau\alpha\frac{\mu}{1-\lambda}(e^{2\tau_0\Delta_{\vec{k}}} - \lambda) - \tau_0\beta\Delta_{\vec{k}} \tag{18}$$

As $k \rightarrow 0$, we find

$$\omega_{\vec{k}} \rightarrow \tau\alpha\mu + \tau_0(\beta - 2\tau\alpha\frac{\mu}{1-\lambda})k^2 \tag{19}$$

Now, $\lambda = e^{-2\mu\tau}$ so, for τ small we get

$$\omega_{\vec{k}} \rightarrow \tau\alpha\mu + \tau_0(\beta - \alpha)k^2 \tag{20}$$

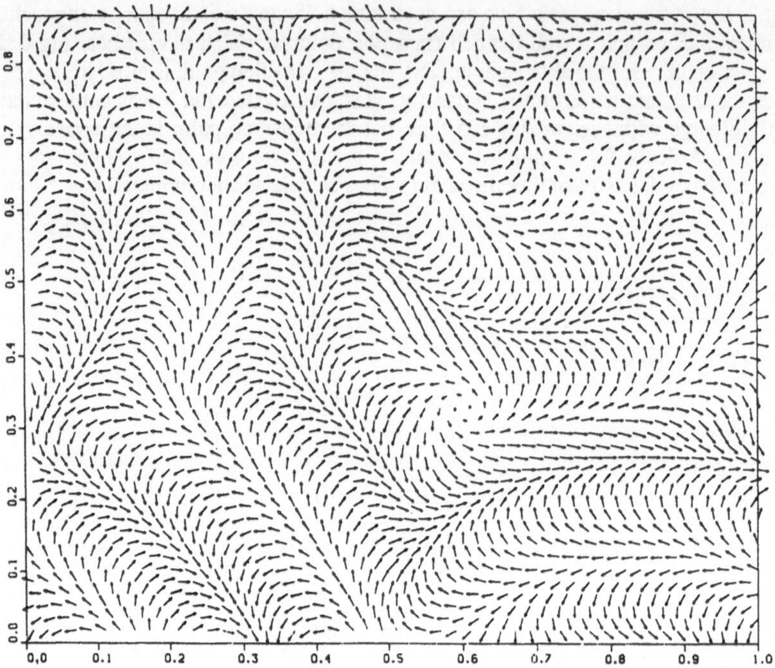

Fig.3. An entangled endstate with one vortex and one antivortex in a system with periodic boundary conditions. The antivortex has completely outgrown the vortex and its spiral arms have acquired several windings with a well-defined wavelength. The picture shows the 1500'th iterate of a random state and the parameters are as in Fig.2 except for α=0.55.

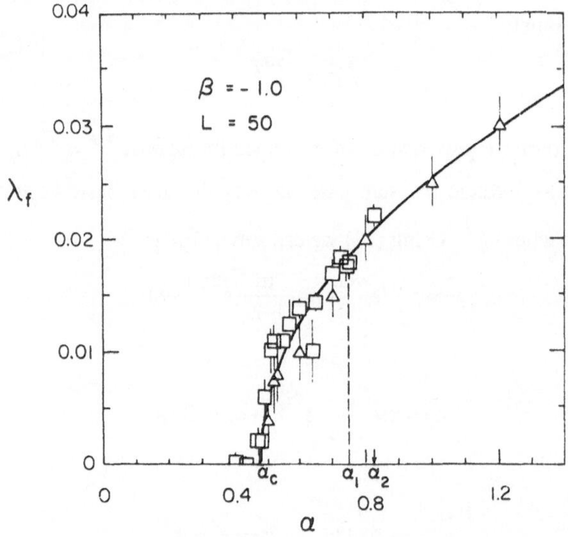

Fig.4. The (finite time) Lyapunov exponent as a function of the parameter α. The squares and the triangles were computed for hexagonal and square lattices, respectively. The true Lyapunov exponent is zero below α_1 as shown by the dotted line. The curve is a power law fit around α_c=0.48 with exponent 0.5. The threshold of linear instability for the uniform state is at α_2=0.82. From [4].

which corresponds to the dispersion relation for plane waves in (2) when $\tau = \tau_0$ and we note that $\omega \cdot n \to \omega \cdot t / \tau$.

LINEAR STABILITY OF PLANE WAVES

To check the stability of these periodic states we consider a small perturbation of (14) in the form

$$A_n(\vec{x}) = (R_{\vec{k}} e^{i\vec{k}\cdot\vec{x}} + u_n(\vec{x}))e^{-i\omega_{\vec{k}} n} \tag{21}$$

Iterating (21) by use of (11) we get to lowest order in u

$$u_{n+1} \approx e^{i\omega(n+1)} \left[e^{-i\omega n} \frac{\partial F}{\partial z} e^{\tau_0(1+i\beta)\Delta} u_n + e^{i\omega n} \frac{\partial F}{\partial \bar{z}} e^{\tau_0(1-i\beta)\Delta} \overline{u_n} \right] \tag{22}$$

where the bar denotes complex conjugation. Here and in the following we shall suppress the index $\vec{k}$ on ω and R. The derivative of F must be evaluated at

$$z = \tilde{A} = Re^{\tau_0(1+i\beta)\Delta_{\vec{k}} + i\vec{k}\cdot\vec{x} - i\omega n} \tag{23}$$

The partial derivatives of F can be written as

$$\frac{\partial F}{\partial z} = \frac{F(z)}{z} \left[\left(\frac{f'(p)}{f(p)} - i\tau\alpha\right)p + 1 \right] \tag{24}$$

$$\frac{\partial F}{\partial \bar{z}} = zF(z) \left[\frac{f'(p)}{f(p)} - i\tau\alpha \right] \tag{25}$$

where

$$p = z\bar{z} = R^2 e^{2\tau_0 \Delta_{\vec{k}}} \tag{26}$$

and the functions F and f are given by (12-13). Specifically

$$\frac{f'(p)}{f(p)} = \frac{d\log f(p)}{df(p)} = -\frac{1-\lambda}{2(\mu\lambda + (1-\lambda)p)} \tag{27}$$

$$\frac{F(\tilde{A})}{\tilde{A}} = e^{-\tau_0\Delta_{\vec{k}} - i(\omega + \tau_0\beta\Delta_{\vec{k}})} \tag{28}$$

$$\tilde{A}F(\tilde{A}) = pe^{-\tau_0\Delta_{\vec{k}} + i(\tau_0\beta\Delta_{\vec{k}} + 2\vec{k}\cdot\vec{x} - (2n+1)\omega)} \tag{29}$$

Substituting into (22) we find to linear order in u_n and $\overline{u_n}$

$$u_{n+1}(\vec{x}) = e^{-\tau_0(1+i\beta)\Delta_{\vec{k}}} M(p) e^{\tau_0(1+i\beta)\Delta} \cdot u_n(\vec{x}) + e^{\tau_0(-1+i\beta)\Delta_{\vec{k}}} N(p) e^{2i\vec{k}\cdot\vec{x}} e^{\tau_0(1-i\beta)\Delta} \cdot \overline{u_n(\vec{x})} \tag{30}$$

where (using (17))

$$M(p) = \frac{\mu\lambda}{2(\mu\lambda + (1-\lambda)p)} + \frac{1}{2} - i\tau\alpha p = \tfrac{1}{2}(\lambda e^{2\tau_0\Delta_{\vec{k}}} + 1) - i\tau\alpha p \tag{31}$$

and

$$N(p) = \frac{\mu\lambda}{2(\mu\lambda + (1-\lambda)p)} - \frac{1}{2} - i\tau\alpha p = \tfrac{1}{2}(\lambda e^{2\tau_0\Delta_{\vec{k}}} - 1) - i\tau\alpha p \tag{32}$$

Substituting in (30) the Fourier expansions $u_n(\vec{x}) = \sum_{\vec{q}} u_n(\vec{q}) e^{i\vec{q}\cdot\vec{x}}$ and $\overline{u_n(\vec{x})} = \sum_{\vec{q}} \overline{u_n(\vec{q})} e^{-i\vec{q}\cdot\vec{x}}$

we get

$$u_{n+1}(\vec{q}) = G_k e^{\tau_0(1+i\beta)\Delta_{\vec{q}}} u_n(\vec{q}) + H_k e^{\tau_0(1-i\beta)\Delta_{\vec{q}-2\vec{k}}} \overline{u_n(2\vec{k}-\vec{q})} \qquad (33)$$

where we use the notation $G_k = M(p) e^{-\tau_0(1+i\beta)\Delta_{\vec{k}}}$ and $H_k = N(p) e^{\tau_0(-1+i\beta)\Delta_{\vec{k}}}$ and one should note that $u_n(\vec{q}) = \overline{u_n}(-\vec{q})$). By complex conjugation and shift of $\vec{q} \to 2\vec{k}-\vec{q}$ in the lower row we get

$$\begin{bmatrix} u_{n+1}(\vec{q}) \\ \overline{u_{n+1}(2\vec{k}-\vec{q})} \end{bmatrix} = \begin{bmatrix} A(\vec{k},\vec{q}) & \overline{B(\vec{k},2\vec{k}-\vec{q})} \\ B(\vec{k},\vec{q}) & \overline{A(\vec{k},2\vec{k}-\vec{q})} \end{bmatrix} \begin{bmatrix} u_n(\vec{q}) \\ \overline{u_n(2\vec{k}-\vec{q})} \end{bmatrix} \qquad (34)$$

where

$$A(\vec{k},\vec{q}) = G_k e^{\tau_0(1+i\beta)\Delta_{\vec{q}}} = M(p) e^{\tau_0(1+i\beta)(\Delta_{\vec{q}} - \Delta_{\vec{k}})} \qquad (35)$$

and

$$B(\vec{k},\vec{q}) = \overline{H_k} e^{\tau_0(1+i\beta)\Delta_{\vec{q}}} = \overline{N(p)} e^{\tau_0(1+i\beta)(\Delta_{\vec{q}} - \Delta_{\vec{k}})} \qquad (36)$$

The eigenvalues of the matrix

$$\mathbf{T} = \begin{bmatrix} A(\vec{k},\vec{q}) & \overline{B(\vec{k},2\vec{k}-\vec{q})} \\ B(\vec{k},\vec{q}) & \overline{A(\vec{k},2\vec{k}-\vec{q})} \end{bmatrix} \qquad (37)$$

appearing in (34) determines the stability. We remind the reader that the vector $\vec{k}$ is the wavenumber of the original harmonic state which also determines p through (17), whereas the vector $\vec{q}$ is the wavenumber of the perturbation.

Stability is determined by the eigenvalues of **T**: whether they lie within the unit circle. To compute them we need

$$tr\,\mathbf{T} = A(\vec{k},\vec{q}) + \overline{A(\vec{k},2\vec{k}-\vec{q})} \qquad (38)$$

and

$$\det \mathbf{T} = A(\vec{k},\vec{q})\overline{A(\vec{k},2\vec{k}-\vec{q})} - B(\vec{k},\vec{q})\overline{B(\vec{k},2\vec{k}-\vec{q})} \qquad (39)$$

which determine the eigenvalues $\chi(\vec{k},\vec{q})$ through the characteristic equation

$$\chi^2 - (tr\,\mathbf{T})\chi + \det \mathbf{T} = 0 \qquad (40)$$

with the solution

$$\chi(\vec{k},\vec{q}) = \tfrac{1}{2}(tr\,\mathbf{T} \pm \sqrt{(tr\,\mathbf{T})^2 - 4\det \mathbf{T}}) \qquad (41)$$

LINEAR STABILITY OF THE UNIFORM STATE

The simplest case is the stability of the uniform state $\vec{k}=0$. Here the stability matrix takes the form

$$\mathbf{T} = e^{\tau_0\Delta_{\vec{q}}} \begin{bmatrix} (\tfrac{1}{2}(\lambda+1) - i\tau\alpha\mu)e^{i\Phi} & (\tfrac{1}{2}(\lambda-1) - i\tau\alpha\mu)e^{-i\Phi} \\ (\tfrac{1}{2}(\lambda-1) + i\tau\alpha\mu)e^{i\Phi} & (\tfrac{1}{2}(\lambda+1) + i\tau\alpha\mu)e^{-i\Phi} \end{bmatrix} \qquad (42)$$

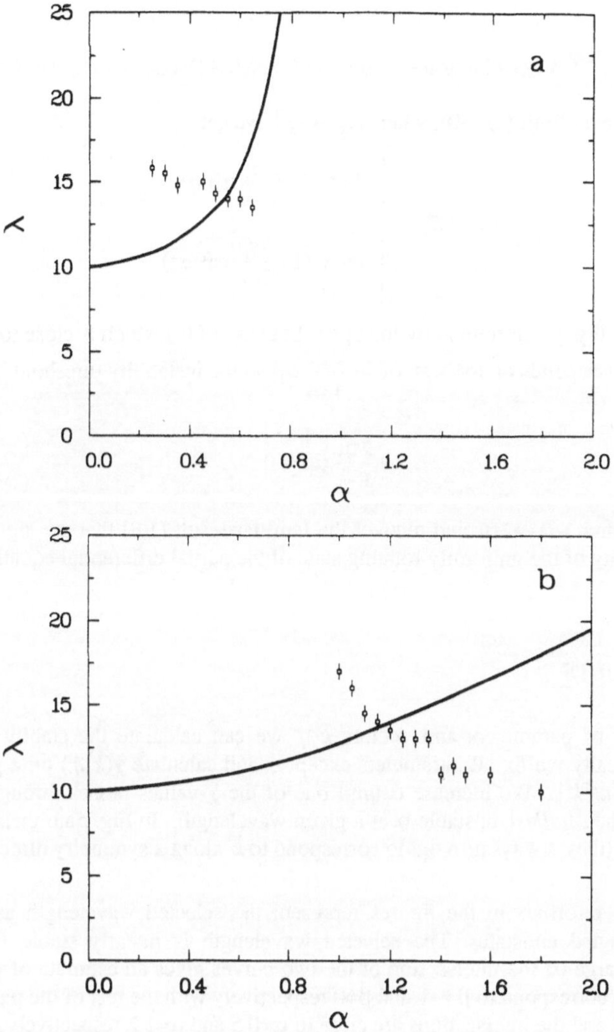

Fig.5. Sideband instability threshold and characteristic vortex wavelength. The full curve shows the wavelength and the corresponding critical α as explained in the text. The boxes represent the characteristic wavelength of entangled endstates. The crossing between the graphs gives an estimate of the transition to turbulence. a): β=−1.0 ; b): β=0. The other parameters are as in Fig.2.

where $\Phi=\tau_0\beta\Delta_{\vec{q}}$ and $p(\vec{k}=0)=\mu$. Now

$$tr\,\mathbf{T} = ((\lambda+1)\cos\Phi + 2\tau\alpha\mu\cdot\sin\Phi)e^{\tau_0\Delta_{\vec{q}}} \tag{43}$$

and

$$\det\mathbf{T} = \lambda e^{\tau_0\Delta_{\vec{q}}} \tag{44}$$

Thus

$$\chi(\vec{k}=0,\vec{q})\cdot e^{-\tau_0\Delta_{\vec{q}}} = \tfrac{1}{2}(\lambda+1)\cos\Phi + \tau\alpha\mu\cdot\sin\Phi \pm \tfrac{1}{2}\sqrt{((\lambda+1)\cos\Phi + 2\tau\alpha\mu\cdot\sin\Phi)^2-4\lambda} \tag{45}$$

and in the long wavelength limit ($q\to0$), where $\Delta_{\vec{q}}\to-q^2$, we get

$$\chi(\vec{k}=0,\vec{q}) \approx \left[\begin{array}{l} 1-\tau_0(1+\frac{2\tau\mu}{1-\lambda}\alpha\beta)q^2 \\ \\ \lambda(1-\tau_0(1-\frac{2\tau\mu}{1-\lambda}\alpha\beta)q^2) \end{array}\right. \tag{46}$$

Since $\lambda=e^{-2\mu\tau}<1$, stability is determined by the upper branch of (46) which is close to 1. Whether it is smaller or larger than 1 depends on the sign of $1+\frac{2\tau\mu}{1-\lambda}\alpha\beta$ so the instability threshold is determined [3] by

$$1+\frac{2\tau\mu}{1-\lambda}\alpha\beta = 0 \tag{47}$$

In the limit $\tau\to0$ we find $1-\lambda\to2\tau\mu$ and recover the familiar result [1,8] that the sign of $1+\alpha\beta$ determines the linear stability of the uniformly rotating state of the partial differential equation (2).

NUMERICAL RESULTS

For a given set of parameters and vectors $\vec{k},\vec{q}$ we can calculate the stability eigenvalues χ through (41). Numerically we fix all parameters except α and calculate $\chi(\vec{k},\vec{q})$ on a grid of $\vec{k},\vec{q}$ with fixed wavelength $\lambda=2\pi/|\vec{k}|$. We increase α until one of the χ-values passes through the unit circle and find in this way the *smallest* unstable α at a given wavelength. In Fig. 5a,b these critical α's are plotted as a full curve (they always turn out to correspond to $\vec{k}$ along a symmetry direction).

The boxes with errorbars in the figures represent the selected wavelength as function of α, estimated from entangled endstates. The selected wavelength is linearly stable for small α and becomes unstable at large α; the intersection of the two curves gives an estimate of the critical value for α. The two figures correspond to $\beta=-1$ and $\beta=0$ respectively with the rest of the parameters fixed at $\mu=0.2$, $\tau=1$ and $\tau_0=0.2$ and the intersections are close to $\alpha=0.5$ and $\alpha=1.2$ respectively.

For $\beta=-1$ the transition to transient turbulence is shown in Fig.4 and is within our accuracy equal to the linear stability threshold $\alpha=0.5$. For $\beta=0$ we have much less accurate data for the transition to turbulence. At $\alpha=2.0$ we believe that the system is truly turbulent: we have at been able to follow the turbulent state for more than 20.000 iterates. Between 1.0 and 1.5 we begin to see clear signals of transient turbulent states, but we find large fluctuations with initial conditions and better statistics is needed. At present we can only say that the linear stability threshold $\alpha=1.2$ is not inconsistent with our data. One should note that we find entangled endstates also for $\beta=0$. However, in our simulations of the dynamics of a symmetric vortex-antivortex pair [3], we never see entanglement, in agreement with the absence of bound states at $\beta=0$ in Ref. [5].

Within the accuracy available to us at this time we can thus say that the sideband instabilities give good estimates for the critical parameter values -at least if we choose the transition to *transient* turbulence which, as argued in [4], might describe the transition to turbulence in the infinite system.

ACKNOWLEDGEMENTS

We would like to thank A.C.Newell for the discussions and correspondence that provoked the stability analysis presented here, E.Bodenschatz and L.Kramer for stimulating discussions and C.Elphick and E.Meron for sending the preprint [5]. AWP is grateful to The Carlsberg Foundation for a Scholarship.

REFERENCES

1. A.C.Newell, Killarney Golfing Gazette (1906).
2. A.N.Zaikin and A.M.Zhabotinsky, Nature 225 , 535 (1970). A.T.Winfree, "The Geometry of Biological Time" (Springer 1980).
3. T.Bohr, M.H.Jensen, A.W. Pedersen and D.Rand, to appear in "New Trends in Nonlinear Dynamics and Pattern Forming Phenomena" ed. P.Coullet and P.Huerre (Plenum 1989).
4. T.Bohr, M.H.Jensen and A.W. Pedersen:"Transition to turbulence in a discrete complex Ginzburg-Landau model", preprint (1989).
5. C.Elphick and E.Meron, preprint (1989).
6. A.C.Newell, private correspondence.
7. P.S.Hagan, Siam, J. Appl. Math. 42 ,762 (1982).
8. A.C.Newell and J.A.Whitehead, J.Fluid Mechanics 38 , 279 (1969); A.C.Newell in *Lectures in Applied Mathematics,* vol. 15, Am. Math. Society, Providence (1974).
9. Y.Kuramoto, *Chemical Oscillations, Waves and Turbulence* Springer, Berlin (1980).

ON THE STABILITY OF PARAMETRICALLY EXCITED STANDING WAVES

Hermann Riecke

Department of Engineering Sciences
and Applied Mathematics
Northwestern University
Evanston, Ill. 60201

A recent theoretical analysis of parametrically excited standing waves has shown that in systems with large aspect ratio they can exhibit very interesting spatial as well as dynamical behavior[1]. Under suitable conditions spatially periodic waves were found to be unstable with respect to spatial modulations of the wave number which do *not* induce the well-known Eckhaus instability. Thus the eventual number of cells in the system does not change, instead the instability leads to the formation of coexisting patches with different wave numbers ("wave-number kinks"). For other parameter values the band of stable wave numbers is found to disappear above a critical value of the external driving. This gives access to a well-controlled transition to a dynamic state which presumably shows chaotic behavior. In the present communication the conditions for the appearance of these two phenomena are discussed in more detail. In particular, the dependence of the phase diffusion coefficient on external as well as internal parameters is presented.

The starting point for the discussion is a supercritical Hopf bifurcation in an extended system, which leads to traveling as well as standing waves. In the present work it is assumed that the traveling waves are stable and the standing waves unstable. To obtain the parametrically excited waves of interest, the control parameter (Rayleigh number $\mathcal{R}$, say) is modulated periodically in time with a frequency ω_e that is close to twice the frequency ω_h of the traveling waves[1-4]. For small amplitudes of the waves as well as the modulation (driving) this system can be described by two coupled complex Ginzburg-Landau equations for the amplitudes η and ζ of the left- and right-traveling wave components,

Nonlinear Evolution of Spatio-Temporal Structures in Dissipative Continuous Systems
Edited by F.H. Busse and L. Kramer
Plenum Press, New York, 1990

$$\partial_t\eta + s\,\partial_x\eta = d\,\partial_x^2\eta + a\,\eta + b\,\zeta + c\,\eta\,(|\eta|^2 + |\zeta|^2) + g\,\eta\,|\zeta|^2 + \dots$$

$$\partial_t\zeta - s\,\partial_x\zeta = d^*\partial_x^2\zeta + a^*\zeta + b\,\eta + c^*\zeta\,(|\eta|^2 + |\zeta|^2) + g^*\zeta\,|\eta|^2 + \dots \qquad (1)$$

The external control parameters enter the linear coefficients $a \equiv a_r + ia_i$ and b in the following way: $\mathcal{R}=\mathcal{R}_c + Aa_r + bB\cos(\omega_e t)$ and $a_i = \omega_h + Ca_r - \omega_e/2$. Here $\mathcal{R}_c$ is the critical Rayleigh number without modulation and A, B and C are O(1)-numbers. The coefficient a_i gives the detuning of the external frequency ω_e with respect to the linear frequency $\omega_h + Ca_r$ of the traveling waves. The spatial derivatives involve the group velocity $s=\partial\omega_h/\partial q$ and the complex dispersion coefficient $d=\partial^2\sigma/\partial q^2$ with σ being the complex growth rate as a function of the wave number q. The analysis assumes that a and b are small. Strictly speaking, consistency requires that s be also small ($s^2=O(a,b)$).

The present discussion focusses on the parametrically excited standing waves (PSW). They bifurcate from the trivial state along $|a-isq-dq^2|^2=b^2$ and appear in fact already for $\mathcal{R}<\mathcal{R}_c$. Therefore (1) also describe parametrically driven systems which do not undergo a Hopf bifurcation, if they exhibit traveling-wave modes which are *weakly* damped: these systems can be viewed as being "close" to a (fictitious) Hopf bifurcation. This approach has been used effectively in various treatments of parametrically driven surface waves[5]. In this general case the group velocity s would have to be allowed to be complex, the investigation of which will be left for later work.

The stability of PSW with respect to long-wavelength perturbations is investigated using the appropriate phase diffusion equation[10]. It is obtained using a WKB-expansion $\eta=S_0 e^{i(\psi+\varphi/\varepsilon)}\left[1+\varepsilon(S+L)\right]$, $\zeta=S_0 e^{i(\theta+\varphi/\varepsilon)}\left[1+\varepsilon(S-L)\right]$, with S_0 being real and the perturbations S and L complex. All quantities depend on the slow scales $X=\varepsilon x$ and $T=\varepsilon^2 t$. The translation symmetry of the system yields at $O(\varepsilon)$ the phase diffusion equation,

$$\partial_T\varphi = D(q)\,\partial_X^2\varphi, \qquad (2)$$

which is also valid for $a_r>0$. The analytic expression for $D(q)$ was given in ref.1. Wave-number kinks arise if the phase diffusion coefficient $D(q)$ becomes weakly negative by developing a minimum with $D(q_{min})<0$, which splits the stable band into two sub-bands[1]. A pattern exhibiting such wave-number kinks is shown in Fig.1, where they mediate between patches with wave number q=-0.5 and patches with q=+0.3. Note that the sign change in q is reflected in the phase shift between η_r and η_i. For sufficiently small wave-number gradients these kinks can be described by an extended phase diffusion equation[1,6]. They

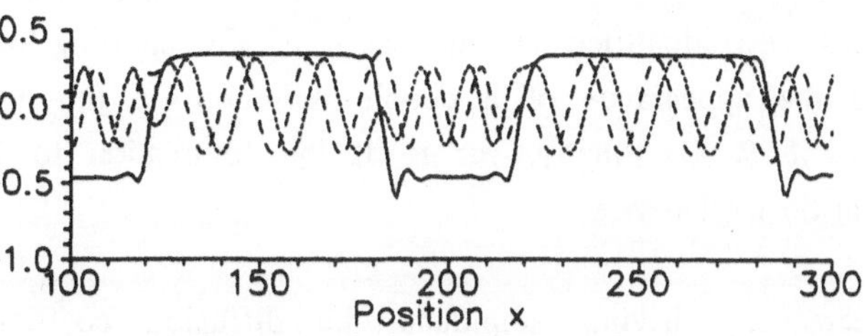

Fig.1 Wave-number kink connecting q=-0.5 with q=+0.3.

—— : q, – – : Re(η), ········ : Im(η)

interact with each other via an attractive potential which decays exponentially for large relative distances[7]. Beyond the phase approximation the potential can also have an oscillatory part. This can lead to a locking of adjacent kinks which then may form a spatially chaotic pattern[8,9]. This appears to be possible in the present case. Under certain conditions *D* becomes negative for all *q*, which eliminates the stable band altogether (see Fig.3a below). In the following the salient features that lead to these two kinds of behavior are discussed.

To get an overview of the behavior of *D(q)* it is useful to consider the special case of vanishing group velocity, *s*=0, and investigate the dependence of *D*(*q*=0) on the relevant parameters. For *s*=0 the condition *D*(*q*=0)=0 is equivalent to

$$2R_nR_d + (2|n|^2R_d+4c_dR_n)S_0^2 + 4|n|^2c_dS_0^4 = 0, \tag{3}$$

with

$$R_n=a_rn_r+a_in_i,\quad R_d=a_rd_r+a_id_i,\quad c_d=c_rd_r+c_id_i,\quad n=2c+g. \tag{4}$$

This leads to

$$S_0^2=S_1^2 \equiv -R_n/|n|^2 \quad \text{or} \quad S_0^2 = S_2^2 \equiv -R_d/2c_d. \tag{5}$$

In addition, one has $D(q)\to c_d/c_r$ for large S_0, implying instability for $c_d>0$ ($c_r<0$ for a supercritical bifurcation to traveling waves). Note that $c_d>0$ is exactly the condition that the traveling waves are Benjamin-Feir unstable in the absence of the temporal modulation (Newell criterion). This instability at large amplitudes was identified as one cause of the splitting of the band. The other cause is the appearance of a double minimum in the neutral curve. For s=0 this arises if R_d>0. Finally, for R_n<0 the bifurcation to PSW is subcritical and supercritical otherwise.

With decreasing driving amplitude the diffusion coefficient D changes sign at $S_0^2=S_1^2$ and $S_0^2=S_2^2$, if these values are positive. At $S_0^2=S_1^2$ the PSW reaches a saddle-node bifurcation and no solution exists for smaller values of b. For s=0 one has therefore the following possibilities in the middle of the wave-number band. In the Benjamin-Feir unstable case (c_d>0) the PSW are unstable at large amplitudes. If the neutral curve has a single minimum (R_d<0) they can become stable for small amplitudes and the two stable sub-bands merge for sufficiently small b (Fig.2a). In the vicinity of this merging wave-number kinks arise. If, however, the neutral curve is double-humped (sufficiently non-convex for s≠0) the two sub-bands remain separate for all b (Fig.2b). In the Benjamin-Feir stable case instability (at q=0 with s=0) arises only if the PSW bifurcate supercritically from a neutral curve with a double minimum.

With increasing group velocity the regions of instability decrease. Thus, the condition for a double-humped neutral curve becomes

$$a_r d_r + a_i d_i \geq 3|d|^2 (s a_i/4|d|^2)^{2/3} + O(s^{4/3}). \tag{6}$$

As long as b can be chosen large as compared to s^2, one obtains useful information from a large-S_0 expansion in the vicinity of the Newell-criterium. Expanding $d_i=(d_{bf}\varepsilon^2-c_r d_r)/c_i$ and $b=b_2/\varepsilon^2$ one finds at leading order in ε that D>0 is equivalent to

$$s^2 c_i^2 (c_r n_r - c_i n_i) - qsc_i d_r (4n_r|c|^2 + c_r|n|^2) + d_r^2 q^2 |c|^2 \left(4(c_r n_r + c_i n_i) + |n|^2\right) +$$

$$+ d_r c_i |n|^2 (a_i c_r - a_r c_i) - 2c_i^2 |n| d_{bf} b_2 > 0. \tag{7}$$

Since q (and a) remain finite in this expansion this expression gives the behavior of $D(q)$ in the middle of the band. From (7) one concludes that $D(q)$ has a maximum there (as a function of q) if

$$(g_i + 4c_i)^2 < c_i^2 - 3c_r^2 - 2c_r g_r - g_r^2/4 \tag{8}$$

and a minimum otherwise. In the latter case (Fig.2b) the stable band is split up into two sub-bands for large modulation as discussed before. Fig.3a shows the stability limits for the same parameters as in Fig.2b, now varying the

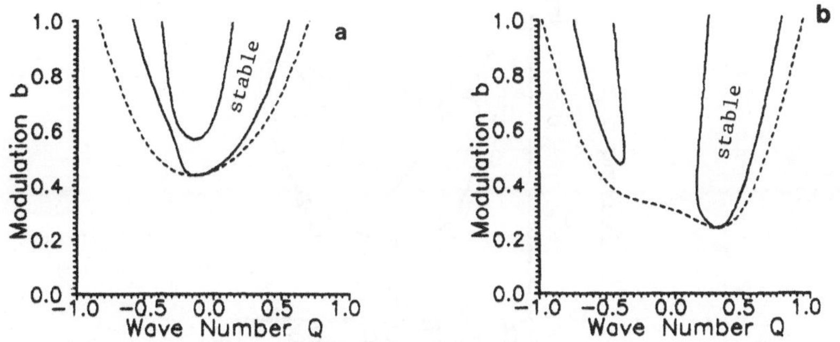

Fig.2 Neutral curve (- -) and D(q)=0 (—) for a_r=-0.05, s=0.2, d=1+0.5*i*, c=-1+4*i*. g=-1-7.5*i*. a) a_i=-0.45 b) a_i=+0.3

detuning a_i with fixed modulation amplitude (b=0.5). Thus, a cut along b=0.5 in Fig.2b is equivalent to a cut along a_i=0.3 in Fig.3a. The neutral curve (short dashes) as well as the line $D(q)$=0 (solid) are closed since both positive and negative detuning shift the driving off the resonance of the waves. Note the separate island of stability, which grows with increasing g_i and eventually merges with the other stability region for negative a_i, whereas for decreasing g_i it disappears altogether. For still more negative g_i the other region starts shrinking as well and one obtains the situation shown in Fig.3b. With (8) satisfied, $D(q)$ has now a maximum at a q_{max} with $D(q_{max})<0$ for suitably chosen d_{bf}. Then the diffusion coefficient is negative for all values of q in the vicinity of q_{max} for large S_0 and, in fact, there is a range in detuning for which no stable PSW exist (a_i>0). In Fig.2b such a change in g_i would be reflected by shifting the left region of stability to

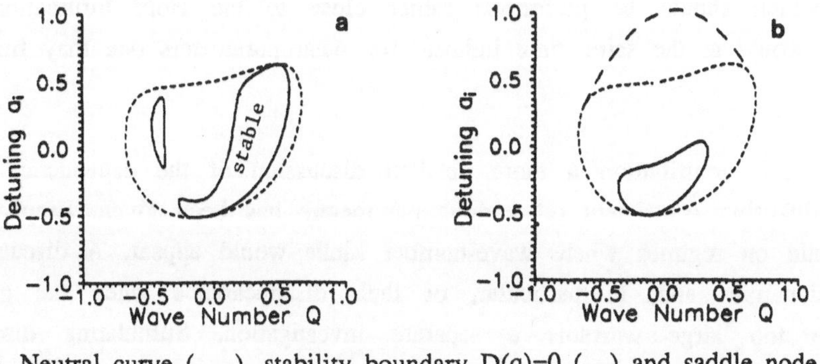

Fig.3 Neutral curve (– –), stability boundary D(q)=0 (—) and saddle node (- -) a_r=-0.05, b=0.5, d=1+0.5*i*, s=0.2 c=-1+4*i*, g_r=-1. a) g_i=-7.5 b) g_i=-14

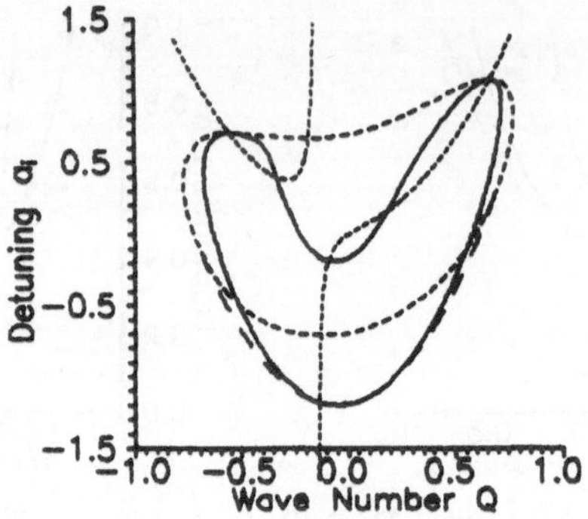

Fig.4 Neutral curve (– –), stability boundary D(q)=0 (—), saddle node (- -) and q_c (—) for a_r=-0.1, b=0.7, d=1+*i*, s=0.3 c=-1+2*i*, g_r=-1+*i*.

larger modulation amplitudes and lowering the upper limit of the right-hand region. Eventually a gap in modulation amplitude with no stable PSW opens between the two regions. Under these conditions numerical simulations exhibit persistent dynamical behavior.

Eq.(6) shows that for sufficiently small group velocity the detuning can always be chosen such that the neutral curve will be non-convex, which will lead to wave-number kinks if the bifurcation is supercritical. Experimentally it would therefore be of interest to have a simple diagnosis whether such conditions can be found. Measuring the full neutral curve is rather cumbersome, if not impossible. Measurements of the critical wave number q_c as a function of the detuning (and thus also of the modulation amplitude) give, however, a good indication whether the detuning can be chosen large enough to make the neutral non-convex. This is shown in Fig.4, where in addition to the neutral curve, the stability limit and the saddle-node line the critical wave number q_c is displayed (dotted line). Although no bifurcation from the single minimum to the double minimum (plus maximum) occurs when changing the detuning, the characteristic behavior should be sufficiently easy to detect. Such a scan, which should be performed rather close to the Hopf bifurcation (a_r<0 small), would at the same time indicate for what parameters one may find these kinks.

In this contribution a more detailed discussion of the dependence of the phase diffusion coefficient on various parameters has been given. Emphasis has been laid on regimes where wave-number kinks would appear. A discussion of their dynamics and, in particular, of their disappearance when the gradients become too large warrants a separate investigation. Stimulating discussions with I. Rehberg are gratefully acknowledged.

References

1. H. Riecke, Europhys. Lett. (to appear).
2. H. Riecke, J. D. Crawford and E. Knobloch, Phys. Rev. Lett **61**, 1942 (1988).
3. D. Walgraef, Europhys. Lett. **7**, 485 (1988).
4 H. Riecke, J.D. Crawford and E. Knobloch in "New Trends in Nonlinear Dynamics and Pattern Forming Phenomena: The Geometry of Nonequilibrium", Eds. P. Coullet and P. Huerre, NATO ASI Series, Plenum Press, 1989.
5. A.B. Ezerskii, M.I. Rabinovich, V.P. Reutov and I.M. Starobinets, Sov. Phys. JETP **64**, 1228 (1986); E. Meron and I. Procaccia, Phys. Rev. **A34**, 3221 (1986).
6. H.R. Brand and R.J. Deissler, Phys. Rev. Lett. **63**, 508 (1989) and this volume.
7. K. Kawasaki and T. Ohta, Physica **116A**, 573 (1982).
8. P. Coullet, C. Elphick and D. Repaux, Phys. Rev. Lett. **58**, 431 (1987).
9. E. Bodenschatz, M. Kaiser, L. Kramer, W. Pesch, A. Weber and W. Zimmermann, in "New Trends in Nonlinear Dynamics and Pattern Forming Phenomena: The Geometry of Nonequilibrium", Eds. P. Coullet and P. Huerre, NATO ASI Series, Plenum Press, 1989.
10. A general linear stability analysis for standing waves with q=0 has been performed by W. Zimmermann (priv. comm.).

DIRECTIONAL SOLIDIFICATION OF TRANSPARENT EUTECTIC ALLOYS

S. de Cheveigné, G. Faivre, C. Guthmann, and P. Kurowski

Groupe de Physique des Solides de
l'Université Paris VII
(Laboratoire associé au C.N.R.S)
Tour 23, 2 Pl. Jussieu, 75251 Paris Cedex 05, France

When an alloy at eutectic concentration is solidified, two distinct phases, α and β, of different concentrations separate. In directional solidification, with thin (50 m) samples of the transparent eutectic mixture CBr_4 - 8.4 % C_2Cl_6, one obtains a periodic structure with lamellae perpendicular to the solidification interface[1] (left-hand side of Fig 1). From the point of view of the fundamental physicist, this is an interesting case of pattern formation[2]. From a more practical point of view, eutectic structures are met during the solidification of a number of metal alloys and it is useful to be able to study their dynamics on a transparent model. The eutectic pattern appears without a threshold, however low the pulling velocity (there is no analogue to the Mullins and Sekerka threshold in cellular growth of dilute alloys). Figure 1 shows one onset

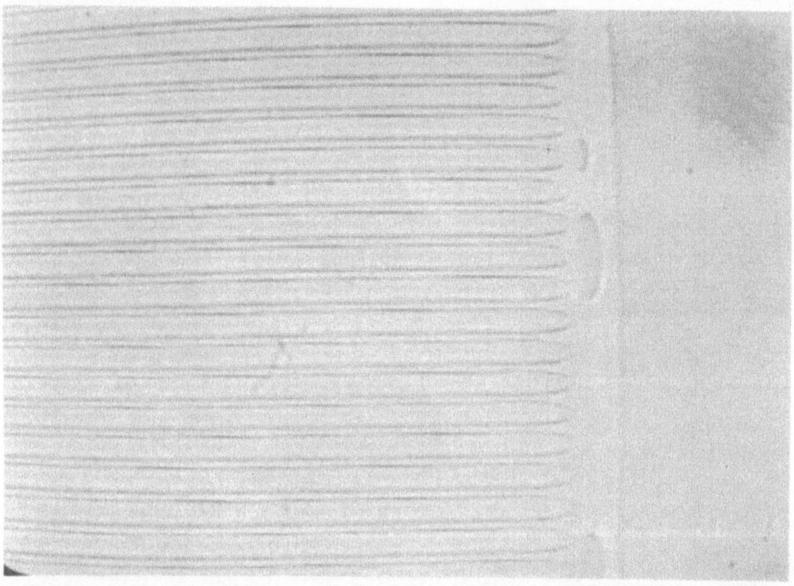

Fig. 1. Directionally solidified CBr_4-C_2Cl_6 at eutectic concentration ($\lambda = 15\ \mu$, $v = 1.5\ \mu$/s). Pulling had been stopped, creating a narrow single-phased layer at the interface. When it resumes one observes nucleation of the second phase.

Nonlinear Evolution of Spatio-Temporal Structures in Dissipative Continuous Systems
Edited by F.H. Busse and L. Kramer
Plenum Press, New York, 1990

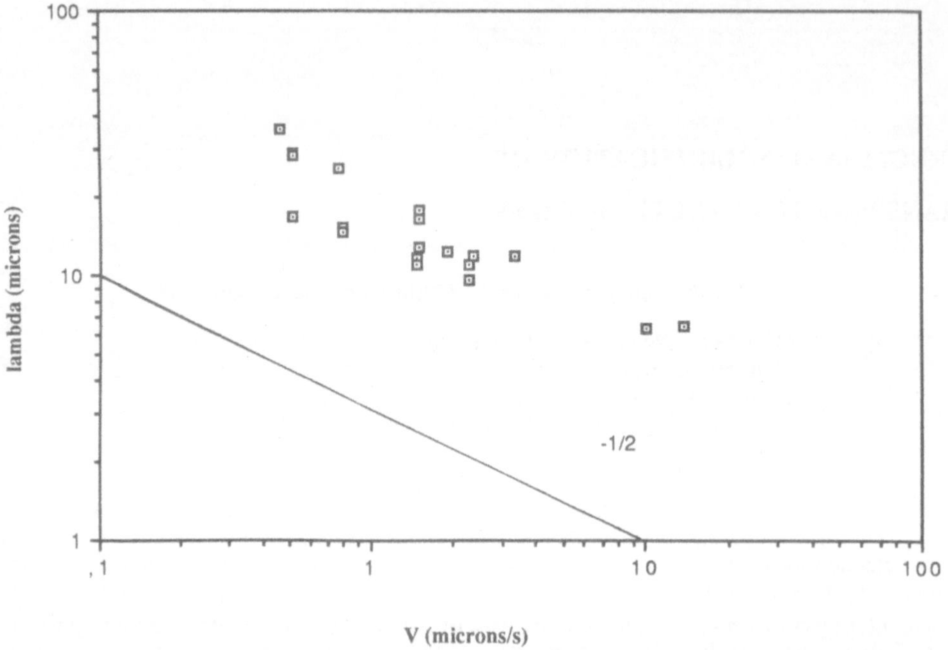

Fig.2. Wavelength vs velocity (G = 100°/s).

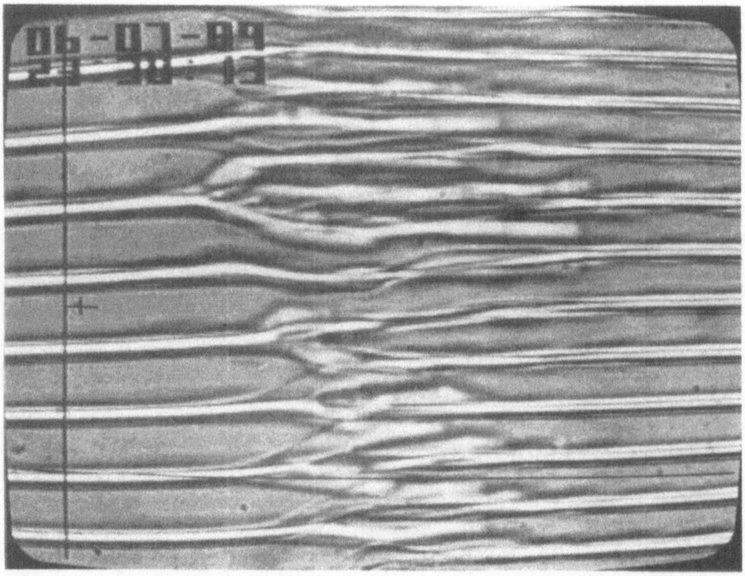

Fig. 3. Wavelength readjustment after a velocity change from 0.8 to 1.5μ/s.

mechanism. In this case the sample had been pulled, creating the pattern to the left. Pulling was then stopped and a narrow single phased band developed at the interface, due to the sample being slightly off-eutectic. When pulling was resumed, the other phase nucleated locally, at the tips of previous lamella. One can also observe propagation of the two phase pattern, not yet well organized, along the interface as reported by Seetharaman and Trivedi[3].

Once a steady-state is reached (and this may require a certain length of time, see below), the system adopts a well-defined wavelength λ (to within about 10%), function of the experimental mechanisms. When the pulling speed is modified, the wavelength should readjust since the steady state wavelength varies with the velocity V. (Fig 2 : $l \alpha V^{-1/2}$). Large velocity changes

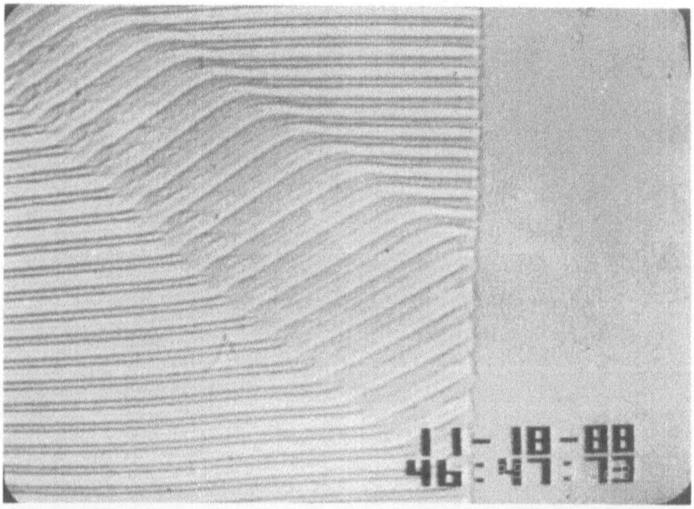

Fig 4. Wavelength readjustment by emission of a tilt wave, after a velocity change from 1,5 to 1,9 μ/s The wave moves down the picture. Before the wave λ = 15 μ, after λ = 13 μ.

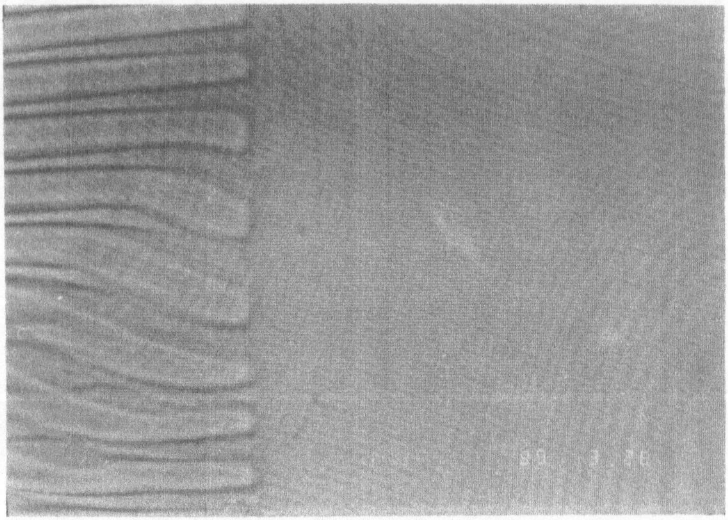

Fig.5. Close-up view of the solid-liquid interface in a tilt wave.

provoking wavelength changes of the order or above a factor 2 are known to cause fusion or splitting of lamella[1]. For small increases, at low pulling speeds (below about 1.7 μ /s for a temperature gradient of 100°/cm), this readjustment takes place in a disordered manner (Fig 3). Individual lamella loose their identity, mix within the thickness of the sample, and reappear with a larger periodicity. The process occurs rapidly following the velocity change.

At higher speeds, the wavelength adjustment after an increase in velocity takes place by the passage of solitary waves, which we call "tilt waves"[3]. The first ones rapidly adjust the wavelength (hence their spreading). In Fig. 4 $\lambda = 15\ \mu$ before the wave and $13\ \mu$ after. These waves subsist long after the main wavelength adjustment has taken place - they then have parallel edges - and reduce the dispersion in wavelength. Figure 5 gives a close-up view of the

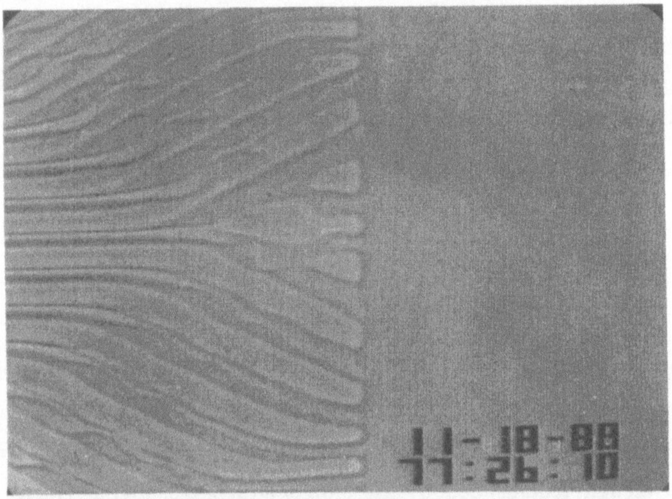

Fig 6. Close-up view of the solid-liquid interface in the collision of two tilt waves.

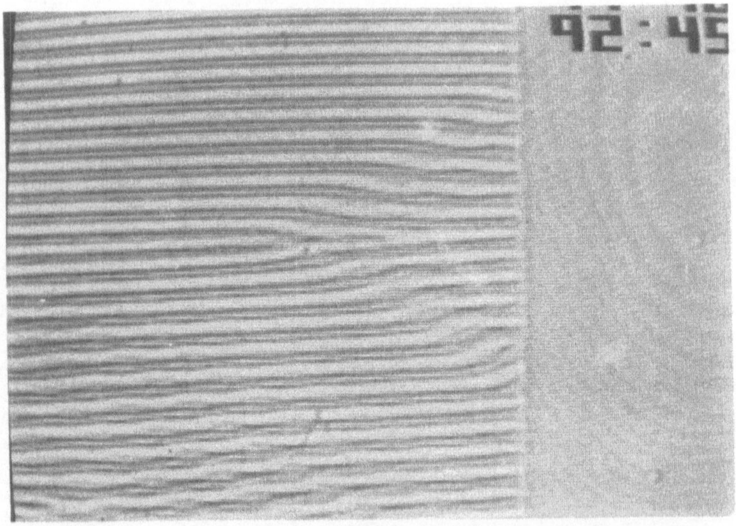

Fig. 7. Birth of two tilt waves on a dust particle. Note that two units have disappeared in the process.

interface on the passing of a tilt wave. Note that the lamella tilt in the direction opposite to the direction of propagation of the wave. The angle between the normal to the interface and the wave envelope is equal and opposite to the angle between the normal to the interface and the tilted lamellae. We generally observe one particular tilt angle (± 26° ±3°) under the conditions described here. Waves propagating in opposite directions interact and can annihilate each other. Figure 6 gives a close-up view of a collision.

The waves appear when the width of a lamella increases beyond a certain threshold, either due to a localized defect (Fig. 7) or to the presence of an inclined grain boundary (Fig 8). In the later case quasi-periodic wave-trains are produced (Fig 9).

When the pulling velocity is decreased, the wavelength is expected to increase but tilt waves are not observed. The wavelength remains frozen and eventually adjusts by fusion of lamellae (Fig 10) when the required adjustment is sufficiently large.

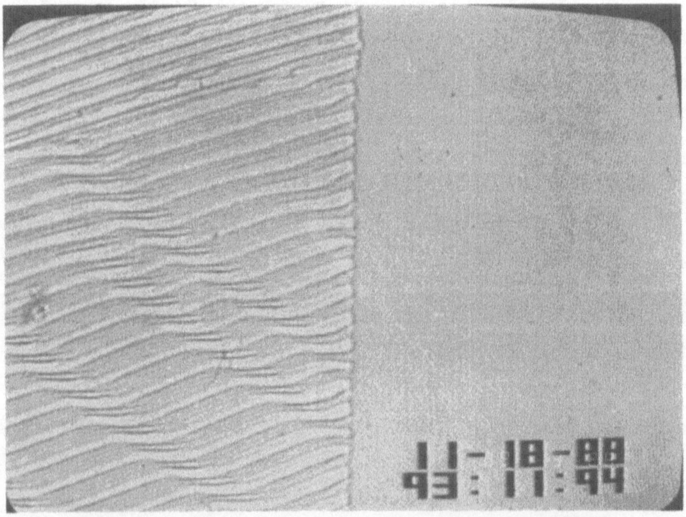

Fig. 8. Birth of tilt waves on a grain boundary.

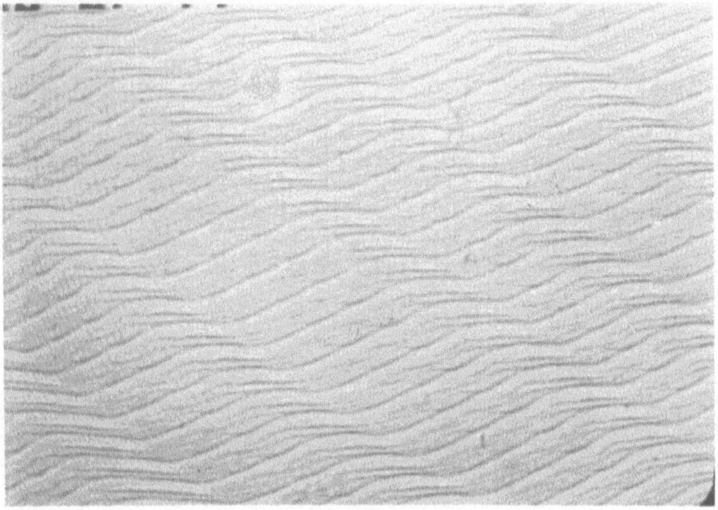

Fig. 9. Quasi-periodic trains of tilt waves.

The features of the tilt waves described here are remarkably similar to those described during the conference by Simon and Libchaber, Dubois, or Couder (see their contributions to the present volume). A unifying phenomenological description based on the breaking of symmetry occuring in each experiment has been proposed by Coullet et al[5]. It is too soon to draw definite conlusions as to the nature of the phenomena observed, but the apparent underlying unity is particularly exciting.

Acknowledgements : The present work was partially funded by the Centre National d'Etudes Spatiales.

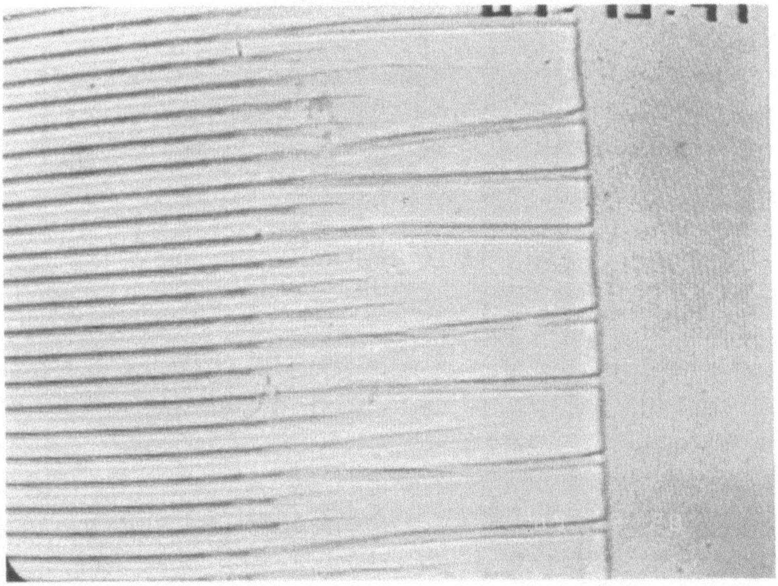

Fig. 10. Wavelength readjustment from 12.7 to 24 μ on average by fusion of cells after a velocity change from 1.1 to 0.5μ/s

References

1 J.D. Hunt and K.A.Jackson, Trans. AIME **236** (1966) 843.
2. K.A.Jackson and J.D.Hunt , Trans. AIME **236** (1966) 1129, J.S. Langer, Phys. Rev. Lett. **44** (1980) 1023, A. Karma, Phys. Rev. Lett. **59** (1987) 71.
3. V.Seetharaman and R. Trivedi, Met. Trans **19A** (1988) 2955.
4. G. Faivre, S. de Cheveigné, C. Guthmann, P. Kurowski, Europhys. Lett **9** (1989) 779.
5. P. Coullet, R.E. Goldstein and G.H. Gunaratne, Phys. Rev. Lett. **63** (1989) 1954.

DIRECTIONAL SOLIDIFICATION OF A FACETED CRYSTAL

B. Caroli*, C. Caroli, and B. Roulet

Groupe de Physique des solides
associé au Centre National de la Recherche Scientifique
Université Paris VII, 2 place Jussieu, 75251 Paris Cedex 05, France
*Also Département de Physique, U.F.R. Sciences Fondamentales et Appliquées, Université de Picardie, 33 rue St Leu
80039 Amiens Cedex, France

When a dilute binary mixture is pulled at constant velocity V in an external thermal gradient G, the solid-liquid interface exhibits a morphological instability when V becomes larger than a threshold velocity $V_c(G)$. For $V < V_c$ the stationary interface is planar, for $V > V_c$ it develops a (quasi)periodic cellular structure. This instability, first analyzed by Mullins and Sekerka(1), results from the competition between the destabilizing effect of solute diffusion and the stabilization due to the thermal gradient and to capillarity.

Most experiments and calculations have been performed in the case where the material has an isotropic surface tension and a microscopically rough solid-liquid interface. The attachment kinetics is then fast enough for the interface to be practically at local thermodynamic equilibrium.

However, faceted cellular fronts have been observed in Bismuth alloys(2) and in semiconductors growing from the melt(3). In these faceted materials interface kinetics and surface tension anisotropy play an important part. Indeed, the kinetics of atomic attachment on a facet, which is controlled by 2-D nucleation or by dislocations, is slow, and the liquid-solid interface energy $\sigma(\theta)$ exhibits a cusp for each facet orientation (see Figure 1).

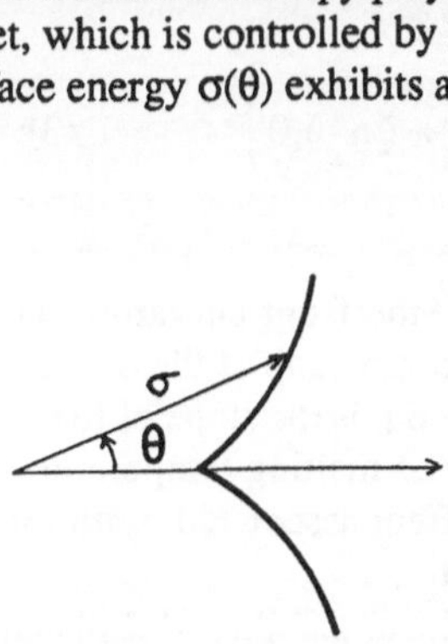

Fig. 1 . Schematic shape of the Wulff plot of the surface energy $\sigma(\theta)$ around the facet direction $\theta = 0$.

In the following we first give the equations describing directional solidification in the presence of a facet. We then show that, when the basic planar front is a facet, there exists,

Nonlinear Evolution of Spatio-Temporal Structures in Dissipative Continuous Systems
Edited by F.H. Busse and L. Kramer
Plenum Press, New York, 1990

beyond the standard velocity threshold $V_c(G)$, an infinite family of deformed periodic small amplitude stationary fronts with crenellated shapes. The analysis of their linear stability shows that these crenellated solutions are unstable against amplitude fluctuations, thus defining an amplitude threshold for the usual cellular instability.

1 - GROWTH EQUATIONS

We describe the two phase system by the usual one-sided model in which solute diffusion in the solid is neglected. As compared with chemical diffusion, thermal diffusion can be considered instantaneous. We assume for simplicity that the two phases have equal thermal diffusivities. The thermal profile is then unaffected by the solidification process and defined by the externally imposed thermal gradient G//Oz. We limit ourselves to 2-D front deformations and call $\zeta(x,t)$ the position of the front along the z-axis.

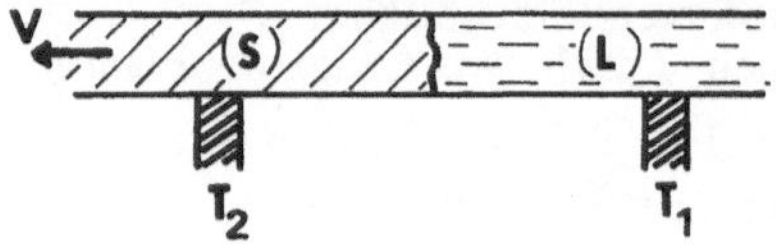

Fig. 2 . Schematic directional solidification set-up.

In the laboratory frame, the solidifying system, pulled along Oz at velocity V (Figure 2) is then described by the following set of equations[4] :

(i) In the liquid ($z > \zeta(x,t)$), the solute concentration $C(x,z,t)$ obeys :

$$\frac{\partial C}{\partial t} = D\nabla^2 C + V\frac{\partial C}{\partial z} \quad (1)$$

(ii) On the front ($z = \zeta(x,t)$), solute conservation is expressed by :

$$- D\vec{n} \cdot \vec{\nabla} C = (\Delta C)(V + \dot{\zeta} n_z) \quad (2)$$

and the generalized Gibbs-Thomson equation reads :

$$T(x, \zeta(x), t) = T_0 + G\zeta$$

$$= T_M + m_L C - \frac{T_M}{L}\left(\sigma + \frac{d^2\sigma}{d\theta^2}\right)K - F\left\{(V + \dot{\zeta} n_z),\theta\right\} \quad (3)$$

$\vec{n}$ is the unit vector normal to the front, $K = - \zeta''/(1 + \zeta'^2)^{3/2}$ the front curvature. D is the solute diffusion coefficient in the liquid, ΔC the solute concentration gap of the mixture (here taken to be constant, a simplifying but unessential assumption). m_L is the slope of the liquidus line on the phase diagram, L the specific latent heat and T_M the melting temperature of the pure solvent. Finally, $F(v,\theta)$ accounts for the undercooling effect associated with interface dissipation, i.e. with the non-instantaneity of atomic attachment.

For rough interfaces, $F \cong 0$ and σ is smooth. The linear stability of a planar front can then be studied by standard bifurcation analysis. On the contrary, for a material faceting in, e.g., the $\theta = O$ direction, $F(v,O)$ is finite. Moreover, the surface stiffness ($\sigma + d^2\sigma/d\theta^2$) contains a singular ($\delta(\theta)$) contribution which accounts for the fact that a deformation of a faceted front always costs a finite energy due to step formation. In other words, equilibrium on a facet is no longer a local, but a global property expressing the fact that a facet can only climb as a whole. Mathematically, as shown by Ben Amar and Pomeau[5], the corresponding global condition is obtained by integrating eq.(3) along the facet. If s is the curvilinear coordinate along the front, the integration of the capillary term from one end of the facet ($s = s_0$) to the other ($s = s_1$) gives, since $K = - d\theta/ds$:

$$\int_{s_0}^{s_1} ds\,(\sigma + d^2\sigma/d\theta^2)\,K(s) = \left[\frac{d\sigma}{d\theta}\right]_{0^+} - \left[\frac{d\sigma}{d\theta}\right]_{0^-} = \varepsilon\,\Delta\sigma' \qquad (4)$$

$\varepsilon = +1$ (resp. -1) for a "convex" (resp. "concave") solid facet (Figure 3).

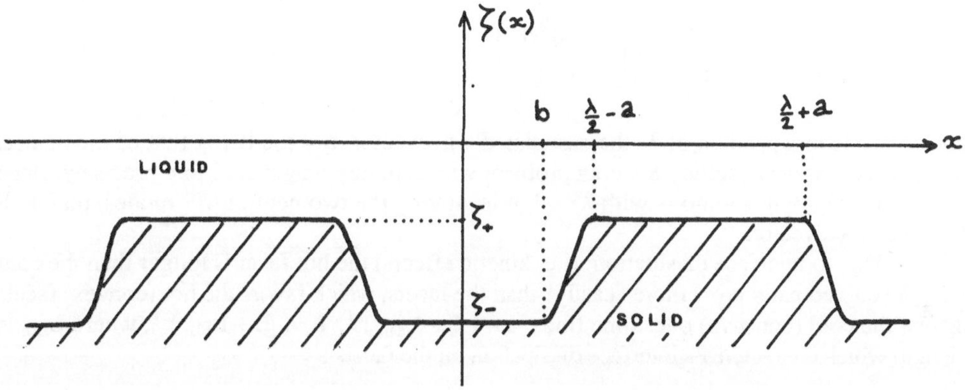

Fig. 3 . Periodic crenellated front. The hot planar region ($\frac{\lambda}{2} - a < x < \frac{\lambda}{2} + a$) is what we call a "convex" facet, the cold one ($-b < x < b$) corresponds to a "concave" facet.

So, when studying a partly faceted front, the Gibbs-Thomson equation (3) valid along the rough parts must be replaced, along each faceted part, by its integrated version, obtained with the help of eq.(4). The matching condition between rough curved parts and facets is obtained by imposing equilibrium at each facet end,the positions of which are themselves to be determined. We assume here that there are no missing orientations on the equilibrium shape of the crystal ($(\sigma + \sigma'')_{\theta=0+} > 0$). The curved parts of the profile then match tangentially with the facets.

2 - STATIONARY CRENELLATED FRONTS[6]

We start from an initial planar front $\zeta = 0$ corresponding to an infinite facet normal to the thermal gradient.

First of all, let us remark that the possible deformations of such a front cannot be studied by a standard linear stability analysis. Indeed, in the presence of a cusp in the Wulff plot (and thus of a $\delta(\theta)$ singularity in the surface stiffness $\sigma + \sigma''$) in the direction of the unperturbed interface, the front equations can no longer be expanded in powers of the deformation amplitude - i.e. of $\theta = \tan^{-1}(d\zeta/dx)$. Since we are unable for this reason to study the dynamic evolution of deformations of the planar front, we limit ourselves to investigating whether there exist stationary front shapes which are non-planar, periodic, with a small but finite deformation amplitude (Figure 3). Moreover, we restrict our investigation to the highly non-local limit where the wavelength λ of these structures is much smaller than the diffusion length $l = D/V$. Note that in usual experimental situations, where $V \approx 1 - 10\ \mu m/sec$, $l \approx 1 - 10^{-1}$ mm. In this $\lambda << l$ limit, the concentration field in the presence of the actual distorted front can be approximated by that associated with its planar average[6]. The problem can then be solved easily, with the following results:

(i) At small velocity, more precisely for $l/l_T > 1$, where the thermal length $l_T = m_L(\Delta C)/G$, there are no such crenellated solutions.

(ii) For $l/l_T < 1$ there is one such solution for each value of λ The lengths a and b of the two (hot and cold) types of facets, and their heights ζ_+ and ζ_- are uniquely defined for a given λ. At fixed λ, the deformation amplitude ($\zeta_+ - \zeta_-$) decreases when V increases beyond the threshold. At fixed velocity, it is a decreasing function of λ.

Hot and cold facets are linked by rough parts, each of which is a half-arch of a sinusoid, with an x-extension $c = \pi\,(d_0 l/(1- l/l_T))^{1/2}$, where d_0 is the capillary length of the problem :

$$d_0 = \frac{T_M}{L m_L(\Delta C)} \left[\sigma + \frac{d^2\sigma}{d\theta^2} \right]_{\theta=0^+} \tag{5}$$

Note that c is independent of λ, the associated wavevector $q = \pi/c$ being that of the neutral mode of the classical Mullins-Sekerka problem with capillary length d_0. More precisely, since we are investigating solutions with $\lambda \ll l$, it is, among the two neutral MS modes, that with the larger wavevector.

Due to interface dissipation, (i.e. kinetic effects) the hot facet is longer than the cold one : the curved parts grow more "easily" than the facets, which favors the hot (convex) facets against the cold (concave) ones. one finds $a \approx (\zeta_+ - L_{kin})^{-1}$, $b \approx (\zeta_- + L_{kin})^{-1}$, where L_{kin} is a length which characterizes interface dissipation on the facets :

$$L_{kin} = F(V, \theta = 0)\, l\, (1 - l/l_T)^{-1} \tag{6}$$

Let us finally remark that the condition of existence of such solutions ($l/l_T < 1$) corresponds precisely to $V > V_c(G)$, where V_c is the standard Mullins-Sekerka threshold :

$$V_c(G) = D / l_T \tag{7}$$

However, our non-local approximation ($\lambda/l \ll 1$) is only valid when :

$$\frac{|V-V_c|}{V_c} \gg (d_0/l_c)^{1/3} \tag{8}$$

where $l_c = D / V_c$. So we cannot predict exactly the position of the threshold, which is only defined, in our approximation, to within an uncertainty measured by $(d_0/l_c)^{1/3}$. In practice, d_0 does not exceed a few hundred angströms, while $V_c \approx 1$ -10 µm/sec ; $(d_0/l_c)^{1/3}$ is then of order 10^{-1} at most, so the restriction implied by condition (8) should be of little consequence for the analysis of experiments.

3 - LINEAR STABILITY OF CRENELLATED FRONTS [7]

We then analyze the stability of the continuum of solutions determined above with respect to small and slowly varying perturbations.

In order to perform this analysis, one cannot simply, as for the stationary case, approximate the difusion field by that of the average planar front, which would leave us with one single degreee of freedom, corresponding to global relaxation in the thermal gradient.

For this reason, we assume the facet lengths a and b to be much larger than the x-extension, c, of the "risers" (rough parts). This enables us to approximate the concentration field associated with the crenellated profile by that of a perfect crenel (c = 0). A small deformation of this stationary profile is then defined by four parameters in each cell (see Figure 4).

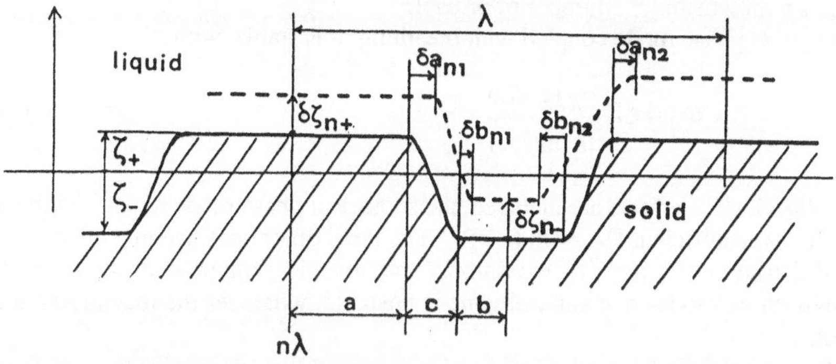

Fig. 4 . Full line : stationary periodic crenellated front. Dashed line : displaced crenellated profile.

Taking advantage of Floquet's theorem, we look for solutions of the form :

$$\delta\zeta_{n\pm}(t) = \delta\zeta_{\pm}\, e^{inp\lambda}\, e^{\omega t}$$

$$\delta a_{n_{1,2}}(t) = \delta a_{1,2}\, e^{ipn\lambda}\, e^{\omega t} \tag{9}$$

Plugging this into the growth equations linearized about the stationary crenellated profile, we obtain four homogeneous linear equations for the four unknowns $\delta\zeta_+$, $\delta\zeta_-$, δa_1, δa_2, which yield the spectrum of the long wavelength front deformation modes. The four corresponding eigenmodes can be analyzed in terms of four basic amplitude and phase types of deformations: climbing ($\delta\zeta_{n_+} \approx \delta\zeta_{n_-}$) and breathing ($\delta\zeta_{n_+} \approx -\delta\zeta_{n_-}$) modes; acoustical ($\delta a_{n_1} \approx \delta a_{n_2}$) and optical ($\delta a_{n_1} \approx -\delta a_{n_2}$) phase ones.

The results are most easily analyzed when kinetic effects are neglected (F = 0) In this simple limit one finds (for modes with very small wavevectors p) :

(i) A pure acoustical phase mode, with a growth rate :

$$\omega_1 = -\mathbf{D}_{\varphi} p^2 \tag{10}$$

where the phase diffusion coefficient :

$$\mathbf{D}_{\varphi} = \frac{\pi\lambda D}{8l}\,\frac{(1-l/l_T)}{\ln(\lambda/c)} \tag{11}$$

is positive for all λ. This mode is stable: the usual Eckhaus instability is blocked by the presence of the facets.

(ii) A pure stable climbing mode :

$$\omega_2 = -D/ll_T \tag{12}$$

It corresponds to global relaxation in the thermal gradient.

(iii) A pure unstable breathing mode :

$$\omega_3 = D\,\pi^2\,(1 - l/l_T)/\Phi_0 l\lambda \qquad (13)$$

where Φ_0 is a dimensionless number of order 1.

(iv) An optical phase mode coupled with breathing. it is stable, with :

$$\omega_4 = -D\,\frac{2\pi}{l\lambda}\,\frac{(1-l/l_T)}{\ln(\lambda/c)} \qquad (14)$$

When kinetic effects are included, the acoustical phase mode remains pure and stable, with a diffusion coefficient $D_\varphi \approx (\ln(\lambda/c))^{-1}$ The three other ones become mixed modes with various admixtures of optical phase, climbing and breathing components. One can show that, among these three modes one and only one is unstable, whatever the magnitude of interface dissipation.

So, crenellated front profiles are always unstable with respect to a mode involving breathing. They can therefore be viewed as defining a finite threshold of deformation amplitude of the planar front to be "jumped over", for $V > V_c$, for the front profile to reach a stable deformed configuration, which can reasonably be guessed to be a partly faceted cellular profile. So, the transition should exhibit an anomalously large velocity hysteresis.

One last remark can be made: it appears from the expressions (11-14) of the growth rates that diffusion slows down the lateral motion of a steep riser by a factor $(\ln(\lambda/c))^{-1}$, where c is the riser width. This slowing down factor is most probably characteristic of all situations involving motion of the so-called "macrosteps"[8] observed on faceted fronts. This is for example the case of periodic vicinal fronts in directional solidification.

REFERENCES

(1) W.W.Mullins, R.F.Sekerka : Stability of a planar interface during solidification of a dilute binary alloy, J.Appl.Phys., 35 : 444 (1964)

(2) B.Billia : Recherches sur la morphologie du front de solidification lors de la croissance unidirectionnelle d'alliages binaires, Thèse, Aix-Marseille (1982)

(3) see for example : D.K.Shangguan, J.D.Hunt: Modelling of cellular growth of faceting crystals, in : Proceedings of the Conference on Solidification Processing (Sheffield, 1987)

(4) J.S.Langer : Instabilities and pattern formation in crystal growth, Rev.Mod.Phys., 52 : 1 (1980)

(5) M.Ben Amar, Y.Pomeau : Growth of faceted needle crystals : Theory, Europhys.Lett., 6 : 609 (1988)

(6) R.Bowley, B.Caroli, C.Caroli, F.Graner, P.Nozières, B.Roulet : On directional solidification of a faceted crystal, J. de Physique 50 : 1377 (1989)

(7) B.Caroli, C.Caroli, B.Roulet : Directional solidification of a faceted crystal II - Phase dynamics of crenellated front patterns, J. de Physique, to be published

(8) A.A.Chernov, T.Nishinaga : Growth shapes and their stability at anisotropic interface kinetics: Theoretical aspects for solution growth, in : "Morphology of Crystals", Part I, I.Sunagawa ed. (Terra Scient. Publ. Corp.) 1987

ROUTES TO CELL FORMATION AND HIDDEN RAMPS

IN DIRECTIONAL SOLIDIFICATION

P.E. Cladis,* P.L. Finn* and J.T. Gleeson†*

* *AT&T Bell Laboratories, Murray Hill, NJ 07974, USA*
† *Physics Department, Kent State University, Kent, OH 44242, USA*

Abstract: Precise measurements of interface position as a crystal grows in a temperature gradient (directional solidification) show that while cellular interfaces are usually steady, stationary planar interfaces are more difficult to achieve. Since the interface's position is directly related to its temperature that is in turn related to impurity concentration at the interface, we found *directional melting* provided a sensitive tool to probe the quality of the crystal grown and obtain a measure of the partition coefficient, k, that determines the nature of the planar-cellular bifurcation.

In directional solidification, the allowed band of linearly unstable wavelengths for cellular patterns without boundary effects extends over 3 orders of magnitude, *i.e.* is much broader than "canonical" hydrodynamic pattern forming systems. In stark contrast, the observed band is much less than one order. We present experimental observations suggesting that soft boundary conditions intrinsic to the experiment are responsible for the dramatic discrepancy between theoretical expectations and experimental observations. In particular, we found that at onset, the interface first develops structure in the vertical, shorter dimension because of curvature at the liquid-solid-glass contact. We use this information to install a slowly varying "ramp", where the control parameter varies spatially from above to below threshold, by inducing a concentration gradient parallel to the solid-liquid interface in the longer direction and find that this leads to further collapse of the allowed band.

1. Introduction

Visual observation of the solid – liquid interface of a transparent crystal as it grows in a temperature gradient (directional solidification) was pioneered by Jackson and Hunt.[1] This opened a window on a technologically important process[2] as well as introduced a pattern forming system[3] that has recently been the subject of much interest.[4] Fig. 1 is a schematic of a directional solidification experiment. The "Hot" and "Cold" temperatures are chosen so that the freezing temperature (and hence, the interface, shown as a wavy line) lies in the field of view. In this way, temperature is mapped onto a spatial dimension, and patterns exhibited by a moving interface, created by pulling the material at a speed, v, towards "Cold", may be observed for arbitrarily long times.

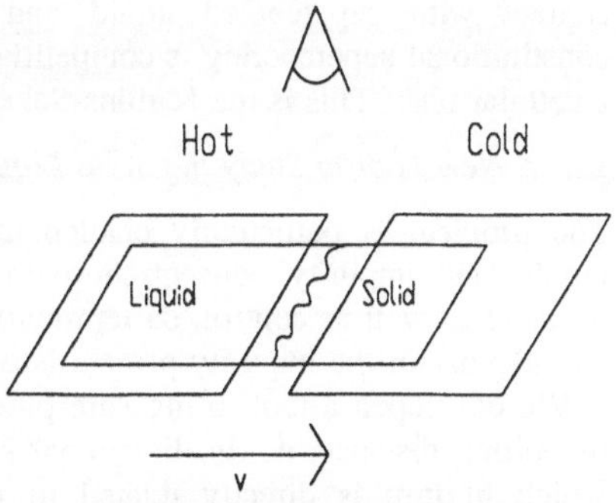

Fig. 1. Directional Solidification

Our objective is to find ways to characterize quantitatively (and eventually control) the physical conditions responsible for the onset of patterns at the solid-liquid interface. To achieve this, we make use of information obtained using high speed image processing and computer controlled data acquisition.

Nonlinear Evolution of Spatio-Temporal Structures in Dissipative Continuous Systems
Edited by F.H. Busse and L. Kramer
Plenum Press, New York, 1990

1.1 *Why Patterns Form at the Solid-Liquid Interface*

To understand why patterns form at a solid-liquid interface during directional solidification, we consider an equilibrium temperature-concentration phase diagram for a binary mixture (Fig. 2). This phase diagram is characterized by a two phase region that broadens with increasing impurity concentration. The partition coefficient, k, the parameter that describes this broadening, is defined as the ratio of the solid to liquid equilibrium impurity concentrations. When the concentration dependence of the liquidus and solidus are linear, as shown in Fig. 2, k is the ratio of the liquidus to solidus slopes.

The dashed line represents the concentration profile at zero speed, the horizontal portion being the discontinuity in concentration at the interface. Since impurity diffusion is faster in the liquid than in the solid (where, in the one-sided approximation, the diffusion constant is assumed infinite), the concentration everywhere in the liquid is taken to be the equilibrium concentration, C_∞. Thus, at rest and in a temperature gradient, the position of the solid-liquid interface is at T_L.

As material freezes, it expels impurity leading to impurity build-up ahead of the solidification front. The dot-dash line represents the uniform speed, steady state concentration profile assumed to be constant, C_∞, in the solid near the interface and exponentially decaying to C_∞ over a length $\ell_D = D/v$ in the liquid. D is the diffusion constant in the liquid (~ $10^{-5}cm^2/sec$ for succinonitrile) and v is the pulling speed. Thus, the steady state interface position at constant speed is at T_S, a distance $\ell_T = \Delta T/G$ from its rest position (G is the temperature gradient and ΔT is the width of the two phase region at c_∞). Since ℓ_D decreases with increasing speed, at a critical speed, v_c, where $\ell_D \sim \ell_T$, the concentration gradient at the interface becomes tangent to the liquidus slope. For $v > v_c$, it dips below the liquidus into the two phase region (*constitutional supercooling.*)[5] The control parameter, the analogue of $\varepsilon = (T - T_c)/T_c$ for critical phenomena, is $(v - v_c)/v_c$.

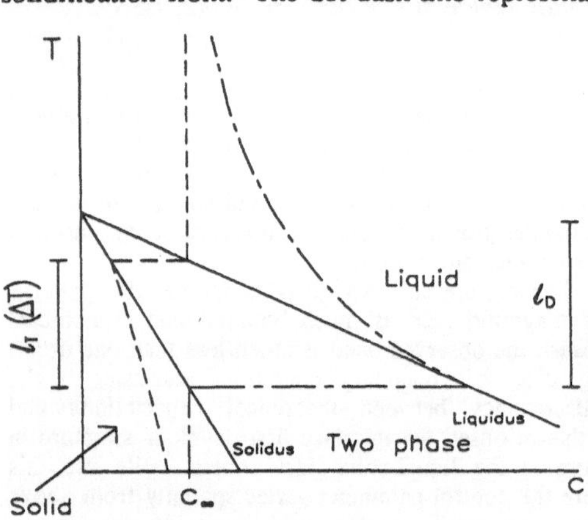

Fig. 2. Binary Mixture Phase Diagram.

"Constitutional" refers to the fact that composition is responsible for the supercooling. Some material on the liquid side of the interface is now at a lower temperature than its local equilibrium, freezing temperature. Thus, perturbations of the solid into the liquid will be in contact with supercooled liquid and may grow. When the driving force provided by constitutional supercooling is competitive with surface forces, the planar interface transforms to a cellular one. This is the Mullins-Sekerka instability (MS).[6]

1.2 *A New Tool to Study the Solid-Liquid Interface*

The problem is particularly challenging to quantify because the driving mechanism is the gradient in impurity concentration on the liquid side of the interface. This is difficult to measure as well as control, so reproducibility can be hard to achieve. Furthermore, the amount and identity of the impurity present is often unknown.

We developed a tool to measure precisely the interface position, z, as a function of time, t, or pulling distance, d. In directional solidification, interface position is directly related to T_L which in turn is directly related to the impurity concentration on the liquid side of the interface. z(d) is sensitive to changes in pulling speed and temperature fluctuations.

We are unaware of any theory for directional melting. We argue that melting is an equilibrium process controlled only by the impurity content and strains in the solid phase. Thus, in directional melting, z is directly related to T_S. We obtain feedback on the quality of the crystal grown by measuring z(d) in directional melting. In particular, when the initial conditions correspond to those of an equilibrium binary mixture phase diagram (frequently not the case), it is possible to extract an independent, dynamical measure of the partition coefficient, k. Since the nature of the planar-cellular bifurcation depends on the magnitude of k, it is an important parameter to know. Caroli, Caroli and Roulet[7] have shown that when $k < 0.45$, the bifurcation is inverted or discontinuous and when $k > 0.45$ it is forward or continuous.

z(t) provides an evaluation of whether patterns are formed under stationary or transient growth conditions. This is important because most theories rely on the system achieving a steady state. In reality, steady state is frequently difficult to achieve and directional solidification provides many interesting examples of time dependent patterns. Another use for z(t) is to measure relaxation times. In particular, when $v \to 0$, the interface does not relax to T_L with a unique time constant. Again, we are unaware of any theory that treats this but it obviously gives useful information about the various physical processes involved in the complex problem of crystal growth.

1.3 *Wavelength Selection in Directional Solidification*

The understanding of whether wavelength selection is an intrinsic property of the system or rather depends on boundary effects is a fascinating question of pattern formation that continues to motivate extensive analytical and numerical investigations. In directional solidification, although the allowed band of linearly unstable wavelengths for cellular patterns without boundary effects extends over 3 orders of magnitude, i.e. is much broader than "canonical" hydrodynamic pattern forming systems, the observed band is much less than 1 order of magnitude.

The onset of the interface instability for an extended system was first analyzed by Mullins and Sekerka.[6] In the weakly non-linear regime, owing to the translational and rotational symmetry of the interface, periodic states can exist near threshold with a continuous band of wave numbers.[8] Far from threshold, Karma[9] suggested that wavelength selection might emerge by requiring a smooth shape at the tip in the limit of infinitely long cells. However, Dombre and Hakim[10] and Ben-Amar and Moussallam[11] showed that a band of physically admissible solutions still survives, leaving wavelength selection an open question.

In a recent paper, Eshelman and Trivedi[12] dramatically demonstrated the magnitude of the discrepancy in directional solidification. In their paper, they compared the theoretical dependence of the critical speed on wave number for both the marginal stability and the fastest growing wave number with their observations. On scales set by theory, all of their data, spanning more than one decade in pulling speed lie on a nearly vertical line. In agreement, our observations are that although there is rarely long range periodicity to the cell pattern, the band of observed wavelengths is certainly not more than a factor of 2 or 3 except in very thin cells.[13]

Recently, Misbah[14] has pointed out that a variable temperature gradient parallel to the interface coupled to a greater interface speed where the gradient was weaker, such as in the rotating solidification experiment (record-player geometry) of de Cheveigne et al.[15] might provide an adequate ramp or "soft" boundary condition to account for wavelength selection in directional solidification for a forward planar-cellular bifurcation. Kramers et al.[16] had previously shown that an infinitely slow spatial variation of the control parameter from above to below threshold results in a collapse of the allowed band to a single wave number. They termed this spatial connection a ramp.

The effect of a ramp on wavelength selection was first elegantly verified in Taylor-Couette.[17] Taylor cells are observed in the Couette geometry but with an inner cylinder tapered at one end. In this way, at a constant rotation of the inner cylinder, above threshold conditions prevail for Taylor cells in part of the container and below threshold in the tapered region where the space between the inner and outer cylinders is wider.

Here, we account for the observed difference in the cell patterns with sample thickness by the existence of a weak ramp created by an induced (by the liquid-solid-glass contact angle) concentration gradient in the shorter, vertical direction parallel to the interface. Furthermore, we use this idea to install a more effective (but still slowly weakening) ramp again induced by a concentration gradient parallel to the interface in the longer, horizontal direction to further collapse the allowed band.

1.4 *Outline*

In this paper, we follow a brief description of our experimental set-up with examples of the information we can obtain by studying the interface position in a gradient as the crystal either grows or melts. In particular, we propose a new way to obtain the partition coefficient, k, from interface position measurements during melting.

Next, we present novel experimental observations of the onset of the cellular pattern at interfaces with a 20:1 or 10:1 aspect ratio suggesting that soft boundary conditions intrinsic to the experiment are responsible for the dramatic discrepancy between theoretical expectations and experimental observations of the allowed band of wavelengths. In particular, we found that the cell pattern in directional solidification of succinonitrile is a 3-dimensional phenomenon so that sample thickness plays an important role in pattern selection. We use this information to install a slowly varying ramp by inducing a concentration gradient parallel to the solid-liquid interface. A concentration gradient parallel to the interface effectively breaks even symmetry and leads to further collapse of the allowed band. The existence of such hidden ramps may contribute to the still not understood thickness dependence of v_c observed by de Cheveigne et al.[18] at an inverted bifurcation.

2. Experimental Setup

Our temperature gradient stage[13] differs from Jackson's original set-up[1] in some respects.

— First, the stepping motor is computer controlled and provides a measure of the pulling distance to 0.1μm. The total distance pulled is a maximum of 2.5 cm. The stepping rate is determined by a frequency generator that can be set with a precision of 1μHz. Thus, in principle, we can control the pulling speed to 1 part in 10^6. In reality, the planar interface frequently selects a growth speed 10-15% less than the pulling speed. Typical pulling speeds for succinonitrile that we have used are 0.5 – 25μm/s.

— Second, the data is acquired and processed using high speed image analysis. In this paper, for example, many of the pictures of the patterns have been enhanced by filtering the digitized image with an edge detector. The effect of the filter is to make regions of uniform intensity black and edges bright. The interface appears as two sharp white lines that result from each of its edges. A comparison of the two kinds of images - enhanced and not are shown in Fig. 3. It is clearly easier to make measurements from an enhanced image.

— Finally, a gentle but firm pressure is exerted on the ends of the sample to improve the thermal contact with the two stages determining the temperature gradient. Care is taken that this does not impede the smooth motion of the sample through the gradient.

The temperature gradient for the measurements that we describe is 7.5±0.1°C/mm. Although the gradient may differ from one run to another, during a single run that can be as long as 48 hours, it has been monitored and found not to vary by more than ±0.005°C/mm.

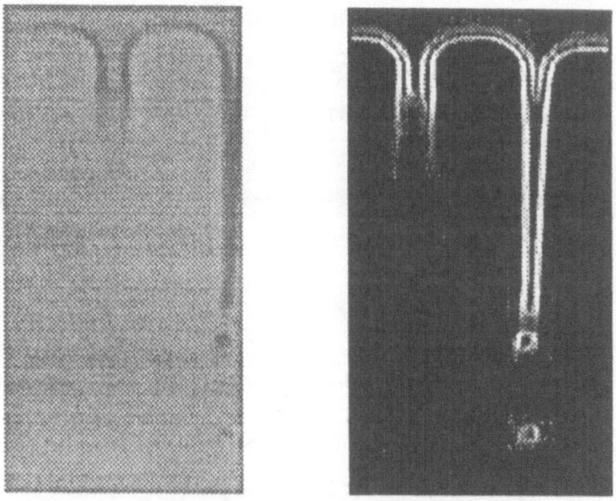

Fig. 3. Image before and after enhancement

2.1 *Succinonitrile*

The experiments we discuss here are on the interface between the cubic plastic crystal and liquid phases of succinonitrile.[19] The motivation for this choice is that it has been extensively studied by others[20] as a pattern forming system as well as a useful model system of the solid-liquid interface of metals and semiconductors. Succinonitrile is a robust material that may be heated to 200°C for 72 hours without decomposition.[19] In comparison, CBr_4, another compound much studied in directional solidification, decomposes at a temperature close to its melting temperature.[21] Consequently, the relevant parameters needed to quantify directional solidification experiments are difficult to measure.

The succinonitrile as received (from Fluka chemical) was judged to be too impure for our experiments. Moreover, the impurity(ies) were not known. Our purification procedure was primarily by sublimation then condensation onto a cold finger while under vacuum resulting in an increase of ~ 2.2°C in the melting temperature and a decrease of ~ 1°C in the width of the two phase region. Both quantities were determined by differential scanning calorimetry.

After purification, the material is vacuum loaded into rectangular (100μm×2mm or 100μm×1mm) 50 mm long glass capillaries (Vitro Dynamics) and the ends sealed with Torr-Seal epoxy (Varian). Our experience has been that Torr-Seal does not contribute to material degradation evidenced by drifts in transition temperatures. The capillaries provide samples of uniform thickness and the same boundary condition with a glass interface on all boundaries of the interface. The entire width of the capillary is in the field of view with a 3.5X objective in the case of wider, and a 10X long distance objective in the case of narrower ones. About 88% of the interface is actually visible. The remaining 12% is obscured by the curvature of the capillaries at each edge.

3. Interface Position Measurements for

3.1 *Directional Solidification*

In these experiments, the sample is displaced towards the cold contact at a constant speed $v < v_c$. Figs. 4a and b, show the interface position, z, as a function of time, t, (or pulling distance d = vt) for two kinds of planar interfaces. In Fig. 4a, succinonitrile is nominally pure whereas in Fig. 4b, about 0.8% acetone has been added. As the crystal grows, the interface moves to lower temperatures (our convention is that a positive displacement is towards the cold contact). Eventually, the succinonitrile-acetone alloy reaches a credible steady position or under-cooling (Fig. 4b) and stayed there for several hours. In contrast, the nominally pure material continued to move to greater undercooling throughout the travel of the stepping motor (Fig. 4a).

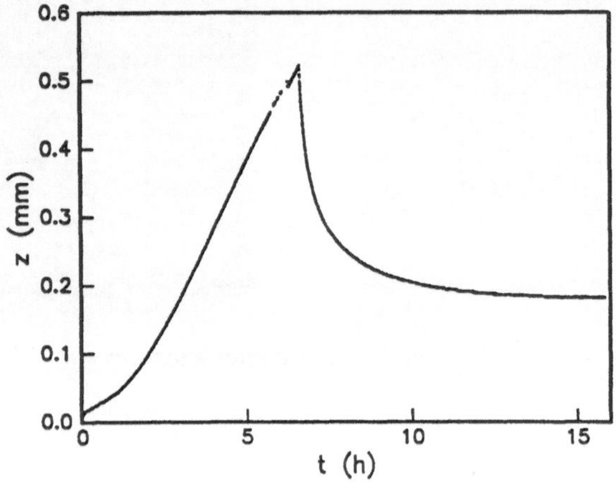

Fig. 4a. Pure Succinonitrile

After the motor stopped (the peak in the λ-like curve of Fig. 4a), the interface relaxed to its equilibrium position at T_L. However, the relaxation to equilibrium, z(t) in Fig. 4a, is not a simple exponential. A power law can be fitted to the data but there is, as yet, no theoretical justification for such behavior.

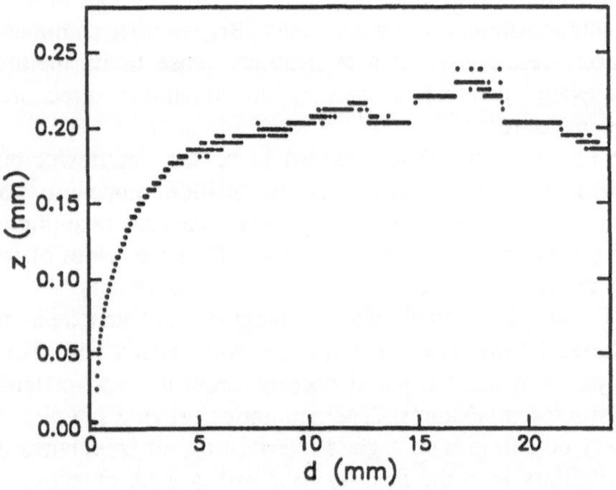

Fig. 4b. Succinonitrile Alloy

Our interpretation of the difference between Figs. 4a and 4b is that the nominally pure material has tiny amounts of some impurities present that are continuously being zone refined out of the system. In contrast, the acetone-succinonitrile mixture more closely realized the idealization of a binary phase diagram and constitutional supercooling. Indeed, after about four runs of the alloy, it also became "too pure" and showed behavior similar to Fig. 4a.

3.2 *Directional Melting of an Alloy*

In this experiment, a succinonitrile-acetone sample is grown first at 1µm/s until cells formed then left to relax at zero speed for 13 hours. Fig. 5 shows that 13 hours was adequate for the interface to reach its original position in the gradient. Since the data acquisition algorithm does not always assign a unique position for a cellular interface, an increase in the scatter in the data is seen in Fig. 5 after the interface transforms from planar to cellular.

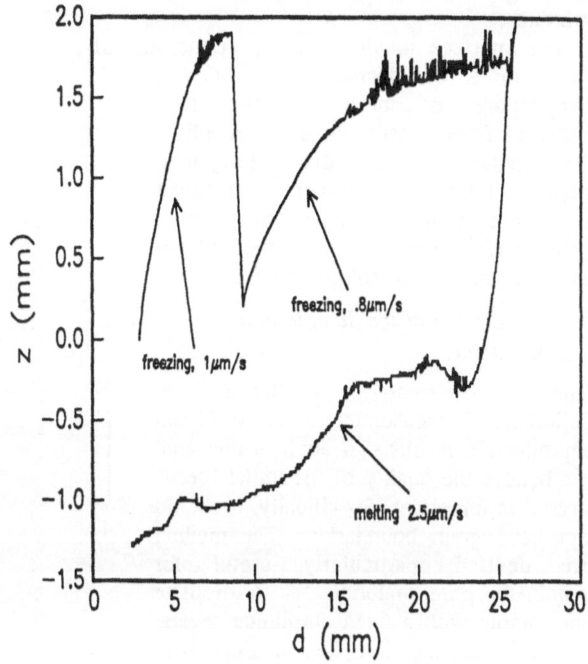

Fig. 5. Solidification followed by Melting

Next, we started crystallizing again but this time at a pulling speed, $v = 0.8\mu m/s$. Once more the interface became unstable.

At the end of the stepping-motor travel, the sample was now displaced towards the hot plate at 2.5µm/s. The result is shown as the lower curve in Fig. 5. These data have been multiplied by 3 to magnify the differences that are small compared to the larger displacements during crystal growth. The quality of the crystal grown can now be evaluated by a direct comparison of the interface position at each value of the stepping motor read-out.

The crystal grown from a cellular interface, melts at a lower temperature (the interface is closer to the cold contact) and the melting temperature is nearly uniformly low. Thus, as expected, the cellular interface does not zone refine. The melting temperature of the crystal grown during growth from a planar, but unsteady interface is clearly seen as an exponential rise in the melting temperature (the interface is displaced by greater amounts towards the hot stage).

Practically no improvement is observed in the melting temperature at the "seam" where the crystal was left during the 13 hour waiting period. This is odd because our understanding is that, at rest, the position of the interface is at T_L. Once crystallization begins, the growing crystal expels impurities into the liquid at a faster rate at smaller pulling distance as the interface shifts towards T_S. So, the exponential behavior observed here is not as expected. Nevertheless, we fitted it to an exponential to find a single characteristic length, $\ell = 12.5$cm. Assuming this length is kv/D and using $v = 0.8\mu m/sec$, $D = 1\times10^{-5} cm^2/s$, we find $k = 0.1$ in good agreement with careful measurements of Esaka and Kurz[22] for succinonitrile-acetone mixtures.

The interface should be at T_L at the start of solidification and it does not appear to be. Furthermore, the decrease in the crystal temperature is more dramatic than expected from constitutional supercooling. We conclude from this that waiting long periods of time in a temperature gradient does not achieve the initial starting conditions prescribed by the binary mixture scenario for solidification.

3.3 *Oscillation of the Sample in a Gradient*

The experimental challenge in solidification measurements is to obtain reproducible results. To achieve this end, we believe the quality of the initial "seed" crystal is important. Specifically, it should not have grain boundaries. The method we devised, particularly useful for nominally pure materials, is to oscillate the sample with a fixed amplitude several times.[23] We mention this method here because it has been used to obtain the grain boundary free interface necessary for the patterns discussed in the next section.

Fig. 6 shows a space time plot[24] of the interface moving in a gradient. The picture is obtained by digitizing along a video-line perpendicular to the interface at fixed time intervals. Each line is then replotted to give a picture of the z-t time evolution that is made quantifiable by the data such as shown in Figs. 5 and 7. Fig. 6 shows that the interface response to a uniform growth rate followed by a uniform melt rate is similar to a capacitor first charging (during growth), then discharging (during melt). The backlash in these experiments is too small to be measured (i.e. is less than 0.1μm).

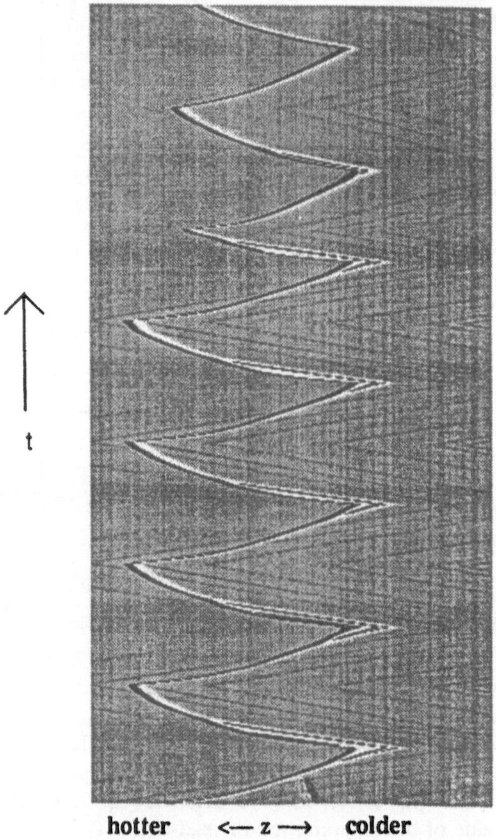

Fig. 6. z-t plot of Oscillating Interface

Fig. 7 shows the interface position as a function of pulling distance, d, for an oscillatory experiment. We could equally well have plotted this data as a function of time to obtain the same pattern shown in Fig. 6. The data shown here are of the last 20 or so oscillations at 2.5μm/s. The oscillation amplitude is about 6mm and the material is nominally pure. The "glitch" in Fig. 7 is a dust speck that is pushed along by the interface a few μm before it is incorporated into the crystal. The final stages of the melting curve are an example of the expected growth transient. Fitting these data points to a simple exponential as was done for Fig. 5, we obtain k = 0.6. Thus, based on the Caroli, Caroli and Roulet criterion,[7] the cellular-planar bifurcation is forward (continuous) for succinonitrile prepared in this way.

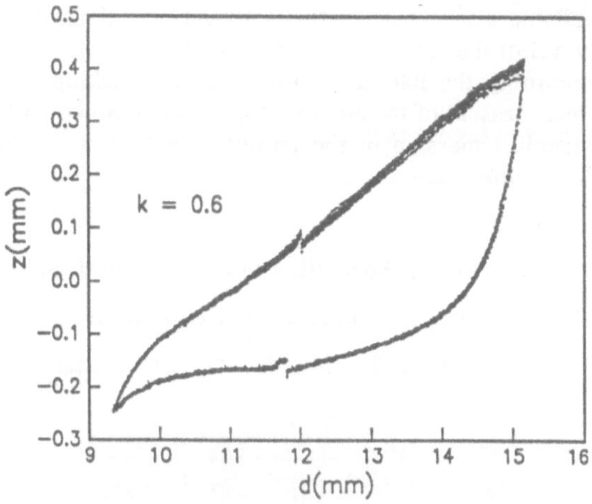

Fig. 7. z-d plot of Oscillating Interface

3.4 *Rapid Interface Response to Changes in Speed*

In Fig. 8, we plot the stepping motor speed (obtained by differentiating the pulling distance with respect to time) vs pulling distance, d. The sample had been previously prepared with the oscillating method outlined above so a stationary planar interface is achieved at 2μm/s. The speed was then increased to 2.5μm/s and the interface responded immediately. Eventually cells formed at this speed. The rapid response of the interface to changes in pulling speed is further illustrated when the speed is dropped briefly to 1μm/s.

Although the interface initially responds in seconds to changes in pulling speed, the time to reach steady state is the order of hours. Furthermore, there is some suggestion, that bears further investigation, that an acceleration pulse, such as the momentary decrease to 1μm/s, may have contributed to the loss of stability of the planar interface.

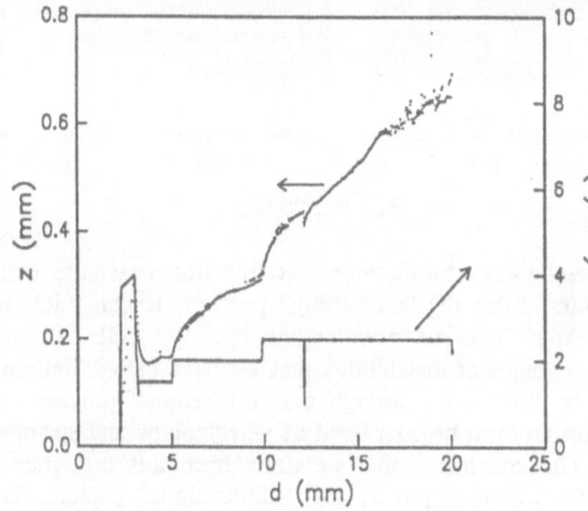

Fig. 8. Effect of acceleration on interface position

We conclude from the above that monitoring the interface position in a gradient provides a useful, new tool for studying crystal growth in directional solidification. The tool would not be available were it not for the powerful, real time, data acquisition provided by direct computer control of the experiment. We have also pointed out that useful information can be obtained by monitoring the interface position during melting. Specifically, the melting temperature is a direct measure of the quality of the crystal grown and its variation with pulling distance gives a dynamical measure of the partition coefficient, k, that determines the nature of the cellular-planar bifurcation.

4. Routes to Cell Formation and Wavelength Selection

4.1 *Thickness Dependence of the Cell Pattern*

As mentioned in the introduction, an outstanding problem in pattern formation during directional solidification is the identification of mechanisms that contribute to wavelength selection of the cellular pattern. The current theoretical prediction for an extended system is a broad band of wavelengths whereas the band of observed lengths is certainly not more than a factor of 2 or 3 except in very thin cells.

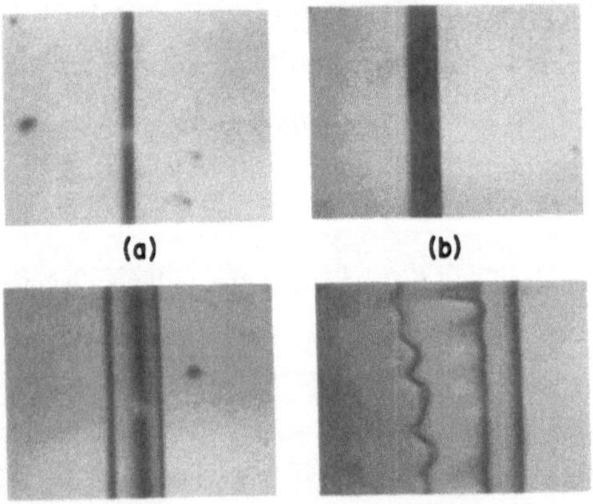

(a) (b) (c) (d)

Fig. 9a. 200μm

We found that the cell pattern depends sensitively on sample thickness.[13] For example, in Fig. 9a, the interface in a 200μm thick sample first "thickens" then develops several lines within the "thickened region" and finally forms two layers of cells. A 140μm thick sample undergoes the same "thickening" process then finally coarsens (Fig. 9b) to a single layer of cells with a narrow band of wavelengths. Furthermore, at this thickness, the cell morphology was stationary for several hours. After the "thickening" process, 100μm thick samples formed one layer of cells with a broader band of wavelengths and the cell pattern was weakly time dependent with the appearance of instabilities that we have called "mitten" and "cell fission".[13] Finally, 15μm thick cells did not go through the "thickening" process but became unstable in the long direction with an even broader band of wavelengths and stronger time dependence.

The conclusion that we draw from this sequence of observations is that the succinonitrile cells we have grown are 3-dimensional objects. Their stability depends on a fine balance between the radius of curvature in both directions perpendicular to the growth direction. When one or the other becomes too large, an MS instability is triggered and a cell fission or mitten instability[13] is generated. When they are both just right (nearly equal), as they appear to be in 140μm thick samples, a stationary cell pattern with a narrow band of wavelengths is stable. A sample only 50μm thicker than this can accommodate two layers of cells and so it does. It does not choose a single row of wider cells. Samples 100μm thick acquire the "wrong" aspect ratio: the radius of curvature in the vertical direction is now constrained to be smaller by the contact angle at the glass boundaries than in the horizontal direction where it tends to be time dependent. The band of wavelengths never varies by more than factors of 2 or 3 but the pattern is always changing with the unrelenting appearance of mitten and cell fission instabilities.[13]

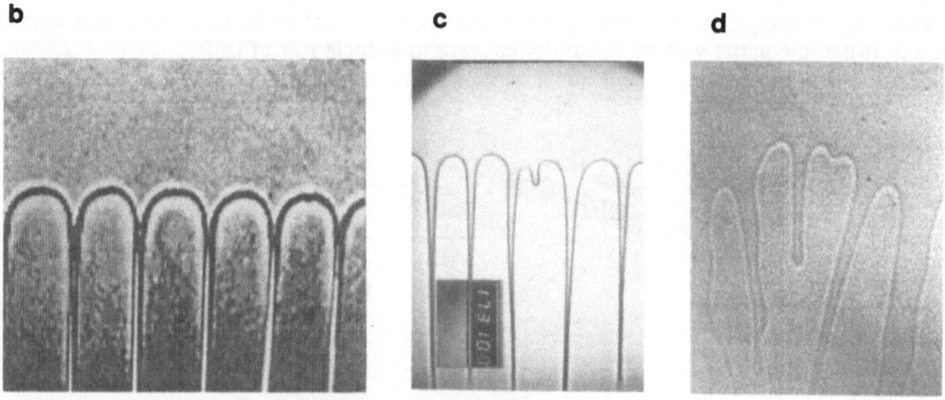

Fig. 9b (140μm*), 9c (*100μm*) and 9d (*15μm)

Finally, in the thin limit, the curvature in the vertical direction is sharpest and the tips first balloon outward then split continuously.

What then is the origin of the "good" length? Of the many lengths in this problem, we pick the diffusion length, $\ell_D = v/D_c \sim 0.1$cm in our observations, and the chemical capillary length, $d_o = 240$Å for succinonitrile.[25] The cell pattern results primarily from a competition between these two,[3] so the characteristic length that emerges is the geometric mean of the two, or $5\mu m \sim \lambda_c/2\pi$ for 2-d cells.[3] Theoretical[3] and experimental[26] results suggest that the length that survives is the fastest growing one given by $\sqrt{3}\,\lambda_c = \lambda_s = 53\mu m$, about half the observed spacing in, for example, Figs. 9b and c.

4.2 *Formation of Cells in a weakly varying Vertical Ramp*

As the sample is pushed in the -z direction, the crystal grows in the +z direction. Initially, then, the interface is in an x-y plane (say) except in two narrow regions where it curves to approach the glass boundaries with the correct contact angle. Call y the viewing direction that will also be referred to as the vertical direction. The distance over which the interface curves is initially small because the capillary length associated with the chemical potential is small. Thus, it is a constant temperature surface everywhere except in the small boundary layers near the glass where it is at a slightly greater undercooling. Thus, according to Fig. 2, the concentration is slightly greater in the boundary layer than elsewhere and, during crystal growth, the impurity flux into the boundary layer is greater than elsewhere. The net effect is to create a gradient parallel to the interface with the regions near the capillary walls growing more slowly than the region in the center of the sample. As the interface becomes more curved, the planar region gets thinner and the boundary layer thicker. This accounts for the optical "thickening" observed in the viewing direction (Fig. 9a and Fig. 10).

In addition, constitutional supercooling may be achieved first in the ever widening boundary layer before the rest of the interface. 2-dimensional cells can then propagate into the center of the sample. These are observed as the many lines that appear after "thickening". Since constitutional supercooling has been achieved, these cells are also unstable in their long direction. Eventually, the cell pattern with mitten and cell fission is observed depending on sample thickness. This is consistent with the observation that each of the several lines that appear in the "thickened" interface undergoes the MS instability initially independent of the others. If the sample is thick enough to support two layers of cells, then two layers of cells result. If not, the initial vertical instability coarsens to a single row of cells.

Thus, we propose that a small initial concentration gradient parallel to the interface leads to

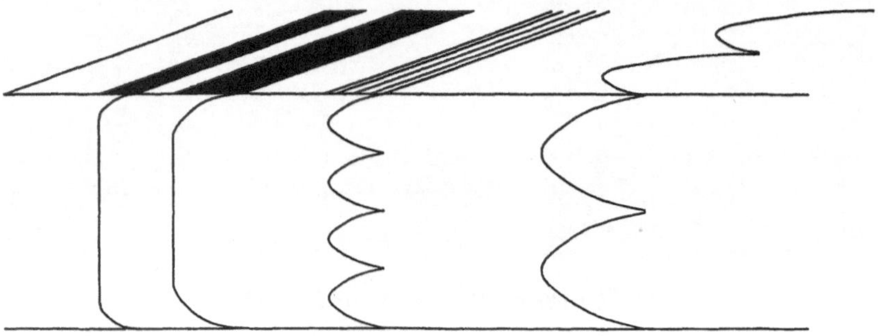

Fig. 10. Vertical Ramp Scenario to account for Interface Contrast

a weak ramping mechanism. As the 2-d cells propagate along the interface, the ramp becomes weaker with correspondingly weaker wavelength selection. This is reminiscent of the propagation of the cell pattern along the interface from grain boundaries first described by Schaeffer and Glicksman.[27] A grain boundary can also result in a concentration gradient parallel to the interface. The analogue of the grain boundary in our case are the glass walls. The liquid-glass-solid contact is an intrinsic grain boundary attached to the sample container and moves with it - not the rest of the interface - leading to its "thickening".

It is worthwhile noting that in our case, 2-d cells appear as transients on the way to full 3-d cell formation. That is, once the interface thickening contrast starts it doesn't stop until cells form in the horizontal direction. Metallurgists have found that in cubic crystals,[28] 3-d cells are associated with the <100> growth direction.

The 4-fold symmetry of <100> planes cannot distinguish between the long or short axis of the interface so 3-d cells result with the initial perturbation oriented by the close-packed, slow growing <111> faces that have 4-fold cubic symmetry. With a much smaller vertical (compared to horizontal) distance to travel, we observe the vertical one first. A 4-fold symmetry of the interface could account for the circular symmetry of our cells.

On the other hand, a <110> oriented crystal has only 2-fold symmetry i.e. two (not four) slow growing planes "to guide the interface morphology to a furrowed structure"[28] and 2-d cells. This would indeed account for the difference in interface patterns between succinonitrile as we have prepared it and, for example, the beautiful patterns observed by others in CBr_4[18] and succinonitrile-acetone alloys.[29] We plan to check this by growing <110> interfaces in succinonitrile using our oscillation technique. However, it remains to be understood how crystallographic anisotropy influences crystal growth patterns at a rough interface.

Returning now to the ramp discussion, the even symmetry of the vertical ramps about the center of the sample (i.e. there are really two ramps here, one from the top and the other from the bottom) hampers good wavelength selection. As cells grow in from the top and the bottom, they eventually meet in the middle and they need not do this exactly in phase. On the other hand, a vertical temperature gradient would break the even symmetry resulting in a more regular cell pattern. The vertical temperature gradient has been made purposely small so our vertical ramp is observed to have even symmetry: by focussing it is possible to observe that initially, the two lines closest to the cold contact are above and below the line closest to the hot contact. In the next section, we describe how a ramp with odd symmetry, an asymmetric ramp, may be installed in the wide direction of the capillaries when they are loaded at a small angle to the edges of the hot and cold contacts but keeping the pulling velocity perpendicular to these contacts.

4.3 *Cell Formation in a weakly varying Horizontal Ramp*

When capillaries are placed at a small angle (~ 2 – 12°) to the edges of the hot and cold contacts, at rest, the interface is perpendicular to the capillary walls in the 10:1 samples and nearly so in 20:1 samples. It does this to minimize its surface energy. Since it does not lie parallel to the edges of the hot and cold contacts, one end, the -x end, say, is at a lower temperature than the other end. Thus, a small concentration gradient parallel to the interface is established. As the sample is pulled with velocity, v, perpendicular to the edge of the hot and cold contacts, there is a component of the impurity flux parallel to the interface towards +x so it eventually moves to lower temperatures. The interface is now obviously at a small angle to the capillary walls (see Fig. 11). When constitutional supercooling is achieved at the +x end, the difference in the interface position in the gradient, Δz, at the two ends of the cells is maximum and cells start to form and propagate towards -x.

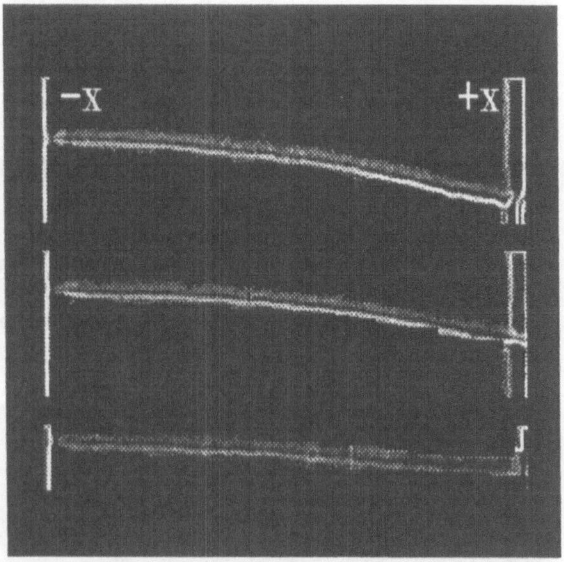

Fig. 11. Ramp Onset

As cells propagate along the interface, impurity build up ceases in the liquid in front of cells and the ramp becomes shorter and shallower. Eventually, Δz decreases so that the interface, now defined by a line tangent to the cell tips, is perpendicular to the pulling velocity. When this happens, the interface is parallel to the hot and cold contacts and the ramp has gone to zero. Grain boundaries disrupt the formation of a uniform concentration gradient parallel to the interface so for these experiments, a grain boundary free interface is essential.

We checked the following points.

— We can control the direction of the ramp by adjusting the capillary angle relative to the temperature gradient. After several zone refining passes, the ramp amplitude decreases showing that it is dependent on impurity flux parallel to the interface.

— The maximum angle the interface makes with the capillary walls increases with decreasing v, again consistent with increasing impurity flux parallel to the interface. This is reasonable because the maximum angle occurs as cells start to move along the interface. Slower speeds mean longer times for the concentration gradient to build.[18]

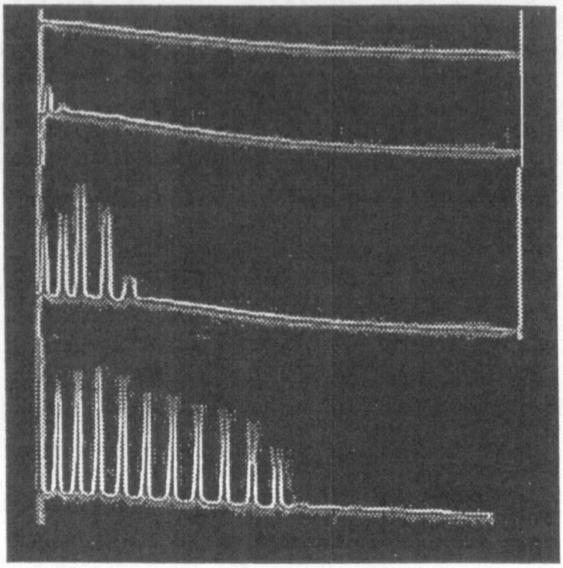

Fig. 12. Cell Propagation along a Ramp

— When growth is stopped, the interface relaxes back to lie normal to the capillary walls and remains so on directional melting. Therefore, the effect is not a peculiarity in the temperature gradient.

Ramping occurs only at low pulling speeds. How low depends on C_∞. For our succinonitrile,

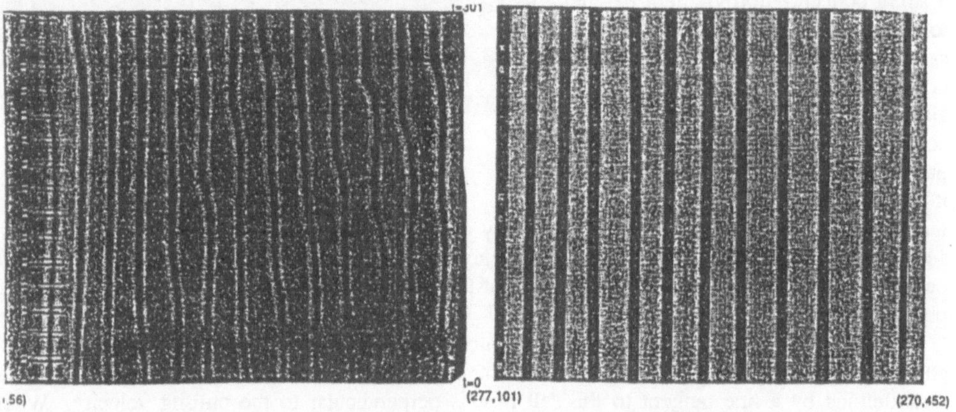

Fig. 13. x-t plot of Unramped (r) and Ramped (l) Cell Pattern

ramping occurs up to about 2.5μm/s ~ $2v_c$ before zone refining and below 2μm/sec < $2v_c$ after. Above this speed, the impurity gradient parallel to the interface does not have enough time to build, and only the usual "thickening" process is observed uniformly along the interface leading eventually to the appearance of cells all at once on the interface.

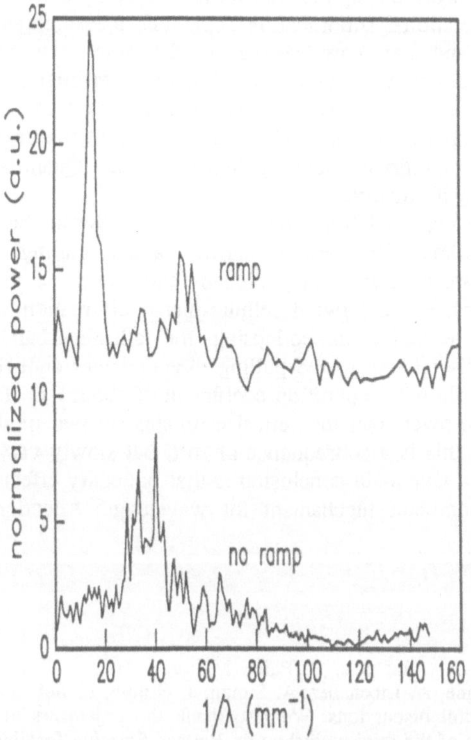

Fig. 14. fft for Ramped (t) and Unramped (b) Pattern

The difference between the two patterns, one grown in a horizontal ramp and the other without, is strikingly shown by the x-t plot, Fig. 13. Fig. 13 was taken 50μm behind the cell tips as parallel to the interface as possible. The speed in both plots is 2.5μm/s but C_∞ in the ramped sample is slightly smaller than in the unramped one.

Clearly the ramped sample shows better wave length selection than the unramped one. In the unramped case, the cell pattern is time dependent. The consequences of mitten and cell fission instabilities are clearly seen as the lines corresponding to the boundary between two cells meander, fuse and split. In the ramped case, these lines are uniformly parallel to each other *and* to the capillary axis! The translational order is not perfect but the orientational order is.

Given that the interface is at a small angle to the capillary axis whereas the cell axis is parallel to it, the symmetry of the ramped cell pattern is similar to a smectic C liquid crystal where the orientational order is at an angle to the direction of translational order i.e. molecules (*cf* cell grooves) are on equidistant layers (*cf* interface) but tilted relative to the layer normal (*cf* pulling direction).

Fig. 14 is the fft obtained for the two cases, ramped and unramped. The fft's for the unramped cells are much broader with a peak not even at the same inverse wavelength. The peak in the ramped case shown in Fig. 14 is at ~70μm, corresponding to about twice the predictions of marginal stability for the length at v_c for 2-dimensional cells and about twice the length of the broad peak in the unramped case. Fig. 14 is an impressive demonstration of the collapse of the allowed band of wavenumbers by "soft" boundary conditions.

5. Conclusions

Directional solidification raises many interesting questions relevant to pattern formation and crystal growth. An impressive variety of patterns are exhibited by a growing crystal interface. Spectacular time dependent patterns can also be observed. A serious study of these may shed light on the kinetics of the many processes at work during non-equilibrium crystal growth.

With the aid of a computer-controlled directional solidification apparatus, measurement of the interface position both as a function of t and d are now feasible. Furthermore, they extend the domain of these experiments to include directional melting and directional oscillation. In this way, we have shown that quantitative information may be gained about the time dependence of the growth process when patterns form, feedback on the quality of the crystals grown under well-defined initial conditions and grain boundary free interfaces. Quantitative data can be obtained and compared precisely with theory.

The discrepancy between the theoretically expected broad band of wavelengths in the cell pattern of an extended system and the experimentally observed narrow band is glaring. We have seen that in these experiments there are probably many hidden ramps intrinsic to the experiment that contribute to this discrepancy. For a forward cellular-planar bifurcation, such soft boundary conditions are particularly effective in collapsing the allowed band of wavelengths. Merely tilting the capillary axis relative to the pulling direction was enough to generate a slowly varying ramp in succinonitrile with a partition coefficient of about 0.6. Even without the horizontal ramp, selection is narrower than theoretical estimates by two or three orders of magnitude. We have proposed that this is a consequence of soft, but slowly varying, boundary conditions in the vertical direction. Our main conclusion is that boundary effects, in particular soft boundary effects, are the dominant mechanism for wavelength selection in directional solidification of succinonitrile.

ACKNOWLEDGEMENTS

We thank W. van Saarloos, J. D. Weeks, K. A. Jackson, P. Palffy-Muhoray, Helmut Brand, A. Zangwill, B. Shraiman, C. Caroli, S. de Cheveigne, A. Libchaber, A. Simon, J. Gollub, R. Sekerka, Y. Pomeau and L. Kramer for stimulating and useful discussions. We also thank the organizers of this Advanced Research Workshop and the organizers of the mini-workshop on *Pattern Forming Instabilities in Liquid Crystals* for their kind and generous hospitality and support. J.G. acknowledges support from the Natural Sciences and Engineering Research Council of Canada.

REFERENCES

1. K.A. Jackson and J.D. Hunt, *Acta Metall.*,**13**,1212 (1965).
2. *Crystal Growth*, edited by B.R. Pamplin (Pergamon, Oxford, 1980).
3. J.S. Langer, *Rev. Mod. Phys.*, **52**, 1 (1980) and Science **243**, 1150 (1989).
4. *Dynamics of Curved Interfaces*, Pierre Pelce (ed) (Academic Press, Inc., New York, 1988); J.D. Weeks & W. van Saarloos, Phys. Rev., **A39**, 2772 (1989); G.J. Merchant & S.H. Davis, Phys. Rev. Lett., **63**, 573 (1989); J. Bechhoefer and A. Libchaber, Phys. Rev. **B35**, 1393 (1987) and P. Oswald, J. Physique, **49**, 1083 (1988); *ibid*, **50**, C3-127 (1989).
5. W. A. Tiller, K. A. Jackson, J. W. Rutter, B. Chalmers, Acta. Metall **1**, 428, 1953.
6. W.W. Mullins and R.F. Sekerka, *J. Appl. Phys.*, **35**, 444 (1964).
7. B. Caroli, C. Caroli and B. Roulet, J. Physique, **43**, 1767 (1982).
8. D. J. Wollkind and L. A. Segel, Philos. Trans. R. Soc. London, **51**, 268 (1970); J. S. Langer and L. A. Turski, Acta. Met. **24**, 1113 (1977); J. S. Langer, Acta. Met. **25**, 1121 (1977); G. Dee and R. Mathur, Phys. Rev. **27**, 7073 (1983).
9. A. Karma, Phys. Rev. Lett. **57**, 858 (1986).

10. T. Dombre and V. Hakim, Phys. Rev. **A36**, 2811 (1987).

11. M. Ben-Amar and P. Moussallam, Phys. Rev. Lett. **60**, 317 (1988).

12. M. A. Eshelman and R. Trivedi, Acta Metall. **35**, 2443 (1987).

13. P. E. Cladis, J. T. Gleeson and P. L. Finn, to be published in *Defects, Patterns and Instabilities*, D. Walgraef (ed.), Kleuwer Academic Publishers (1990).

14. Chaouqi Misbah, J. Physique **50**, 971 (1989).

15. S. de Cheveigne and C. Guthmann, *Les Houches 1987, Propagation in Systems Far from Equilibrium*, J. E. Wesfried, H. R. Brand, P. Manneville, G. Albinet and N. Boccara (eds) (Springer Verlag).

16. L. Kramer, E. Ben-Jacob, H. R. Brand and M. C. Cross, Phys. Rev. Lett. **49**, 1891 (1982). See also: P.C. Hohenberg, L. Kramer and H. Riecke, Physica 15 D, 402, (1985).

17. D. S. Cannell, M. A. Dominguez-Lerma and G. Ahlers, Phys. Rev. Lett. **50**, 1365 (1983); M. A. Dominguez-Lerma, D. S. Cannell and G. Ahlers, Phys. Rev. **A34**, 4956 (1986); I. Rehberg, E. Bodenschatz, B. Winkler and F. H. Busse, Phys. Rev. Lett. **59**, 282 (1987).

18. S. de Cheveigné, C. Guthmann & M.M. Lebrun, *J. Physique*, **47**, 2095 (1986).

19. *The Merck Index* edited by M. Windholz, entry no. 8746.

20. See for example: S.C. Huang & M.E. Glicksman, *Acta Metall.*, **29**, 701 (1981); M.A. Eshelman, V. Seetharaman & R. Trivedi, *Acta Metall.*, **36**, 1165 (1988); H. Chou & H.Z. Cummins, *Phys. Rev. Lett.*, **61**, 173 (1988).

20. *see for example*: H. Chou, PhD thesis, CCNY (1988).

21. We are indebted to A. Dougherty for drawing our attention to this point.

22. H. Esaka and W. Kurz, J. Cryst. Growth, **72**, 578 (1985).

23. J. T. Gleeson, P. L. Finn and P. E. Cladis (unpublished).

24. Marc Rabaud and Yves Couder, to be published in *Defects, Patterns and Instabilities*, D. Walgraef (ed.), Kleuwer Academic Publishers (1990).

25. S. de Cheveigne, C. Guthmann, P. Kurowski, E. Vicente and H. Biloni, J. Cryst. Growth, **92**, 616 (1988).

26. A. Buka and P. Palffy-Muhoray, Phys. Rev. **A36**, 1527 (1987).

27. R. J. Schaeffer and M. E. Glicksman, Met. Trans. **1**, 1973 (1970).

28. see for example: Merton C. Flemings, *Solidification Processing*, McGraw-Hill Book Company, New York (1974)p 67.

29. M. A. Eshelman, V. Seetharaman and R. Trivedi, Acta. metall. **36**, 1165 (1988).

INFLUENCE OF AN EXTERNAL PERIODIC FLOW ON DENDRITIC CRYSTAL GROWTH

Ph. Bouissou, B. Perrin, and P. Tabeling

Ecole Normale Supérieure
24, rue Lhomond, 75231 Paris (France)

1 - INTRODUCTION

The problem of dendritic shape selection in crystal growth processes has been the subject of many theoretical and experimental studies these last years(1-8). Experimentally, dendritic side-branching has received less attention despite the significant number of theoretical and numerical studies devoted to this problem(9-12). A recent experimental study(13) shows results consistent with the assumption that side-branching originates into the selective amplification of microscopic noise at the crystal tip. Indeed, for testing such an assumption, it would be necessary to control the noise in the experimental system. Such a control has been achieved in the case of anomalous Saffman-Taylor fingers(14). In the present crystal growth experiment, we achieve a control of the side-branching by imposing a time dependent flow during the growth process. We report herein a first quantitative study of the effect of a periodic external flow on dendritic growth.

2 - EXPERIMENTAL SETUP

The system from which crystals grow is a mixture of pivalic acid and ethanol alcohol (PVA/AL). The crystallographic structure of the solid phase (which is essentially pivalic acid) is fcc(4). In all the experiments, the alcohol concentration is 0.7 % in volume. This corresponds to a melting temperature for the mixture of 25 °C. Because of the presence of alcohol which plays the role of impurity, the growth is controlled by chemical diffusion.

The quartz cell is a parallelepiped of dimensions 45 x 5 x 1 mm. In order to obtain a longitudinal flow of the PVA/AL system, an open hydraulic circuit is formed, including the cell, small diameter tubes and a reservoir. The time dependent flow is produced by varying periodically the position of the reservoir, at a frequency $\nu_{exc.}$ ranging between 5 and 500 mHz. The maximum velocity amplitude of the resulting flow is fixed at the value 19.6 μm/s for all the experiments ; there is no mean flow.
In a similar way as in a previous work(15), the cell is entirely immersed in a circulation of thermally regulated water maintained at temperature T_1 ; by this way, a regulation better than 20 mK for the operating temperature is achieved. On the other hand, the tubes forming the hydraulic circuit are in thermal contact with another circulation of thermally regulated water ; such tubes can therefore be held at a temperature T_2 slightly above T_1. Most of the experiments are done with an overheating $\Delta T = T_2 - T_1 \approx 0.2°C$, which is sufficient to avoid cristallization in the tubes and small enough to leave the growth process unperturbed by any temperature gradient effect.

Nonlinear Evolution of Spatio-Temporal Structures in Dissipative Continuous Systems
Edited by F.H. Busse and L. Kramer
Plenum Press, New York, 1990

The dendrites are observed with a microscope which allows magnification up to 750. The images are sent to a video camera and digitized into 512 x 512 pixels. The resulting overall image separation ranges from 0.6 to 0.9 μm/pixel.

3 - EXPERIMENTAL PROCEDURE

The experimental procedure is as follows : starting from a crystallized state, the cell is heated up at temperature T_1 slightly below equilibrium temperature T_e so that only a single crystal remains in the cell. Then the periodic flow is turned on and after a few minutes the cell is cooled down at a temperature T_o smaller than T_1 by an amount ranging from 0.3 °C to 1 °C. After a short transient of a few minutes (which corresponds to the thermal time constant of the system), dendrites grow from the germ in the <100> directions. We choose one isolated dendrite and follow its temporal evolution. In all the experiments, the flow is directed along the growth direction of the dendrite.

4 - RESULTS

A typical dendrite obtained in the presence of an external periodic flow is shown in Pic. 1. In this case, the amplitude of the flow is 19.6 μm/s, the imposed frequency 15 mHz, the growth velocity 0.9 μm/s and the tip radius 8.0 μm. Also

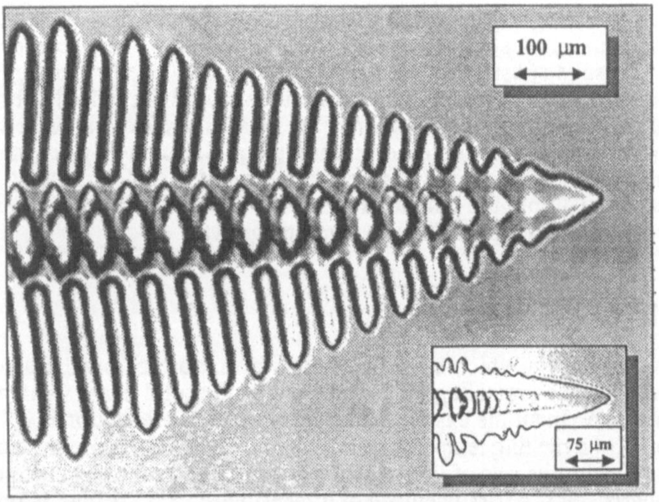

Picture 1 . A typical dendrite growing in the presence of an external periodic flow (in insert : a natural dendrite growing at 1.6 μm/s with a curvature tip radius of 8 μm).

shown, on the lower right of the picture, a "natural" dendrite of pivalic acid, *i.e.* a dendrite growing out of a solution at rest. One obtains, when the periodic external flow is applied, a spectacular regularity of the spatial structure of the side-branching. The wavelength of the side-branching is well defined in this case, and the branches appear to be correlated over large distances – typically one hundred times the radius of curvature at the tip – ; this contrasts with natural dendrites where coherence rapidly decays as the distance from the tip increases[13]. The apparent wavelength of the side-branching can be estimated to 65 μm in the case corresponding to Pic. 1. This is close to the imposed wavelength $\lambda_{exc.} = V / \nu$, which is about 60 μm in this case, so that one can consider that the actual wavelength of the side-branches is imposed by the forcing.

In order to perform a quantitative study of the influence of the periodic flow, we have studied the time dependent deviation w_z (t) of the dendrite (from the center to one side) at a fixed distance z behind the tip. Fig. 1 shows the power spectrum P_z (ν) of the signal w_z (t) at a distance of 26 tip radius ρ from the nose of the dendrite, for a forced frequency of 20 mHz. The existence of a sharp peak at 20 mHz shows that the system is locked at the imposed frequency.

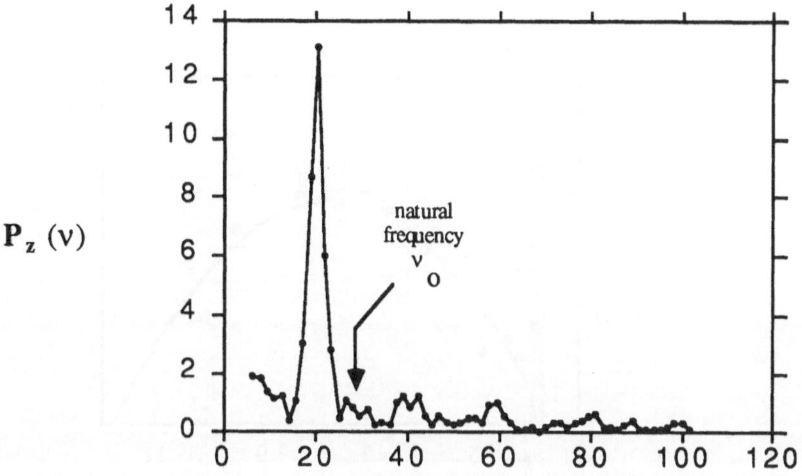

Figure 1 . Power spectrum P_z (ν) of half distance w_z (t) (at a fixed distance z = 26 ρ behind the tip) of a dendrite forced periodically at the frequency 20 mHz.

We have studied the response of the system *i.e.* the amplitude A_z of the side-branching instability as a function of the reduced imposed wavelengh $\Lambda = \lambda_{exc.} / \rho$, at a given location z from the tip. Fig. 2 shows the dependance A_z with Λ, for z = 26 ρ. The curve looks like a resonance curve with a maximum response for Λ ~ 7. This value turns out to be the mean wavelength of the side-branching in the case of the natural dendrite, *i.e.* when no forcing is applied. For Λ < 4, the system ceases to be influenced by the forcing, and the dendrite looks "natural".

5 - DISCUSSION

We can build a rough model for the influence of a forcing on dendritic side-branching by using a simplified version of Zel'dovich theory[18]. Let us recall the expression of the amplification rate σ for the linear instability of a plane crystal interface growing at velocity V out of a saturated solution :

$$\sigma(\Lambda) = 2\pi \frac{V}{\rho} \frac{1}{\Lambda} \left(1 - \frac{\Lambda_0^2}{3\Lambda^2} \right) , \quad (1)$$

where $\Lambda = \lambda / \rho$ and $\Lambda_0 = 2\pi \sqrt{\frac{3 d_0 D}{\rho^2 V}}$ (for the symmetric model),

(here, d_0 is the capillary lengh and D the chemical diffusivity). Relation (1) gives $\Lambda = \Lambda_0$ for the maximum growth rate $\sigma_{max.}$; on the other hand, σ is negatif when $\Lambda < \Lambda_{min} = 3^{-1/2} \Lambda_0$ which thus defines the lower boundary of the unstable range of wavelenghts.

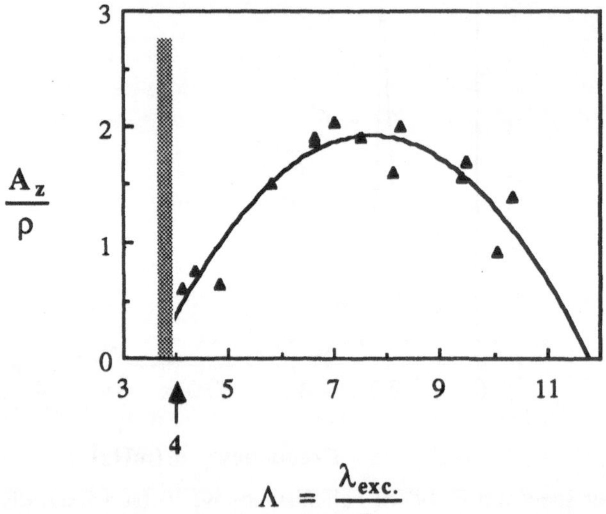

Figure 2. Dependance of the amplitude A_z of the side branching instability at a fixed distance behind the tip with the reduced wavelengh $\Lambda = \lambda_{exc.} / \rho$. For $\Lambda < 4$, the dendrite looks natural and the system is no more locked.

In the Zel'dovich picture, the perturbation starts growing close to the tip - where the amplification rates are larger- and is further advected along the flanks. Let us assume a localized disturbance of amplitude A_0 at the tip, and suppose - as a rough model - that it grows exponentially with an amplification factor given by (1). In this model, the role of the external periodic flow is to impose the magnitude A_0 and the wavelenght λ of the initial perturbation.

Under these assumptions, the amplitude A of the disturbance at distance z is given by :

$$A = A_0 \exp\left(\frac{z}{\rho} F\left(\frac{\Lambda}{\Lambda_0}\right)\right) ,$$

$$\text{where } F\left(\frac{\Lambda}{\Lambda_0}\right) = \frac{2\pi}{\Lambda}\left\{1 - \frac{\Lambda_0^2}{3\Lambda^2}\right\} .$$

When $\Lambda < \Lambda_{min}$, F is negative and the forced wavelenght is not amplified. In a real experiment, this should correspond to a case where the forcing is ineffective, so that the side-branching is controlled only by the natural noise of the experiment. This is precisely what we observe in the range of small wavelengths.
When Λ becomes larger than Λ_{min}, perturbations grow. Their amplitude increases with Λ, reaches a maximum for $\Lambda = \Lambda_0$, and further decreases. The shape of the theoretical curve thus corresponds the response curve obtained experimentally.
This rough model is thus in good qualitative agreement with the experiment. Actually, on a more quantitative level, it turns out to provide reasonably good estimates for various quantities found experimentally to characterize the system : if one admits that the preferred excited wavelenght is $\Lambda_0 = 7$ (according to experimental results(16-17)), one has $\Lambda_{min} = 4$ which is precisely the experimental value under which the system becomes insensitive to the forcing.

6 - CONCLUSION

We have shown that it is possible to force the side-branching instability by using a time periodic modulation produced by an externally controlled flow. The effect of the forcing is correctly described within the framework of Zel'dovich approach. This experiment thus provides an additional support for this theory.

ACKNOWLEDGMENTS

We thank P. Pelcé, V. Croquette, D. Bensimon and C. Caroli for many helpful discussions related to this experiment.

REFERENCES

(1) J. S. Langer (1986), *Phys. Rev. A* **33**, 435
(2) S.C. Huang, M.E. Glicksman (1981), *Acta Metall.* **29**, 717
(3) P. Pelcé, Y. Pomeau (1986), *Stud. Appl. Math.* **74**, 245
(4) D. A. Kessler, H. Levine (1986), *Phys. Rev. Lett.* **57**, 24
(5) M. E. Glicksman, N. B. Singh (1986) in "Rapidly Solidified Powder Aluminium Alloys"
(6) A. Dougherty, J. P. Gollub (1988), *Phys. Rev. A* **38**, 3043
(7) M. Ben Amar, Y. Pomeau (1986), *Europhys. Lett.*, **2**,307
(8) For a recent review, see P. Pelcé, *Dynamics of curved fronts*, Perspectives in Physics (Academic New York, 1988)
(9) P. Pelcé, P. Clavin (1987), *Europhys. Lett.* **3**, 907
(10) D. A. Kessler, H. Levine (1987), *Europhys. Lett.* **4**, 215
(11) M. N. Barber, A. Barbieri, J. S. Langer (1987), *Phys. Rev. A* **36**, 3340
(12) B. Caroli, C. Caroli, B. Roulet (1987), *J. Phys.* **48**, 1423
(13) A. Dougherty, P. D. Kaplan, J. P. Gollub (1987), *Phys. Rev. Lett.* **58**, 1652
(14) M. Rabaud, Y. Couder, N. Gerard (1988), *Phys. Rev. A* **37**, 935
(15) Ph. Bouissou, B. Perrin, P. Tabeling (1989), *Phys. Rev. A* **40**, 509
(16) S. C. Hung, M. E. Glicksman (1981), *Acta Metall.* **29**, 701
(17) Ph. Bouissou (1989), Université P. et M. Curie Thesis "Etude de l'influence d'un écoulement sur la croissance dendritique : aspect théorique et expériemental"
(18) Ya. B. Zel'dovich, D. A. Frank-Kamenetskii (1938), Acta Physicochimica *USRR* **2**, 341

PATTERN GROWTH : FROM SMOOTH INTERFACES TO FRACTAL STRUCTURES

A. Arnéodo*, Y. Couder**, G. Grasseau*, V. Hakim** and M. Rabaud**

* Centre de Recherche Paul Pascal, Chateau Brivazac, 33600 Pessac , France
** Laboratoire de Physique Statistique, 24 rue Lhomond, 75231 Paris Cedex 05, France

INTRODUCTION

Fractal structures can be obtained in several growth experiments, the resulting patterns being fractal only in a limited range between a smallest and a largest length scale. In this regard they differ from mathematically defined fractals which cover an infinite range of length scales. In the systems that we will discuss the smallest scale l_{min} can be linked with the instability process or with the computational technique, the largest scale l_{max} can be imposed by either geometrical boundary conditions or, in free growth, the overall size reached by the pattern. We can therefore tune the extent of the fractal range and go from compact objects (when the two length scales are close to each other) to structures with a large fractal range (when the two scales are far from each other). Limiting ourselves here to patterns grown in Laplacian fields, we investigate their statistical properties and show that, as the fractal range opens up, some properties of the compact forms are retained by the fractal structure. We use two isotropic Laplacian pattern forming systems : The Saffman Taylor instability[1,2] and the formation of clusters in the numerical model of Diffusion Limited Aggregation[3] (D.L.A.). These two systems are very different in nature but their ruling laws have strong similarities.

The instability giving rise to Saffman Taylor viscous fingering occurs at the interface between two fluids moving between narrowly spaced solid plates. The interface is unstable when the less viscous fluid forces the most viscous fluid to recede. The flow of the fluids is dominated by the viscous dissipation on the plates and the mean velocity in the cell's plane is proportional to the pressure gradient $\mathbf{V}=-(b^2/12\mu)\nabla p$. Because of the incompressibility of the fluids, the pressure field obeys a Laplace law : $\nabla^2 p = 0$. Surface tension has a stabilizing influence, this is taken into account by adding a boundary condition for the pressure jump at the interface. The linear analysis[2] of the stability of a plane interface moving at velocity V gives the wavelength of maximum instability : $l_c = \pi b(T/\mu V)^{0.5}$. where b is the cell thickness, μ the viscosity of the most viscous fluid and T the surface tension. The capillary length, l_c, is the characteristic small scale of the viscous fingering.

In the numerical model of aggregation of randomly walking particles introduced by Witten and Sander[3], the probability P of visit of a site is given by $\nabla^2 P = 0$ and the normal velocity of growth of a region of the interface is $V_n = (\nabla P)_n$. As was first noted by Paterson[4] the equations of growth of DLA are similar to those of Saffman-Taylor fingering in the limit of zero surface tension. In DLA it would thus appear that all length scales are unstable ; however the computation of DLA clustering is usually done on a lattice and the lattice-mesh size introduces a smallest length scale l_{min} .

Laplacian fields are defined by the boundary conditions. In growth problems (i.e. Stefan problems) at least one of the boundaries is the moving front itself and the others depend on the chosen experimental configuration which is therefore crucial[1,5,6]. We will concentrate here on geometries in which stable smooth solutions are known.

In the configuration initially chosen by Saffman and Taylor[1] the fluids move in a very long linear channel of width W. These boundaries simplify the problem by imposing

Nonlinear Evolution of Spatio-Temporal Structures in Dissipative Continuous Systems
Edited by F.H. Busse and L. Kramer
Plenum Press, New York, 1990

translational invariance. The control parameter is usually defined as: $1/B \approx 12\pi^2(W/l_c)^2$. Experiments show that the two length scales can be identified as $l_{min} = l_c$ and $l_{max} = W$. The control parameter represents their ratio as $l_{max}/l_{min} \approx (118\ B)^{-0.5}$. If the experiment is done in a given cell so that l_{max} is constant, l_{min} can be chosen at will by changing the velocity. For $l_{max}/l_{min} < 8$ a single stable finger is observed, scaled on the cell width. In the upper part of its stability range the relative width λ of the finger tends asymptotically towards 0.5. Saffman and Taylor[1] had found a family of solutions for the interface shape at T = 0. The selection of the observed asymptotic finger width λ=0.5 was understood a few years ago[7]. It results from the action of surface tension acting as a singular perturbation.

A variant of this geometry was introduced[6] (for viscous fingering) where the cell has a sector shape of angle θ_0. The growth can be started either from the apex of the cell (divergent cell $\theta_0>0$) or from its periphery (convergent cell $\theta_0<0$). For a Saffman Taylor front moving at a constant velocity, l_c remains constant. The other length scale, the local width $W(\rho)=\theta_0\rho$ of the cell is a function of ρ the distance of the front to the apex. In this geometry the ratio l_{max}/l_{min} changes spontaneously with the distance of the active front to the cell's apex. Previous investigation showed that near the apex there is a region where the finger is stable and tends to occupy a finite fraction $\lambda(\theta_0)$ of the cell's angular width.

Here we will investigate the very unstable fingers observed at large velocities in these two types of cells. We will also study DLA clusters in similar configurations. This is possible if the cluster grows between walls which reflect the random walkers. These walls can either be parallel to each other at a distance W, or form an angle θ_0. The main results of this study have been announced in Ref. 8

EXPERIMENTAL SET-UP

We performed the viscous fingering experiments in a linear channel and four sector shaped cells (θ_0= 23°, 45°, 60° and 90°). All were made of float glass 19 mm thick. The cells thickness b was either 0.25 mm or 0.125 mm, the measurements were done in the region of width W~10 cm. We used a silicon oil Rhodorsyl 47 V 500 with $T = 21\ 10^{-3}$ N/m and μ=0.48 kg/m.s. In both the linear and the divergent cells, the air was injected at a controlled pressure. For convergent fingers oil was pumped out of the apex.

We wanted, after a large number N of independent runs of the same experiment, to measure the mean occupancy of each site of the channel. The results are reported with the following conventions. In the linear cell we choose Ox along the cell's axis and Oy across the cell. For sector shaped cells the origin is at the apex and the angles are measured from the cell's bisector. In the case of viscous fingering the complete processing of a large number of complex images was beyond our means, so we limited ourselves to the measurement of the mean occupancy across the cell. We chose a section of the cell where the pattern had finished its evolution and built a transverse histogram of the occupancy by air in all the runs. A division by N gave the mean occupancy r(y) in this section.

The DLA clusters were computed in strip geometries of width W = 32, 64 or 128 lattice units. We used an on-square lattice algorithm and reflecting lateral walls. As in the experiments, a numerical trick was used to initiate the growth from the cell axis : the very first random walking particle sticks onto a needle (a few lattice units long) centered on the cell axis. Similar DLA simulations were conducted in a sector geometry (the cell boundaries are approximated by a staircase structure and the particles are locally reflected normally to these boundaries). A more extensive statistical investigation could be carried out in this case. In a given strip we grew N aggregates with the same total number M of particles. We then counted for each point of the grid how many times it had been occupied by a particle of an aggregate. This number, divided by N, gave r(x,y), the mean occupancy of this point.

RESULTS

Linear cells

Figure 1 shows individual patterns obtained respectively for various width of the range of fractality. For $l_{max}/l_{min}<8$ the stable Saffman Taylor finger is observed (Fig. 1a), for l_{max}/l_{min} 50 we get very unstable fingers (Fig. 1b). Finally a DLA pattern obtained in a linear strip of width 128 is shown on Fig. 1c. For each of these situations we measured the transverse occupancy profile in a region where the pattern did not grow anymore (Fig. 2). In

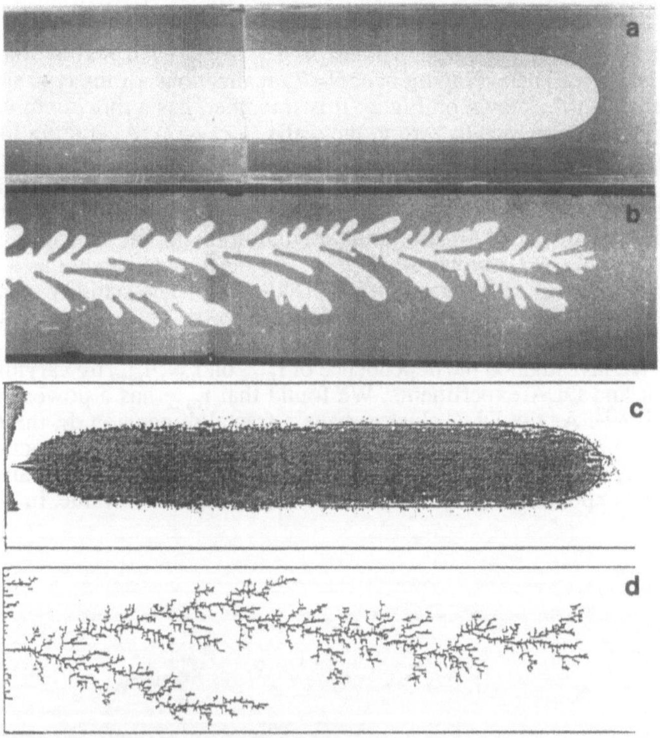

Fig. 1.(a) A Stable Saffman Taylor finger λ =0.5
(b) An unstable Saffman Taylor finger for l_{max}/l_{min}=50
(c) A DLA cluster with 6000 particles grown in a strip of width 128
(d) Region of the strip in (c) with mean occupancy above average ($r > r_{max/2}$). 510 aggregates of the type shown in (c) were grown to obtain this repartition.

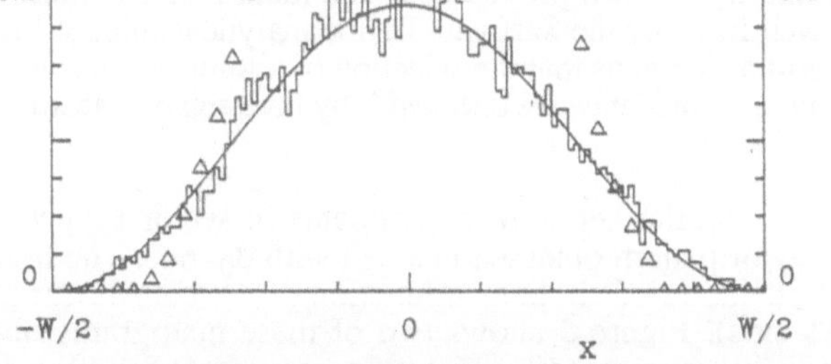

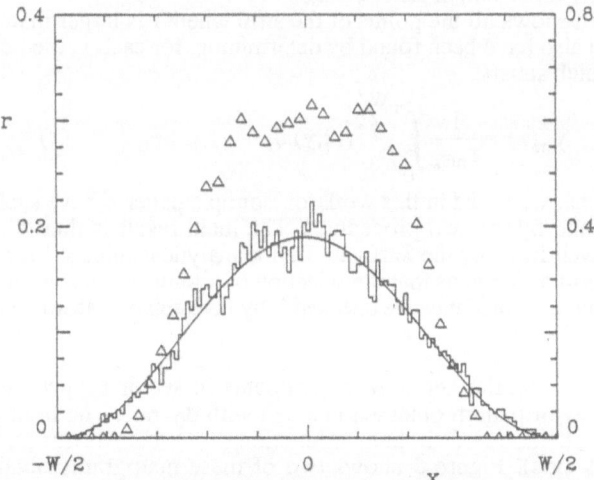

Fig. 2. Histograms showing the mean occupancy across a section of the linear cell for 75 unstable Saffman Taylor fingers at l_{max}/l_{min}=50 (scale of r(x,y) on the right) and N=510 DLA clusters of mass M=6000 grown in a strip W=128 and its fit by $r(x,y)=r_{max}\cos^2(\pi y/W)$ (scale on the left).

the stable case it is a square profile with probability of occupation 1 in the central half of the channel and 0 elsewhere. For larger values of l_{max}/l_{min} each realization of the pattern becomes very unstable. The averaging over N= 75 realizations of the type shown on Fig. 1b gives a transverse profile shown on Fig 2 ; it is smoother, has a maximum value $r_{max} <1$ at the center (y=0) and decreases to zero at the walls (y=±W/2). Averaging in the DLA case requires a larger number of realizations : Fig 1c shows one out of a series of 510 DLA simulations of clusters with the same mass M=6000. The transverse occupancy is now well fitted (Fig.2) by $r(x,y)=r_{max} \cos^2(\pi y/W)$. But the computational technique lends itself to the measurement of r(x,y) in all the points of the domain. We can thus also obtain the longitudinal profile r(x,y=0)shown on Fig.3 . The mean occupancy has a constant value r_{max} along the cell's axis and a region of fall-off at the tip. The stability of r_{max} shows that the translational invariance of the cell imposes itself to the mean properties of the cluster. Even though fractal at small scales, the cluster occupies a finite fraction of the space on scales larger than W. We investigated the dependence of r_{max} on l_{max}/l_{min} by varying W in both the Saffman Taylor and DLA experiments. We found that r_{max} has a power law dependance $(l_{max}/l_{min})^{-0.38\pm0.02}$. As the DLA clusters have a fractal dimension d_f, their sections by a line have a fractal dimension d_f-1. The mean transverse density (the fraction of the cell's width occupied by the cluster) should scale[10] as $\lambda r_{max}= 1/2r_{max}$ i.e. with an exponent d_f-2. The experimental exponent -0.38 corresponds to d_f=1.62±0.02, a value in good agreement

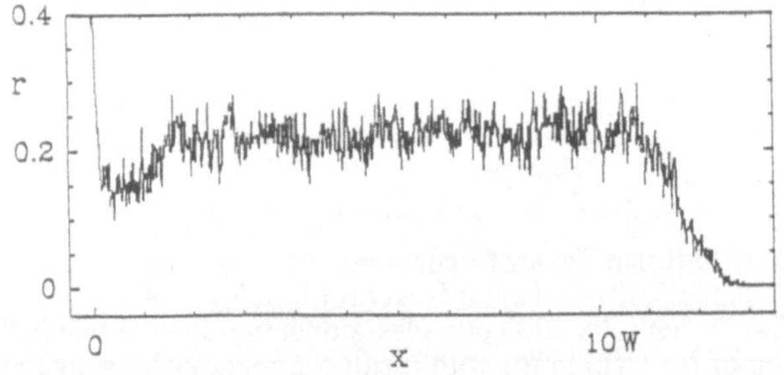

Fig. 3.Histogram of the mean occupancy along the axis of the linear strip.N=255 DLA clusters of mass M=6000 were grown in a strip of width W=64

with the usual fractal dimension of DLA clusters[11].

Finally Fig. 1d shows all the points of the strip where r is larger that $r_{max}/2$. The limit of this region could also have been found by determining, for each x, the points on each side of the cell's axis which satisfy

$$y_m^{\pm}(x) = \frac{1}{r_{max}} \int_0^{\pm W/2} r(x,y)\, dy \qquad (1)$$

All the histograms that we found in this work for isotropic patterns have such profiles that the same points are found by the two procedures. The main result is that this region of mean occupancy is very well fitted by the Saffman-Taylor analytical solution[1] for λ = 0.5. This is a rather surprising result as it means that the selection of a solution survives its instability. As it could have been a mere coincidence we checked it by investigating sector shaped geometries.

Sector-shaped cells

We performed a similar series of experiments in sector shaped cells[6,12]. Figure 4 shows a Saffman Taylor pattern obtained in a cell with θ_0=60°. The transverse histograms are now measured on an arc of circle as a function of the curvilinear position s (from an origin on the cell's axis). Figure 5 shows two of these histograms obtained for viscous fingering respectively in a divergent (θ_0=45°) and a convergent cell (θ_0=-45°) cell. They have a different shape : a large plateau at r_{max} with a rather abrupt fall-off to zero in the first case ; a narrow peak with a large region on each side where r=0 in the second case. For all angles $-90°<\theta_0<90°$, the width at mid-height is best fitted by a linear relation (Fig. 6) :

$$\lambda_m= 0.5 + (3.4\pm 0.2)\ 10^{-3}\, \theta_0 \qquad (2)$$

For all the convergent channels and for the divergent ones with $\theta_0<20°$, these widths,

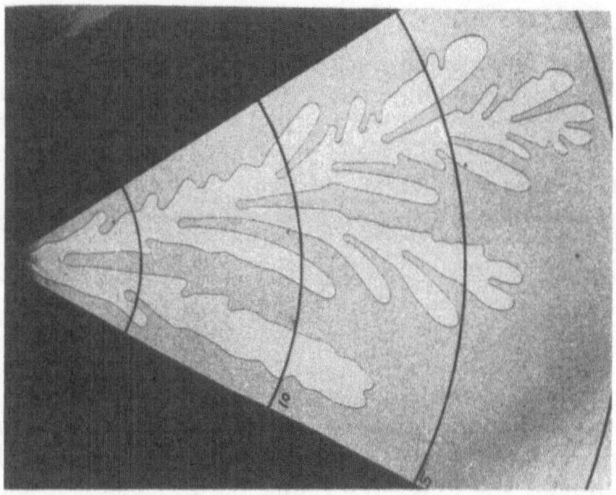

Fig. 4. Unstable viscous fingers obtained in an angular cell with θ_0= 60°.

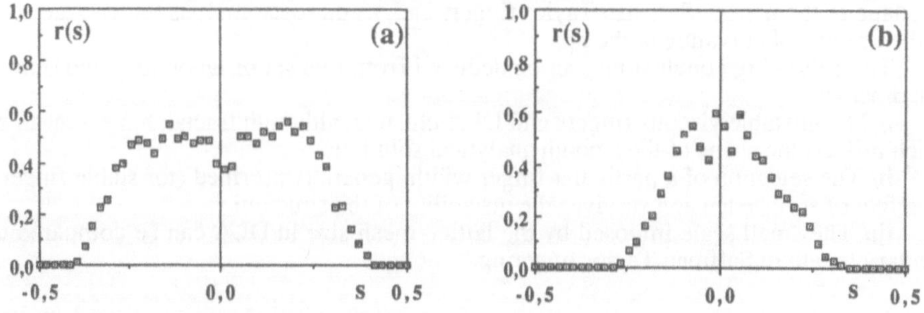

Fig 5.Histograms showing the mean occupancy across a section of an angular cell of angle θ_0=45° for (a) N=75 patterns grown in the divergent direction, (b) N=55 patterns grown in the convergent direction.

coincide with the asymptotic widths of stable fingers[6]. For θ_0>20° the same identity is not observed because the stable fingers destabilized before reaching their asymptotic width.

For DLA clusters a statistics for the whole domain can be obtained. A radial histogram along the bissector of the cell now gives a r_{max} which is a decreasing function of ρ the distance to the apex. This is due to the increase of the ratio l_{max}/l_{min} with the distance of the front region to the cell's apex. In order to find the mean finger shape, one can actually use the

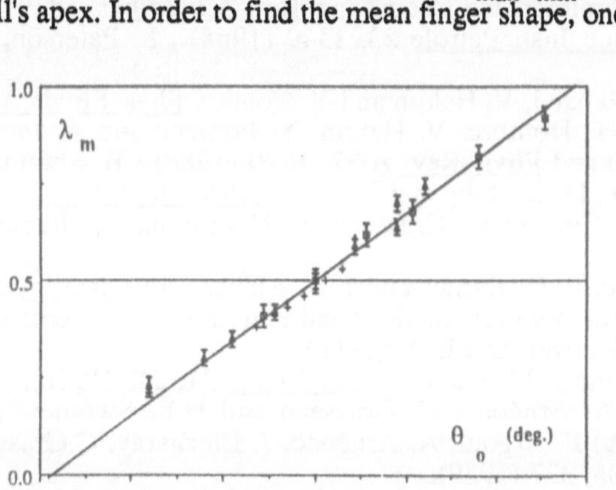

Fig. 6. Width of the region of maximum occupancy in sector shaped cells as a function the angle of the cell. (▲) Unstable Saffman Taylor fingers, (⊡) DLA aggregates, (+) Recall of the minimum width observed for stable fingers in these cells. (-----------) Fit of the asymptotic values by relation (2).

procedures defined previously (either the cut at mid-height or relation (1) ; they are still valid but according to the distance ρ from the apex where we define $r_{max}(\rho)$, we get the same mean finger of constant relative width λ but at different stages of growth.

Divergent cells with θ_o=90° are of particular interest because a family of analytical solutions is known[6], giving the finger shape in the absence of surface tension. These fingers form a self-similar counterpart to the Saffman Taylor solutions. The region of large occupancy has an angular width λ_m=0.82±0.02. Its shape is very well fitted by the corresponding analytical solution[8].

CONCLUSIONS

We have also extended this work to anisotropic patterns[13]; in particular we have investigated the unstable region of a dendrite, far away from the tip. For dendrites growing at small Peclet numbers in a nearly two dimensional cell, the competition between different side branches creates a fractal structure. We found its dimension to be 1.58±0.03. We also showed that in this unstable region the dendrite occupies a mean profile which is the same parabola that is stable at the tip. Similarly noise-reduced DLA clusters which exhibit the anisotropy of the underlying lattice were grown in a linear channel. Their mean profiles have the shape of the narrow Saffman Taylor fingers and, as these anomalous fingers, are selected by their radius of curvature at the tip.

The following conclusions can be deduced from this set of laboratory and numerical experiments:

i/ The unstable viscous fingers and DLA clusters, although fractal, have a mean shape which reflects the shape of the smooth analytical solutions.

ii/ The selection of a particular finger width, generally ascribed (for stable fingers), to the effect of surface tension survives the instability of the structure.

iii/ The small scale imposed by the lattice-mesh size in DLA can be compared to the capillary length in Saffman Taylor fingering.

REFERENCES

1. P. G. Saffman and G. I. Taylor, Proc. Roy. Soc. London, Ser. A 245, 312 (1958).
2. R. L. Chuoke, P. Van Meurs and C. Van der Pol, Tr. AIME 216,188, (1959).
3. (a) T. Witten and L. M. Sander, Phys. Rev. Lett 47, 1400 (1981) and Phys. Rev. B **27**, 5686 (1983) ; (b) P. Meakin, in "Phase Transitions and Critical Phenomena", Vol. 12, edited by C. Domb and J.L. Lebowitz (Academic Press, Orlando, 1988) and references therein.
4. (a) L. Paterson, Phys. Rev. Lett. 52, 1621 (1984). (b) D. Bensimon, L. P. Kadanoff, S. Liang, B.I. Shraiman and Chao Tang, Rev. Mod. Phys. 58, 977 (1986).
5. J. Bataille, Revue Inst. Pétrole 23, 1349 (1968) ; L. Paterson, J. Fluid Mech. 113, 513 (1981).
6. H. Thomé, M. Rabaud, V. Hakim and Y. Couder, Phys. Fluids A 1, 224 (1989).
7. R. Combescot, T. Dombre, V. Hakim, Y. Pomeau and A. Pumir, Phys. Rev. Lett. 56, 2036 (1986) and Phys. Rev. A 37, 1270 (1988) ; B. Shraiman, Phys. Rev. Lett. 56, 2028 (1986) ; D. C. Hong and J. Langer, Phys. Rev. Lett. 56, 2032 (1986).
8. A. Arnéodo, Y. Couder, G. Grasseau, V. Hakim and M. Rabaud, Phys. Rev. Lett., 63, 984 (1989).
9. (a) S. N. Rauseo, P. D. Barnes and J. V. Maher, Phys. Rev. A 35, 1245 (1987), (b) Y. Couder, in "Random Fluctuations and Pattern Growth", edited by H.E. Stanley and N. Ostrowsky (Kluwer, Dordrecht,1988).
10. A.R. Kopf-Sill and G.M. Homsy, Phys. Fluids 31, 242 (1988).
11. (a) F. Argoul, A. Arnéodo, G. Grasseau and H.L. Swinney, Phys. Rev. Lett. 61, 2558 (1988) ; (b) F. Argoul, A. Arnéodo, J. Elezgaray, G.Grasseau and R. Murenzi, Phys. Lett. A 135, 327 (1989).
12. It was brought to our attention that experiments on DLA growth in sector shaped cells had been done previously by H. Kondo, M. Matsushita and S. Ohnishi, "Science on form: Proceedings of the first International Symposium for Science on form", edited by S. Ishizaka (K.T.K.Scientific Publishers, Tokyo,1986).
13. Y. Couder, F. Argoul, A. Arnéodo, J. Maurer and M. Rabaud, Preprint.

THE PRINTER'S INSTABILITY :

THE DYNAMICAL REGIMES OF DIRECTIONAL VISCOUS FINGERING

Y. Couder, S. Michalland, M. Rabaud and H. Thomé

Laboratoire de Physique Statistique, Ecole Normale Supérieure
24 rue Lhomond, 75231 Paris Cedex 05

INTRODUCTION

Important experimental and theoretical efforts have been devoted recently to the dynamical behaviours of extended systems (reviews can be found in References 1 and 2). In this spirit, experiments were done in the Rayleigh-Bénard instability aimed at the study of the onset of chaos in periodic structures with a large number of cells[3,4]. On the other hand electroconvection of liquid crystals[5,6], convection in binary fluid mixtures[7,8] and the Faraday experiment[9,10] show a rich dynamics due to the presence of two independent control parameters. In particular, propagative modes are observed, sometimes confined to limited domains. Propagation also exists in Taylor-Couette flows where it creates spiral structures[11-13]. For the sake of simplicity and comparison to theoretical models, one-dimensionality has been sought in extended systems. This is obtained by confinement in the above-mentioned convection experiments[3,4] Another way to reach one-dimensionality is to study a boundary. This boundary can be the limit between two convective rolls affected by the oscillatory instability[14] or an interface between two-dimensionally confined media. In the latter case, if a front instability generates a large number of cells, their dynamics is one dimensional to a good approximation. Such situations are obtained in directional phase transition of liquid crystal[15], eutectic crystallisation[16] and directional viscous fingering[17a,b]. We presented previously[17a] results on the instability onset of the front of a viscous fluid in a widening gap. The present article reports the development of our work on the nonlinear regimes of this front instability.

Theoretical progress on the dynamics of extended systems has been linked with the use of the formalism of amplitude and phase equations to describe the secondary instability of a regular cellular structure [1,2]. Model phase equations, the Kuramoto-Sivashinsky (KS) equation[18] and its variants[19] were widely investigated numerically[20]. These equations exhibit a specific behaviour, the spatiotemporal intermittency as a precursor to a fully chaotic state. Though these equations were shown to be relevant to the dynamics of the fronts of directional solidification[21] in which they should describe the onset of chaos, spatio-temporal intermittency had not been observed yet in experiments of this type, but only in Rayleigh Bénard convection experiments[3,4]. The more general case of coupled complex Ginzburg-Landau equations has been recently investigated[22-24] in order to explain the propagation of localized structures and to explore the dynamics of the fronts due to subcritical transitions.

Directional viscous fingering

In the present experiment the interface between air and oil destabilizes to form a one -dimensional array with a large number of identical cells. The basic instability is not new ; printers have known for a long time that the interface of a lubricating film in a widening gap is unstable at large velocities. Previous experiments had been done in various configurations ; between a plane and a cylinder[25], between two rotating solid rollers[26-29], in the narrow passages of journal bearings[30], and in the peeling of an adhesive tape[31]. In our previous article[17a] we

Nonlinear Evolution of Spatio-Temporal Structures in Dissipative Continuous Systems
Edited by F.H. Busse and L. Kramer
Plenum Press, New York, 1990

showed that this cellular instability could be understood as being a Saffman-Taylor type of instability : in the frame of reference of the moving boundaries a low viscosity fluid (e.g. air) is penetrating one of high viscosity (e.g. oil). The front is dynamically confined to a region of non constant thickness. The thickness gradient adds a stabilizing factor to the linear analysis. This factor has two effects. While the usual Saffman Taylor front is always unstable, there is here a finite onset to the instability. It also creates non-linear terms which saturate the instability so that steady regimes are reached. This experiment has the same relation to viscous fingering as directional solidification has to free dendritic growth[32,33]. We showed that the ruling equations of the front are similar to those of directional solidification in the limit of the small Péclet number, the pressure field here playing the role of the impurity concentration. The shape and dynamics of the cells are also analogous in the two types of experiments.

But the behaviour of this instability as a function of the two velocities of the solid boundaries had not been explored. The present experiment has been designed so that they can be tuned independently. We will describe the behaviour of the front as a function of these velocities and show that they form independant control parameters. A rich phenomenology of non linear dynamical behaviors results from the existence of two control parameters, illustrating various processes typical of one dimensional extended systems.

EXPERIMENTAL SET-UP

The cell

The cell is formed of two horizontal plexiglass cylinders of different radii ($R_1 = 49$ mm and $R_2 = 71$ mm), placed horizontally, one inside the other and off-centered so as to be separated only by a small gap on one generatrix at the bottom of the apparatus (Fig.1). This gap (of minimum thickness b_o), can be adjusted between 100 µm and 1 mm with a resolution of 15 µm. The two cylinders rotate independently with tangential velocities V_1 and V_2, (with a resolution of 1mm/s). A small amount of oil is introduced between the cylinders and fills the lower part of the gap. As the oil wets the plexiglass, when rotation is set two films coating the cylinders are formed. The quantity of oil has to be sufficient to form these films but not too large. In these limits the experimental results do not depend on the quantity of oil set in the cell. We used a silicon oil, Rhodorsyl 47V100, with dynamical viscosity $\mu = 96.5\ 10^{-3}$ Kg/m.s and surface tension $T = 20.9\ 10^{-3}$ N/m at 25°C.

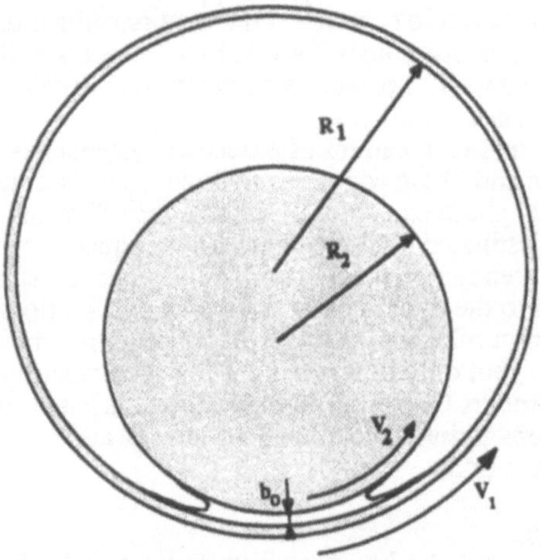

Fig. 1. Sketch of a section of the experimental cell showing the position of the cylinders and the two menisci.

The boundary conditions

The inner roller is shorter than the outer one (L_1= 200 mm and L_2= 210 mm), so there is a 5 mm zone at each end of the inner roller where an important quantity of oil is present. Due to recirculation in these regions, the interface is perturbed. The boundary conditions are then intermediate between solid and free conditions ; there is no phase pinning of the pattern at the ends, but the cells amplitude is changed by the presence of the boundary.

We will limit ourself to the case of wetted cylinders. In the case of dry cylinders (first rotation after the cleaning of the cell) the non linear dynamics of the interface is different.

Visualization

The oil only fills a small region at the bottom of the cell. Seen from below through the transparent outer cylinder, the two menisci separating oil from air appear as two dark lines. The shapes of these interfaces are recorded by a CCD video camera. The video pictures can be digitalized on a Macintosh IIx computer. We also constructed spatiotemporal pictures of the evolution of the front. We selected one single video sweep line intersecting the cells and, by collecting 512 of these lines separated from the next by a chosen delay, generated a new numerical image.

EXPERIMENTAL RESULTS

As will be shown below, the threshold value and the nonlinear dynamics of the unstable front are controlled by the two velocities V_1 and V_2 of the two cylinders, which act as two independent control parameters. Two symmetries of our experiments are never broken. First, the left interface for velocities V_1 and V_2 is in the same state as the right interface for velocities $-V_1$ and $-V_2$. Secondly, the state of a given interface is the same if we permute V_1 and V_2. The two cylinders play the same role showing that in the experimental range of velocities, centrifugal forces can be neglected. This could be foreseen, the Taylor number in these experiments being very small. ($Ta < 10^{-3}\,Ta_c$)

We will first describe the onset of the instability, then successively the various non linear behaviours observed when only one cylinder is rotating, when the two cylinders are contra-rotating and when they are co-rotating.

Instability threshold

For each sign of (V_1+V_2) the meniscus in the convergent region is stable and the other can be unstable. We limit ourselves to the study of one meniscus $(V_1+V_2>0)$. For low velocities of the cylinders this interface is stable and observed as a straight line. Over a velocity onset it becomes wavy. Fig.2 shows the stability limit of the linear interface. For contra-rotating cylinders this limit is well fitted by a linear relation $V_1+V_2=V_c$ (with $V_c \approx 47$ mm/s for $b_o = 0.37$

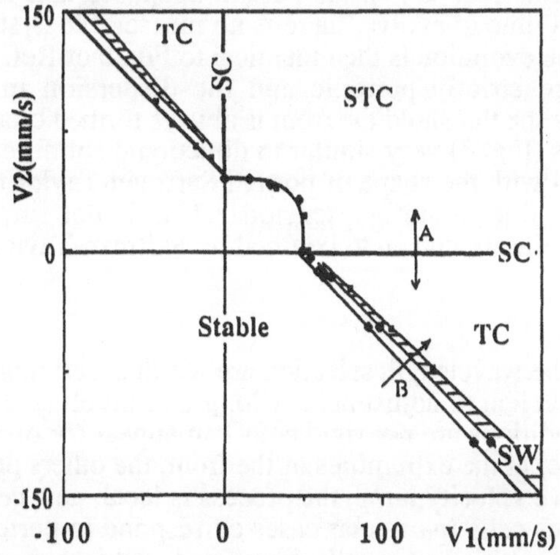

Fig. 2. State diagram of one interface in the (V_1, V_2) plane measured with b_o=0.37mm. The different regions are labelled : SC, Stationnary cells, TC, Travelling cells, SW Solitary waves, STC Spatio-temporal chaos.

mm). For co-rotating cylinders (V_1 and V_2>0) the front remains stable up to larger values of the mean rotation. Everywhere along this transition limit (as in the plane-cylinder experiment [17a]) the instability appears as a supercritical process in which the front becomes a sinusoid of wavelength λ_c. The amplitude of the sinusoid grows as the square root of the distance to the onset. The value of this critical wavelength is almost constant on all the threshold line. The linear analysis presented in Ref. 17a, in the particular case where only one cylinder is rotating, gives reasonable predictions for the values of λ_c. However for two cylinders this treatment also suggests that the state of the front depends essentially on (V_1+V_2) only. This is not what is observed experimentally. The existence of a bump in the onset curve and the non linear behaviour show that V_1 and V_2 form two independent control parameters. The reason of this discrepancy is certainly related to drastic changes (that our theoretical treatment did not include) in the small scale velocity field in the immediate vicinity of the meniscus.

Rotation of only one cylinder

If one of the cylinders is motionless (the states of the system represented by the points of the axes of Fig. 2), the non linear evolution is apparently simple. Above the threshold, the amplitude of the deformation grows and the front departs from its sinusoidal shape, the wavelength λ of the pattern decreases rapidly and tends to saturate for large values of the constraint. At all the velocities that we investigated (up to 50 times the initial threshold) the

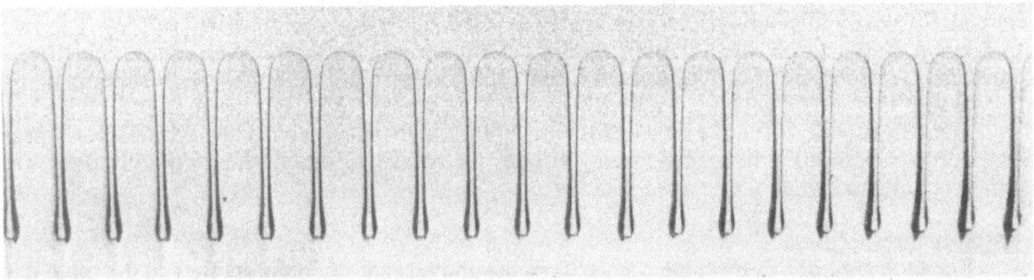

Fig. 3. A part of a stable pattern of stationary cells (V_2=0) A front can have up to several hundred of these cells.

pattern reaches a stable, regular state of identical motionless cells and no chaos is observed (Fig.3). For all velocity changes, after a long transient (typically of the order of the viscous diffusion time along the interface L^2/ν), the pattern adapts to take an optimum wavelength. Provided the structure has time to evolve, there is no measurable hysteresis in the wavelength versus velocity curve. (The evolution is then identical to Fig. 6 of Ref. 17a). Furthermore in the steady state the cells are strictly periodic and the dispersion in their widths is below measurability. Except near the threshold the front is always formed of a series of parallel fingers separated by thin oil walls (Fig. 3) very similar to directional solidification cells. The shape of these fingers is well fitted with the shape of normal Saffman-Taylor fingers of large λ, taking into account a correction due to surface tension[34].This is not surprising as the shapes of directional solidification cells are known to be fitted by Saffman-Taylor solutions in the limit of small Péclet number[35].

Transients

Some transients of the wavelength selection are worth describing. For very small velocity jumps, the process of wavelength adjustment is long and involves the whole array. Usually, because the boundary conditions are not rigid cells can appear (or disappear depending on the sign of the velocity change) at the extremities of the front, the others progressively adjusting by diffusion. For large positive velocity jump, the process is local, new cells are created locally by tip splitting of the previous cells (particular cases correspond to perfect halving of the spatial period). For large negative velocity jump cells disappear locally when pinched off.

Specific transients are observed for jumps in velocity of intermediate amplitude. The reorganization of the structure (of initial regular wavelength λ), occurs by a compression wave propagating along the front. One particular cell, often at the front boundary, adapts to the new

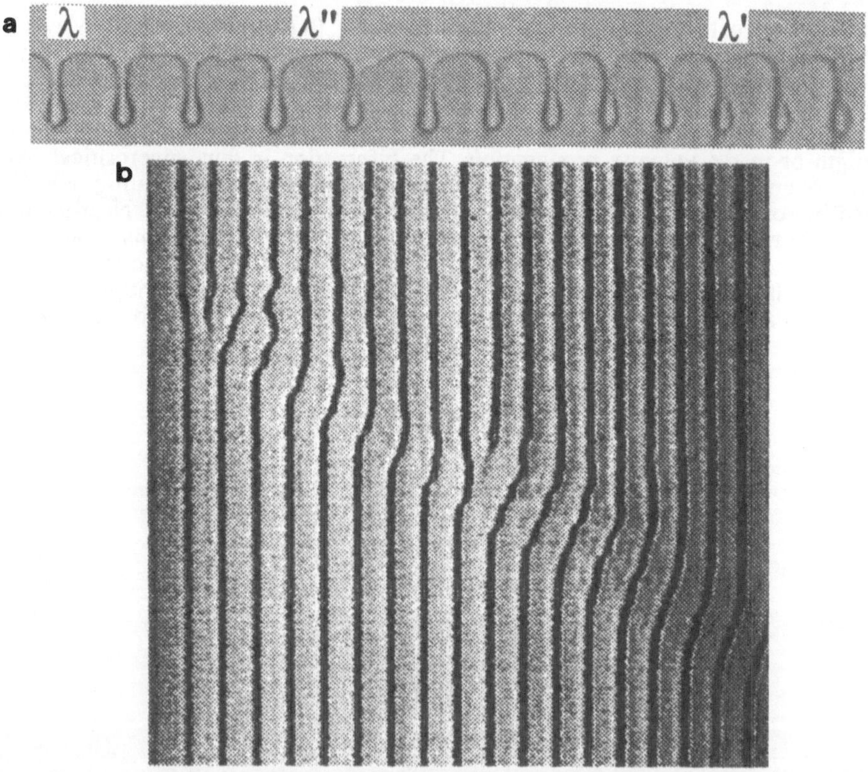

Fig.4. The compression wave generated after a jump in velocity. (a) Photograph of a part of the interface. λ and λ' the wavelengths before and after compression. (b) Spatio -temporal evolution shown by the image processing, time ellapses from bottom to top.

shorter wavelength λ'; its neighbour is then first dilated to λ'', then shifts, and shrinks to λ' (Fig. 4a). The process repeats itself along the front so that, transitorily, the interface is formed of a domain of a few shifting cells separating two regions of different wavelengths. The shifting cells propagate towards the cells of smaller wavelength (phase velocity) but the boundaries of their domain propagate in the opposite direction (group velocity). These waves appear to be of the type observed in liquid crystal directional phase transition[15] and in eutectics growth[16] after a jump in the velocities. As has been described in the case of eutectic[16], the domain of shifting cells spreads with time. There is a difference however ; in our experiment the domain widening is limited: when the drift of a cell becomes larger than one wavelength (Fig. 4b) a cell splits and a new cell is formed. The compression wave starts propagating again until the phase mismatch becomes again larger than 2π. The whole process has been observed easily for positive velocity jumps. For jumps to smaller velocities the trend is to increase the wavelength. the destruction of a cell can sometimes be followed by a localized dilatation wave with a shrinking length. These processes are clearly subcritical as they start only in regions which are locally disturbed, usually at the end of the front, sometimes at imperfections of the solid surfaces. The compression wave has been recently associated with a subcritical process of breaking of the parity symmetry [22]. It can also be noticed that when the intermediate cells have been stretched, λ'' is so large that its first harmonic λ''/2 is also unstable. This has been shown to induce propagation of the cells[23].

Contra-rotating cylinders

As shown on Fig. 2, four different regimes exist in the contra-rotating domain ; one where the interface remains stable ($V_1+V_2<V_c$), the second where it destabilizes into a static cellular pattern, a third where cells propagate and finally a small domain in which there is a steady regime where localized travelling structures and static cells coexist on the front. In order to describe the transitions between different regimes we will follow two particular paths accross this quadrant of the phase diagram.

In the first path (Fig.2) we start from one axis (e.g. from point A, where $V_2=0$). Increasing slowly the velocity of the other cylinder in contra-rotation (e.g. $V_2<0$) the cells become propagative and start drifting. We could not find a finite lower threshold for the propagation ; as soon as V_2 is non zero, the cells start moving. The velocity of translation is proportional to $(V_2)^{1/2}$. No hysteresis has been found and there is no discontinuity in the wavelength or in the velocity propagation. The bifurcation is thus supercritical. After the transition the unstable interface is formed of domains of left or right propagating cells. They are separated by sources and sinks where cells are formed or vanish (Fig. 5b). Near the threshold the front is formed of domains with a chaotic behaviour ; they form and vanish constantly, their limits move and never stabilize. It is only for larger values of V_2 that the structure tends to stabilize with large regions of left or right travelling waves. It is worth noting that the cells motion is then associated with a slow large scale flow along the front : if one single source is

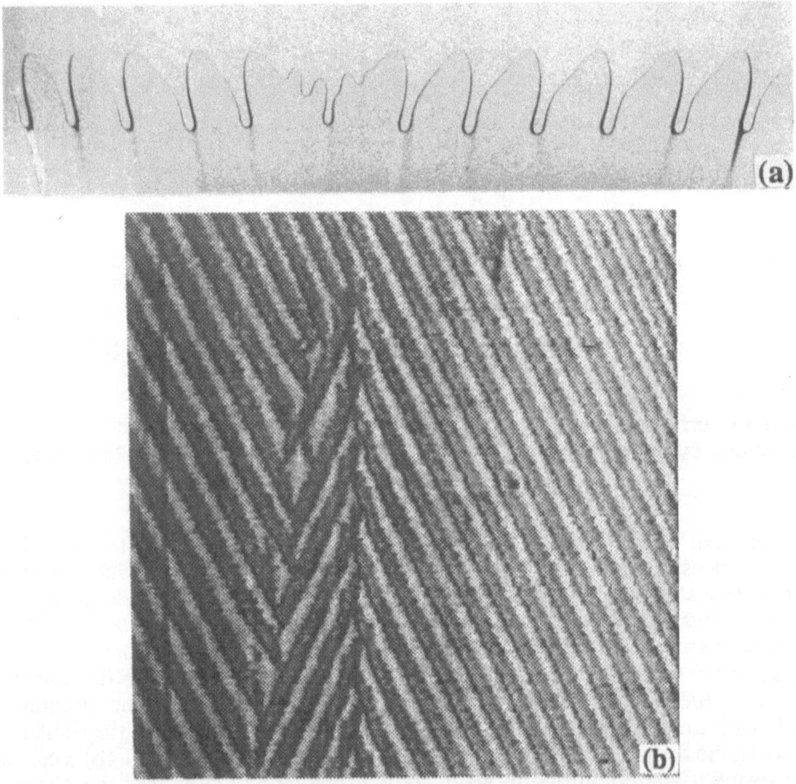

Fig. 5. Propagative state for contra-rotating cylinders. (a) Photograph of a part of an interface showing a source. (b) Spatio-temporal evolution showing the annihilation of a sink with a source. Time elapses from bottom to top.

present at the center, the fluid moves very slowly towards the extremity of the cylinders and accumulates there. Finally the shape of the translating cells is asymmetrical, their backs being sensitive to side branching instability which tends to create cells with a motion in the opposite direction.

The second path we will describe starts from point B in Figure 2. Increasing slowly V_1 and V_2 (following for example a line V_1-V_2= Cste), we first cross the onset of the first instability where a motionless sinusoidal front is formed. Then over a second threshold we begin to see recurrently the passage of a solitary wave moving in either direction along the front. These cells have a large, well determined amplitude and a characteristic shape (Fig. 6a). They usually form after the splitting of a steady cell. Once formed, they travel at constant velocity, locally destroying the motionless sinusoid which reappears afterwards with the same phase. When two

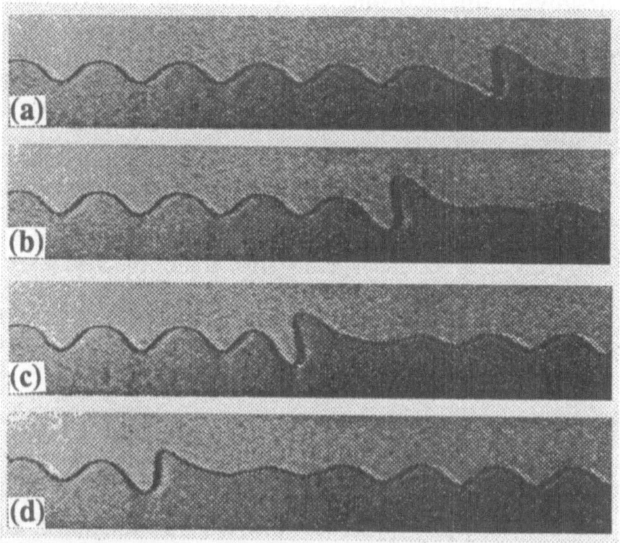

Fig. 6. Successive photographs of a solitary wave propagating along the interface. The underlying stationary structure is destroyed then reappears. The phase of the stationary pattern is unaffected by the passing wave.

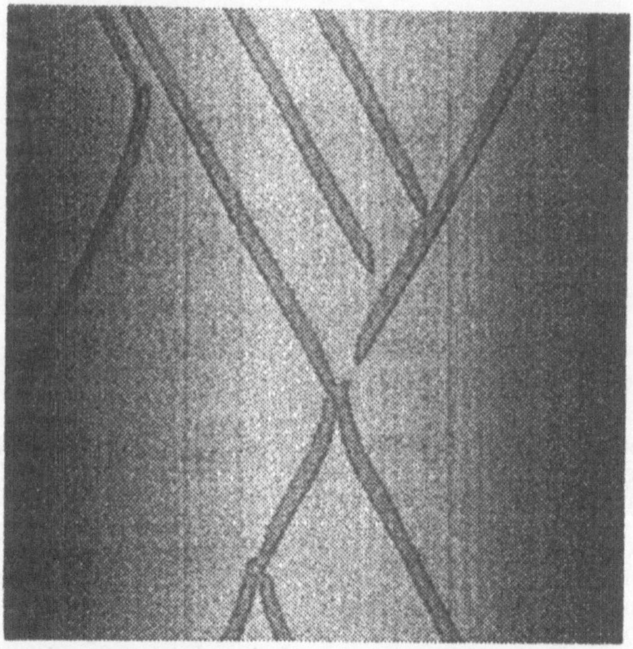

Fig. 7. Interacting solitary waves (SW). The line chosen for the image processing is just above the front so that only the solitary waves are observed. Several types of interactions are observed. Collisions : at the bottom one of the SW is destroyed, in the middle of the picture two SWs cross each other, at the top one of the colliding SWs turns back. A SW is also seen to generate contrapropagating SWs. Time ellapses from bottom to top.

of these structures, moving in opposite directions meet, they can cross each other, but more often one of them is destroyed. Still increasing the velocities, the mean density of these solitary waves increases and they can be generated by packets of several waves following each other. The front is then formed of "bubbles" of propagating structures travelling on a motionless pattern. We must emphasize that these waves are completely different from what we called above the compressive waves. Here the waves and their domain move in the same direction, in compressive waves they move in opposite directions. Increasing V_1 and V_2 further, the number and the size of these bubbles increase and the front becomes entirely formed of travelling cells separated by sources and sinks : the system has reached the travelling cells domain.

Altogether the domain of propagating cells can be entered either by a supercritical transition along the axis or by localized subcritical processes in the wedge shaped region shown on Fig.2. In this region there is an overlap of the propagating state and of the stationary state. One of the border lines of this coexistence region corresponds to the apparition of the first propagating cell, the other to the loss of the last stationary cell. The apex of the wedge is very near the axis in a region where the front behaviour could not be clearly characterized.

Co-rotating cylinders

The last region of Figure 2 ($V_1>0$ and $V_2>0$) corresponds to the situation where the two cylinders rotate in the same direction. We will start again from one of the two axes (e.g. from point A) where a regular stationary pattern has been created. If co-rotation of the other cylinder is set in, temporal and spatial defects appear in the linear array of cells (Fig. 8a). This is manifested by local fluctuations of the wavelength and by splitting and pinching-off of some cells. These processes are not uniformly distributed. Some parts of the front remain strictly ordered, formed of steady, spatially periodic, cells. We will call them quiescent domains. Other parts are

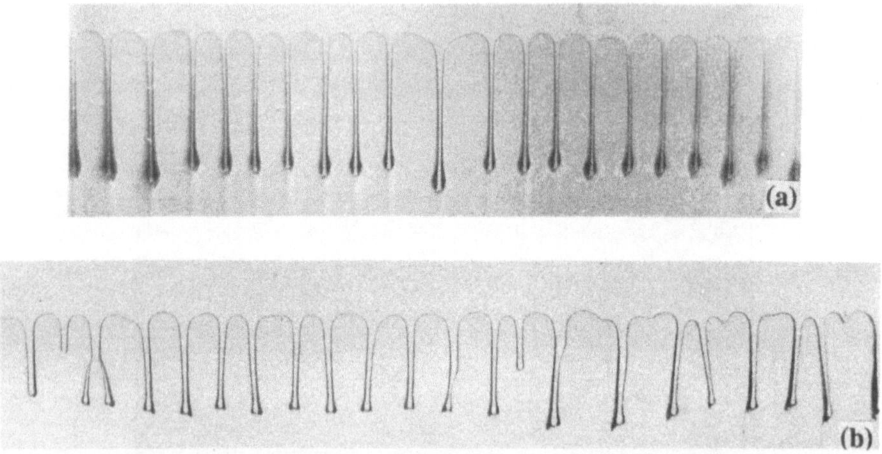

Fig. 8. Co-rotating cylinders : (a) A part of a front in a weakly chaotic state showing cells oscillating in an optical mode and a pair of anomalous cells, (b) A part of a front in a strongly chaotic state.

disordered with strong fluctuations of the wavelength and repeated formation and destruction of cells. We will call them chaotic domains. The difference between the two types of domains and their time evolution are clearly seen after image processing (Fig. 9 a).

There is, along the front, coexistence of these chaotic and quiescent domains. Their borders are not stable : the domains appear, extend, then shrink and vanish. Again we did not find a finite velocity threshold for this transition. Even for very slow co-rotation some disorder is observed (Fig. 8a). As co-rotation is increased, the disorder increases and ultimately the front is chaotic practically everywhere (Fig. 8b and 9b). Qualitatively, for each value of the control parameters a certain fraction of the front is, on the average, occupied by chaotic domains. The whole area of co-rotation on Fig.2 (labeled STC) exhibits this type of spatiotemporal chaos. This behaviour appears very similar to that observed by Chaté and Maneville[20] in numerical simulations of the Kuramoto Sivashinsky (KS) equation and on its variants, and called by them

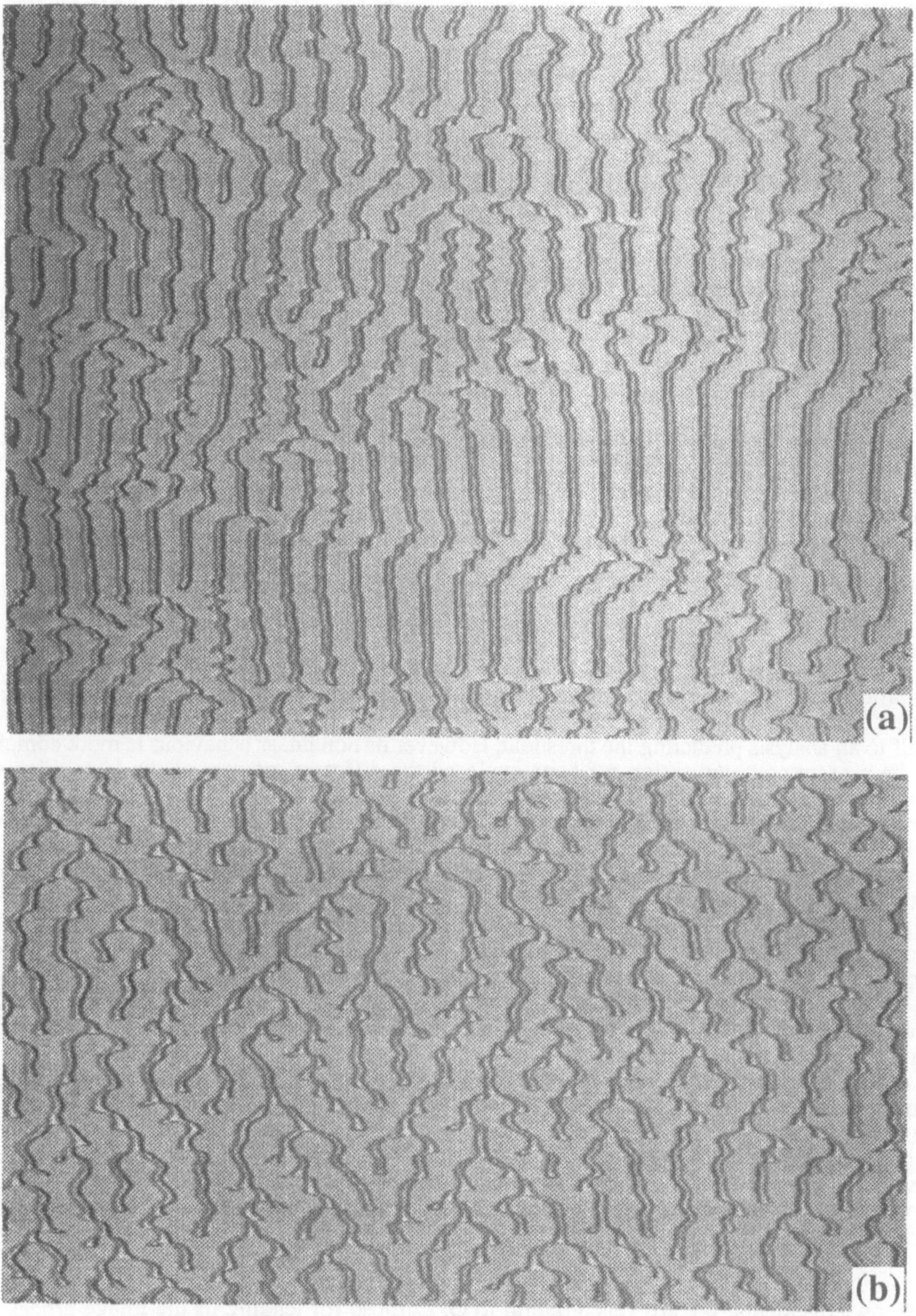

Fig. 9. Co-rotating cylinders :
(a) The spatiotemporal evolution of a part of the front showing the coexistence of quiescent and chaotic domains. The image summarizes a long period of evolution of the front (49s). It shows that the chaotic regions are inter-related by the propagation of phase waves ; these waves often form the uptime limit of the chaotic bursts. Also observed are local oscillations of the cells in an optical mode (b) The spatiotemporal evolution of a fully chaotic front. Time elapses from bottom to top.

spatiotemporal intermittency. A quantitative investigation seems promising because in the present experiment the basic state is strictly periodic and because the co-rotation velocity is a control parameter of the degree of chaos. A statistical investigation of the distribution of the domains' sizes as a function of the control parameter is under progress.

The precision of the visualization of the front permits the observation of the individual processes at work. For instance, after the pinching-off of a cell, the neighbouring quiescent regions are affected by a phase wave (Fig.9a) similar to that described above (Fig.4) propagating along the front and by which the neighbouring cells close the gap. When the phase wave, travelling along the front, reaches an already slightly disturbed region, it can trigger another chaotic burst. We thus observe that the activity in the chaotic domains excites the quiescent regions and that the waves propagating across them can correlate chaotic bursts far away from each other. This characteristic difference between the activity of the two types of domains corresponds to a behaviour previously observed in numerical simulations. Other characteristics of the observed weak chaos points to a similarity of behaviour with the numerical simulations. The lateral displacement due to the phase wave can generate oscillations of the cells in an optical mode ; in these oscillations each cell swells and shrinks in opposition of phase with its neighbours (Fig.8a). We also observe localized metastable anomalies in the periodic array, formed of a pair of adjacent cells which occupy approximately the space of three normal cells. They have a characteristic, asymetrical, shape (Fig 8a), are transient in the strongly chaotic state but long lived near the threshold. These properties correspond exactly to those of the coherent structures observed in simulations of the Kolmogorov-Spiegel-Sivashinsky (KSS) equation[19] (a variant of the damped K.S. equation with an extra non-linear term).

CONCLUSION

The front instability that we call directional viscous fingering is basically similar directional solidification. However it has a certain number of specific characteristics. It is somewhat simpler because the medium is isotropic and because the first instability onset is super-critical and lends itself to an analysis predicting the threshold. However its non linear behaviour is more complex. In the present communication we have explored the different regimes of the instability as a function of the two solid surfaces velocities and shown these to be two independent control parameters. For this reason the dynamics of directional viscous fingering turns out to be very rich. Three main non-linear regimes have been characterized. Stationary periodic fronts are observed which can sustain the propagation of compressive waves. Propagative modes can be obtained on the whole front, where they form domains of cells propagating to the left or to the right separated by sources and sinks. They can also form solitary waves of well defined shape and amplitude. Finally the front can be chaotic with spatio-temporal intermittency. It had been shown[21] that the front of directional solidification is governed by a damped KS equation. The instability of our front is very similar to that of directional solidification ; in the corotating situation, we do find a behaviour which exhibits the main characteristics of the numerical simulations of the KS equations.

The transitions between the different non-linear regimes of the front can be supercritical or subcritical. Particularly surprising is the behaviour near the axes of the phase diagram : on each side of it, for any contra or co-rotation of the other cylinder, there is a supercritical transition to either propagation or chaos. So the two axes appear to form by themselves singular regions sustaining a specific behaviour. More work is needed to understand this oddity, to study the propagative solitary waves, and to achieve a quantitative characterization of the spatiotemporal chaos.

Acknowledgements. We are very grateful to O. Cardoso for lending us the image processing program he designed. We had particularly useful discussions with P. Coullet, S. Fauve, V. Hakim, and J. Lega.

REFERENCES

1. Propagation in Systems Far From Equilibrium, J.E. Wesfreid, H.R. Brand, P. Maneville, G. Albinet and N. Boccara editors (Springer, Berlin, 1984). Reviews on the dynamics of extended systems : A. C. Newell, p.122, H. R. Brand, p. 206, P. Maneville, p. 265.

2. New trends in Nonlinear Dynamics and Pattern Forming Phenomena : The Geometry of Nonequilibrium, P. Coullet and P. Huerre editors, (Plenum Press, New York). To appear in 1990.
3. S. Ciliberto and P. Bigazzi, Phys. Rev. Lett. 60, 286 (1988).
4. F. Daviaud, M. Dubois and P. Bergé. Europhys. Lett. 9, 441 (1989).
5. A. Joëts and R.Ribotta, Phys. Rev. Lett. 60, 2164 (1988).
6. I. Rehberg, S. Rasenat, and V. Steinberg, Phys. Rev. Lett. 62, 756 (1989).
7. P. Kolodner and C. Surko, Phys. Rev. Lett. 61, 842 (1988).
8. E. Moses, J. Fineberg and V. Steinberg, Phys. Rev. Lett. 61, 838 (1988).
9. S. Douadi , S. Fauve and O. Thual, Europhys. Lett. 10, 309 (1989).
10. N.B. Turafillaro, N.B. Ramshankar and J.F. Gollub, Phys. Rev. Lett. 62, 422 (1989).
11. C. D. Andereck, S.S. Liu and H.L. Swinney, J. Fluid Mech. 164, 155 (1986).
12. I. Mutabazi, J.J. Hegseth, C. D. Andereck and J. E. Weisfreid, Phys. Rev. A, 38, 4752 (1988).
13. Li Ning, G. Alhers and D.S. Cannell. Phys. Rev. Lett. 64, 1235 (1990).
14. B. Janiaud, E. Guyon, V. Croquette and D. Bensimon, in this volume.
15. A. Simon, J. Bechhoefer and A. Libchaber. Phys. Rev. Lett. 61, 2574 (1988) and J. Bechhoefer, A. Simon, A. Libchaber and P. Oswald. Phys. Rev. A 40, 2042 (1989).
16. G. Faivre, S. de Cheveigné, C. Guthmann and P. Kurowski. Europhys. Lett. 9, 779 (1989).
17. (a) V. Hakim, M. Rabaud, H. Thomé and Y. Couder. in Ref.2. (b) M. Rabaud, S. Michalland and Y. Couder. Phys. Rev. Lett. 64, 184 (1990).
18. Y. Kuramoto and T. Tsuzuki, Prog. Theor. Phys. 55, 356 (1976), and G.I. Sivashinsky, Acta Astronaut. 4, 1177 (1977).
19. A.N. Kolmogorov, in Seminar Notes edited by V.I. Arnold and L.D. Meshalkin, Uspekhi Mat. Naut. 15, 247 (1960).
20. H. Chaté and P. Maneville, Phys. Rev. Lett. 58, 112 (1987) and H. Chaté and B. Nicolaenko, to appear in Ref. 2.
21. G. I. Sivashinsky, Physica D 8, 243 (1983), A. Novick-Cohen, Physica D, 23, 118 (1986) and A. Novick-Cohen, Physica D 26, 403 (1987).
22. P. Coullet, R. E. Goldstein and G.H. Gunaratne, Phys. Rev. Lett. 63, 1954 (1989).
23. S. Douady, S. Fauve and O. Thual. Europhys. Lett. 10, 309 (1989).
24. W. Van Saarloos and P.C. Hohenberg. Phys. Rev. Lett. 64, 749 (1990).
25. J. R. A. Pearson, J. Fluid Mech. 7, 481 (1960).
26. M. D. Savage, J. Fluid Mech. 80, 743 (1977) ; J. Fluid Mech. 80, 757 (1977) and J. Fluid Mech. 117, 443 (1982).
27. E. Pitts and J. Greiller, J. Fluid Mech. 11, 33 (1961).
28. C. C. Mill, G. R. South, J. Fluid Mech. 28, 523 (1967).
29. J. Greener, T. Sullivan, B. Turner, S. Middleman, Chem. Eng. Commun. 5, 73 (1980).
30. G. I. Taylor, J. Fluid Mech. 16, 595 (1963).
31. A. D. McEwan and G. I. Taylor, J. Fluid Mech. 26, 1 (1966).
32. R. Trivedi and K. Somboonsuk, Acta Metall. 33, 1061 (1985).
33. S. de Cheveigné, C. Guthmann and M. M. Lebrun, J. Physique, 47, 2095 (1986).
34. These profiles were obtained numerically by M. Maashal (private communication).
35. T. Dombre and V. Hakim, Phys. Rev. A, 36, 2811 (1987).

FRONT PROPAGATION INTO UNSTABLE STATES: SOME RECENT DEVELOPMENTS AND SURPRISES

Wim van Saarloos

AT&T Bell Laboratories
Murray Hill, N.J. 07974
U.S.A.

ABSTRACT

I review the differences and similarities between the marginal stability approach to front propagation into unstable states and the "pinch point" analysis for the space-time evolution of perturbations developed in plasma physics. I then briefly discuss the following developments and surprises: (i) the resolution of a discrepancy between the theory and experiments on Taylor vortex fronts; *(ii)* some new results for the regime where front propagation is dominated by nonlinear effects (nonlinear marginal stability regime); *(iii)* ongoing work on fronts and pulses in the complex Ginzburg-Landau equation.

INTRODUCTION

In the research communities represented here at this workshop on "Nonlinear Evolution of Spatio-Temporal structures in Dissipative Continuous Systems" we have in recent years seen a growing interest in front propagation into unstable states (see e.g. Refs. 1,2 and references therein). At the same time, especially due to the shift in attention from the Rayleigh-Bénard instability in a one component fluid to the Rayleigh-Bénard instability in binary mixtures, the awareness of the importance of the distinction between a convective instability and an absolute instability has become appreciated.[3] In a convectively unstable state, a perturbation grows but at the same time is convected away; the effect of the convection is strong enough that the perturbations are found to decay when viewed at a fixed position. At an absolute instability, on the other hand, the instability is strong enough that perturbations do grow when viewed at a fixed position. When a system of finite size exhibits a convective instability, the long term dynamics depend on whether perturbations have a chance to grow sufficiently during the time over which they are convected away from one side of the system to the other, and on the boundary conditions. As a result, the final state often depends on the size of the system. A nice example of these effects within the context of the Rayleigh-Bénard instability is given by the linear behavior as well as the nonlinear "sloshing" states in binary fluid mixtures.[3–5]

The term front propagation is often used to refer to describe situations in which the dynamical evolution of a spatially extended system that starts out in an unstable state, is governed mainly by the propagation of well defined spatially separated fronts or interfaces connecting the unstable region with a region in which the system is in some *nonlinear* (stable) state. Clearly, the dynamical properties of such fronts – e.g. their speed of propagation – determine to a large extent the qualitative behavior of a convectively unstable finite system like those found in binary fluid mixtures.[4]

Nonlinear Evolution of Spatio-Temporal Structures in Dissipative Continuous Systems
Edited by F.H. Busse and L. Kramer
Plenum Press, New York, 1990

Convective instabilities abound in plasma physics[6] and in fluid dynamics[7] (essentially in all cases where a fluid flow past a fixed body becomes unstable). So it should come as no surprise that the *linear* spatio-temporal behavior at convective instabilities has been studied for over thirty years in these fields, and that the essential results are summarized in a standard book like the "Physical Kinetics" volume in the Landau-Lifshitz Course on Theoretical Physics.[8] Nevertheless, the relation between this work and that on front propagation has often been overlooked. Presumably, this is due to the fact that the theoretical work in plasmas and fluid dynamics is normally limited from the start to the linear behavior of perturbations, whereas originally most of the work on moving fronts concentrated on the development of fully nonlinear front solutions right away. That nevertheless some of the essential results of these rigorous approaches could be reformulated in terms of concepts related to the linear stability properties of the equations was recognized, to my knowledge, for the first time by Dee and Langer[9] and by Ben-Jacob et al.[10] However, the the similarity of some of their ideas with the tools already developed to determine the convective or absolute nature of an instability from an analysis of the linear evolution of perturbations, were not immediately recognized.

Most recently, I have obtained a number of new results for the rate of approach to the asymptotic front speed and for the mechanism of velocity selection in the regime dominated by the nonlinearities.[2] As explained above, in many theoretical discussions the effects of nonlinearities on the convective versus absolute nature of instabilities are not considered, even though our analysis indicates that they can be extremely important. Since these results have just appeared in a detailed paper, I will confine myself here to bringing some of the differences and similarities of the two approaches to the reader's attention. Moreover, I will highlight the recent resolution[11] of the longstanding discrepancy between theory and experiment on front propagation in Taylor-Couette vortex flow.[12] Finally, I will briefly draw attention to some of the implications of and surprises from my recent work as well as ongoing work in collaboration with P. C. Hohenberg.[13]

BRIEF SKETCH OF HISTORICAL DEVELOPMENTS

The first investigations of how a class of initial conditions develops into a well-defined front solution propagating into an unstable state goes back to the work of Kolmogorov et al.[14] and Fisher.[15] These authors studied partial differential equation of the type

$$\frac{\partial \phi}{\partial t} = \frac{\partial^2 \phi}{\partial x^2} + F(\phi)\,, \qquad \textit{with}\;\; F(0) = 0, \quad F'(0) = 1\,. \tag{1}$$

The simplest example of such an equation is

$$\frac{\partial \phi}{\partial t} = \frac{\partial^2 \phi}{\partial x^2} + \phi - \phi^3\,. \tag{2}$$

In the latter equation, the homogeneous state $\phi=0$ is unstable, whereas the homogeneous states $\phi=\pm 1$ are the absolutely stable states. The above authors were able to show that for a large class of functions like the one corresponding to (2), initial conditions $\phi(x, t=0)$ for which ϕ decays sufficiently rapidly for $x \to \infty$ develop under the dynamics (1) into a moving front solution $\phi(x-vt)$ with a particular value of the velocity. Subsequent investigations concentrated on equations of type (1), and culminated in the seminal work by Aronson and Weinberger.[16] The results of their rigorous mathematical analysis of (1) were popularized and reinterpreted for the physics community by Dee and Langer[9] and Ben-Jacob et al.[10] They noted that for a large class of functions $F(\phi)$ like the one in (2), the mathematical analysis showed that the asymptotic front velocity v_{as} was in fact *independent* of the details of the nonlinear form of $F(\phi)$ and equal in value to a particular velocity v^* that is singled out by the linear dispersion relation corresponding to Eqs. (1) and (2). That is, if one linearizes these equations and substitutes a Fourier mode of the form $e^{ikx-i\omega t}$ to determine the dispersion relation $\omega(k)$, one observes that the rigorous results for a large class of functions $F(\phi)$ amount to the statement that $v_{as}=v^*$, with v^* given by the solution ω^*, k^* of

$$v^* = \frac{\mathrm{Im}\omega}{\mathrm{Im}k} = \frac{\partial \mathrm{Im}\omega}{\partial \mathrm{Im}k}, \qquad \frac{\partial \mathrm{Im}\omega}{\partial \mathrm{Re}k} = 0 . \tag{3}$$

Stability considerations showed[9,10] that for equations of the type (1), v^* defined by (3) can interpreted as the velocity of the front at which front profiles are *marginally stable.* Since v^* is completely determined by the linear dynamics of small perturbations around the unstable state $\phi=0$, I refer to v^*, ω^* and k^* as the linear marginal stability values.

From the work of Aronson and Weinberger,[16] it is known that only for a certain class of functions $F(\phi)$ the front speed approaches v^* asymptotically. It is also possible to have $v_{as}=v^\dagger>v^*$ in (1). In such cases, Langer and coworkers[9,10] also showed that the point where $v=v^\dagger$ is the point where the stability of the uniformly moving profiles $\phi(x-vt)$ changes, in that front profiles with velocity $v>v^\dagger$ are stable while those with $v<v^\dagger$ are unstable. However, since in this case the stability can not be inferred from the linearized dynamical equation, but depends strongly on the whole nonlinear behavior of the profiles $\phi(x-vt)$, we refer to this case as nonlinear marginal stability.

Numerical studies demonstrated that the asymptotic speed of nonlinear fronts in several more complicated equations that do not admit uniformly translating front solutions, is in fact correctly predicted by the *linear* marginal stability value (3). This led me to reformulate some of the marginal stability ideas.[1,2] This reformulation, based on an analysis of the dynamics in the "leading edge" where perturbation are small enough that their dynamics is essentially governed by the linearized equations, not only leads to a better understanding of the physical mechanism that drives the front velocity either to the linear marginal stability value v^* or to the nonlinear marginal stability value $v^\dagger$, but also leads to explicit predictions for the rate of approach of the front velocity to v^*. We will give examples of this in the next sections.

As the above sketch indicates, the above line of research started with a fully nonlinear analysis of a particular class of relatively simple partial differential equations. From the very first investigations on, it was clear that the propagation of fronts into unstable states can depend critically on the nonlinearities in the dynamical equation. However, the recent realization that the linear marginal stability prediction for v^* in terms of the linear dispersion relation $\omega(k)$ appears to be correct for a large class of equations, has focussed attention on the fact that to a large extent the dynamical selection takes place in the "leading edge" governed by the linearized equations.

The line of research that developed independently since the late fifties in plasma physics[6,8] and fluid dynamics started almost at the other end: here the attention was confined right away to the linear dynamical evolution of perturbations, with the advantage that the arguments were general enough that no specific assumption concerning the specific form of the equations needed to be made. On the other hand, even to date the effect of nonlinearities is hardly ever discussed explicitly in these approaches, even though some of the most interesting applications are to cases in which the perturbations grow sufficiently large that nonlinear dynamical patterns, e.g. vortices,[17] are generated. Moreover, the argument is commonly formulated in such a way that it only applies in the limit $t\to\infty$, x fixed. When perturbations around an absolutely unstable state are investigated, this is not the proper limit: to make contact with the work on front propagation, we should then analyze the equation in a frame moving with speed v'. This speed should be chosen self-consistently so that it agrees with the front speed predicted by the analysis, since only then we stay within the front region with an analysis based on keeping the co-moving coordinate $x'=x-vt$ fixed. In other words, if the analysis predicts the selection of a mode $e^{-i\omega t+ikx}$ in the lab frame, we need to choose

$$v' = \frac{\text{Im}\omega}{\text{Im}k} \,. \tag{4}$$

Consider now a system whose reference state is spatially homogeneous in the x-direction. If we linearize the dynamical equation about this state then we get upon Fourier transformation in space-time an expression for the Green's function $G(x', t)$ in the comoving frame $x' = x - v't$ of the form[6]

$$G(x', t) = \int_{C_\omega} d\omega' \int_{C_k} dk' \, \frac{e^{-i\omega' t + ik'x'}}{D(\omega', k')} \,. \tag{5}$$

Here the contour C_ω in the ω' plane is dictated by causality considerations, while the k' integration is along the real k' axis. The zeroes of $D(\omega', k')$ determine the dispersion relation $\omega'(k')$ which is related to the dispersion relation $\omega(k')$ in the fixed frame according to

$$\omega'(k') = \omega(k') - v'k' \,. \tag{6}$$

To extract the long-time behavior of the Green's function (5), we use the fact that the integration contours in (5) can be deformed continuously in an suitable manner. For an arbitrary fixed complex value of ω', $D(\omega', k')$ will have a number of zeroes in the complex k' plane. This situation is sketched in the rightmost column of Fig. 1. When the contour C_ω is deformed, the poles in the k' plane move, but we can deform the contour C_k continuously around the poles without changing the value of the integral *until* two poles "pinch off" the k' contour – see Fig. 1. As a result, the asymptotic $t \to \infty$ behavior of (5) is determined by points in the complex k' plane where two poles arrive from opposite sides of the real axis and pinch off the C_k contour.

The location of the pinch point where two roots coincide is given by[6,8]

$$\frac{d\omega'(k')}{dk'} = 0 \,. \tag{7}$$

Upon using (6) and the consistency requirement (4), this can be written as

$$\frac{d\omega(k')}{dk'} = v' = \frac{\text{Im}\omega(k')}{\text{Im}k'} \,. \tag{8}$$

From the real and imaginary part of this equation one recovers the linear marginal stability equations (3). Hence, *provided* one analyzes the evolution of perturbations in a comoving frame, the pinch point coincides with the linear marginal stability point, and $v' = v^*$.

An advantage of the pinch point analysis is that it can also be used to study perturbations in the lab frame or in any other frame moving with a nonzero velocity. Such an analysis is, of course, only useful as long as perturbations are still so small that a nonlinear front has not developed yet. Here, we focus on the aspects most relevant for a comparison with front propagation.

Fig. 1 illustrates the various differences between the marginal stability analysis[1,2] and the pinch point analysis.[6-8] *(i)* The former is conveniently formulated in terms of an analysis of the local wavenumber $q \equiv -i\,\partial(ln\phi)/\partial x$. The dynamical equation for q supports not only the intuitive interpretation of the dynamical velocity selection, but can also be used to derive the rate of approach to the asymptotic velocity v^*. The pinch point analysis, on the other hand, is a straightforward asymptotic analysis of the Green's function with the aid of contour integra-

	MARGINAL STABILITY	PINCH POINT
type of analysis	leading edge dynamics $\phi = e^{iu(x,t)}$, $q \equiv u_x$ $q_t = (f' - f_q q')\,q_x + \mathcal{L}q$	contour integration 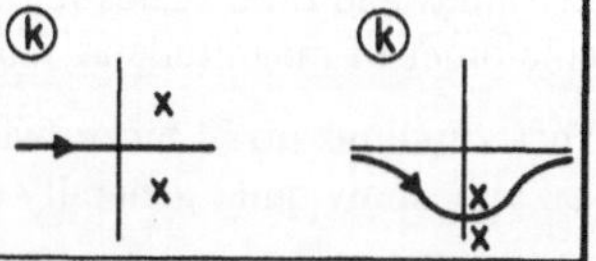
related conditions	maximum growth rate 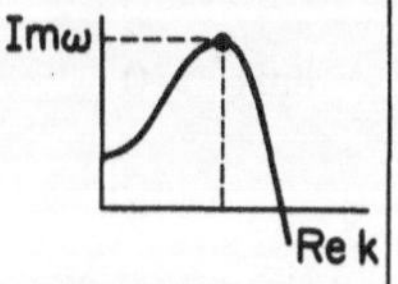	integration contour has to be "pinched" off by two poles
initial conditions	can be important	often ignored
importance of nonlinearities	linear marginal stability = ⇕ nonlinear marginal stability	pinch point ? (nonlinearities not considered)

Figure 1. Illustration of the difference between the marginal stability approach and the pinch point analysis.

tion. *(ii)* The requirement that the pinch point corresponds to the coalescence of two poles from *opposite* sides of the real axis translates in the linear marginal stability analysis into the requirement that in solving the condition $\partial \mathrm{Im}\omega / \partial \mathrm{Re}k = 0$, we always pick the solution that corresponds to an (absolute) maximum of $\mathrm{Im}\omega$ as a function of $\mathrm{Re}k$. *(iii)* In the marginal stability analysis, the fact that the initial conditions have to be sufficiently localized is recognized immediately, whereas in the pinch point analysis this is often overlooked (see, however, Lifshitz and Pitaevskii[8]). *(iv)* Finally, in the pinch point analysis, no attempt is made to consider the effects of nonlinearities, so that there is no regime equivalent to nonlinear marginal stability.

RESOLUTION OF DISCREPANCY BETWEEN THEORY AND EXPERIMENT

One of the first experimental tests of linear marginal stability in a convection experiment was done by Ahlers and Cannell.[12] They studied the velocity of a propagating Taylor vortex front close to the instability threshold. In that region, the dynamical behavior is described by an amplitude equation similar to (2). Surprisingly, however, the front velocity Ahlers and Cannell[12] observed was substantially smaller than predicted theoretically on the basis of the asymptotic front speed v^*. This discrepancy was only resolved recently by Niklas et al.[11], who noticed in numerical simulations that the approach of the front speed to the asymptotic value v^* is relatively slow, and that v^* is approached from *below*. Moreover, for parameter values and time scales relevant to the experiments, they observed deviations from v^* of the same order of magnitude as found numerically. Fig. 2 shows some of their results.

These findings tie in nicely with some of my own results on the rate of approach to v^*. Indeed, on may show quite generally that in the leading edge of the profile[2]

$$v(t) = v^* - \frac{3}{2\mathrm{Im}k^* t} + \cdots, \qquad \mathrm{Im}k^* > 0 . \tag{9}$$

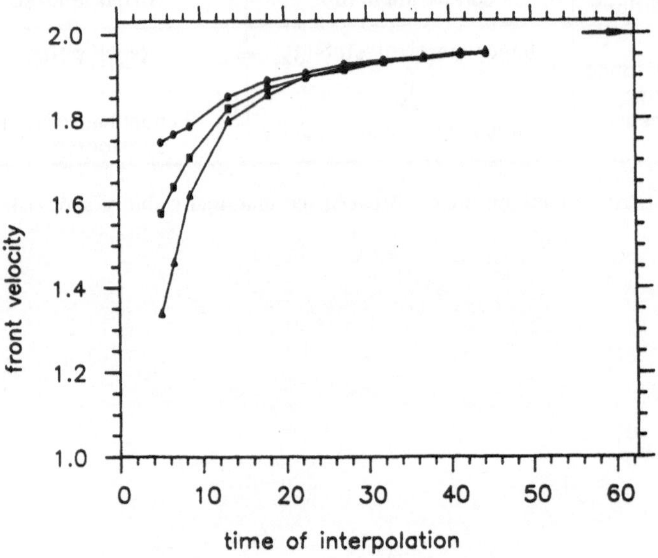

Figure 2. Numerical simulation results of Niklas et al.[11] for the dimensionless front velocity as a function of time for a propagating Taylor vortex front. The marginal stability value $v^* = 2$ is indicated by an arrow. Cirlces, squares and triangles denote the front velocity measured at 10%, 25% and 50% of the maximum amplitude.

Here v^* and k^* are determined from Eq. (3). This result shows that v^* is *generally* approached from *below*, and that the convergence is not exponentially fast, but with a power law. It is interesting to note that our result is different from the expression found in the pinch point literature, which corresponds to

$$v(t) = v^* - \frac{1}{2\mathrm{Im}k^* t} + \cdots \tag{10}$$

In Fig. 3 we show some of our results from numerical simulations of Eq. (2), which clearly support our expression (9) rather than (10). For a more extensive discussion of this and of how (9) might arise from the pinch point analysis, we refer to Ref. 2.

NONLINEAR MARGINAL STABILITY

As stated earlier, from studies of Eq. (1) it has been known since long that there are situations in which the nonlinearities drive the selected front speed to some value $v^\dagger > v^*$. To my knowledge, however, the first examples of this in model systems more complicated than (1) were given only recently in Ref. 2. In this paper, I formulated a general picture for this so-called nonlinear marginal stability mechanism. I will not discuss my arguments in detail here, but just note that although in the nonlinear marginal stability regime the velocity $v^\dagger$ is selected by the nonlinear terms in the equation, the selected profile in this regime is nevertheless argued to obey the following nontrivial constraint between $\mathrm{Re}k^\dagger$ and. $\mathrm{Im}k^\dagger$ obtained from the linear dispersion relation,

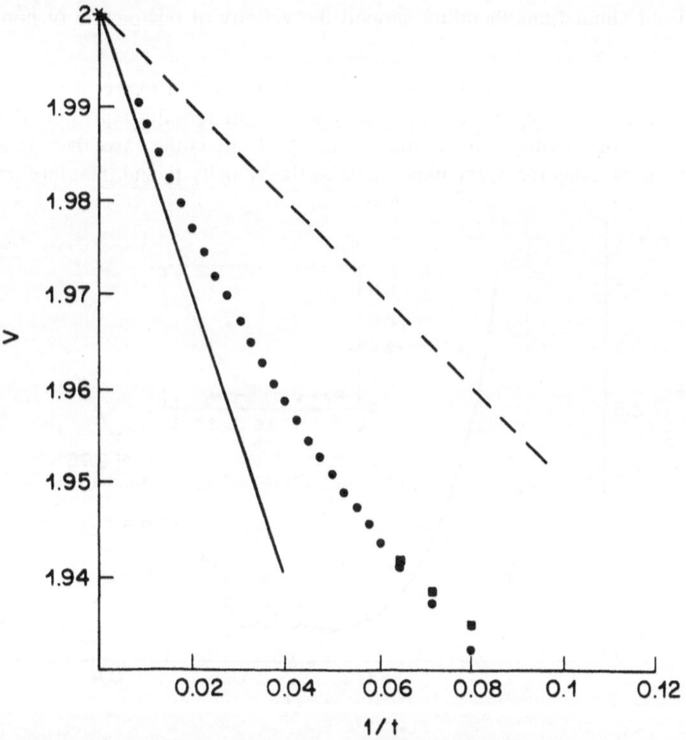

Figure 3. Front velocity as a function of $1/time$ for Eq. (2) with Gaussian initial conditions. The circles denote the velocity measured at $\phi = .05$, and the squares the velocity at $\phi = .5$. For $t^{-1} < .06$ the squares are not shown but the data fall on top of each other. The solid line indicates our result (9) and the dashed line expression (10).

$$\left.\frac{\partial \mathrm{Im}\omega}{\partial \mathrm{Re}k}\right|_{k=k^{\dagger}} = 0 . \qquad (11)$$

The above equation can be used to write e.g. Rek as a function of Imk, and according to our arguments, the solution relevant for the nonlinear marginal stability regime is the one corresponding to the root with the second smallest value of Imk. Together with the usual relation $v^{\dagger} = \mathrm{Im}\omega/\mathrm{Im}k^{\dagger}$, our arguments then imply that $v^{\dagger}$ and Imk should obey a particular relation in the nonlinear marginal stability regime.

We stress that while my explicit calculations of $v^{\dagger}$ for pattern forming equations that are not of the form (1) all rely on perturbative amplitude expansions, Eq. (11) is a nonperturbative result that can be tested even far into the nonlinear marginal stability regime. I have done so for the following extension of the Swift-Hohenberg equation,[18]

$$\frac{\partial \phi}{\partial t} = -2\frac{\partial^2 \phi}{\partial x^2} - \frac{\partial^4 \phi}{\partial x^4} + (\varepsilon - 1)\phi + b\,\phi^2 - \phi^3 . \qquad (12)$$

Fig. 4 presents some of our data for the velocity of fronts propagating into the unstable state $\phi = 0$ for $\varepsilon = 1/4$. The full line in the figure denotes the branch of solutions of (11) corresponding to the smallest root Imk, and the dashed branch the next smallest one. The minimum of the curve corresponds to the point v^*, Imk^*. The open circles are the data points from our simulations for various values of b as indicated, and the inset shows the measured value of the velocity as a function of b. Clearly, to within our numerical accuracy, the data lie on the dashed branch, and our simulations therefore support the validity of our picture of nonlinear marginal stability for Eq. (12).

We remark that although this was not noted in Ref. 2, I expect my analysis of the nonlinear marginal stability regime to apply only to pattern forming equations that admit a two-parameter family of moving front solutions. Indeed, if this is the case, there is a one-parameter family of front solutions for every fixed value of the velocity v, and therefore one expects there

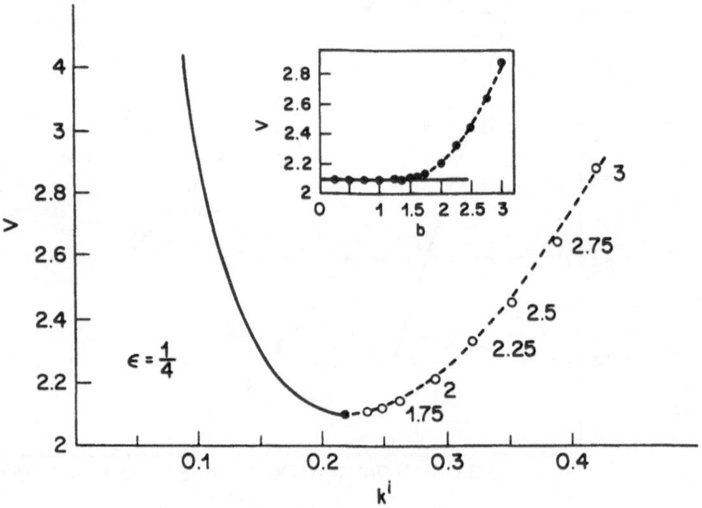

Figure 4. The measured front velocity v versus $k^i = \mathrm{Im}k$ for Eq. (12) with $\varepsilon = 1/4$. Open circles are data points for $b > 1.5$ with the value of b as indicated. The inset shows the velocity as a function of b. For $b \lesssim 1.5$, we have linear marginal stability, and for $b \gtrsim 1.5$ nonlinear marginal stability.

to be a particular moving front solution that satisfies the condition (11) at $v=v^{\dagger}$. For the Swift-Hohenberg equation, it is indeed known[19] that there is a two-parameter family of moving front solutions, and our picture appears to be consistent. I do not know, however, whether the nonlinear marginal stability analysis can be extended to equations that admit only a one-parameter family of solutions. Further study of this question appears necessary.

COMPLEX GINZBURG-LANDAU EQUATION NEAR A SUBCRITICAL BIFURCATION

Recently, Hohenberg and I[13] have obtained a number of exciting new results for the existence and stability of fronts and pulses in the complex Ginzburg-Landau equation near a critical bifurcation,

$$\frac{\partial A}{\partial t} = (1+c_1)\frac{\partial^2 A}{\partial x^2} + \varepsilon A + (1+c_3)\,|A|^2 A + (-1+c_5)\,|A|^4 A\,. \qquad (13)$$

In this investigation, both front propagation into an unstable state $\varepsilon>0$ and front propagation into a metastable state for $\varepsilon<0$ was analyzed. An extremely surprising finding of our work is that an exact analytic front solution of this equation can be found that allows one to compute $v^{\dagger}$ both for $\varepsilon>0$ and for $\varepsilon<0$ analytically. With the aid of this exact solution, a number of new and interesting additional results are obtained: *(i)* For fixed c_1, c_3 and c_5, there usually is a range of values of $\varepsilon<0$ where stable periodic nonlinear pulse solutions with stationary envelope are found. The upper edge of this range is determined by the condition $v^{\dagger}(\varepsilon)=0$. *(ii)* In a co-dimension one subspace of parameter space, we have obtained exact analytic nonlinear pulse solutions. The analytic form we find has subsequently been used by Niemela et al.[20] to successfully fit pulse shapes found in convection experiments. *(iii)* In large regions of the c_1, c_3, c_5 parameter space, nonzero initial conditions never dynamically develop into a front solution that propagates into the $A=0$ state for any $\varepsilon<0$. In this case, stable pulses continue to exist all the way up to $\varepsilon=0$. *(iv)* In this parameter range where pulses exist for all $\varepsilon \to 0^-$, all fronts propagating into the unstable state $A=0$ for $\varepsilon>0$ propagate with the linear marginal stability velocity v^*. This contradicts my earlier speculation that for arbitrarily small but positive ε, one would always expect fronts to propagate with $v^{\dagger}$ rather than v^* near a subcritical bifurcation.[2]

CONCLUSION

The results summarized in the last three section illustrate that important progress is being made concerning front propagation into unstable states. Moreover, as our recent work on the complex Ginzburg-Landau equation indicates, the distinction between front propagation into unstable states and metastable states may be less large than we originally believed. I expect these type of problems to remain a fruitful area for research in the next few years.

REFERENCES

1. W. van Saarloos, Phys. Rev. *A37*, 211 (1988).
2. W. van Saarloos, Phys. Rev. *A39*, 6367 (1989).
3. See e.g. P. Kolodner, this volume, and references therein.
4. M. C. Cross, Phys. Rev. Lett. *57*, 2935 (1986); Phys. Rev. *A38*, 3593 (1988).
5. See e.g. P. Kolodner, A. Passner, C. M. Surko and R. W. Walden, Phys. Rev. Lett. *56*, 2621 (1986).

6. A. Bers, in: *Handbook of Plasma Physics*, M. N. Rosenbluth and R. Z. Sagdeev, eds. (North-Holland, Amsterdam, 1983).

7. P. Huerre, in: *Propagation in Systems far from Equilibrium*, J. E. Wesfreid, H. R. Brand, P. Manneville, G. Albinet, and N. Boccara, eds. (Springer, New York, 1988).

8. E. M. Lifshitz and L. P. Pitaevskii, *Physical Kinetics*, Course of Theoretical Physics, (Pergamon, New York, 1981), vol. 10, chap. VI.

9. G. Dee and J. S. Langer, Phys. Rev. Lett. *50*, 383 (1983).

10. E. Ben-Jacob, H. R. Brand, G. Dee, L. Kramer and J. S. Langer, Physica *14D*, 348 (1985).

11. M. Niklas, M. Lücke and H. Müller-Krumbhaar, Phys. Rev. *A40*, 493 (1989).

12. G. Ahlers and D. S. Cannell, Phys. Rev. Lett. *50*, 1583 (1983).

13. W. van Saarloos and P. C. Hohenberg, to be published.

14. A. Kolmogorov, I. Petrovsky and N. Piskunov, Bull. Univ. Moskou, Ser. Internat., Sec. A *1*, 1 (1937), reprinted in: *Dynamics of Curved Fronts*, P. Pelcé, ed. (Academic, San Diego, 1988).

15. R. A. Fisher, Ann. Eugenics *7*, 355 (1937).

16. D. G. Aronson and H. F. Weinberger, Adv. Math. *30*, 33 (1978).

17. See e.g. G. S. Triantafyllou, K. Kupfer and A. Bers, Phys. Rev. Lett. *58*, 1914 (1987).

18. J. Swift and P. C. Hohenberg, Phys. Rev. *A15*, 319, (1977).

19. P. Collet and J. P. Eckmann, Commun. Math. Phys. *107*, 39 (1986).

20. J. J. Niemela, G. Ahlers and D. S. Cannell, to be published.

COLLAPSING SOLUTIONS IN THE 3-D EULER EQUATIONS

Alain Pumir[1] and Eric D. Siggia[2]

[1] L. P. S., E.N.S. 24, rue Lhomond, 75231 Paris Cedex, France
[2] LASSP, Cornell University, Ithaca, N.Y. 14853, USA

Intermittent fluctuations in a turbulent flow constitute one of the most original, and also challenging problems in fluid mechanics[1]. They have been observed in many experimental situations, and it has been established that they play a very important role in many processes of great practical importance. As an example, it has been shown experimentally that about a half of the Reynolds stress in turbulent boundary layer flows is produced during the 'bursting' events[2]. These events are associated with the generations of small scales in the flow. Indeed, variations of the velocity field over extremely small length scales (even smaller than the Kolmogorov length !) have been reported in turbulent boundary layer flows[3]. Numerical simulations of turbulent channel flow at much more modest Reynolds numbers have revealed similar features[4].

So far, a detailed understanding of these phenomena remains a challenge. In a previous paper[5], it has been suggested that singularities in the fluid equations may provide a paradigm for many properties of turbulent flows. Mathematically, one is faced with a deep and still unsolved problem, namely the problem of regularity of the solution of the 3-d Euler and Navier-Stokes equations[6,7]. The nonlinear term in the fluid equations plays a crucial role, and very little is known about the mechanisms it induces. In particular vortex stretching is expected to make the vorticity to grow, possibly in an unbounded way.

As it is clear when one writes the Euler equations in term of the vorticity field :

$$\frac{D\omega_i}{Dt} \equiv \partial_t\omega_i + (\vec{u}.\nabla)\omega_i = s_{ij}\omega_j$$

where D/Dt denotes the Lagrangian derivative, ω_i is the ith component of the vorticity and $s_{ij} = 1/2\ (\partial_i v_j + \partial_j v_i)$ is the (symmetric) rate of strain tensor. At each point, the vorticity field can be decomposed according to the eigenvectors of the rate of strain tensor. Obviously, the components corresponding to eigendirections with positive eigenvalues will be amplified. The rate of vorticity stretching thus depends on the eigenvalues of s, and on the angles between the vorticity vectors and the eigenvectors of s. It is therefore of fundamental importance to understand the correlations between vorticity and the rate of strain tensor.

The Kolmogorov analysis of turbulent flows suggests that in the inviscid limit, singularities should exist. Actually the question has not been settled mathematically, despite more than 50 years of effort. In fact very little is known concerning the strictly inviscid case of the Euler equations. It has been proven by Beale et al.[8] that a solution that diverges in a finite time must lead to unbounded values of vorticity, hence to infinite gradients and to the generation of very small scales. The difficulties come from both the nonlinear and non local character of the 3-d fluid equations. Numerical analysis can provide new insight on these questions, like in many other fields in nonlinear sciences. In order to make some progress, we have studied simple flow configurations that can lead to the formation of very large gradients (very small scales). By following their evolution, one can try to understand the relevant physical mechanisms. We summarize here the results of a numerical attempt to simulate collapsing solutions of the fluid equations. A more detailed report will be published elsewhere[9].

In order to simulate solutions focusing on vanishingly small scales, one has to devise appropriate numerical methods. Our previous experience with a fixed mesh, spectral code has taught us that it is hard to track reliably solutions generating very small spatial scales [10]. Our overriding concern here is to simulate a large range of scales. In order to follow the solution as it goes to smaller and smaller scales, the number of modes should not increase significantly as the collapse proceeds. In addition an accurate way is needed to reallocate grid points so as to always have the flow fully resolved. Essentially we require that the dynamics is sufficiently local. For a singularity to occur requires real nonlinearity : if the small scale vorticity containing

Nonlinear Evolution of Spatio-Temporal Structures in Dissipative Continuous Systems
Edited by F.H. Busse and L. Kramer
Plenum Press, New York, 1990

modes are acted upon by a strain field which is dominated by the large scales, only exponential growth will result. One could however imagine weak singularities which arise when octave of scales, $2^{-n-1} < \lambda < 2^{-n}$ make a constant contribution to the strain rate. Examination of our flow field suggests that this is not occuring. In order to simulate more scales than is possible with a fixed mesh code, some approximation has to be made.

The above considerations led us to write the following code. Separable rectangular coordinates are emloyed. The real line is mapped onto the interval $\xi \in (-1, 1)$ with a suitable function and a uniformly spaced grid is laid down in ξ. Derivatives are computed with centered differences (second order accurate) which are so arranged that the discrete equations of motion explicitly respect incompressibility and conserve energy and momentum. Since we are ultimately interested only in small scales we chose boundary conditions with velocity and vorticity zero at infinity. The time stepping is also second order accurate.

When gradients became too large we remeshed the flow by adjusting the scales in the coordinate transformation, which can be done separately in the x,y and z direction. Interpolation from one mesh to another is done with cubic splines. Vorticity, strain rate and velocity are monitored before and after to assess errors. For the Euler equations, we had no difficulty in maintaining good resolution around the points of maximum vorticity.

Our study of the Biot-Savart model for one or several vortex filaments [11] showed that a singularity would occur when two pieces of filaments paired antiparallely. The collapse that followed was universal i.e., independent of initial conditions. We therefore initialized our Euler code with paired vortex filaments. The range of parameters over which we were able to follow the flow is shown in table 1.

Table 1. Ranges of values obtained in our simulation

	T = 0.	T = 8.
x-scale	0.5	$1.\ 10^{-3}$
y,z-scale	1.0	1.10^{-2}
max vel.	0.5	1.0
$\lvert\omega\rvert_{max}$	3.	100.
$\omega.e.\omega_{max}$	2.5	$6.\ 10^{+3}$

The time dependence of the inverse maximum vorticity and normalised strain rate along the vorticity are shown in Fig.1a,b.

Several comments are in order. Firstly it is apparent that the vorticity initially grows faster than exponential, as is necessary for a finite time singularity, but then crosses over and becomes exponential. This is obvious from Fig.1b, which shows the rate of growth of vorticity, namely $\omega.s.\omega_{max}/\omega^2_{max}$. The latter quantity must diverge if there is to be a finite time singularity. Instead, a saturation is observed (the decrease at late stages of $\omega.s.\omega_{max}/\omega^2_{max}$ is due to the increasing separation of the locations of the two maxima). Although we find no evidence of a singularity here, the range of scales explored is very large. If an adaptative mesh scheme was not used, a fixed mesh of order $6.10^4 \cdot 5.10^3 \cdot 5.10^3$ would have been necessary, which of course is absurdly large. In practice we were able to do the entire run with approximately a 100^3 mesh.

The limited growth in velocity strongly suggests that the solution would not lead to singularities in the Navier-Stokes equations. Comparison of the mean square strain with its component along the vorticity direction suggests that the eigenvalues of the strain matrix are proportional to $(1, \varepsilon, -1-\varepsilon)$, $0 < \varepsilon << 1$ and that the vorticity is parallel to the direction of weak stretching. Similar results have been obtained from the numerical solution of a turbulent, fully resolved 3 dimensional flow [12]. The strongest stretching is thus perpendicular to the vorticity. The resulting enstrophy production is therefore rather weak.

In figure 2, we show 3 dimensional surface plots of the regions within which the vorticity is larger than some threshold. The vortex tubes have become paired vortex sheets. Figure 3 shows the vorticity vectors in regions where the vorticity is most intense. The alignment of vorticity with the z-axis is obvious. Also, the wide difference between the x, y and z scales suggest that the solution is mostly 2 dimensional. Our simulations also reveal that the most amplified regions of the flow contain less and less circulation as times goes on. As a result, the nonlinearity induced by the small, fast scales of the motion are becoming less and less efficient, and are eventually dominated by the nonlinearities originating from the large, slow scales of the flow. The crossover to exponential growth can be understood fluid mechanically as the concentration of the region of maximum vorticity into a thin sheet which contains negligible circulation. Hence, the maximum

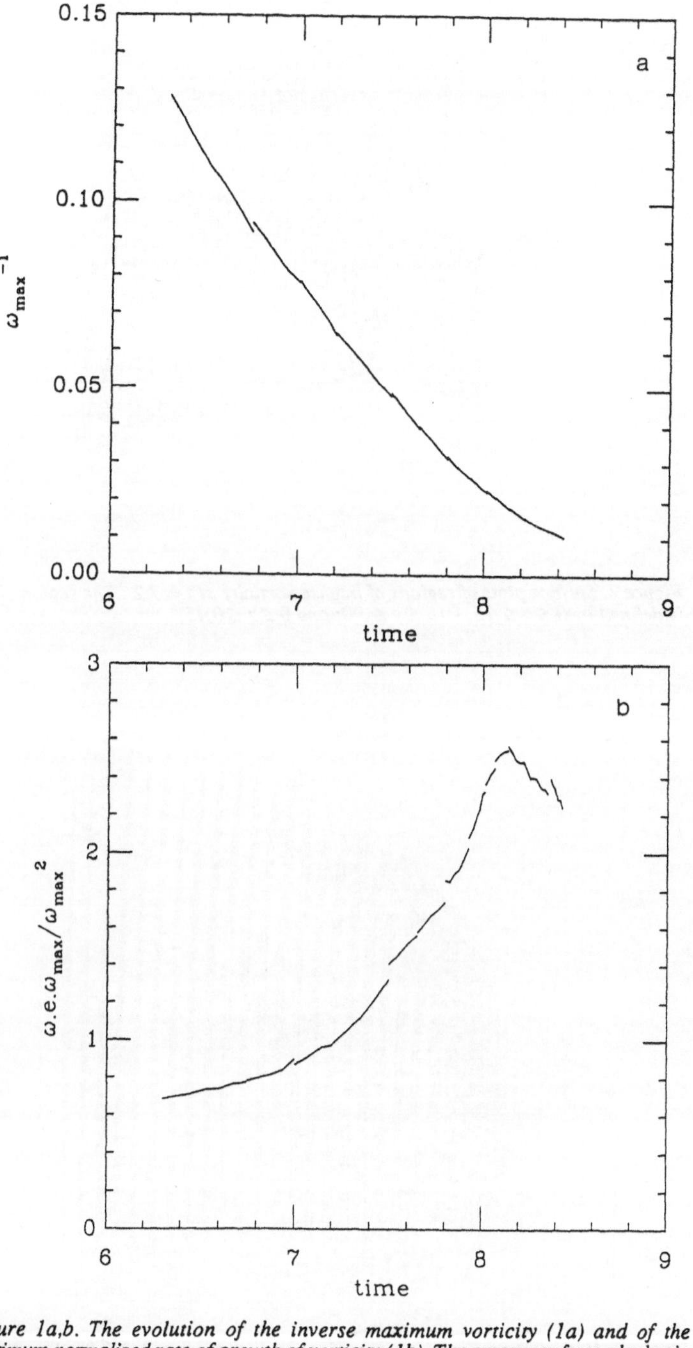

Figure 1a,b. The evolution of the inverse maximum vorticity (1a) and of the maximum normalized rate of growth of vorticity (1b). The crossover from algebraic to exponential growth is obvious from the decrease of the slope in fig. 1a, and from the saturation in fig. 1b.

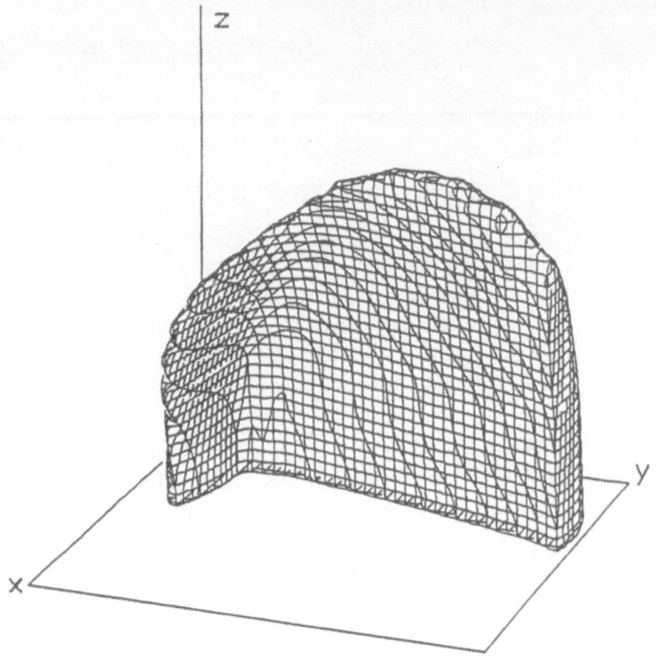

Figure 2. Surface plots of regions of intense vorticity at $t = 7.2$. The region shown is defined by $0 \leq x \leq .15$; $-.15 \leq y \leq .90$; and $0 \leq z \leq 0.45$.

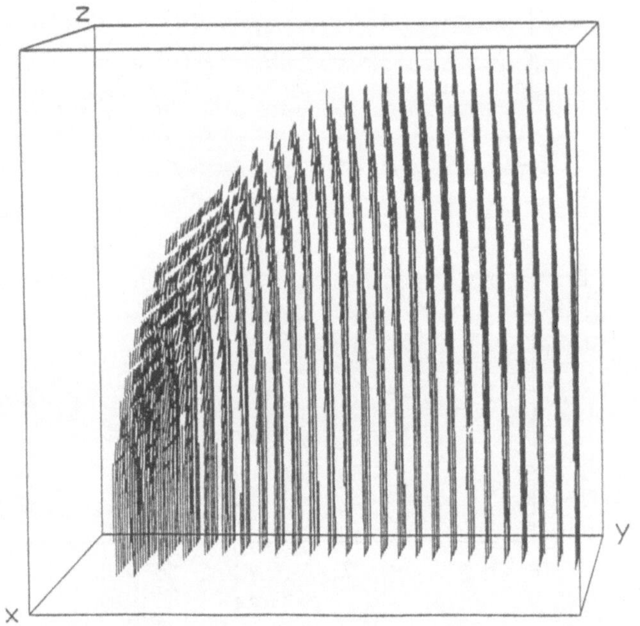

Figure 3. Vorticity vectors in regions of intense vorticity at $t = 7.2$. The region shown is defined by $0 \leq x \leq .15$; $-.15 \leq y \leq .90$; and $0 \leq z \leq 0.45$. The alignment of the vorticity field with the z axis is obvious.

vorticity eventually undergoes passive stretching and whether the amplification in the region containing most of the circulation itself eventually diverges is unclear.

The paired sheets never become unstable because there is always some self induced stretching present. The time of integration is many times the spacing divided by velocity so any instabilities present would certainly have time to develop.

Although we have succeeded in simulating very large gradients and a very dramatic contraction of spatial scales, we have not found any evidence that our solution must collapse in a finite time. Of course, our calculation does not disprove the existence of such solutions. There may very well be unstable collapsing solutions (as is suggested by the initial rapid growth of vorticity, followed by the exponential amplification). Tracking such solutions would require a better understanding of the phase space. The weakening of the effective nonlinearity in our flow field is itself of great interest. Depression of nonlinearity has been reported in other circumstances[13,14]. In our case, it is due to the fact that the stretching when amplifying the vorticity field, also concentrates the circulation on (locally) 2-dimensional sheets, that do not generate self stretching.

Other initial conditions have been studied as well, details can be found in reference 9.

References

1) A.A. Townsend, Proc. Roy. Soc. **A208**, 534 (1951)
2) W. Willmarth and S. Lu , J. Fluid Mech. **48**, 775 (1973)
3) W. Willmarth and T. Bogar, Phys. Fluids **20**, S9 (1977)
4) J. Kim, Phys. Fluids **28**, 52 (1984)
5) A. Pumir and E.D. Siggia, Physica **D37**, 539 (1989)
6) J. Leray, Acta Math. **63**, 193 (1934)
7) O. Ladyzhenskaya, The mathematical theory of viscous incompressible flows, Gordon and Breach (1963)
8) J. T. Beale, T. Kato and A. Majda, Commun. Math. Phys. **94**, 61 (1984)
9) A. Pumir and E.D. Siggia, to be published in Phys. Fluids.
10) A. Pumir and R.M. Kerr, Phys. Rev. Lett. **58**, 1963 (1987)
11) E.D. Siggia and A. Pumir, 1985, Phys. Rev. Lett. **55**, 1749 (1985) and Phys. Fluids **30**, 1606 (1987)
12) W. Ashurst, A. Kerstein, R. Kerr and C. Gibson, Phys. Fluids **30**, 2343 (1987)
13) R.H. Kraichnan and R. Panda, Phys. Fluids **31**, 2395 (1988)
14) A. Tsinober, oral presentation at the conference 'Turbulence 89' held in Grenoble (France), September 1989

LARGE MAGNETIC REYNOLDS NUMBER DYNAMO ACTION IN STEADY SPATIALLY PERIODIC FLOWS

A.M. Soward[1] and S. Childress[2]

[1] Department of Mathematics and Statistics
The University
Newcastle upon Tyne, NE1 7RU, U.K.

[2] Courant Institute of Mathematical Sciences
New York University
251 Mercer Street
New York, NY 10012, U.S.A.

ABSTRACT

Dynamo action at large magnetic Reynolds number R is very sensitive to the streamline topology. A starting point for a systematic study is the 2-dimensional steady spatially periodic flow $\mathbf{u}' = (\partial\psi'/\partial y, -\partial\psi'/\partial x, K\psi')$, $\psi' = \sin x \sin y + \delta \cos x \cos y$, where δ, K = constants. Slow dynamo action is possible within the spiralling vortices, and individual modes are localised in the neighbourhood of particular streamsurfaces. When $\delta = 0$, almost fast dynamo action occurs in boundary layers of width $R^{-1/2}$ containing the streamline separatrices. These connect the stagnation points and bound adjacent vortices. This streamline pattern is structurally unstable, and so, when $0 < \delta << 1$, channels emerge between the vortices. With the addition of mean motion $\bar{\mathbf{u}} = (M,N,0)/(M^2+N^2)^{1/2}$, where M,N = relatively prime integers, $\epsilon << 1$, but $\delta = 0$, the streamlines connected to the X-type stagnation points bound channels with multiplicity of order $L = M + N$, and are dense in the *irrational* limit $L \to \infty$. Dynamo action in all these systems is discussed. Results for the latter case shed light on dynamo processes, which may occur in the fully 3-dimensional ABC-flows. There the flow downstream of stagnation points on certain 2-dimensional manifolds provides the sites of possible fast dynamo action. The manifolds form dense subregions, where the fluid particles paths are chaotic.

Nonlinear Evolution of Spatio-Temporal Structures in Dissipative Continuous Systems
Edited by F.H. Busse and L. Kramer
Plenum Press, New York, 1990

1. INTRODUCTION

In dimensionless units based on the typical velocity U, length L, and convective time $T_C(=L/U)$, the generation of magnetic field **B** in an electrically conducting fluid (magnetic diffusivity η) moving with velocity **u** is governed by the magnetic induction equation

$$\partial \mathbf{B}/\partial t = \nabla \times (\mathbf{u} \times \mathbf{B}) + R^{-1}\nabla^2\mathbf{B}, \qquad \nabla\cdot\mathbf{B} = 0, \tag{1.1a}$$

where $R(=UL/\eta)$ is the magnetic Reynolds number. When R is large,

$$R \gg 1, \tag{1.1b}$$

the magnetic diffusion time scale $T_D(=L^2/\eta)$ is much larger than the convective time $T_C(=T_D/R)$. In the case of steady flow, magnetic modes proportional to exp(pt) may be sought. Those modes which grow on the convective time scale, $pT_C = O(1)$, are said to be fast. Otherwise, when $0 < pT_C = O(1)$, they are slow. A central issue in dynamo theory is whether magnetic field can be systematicaly generated on a convective time scale as necessary in many astrophysical applications.

A balance of convection and diffusion can only occur when the magnetic field varies on a length scale of order $R^{-1/2}$. Typically this balance is met in boundary layers. Elsewhere, convection dominates and at lowest order the magnetic field is frozen to the fluid. This we will refer to as the mainstream. In steady flows, to which we restrict attention throughout this article, the formation and location of boundary layers is a delicate matter linked to the streamline topology. Moffatt and Proctor[1] have argued that field generation on an $R^{-1/2}$-length scale is essential for fast dynamo action. Consequently the presence of the boundary layers is an essential ingredient and the raison d'étre for any fast dynamo. A dynamo with only mainstream features like the slow nearly axially symmetric dynamo of Braginsky[2] and the screw dynamo of Ruzmaikin et al.[3], and Gilbert[4] must be slow. Both these models operate on length scales long compared to $R^{-1/2}$.

Some of the important topological aspects of the dynamo are apparent even in the simple case of 2-dimensional flows

$$\mathbf{u} = \mathbf{u}_H + w\hat{\mathbf{z}}, \qquad \mathbf{u}_H = \nabla\psi \times \hat{\mathbf{z}}. \tag{1.2}$$

Here $\mathbf{u}_H$ defines motion in a horizontal x,y-plane, while $w(x,y)$ is the velocity in the vertical z-direction; the hat is used to denote a unit vector. All motion is confined to streamsurfaces $\psi(x,y)$ = constant. Throughout this article we focus attention on such flows and demand in addition that they are spatially periodic. The topology of the flow is determined by the X-type stagnation points of the horizontal motion, when they exist, and the streamlines that pass through them. For our spatially periodic flows, there is at least one closed streamline connected to a stagnation point. It defines the boundary of a cell inside which all motion is on closed streamsurfaces. In addition to the closed eddies there may be open channels in which motion is not recirculating but proceeds indefinitely through the array of eddies. The boundaries of the channels are defined by the streamlines through the X-type stagnation points. When $w[=w(\psi)]$ is constant on streamsurfaces, slow dynamo modes are sometimes possible, which are localised near a particular closed streamsurface in the eddy interiors. This non axisymmetric extension[5] of the screw dynamo[3,4], is described in Section 3 below. Essentially helical field lines move with the flow on this resonant streamsurface and are eliminated elsewhere by the flux expulsion mechanism.

Roberts[6] studied the generation of magnetic field by the particularly symmetric flow characterised by

$$w = K\psi', \qquad \psi = \psi' = \sin x \sin y, \tag{1.3}$$

with K = constant. The eddies fill squares $\Pi_{m,n}=[m\pi,(m+1)\pi] \times [n\pi,(n+1)\pi]$, there are no channels and the cell boundary streamline consists of a heteroclinic orbit through four X-type stagnation points. A boundary layer approach to this problem was developed by Childress[7] and refined by Soward[8] to show that almost fast dynamo action is possible. In these models magnetic flux is expelled from the mainstream eddies by the process illustrated numerically by Weiss[9].

Childress and Soward[10] applied the boundary layer method[7] to the flow (1.2) with

$$w = K\psi', \qquad \psi' = \sin x \sin y + \delta \cos x \cos y, \qquad \delta \ll 1, \tag{1.4}$$

where K and δ are constants. The inclusion of the term proportional to δ breaks a symmetry of the Roberts flow (1.3). Now the heteroclinic orbit containing the eddies only passes through two X-type stagnation points and the other two give way to a narrow ($\delta \ll 1$) channel region. The resulting cat's-eye configuration is illustrated in Figure 1. Another periodic motion discussed by Soward and Childress[11] invokes the addition of slow mean motion

$$\bar{u}_H = (\bar{u}_x, \bar{u}_y), \qquad \bar{w} = 0, \qquad |\bar{u}_H| \equiv \epsilon \ll 1 \qquad (1.5)$$

to the Roberts flow (1.3). Now the eddies are generally bounded by homoclinic orbits, which begin and end at an X-type stagnation point. The other two streamlines connected to the X-type stagnation point, generally separate two neighbouring channels (see Figures 2 and 3). For both the cat's-eye and mean flow models the X-type stagnation points trigger boundary layers, which contain the streamlines through them, and they provide the sites of any fast dynamo activity. These models are described in Section 4. They are characterised by the flow (1.2) with

$$w = K(\psi - \bar{\psi}), \qquad \psi = \psi' + \bar{\psi}, \qquad \bar{\psi} = y\bar{u}_x - x\bar{u}_y, \qquad (1.6)$$

where ψ' is given by (1.4).

The topology of our spatially periodic 2-dimensional flows is very simple with all motion lying on streamsurfaces. For fully 3-dimensional flows like the Beltrami ABC-flow

$$u = (A \sin z + C \cos y,\ B \sin x + A \cos z,\ C \sin y + B \cos x), \qquad (1.7)$$

the streamline structure is more complicated. Parts of the flow still lie on streamsurfaces, but there are other regions, in which the particle paths are chaotic[12]. There exponential stretching occurs, just as in our 2-dimensional cases, on the streamlines through the X-type stagnation points. It is widely believed that such flows will yield fast dynamos[13,14]; though it is not yet established. The importance of exponential stretching for the fast dynamo process was first demonstrated by simple rope models[15,16]. These fundamental ideas form the basis of the recent theoretical abstractions[17,18] which employ Baker's maps[19].

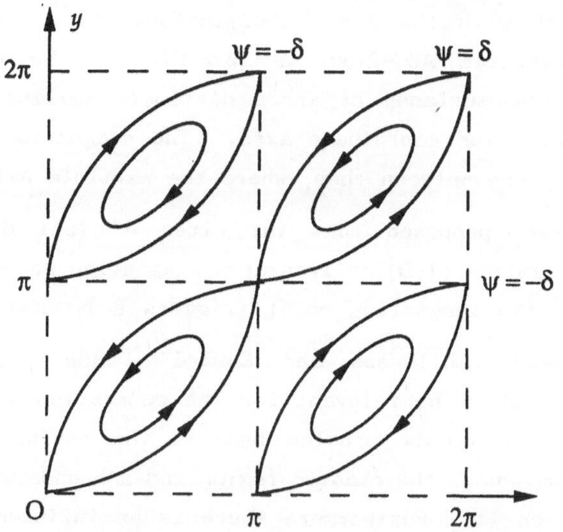

Fig. 1. The cat's-eyes. The values of ψ on the separtrices are indicated.

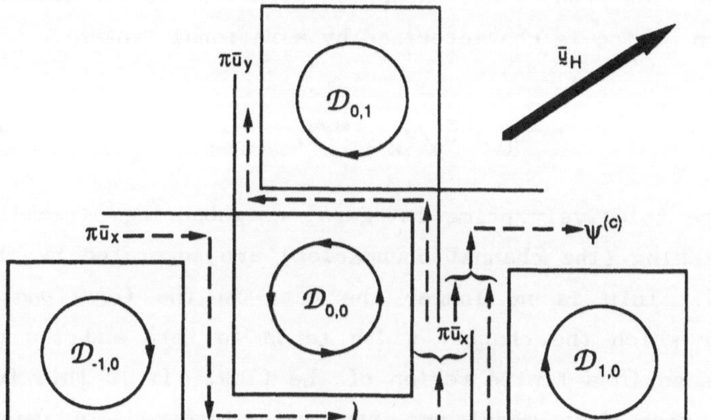

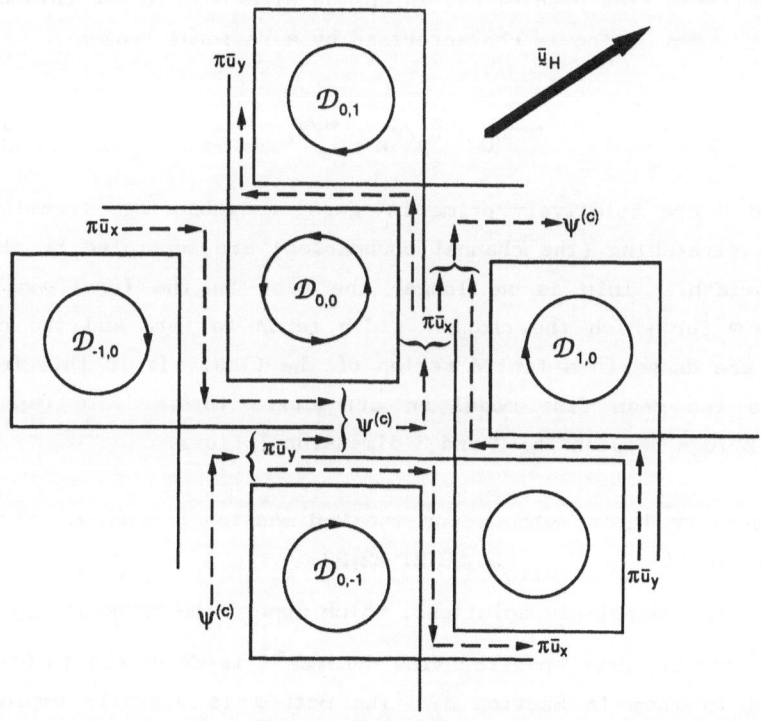

Fig. 2. Mean flow $\bar{u}_H$. The fluid fluxes between the streamlines through the X-type stagnation points of the eddy $\mathcal{D}_{0,0}$ and its immediate neighbours are indicated.

To form a link with the 2 and 3-dimensional flows, we outline the nature of the symmetric ABC-flow, $A = B = C = 1$. Motion is largely confined to the streamsurfaces of three distinct families of spiralling vortices aligned with the coordinate axes. The stagnation points of the flow lie in the regions between them, where the particle paths are largely chaotic. Childress[7] proposed that the sites of fast dynamo activity occured when a streamline (1-D) arrives at a stagnation point and leaves on a surface (2-D). The stagnation point triggers a boundary layer on the downstream surface. Childress and Soward[20] made estimates of an alpha-effect, which might be relevant for the generation of mean magnetic field. This is a difficult problem because the surface spreads in a complicated way throughout the chaotic region and a boundary layer solution is required valid on it. Furthermore there is an infinite but countable number of these surfaces and they are likely to be dense in a substantial part of the chaotic region. One boundary layer will interact to a greater or lesser extent with another depending on the downstream distance from the corresponding stagnation point. Curiously this is a feature that the 2-dimensional mean flow problem has in common with it. To see this we note than when the mean motion is characterised by a rational tangent

$$\tan\theta_0 = \bar{u}_y/\bar{u}_x = N/M, \tag{1.8}$$

where M and N are relatively prime integers, neighbouring streamlines of exponential stretching (the channel boundaries) are separated by channels of finite width. This is no longer the case in the *irrational* limit $L = M + N \to \infty$ for which the channel width tends to zero and the channel boundaries are dense in a finite region of the flow. It is this feature, which makes the mean flow model an attractive intermediate case. It provides a bridge between the 2 and 3-dimensional flows.

The boundary layer methods for the 2-dimensional problems have been developed[7,8,10,11] and reviewed[20,21,22,23] elswhere. Here we devote attention to the mainstream solutions, which append the boundary layers. A technique[5] recently developed for slow dynamos[24] is described in Section 2 and applied to them in Section 3. The method is usefully employed in Section 4 to find the mainstream eddy and channel solutions. The magnetic fields in the channels, fixed by jump conditions across their boundaries, can be used to calculate the alpha-matrices (see (4.35), (4.38) below), which determine the growth of mean magnetic field.

2. MATHEMATICAL FORMULATION AND AVERAGING PROCEDURES

The basic idea underlying our mainstream solutions is that the magnetic field is almost frozen to the flow. The method employed here to study magnetic induction by the 2-dimensional steady flow (1.2) is a special case of a more general procedure[5]. In our applications, we define a *primary* solenoidal flow

$$\mathbf{U} = \nabla\psi \times \nabla\xi = \mathbf{u}_H + v\hat{\mathbf{z}}, \tag{2.1a}$$

in which

$$\xi = z - \theta(\mathbf{x}_H), \qquad v = \mathbf{u}_H\cdot\nabla\theta, \tag{2.1b}$$

where θ, chosen later at our convenience, is a function of the horizontal coordinates $\mathbf{x}_H = (x,y)$. The remaining part of the motion

$$\tilde{\mathbf{u}} = \tilde{w}\hat{\mathbf{z}}, \qquad \tilde{w}(\psi) = w - v, \tag{2.2a}$$

defines the *secondary* flow and as indicated will be a function of ψ alone. The complete flow is

$$\mathbf{u} = \mathbf{U} + \tilde{\mathbf{u}}. \tag{2.2b}$$

The magnetic field is resolved parallel and perpendicular to the *primary* velocity in the form

$$\mathbf{B} = B_{\parallel}\mathbf{U} + |\mathbf{U}|^{-2}(B_\psi\nabla\xi - B_\xi\nabla\psi) \times \mathbf{U}, \tag{2.3}$$

where

$$B_{\parallel} = \mathbf{B}\cdot\mathbf{U}/|\mathbf{U}|^2, \qquad B_\psi = \mathbf{B}\cdot\nabla\psi, \qquad B_\xi = \mathbf{B}\cdot\nabla\xi. \tag{2.4}$$

The velocity decomposition is not unique. Provided, however, that we arrange the *secondary* velocity to be small ($|\tilde{w}| \ll |\mathbf{U}|$), we expect that the component $B_{\parallel}\mathbf{U}$ to dominate in (2.3) and so we call it the *primary* magnetic field. The remaining perpendicular components define the *secondary* magnetic field.

We assume that each *primary* streamline $C(\psi,\xi)$ is periodic and invariant under the same constant coordinate shifts

$$\mathbf{X} = \mathbf{X}_H + Z\hat{\mathbf{z}}. \tag{2.5}$$

The corresponding periodicity sections $\overline{C}(\psi,\xi)$ are either closed loops when $\mathbf{X} = 0$, or open arcs when $\mathbf{X} \neq 0$. For any periodic function ϕ invariant under the shift $\mathbf{X}$, we define the average

$$\langle\phi\rangle(\psi,\xi,t) = T^{-1}\int_{\overline{C}} \phi|\mathbf{u}_H|^{-2}\mathbf{u}_H\cdot d\mathbf{x}, \qquad T(\psi) = \int_{\overline{C}}|\mathbf{u}_H|^{-2}\mathbf{u}_H\cdot d\mathbf{x}, \tag{2.6a}$$

where T is the transit time of $\overline{C}$. We also note that the element of area $\delta\Sigma$ between the horizontal projections of the two neighbouring sections $\overline{C}(\psi+\delta\psi, \xi)$ and $\overline{C}(\psi,\xi)$ is related to T by

$$d\Sigma/d\psi = T. \tag{2.6b}$$

We introduce the circulation integral

$$\gamma(\psi) = T\langle|\nabla\psi|^2\rangle = \int_{\overline{C}} \mathbf{u}_H\cdot d\mathbf{x} \tag{2.7a}$$

and note that it satisfies

$$d\gamma/d\psi = T\langle\nabla^2\psi\rangle. \tag{2.7b}$$

We define the helicity

$$-\Gamma = \mathbf{U}\cdot\nabla \times \mathbf{U} = -v\nabla^2\psi + \nabla v\cdot\nabla\psi \tag{2.8a}$$

and note that, when v is a function of ψ alone, (2.7) gives

$$-T\langle\Gamma\rangle = -v(d\gamma/d\psi) + \gamma(dv/d\psi), \qquad v = v(\psi). \tag{2.8b}$$

In all our applications, the lowest order approximations to the magnetic induction equation yield

$$(\mathbf{U}\cdot\nabla)\,(B_{\parallel},B_{\psi},B_{\xi}) = 0. \tag{2.9}$$

Since each component of magnetic field is almost constant on $\overline{C}$-contours, it is convenient to define the average magnetic field by its component average. The result for the solenoidal magnetic field is

$$\langle \mathbf{B} \rangle = \langle B_{\parallel} \rangle \mathbf{U} + T^{-1} |\mathbf{U}|^{-2} \nabla (T \langle \chi \rangle) \times \mathbf{U}, \tag{2.10}$$

where $\langle \chi \rangle$, like $\langle B_{\parallel} \rangle$, is a function of ψ, ξ, t alone and it is the scalar potential which defines the components $\langle B_{\psi} \rangle$, $\langle B_{\xi} \rangle$. Equations governing the evolution of $\langle B_{\parallel} \rangle$ and $\langle \chi \rangle$ are obtained by resolving the magnetic induction equation parallel and perpendicular to **U**. The latter equation, which determines the weak *secondary* magnetic field, is more interesting. It is

$$D\langle \chi \rangle / Dt = R^{-1} \langle -(\nabla \times \mathbf{U}) \cdot \mathbf{B} + \nabla \cdot (\mathbf{U} \times \mathbf{B}) \rangle, \tag{2.11}$$

where the material derivative is

$$D/Dt = \partial/\partial t + \tilde{w}\, \partial/\partial \xi \tag{2.12a}$$

and

$$\langle B_{\psi} \rangle = \partial \langle \chi \rangle / \partial \xi . \tag{2.12b}$$

This concludes the summary of the mathematical apparatus[5] required in the subsequent sections.

3. SLOW MODES

In this section we restrict attention to flows of the type (1.3) and (1.4), for which the vertical velocity w is constant on stream surfaces. We consider the possibility that magnetic field can be generated in the neighbourhood of some streamsurface $\psi = \psi_0$ in the interior of a closed eddy. We anticipate that the *primary* magnetic field is helical, periodic under the shift

$$\mathbf{X} = Z\hat{\mathbf{z}} \qquad (\mathbf{X}_H = 0), \tag{3.1}$$

and propagates as a wave with constant velocity W parallel to the z-axis. To apply the mathematical apparatus of Section 2 we must work in the moving frame in which the field is stationary and the vertical velocity is

$w(\psi) - W$. The corresponding component of U in (2.1a) is

$$v(\psi) = Z/T. \tag{3.2}$$

Values on $\psi = \psi_0$ are denoted by the subscript 0. To minimise the *secondary* velocity (2.2a) in its vicinity we demand that

$$\tilde{w}_0 = 0, \qquad (d\tilde{w}/d\psi)_0 = 0. \tag{3.3}$$

Given ψ_0, these conditions with (2.2a) and (3.2) fix the values of W and Z by the formulas

$$W = w_0 - v, \qquad (dT/d\psi)_0 = -(T_0{}^2/Z)(dw/d\psi)_0. \tag{3.4}$$

It follows that close to $\psi = \psi_0$, the *secondary* velocity is small and given at leading order by

$$\tilde{w} = \tfrac{1}{2}(d^2\tilde{w}/d\psi^2)(\psi-\psi_0)^2. \tag{3.5}$$

The equations governing the slow evolution of $\langle\chi\rangle$ and $\langle B_{\parallel}\rangle$ on short ψ and ξ length scales are given approximately by

$$D\langle\chi\rangle/Dt = R^{-1}\langle\Gamma\rangle_0\langle B_{\parallel}\rangle + R^{-1}\mathcal{L}_0(\langle\chi\rangle), \tag{3.6a}$$

$$D\langle B_{\parallel}\rangle/Dt = \langle S\rangle_0\partial\langle\chi\rangle/\partial\xi + R^{-1}\mathcal{L}_0(\langle B_{\parallel}\rangle) \tag{3.6b}$$

in which $\langle\Gamma\rangle$ and $\langle S\rangle$ are defined by (2.8b) above and (3.9) below respectively, the material derivative is (2.12a) with $\tilde{w}$ approximated by (3.5), and the diffusion operator $\mathcal{L}_0$ is

$$\mathcal{L}_0 \equiv \langle|\nabla\psi|^2\rangle_0\, \partial^2/\partial\psi^2 + \langle|\nabla\xi|^2\rangle_0\, \partial^2/\partial\xi^2 \tag{3.6c}$$

where ξ is normalised such that $\langle\nabla\psi\cdot\nabla\xi\rangle$ vanishes. Localised solutions of (3.6) exist proportional in amplitude to

$$\{\exp[-\beta(\psi-\psi_0)^2]\}e^{pt + in\xi}, \tag{3.7}$$

in which p is the complex growth rate, nZ is an arbitrary integer multiple of 2π, and β is a constant fixed by the coefficients in (3.6). Solutions analagous to this have been found[3] for the axisymmetric screw dynamo. The fastest growing modes[4] have

$$n = O(R^{1/3}), \qquad \beta = O(R^{2/3}) \tag{3.8a}$$

with

$$p = O(R^{-1/3}). \tag{3.8b}$$

This slow dynamo operates on length scales of order $R^{-1/3}$ and grows on the long time scale $R^{1/3}T_C(=R^{-2/3}T_D)$.

The *primary* magnetic field $B_{\parallel}\mathbf{U}$ is strongly helical. Its only contribution to the diffusive term on the right of (2.11) is $\langle\Gamma\rangle\langle B_{\parallel}\rangle/R$. This provides the alpha-effect in (3.6a) responsible for the generation of *secondary* magnetic field in planes normal to **U**. This type of alpha-effect has its counterpart in the hybrid Eulerian-Lagrangian[24] approach to the Braginsky[2] dynamo. *Primary* magnetic field is maintained by the omega-effect term $\langle S\rangle_0\partial\langle\chi\rangle/\partial\xi$ in (3.6b). The coefficient

$$\langle S\rangle_0 = -\, T_0^{-1}(dT/d\psi)_0 \tag{3.9}$$

is analagous to the angular velocity gradient in an axisymmetric system. The effect vanishes when the time taken to complete neighbouring periodicity sections $\overline{C}(\psi+\delta\psi,\xi)$ and $\overline{C}(\psi,\xi)$ are the same, $(dT/d\psi)_0 = 0$, just as in the case of rigid rotation.

4. FAST DYNAMO MECHANISMS

4.1 Generalised alpha-effect

One of the main advantages of considering the dynamo action caused by the 2-dimensional flow (1.2) is that it is possible to seek separable solutions of (1.1) in the form

$$B(x,t) = \mathrm{Re}[\hat{B}(x_H)e^{pt + inz}], \qquad \hat{B} = \overline{B}_H + B'. \tag{4.1}$$

Here the complex vector $\hat{B}$ is separated into $\overline{B}_H$, its constant horizontal average over the x,y-plane, and B', its remaining fluctuating part. Note that the horizontal and periodicity section averages are distinguished by the bracket and bar notation respectively. The slow mode (3.7) has the form (4.1) but, because of the helical nature of the field, has zero horizontal average, $\overline{B}_H = 0$. Throughout this section we are concerned only with field possessing non-zero horizontal mean, but no mean vertical component, see (4.3a) below.

Of particular interest is the mean EMF

$$\overline{u \times \hat{B}} = \overline{\mathcal{E}}_H - \overline{E}\hat{z}, \qquad \overline{\mathcal{E}}_H = (\overline{u' \times B'})_H, \qquad -\overline{E} = (\overline{u}_H \times \overline{B}_H)_z, \tag{4.2}$$

required to maintain the mean field $\overline{B}_H$. The average of the induction equation yields

$$(p+R^{-1}n^2)\overline{B}_H = in\hat{z} \times \overline{\mathcal{E}}_H, \qquad \hat{z}\cdot\overline{B} = 0, \tag{4.3a}$$

which implies that $\overline{\mathcal{E}}_H$ is linearly related to $\overline{B}_H$. Accordingly the solution of (4.3a), namely

$$\overline{\mathcal{E}}_H = \alpha\cdot\overline{B}_H, \tag{4.3b}$$

defines a generalised alpha-matrix $\alpha(R,n)$. Classical mean field theory is based on the assumption of long z-length scale. In that small n-limit, α has the Taylor series expansion

$$\alpha = \alpha_0 + (\partial\alpha/\partial n)_0 n + O(n^2). \tag{4.4}$$

The first term $\alpha_0[\equiv \alpha(R,0)]$ is the usual alpha-effect and the second term, also evaluated at n = 0, is linked to turbulent diffusivity. When n is finite, $\alpha(R,n)$ and p(R,n) can only be determined simultanously as part of the complete solution. On the other hand, when n is small, we can first solve for $\hat{B}(x_H)$ by ignoring the z and t dependences. That gives α_0 and hence the growth rate p is determined from (4.3).

The symmetries of the flow (1.3) lead to an isotropic alpha-matrix $\alpha(R,n)I$, which make it a particularly attractive starting point for a fast dynamo study. The value of $\alpha_0[=\alpha(R,0)]$ is calculated by considering the steady 2-dimensional field $B(x_H)$, which ensues when a uniform horizontal field is applied parallel to one of the principle axes. Magnetic flux is expelled from the cells into sheets of width $R^{-1/2}$. There flux is pulled out into tongues by the flow which leaves the stagnation points in the perpendicular direction. This exponential stretching is the key ingredient of the fast dynamo process. The upward and downward motion in the cells enables this flux to be used to generate mean field in the perpendular direction. The predicted size of α_0 is of order $R^{-1/2}$, which would lead to a fast dynamo if the z-length scale was $R^{-1/2}$ like the boundary layer thickness. In fact at such large values of n, the complete form of $\alpha(R,n)$ must be considered[8,23]. This shows that there is a large n cut off at $n = O(R^{1/2}/\ell n\ R)$, from which we deduce that fast dynamo action is not quite possible. Despite the refinements to the theory necessary for large n, the main physical processes are adequately accounted for by considering α_0 alone. For this reason we only consider the matrix α_0 here. The limitations of the range of its validity have been carefully estimated in the original papers[10,11].

4.2 Closed eddies, open channels and boundary layers

For the general flow (1.4) and (1.5), closed edies $\mathcal{D}_{m,n}$ continue to virtually fill the squares $\Pi_{m,n}$ provided that δ and ϵ are small. The flow velocity remains dominated by the Roberts motion (1.3), for which the circulation integral $\int \mathbf{u}_H \cdot d\mathbf{x}$ for each edge of square is 2. Thus the circulation is

$$\gamma = 8 \qquad \text{about} \qquad \mathscr{C}_{m,n} \qquad (\ell=m+n), \tag{4.5}$$

namely the closed streamline consisting of four edges bounding the square. The sense of circulation, is identified by whether ℓ is even or odd.

For the cat's-eye problem defined by the flow (1.4), narrow channels $|\psi'| < \delta(\ll 1)$ emerge. There the u_H-streamlines are periodic under the shift

$$X_H = \pi \mathbf{i}, \qquad \mathbf{i} = (1,1) \tag{4.6a}$$

with circulation $\qquad \gamma = 4 \quad \text{on } \overline{C}(\psi),$ (4.6b)

namely the periodicity section composed of two sides of the square. To calculate the alpha-matrix α_0, we consider the effect of the motion on an applied uniform horizontal magnetic field. The magnetic flux is expelled from the eddies. The X-type stagnation points at the vertices of the cat's-eyes trigger boundary layers of width $R^{-½}$ which bound the separatixes. When the mean field is aligned with the channel [$\overline{\mathbf{B}}_H=\mathbf{i}$, see (4.6a)] the flux simply aligns itself with the channel motion and there is little inductive effect. On the other hand, when there is a component perpendicular to $\mathbf{i}$, flux is pulled out into long tongues in the channel and intensified considerably. This leads to the dominant contribution to the alpha-effect. When the channels are wide, $\delta R^{½} \gg 1$, in the sense that the boundary layers on two neighbouring separatrixes $\psi = +\delta, -\delta$ no longer overlap, we may derive mainstream channel solutions. These give the dominant contribution to the alpha-effect which is given by (4.35) below. These analytic solutions together with numerical solutions of the boundary layer equations valid when $\delta R^{½} = O(1)$ were derived by Childress and Soward[10].

For the mean flow problem (1.5), the streamline topology is more complicated and linked to $\tan\theta_0 = N/M$ (see (1.8)). We, therefore, write

$$\overline{\mathbf{u}}_H = \{(\Delta\psi)/\pi\}(M,N), \qquad (\Delta\psi)/\pi = \epsilon/(M^2+N^2)^{½}, \qquad \epsilon \ll 1. \tag{4.7}$$

Each $\mathcal{D}_{m,n}$-eddy is bounded by a $\mathcal{C}_{m,n}$-streamline with a single X-type stagnation point located at $\mathbf{x}_{m,n}$, where at lowest order

$$\mathbf{x}_{m,n} = \pi(m+½,n+½) + (-1)^{\ell}\pi(-½,½) \qquad (\ell=m+n). \tag{4.8a}$$

There, correct to order ϵ, ψ and $\overline{\psi}$ take the same values

$$\psi_{m,n} = [M(n+½) - N(m+½)](\Delta\psi) + ½(-1)^{\ell}\psi^{(c)}, \tag{4.8b}$$

where $\qquad \psi^{(c)} = L(\Delta\psi), \qquad L = M + N.$ (4.9)

The streamline leaving the X-type stagnation point $\mathbf{x}_{-1,0}$ close to the origin arrives at a similar stagnation point $\mathbf{x}_{m,n}$ after a shift

$$X_H = (2\pi/\tau)(M,N) \qquad (m+n+1 = 2L/\tau, \text{ even}), \tag{4.10a}$$

where
$$\tau = 2 \quad (L \text{ even}), \qquad \tau = 1 \quad (L \text{ odd}). \tag{4.10b}$$

The shift defines the periodicity section $\overline{C}(\psi)$ of streamlines within the channel regions. The channel boundaries are defined by the streamlines through the X-type stagnation points. For small ϵ, they are

$$C_k: \quad \psi = \psi_k = k(\Delta\psi) \qquad (k \text{ integer}), \tag{4.11a}$$

which bound channels

$$D_k: \quad \psi_k < \psi < \psi_{k+1}. \tag{4.11b}$$

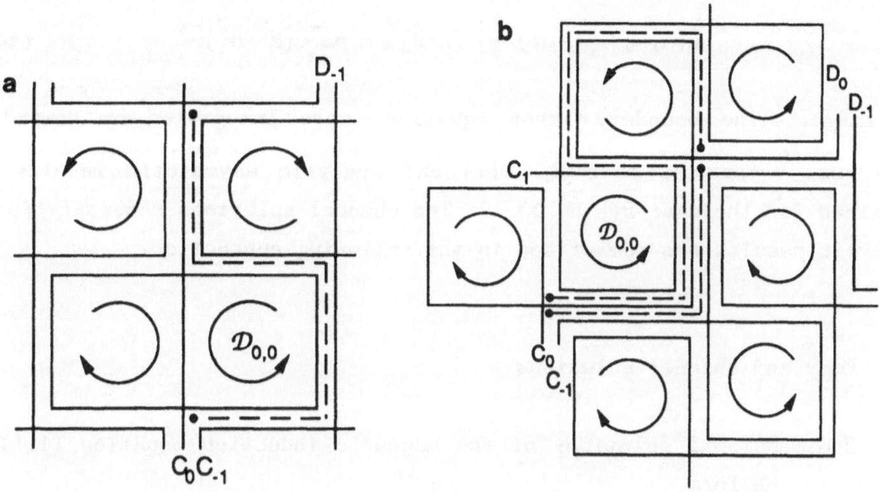

Fig. 3. Eddy and channel structure. The periodicity channel sections are indicated by the broken lines.
(a) (M,N) = (0,1), L odd, (b) (M,N) = (1,1), L even.

When L is odd all channel periodicity sections are similar (see eg L = 1 in Figure 3(a) but note that it is exceptional in that two neighbouring eddy boundaries coincide). When L is even, on the other hand, the channels are of two different types depending on whether k is odd or even (see eg L = 2 in Figure 3(b)). The number of eddy sides on D_k is

$$S_k = (4/\tau)\Delta_k L, \qquad \Delta_k = 1 + (-1)^k(\tau-1)/L. \tag{4.12}$$

so that the corresponding circulation is $2S_k$.

As in the case of the cat's-eye problem the complete magnetic field is determined when an applied uniform horizontal field $\overline{B}_H$ is maintained. The gap between neighbouring eddies is comparable to the eddy boundary layer width when

$$\beta = \epsilon R^{1/2} \tag{4.13a}$$

is of order unity. For the channels the boundary layer width increases downstream over a periodicity section of circulation $2S_k$ of order L to a width of order $(L/R)^{1/2}$. This means that neighbouring channel boundary layers are non intersecting only when

$$\beta(M,N) = [(\Delta\psi)/\pi]/[L/R]^{1/2} = \beta L^{-1/2}(M^2+N^2)^{-1/2} \tag{4.13b}$$

is large. The boundary layer equations were integrated by Soward and Childress[11] for the case $\beta = O(1)$ and analytic asymptotic results were obtained for the case $\beta(M,N) \gg 1$. The channel solutions necessary for the analytic results are summarised in the following subsection.

4.3 Eddy and channel solutions

2-dimensional solutions of the magnetic induction equation (1.1) are sought in the form

$$B = \overline{B}_H + KB\hat{z}, \qquad B_H = \nabla A \times \hat{z} \tag{4.14}$$

in the presence of the applied uniform horizontal field

$$\overline{B}_H = \nabla\overline{A} \times \hat{z}, \qquad \overline{A} = y\overline{B}_x - x\overline{B}_y, \qquad \overline{B} = 0, \tag{4.15}$$

Accordingly A and B satisfy the inhomogeneous heat conduction equations

$$(u_H \cdot \nabla)A - R^{-1}\nabla^2 A = \bar{E}, \tag{4.16a}$$

$$(u_H \cdot \nabla)B - R^{-1}\nabla^2 B = B_H \cdot \nabla(\psi - \bar{\psi}), \tag{4.16b}$$

where $-\bar{E}$ and $K(\psi-\bar{\psi})$ are the vertical mean electric field (4.2) and vertical velocity (1.6) respectively. Equations (4.16) are solved successively and the results are used to calculate the mean horizontal EMF (4.2), which reduces to

$$\bar{\mathcal{E}}_H = K\overline{[(\psi-\bar{\psi})\nabla A - B\nabla\psi]}. \tag{4.17}$$

The channel and eddy solutions for the flows (1.4), (1.5) with universal representation (1.6) are now outlined briefly. In the formalism of Section 2, we introduce the *primary* and *secondary* flows (2.1) and (2.2) with

$$v = -K(\bar{\psi}-\langle\bar{\psi}\rangle), \qquad \tilde{w} = K(\psi-\langle\bar{\psi}\rangle). \tag{4.18}$$

The former ensures that the vertical Z-shift about a closed eddy streamline or open channel periodicity section vanishes [$T\langle v\rangle=0$], as necessary for the direct application of the results. The magnetic field components (2.4) are

$$B_{\parallel} = (\nabla\psi \cdot \nabla A + Kv B)/(|\nabla\psi|^2 + |v|^2), \tag{4.19a}$$

$$B_\psi = -(u_H \cdot \nabla)A, \qquad B_\xi = KB - B_H \cdot \nabla\theta. \tag{4.19b}$$

The vertical electric field $-\bar{E}$, like $\bar{u}_H$, is of order ϵ and like the diffusion terms in (4.16) can be neglected at lowest order. The successive solution of (4.16a,b) yields the results that A and B_ξ are constant on streamlines:

$$A = A(\psi), \qquad B_\xi = B_\xi(\psi). \tag{4.20}$$

Further simplifications to (4.19) follow, on the basis that $|\nabla\psi| = O(1)$, $v = O(\epsilon)$. They give with the help of (2.1b) the approximate expressions

$$B_{\parallel} = dA/d\psi, \qquad B_{\xi} = KB - v(dA/d\psi). \tag{4.21}$$

The former measures the strength of the aligned *primary* magnetic field, while the latter provides a measure of the z-directed *secondary* magnetic field, which is not aligned to the *primary* flow $U = u_H + v\hat{z}$. Note that their periodicity section averages are

$$\langle B_{\parallel}\rangle = dA/d\psi, \qquad \langle B_{\xi}\rangle = K\langle B\rangle, \tag{4.22}$$

since v has zero average.

The average of the A-equation (4.16a) yields

$$[A]_{\overline{C}} = -T\langle B_{\psi}\rangle = T\overline{E} + R^{-1}d[\gamma(dA/d\psi)]d\psi, \tag{4.23a}$$

where
$$[A]_{\overline{C}} = -(X_H \times \overline{B}_H)_z \tag{4.23b}$$

is the jump in A under the shift X_H between the two ends of a periodicity section. In the eddy interiors X_H vanishes and so with (2.6b) integration gives

$$\gamma(dA/d\psi) = R[(-1)^{\ell}\pi^2 - \Sigma(\psi)]\overline{E} \qquad \text{on } \mathscr{D}_{m,n}, \tag{4.24}$$

where we have noted that $\gamma = 0$, $\Sigma = (-1)^{\ell}\pi^2$ at the O-type stagnation point at the eddy centre. Furthermore we measure Σ from the bounding streamline, where $\gamma = 8$ (see (4.5)) implying

$$dA/d\psi = (-1)^{\ell}R(\pi^2/8)\overline{E} \qquad \text{on } \mathscr{C}_{m,n}. \tag{4.25}$$

In the channels, on the other hand, $\overline{E}$ is small compared to $[A]_{\overline{C}}$ and can be neglected. Integration of (4.23a) yields

$$dA/d\psi = -(R/\gamma)(X_H \times \overline{B}_H)_z(\psi - \psi_M), \tag{4.26}$$

where γ is assumed to be constant on each channel and ψ_M is the value of ψ on the middle streamline. This choice of ψ_M ensures that the total flux in the large tongue of magnetic field in the channel vanishes. For the cat's-eye problem, we have no vertical electric field ($\bar{E}$=0) and

$$\psi_M = 0, \qquad \gamma^{-1}(X_H \times \bar{B}_H)_z = (\pi/4)(i \times \bar{B}_H)_z \tag{4.27}$$

(see (4.6a)), while for the mean flow problem the corresponding values are

$$\psi_M = \tfrac{1}{2}(\psi_k+\psi_{k+1}), \qquad \gamma^{-1}(X_H \times \bar{B}_H)_z = -(\pi^2/4)\bar{E}/(\Delta_k\psi^{(c)}) \tag{4.28}$$

(see (4.2), (4.10a), (4.12)). In the former case the tongue field strength is of order $R\delta$, while in the latter case it is of order $R\epsilon/L$ tending to zero in the *irrational* limit $L(=M+N) \to \infty$.

The derivation of an equation of the weaker *secondary* magnetic field B_ξ is a more delicate matter, which is best obtained directly from (2.11) and (2.12). Correct to leading order they give

$$\text{const.} + \tilde{w}[A]_{\bar{C}} = -R^{-1}T\langle\Gamma\rangle(dA/d\psi) + R^{-1}\gamma(dB_\xi/d\psi). \tag{4.29}$$

Here we note that $\partial\langle\chi\rangle/\partial t$ in (2.11) is a constant, which may take different values in different eddies and channels. In the evaluation of the mean helicity $\langle\Gamma\rangle$, we note that, whereas v is of order ϵ, its gradient may be significantly larger and dominated by $K(d\langle\bar{\psi}\rangle/d\psi)\nabla\psi$. This means that the helicity coefficient (2.8b) is

$$-T\langle\Gamma\rangle = K\gamma(d\langle\bar{\psi}\rangle/d\psi). \tag{4.30}$$

On the eddies, the symmetries about the cell centre implies that $\langle\bar{\psi}\rangle$ takes the cell centre value everywhere. Hence its derivative vanishes as does the helicity contribution (4.30). Since $[A]_{\bar{C}}$ vanishes also, and $\gamma = 0$ at the eddy centre, the constant in (4.29) must be zero. Trivial integration gives

$$(K\langle B\rangle =)\ B_\xi = (-1)^\ell KB_0 \qquad \text{on } \mathcal{D}_{m,n}, \tag{4.31}$$

where B_0 is a constant. By this choice the horizontal average of B over two neighbouring eddies vanishes as required by (4.15). In the channels, the helicity effect is present but its role is illusory. That is because if we multiply (4.23a) by $K(\langle\overline{\psi}\rangle-\psi_M)$ and add the result to (4.29) we obtain

$$\text{const.} + K(\psi-\psi_M)[A]_{\overline{C}} = R^{-1}\gamma d[B_\xi+K(\langle\overline{\psi}\rangle-\psi_M)(dA/d\psi)]/d\psi, \qquad (4.32a)$$

where
$$B_\xi + K(\langle\overline{\psi}\rangle-\psi_M)(dA/d\psi) = K[B+(\overline{\psi}-\psi_M)(dA/d\psi)], \qquad (4.32b)$$

and $\overline{E}$ and derivatives of γ have been neglected. The mean EMF (4.17) becomes

$$\overline{\mathcal{E}}_H = K\overline{\mathcal{E}(\psi)\nabla\psi}, \qquad \mathcal{E} = \psi(dA/d\psi) - \mathcal{B}, \qquad (4.33a)$$

where
$$\mathcal{B} = B + \overline{\psi}(dA/d\psi). \qquad (4.33b)$$

Soward and Childress[11] consider the nature of the channel and eddy boundary layers. They conclude that $\gamma(d\mathcal{B}/d\psi)$ is continuous across channel boundaries. In addition they argue that at lowest order B is continuous in the neighbourhood of the X-type stagnation points across the streamlines downstream of them. This condition is met when $\mathcal{E}$ is continuous across both eddy and channel boundaries. Together with (4.31) these boundary conditions fix the solutions of (4.32) for B.

For the cat's eye problem B, like A, vanishes inside the eddies, with the consequence that $\mathcal{E} = 0$ also. In the channels we must apend to the solutions (4.26), (4.27) for $dA/d\psi$, the results

$$B = -R(\pi/8)(\mathbf{i} \times \overline{\mathbf{B}}_H)_z(\psi^2-\delta^2), \qquad \mathcal{E} = -R(\pi/8)(\mathbf{i} \times \overline{\mathbf{B}}_H)_z(\psi^2+\delta^2). \qquad (4.34)$$

Here the vertical field KB is of order $R\delta^2$, small compared to the horizontal field of order $R\delta$. The mean EMF (4.33a) is computed from (4.34) on the basis that one channel periodicity section gives the contribution from one square of area π^2. The result is

$$K\overline{\mathcal{E}\nabla\psi} = -(K/\pi)\left[\int_{-\delta}^{\delta}\mathcal{E}(\psi)d\psi\right]\mathbf{i} \times \hat{\mathbf{z}}, \qquad (4.35a)$$

which gives the alpha-matrix in the diadic form

$$\alpha_0 = -(2K/3)R\delta^3[I - \tfrac{1}{2}\mathbf{i}\mathbf{i}]. \qquad (4.35b)$$

For the mean flow problem, it is readily verified that the solution of (4.31) and (4.32), which satisfies the boundary conditions stated, is

$$B - B^{(s)} = \begin{cases} (-1)^{\ell}(\pi^2/8)R\bar{E}(\psi_{m,n}-\bar{\psi}) & \text{on boundary of } \mathcal{D}_{m,n}, \quad (4.36a) \\ (\pi^2/8)R\bar{E}[(\psi-\bar{\psi})^2-(\psi_{k+1}-\bar{\psi})(\psi_k-\bar{\psi})]/(\Delta_k\psi^{(c)}) & \text{on } D_k, \quad (4.36b) \end{cases}$$

$$\mathcal{E} + B^{(s)} = (\pi^2/8)R\bar{E}(\psi-\psi_{k+1})(\psi-\psi_k)/(\Delta_k\psi^{(c)}) \quad \text{on } D_k, \tag{4.36c}$$

where
$$B^{(s)} = -(\pi^2/16)R\bar{E}\psi^{(c)}. \tag{4.36d}$$

The interesting feature of this result is that $B^{(s)}$, the X-type stagnation point value of B, is of order $R\epsilon^2$, whereas the channel values of $\mathcal{E} + B^{(s)}$ are of order $R\epsilon^2/(M^2+N^2)$. Now the eddy contributions of $\mathcal{E}$ to the EMF vanish on averaging and so only the channel contributions (4.36c) count. Thus in the *irrational* limit $L \to \infty$, we see that $\mathcal{E}$ take the constant value $-B^{(s)}$ everywhere in the channels. Finally to evaluate $\bar{\mathcal{E}}_H$ we note that when L is even (odd) the two (one) distinct channel periodicity sections provide the contribution of two squares area $2\pi^2$. Both cases are catered for by the expression

$$K\overline{\mathcal{E}\nabla\psi} = -(K/2\pi)\left[\int_{\psi_{k-1}}^{\psi_{k+1}} \mathcal{E}(\psi)d\psi\right](N,-M).$$

Use of (4.36c) yields the result

$$K\overline{\mathcal{E}\nabla\psi} = (K\pi^2/16)R[1-(3\Lambda L^2)^{-1}]\psi^{(c)}\bar{u}_H \times (\bar{u}_H \times \bar{B}_H), \tag{4.37a}$$

where
$$\Lambda = 1 - (\tau-1)L^{-2}. \tag{4.37b}$$

The corresponding alpha-matrix is

$$\alpha_0 = -\alpha_0[I - \epsilon^{-2}\bar{u}_H\bar{u}_H], \tag{4.38a}$$

where
$$\alpha_0 = [1-(3\Lambda L^2)^{-1}]\alpha_0^{(irr)}, \tag{4.38b}$$

$$\alpha_0^{(irr)} = KR\epsilon^3(\pi^3/16)[|\sin\theta_0|+|\cos\theta_0|]. \tag{4.38c}$$

Here $\alpha_0^{(irr)}$ is the limiting value of α_0 ($\tan\theta_0 = N/M$) in the *irrational* limit $L \to \infty$.

For both the cat's-eye and mean flow problem there is no EMF at the order calculated when the mean field is aligned with the mean channel directions i and $\bar{u}_H$ respectively. To calculate that smaller contribution, the boundary layers must be considered in detail. The calculation for the cat's-eye problem[10] shows that there is an additional contribution of order $R^{-3/2}\delta^{-2}$ to the alpha-matrix (4.35b). For the dynamo to operate this contribution is essential because, without it, the determinant of α_0 vanishes, there is no generation of mean magnetic field perpendicular to i and the dynamo fails. With it, the determinant is of order β/R, which increases with β from the order R^{-1} values (see Section 4.1 above) appropriate to the symmetric Roberts flow (1.3). This means that opening up the channels improves the alpha-effect. It does not, however, imply faster growth rates because the z-length scale of modes admissible to the theory also increases unfavourably with β. Similar results are to be expected for the mean flow problem but no such comparable calculation has been undertaken because of its complexity.

5. DISCUSSION

The large R-asymptotics presented here draw attention to a number of different aspects of magnetic induction and their relation to the growth of magnetic fields. The slow dynamo example of Section 3 stresses the fact that motion constrained to lie on smooth closed streamsurfaces is unlikely to provide a source of fast dynamo action. As equations (3.6) suggest stretching only intensifies magnetic field on the streamsurface. There is no convective mechanism available to create magnetic field normal to the surface. Instead that normal component only emerges from diffusion of the strong streamsurface field. The resulting $\alpha\omega$-dynamo is slow and for the model considered here it grows on the hybrid time scale $(T_D^{1/3}T_C^{2/3})$. This result has the further implication that the magnetic field of a fast dynamo mode must be expelled from the regions, where motion is constrained as here to lie on smooth closed streamsurfaces.

Though the motion considered in Section 4 also lies on streamsurfaces, it contains the essential ingredient necessary for fast dynamo action, namely the exponential stretching at the stagnation points where the streamsurface is not smooth. This ingredient is clear in the case of the Roberts dynamo associated with the symmetric motion (1.3). There the mean magnetic field along one of the principle axes (say Ox) is plucked into tongues along the other (say Oy). This provides a convective mechanism which creates magnetic field in two mutually orthogonal directions, one from the other. This overcomes the slow dynamo need to invoke diffusion for the generation of one of the two crucial components of field. Of course, the detailed structure of the boundary layers, which contain the field, depends on diffusion but, unlike the slow dynamo, diffusion does not provide the field source. From that point of view, the development in Section 4 has sidestepped all the key questions linked to the nature of the diffusive boundary layers. Instead we have highlighted the mainstream aspects of the problem to establish the similarities and differences with the slow dynamo of Section 3.

The cat's-eye flow (1.4) provides the natural extension of the Roberts dynamo. The mean flow problem (1.5), on the other hand, introduces a fundamentally new ingredient. For as L(=M+N) increases the number of boundary layers increases so that in the *irrational* limit $L \to \infty$ they become dense in a finite region of the flow rather than on a set of measure zero. Interestingly the strength of the horizontal magnetic field (4.26) in the channels is of order $R\epsilon/(M^2+N^2)^{1/2}$ and so its magnitude decreases with increasing L. The vanishing of this channel field in the limit R and ϵ fixed, $L \to \infty$ may be regarded as a flux expulsion mechanism which operates because of the ever decreasing channel width. In this limit it is the horizontal eddy field, which provides the source of vertical field $KB^{(s)}$ residing in the channels (see (4.36)) necessary to yield the alpha-effect.

The dense feature, which we refer to, is common to the fully 3-dimensional Beltrami flows as we explained in the Introduction. There some flux explusion is also likely to occur in the chaotic regions but the changing orientations of the surfaces throughout the flow may lead to other constructive processes not present in our 2-dimensional flows. The importance of our mean flow problem lies in the fact that it provides a bridge between the 2-dimensional flows, which probably, like the Roberts dynamo, only just fail to be fast dynamos, and the fully 3-dimensional flows, for which we reasonably expect some to yield fast dynamos.

As a tentative preliminary step towards 3-dimensional flow, it is of some interest to consider the effect of rotating the direction of the mean flow with height z, say by setting $\theta_0 = kz$, where k is the z-wave number. The resulting mean motion now has helicity. Provided for sufficiently small k the alpha-matrix (4.38) remains valid, the vanishing of its determinant is no longer a serious matter as the generation of field in the slowly varying direction $\bar{u}_H$ is extremely potent. Indeed estimates[10] show that growth rates on the convective time scale T_C are possible when the z-length scale k^{-1} is long of order KR. On these long length scales, the approximations, upon which the present results are based, appear to be valid. On shorter length scales large mainstream fluxes are expelled from the eddy interiors, because of the z-variations and our results no longer apply. That does not imply that fast dynamo action is not possible at fixed k, K, ϵ as $R \to \infty$. Rather it means that a more detailed analysis is necessary which takes into account the small regions of chaotic motion occuring close to the separatices of the unperturbed 2-dimensional flow.

REFERENCES

1. H. K. Moffatt and M. R. E. Proctor, Topological constraints associated with fast dynamo action, J. Fluid Mech. 154:493 (1985).
2. S. I. Braginsky, Self-excitation of a magnetic field during the motion of a highly conducting fluid, Zh. Eksp & Teor. Fiz. 47:1084 (1964).
3. A. Ruzmaikin, D. Sokoloff and A. Shukurov, Hydromagnetic screw dynamo, J. Fluid Mech. 197:39 (1988).
4. A. D. Gilbert, Fast dynamo action in the Ponomarenko dynamo, Geophys. Astrophys. Fluid Dynamics 44:241 (1988).
5. A. M. Soward, A unified approach to a class of slow dynamos, Geophys. Astrophys. Fluid Dynamics (submitted, 1989).
6. G. O. Roberts, Dynamo action of fluid motions with two-dimensional periodicity, Phil. Trans. Roy. Soc. Lond. A271:411 (1972).
7. S. Childress, Alpha-effect in flux ropes and sheets, Phys. Earth Planet. Int. 20:172 (1979).
8. A. M. Soward, Fast dynamo action in a steady flow, J. Fluid Mech. 180:267 (1987).
9. N. O. Weiss, The expulsion of magnetic flux by eddies, Proc. Roy. Soc. Lond. A293:310 (1966).
10. S. Childress and A. M. Soward, Scalar transport and alpha-effect for a family of cat's-eye flows, J. Fluid Mech. 205:99 (1989).
11. A. M. Soward and S. Childress, Large magnetic Reynolds number dynamo action in a spatially periodic flow with mean motion, Phil. Trans. Roy. Soc. Lond. A (accepted, 1989).

12. T. Dombre, U. Frisch, J. M. Greene, M. Hénon, A. Mehr and A. M. Soward, Chaotic streamlines in the ABC flows, J. Fluid Mech. 167:353 (1986).

13. D. J. Galloway and U. Frisch, A numerical investigation of magnetic field generation in a flow with chaotic streamlines, Geophys. Astrophys. Fluid Dynamics 29:13 (1984).

14. D. J. Galloway and U. Frisch, Dynamo action in a family of flows with chaotic streamlines, Geophys. Astrophys. Fluid Dynamics 36:53 (1986).

15. H. Alfvén, Tellus 2:74 (1950).

16. S. I. Vainshtein and Ya. B. Zeldovich, Origin of magnetic fields in Astrophysics, Sov. Phys. Usp. 15:159 (1972).

17. B. Bayly and S. Childress, Construction of fast dynamos using unsteady flows and maps in three dimensions, Geophys. Astrophys. Fluid Dynamics 44:211 (1988).

18. J. M. Finn and E. Ott, Chaotic flows and fast magnetic dynamos, Phys. Fluids 31: 2992 (1988).

19. J. M. Ottino, "The kinematics of mixing: stretching, chaos and transport," Cambridge University Press, Cambridge (1989).

20. S. Childress and A. M. Soward, On the rapid generation of magnetic fields, in "Chaos in Astrophysics," J. R. Buchler, J. M. Perdang and E. A. Spiegel, ed., NATO ASI Series C: Mathematical and Physical Sciences 161:223 (1985).

21. A. M. Soward and S. Childress, Analytic theory of dynamos, Adv. Space Res. 6:7 (1986).

22. D. R. Fearn, P. H. Roberts and A. M. Soward, Convection, stability and the dynamo, in "Energy, Stability and Convection," G. P. Galdi and B. Straughan, ed., Pitman Research Notes in Mathematics Series 168:60 (1988).

23. A. M. Soward, On dynamo action in a steady flow at large magnetic Reynolds number, Geophys. Astrophys. Fluid Dynamics 49:3 (1989).

24. A. M. Soward, A kinematic theory of large magnetic Reynolds number dynamos, Phil. Trans. Roy. Soc Lond. A272:431 (1972).

ON SPATIAL STRUCTURE OF KOLMOGOROV SPECTRA OF WEAK TURBULENCE

G.E. Fal'kovich and A.V. Shafarenko

Institute of Automation and Electrometry
Siberian Branch of Academy of Science
630090 Novosibirsk, U.S.S.R.

This lecture is devoted to the stability problem for the Kolmogorov spectrum of weak turbulence, when the characteristic time of the nonlinear interaction between the waves is much greater than that of the dispersion. This condition makes at possible to get a closed analytical description in terms of the amplitude pair correlator n(k), defined as $n(k)\ (k-k') = a(k)a^*(k)$ with a(k) being a canonical variable of the wave field 3 :

$$\partial n_k / \partial t = \int W(k,k_1,k_2)\ (n(k_1)n(k_2)-n(k)n(k_1)-n(k)n(k_2)) \times$$
$$\times\ \delta(k-k_1-k_2)dk_1dk_2 - 2\int W(k_2,k,k_1)(n(k)n(k_2)-n(k_1)n(k)-$$
$$-n(k_1)n(k_2))\ \delta(k_1-k-k_2)\ dk_1dk_2 + \Gamma(n(k),k) - D(n(k),k). \qquad (1)$$

Here $W(k,k_1,k_2) = |V_{k,k_1k_2}|\ \delta(\omega(k) - \omega(k_1) - \omega(k_2))$, and $V_{kk_1k_2}$: the matrix element of three-wave interaction. The terms $\Gamma(n(k),k)$, and $D(n(k),k)$ describe external pump and damping, respectively.

Eq. (1) is applicable when the dispersion law is such that it allows decay processes.

Kolmogorov and Obukhov [1,2] suggested that if a turbulence is local (i.e. effectively interacting are the waves with comparable wavelengthes only), scale-invariant (i.e. there is no characteristic scale in the medium) and isotropic, the energy distribution in k-space will obey a power law $n(k)=k^{\varkappa}$ for all k distant from the energy source and from the dissipation domain. For the weakly turbulent system, this hypothesis got its first confirmation when Zakharov [3] found that Eq. (1) admits a power stationary solution

$$n^0(k) = p^{1/2}k^{-\alpha-d} \qquad (2)$$

where α is the uniformity index of the matrix element

Nonlinear Evolution of Spatio-Temporal Structures in Dissipative Continuous Systems
Edited by F.H. Busse and L. Kramer
Plenum Press, New York, 1990

$V(\lambda k, \lambda k_1, \lambda k_2) = \lambda^{\alpha} V(k, k_1, k_2)$, p is the energy flux, and d the spatial dimensionality. Solution (2) is, however, purely formal: it has a singularity at k=0 and contains infinite energy. To give it a meaning, one has to prove its global stability, in other words, the fact that it is a spectrally intermediate asymptotic of the steady state to where the evolution comes in a long time. So, an initial value problem should be solved, with a specific pump and damping mechanism and with some arbitrary initial data, which is only possible by means of numerical simulation so far.

The most vulnerable point of the hypothesis is the influence the spectral shape of the pump may exert on the spectrum in the inertial range. The Kolmogorov concept implicitly involves the supposition that in the process of the energy stepwise transfer from the pump on, the spectral characteristics of the pump are gradually becoming less and less determinant for the spectrum. So they contribute to the macroscopic parameter only, viz to the energy flux. This is essentially the contents of the "universality hypothesis".

Being very important for the Kolmogorov picture of turbulence, the universality hypothesis is first to be proved. The problem is naturally divided on two issues:

(i) whether the spectrum is sensitive to the isotropic characteristics of the turbulence source (i.e. to the width of the pump in the k-space), and

(ii) whether the spectrum becomes essentially angular dependent if the pump is made slightly anisotropic in the k-space.

The first issue was treated by us in our paper [4]. We found that if the (isotropic) pump width is much smaller than its characteristic wavenumber, a large intermediate region appeared between the pump and the Kolmogorov spectrum, where the solution had a peak sequence shape, with a limited number of the peaks whose characteristic frequencies were multiples of that of the pump, and whose spectral widthes were proportional to those frequencies (see fig. 1).

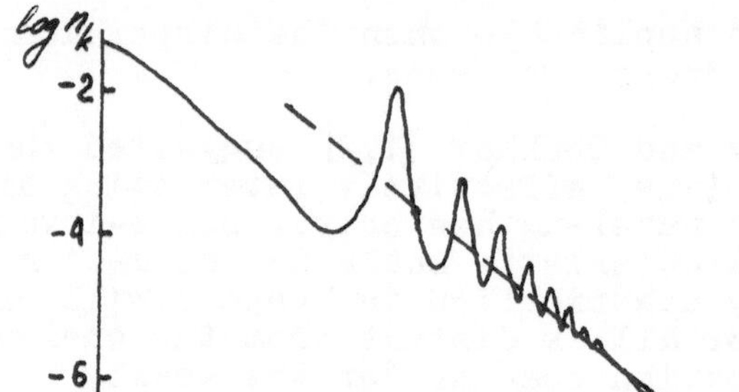

Fig. 1. The spectrum of weak acoustic turbulence excited by a narrow pump (k_o=20, Δk=4, k_{max} = 400, d = 2).

The results show that the Kolmogorov regime may generally be implemented a rather long way from the pump, where peak widths are greater than the spacing between them and, consequently, where the peak structure is only vaguely expressed. But still the Kolmogorov asymptotic retains its universal form (2), with the flux P being the only parameter affected by the pump spectral properties:

$$P = k_0^{2(\alpha - m)},$$

where k is the pump location and m the uniformity index of the dispertion law: $\omega(\lambda k) = \lambda^m \omega(k)$.

The situation is entirely different in the case (ii). It turned out in our study [5], that even small anisotropy of the pump may cause significant angular modulation of the spectrum in the inertial range, which may, in principle, lead to a multiparametric inertial-range asymptotic , i.e. to the lack of simple universality. Because of great technical difficulties in numerically solving Eq. (1), we restricted ourselves to the simplest case of 2d acoustic turbulence, with the dispertion law $\omega(k)=ck(1+\beta k^2)$ close to the scale invariant as $\beta \ll k_{max}^{-2}$, matrix element $V=(kk_1k_2)^{1/2}$, slightly anisotropic pump with angular dependence like $1+\mathcal{E}\cos m\varphi$, $\mathcal{E} \ll 1$, and some (model) strong damping mechanism switched on at large k. We managed to perform numerical solution of the Eq. 1 in the discrete form

$$\partial n_k/\partial t = \sum_{l=1}^{k} l(k-l)\,[n_l n_{k-l} - n_k(n_l + n_{k-l})] -$$

$$-2\sum_{l=k} l(l-k)\,[n_k n_{l-k} - n_l(n_k + n_{l-k})] + \gamma_k n_k -$$

$$-2n_k \sum_{l=k_m}^{k_m+k} l(l-k)\,n_{l-k},$$

where $n_k=n_k(\theta)$, $n_l=n_l(\theta \pm ak \mp al)$,

$n_{k-l}=n_{k-l}(\theta \mp al)$

in the first sum and

$n_k=n_k(\theta)$, $n_l=n_l(\theta \pm al \mp ak)$,

$n_{l-k}=n_{l-k}(\theta \pm al)$

in the second and third ones, using the grid which had 100 points in k and 32 in φ. The results are summarized in fig. 2, where the k-dependence of the relative angular modulation of the spectrum versus k

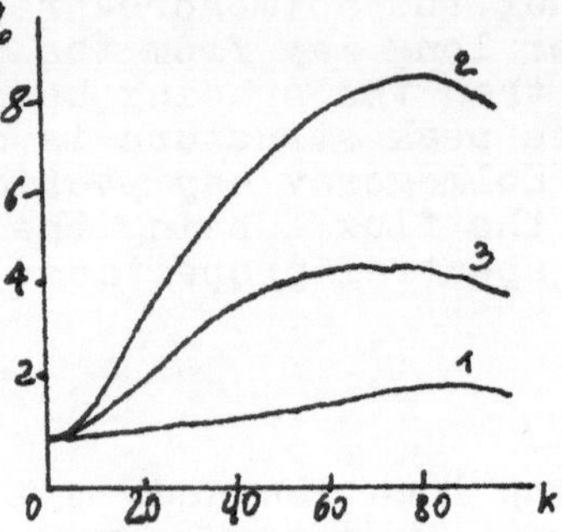

Fig. 2. Dependence of the relative angular modulation on k for different angular harmonics m and dispersion parameter a: 1) m=1, a= $\pi/128$; 2) m=1, a= $\pi/256$; 3) m=2, a = $\pi/256$.

is presented. It is clearly seen that the modulation increases with k, reaching in one case the value ten times greater than in the pump region.

These results show the need for a more accurate formulation of the universality hypothesis. Namely: the stationary spectrum in the inertial range is probably defined by a number of the parameters equal to the number of intergals of motion. There is many integrals ($L \approx (Bk)^{-1}$) in the discussed case of weak acoustic turbulence because of the fact that this case is close to the degenerated one with β =0 and with infinite set of integrals.

References

1. A.N. Kolmogorov, Doklady Ac.Sc. USSR 30 (1941) 259.
2. A.M. Obukhov, Izv. Ac.Sc.
3. V.E. Zakharov, in: Handbook of Plasma Physics, vol. 2 (Noth-Holland, Amsterdam, 1984).
4. G.E. Fal'kovich, A.V. Shafarenko, JETF, vol.94, N 7,1988.
5. G.E. Fal'kovich and A.V. Shafarenko, Physica 27D (1987) 399-411.

SPATIOTEMPORAL INTERMITTENCY

Paul Manneville

Institut de Recherche Fondamentale
DPh-G/PSRM, Centre d'Etudes Nucléaires de Saclay
91191 Gif-sur-Yvette Cedex, France

INTRODUCTION

As a first step toward the understanding of the nature of turbulence, the study of the emergence and destabilization of dissipative structures in out-of-equilibrium systems has attracted considerable attention. During this study, confinement effects have been shown to play a major role in controlling the nature of the nonlinear effective dynamics driving the unstable modes. In *confined systems*, the spatial structure of the modes remains frozen, so that we are left with a small number of degrees of freedom well accounted for by the theory of dissipative dynamical systems. Intrinsic stochasticity is then mainly synonymous of temporal chaos. In *extended systems* confinement effects are weak and spatial modulations are allowed, leading to the formation of patterns. A much larger number of degrees of freedom is then involved in the dynamics, which raises the problem of the transition to *weak turbulence*.

Two cases can be distinguished according to whether the primary instability is *super*-critical or *sub*-critical. When it is super-critical, the primary modes gently saturate beyond threshold. In weakly confined systems, the most "dangerous" secondary modes then appear connected to the *phase* of some complex envelope. The transition to turbulence can be understood as a spatiotemporal process associated with the emergence of temporal chaos for the phase variable. When the primary instability is sub-critical, different attracting states can *coexist* in phase space and a "competition" between the corresponding regimes occurs. Following Pomeau's ideas,[1] we understand easily that, when modulations are allowed, the coexistence of attractors in phase space can be converted into a coexistence in physical space of homogeneous domains of different kinds separated by *walls* or *fronts*; hence a possible different transition to turbulence with a novel intrinsic spatiotemporal meaning.

When one of the two competing regimes is chaotic ("turbulent") and the other regular ("laminar"), the system can fluctuate in both space and time giving rise to a specific turbulent regime called *spatiotemporal intermittency* (STI). In that case, Pomeau further conjectured that the front propagation would be randomized by the intrinsic stochasticity of the turbulent domains, yielding a *contamination* process akin

Nonlinear Evolution of Spatio-Temporal Structures in Dissipative Continuous Systems
Edited by F.H. Busse and L. Kramer
Plenum Press, New York, 1990

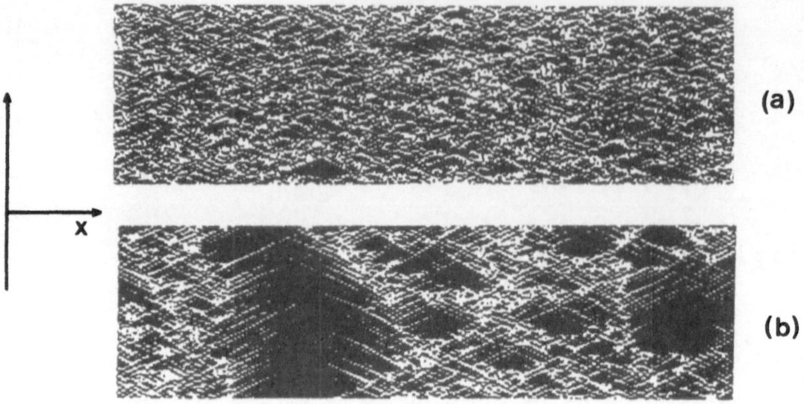

Fig. 1. STI in model (1): "laminar" → "black", "turbulent" → "white"; (a) deep inside the turbulent regime; (b) very close to the STI treshold.

to *directed percolation* (DP). STI was then presented not only as a specific turbulent regime but also as a new "generic" transition scenario valid for sub-critically unstable weakly confined systems. A great interest of this last viewpoint lies in the fact that it introduces *critical phenomena* concepts in the field of turbulence in a novel way. Below, I will first present recent results on STI obtained in collaboration with H. Chaté at Saclay.[2] I will then review some experimental situations where our approach has already received some application or could be of direct relevance.

ILLUSTRATION OF SPATIOTEMPORAL INTERMITTENCY

Continuous systems of interest in physical applications are usually described in terms of partial differential equations (PDE). Accordingly, before turning to a discussion of discrete-space discrete-time systems for which most of the results have been obtained, I will begin my presentation by illustrating STI and the associated scenario using the modified Swift-Hohenberg model[4]

$$\partial_t v + v\partial_x v = rv - \left(\partial_{xx} + 1\right)^2 v \tag{1}$$

Numerical simulations of this model were performed on an interval of length $L \gg \lambda_c = 2\pi$. Since the evolution of a large number of cells, typically a few hundreds, had to be followed, a *black-and-white* reduced description of the system was developed. As seen in Fig. 1, the whole solution $v(x,t)$ at given t could be decomposed into locally periodic structures (in black) and strongly distorted domains (in white), the evolution of which was followed along time. The turbulent regime obtained for r sufficiently large could then be resolved in terms of a *space-time fluctuating* mixture of laminar and turbulent domains. For such a regime, at a given fixed point in space, the state of the system can be analyzed as an intermittent alternation of laminar intermissions and turbulent bursts but this is clearly due to a process that develops in both space and time, hence the term *spatiotemporal intermittency*.

When r is decreased from 1 down to 0, a transition from a fully intermittent regime to a fully laminar regime is observed. When $r = 1$, model (1) is equivalent to the well-known Kuramoto–Sivashinsky equation and, for L large, a "fully intermittent" regime made of relatively small laminar domains is observed. The sizes of

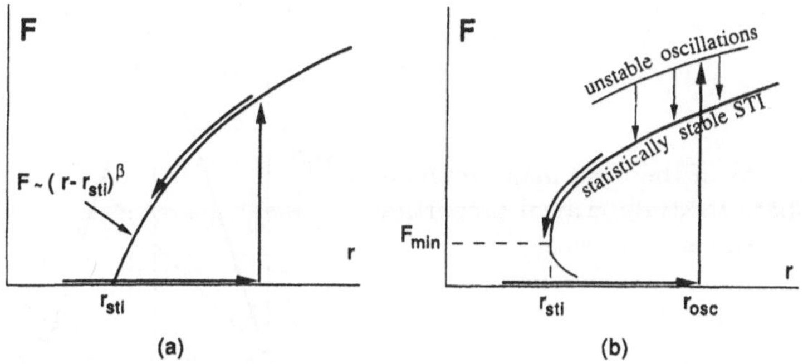

Fig. 2. Conjectured behavior of the turbulent fraction F for STI in model (1); the transition can be either "continuous" (a) as for DP, but more probably weakly "discontinuous" as in (b); experimentally, the branch corresponding to the sub-critical oscillations (threshold r_{osc}) seems unstable and therefore unreachable.

laminar domains increase when r is lowered (Fig. 1a) but for r sufficiently large, an exponentially decreasing distribution of sizes is obtained. Closer to the threshold (r_{sti}), the sizes of laminar domains becomes "macroscopic" (Fig. 1b) and their distribution decays slower than exponentially. Finally, below the threshold, the intermittent state recedes and, at the end of a long transient, the system becomes "fully laminar".

If the threshold is approached from below, the result depends on the nature of the initial condition, either *regular* or *distorted*. Regular initial conditions can be maintained above STI threshold (laminar state is metastable) up to a point r_{osc}, where a secondary instability takes place. Corresponding oscillations can be observed without ambiguity only thanks to confinement effects for moderate L. At large L, they immediately decay into turbulence. For disordered initial conditions, the STI regime is reached as soon as $r > r_{sti}$ whereas for $r < r_{sti}$, STI is only transient though with transients of increasing duration when r approaches r_{sti}.

Reliable statistical results close to the threshold are difficult to obtain owing to size effects. In terms of the turbulent fraction F (the time average of the fraction of the system occupied by turbulent domains), one of the two behaviors displayed in Fig. 2 can be conjectured. Keeping in mind the analogy with DP—a second-order transition—one can guess a continuous transition with a well-defined critical behavior at threshold (Fig. 2a) but small turbulent fractions seem difficult to stabilize, so that a discontinuous transition of the type sketched in Fig. 2b is equally plausible. Such ambiguities, unfortunately typical of experimental results to be discussed later, should be kept in mind while building models of STI.

MODELING SPATIOTEMPORAL INTERMITTENCY

Directed percolation presents itself as the simplest stochastic process describing propagation in space by contamination. Let us take the concrete example of a flow through a porous medium viewed as a network of pores that can be *open* or *closed* with probability p. Having poured some fluid at one end of the system, we then try to know whether pores at some level are *wet* or *dry*. A two-state probabilistic cellular automaton (PCA) has been defined implicitly. The dry state is called *absorbing* since a given pore with dry "parents" can only remain dry. The wet state is usually called *active* and when $p > p_c$, some "activity" is maintained down to the other end of the system; the fluid percolates through the medium.

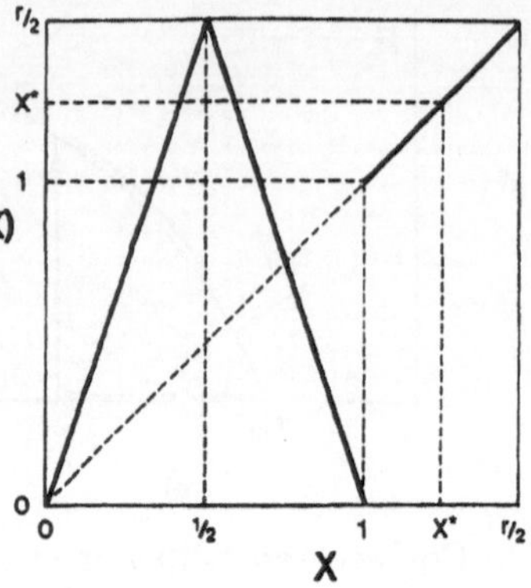

Fig. 3. Shape of the local map proposed by H. Chaté to study critical properties of STI and the analogy with DP.

In building the analogy with the transition to turbulence *via* STI, we are obviously led to identify *laminar* with absorbing and *turbulent* with active. The problem is then to understand in which way stochasticity enters the *a priori* deterministic extended system to make a fully probabilistic process such as DP relevant. An interesting intermediate class of systems is offered by so-called *coupled map lattices* (CML)[5]:

$$X_j^{n+1} = \sum_{j' \in \mathcal{V}_j} W_{jj'} f(X_{j'}^n) ,$$

where subscript j and superscript n denote space and time, respectively. The variable X (not necessarily one-dimensional) characterizes each local sub-system and $W_{jj'}$ describes the coupling to other sub-systems belonging to the neighborhood $\mathcal{V}_j$ of a given site j. (Notice that the decomposition of the system into sub-units interacting locally at discrete times is more or less justified by the fact that, for a sub-critical instability, no divergence of the time and space scales is expected.)

To define a concrete CML we have to specify the analytic expression of the local map and the structure of the coupling. The main problem is to avoid spurious features linked with the local dynamics chosen. In most of our studies we took the *minimal model* proposed by H. Chaté[6] and displayed in Fig. 3. It describes a system with a local phase space split into two pieces. In the first piece, $X \in [0,1]$, a chaotic but transient dynamics is generated by a tent map with slope $\alpha > 2$. The second piece, $X > 1$, corresponds to the laminar state and is simply accounted for by $f(X) = X$. For the coupling, simple diffusion is assumed, which, for a one-dimensional system with a three-site neighborhood, yields:

$$X_j^{n+1} = f\left(X_j^n\right) + g\left(f\left(X_{j-1}^n\right) - 2f\left(X_j^n\right) + f\left(X_{j+1}^n\right)\right)$$

The transition to turbulence *via* STI can then be studied by varying the coupling g at given slope α and given space dimension d. A critical value of the coupling $g_{\rm sti}$ is found such that, for $g < g_{\rm sti}$, the STI regime is *transient* and for $g > g_{\rm sti}$, STI is *sustained*. Figure 4 and 5 give illustrations of the transition for $\alpha = 2.1$ and $\alpha = 3$.

This example, where spatial coupling is seen to convert local transient chaos into global sustained turbulence, is particularly well suited to analyze the relation between STI and critical phenomena, as reviewed in the next section. However the modeling of STI is still a rather open problem since there is no systematic way to pass from the fully continuous space-time formulation of the initial physical problem to a fully discrete space-time description in terms of CMLs.

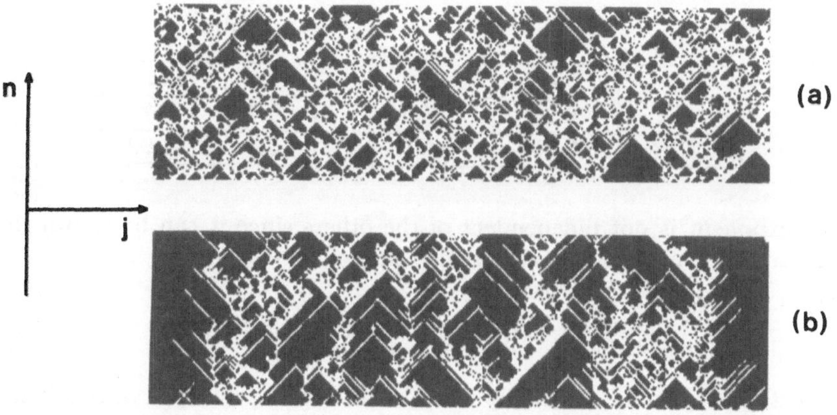

Fig. 4. STI in Chaté's model for $\alpha = 3$; (a) above threshold; (b) "at" threshold.

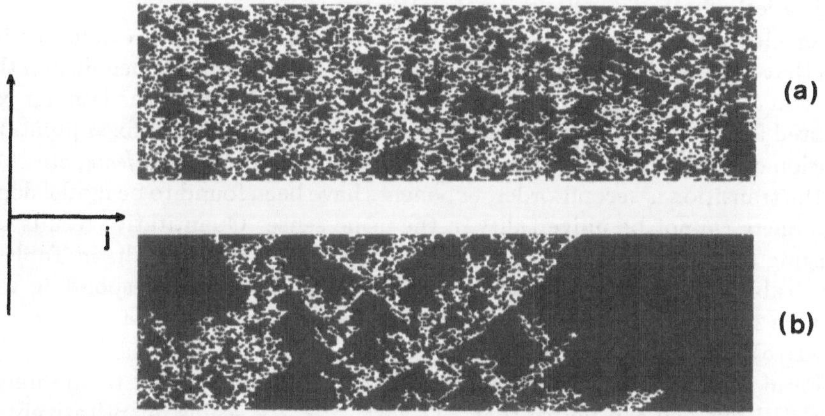

Fig. 5. STI in Chaté's model for $\alpha = 2.1$; (a) above threshold; (b) "at" threshold.

STI AND CRITICAL PHENOMENA

The DP transition displays a *critical behavior* typical of a second order transition. For example, the natural order parameter is the fraction F of active sites that varies as

$$F \propto (p - p_c)^\beta$$

above p_c; hence the definition of the critical exponent conventionally called β. In the same way, the size of active clusters ξ varies as

$$\xi \propto (p_c - p)^{-\nu}$$

below p_c; two directions can be defined, either transverse or parallel to the direction of propagation. Accordingly, two coherence lengths $\xi_\perp$ and $\xi_\parallel$ and two critical exponents $\nu_\perp$ and $\nu_\parallel$ can be defined. In one space dimension, a further scaling quantity is of interest, the distribution $\mathcal{N}$ of the length l_{abs} of absorbing domains at the critical point, from which a new critical exponent, here denoted ζ, can be obtained:

$$\mathcal{N}(l_{\text{abs}}) \propto l_{\text{abs}}^{-\zeta}$$

Table 1.

d	1	2
STI	0.25	0.50
DP	0.28	0.60

Table 2.

α	ζ
2.1	1.78
3.0	1.99

This last exponent is not independent of the others since it can be shown that, d_f being the fractal dimension of set of active sites at threshold, one has

$$\zeta = 1 + d_f \qquad \text{with } d_f = 1 - \beta/\nu_\perp .$$

Directed percolation is usually thought to be the prototype of the *universality class* of elementary two-state PCA with one absorbing state. In checking the critical behavior of STI we have to determine the order of the transition and, when it is second-order, whether it belongs to the DP class.

Results obtained with the family of CMLs defined in the previous section have shown that [7] the transition can be either 1st order or 2nd order, depending on the exact expression of the local map chosen (and especially the fact that the laminar regime is accounted for by a single attracting fixed point or a continuum of fixed points), on the dimension of space, on the existence of localized solutions called *defects*, etc. Moreover when the transition is second-order, exponents have been found to be model dependent, so that there cannot be universality in the usual sense. Quantitative results obtained by varying the slope α are presented and compared with DP results in Table 1 and 2 above. Table 1 gives β for $\alpha = 3$ and $d = 1$ and 2 with the corresponding values for DP. Table 2 displays ζ in the one-dimensional case for two values of α (for comparison $\zeta_{\mathrm{DP}} = 1.75$).

From these results it seems clear that STI does not belong to the universality class of DP but one can notice that STI with $\alpha = 2.1$ seems quantitatively "closer" to DP than STI with $\alpha = 3$. This fact could have been anticipated at the qualitative level from the consideration of the aspect of turbulent "clusters" in Fig. 4 and 5. For $\alpha = 2.1$ they appear similar to those generated by a DP process whereas for $\alpha = 3$ they seem to come out from the simulation of a *deterministic cellular automaton* (DCA). This points to a likely origin of non-universality: the role of *local dynamics* cannot be evacuated so easily.

To make this role more apparent, we performed a further step reducing the CML to a multi-state DCA.[8] The problem of passing from a local continuous phase space to a discrete phase space was solved by approximating the local map by a *step function* and defining the states of the automaton by a backward iteration of the absorbing region corresponding to the laminar state. This procedure, sketched in Fig. 6, preserves the most relevant feature of the local dynamics: the duration of chaotic transients. The evolution rule, directly derived from the diffusive coupling, then depend on α and g.

The approximation level being fixed, a path is followed in the space of DCA with a given number of states when g is varied. When α is large, this path leads from *trivial* dynamics (class I or II) for g small to *complex* dynamics (class III) for g large and the change occurs in a well-defined transition region where class IV automata are observed. By contrast for α close to 2, only *trivial* dynamics is observed in the reduced DCA and the transition displayed by the CML can be explained only by some randomization associated with the divergence of local trajectories during the long chaotic transients in the vicinity of the local map's crisis point. The presence of such local "random number generator" then makes the situation similar to that in DP.

Fig. 6. Reduction of the local map into a step-function defining a multi-state DCA and preserving the duration of chaotic transients, which accounts for the difference between cases $\alpha = 2.1$ and $\alpha = 3$.

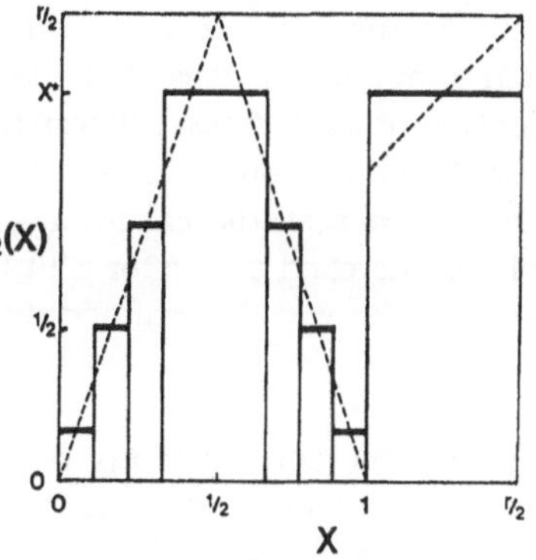

EXPERIMENTAL OBSERVATIONS OF STI

One of the most clear-cut example of spatiotemporally intermittent behavior has been observed in *quasi-one-dimensional convection* experiments [9]. Recent findings of the Saclay group [10] in annular geometry at moderate Prandtl have been presented by M. Dubois at this conference. Difficulties in the interpretation arise from the fact that the laminar state does not seem strictly absorbing so that the transition is "imperfect" but many features predicted form the analogy with DP and seen on models are recovered, e.g., the behavior of the distributions of size of laminar domains.

Another possible candidate is the so-called *printer instability* [11] discussed by Y. Couder. This instability of the interface between air and a fluid lubricating the interval between two rollers could probably be accounted for by a PDE similar to model (1).

In the field of hydrodynamic instabilities, the much studied *Taylor–Couette* problem [12] also offers situations of interest in the perspective of STI, especially a regime called "interpenetrating spirals" for which a model in terms of two coupled PDEs analogous to the Kuramoto–Sivashinsky equation with appropriate symmetries for two phase variables should be relevant.

Spatiotemporal intermittency, as invoked up to now, only involves domains of *small-scale turbulence* in competition with a supposedly well-identified *laminar flow*. Still in the Taylor–Couette flow, the regime called "spiral turbulence" offers a good opportunity to study the stability and the evolution of the front between such domains.[13] A similar but probably more complicated situation occurs in parallel flows that are stable at the inviscid limit,[14] e.g., the plane Poiseuille flow or the Blasius boundary layer flow. In such cases, the transition to turbulence generally involves *turbulent spots* with well-defined boundaries slowly spreading by contamination of the surrounding laminar flow.[15] In pipe flows, the growth of *turbulent slugs* at the origin of the intermittency first observed by Reynolds (1883) also involves the same kind of processes.

A related problem involves so-called *coherent structures* (CS), i.e., structures that can be understood as special nonlinear localized solutions of the primitive equations "at finite distance" from the solution corresponding to the considered basic flow. In some CMLs displaying STI, structures called *defects* play the role of these CSs. However, the relevance of the Kolmogorov–Spiegel–Sivashinsky equation [16] to the problem of the interaction between STI and CSs is probably greater since this equation is a PDE similar to model (1), therefore seemingly more appropriate to deal with the properties of turbulent continuous media.

To conclude, I would like to suggest an tentative application of STI to more fully developed turbulent regimes, especially in boundary layers. After the merging of turbulent spots, late stages of the transition to turbulence in such systems [17] involve the formation of *wall streaks*. These finite life-time unsteady counter-rotating rolls parallel to the flow break down in an intermittent fashion and it would perhaps be interesting to search for a statistical interpretation of the so-called *bursting phenomenon* along the lines sketched above for STI that, in any case, will remain an original scenario at least valid for weakly confined dissipative structures.

Acknowledgements. It is my pleasure to thank H. Chaté for his enthusiastic collaboration, F. Daviaud and M. Dubois at Saclay, Y. Couder and M. Rabaud at the ENS, and of course Y. Pomeau, for many enlightening discussions.

REFERENCES

1. Y. Pomeau, Physica D **23**(1986)3.
2. H. Chaté and P. Manneville, *Transition to turbulence via spatiotemporal intermittency: modeling and critical properties* in Ref. 3 below.
3. P. Coullet and P. Huerre, eds., *New trends in nonlinear dynamics and pattern forming phenomena*, Proceedings Cargèse'88, (Plenum Press, to appear).
4. H. Chaté and P. Manneville, Phys.Rev.Lett. **58**(1987)112.
5. K. Kaneko, Prog. Theor. Phys. **74**(1985)1033.
6. H. Chaté and P. Manneville, Physica D **32**(1988)409.
7. H. Chaté and P. Manneville, Europhys. Lett. **6**(1988)591.
8. H. Chaté and P. Manneville, J. Stat. Phys. **56**(1989)357.
9. P. Bergé, Nucl. Phys. B (Proc.Suppl.) **2**(1987)247; S. Ciliberto and P. Bigazzi, Phys. Rev. Lett. **60**(1988)286.
10. F. Daviaud, P. Bergé, and M. Dubois, Europhys. Lett. **9**(1989)441; M. Dubois, these proceedings.
11. M. Rabaud, S. Michalland, and Y. Couder, preprint ENS Paris, 1989; Y. Couder, these proceedings.
12. C.D. Andereck, S.S. Liu, and H.L. Swinney, J. Fluid Mech. **164**(1986)155.
13. J.J. Hegseth, C.D. Andereck, F. Hayot, and Y. Pomeau, Phys. Rev. Lett. **62**(1989)257.
14. see the review by D.J. Tritton, *Physical fluid dynamics* 2nd edition (Clarendon Press, Oxford, 1988).
15. S.E. Widnall, "Growth of turbulent spots in plane Poiseuille flow, in *Turbulence and Chaotic Phenomena in Fluids*, T. Tatsumi, ed. (Elsevier, 1984); J.J. Riley and M. Gad-el-Hak, "The dynamics of turbulent spots", in *Frontiers in fluid dynamics*, S.H. Davis and J.L. Lumley, eds (Springer, 1985).
16. H. Chaté and B. Nicolaenko, in the Cargèse'88 Proceedings, Ref. 3.
17. for a brief but recent description of late stages of the transition to turbulence in boundary layers see: D.M. Bushnell, C.B. McGinley, "Turbulence control in wall flows," Annual Review of Fluid Mechanics, **21**(1989)1.

ON THE SCALE-INVARIANT THEORY OF DEVELOPED HYDRODYNAMIC TURBULENCE KOLMOGOROV TYPE

Victor S. L'vov

Institute of Automation and Electrometry
Siberian Branch of Academy of Science
630090 Novosibirsk, U.S.S.R

The modern statistical theory of hydrodynamic turbulence goes back to the papers by Kraichnan and Wyld [1,2], who suggested to simulate excitation of stationary space-homogeneous developed hydrodynamic turbulence with the help of a space-distributed variable force $\vec{f}(\vec{r},t)$. According to the Kolmogorov-Obukhov's universality hypothesis [3,4] one can believe that in the limit of a large Reynolds number, the properties of fine-scale part of the turbulence (in the inertial range) will not depend on the way of turbulence excitation, i.e. of the type of the boundary conditions for the liquid flow or characteristics of the exciting force $\vec{f}(\vec{r},t)$. Therefore, one can suppose that the force $\vec{f}$ is a random force with a Gaussian statistics, it does not excite the mean flow: $\langle\vec{f}(\vec{r},t)\rangle = 0$, and its pair correlator D depends only on the coordinates and time difference:

$$\langle f_i(\vec{r}_1,t_1) f_j(\vec{r}_2,t_2)\rangle = D_{ij}(\vec{r}_1-\vec{r}_2,t_1-t_2)\ . \quad (1)$$

This formula is the condition for the turbulence excitation to be stationary and homogeneous. On the other hand, one should warn against arbitrary assumptions on the properties of $D(R,\tau)$: if follows from physical consideration that there is no forced excitation of the turbulence in the inertial range. Therefore, the correlator $D(R,\tau)$ must be concentrated in the region $R > L, \tau > L/V$. There L, that is the external turbulence scale, is a characteristic scale of energy-containing vortices, $V_T = \langle|v(\vec{r},t)|^2\rangle^{1/2}$ is the mean square turbulence velocity. Provided that (as it is in a number of papers [5-7]) D is the power function of R proportional to $\delta(\tau)$ (which means that it is not localized in the frequency dom.in at all), we come to a curious problem of behavior of the velocity field with that exciting force. The reference of this problem to "natural" hydrodynamic turbulence is not, however, clear.

The next principal step, conventional for researchers (see monograph [4]), is to come over to $\vec{k}$-representation, that is to expande the turbulent liquid velocity field in

Nonlinear Evolution of Spatio-Temporal Structures in Dissipative Continuous Systems
Edited by F.H. Busse and L. Kramer
Plenum Press, New York, 1990

plane waves. Such expansion is a long way from intuitive knowledge of the liquid turbulence as a system of interacting well-localized vortices. On the other hand, it will enable one to use detailed and rather powerful techniques for the diagram analysis of the perturbation theory serieses in the $\vec{k}$-space.

In the $\vec{k}$,t-representation the Navier-Stokes equations for incompressible liquid with an exciting force $\vec{f}$ in the right-hand side can be rendered in the form [1,2,4]:

$$\partial v_i(t,\vec{k})/\partial t+\nu k^2 v_i(t,\vec{k})=(i/2)\int \gamma_{ijl}(\vec{k},\vec{1},\vec{2}) *$$

$$* v_j(t,\vec{k}_1) v_l(t,\vec{k}_2)\, d\vec{k}_1 d\vec{k}_2 + f_i(t,\vec{k}) .$$

$$\gamma_{ijl}(\vec{k},\vec{1},\vec{2})=\Delta_{im}(\vec{k})\gamma_{mjl}(\vec{k})\,\delta(\vec{k}-\vec{k}_1-\vec{k}_2) , \tag{2}$$

$$\gamma_{ijl}(\vec{k})=(k_j\,\delta_{il}+k_l\,\delta_{ij}),\ \Delta_{ij}(k)=\delta_{ij}-k_i k_j/k^2.$$

Here $\gamma_{ijl}(k,1,2)$ is the <u>complete Euler vortex</u>, $\Delta_{ij}(k)$ is the transverse projector. For statistical description of the turbulence, we define in a usual manner the <u>different time pair correlator</u> (PC) of the velocity $\hat{F}(\vec{k},\tau)$ and its Fourier image $F(\vec{k},\omega)$:

$$2\pi F_{ij}(\vec{k},\omega)\,\delta(\omega-\omega')\,\delta(\vec{k}-\vec{k}')=\langle v_i(\vec{k},\omega)v_j^*(\vec{k}',\omega')\rangle ,$$

$$\hat{F}(\vec{k},\tau)=\int\hat{F}(\vec{k},\omega)\exp(-i\omega\tau)d\omega/2\pi , \tag{3}$$

$$\hat{F}(\vec{k})=\int\hat{F}(\vec{k},\omega)d\omega/2\pi .$$

Note that in the theory of weak wave turbulence the socalled kinetic equation can be constructed. It describes time evolution of simultaneous PCs of wave field amplitudes and is closed. Kraichnan and Wyld [1,2] has demonstrated that this approach is impossible in the theory of developed hydrodynamic turbulence: $F(\vec{k})$ proves to be a functional not of simultaneous PC $F(\vec{k}')$, but of the different time velocity PC $\hat{F}(\vec{k}',\omega')$ and the <u>Green functions</u> (GF) $\hat{G}(\vec{k}',\omega')$:

$$\hat{G}_{ij}(\vec{k},\omega)\,\delta(\vec{k}-\vec{k}')\delta(\omega-\omega')=\langle\delta v_i(\vec{k},\omega)/\delta f_j(\vec{k}',\omega')\rangle . \tag{4}$$

These values satisfy the Dyson-Wyld equation [2]:

$$\hat{G}(q)=\hat{G}_0(q)+\hat{G}_0(q)\hat{\Sigma}(q)\hat{G}(q),$$

$$\hat{G}_{0,ij}(q)=\Delta_{ij}(\vec{k})[\omega+i\nu k^2]^{-1}, \tag{5}$$

$$\hat{F}(q)=\hat{G}(q)[\hat{D}(q)+\Phi(q)]G^*(q),\quad q=\vec{k},\omega .$$

where $\hat{G}_0$ is the bare GF of noninteracting field, $\hat{D}(q)$ is the Fourier image of the random force correlator (1), the <u>mass operators</u> $\Sigma(q)$, $\Phi(q)$ are expressed through $G(q)$ and $F(q)$ according to the Feynman's rules [4] with the vertex of a single type:

$\gamma_{ijl}(q_1 q_2 q_3) =$ 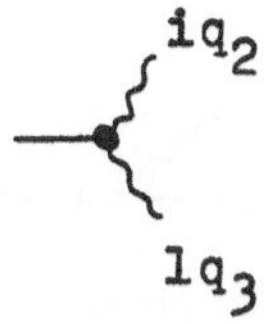 $= \gamma_{ijl}(\vec{k})\delta(q - q_1 - q_2)$, (6)

provided that we take "natural" graphic notation for $\hat{G}$ and $\hat{F}$ [5,6]:

$G(q) =$ , $G^*(q) = G(-q) =$ ,

$F(q) = F^*(+q) = F(-q) =$  . (7)

Let us write first diagrams for $\Sigma(q)$ and $\Phi(q)$:

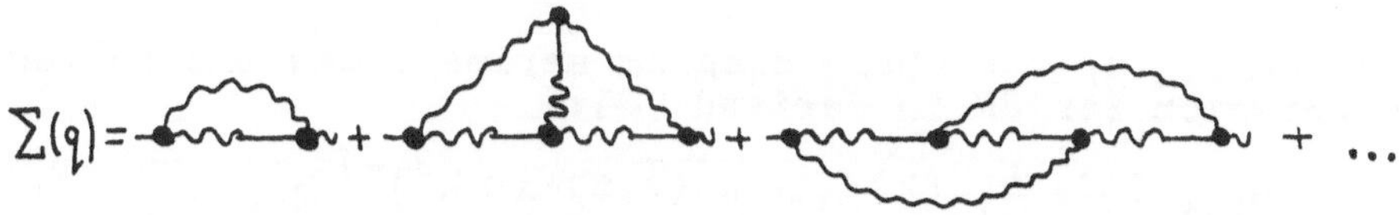

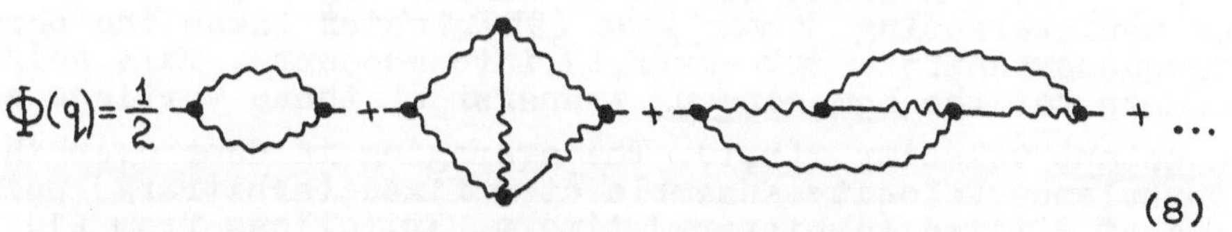

(8)

As Kraichnan [1] and Wyld [2] mentioned, the integrals in the momenta q of the velocity PC in the above diagrams diverge in the region of small q. They related these divergences to the effect of small k-vortices transfer (i.e. the vortices with characteristic size 1/k) by almost homogeneous velocity field of large L-vortices with the energy-containing size L. In 1977 the author suggested the homogeneous transfer approximation [8], that enables one to extract the most diverging part from each diagram (8). Given this approximation, in the diagrams for $\Sigma(q)$ one should neglect the PC momenta q_j compared to the outer momentum q in the arguments of vetices and backbone GF (i.e. in the way along GF from the entry to the exit of the diagram). Formally, the homogeneous transfer approximation can be obtained if the complete Euler vertices (2) and (6) in the backbone are replaced by the so-called kinematic vertices:

$\gamma_{ijl}(qq_1q_2) =$ 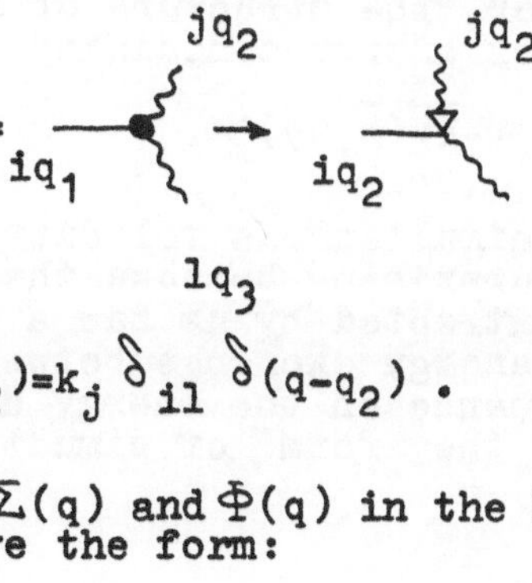 $=$ (9)

$= \tilde{\gamma}_{ijl}(qq_1q_2) = k_j \, \delta_{il} \, \delta(q - q_2)$.

The diagram for $\Sigma(q)$ and $\Phi(q)$ in the homogeneous transfer approximation have the form:

$\Sigma_t(q)=$ 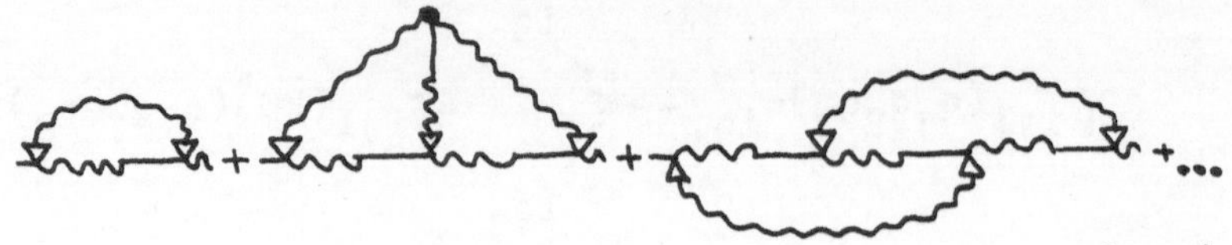

(10a)

$\Phi_t(q)=$ 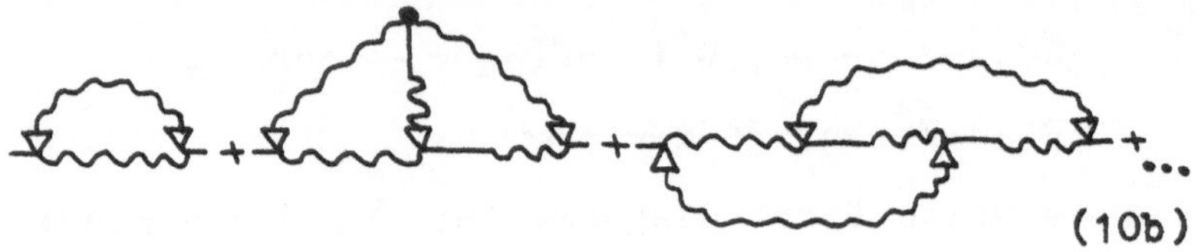

(10b)

By summing up the whole diagram series (10a) the following expression for GF is derived [8]:

$$G_{t,ij}(q)= \Delta_{ij}(\vec{k})\langle[\omega-\vec{k}\vec{v}(\vec{r},t)+i\nu k^2]^{-1}\rangle . \tag{11}$$

Its physical meaning is fairly simple: G_t represents the GF of noninteracting k-vortices (9b), which takes the Doppler frequency shift $\omega \Rightarrow \omega-\vec{k}\vec{v}(\vec{r},t)$ into account . This shift accounts for the <u>homogeneous</u> transfer of these vortices at <u>constant</u> velocity v(r,t). The averaging is done over the turbulence velocity ensemble at a fixed (arbitrary) point $\vec{r}$ and at a fixed (arbitrary) time t. It follows from (1) that in the inertial range (where $kv \gg \nu k^2$) the GF has the form:

$$G_{t,ij}(q)= \frac{\Delta_{ij}(\vec{k})}{kV_T}\, q(\frac{\omega}{kV_T}). \tag{12}$$

Expressions (11), (12) in the theory of turbulence for the inertial scale range should be considered as final: $G_t(q)$ is expressed through external (for this theory) characteristics of turbulence in the energy-containing range: $V_T \simeq \varepsilon^{1/3}L^{1/3}$, where ε is the energy flux along the spectrum, and statistical properties of the velocity field determining the form of the structural function q respectively.

Also, the author has managed [8] to analyse the series (10b) for Φ_t and to demonstrate that in the approximation of the homogeneous transfer, the structure of $F(k,\omega)$ in ω is completely determined by energy-containing vertices:

$$\hat{F}_t(\vec{k},\omega)=\hat{F}(\vec{k})\langle\delta(\omega-\vec{k}\vec{v}(\vec{r},t))\rangle . \tag{13}$$

At the same time, the simultaneous velocity PC F(k) remains arbitrary. It is not surprising because the homogeneous part of the transfer extracted by us has a kinematic character. It results in no energy exchange between the vortices and, thus, has no influence on the energy distribution over scales, that is on the form of simultaneous velocity PC F(k).

It is clear that the velocity field of large vortices is indeed inhomogeneous. This leads to some deformation of k-vortices while these are being transfered. The deformation is the more essential the less is the contrast in the scales of interacting vortices. It is this dynamic interaction (but not the kinematic effect of the homogeneous transfer) that must determine the energy scale distribution. Thus the problem is to distinguish the relatively low dynamic interaction against the masking homogeneous transfer in the formal technique of the theory. A natural way to solve this problem is to use the variables for which the homogeneous transfer is absent. For this propouse Kraichnan [9] suggested to use Lagrange variables $u(r_1, t_0, t)$, i.e. the velocity at the time t of a liquid particle which was at the initial point r_1 at the time t_0. The particle trajectory can easily be found from the Lagrange velocity

$$\vec{r}(t) = \vec{r}_1 + \int_{t_0}^{t} \vec{u}(\vec{r}_1, t_0, \tau) d\tau. \tag{14}$$

It is evident, that the Lagrange velocity $\vec{u}(\vec{r}_1, t_0, t)$ coincides with the Euler velocity $\vec{v}(\vec{r}, t)$, if $\vec{r}(t)$ is taken for r on the trajectory (14):

$$\vec{u}(\vec{r}_1, t_0, t) = \vec{v}\left[\vec{r}_1 + \int_{t_0}^{t} \vec{u}(\vec{r}_1, t_0, \tau) d\tau, t\right]. \tag{15}$$

Unfortunately, the diagram technique (DT) for the Lagrange velocity proved to be very sophisticated, and Kraichnan had advanced the theory until the approximation of direct interactions only [9]. The fact is that Lagrange variables exclude the transfer in the whole volume of turbulent liquid, which is unnecessary for physical consideration of the problem. Indeed, if one whishes to describe the evolution of a certain k-vortex placed whithin a certain volume of the order 1/k, it is sufficient to exclude the homogeneous part of the transfer exactly from that volume. However, to have something good out of it in the formal technique of the theory, one have to give up the purpose of describing the turbulence in the whole $\vec{r}$-space. To do so, the functions G(q) and F(q) (which describe global properties of the turbulence) must be replaced in the new theory by the local characteristics $G(\vec{r}, q)$ and $F(\vec{r}, q)$ describing the properties of the turbulence at a specific point $\vec{r}$. If the new theory should be adequate to physical consideration of the problem then the functions $G(\vec{r}, q)$ and $F(\vec{r}, q)$ in its formalism will be functionals of $G(\vec{r}', q')$ and $F(r', q')$ defined in the bounded domains $k|\vec{r} - \vec{r}'| \simeq 1$ and $|q - q'| \simeq q$. In 1986, Belinicher and L'vov [10] constructed the theory of turbulence, which possessed those properties. The theory used a quasi-Lagrange velocity $\vec{u}(\vec{r}_0, \vec{r}, t_0, t)$. The expression for this velocity can be derived from (15). For this purpose, one should replace in the integrand the velocities $\vec{u}(\vec{r}_1, \vec{r}_0, t)$ of the particles being contained in the volume of interest by the velocity of one of them. As such one should take, for example, the velocity $u(r_o,r_o,t)$:

$$\vec{v}(\vec{r},t)=\vec{u}\left[\vec{r}_0,\vec{r}-\int_{t_0}^{t}\vec{u}(\vec{r}_0,\vec{r}_0,\tau)d\tau\,,t_0,t\right]. \qquad (16)$$

The argument r_o is the coordinate of the above-mentioned point $\vec{r}_o$. It should be emphasized that formula (16) containes no approximations. It is an exact relation between the Euler and quasi-Lagrange velocities. Also note that the quasi-Lagrange velocity is a physical reality just like Euler or Lagrange velocities. It can be experimentally measured, successfully used for numerical simulation of turbulence, and the theory of turbulence can be adequately constructed in its terms [10].

The equations of motion for the quasi-Lagrange velocity coincide in form with the Navier-Stokes equation, but with a different vertex V:

$$[\partial/\partial t+\nu k^2]u_i(\vec{r}_0,\vec{k},t)=\int V_{ijl}(\vec{r}_0,\vec{k},\vec{1},\vec{2})u_j(\vec{r}_0,\vec{k}_1,t)*$$

$$*u_l(\vec{r}_0,\vec{k}_2,t)d\vec{k}_1d\vec{k}_2+f_i(\vec{r}_0,\vec{k},t), \qquad (17a)$$

$$V_{ijl}(\vec{r}_0,\vec{k},\vec{1},\vec{2})=\{k_{2j}\,\delta_{il}[\delta(\vec{k}-\vec{k}_1-\vec{k}_2)-\delta(\vec{k}-\vec{k}_2)\exp(i\vec{k}_1\vec{r}_0)]+$$

$$+k_{il}\,\delta_{ij}[\delta(\vec{k}-\vec{k}_1-\vec{k}_2)-\delta(\vec{k}-\vec{k}_1)\exp(i\vec{k}_2\vec{r}_0)]\}, \qquad (17b)$$

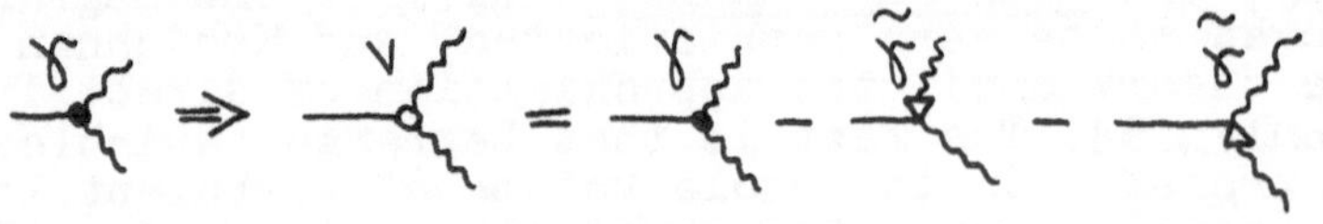

(17c)

The expression for the vertex V is the difference between complete Euler vortex γ (2) and two kinematic vertices (9) describing the homogeneous transfer. Therefore, it is clear that V describes the dynamic vertices interaction only. So, the vertex V can be naturally referred to as dynamic. The principal difference between the quasi-Lagrange equations (17) and Navier-Stokes one (2) is that the momentum $\vec{k}$ in the dynamic vertex V is not conserved. Needless to say, this accounts for to the fact that the point $\vec{r}_o$, where transfer is excluded, is fixed while coming over to quasi-Lagrange variables. As a result, the GF and PC in quasi-Lagrange variables are nondiagonal in $\vec{k}$. The diagonal form in ω, however, remaines unchanged:

$$F_{ij}(\vec{r}_0,\vec{k}_1,\vec{k}_2,\omega_1)\,\delta(\omega_1-\omega_2)=\langle u_i(\vec{r}_0,\vec{k}_1,\omega_1)u_j^*(r_0,k_2,\omega_2)\rangle,$$

$$(18)$$

$$G_{ij}(\vec{r}_0,\vec{k}_1,\vec{k}_2,\omega_1)\,\delta(\omega_1-\omega_2)=\langle\delta u_i(\vec{r}_0,\vec{k}_1,\omega_1)/\delta f_j(\vec{r}_0,\vec{k}_2,\omega_2)\rangle.$$

To physically interpret the theory, the transition to a mixed $(\vec{k},\vec{r})$-representation is extremly useful:

$$L(\vec{r}_0-\vec{r},\vec{k},\omega)=\int \exp(i\vec{s}\vec{r})L(\vec{r}_0,\vec{k}+\vec{s}/2,\vec{k}-\vec{s}/2,\omega)d\vec{s}\ , \qquad (19)$$

Here L is any of the functions G and F . In this way of coming over to the $\vec{r}$-representation (dictated by the quasi-classical approach), the functions $G(\vec{r},\vec{k},\omega)$ and $F(\vec{r},\vec{k},\omega)$ have the physical meaning mentioned above, that is, the GF and PC near the point r, respectively. Note that G and F in formula (19) do not depend on $\vec{r}$ and $\vec{r}_0$ separately, but on their difference $\vec{r}-\vec{r}_0$. The reason for that is that the homogeneous turbulence is translationally symmetric.

In discussion on the theory, it is of principal importance to study the dependence of the velocity PC and the GF on their arguments k_1, k_2 and ω. One can naturally suppose that they contain term diagonal in (k_1-k_2), that is, then can be represented in the form:

$$\hat{F}(\vec{k}+\vec{s}/2,\vec{k}-\vec{s}/2,\omega)=\hat{F}_1(\vec{k},\omega)\,\delta(\vec{s})+F_2(\vec{k}-\vec{s}/2,\vec{k}-\vec{s}/2,\omega),$$
$$\hat{G}(\vec{k}+\vec{s}/2,\vec{k}-\vec{s}/2,\omega)=G_1(\vec{k},\omega)\,\delta(\vec{s})+G_2(\vec{k}+\vec{s}/2,\vec{k}-\vec{s}/2,\omega). \qquad (20)$$

Here the functions F_2 and G_2 are either regular at $s\to 0$, or have an integrable singularity. Belinicher and L'vov[10] had assumed with practically no discussion that $F_1=G_2=0$, and then they found the functions F_2 and G_2. It was stated in [10] that F_2 and G_2 do not depend on the external turbulence scale L^2 and they describe the <u>Kolmogorov type</u> of turbulence, i.e. the one for which the simultaneous velocity PC $F(k)\propto k^{-11/3}$. Later on, in Ref. 11, this point of view was questioned and a quite different statement was suggested: the Kolmogorov turbulence is described by the functions F_1 and G_1, independent of L and having the form:

$$\hat{G}_1(\vec{k},\omega)=\hat{g}_1\,(\omega/\varepsilon^{1/3}k^{2/3})/\varepsilon^{1/3}k^{2/3},$$
$$\hat{F}_1(\vec{k},\omega)=\varepsilon^{1/3}f_1(\omega/\varepsilon^{1/3}k^{2/3})/k^{2/3},\ F(\vec{k})=\varepsilon^{2/3}k^{-11/3}, \qquad (21)$$

where ε is the energy spectrum flux, with dimensionless numerical coefficients ommitted. The authors of Ref. 11 assumed the functions F_2 and G_2 to describe a non-Kolmogorovian part of the turbulence associated with intemetency.

I do not agree with the statements proposed in Ref.11, and suppose them to be erroneous. First of all, the dependence of $\hat{G}(\vec{r},\vec{k},\omega)$ and $\hat{F}(\vec{r},\vec{k},\omega)$ on $\vec{r}$ (or, which is the same, the nondiagonality of $G(k_1,k_2,\omega)$ and $F(k_1,k_2,\omega)$ in $(\vec{k}_1-\vec{k}_2)$) is a formal consequence of the fact that a certain $\vec{r}$-coordinate is fixed in the quasi-Lagrange formulation of the problem. Likewize, the procedure of "zero momenta subtraction" suggested in Ref. 12, 13 and used in Ref. 11 is equivalent to the just mentioned one. This fact is somehow disquised in Refs. 11, 13 by absence of the letter r_o in the there presented formulae, because from the very beginning r_o was set to zero. So, as opposed to the statements in Ref. 11, we believe the nondiagonality in $(\vec{k_1}-\vec{k_2})$ to bear no relationship to physical problems, such as fluctuations of the energy spectrum flux, intermetency, etc.

It is accepted to understand the intermetency as the following property of hydrodynamic turbulence: at $Re \to \infty$ the fraction

$$K_n = \langle (\partial V/\partial r)^n \rangle / \langle (\partial V/\partial r)^2 \rangle^{n/2} \tag{22}$$

tends to infinity proportionaly to a certain power of Re. This property we shall term a <u>strong intermetency</u>. This phenomenon if exists at all, can not be found in a superficial insight into the turbulent theory which uses the diagram technique of both for the cannonical Klebsh variables [14, 8, 11-13] and the Euler velocity [2, 5-7, 10]. The intermetency can be caused by the onset of a Kolmogorov's solution instability with respect to perturbations which violate the spatial homogeneity of the turbulence. The new solution with the intermetency will then be concerned with a <u>nonperturbative</u> (i.e. giving zero in serieses of the perturbation theory) contribution to values to be measured in the experiment. Note that it might be that the strong intermetency does not exist in hydrodynamic turbulence. If so, the experiments [4] could be explained by a <u>weak intermetency</u>, when the value K (see Eq. (22)) at $Re \to \infty$ is finite, although numerically high.

Now again we consider the problem on the diagonal parts of GF and PC. Substituting functions (20) in the diagram serieses for mass operators of the quasi-Lagrange DT (they differ from (8) by replacement of γ by V), one can easily see that the terms G_1 and F_1 form the system of equation, which does not contain G_2 and F_2. In these serieses, the diagrams diverge in the infra-red region exactly in the same way as they do in DT for the Euler velocity. All integrals in these diagrams one can regularize by putting $F(k_j)=0$ at $Lk<1$. The serieses obtained can be summed up in the framework of homogeneous transfer approximation. The method for this summing was developed in [8] and described in detail in [10]. As a result, we have (V.I.Belinicher and V.S.Lvov, to be published):

$$\hat{G}_1(\vec{k},\omega)=\langle[\omega-\vec{k}(\vec{V}_1-\vec{V}_2)+i\nu k^2]^{-1}\rangle_{V_1V_2}\ ,$$
$$\hat{F}_1(k,\omega)=\hat{F}_1(\vec{k})\langle\delta(\omega-\vec{k}(\vec{V}_1-\vec{V}_2))\rangle_{V_1V_2}\ . \tag{23}$$

Here $\langle\ \rangle_{V_1V_2}$ - stands for avergaging over two statistically independent ensembles of the velocities V_1 and V_1, where $V_j^n = \langle v(r,t)^n\rangle$, j=1,2. To understand the physical meaning of these formulae we note that according to (19) and (20) $\hat{G}_1(\vec{k},\omega)$ and $\hat{F}_2(\vec{k},\omega)$ are the limits at $r\to\infty$ of the values $G(\vec{r}-\vec{r}_o,\vec{k},\omega)$ and $F(\vec{r}-\vec{r}_o,\vec{k},\omega)$, respectively. Also note that the transfer is excluded at the point $\vec{r}_o$, that is, the values G_1 and F_1 have been calculated in the reference system moving at the velocity $\vec{v}(\vec{r}_o,t)$. It is clear now, that in formula (23) $\vec{V}_1=\vec{v}(\vec{r}_o,t)$, $\vec{V}_2=\vec{v}(\vec{r},t)$. These velocities do not correlate, because the distance between the points r and r_o exceeds the scale L of largest vortices. The structure of expression (23) corresponds to expression (11) and (13) for the GF and PC velocities in the homogeneous transfer approximation in the DT for the Euler velocity. Particularly, the following formulae, similar to (12), follow from Eq.(23):

$$\hat{G}_1(\vec{k},\omega)=\frac{1}{k\tilde{V}_T}\,g_1\left(\frac{\omega}{k\tilde{V}_T}\right),\quad \hat{F}_1(\vec{k},\omega)=\frac{F_1(k)}{k\tilde{V}_T}\,f_1\left(\frac{\omega}{k\tilde{V}_T}\right).$$

$$\tilde{V}_T^2=2V_T^2,\quad V_T^2=\langle[v(r,t)]^2\rangle\simeq\varepsilon^{2/3}L^{2/3}. \tag{24}$$

This expression comprises the external scale L. Which is a fundamental distinction from the Eq. (21) of the Ref.11.

The general structure of expressions (20) for the GF and velocity PC can be fairly well understood in the $(\vec{k},\vec{r})$-representation from the results given in Ref. 20 and formulae (22), (23). Namely:

$$\hat{G}(\vec{r},\vec{k},\omega)=\hat{g}\,[\omega/\Omega(k,r)]/\Omega(k,r),$$

$$\hat{F}(r,k,\omega)=\varepsilon^{2/3}k^{11/3}\hat{f}\,[\omega/\Omega(k,r)]\,\Omega(k,r). \tag{25}$$

At $r\gg L$ according to (24), $\Omega=\sqrt{2}\,kV_T$. At $L>r>1/k$ the formulae for G and F have the form (24), but the expression for V_T has no contribution of vortices with $l>r$. Besides, $V_T(r)\simeq r^{1/3}\varepsilon^{1/3}$ [20]. Thus in this region, $\Omega=(\varepsilon r)^{1/3}k$.

Finally, at $kr\ll 1$ the transfer is completely excluded and Ω has the Kolmogorov value $\Omega=\varepsilon^{1/3}k^{2/3}$. For illustration, an interpolation formula for Ω can be suggested, which provides a qualitively valid representation of all asymptotics:

$$\Omega(k,r,L)=\varepsilon^{1/3}k^{2/3}\,\frac{1+(kr)^{1/3}}{1+(r/L)^{1/3}}. \tag{26}$$

Formulae (25), (26) give an idea of how the external turbulence scale L occurs in the structural function g and f for GF and velocity PC, respectively. At $L\to\infty$ the expression for Ω depends no longer on L. And expression (25) for G and F approaches zero, if $r\to\infty$. This means that $G_1=F_1=0$ in the limit $L\to\infty$. It is exactly that limit that was derived in Ref. 20. If L is finite, then $G_1=0$ and $F_1=0$ are given by Eqs. (22), (23). The expressions for G_2 and F_2 can be taken from Eqs. (25), (26), if one considers momentum representation in the variable $\vec{r}$. The above expressions can easily be analyzed, so we shall not do all that. Still note, that the integrals over ω in (25) for $G(\vec{r},\vec{k},\omega)$ and $F(\vec{r},\vec{k},\omega)$ are independent of $\vec{r}$, and thus

$$\int d\omega F_2(\vec{k}+\vec{s}/2,\vec{k}-\vec{s}/2,\omega)=F_2(\vec{k})\delta(\vec{s}),$$
$$\int d\omega G_2(\vec{k}+\vec{s}/2,\vec{k}-\vec{s}/2,\omega)=G_2(\vec{k})\delta(\vec{s}). \quad (27)$$

This fact is a consequence of the general statement that the simultaneous correlators of Euler and quasi-Lagrange velocities coincide.

Later on, we shall give detailed analysis of how the scale L influences upon some Kolmogorov turbulence properties in the inertial range.

References

1. Kraichnan R.H. - J.Fluid Mech., 1959, v. 6, p. 497.
2. Wyld H.W. - Ann. Phys., 1961, v.14, p.143.
3. Kolmogorov A.N. - DAN SSSR, 1941, v.30, p.229; DAN SSSR, 1941, v.32, p.19. Obukhov A.M. DAN SSSR, 1941, v.32, p.22; AN SSSR, Ser.geograph. i geophizika, 1941, v.5, p. 443.
4. Monin A.S., Yaglom A.M. Statisticheskaya gidromekhanica, part 2, M.: Nauka, 1967.
5. Foster D., Nelson D.R., Stephen M.J. - Phys. Rev., 1977, v.A16, p.732.
6. Martin P.C., Siggia E.D., Rose H.A. - Phys. Rev. A,1972, v.8, p.423; DeDominics C., Martin P.C. - Phys. Rev., 1979, v.A19, p.419.
7. Adzhemyan L.Ts., Vasiljev A.N., Pis'mak Yu.M. - TMF, 1983,v.57, N 2, p.268-281.
8. L'vov V.S. To the theory of developed hydrodynamic turbulence.Preprint IA&E SO AN SSSR,1977,N 53 (in Russian).
9. Kraichnan R.H.-Proc.Symp.Dynamics of Fluids and Plasma. N.Y.: Ac.Press,1967, J.Fluid Mech.,1977, v.83,p.349; Phys.Fluid, 1965, v.8, p.575, 1966, v.9, p.1728.
10. Belinicher V.I., L'vov V.S. Scale-invariant theory of developed hydrodynamic turbulence.Preprint IA&E SO Ac.Sci. USSR, N 333,1986 (in Russian),N 358,1987 (in English), Zh.Eksp.Teor.Fis., 1987, v.93,N 2(8),pp.533-551.
11. A.V.Tur,V.V.Yanovsky. On the theory of developed turbulence in incompressible fluid.Preprint of Space Researched Institute USSR Ac.Sci. N1203, Moscow 1986.
12. Tur A.V.,Yanovsky V.V., Moiseev S.S. Proceeding of the International working group on nonlinear physics and turbulence, Kiev, 1983, p.1693; DAN SSSR, 1984,v. 279, p.96.
13. Sagdeev R.Z., Moiseev S.S., Tur A.V., Yanovsky V.V. - In "Nonlinear phenomena in plasma phys. and hydrodinamics", ed. R.Z. Sagdeev, Mir Publ. 1986, p.137.
14. Zakharov V.E., L'vov V.S. - IVUZ, Radiofizika, 1975, XVIII, pp. 1470-1487.

CONTRIBUTORS

Armbruster, D.
Aranson, J.S.
Arnéodo, A.

Banks, W.H.H.
Barten, W.
Bensimon, D.
Bestehorn, M.
Bohr, T.
Bonetti, M.
Bouissou, Ph.
Brand, H.R.
de Bruyn, J.R.
Buka, A.
Bühler, K.
Busse, F.H.
Buzug, Th.

Caroli, B.
Caroli, C.
Cessi, P.
Chen, K.
de Cheveigné, S.
Childress, S.
Chu, X.-L.
Cladis, P.E.
Couder, Y.
Croquette, V.

Dangelmayr, G.
Daviaud, F.
Donnelly, R.J.
Drazin, P.G.
Dubois, M.

Elphick, Ch.

Faivre, G.
Fal'kovich, G.E.
Finn, P.L.
Friedrich, R.
Frisken, B.J.

Ghoniem, N.M.
Gleeson, J.T.
Gorshkov, K.A.
Grasseau, G.
Guthmann, C.
Guyon, E.

Haken, H.
Hakim, V.
Hort, W.
Hughes, D.W.

Ierley, G.R.

Janiaud, B.
Jensen, M.H.
Joseph, D.D.

Kamps, M.
Kessler, D.A.
Knobloch, E.
Kramer, L.
Kurowski, P.

Levine, H.
Lhost, O.
Linz, S.J.
Lomov, A.S.
Lücke, M.
L'vov, V.S.

Malomed, B.A.
Manneville, P.
May, A.D.
Michalland, S.
Miike, H.
Moore, D.R.
Morris, S.W.
Müller, S.C.
Müller, U.

Newell, A.C.
Nepomnyashchy, A.A.

Palffy-Muhoray, P.
Parisi, J.
Passot, T.
Pedersen, A.W.
Perrin, B.
Pfister, G.
Pismen, L.M.
Platten, J.K.
Plesser, T.
Polifke, N.
Pomeau, Y.
Proctor, M.R.E.
Pumir, A.

Rabaud, M.
Rabinovich, M.I.
Rand, D.A.
Reimers, T.
Regev, O.
Rehberg, I.
Roulet, B.

van Saarloos, W.
Schnaubelt, M.
Schöpf, W.
Shraiman, B.I.
Siggia, E.D.
Singh, P.
Souli, M.
Soward, A.M.
Spiegel, E.A.

Tabeling, P.
Thome, H.
de la Torre Juárez, M.

Velarde, M.G.

Walgraef, D.
Weiss, N.O.
Williams, H.
Winkler, B.L.

Young, W.R.
Yuan, H.J.

Zaturska, M.B.
Zimmermann, G.
Zimmermann, W.
Zinemanas, D.

INDEX